Russian Mathematicians in the 20th Century

Russian Mathematicians in the 20th Century

editor

Yakov Sinai

Princeton University, USA

World Scientific

New Jersey • London • Singapore • Hong Kong

Published by

World Scientific Publishing Co. Pte. Ltd.
5 Toh Tuck Link, Singapore 596224
USA office: Suite 202, 1060 Main Street, River Edge, NJ 07661
UK office: 57 Shelton Street, Covent Garden, London WC2H 9HE

British Library Cataloguing-in-Publication Data
A catalogue record for this book is available from the British Library.

The editor and publisher would like to thank the following organizations and publishers of the various journals and books for their assistance and permission to reproduce the selected reprints found in this volume:

American Mathematical Society
C. R. Acad. Sci. Paris
Doklady Akademii Nauk SSSR
Elsevier Science Publisher
Graylock Press
Historia Mathematica
Journal d'Analyse Mathematique
Journal of Indian Mathematical Society
Journal Uspekhi Math Nauk (Russian Math Surveys)
Kluwer Academic / Plenum Publishers
Springer-Verlag
Studia Mathematica
University of St Andrews, Scotland

While every effort has been made to contact the publishers of reprinted papers prior to publication, we have not been successful in some cases. Where we could not contact the publishers, we have acknowledged the source of the material. Proper credit will be accorded to these publications in future editions of this work after permission is granted.

RUSSIAN MATHEMATICIANS IN THE 20TH CENTURY

The biographies and other additional information in this book were prepared entirely at the discretion of World Scientific Publishing.

ISBN 981-02-4390-1
ISBN 981-238-385-9 (pbk)

Desk Editor: Tan Rok Ting
Cover Design: Ng Chin Choon

Printed in Singapore by World Scientific Printers (S) Pte Ltd

Preface

Several years ago, Dr. Sen Hu from World Scientific Publishing Co. approached me with a suggestion to publish a volume of selected papers of Soviet mathematicians that were done in the twentieth century. Certainly, it was a sign of respect and appreciation for Soviet mathematics, and I agreed to participate in the preparation of the corresponding edition. Dr. Hu asked several leading Russian mathematicians for the names of people whose works should be included. After the list was completed, he and I started a long and difficult process of selecting papers from each author. We wanted to maintain some parity, and at the same time, include papers of some interest even nowadays.

This volume is essentially the result of our efforts. There is some difference between the mathematical talent of people working under a totalitarian regime, and those working in a democratic society. Soviet mathematicians much more often discussed "what to do" rather than "how to do". Maybe this difference of styles can be seen in this edition.

There is every reason to expect that this volume will be useful to many researchers who preserve some interest in how mathematics was done in the Soviet Union in the past.

Ya. G. Sinai
February 2003

Contents

Mark Grigorievich Krein

Lazar Aronovich Liusternik

Vladimir Abramovich Rokhlin

Petr Sergeevich Novikov

Nikolai Nikolaevich Bogoliubov

Aleksandr Danilovich Aleksandrov

Leonid Vitalyevich Kantorovich

Anatoly Ivanovich Malcev

Aleksandr Mikhailovich Lyapunov

Born – 3 June 1857 in Yaroslavl, Russia
Died – 3 Nov 1918 in Odessa, Russia

Aleksandr Mikhailovich Lyapunov was born in 1857 at Yaroslavl where his father was director of the Demidovsk Lycée, a general high school. In 1870 his mother, who had been widowed in 1868, moved with her three sons to Gorki, and it was here that Lyapunov received his secondary school education, graduating from the gymnasium in 1876 with a gold medal. In the same year he became a student in mathematics in St. Petersburg University, where he was particularly influenced by the lectures of Professor P. L. Chebyshev. Lyapunov graduated in 1880, again winning a gold medal for an essay on hydrostatics. He stayed on at the university in the Mechanics Faculty, developing his Master's thesis *On the stability of ellipsoidal forms of equilibrium of rotating fluids* which he defended in 1884. This work, which was translated into French in 1904, established his name in Europe.

In the autumn of 1885, Lyapunov moved to Kharkov where he became a "privat-dozent" in the university's Department of Mechanics. In the winter break of 1885/1886 he returned to St. Petersburg to marry his first cousin, Natalia Rafailovna Sechenova. In his work in Kharkov, Lyapunov first had to spend much time preparing lectures in mechanics and notes for his students but in 1888 he started to publish papers devoted to the stability of motion. This work led to his famous doctoral thesis of 1892. *The general problem of the stability of motion.* He defended this thesis in Moscow on 12 October 1892, one of his "opponents" being N. E. Zhukovski (better known as Joukowski — "the father of Russian aviation"; this thesis was republished in French in 1908 and 1949, Russian in 1935, and English. Lyapunov was promoted in the following year 1893 to an ordinary professorship at Kharkov.

Although Lyapunov continued his work on stability problems until 1902, he also turned his attention to other topics, including potential theory and probability theory, particularly the central limit theorem of Laplace. He was also an active participant in activities of the Kharkov Mathematical Society, becoming its president and editor of its *Communications* from 1899 to 1902.

In 1902, Lyapunov moved back to St. Petersburg where he had been appointed Academician and the successor to his former teacher P. L. Chebyshev, who had died in 1894. He was now able to devote himself entirely to research, returning to the problem of forms of equilibrium of bodies of uniformly rotating fluid particles under mutual Newtonian gravitational attraction. In particular, Lyapunov showed that pear-shaped forms of rotating fluid were unstable. This result had important consequences in astronomy, since it disproved the current theories of formation of satellites from bodies of rotating fluid e.g. the Earth–Moon system. G. H. Darwin (son of the famous naturalist Charles Darwin) at Cambridge had proposed such a mechanism at this time which naturally required pear-shaped bodies to be stable. In 1917, J. H. Jeans, in an Adams Prize Essay at Cambridge, was to confirm that Lyapunov's result was correct and that therefore Darwin's theory was incorrect.

In June 1917, Lyapunov left St. Petersburg for Odessa on doctor's orders, since his wife was suffering badly from tuberculosis. However, in Odessa her condition deteriorated, and on 31 October 1918 she died. Lyapunov himself was in poor health and his eyesight was failing. The tragic events of the revolution, the burning of the family estate on the River Volga including a great library built up by his father and grandfather, and the illness of his wife had greatly depressed Lyapunov. On the day his wife died, Lyapunov shot himself in the head with a pistol. He died on 3 November 1918, the day of his wife's funeral. In a suicide note he asked to be buried with her. The inscription on the gravestone reads:

The creator of the theory of stability of motion, the doctrine of equilibrium figures of rotating liquid, methods of the qualitative theory of differential equations, author of the central limit theorem and other deep investigations in areas of mechanics and mathematical analysis.

A NEW CASE OF INTEGRABILITY OF DIFFERENTIAL EQUATIONS OF MOTION OF A SOLID BODY IN LIQUID

"Reports of the Kharkov Mathematical Society,"
Series II, Vol. IV, No. 1 and 2, 1893,
pp 81-85

It is known that the problem of the motion of a solid body in an unlimited liquid in the absence of external forces and under certain assumptions about the nature character of liquid motion depends on integrating a system of differential equations of following form:

$$\frac{dx_1}{dt} = x_2\frac{\partial T}{\partial y_3} - x_3\frac{\partial T}{\partial y_2},$$
$$\frac{dx_2}{dt} = x_3\frac{\partial T}{\partial y_1} - x_1\frac{\partial T}{\partial y_3},$$
$$\frac{dx_3}{dt} = x_1\frac{\partial T}{\partial y_2} - x_2\frac{\partial T}{\partial y_1},$$
$$\frac{dy_1}{dt} = x_2\frac{\partial T}{\partial x_3} - x_3\frac{\partial T}{\partial x_2} + y_2\frac{\partial T}{\partial y_3} - y_3\frac{\partial T}{\partial y_2},$$
$$\frac{dy_2}{dt} = x_3\frac{\partial T}{\partial x_1} - x_1\frac{\partial T}{\partial x_3} + y_3\frac{\partial T}{\partial y_1} - y_1\frac{\partial T}{\partial y_3},$$
$$\frac{dy_3}{dt} = x_1\frac{\partial T}{\partial x_2} - x_2\frac{\partial T}{\partial x_1} + y_1\frac{\partial T}{\partial y_2} - y_2\frac{\partial T}{\partial y_1},$$

where T is some quadratic form of variables x_1, x_2, x_3, y_1, y_2, y_3 with constant coefficients.

These equations in any case allow the three following integrals:

$$T,\quad x_1^2 + x_2^2 + x_3^2 \text{ and } x_1y_1 + x_2y_2 + x_3y_3$$

[which in the considered question will always be independent since the form of T in this question is such that when $x_1 = x_2 = x_3 = 0$ it turns into defined (définie) form of variables y_1, y_2, y_3], and whenever a fourth integral is found which, like those just indicated does not depend on t and does not reduce to a function of them, the principle of last factor, which obviously applicable to the considered system of differential equations, will permit it to be integrated by means of quadratures.

Clebsch indicated three cases where such an integral can be found in the guise of an integral function of values x_1, x_2, x_3, y_1, y_2, y_3 not higher than second degree (Math. Ann., Vol. 3).

All these cases pertain to following type of form T:

$$T = \frac{1}{2}(a_1x_1^2 + a_2x_2^2 + a_3x_3^2) + (b_1x_1y_1 + b_2x_2y_2 + b_3x_3y_3) + \\ + \frac{1}{2}(c_1y_1^2 + c_2y_2^2 + c_3y_3^2),$$

and the first of them is characterized by condition that corresponding elements of any two columns of table

$$\begin{matrix} a_1, & a_2, & a_3 \\ b_1, & b_2, & b_3 \\ c_1, & c_2, & c_3 \end{matrix}$$

are mutually equal.

If this holds true, for instance, in relation to the second and third columns so that

$$a_2 = a_3, \quad b_2 = b_3, \quad c_2 = c_3,$$

then y_1 can serve as the fourth independent integral.

The second of the cases pointed out by Clebsch is characterized by equalities

$$b_1 = b_2 = b_3$$

together with condition that with proper selection of constants a and τ coefficients a_1, a_2, a_3 are expressed by formulas

$$a_1 = a + \tau c_2c_3,$$
$$a_2 = a + \tau c_3c_1,$$
$$a_3 = a + \tau c_1c_2.$$

So that this case may not be included in the preceding the coefficients c_1, c_2, c_3 have to be all different.

Under this condition the fourth independent integral will be function

$$\tau(c_1x_1^2 + c_2x_2^2 + c_3x_3^2) - (y_1^2 + y_2^2 + y_3^2).$$

Finally, the third case of Clebsch is characterized by conditions

$$b_1 = b_2 = b_3,$$
$$c_1 = c_2 = c_3,$$

under which, if there are different ones among the values a_1, a_2, a_3 the fourth independent integral may be function

$$a_2a_3x_1^2 + a_3a_1x_2^2 + a_1a_2x_3^2 - c(a_1y_1^2 + a_2y_2^2 + a_3y_3^2),$$

where c is the total value of coefficients c_1, c_2, c_3.

When a_1, a_2, a_3 are all different this case is not included in the preceding, but, if it is desired, it may be considered as a certain limit to the second, from which conversion to it is possible by changing c_2, c_3, a and τ so that $c_2 - c_1$ and $c_3 - c_1$ became infinitesimal, a and τ became infinitely large, and a_1, a_2, a_3 remained constant. The integral just now pointed out will be then obtained as the limit of some linear combination of the four whole integrals corresponding to the second case.

It may be noted that second and third cases can be replaced by a single one which is characterized by conditions

$$b_1 = b_2 = b_3,$$

$$\frac{a_2 - a_3}{c_1} + \frac{a_3 - a_1}{c_2} + \frac{a_1 - a_2}{c_3} = 0.$$

Recently V. A. Steklov has shown a new case of integrability, in which, like the preceding, the fourth independent integral also can be found in the guise of the entire function of second degree of variables x_1, x_2, x_3, y_1, y_2, y_3 ("Reports of the Khar'kov Math. Society," Vol. 3, and "Math. Ann.," Vol. 42).

This case V. A. Steklov defines by conditions

$$\left.\begin{aligned} b_1 &= \sigma c_2 c_3, \quad a_1 = \sigma^2 c_1 (c_2^2 + c_3^2), \\ b_2 &= \sigma c_3 c_1, \quad a_2 = \sigma^2 c_2 (c_3^2 + c_1^2), \\ b_3 &= \sigma c_1 c_2, \quad a_3 = \sigma^2 c_3 (c_1^2 + c_2^2). \end{aligned}\right\} \tag{1}$$

in fulfillment of which the considered differential equations allow following integral:

$$\sigma^2 S (c_2 - c_3)^2 x_1^2 - 2\sigma S c_1 x_1 y_1 + S y_1^2,$$

where S signifies the summation of three terms, obtained from the written circular permutation of signs 1, 2, 3.

So that these conditions may determine a case not occurring in the preceding, constant σ must not be zero, but coefficients c_1, c_2, c_3 must all be different.

Under this last condition the integral just now indicated surely will not be function of three known.

To the enumerated cases of integrability I can here add one more.

This case is characterized by conditions

$$c_1 = c_2 = c_3 = c,$$

$$a_1 - \frac{(b_2-b_3)^2}{c} = a_2 - \frac{(b_3-b_1)^2}{c} = a_3 - \frac{(b_1-b_2)^2}{c},$$

under which the considered differential equations, may be easily ascertained, allow the following integral:

$$S b_1 [(b_2+b_3)x_1 + cy_1]^2,$$

which does not lead a function of three known, if there are different ones among values b_1, b_2, b_3.

When b_1, b_2, b_3 all are different this case is not included in any of the preceding. But if the case of V. A. Steklov is properly generalized the case just now indicated will be derived from it as the limiting one, just as the third case of Clebsch is derived from the second.

In fact, noting that, whatever the constants A and B, the considered differential equations from replacement of T by

$$T + A(x_1^2 + x_2^2 + x_3^2) + B(x_1y_1 + x_2y_2 + x_3y_3)$$

do not change, we may replace condition (1) by the following:

$$b_1 = b + \sigma c_2 c_3,$$

$$b_2 = b + \sigma c_3 c_1,$$

$$b_3 = b + \sigma c_1 c_2,$$

$$a_1 = a + \sigma^2 c_1 (c_2 - c_3)^2 = a + \frac{(b_2-b_3)^2}{c_1},$$

$$a_2 = a + \sigma^2 c_2 (c_3 - c_1)^2 = a + \frac{(b_3-b_1)^2}{c_2}, \quad a_3 = a + \sigma^2 c_3 (c_1 - c_2)^2 = a + \frac{(b_1-b_2)^2}{c_3},$$

from which the last three, because of the arbitrariness of constant a, are equivalent to

$$a_1 - \frac{(b_2-b_3)^2}{c_1} = a_2 - \frac{(b_3-b_1)^2}{c_2} = a_3 - \frac{(b_1-b_2)^2}{c_3}.$$

In order now to proceed to the case shown by me it must be assumed that $c_2 - c_1$ and $c_3 - c_1$ become infinitesimal, while b and σ become infinitely large, and moreover in such a way that b_1, b_2, b_3 remain constant.

In conclusion will remark that the last two cases can be replaced one which is characterized by conditions

$$a_1 - \frac{(b_2 - b_3)^2}{c_1} = a_2 - \frac{(b_3 - b_1)^2}{c_2} = a_3 - \frac{(b_1 - b_2)^2}{c_3},$$

$$\frac{b_2 - b_3}{c_1} + \frac{b_3 - b_1}{c_2} + \frac{b_1 - b_2}{c_3} = 0.$$

A GENERAL PROPOSITION OF PROBABILITY THEORY

"Comptes rendus de l'Académie des Sciences,"
Vol. CXXXII, Paris, 1901, pp. 814-815

In the note which I published two months ago in "Comptes rendus" (21 January 1901) I indicated a proposition, representing a certain generalization of the theorem which was the subject of my memoir, printed in "News of the Academy of Sciences," 5th Series, Vol. XIII, No. 4, 1900, pp. 359-386. In continuing my investigations in the same direction, I arrived at a result more still general. That result I would like to report in this note.

Let

$$x_1, x_2, x_3, \ldots$$

be an infinite sequence of independent real variables the value of which are determined by the case. Designating by

$$x_1, x_2, \ldots$$
$$a_1, a_2, \ldots,$$
$$d_1, d_2, \ldots$$

the respective mathematical expectations of values

$$x_1, x_2, \ldots$$
$$(x_1 - \alpha_1)^2, (x_2 - \alpha_2)^2, \ldots$$
$$|x_1 - \alpha_1|^{2+\delta}, |x_2 - \alpha_2|^{2+\delta}, \ldots,$$

where δ is any <u>positive</u> number we will have the following proposition:

<u>In all cases where there exist constant numbers</u> δ; <u>for which the ratio</u>

$$\frac{(d_1 + d_2 + \cdots + d_n)^2}{(a_1 + a_2 + \cdots + a_n)^{2+\delta}}$$

tends toward zero when n increases without limit the probability of inequalities

$$z_1 < \frac{x_1 - a_1 + x_2 - a_2 + \cdots + x_n - a_n}{\sqrt{2(a_1 + a_2 + \cdots + a_n)}} < z_2,$$

in which z_1 and $z_2 > z_1$ are any given numbers converges toward limit

$$\frac{1}{\sqrt{\pi}}\int_{z_1}^{z_2} e^{-z^2}\,dz$$

when $n = \infty$ and moreover does so uniformly for all values of z_1 and z_2.

It is obvious that this hypothesis has a much more general nature than that in the preceding note, since on the one hand number $\delta \leq 1$ introduced by me before, and on the other hand instead of the mathematical expectations of the quantities

$$|x_1|^{2+\delta},\ |x_2|^{2+\delta}, \ldots$$

I am now considering the mathematical expectations of quantities

$$|x_1 - a_1|^{2\cdot\delta},\ |x_2 - a_2|^{2+\delta}, \ldots,$$

and moreover introduce the arithmetic mean value into the conditions of the theorem of their largest mathematical expectation from

$$d_1, d_2, \ldots, d_n$$

Deferring proof until publication of a detailed memoir on this subject I will limit myself here only to the remark that the new proposition is established by the same method which I used in the particular case considered in the memoir cited.

Automatica, Vol. 28, No. 5, pp. 865–871, 1992
Printed in Great Britain.

0005–1098/92 $5.00 + 0.00
Pergamon Press Ltd

Alexander Mikhailovitch Lyapunov: On the Centenary of his Doctoral Dissertation on Stability of Motion*

P. S. SHCHERBAKOV†

An account of the life and scientific activity of the mathematician A. M. Lyapunov (1857–1918) is given, a curriculum vitae and list of his publications is included.

Key Words—Control theory; stability; Lyapunov methods.

Abstract—To celebrate the centenary of the publication in 1892 of A. M. Lyapunov's doctoral thesis "On the general problem of the stability of motion" an account of Lyapunov's life and scientific work is given. A *curriculum vitae* of Lyapunov and a bibliography of his published works are appended.

1. INTRODUCTION

THE NAME OF A. M. Lyapunov is well known to control engineers and mathematicians through Lyapunov's second method for investigating the stability of dynamical systems, and the method of characteristic functions to prove various limit theorems in probability theory. The present note gives an account of his life and scientific activity; a list of his works in chronological order is also given. The information has been obtained from obituary notices and biographical essays cited in the list of references.

2. LIFE OF LYAPUNOV

"Educated at first by his father, the colleague of N. I. Lobatchevskii at Kazan University, then among the people close to I. M. Setchenov, who spent his youth among the most educated part of our society of that time, Lyapunov embodied the best type of idealist for 60 years, that may be not understood by everybody at present". These words were said in the obituary speech in 1919 by Academician V. A. Steklov, the disciple and the best friend of Lyapunov. The environment he had grown up and lived in was one of the sources that nourished the culture of Russian and European society of that time. One of his younger brothers—Sergei, was a prominent conductor, pianist and composer of the M. A. Balakirev circle; another—Boris, was a slavist, historian, member of the Petersburg Academy of Sciences. His wife's uncle and the close relative of his mother, the great physiologist I. M. Setchenov took a great part in his education and life. Lyapunov's family was on friendly terms with their close relative Academician A. N. Krylov—the outstanding naval architect, mathematician, and mechanical engineer; their relatives were the famous physicians N. F. Filatov (epidemiologist and paediatrician) and V. P. Filatov (ophthalmologist). This was the environment of the Russian intelligentsia, alien to materialistic claims, participating in the destiny of the world culture. Some eccentricity, originating from the absolute primacy of spiritual values was typical of these people.

Alexander Mikhailovitch Lyapunov belonged to the ancient noble clan (according to some sources descending from Alexander Nevskii) that associated its activities with Kazan University, already famous in the 18th century. His grandfather was a syndic at the University Treasury. His father worked as an astronomer–observer in the University observatory, and later as its director. A. M. Lyapunov was born on 25 May 1857 in Yaroslavl', when his father was principal of Yaroslavl' Lycée—one of the four high general educational institutions in Russia (founded on the wealth of the foundry proprietor, Demidov). In 1865 the family moved to his mother's patrimonial estate Shipilovo in Simbirsk province on the banks of the Volga river. Just there and in the neighbourhood on Setchenov's estate was formed the circle that defined in many respects the personality of Alexander Mikhailovitch. There also he became acquainted with the niece of I. M. Setchenov, Natalia, who was to become his wife in 1886.

Graduating from the gymnasium of Nizhnii

* Received 29 July 1991; revised 13 November 1991; received in final form 24 February 1992. The original version of this paper was not presented at any IFAC meeting. This paper was recommended for publication in revised form by Editor P. C. Parks.

† Institute of Control Sciences, Profsoyuznaya 65, 117806, Moscow, Russia.

Novgorod in 1876 with a gold medal, Lyapunov entered the physical–mathematical department of St. Petersburg University but being inspired by his vocation he changed to the mathematical department within a month. He attended lectures there and made contact with such titans of science as D. I. Mendeleev and P. L. Tchebyshev, who made a great impression on him. A less than modest financial situation and his position as head of the family (his father died when he was 11, and his mother in 1879) led him to give private lessons in mathematics, and in 1882 I. M. Setchenov wished to take his tuition, feeling the necessity to master the exact sciences.

These circumstances determined to some extent the nature of Lyapunov. The circle of his acquaintances was restricted to relatives and a small number of his mathematical colleagues. He did not have many outside interests and went to the theatre only once or twice a year to hear his brother's concerts. Here was a man of fanatic dedication to his work and studies from his college days onwards. And later, when in charge of the faculty at Kharkov University, he often appeared in the auditorium not having had a wink of sleep.

That is why the life of Lyapunov did not abound in bright worldly events and sharp changes. The purposes of his life were his work, his wife and . . . flowers. The whole plant kingdom was on display at his Kharkov apartment. And it was essentially refilled by saplings of exotic palm-trees, rubber plants, araucarias, brought by him from the botanical gardens of Palermo, Messina, Naples. These were the trophies from the 6th International Mathematical Congress of 1908 in Rome, where he personally became acquainted with prominent mathematicians such as representatives of the French school, Picard, Goursat and Hadamard, with whom he had been earlier acquainted, but only through scientific correspondence.

Lyapunov's wife was a woman of poor health and since the 1900s she had been suffering serious attacks of tuberculosis. On her doctor's orders Lyapunov moved with his wife to Odessa in mid-summer of 1917. There they got the news that their estate on the Volga river had burnt down, together with the great library founded by his father and grandfather. The disease of his wife was progressing, the health of Lyapunov himself left much to be desired (his eyesight had deteriorated), the tragic events in the country, and the impossibility to sacrifice all of himself to work had an oppressive effect on Lyapunov. And when on 31 October 1918 Natalia died, he shot himself in his head with a pistol. The wound turned out to be fatal, and on 3 November 1918, the day his wife was buried, A. M. Lyapunov died. In his suicide note he asked to be buried together with his wife.

Thus tragically ended the life of the individual and mathematician, faithfully and deeply devoted to his science.

3. SCIENTIFIC ACTIVITIES

The name of Lyapunov is primarily connected with the so-called "Lyapunov's second method" for investigating the stability of dynamical systems. This was developed in his celebrated doctoral dissertation "The General Problem of the Stability of Motion" in 1892. Lyapunov established a criterion of stability or instability as a function in generalized coordinates (Lyapunov's function) which was not necessarily an energy integral as in the works of Lagrange and Dirichlet. Some regularity of signs of Lyapunov's function and its derivative with respect to time is essential in determining stability. Later this method turned out to be very convenient and a general-purpose tool for investigating stability, although Lyapunov himself attached much more importance to his first method ("solution by first approximation"), which was also introduced in his dissertation and investigated there in detail. Extremely intensive work on the dissertation took four years. The final stage of the work was complicated by an epidemic of cholera in the summer of 1891 on the Volga river at the place his family lived at that time. Not long before the prescribed date of the report one of the thesis opponents died, and the defence procedure had to be postponed. Nevertheless the work was defended with brilliance at Moscow University on 30 September 1892. One of the opponents at the defence procedure was the famous applied mathematician N. E. Zhukovskii, the founder of the Russian school of aerodynamics. Nowadays this thesis is considered to be a classical one and has appeared as a book in four editions. It was translated into French in 1908 and the French edition has been reprinted several times. An English translation of this French edition has just been published (Fuller, 1992).

Essential contributions were made by Lyapunov to the solution of important problems of mechanics. One such problem at that time was the problem of equilibrium figures of a rotating liquid, which was treated by Maclaurin, Jacobi and Laplace years earlier. In 1882 while discussing with young Lyapunov the subject of his Master's dissertation, Tchebyshev had proposed that he should investigate this problem in detail. "If you intend to be a serious mathematician, you should deal with serious problems", he said. Having encountered substantial technical difficulties, Lyapunov had to postpone the solution of this problem. But the work had not gone for nothing and his Master's dissertation was defended on the topic "On the Stability of Ellipsoidal Figures of Equilibrium of Rotating Liquid".

In the late 1890s Lyapunov published a number of papers on potential theory and probability theory, among the latter was the method of characteristic functions developed by him, which expanded the field of application of the central limit theorem and turned out to be an extremely suitable and effective tool to prove the convergence of a series of random variables. The most popular scientific journals of that time, closest to the areas of Lyapunov's interest, were the French ones, and that is why from 1896 onwards he published his works in French, even in Russian journals.

In 1901 being formally introduced by Academician A. A. Markov, Lyapunov was elected an Academician in the Department of Applied Mathematics at St. Petersburg, a vacancy due to Tchebyshev's death. The election was made on account of his works in potential theory, on the Dirichlet problem and probability theory. At that time his fundamental work on the stability of motion had not received proper recognition at the Academy of Sciences.

After moving to St. Petersburg in May 1902, Lyapunov reverted to the problem of equilibrium figures. In 1902 the work of H. Poincaré on equilibrium figures had appeared, and in 1902–1903 the British astronomer G. H. Darwin, the son of the great Charles Darwin, and head of Newton's faculty in Cambridge, also the president of the Royal Astronomical Society, published his papers on the same topic. He proved there the stability of so-called "piriforms"—equilibrium figures in the shape of a pear, a fact that fitted perfectly the cosmogony theory developed by him. However in 1905 Lyapunov proved the mathematical feasibility of pear-shaped figures, but rejected the possibility of their stable existence. This extremely complicated work was done with absolute conscientiousness, for instance, some calculations were made to an accuracy of 14 significant digits! The results of his work on equilibrium figures Lyapunov published in a huge treatise (about 1000 pages, consisting mainly of formulas). Faultless rigour, thoroughness of the arguments, masterful operation with the exquisite mathematical technique peculiar to most of Lyapunov's work, reached their peak in this investigation. But the complexity of the analytical apparatus and the breadth of scope of the problem made the work hard to read. Despite the exhaustive conclusions, scientific dispute continued; Darwin and Lyapunov verified their results and could not find any contradictions. In 1911 Poincaré said: "Piriforms may be stable, but there is no confidence that it is really so. Darwin finds this figure stable, but as Lyapunov contends, it is unstable. In order to resolve the problem finally, one should make calculations once more, but they are extremely complicated". Only in 1917 did the expert opinion of the brilliant British astronomer J. H. Jeans put an end to this dispute, and confirm the correctness of Lyapunov's concept.

Lyapunov's merits were highly appreciated in the scientific community. In addition to honorary degrees from St. Petersburg, Kharkov and Novorossiisk (Odessa) Universities, he was admitted to the Mathematical Society of Palermo, the Academy of Sciences dei Lincei in Rome, and in 1916 he gained Corresponding Membership of the Paris Academy of Sciences.

4. PEDAGOGIC AND PUBLIC ACTIVITIES

Despite the fact that Lyapunov himself affirmed that in giving lectures he was not in his element, his pedagogic activity was quite diverse and fruitful, Krylov (1919). Since 1883, being attached to St. Petersburg University for preparing his Master's dissertation, Lyapunov gave lectures there for about a year as an assistant. After he had defended his thesis in 1885 he was appointed to the Chair of Mechanics in Kharkov University. Here he gave some lecture courses which became very popular later. After the murder of the Tsar Alexander II there was a period of some reaction in Russia, which also affected the University Charter, and caused discontent among the students. That is why the news about the appointment of a new professor was met by the students with some suspicion, which was cleared away only after Lyapunov's first lecture. The level of narration and the demands on the audience were so high, that pupils did not ask questions, being afraid of looking ignorant. An "ambassador" was elected who transmitted to Lyapunov the questions of the whole class and who, consequently, had to blush for everyone. "Why do the students ask me for explanations so seldom, is everything really clear?", Lyapunov naively asked the "ambassador", V. A. Steklov.

For about five years Lyapunov did not have an opportunity to be engaged exclusively in scientific research since he was giving lectures at Kharkov University and preparing texts. These courses on potential theory, analytical mechanics, differential equations and probability theory were published lithographically later, and were popular with several generations of students. At the same time Lyapunov gave lectures in theoretical mechanics at the Kharkov Polytechnic. He managed to give several lectures in Novorossiisk University on the shape of celestial bodies but his death prevented him from finishing the course.

After moving to St. Petersburg Academician Lyapunov took part in the work of the Academy of Sciences with enthusiasm. There remained work on documents concerning his participation in very diverse, sometimes quite unexpected, commissions of the Academy of Sciences: on the publishing of the works of the great Euler; on the reform of the calendar in Russia*; on the problems of teaching mathematics in secondary schools. His report "About the measures for eliminating possible epidemics in case of military actions" (1915) is also widely known. In 1905 among 16 academicians Lyapunov signed the famous "note of 342 scientists"—an appeal of scientists concerning the disastrous situation in the field of enlightenment—that caused immense embarrassment to the Tsarist authorities.

According to the opinions of his contemporaries Lyapunov "· · · belonged to those professors who form the soul of the University, thanks to whom the University lives and prospers, and who bear the ideal of a professor and a scientist".

5. CURRICULUM VITAE OF LYAPUNOV

Born 25 May 1857 at Yaroslavl'
Died 3 November 1918 (new style*) at Odessa

1876 Graduated from the Gymnasium of Nizhnii Novgorod with a gold medal.
1876 Entered St. Petersburg University, Physics and Mathematics Department.
1880 Graduated from the University with a gold medal.
1885 Master's degree in Applied Mathematics.
1885 Appointed to the Department of Mechanics, Kharkov University as a lecturer (Assistant Professor).
1892 Doctoral dissertation in Applied Mathematics.
1893 Professor-in-Ordinary of Mechanics at Kharkov.
1901 Elected Academician-in-Ordinary at St. Petersburg.

* Before 1918 Russia used the Julian calendar giving "old style" dates. In the 19th. century this calendar was 12 days behind, and in the 20th. century 13 days behind, the Gregorian calendar is used elsewhere.

Memberships

1899: Chairman of Kharkov Mathematical Society;
1900, Dec: Corresponding Member of St. Petersburg Academy of Sciences.
1901, Oct: Academician of St. Petersburg Academy of Sciences.
1907, Dec: Membership in Mathematical Society of Palermo, Italy.
1908: Membership in Rome Academy of Sciences (*dei Lincei*).
1911–1913: Honorary Membership in Kharkov and St. Petersburg Universities.
1916: Corresponding Member of Paris Academy of Sciences.
1917: Honorary Membership in Odessa University.

Publications:
More than 40 articles in his lifetime and, after his death, several editions of Selected papers, Lectures in Theoretical Mechanics, Collected papers in five volumes, Doctoral dissertation.

Teaching experience:
Kharkov, Petersburg, Odessa Universities.

Contributions to science:
Founder of contemporary Stability Theory, Mechanics (Equilibrium figures of rotating liquid), Probability Theory (Central Limit Theorem), Mathematical Physics (Dirichlet problem, Potential Theory).

Concepts and terms named after Lyapunov

(1) *Stability Theory and Differential Equations*:
Stability in Lyapunov sense, Asymptotically stable solution, Lyapunov's function (Lyapunov's 2nd method), Lyapunov's characteristic index, Lyapunov's transformation, Lyapunov exponents.
(2) *Probability Theory and Stochastic Processes*:
Lyapunov's theorem (Central Limit Theorem), Method of Lyapunov's characteristic functions, Lyapunov's stochastic function.
(3) *Mechanics and Mathematical Physics*:
Lyapunov's theorem (in potential theory), Lyapunov's surfaces and curves.

Inscription on Lyapunov's gravestone
"The creator of the theory of Stability of Motion, the doctrine of equilibrium figures of rotating liquid, methods of the qualitative theory of Differential Equations, the author of the Central Limit Theorem, and other deep investigations in the areas of mechanics and mathematical analysis".

6. LYAPUNOV'S WORKS

List of scientific papers published during Lyapunov's life (according to Smirnov, 1953), collected papers and special re-issues of his most important works are presented (Tsykalo, 1988). Lithographic lecture courses, various documents and reports are not included.

1881 On the equilibrium of heavy bodies in heavy liquids contained in a vessel of a certain shape. (Candidate dissertation). *Zhurn. Russk. Fiz.-Khim. Obsh.*, **XIII,** 5, *fiz. otd.*, 197–238; 6, 273–307 (in Russian).

1881 On the potential of hydrostatic pressures. *Ibid.*, **XIII,** 8, *fiz. otd.*, 349–376 (in Russian).

1884 On the stability of ellipsoidal figures of equilibrium of a rotating liquid. Treatise of the Master's degree in Applied Mathematics. St. Petersburg, *Akad. Nauk.*, **XV,** 109 (in Russian).

1885 Some generalizations of the Lejeune–Dirichlet formula for the potential function of the ellipsoid to the case of the inner point. *Soobsh. i prot. zased. Mat. Obsh. Khark. Univ.*, 2, 120–130 (in Russian).

1886 On the body of maximum potential. *Ibid.*, 2, 63–73 (in Russian).

1888 On the steady screw motions of a rigid body in a liquid. *Soobsh. Khark. Mat. Obsh.*, *2nd ser.*, **1,** 1–2, 7–60 (in Russian).

1889 On the stability of motion in one particular case of the three-body problem. *Ibid.*, **II,** 1–2, 1–94 (in Russian).

1892 General problem of the stability of motion. Doctoral dissertation. Kharkov Mathematical Society, **XI,** 250 pp. (in Russian).

1893 Investigation of one of the special cases of the problem of the stability of motion. *Mat. sb.*, **XVII,** 2, 253–333 (in Russian).

1893 On a question of the stability of motion. *Zap. Khark. Univ.*, **1,** 99–104 (in Russian).

1893 New case of integration of differential equations of the motion of a rigid body in a liquid. *Soobsh. Khark. Mat. Obsh.*, *2nd ser.*, **IV,** 1–2, 81–85 (in Russian).

1894 On a property of differential equations for the problem of motion of a heavy rigid body with a fixed point. *Ibid.*, **IV,** 3, 123–140 (in Russian).

1895 Some words concerning the paper by G. G. Appel'rot "In connection with a paragraph of the first memoir by S. V. Kovalevskaya "Sur le problème de la rotation d'un corps solide autour d'un point fixé" ". *Ibid.*, **IV,** 5–6, 292–297 (in Russian).

1896 On the series proposed by Hill for representation of the motion of the Moon. *Trudy Otd. Fiz. nauk Obsh. lyubit. yestestvozn.*, **VIII,** 1, 1–23 (in Russian).

1896 On a question concerning second order linear differential equations with periodic coefficients. *Soobsh. Khark. Mat. Obsh.*, *2nd ser.*, **V,** 3–6, 190–254 (in Russian).

1896 Sur une série relative à la théorie des équations différentielles linéaires à coefficients périodiques. *Computes rendus de l'Acad. des sciences,* t., **CXXIII,** 1248–1252.

1897 Sur l'instabilité de l'équilibre dans certains cas où la fonction de forces n'est pas un maximum. *J. de mathém. pures et appl.*, 5e *série,* **III,** 81–94.

1897 Sur le potentiel de la double couche. *Comptes rendus de l'Acad. des sciences,* **CXXV,** 694–696.

1897 Sur le potential de la double couche. *Soobsh. Khark. Mat. Obsh.*, *2nd ser.*, **VI,** 2–3 129–138.

1897 Sur certaines questions se rattachant au problème de Dirichlet. *Comptes rendus de l'Acad. des sciences,* **CXXV,** 808–810.

1898 Sur certaines questions qui se rattachent au problème de Dirichlet. *J. de mathém. pures et appl.* (*J. Liouville*), 3e *série,* **IV,** 241–311.

1899 Sur une équation différentielle linéaire du second ordre. *Comptes rendus de l'Acad. des sciences,* **CXXVIII,** 911–913.

1899 Sur une équation transcendante et les équations différentielles linéaires du second ordre à coefficients périodiques. *Ibid.*, **CXXVII,** 1085–1088.

1900 Sur une proposition de la théorie des probabilités. *Izv. Akad. Nauk, 5th ser.* **XIII,** 359–386.

1900 Sur une série relative à la théorie d'une équation différentielle linéaire du second ordre. *Comptes rendus de l'Acad. des sciences,* **CXXXI,** 1185–1188.

1901 Novelle forme du théorème sur la limite de probabilité. *Zap. Akad. Nauk po Fiz.-Mat. otd.*, *8th ser.*, **XII,** 5, 1–24.

1901 Sur un théorème du calcul des probabilités. *Comptes rendus de l'Acad. des sciences,* **CXXXII,** 126–128.

1901 Une proposition générale du calcul de probabilités. *Ibid.*, **CXXXII,** 814–815.

1901 Reply to P. A. Nekrasov. *Zap. Khark. Univ.*, 3, 51–63 (in Russian).

1902 Sur une série dans la théorie des équations différentielles linéaires du second ordre à coefficients périodiques. *Zap. Akad. Nauk po Fiz.-Mat. otd.*, *8th ser.*, **XIII,** 2, 1–70.

1902 Sur le principe fondamental de la méthode

de Neumann dans le problème de Dirichlet. *Soobsh. Khark. Mat. Obsh., 2nd ser.*, **VII,** 4–6, 229–252.
1903 Recherches dans la théorie de la figure des corps célestes. *Zap. Akad. Nauk po Fiz.-Mat. otd., 8th ser.*, **XIV**, 7, 1–37.
1904 Sur l'équation de Clairaut et les équations plus générales de la théorie de la figure des planètes. *Ibid.*, **XX,** 10, 1–66.
1905 Sur un probléme de Tchébycheff. *Ibid.*, **XVII,** 3, 1–32.
1906 Sur les figures d'équilibre peu différentes des ellipsoides d'une masse liquide homogène douée d'un mouvement de rotation. Premier partie: Étude générale du problème. St. Petersburg, *Acad. Sci.*.
1908 Problème de minimum dans une question de stabilité des figures d'équilibre d'une masse fluide en rotation. *Zap. Akad. Nauk po Fiz.-Mat. otd., 8th ser.*, **XXII,** 5, 1–140.
1909 Sur les figures d'équilibre peu différentes des ellipsoides d'une masse liquide homogène douée d'un mouvement de rotation. Deuxième partie: Figures d'équilibre dérivées des ellipsoides de Maclaurin. St. Petersburg, *Acad. Sci.*.
1909 Sur une classe de figures d'équilibre d'un liquide en rotation. *Ann. scientif. de l'Éc. norm. supér., 3e série* **XXVI,** 473–483.
1912 Sur les figures d'équilibre peu différentes des ellipsoides d'une masse liquide homogène douée d'un mouvement de rotation. Troisième partie: Figures d'équilibre dérivées des ellipsoides de Jacobi. (Recherches relatives à la vitesse angulaire et au moment des quantités de mouvement). St. Petersburg, *Acad. Sci.*.
1914 Sur les figures d'équilibre peu différentes des ellipsoides d'une masse liquide homogène douée d'un mouvement de rotaton. Quatrième partie: Nouvelles formules pour la recherche des figures d'équilibre. St. Petersburg, *Acad. Sci.*.
1915 Sur les séries de polynômes. *Izv. Akad. Nauk, 6th ser.* **IX,** 17, 1857–1868.
1916 Sur les équations qui appartiennent aux surfaces des figures d'équilibre dérivées des ellipsoides d'un liquide homogène en rotation. *Ibid.*, **X,** 3, 139–168.
1916 Nouvelles considérations relatives à la théorie des figures d'équilibre dérivées des ellipsoides dans le cas d'un liquide homogène. 1–2. *Ibid.*, **X,** 7, 471–502; 8, 589–620.
1917 Sur une formule d'analyse. *Ibid.*, **XI,** 2, 87–118.
1925 Sur certain séries de figures d'équilibre d'un liquide hétérogène en rotation. Premier partie. Leningrad, *Acad. Sci.*.
1927 Sur certain séries de figures d'équilibre d'un liquide hétérogène en rotation. Deuxième partie. Leningrad, *Acad. Sci.*, 225–437.
1930 On the shape of celestial bodies. *Izv. Akad. Nauk SSSR, 7th ser., Otd. Fiz.-Mat. nauk*, 1, 25–41 (in Russian).
1935 General problem of the stability of motion. (Dissertation and papers). 2nd ed. M.-L., Ob'ed. *Nauchno-Tekhn. Izd.*, 386 (in Russian).
1948 Selected papers. ed. Acad. V. I. Smirnov, M., *Akad. Nauk SSSR*, 540 (in Russian).
1949 Works on potential theory. (Translation from French by O. D. Kanevskaya, N. O. Rakhlenko), ed. N. I. Akhiezer, G. I. Drinfel'd, M.-L., *Gos. Izd. Tekhn.-Teor. Lit.*, 178 (in Russian).
1950 General problem of the stability of motion. 3rd ed. M.-L., *Gos. Izd. Tekhn.-Teor. Lit.*, 471 (in Russian).
1954 Collected works in 5 volumes. M., Acad. Sci., 1954–1965 (in Russian).
1982 Lectures on theoretical mechanics. Kiev. *Naukova Dumka*, 632 (in Russian).

There are several of Lyapunov's works which are still unpublished. One can obtain some information about them from the following papers:

(1) Smirnov, V. I. (1956). Scientific archives of A. M. Lyapunov on stability and the theory of ordinary differential equations. *Proc. 3rd All-Union Math. Congress, M.*, **1,** (in Russian).
(2) Sretenskii, L. N. (1958). Unpublished manuscripts of A. M. Lyapunov. *Ibid., M.*, **3,** 490–500 (in Russian).
(3) Alexander Mikhailovitch Lyapunov: scientific papers, mainly unpublished and materials for them. *Trudy AAN SSSR* (1959) **4,** 16, 12–14 (in Russian).

Acknowledgement—The author thanks with gratitude the assistance of *Automatica* Editor P. C. Parks in the preparation of this paper.

REFERENCES

Anon. (1915). *Materials for the Biographic Dictionary of the Members of the Imperial Academy of Sciences*, Vol. 1. Petersburg (in Russian).
Fuller, A. T. (1992). Translation from French of Lyapunov's thesis of 1892. *Int. J. Control*, **55,** 581–773.
Krylov, A. N. (1919). Alexander Mikhailovitch Lyapunov (speech at the public session of Russian Academy of Sciences, 3 May 1919). *Izv., Ross. Akad. Nauk*, **13,** 8–11, 389–394 (in Russian).
Lyapunov, B. M. (1930). Brief essay on the life and activities of A. M. Lyapunov. *Izv. Akad. Nauk SSSR*, **1,** 1–24 (in Russian).
Shibanov, A. S. (1985). *Alexander Mikhailovitch Lyapunov*. M.–L., Molodaya Gvardiya, 336 (in Russian).

Smirnov, V. I. (Ed.) (1953). *Alexander Mikhailovitch Lyapunov*. Bibliography. Compiler A. M. Lukomskaya, M.–L., Academy of Sciences of U.S.S.R., 268 (in Russian).

Steklov, V. A. (1919). Alexander Mikhailovitch Lyapunov. (speech at the public session of Russian Academy of Sciences, 3 May 1919). *Izv. Ross. Akad. Nauk,* **13,** 8–11, 367–388 (in Russian).

Tsykalo, A. L. (1988). *Alexander Mikhailovitch Lyapunov*. M., Nauka, 247 (in Russian).

APPENDIX

Gos. Izd. Tekhn.-Teor. Lit. = State techn.-Theor. Publ.
Izv. Akad. Nauk = Proc. Acad. Sci.
M = Moscow.
M.–L. = Moscow–Leningrad.
Mat. sb. = Math. Trans.
Molodaya Gvardiya = Young Guards.
Naukova Dumka (Ukraine) = Scientific Thought.
Ob'ed. Nauchno-Tekhn. Izd. = United Sci.-Techn. Publ.
Soobsh. i prot. zased. Mat. Obsh. Khark. Univ. = Comm. and Proc. of Math. Society of Kharkov University.
Soobsh, Khark. Mat. Obsh. = Comm. of Kharkov Math. Society.
Trudy AAN SSSR = Proc. Archives Acad. Sci. U.S.S.R.
Trudy Otd. Fiz. nauk Obsh. lyubit. yesterstvozn. = Proc. Soc. Natural Sci. Lovers, Sect. Phys. Sci.
Zap. Akad. Nauk po Fiz.-Mat. otd. = Notes Acad. Sci., Phys.-Math. Section.
Zap. Khark. Univ. = Notes of Kharkov University.
Zhurn. Russk. Fiz.-Khim. obsh. = J. of Russian Phys.–Chem. Society.

Nikolai Nikolaevich Luzin

Born – 9 Dec 1883 in Irkutsk, Russia
Died – 25 Feb 1950 in Moscow, USSR

Nikolai Nikolaevich Luzin was born in Irkutsk. His father was a businessman, half Russian and half Buryat. Nikolai was the only son and the family moved to Tomsk when he was about eleven years old so that he could attend the Gymnasium there.

One might expect that Luzin would have shown a special talent for mathematics at the Gymnasium, but this was far from the case. This was because the system of instruction was based on mechanical memory: it was required to learn the theorems by heart and to reproduce their proofs exactly. For Luzin this was torture. His progress in mathematics at the Gymnasium became worse and worse, so that this father was obliged to engage a tutor. Fortunately the tutor was a talented young man who quickly discovered that, despite Luzin's poor performance in mathematics, he could solve hard problems but often using a novel method that the tutor had never seen before. Soon the tutor had shown Luzin that mathematics was not a subject where one had to learn long lists of facts, but a topic where creativity and imagination played a major role.

In 1901 Luzin left Gymnasium and at this time his father sold his business and the family moved to Moscow. There Luzin entered the Faculty of Physics and Mathematics at Moscow University intending to train to become an engineer. At first Luzin lived in the new family home in Moscow, but Luzin's father began to gamble on the stock exchange with the money he had made from the sale of his business. The family soon hit hard times as Luzin's father lost all their savings and the family had to leave their home. Luzin, together with a friend, moved into a room owned by the widow of a doctor. His friend soon became involved with the Revolution and was forced into hiding. Luzin stayed on by himself in the room but he clearly got on well with the owners since he later, in 1908, married the widow's daughter.

At Moscow University Luzin studied under Bugaev, learning from him the theory of functions which was to influence greatly the direction his research would eventually take. However he was only an average student who seemed to show little flair for mathematics. However, although Luzin appeared to lack talent in mathematics, one of his teachers Egorov spotted his great talent, invited him to his home, and began to set him hard problems. It was until 1909 that Luzin seems to have finally committed himself completely to mathematics. Under Egorov's supervision he worked on his master's thesis. In 1910 he was appointed as assistant lecturer in Pure Mathematics at Moscow University. He worked for a year with Egorov and they went on to publish joint papers on function theory which mark the beginnings of the Moscow school of function theory.

In 1910 Luzin travelled abroad visiting Göttingen where he was influenced by Landau. He returned to Moscow in 1914 and he completed his thesis *The integral and trigonometric series* which he submitted in 1915. After his oral examination he was awarded a doctorate, despite having submitted his thesis for the Master's Degree. Egorov was extraordinarily impressed by the work and had pressed for the award of the doctorate. In 1917 Luzin was appointed as Professor of Pure Mathematics at Moscow University just before the Revolution. The Revolution caused Luzin rethink some of the same thoughts as he had done at the time of his crisis. By this stage, however, his mathematical career was extremely successful and the second crisis did not materialize.

Over the ten years Luzin and Egorov built up and impressive research group at the University of Moscow which the students called "Luzitania". The first students included P. S. Aleksandrov, M. Ya Suslin, D. E. Menshov and A. Ya Khinchin. The next students included P. S. Urysohn, A. N. Kolmogorov, N. K. Bari, L. A. Liusternik and N. G. Shnirelman. In 1923 P. S. Novikov and L. V. Keldysh joined the group.

Luzin's main contributions are in the area of foundations of mathematics and measure theory. He also made significant contributions to descriptive set topology. In the theory of boundary properties of analytic functions he proved an important result in 1919 on the invariance of sets of boundary points under conformal mappings. He also studied, together with Privalov, boundary uniqueness properties of analytic functions. From 1917 onwards, Luzin studied descriptive set theory. Much of his work on set theory involved the study of effective sets, that is sets which can be constructed without the axiom choice.

In 1927 Luzin was elected as a member of the Academy of Science of the USSR. Two years later he became a full member of first the Department of Philosophy, then to the Department of Pure Mathematics. He worked from this time until his death in the Academy of Sciences of the USSR. From 1935 he headed the Department of the theory of Functions of Real Variables at the Steklov Institute.

In 1931 Luzin himself turned to a new area when he began to study differential equations and their application to geometry and to control theory. His work in this area led him to study the bending of surfaces.

Luzin always had an interest in the history of mathematics and late in his career he wrote important articles on Newton and Euler.

THÉORIE DES FONCTIONS. — *Sur l'absolue convergence des séries trigonométriques.* Note de M. N. **Lusin**, présentée par M. Appell.

La lecture de la Note de M. Denjoy (*Comptes rendus*, 8 juillet 1912) m'a inspiré l'idée de résumer ici les résultats sur la convergence des séries trigonométriques déjà publiés par moi en russe dans le *Recueil de la Société mathématique de Moscou* (t. XXVIII, 3-24 février 1912), dans le Mémoire intitulé : *Sur l'absolue convergence des séries trigonométriques.*

Je me permets d'abord de reproduire une démonstration du théorème général, qui me paraît plus simple que celle de M. Denjoy.

THÉORÈME. — *Si une série trigonométrique* $\frac{a_0}{2} + \sum_{n=1} a_n \cos nx + b_n \sin nx$ *est absolument convergente pour tous les points* x *d'un ensemble de mesure* non nulle, *les deux séries* $\sum_{n=1}^{\infty} |a_n|$ *et* $\sum_{n=1}^{\infty} |b_n|$ *sont convergentes en même temps.*

Soit, en effet, la série $\frac{|a_0|}{2} + \sum_{n=1}^{\infty} \rho_n |\cos(nx + \alpha_n)|$, $\rho_n = \sqrt{a_n^2 + b_n^2}$ convergente pour tous les points x d'un ensemble E ayant la mesure non nulle. D'après un théorème important de M. Egoroff (*Comptes rendus*, 30 janvier 1911), il existe un ensemble P parfait et non dense et jouissant des propriétés suivantes :

1° La série $\frac{|a_0|}{2} + \sum_{n=1} \rho_n |\cos(nx + \alpha_n)|$ converge *uniformément* pour cet ensemble P;

2° La mesure de P est *non nulle* (soit Mes. P $= p > 0$).

Désignons par $S(x)$ la somme de la série. On voit bien que pour cet ensemble P on peut écrire, les intégrales étant prises au sens de M. Lebesgue,

$$\int_P S(x)\, dx = \frac{|a_0|}{2} p + \sum_{n=1}^{\infty} \rho_n \int_P |\cos(nx + \alpha_n)|\, dx.$$

D'autre part, on a toujours

$$\int_P |\cos(nx+\alpha_n)|\,dx \geqq \frac{p}{2}\sin\frac{p}{8} \qquad (n=1,2,3,\ldots).$$

Par conséquent,

$$\sum_{n=1}^{\infty}\rho_n \leqq \frac{1}{\frac{p}{2}\sin\frac{p}{8}}\int_P S(x)\,dx.$$

Donc la série $\sum_{n=1}^{\infty}\rho_n$ est convergente (C. Q. F. D.).

On peut de plus montrer que *la présence* de points de convergence absolue exerce une influence considérable sur la convergence *non absolue* d'une série trigonométrique.

D'abord, suivant M. Fatou (*Acta mathematica*, t. XXX, p. 398), les points de convergence, de divergence et de convergence absolue sont deux à deux symétriques par rapport aux points de convergence absolue.

D'autre part, on s'assure facilement que tout ensemble mesurable de points situés sur le segment $(0, 2\pi)$ et ayant une infinité de points de symétrie est ou de mesure 0, ou de mesure 2π.

Nous trouvons ainsi ce théorème général :

Toute série trigonométrique ayant dans l'intervalle $(0, 2\pi)$ *un ensemble dénombrable de points de convergence absolue est ou partout convergente ou partout divergente, sauf peut-être les points d'un ensemble de mesure nulle.*

Si la série a *deux* points de convergence absolue dont la différence des arguments est incommensurable à π, on en déduira l'existence de tels points dans tout intervalle. Comme conséquence nous obtiendrons :

Toute série trigonométrique ayant deux *points de convergence absolue dont la distance est incommensurable à* π, *est ou partout convergente ou partout divergente, sauf peut-être les points d'un ensemble de mesure nulle.*

Par exemple, on voit bien que la série $\sum_{n=1}^{\infty} b_n \sin 2^n x$ a toujours l'ensemble dénombrable de points de convergence absolue.

Par conséquent :

Il existe des séries trigonométriques incomplètes qui sont, quels que soient leurs coefficients, ou partout convergentes ou partout divergentes, sauf peut-être les points d'un ensemble de mesure nulle.

C'est pourquoi il serait très intéressant de construire un exemple de série trigonométrique qui soit convergente pour un ensemble de mesure *non nulle* et *inférieure* 2π.

Il semble que, *si une telle série existe*, elle ne sera pas une série de Fourier proprement dite. Il est infiniment probable que toute série de Fourier, correspondante à une fonction à carré sommable, est partout convergente, sauf peut-être les points d'un ensemble de mesure nulle.

ANALYSE MATHÉMATIQUE. — *Sur la convergence des séries trigonométriques de Fourier*. Note de M. N. LUSIN, présentée par M. Appell.

1° Dans la Note présente je me propose de déduire une condition nécessaire et suffisante pour la convergence presque partout de toute série trigonométrique de Fourier d'une fonction à carré sommable.

Considérons deux séries trigonométriques conjuguées

$$\sum_{n=1}^{\infty} a_n \cos nx + b_n \sin nx \tag{1}$$

et

$$\sum_{n=1}^{\infty} - b_n \cos nx + a_n \sin nx, \tag{2}$$

où la série $\sum_{n=1}^{\infty} a_n^2 + b_n^2$ est convergente. D'après un théorème de MM. Fischer et Riesz, ces séries trigonométriques sont les séries de Fourier de deux fonctions à carré sommable $f(x)$ et $g(x)$, ce qui donne lieu aux égalités

$$(3)\quad \begin{cases} a_n = \frac{1}{\pi}\int_0^{2\pi} f(\alpha)\cos n\alpha\, d\alpha = \frac{1}{\pi}\int_0^{2\pi} g(\alpha)\sin n\alpha\, d\alpha, \\ b_n = \frac{1}{\pi}\int_0^{2\pi} f(\alpha)\sin n\alpha\, d\alpha = -\frac{1}{\pi}\int_0^{2\pi} g(\alpha)\cos n\alpha\, d\alpha \end{cases}$$

Donc, en désignant par $S_n(x)$ la somme

$$\sum_{k=1}^{k=n} a_k \cos kx + b_k \sin kx,$$

nous pouvons écrire

$$S_n(x) = \frac{1}{\pi}\int_0^{\pi} [g(x+\alpha) - g(x-\alpha)] \left[\frac{1}{2\operatorname{tang}\frac{\alpha}{2}} - \frac{\cos\left(n+\frac{1}{2}\right)\alpha}{2\sin\frac{\alpha}{2}} \right] d\alpha.$$

De là, *il suffit* pour la convergence presque partout dans $(0, 2\pi)$ de la série (1) de Fourier qu'on ait simultanément :

$$\text{I.}\qquad \frac{1}{\pi}\int_0^{\pi} \frac{g(x+\alpha) - g(x-\alpha)}{2\operatorname{tang}\frac{\alpha}{2}}\, d\alpha = f(x)$$

presque partout, l'intégrale étant définie comme $\lim\limits_{\varepsilon=0}\int_\varepsilon^\pi$;

$$\text{II.}\qquad \lim_{n=\infty}\int_0^{\pi} \frac{g(x+\alpha) - g(x-\alpha)}{\alpha}\cos n\alpha\, d\alpha = 0$$

presque partout.

Nous allons démontrer que ces deux conditions I et II sont en même temps *nécessaires* pour la convergence presque partout de la série (1) de Fourier.

2° Considérons d'abord la nature de la condition I. Supposons pour cela $g(x)$ une fonction continue dans $(0, 2\pi)$. Il est évident que l'existence de l'intégrale $\int_0^{\pi} \frac{g(x+\alpha) - g(x-\alpha)}{2\operatorname{tang}\frac{\alpha}{2}}\, d\alpha$ est équivalente à l'existence de l'intégrale $\int_0^{\pi} \frac{g(x+\alpha) - g(x-\alpha)}{\alpha}\, d\alpha$. En général, $\left|\frac{g(x+\alpha) - g(x-\alpha)}{\alpha}\right|$ ne restera pas bornée lorsque α tend vers zéro. En effet, on peut construire une fonction *continue* $g(x)$ telle qu'on a

$$\int_0^\pi \left| \frac{g(x+\alpha)-g(x-\alpha)}{\alpha} \right| d\alpha = +\infty$$

pour un ensemble de points x de mesure *non nulle*. Nous allons, au contraire, montrer que l'intégrale $\int_0^\pi \frac{g(x+\alpha)-g(x-\alpha)}{\alpha} d\alpha$ existe toujours presque partout et cela quelle que soit la fonction $g(x)$ à carré sommable. Donc cette intégrale doit son existence à quelque sorte de l'*interférence* des valeurs infinies positives et négatives de la fonction $\frac{g(x+\alpha)-g(x-\alpha)}{\alpha}$ dans le voisinage de $\alpha = 0$. On peut considérer cette interférence comme la cause de la convergence de toutes les séries trigonométriques de Fourier.

3° Considérons la fonction harmonique

$$\sum_{n=1}^{\infty} \rho^n (a_n \cos nx + b_n \sin nx).$$

D'après les formules (3) on a identiquement, pour $0 \leqq x \leqq 2\pi$ et $0 \leqq \rho < 1$,

$$\frac{1}{2\pi}\int_0^{2\pi} f(\alpha) \frac{1-\rho^2}{1-2\rho\cos(\alpha-x)+\rho^2} d\alpha = \frac{1}{2\pi}\int_0^{2\pi} g(\alpha) \frac{2\rho\sin(\alpha-x)}{1-2\rho\cos(\alpha-x)+\rho^2} d\alpha.$$

L'intégrale du premier membre est une intégrale de Poisson qui correspond à une fonction à carré sommable; par conséquent, cette intégrale tend vers $f(x)$ presque partout quand ρ tend vers 1. De là, suivant l'analyse de M. Fatou (*Acta mathematica*, t. XXX, p. 360), et en vertu de « C-propriété » de la fonction $g(x)$ (ma Note, *Comptes rendus*, 17 juin 1912), on démontre sans difficulté que l'intégrale $\frac{1}{\pi}\int_\varepsilon^\pi \frac{g(x+\alpha)-g(x-\alpha)}{2\tan\frac{\alpha}{2}} d\alpha$ tend vers $f(x)$ presque partout quand ε tend vers zéro pour toute fonction $g(x)$ à carré sommable.

4° La condition I étant satisfaite pour toute fonction $f(x)$ à carré sommable, on voit bien que la condition II est nécessaire et suffisante pour la convergence presque partout de la série de Fourier (1). Il est infiniment probable que la condition II est de même réalisable pour toute fonction $g(x)$ à carré sommable et, par conséquent, que toute série trigonométrique de Fourier qui correspond à une fonction à carré sommable, est toujours convergente presque partout. Il faut ici signaler une différence importante entre le cas de fonction *sommable* et le cas de fonction *à carré sommable :* en effet, on peut construire une fonction *sommable*

$$f(x) \sim \sum_{n=1}^{\infty} a_n \cos n x + b_n \sin n x$$

telle que la somme de la série conjuguée

$$\sum_{n=1}^{\infty} - b_n \cos n x + a_n \sin n x$$

est une fonction *non sommable* dans aucun intervalle de $(0, 2\pi)$ et, par conséquent, non totalisable au sens de M. Denjoy.

ANALYSE MATHÉMATIQUE. — *Sur une décomposition d'un intervalle en une infinité non dénombrable d'ensembles non mesurables.* Note de MM. **W. Sierpiński** et **N. Lusin**, présentée par M. Hadamard.

Le but de cette Note est de démontrer l'existence d'*une décomposition de l'intervalle* (o, 1) *en une infinité de puissance du continu d'ensembles sans points communs deux à deux et tels que la mesure extérieure de chacun de ces ensembles est égale à* 1. La démonstration de cette existence utilisera le théorème de M. Zermelo, d'après lequel il existe un ensemble bien ordonné de puissance du continu.

Soit Ω_0 le plus petit nombre transfini tel que l'ensemble de tous les nombres transfinis $\alpha < \Omega_0$ ait la puissance du continu. Il existe donc un ensemble bien ordonné du type Ω_0

(1) $$x_1, \quad x_2, \quad x_3, \quad \ldots, \quad x_\omega, \quad \ldots, \quad x_\alpha, \quad \ldots \quad (\alpha < \Omega_0)$$

contenant une et une seule fois tout point x de l'intervalle (o, 1). Or, l'ensemble de tous les ensembles parfaits contenus dans (o, 1) ayant la puissance du continu (et d'après $c.c = c$, c désignant la puissance du continu), il existe un ensemble bien ordonné du type Ω_0

(2) $$P_1, \quad P_2, \quad P_3, \quad \ldots, \quad P_\omega, \quad \ldots, \quad P_\alpha, \quad \ldots \quad (\alpha < \Omega_0)$$

contenant tout ensemble parfait de l'intervalle (o, 1) une infinité de puissance du continu de fois.

Prenons maintenant le premier point p_1 de la suite (1) contenu dans l'ensemble P_1, ensuite le premier point p_2 de la suite (1) contenu dans P_2 et autre que p_1, puis le premier point p_3 de (1) contenu dans P_3 et autre que p_1 et p_2, et ainsi de suite transfiniment. Généralement, α étant un nombre ordinal fini ou transfini donné $< \Omega_0$, désignons par p_α le premier point de la suite (1) contenu dans P_α et autre que tous les points

(3) $$p_1, \quad p_2, \quad \ldots, \quad p_\omega, \quad \ldots, \quad p_\xi, \quad \ldots \quad (\xi < \alpha).$$

[On voit sans peine que, pour tout nombre $\alpha < \Omega_0$ un tel point p_α existe dans P_α, puisque l'ensemble parfait P_α a la puissance du continu et (3) est un ensemble du type $\alpha < \Omega_0$, donc, d'après la définition du nombre Ω_0, de puissance inférieure à celle du continu.]

Nous avons donc ainsi défini un ensemble bien ordonné de points différents de (0, 1)

$$(4) \qquad p_1,\ p_2,\ p_3,\ \ldots,\ p_\omega,\ \ldots,\ p_\alpha \quad (\alpha < \Omega_0)$$

du type Ω_0.

Soit maintenant P un ensemble parfait, situé dans (0, 1). La suite (2) contenant P une infinité de puissance du continu de fois, l'ensemble des indices successifs α satisfaisant à la condition $P_\alpha = P$ sera évidemment du type Ω_0. Soient

$$\alpha_1,\ \alpha_2,\ \alpha_3,\ \ldots,\ \alpha_\omega,\ \ldots,\ \alpha_\beta,\ \ldots \quad (\beta < \Omega_0)$$

ces indices et posons

$$p_{\alpha_\beta} = q_\beta(P) \qquad (\text{pour } \beta < \Omega_0).$$

A tout ensemble parfait P de (0, 1) correspondra donc un ensemble bien ordonné de points différents

$$(5) \qquad q_1(P),\ q_2(P),\ \ldots,\ q_\omega(P),\ \ldots,\ q_\beta(P),\ \ldots \quad (\beta < \Omega_0)$$

du type Ω_0; l'ensemble (5) sera évidemment sous-ensemble de P et aux ensembles P inégaux correspondront toujours des ensembles (5) sans point commun.

Posons maintenant, pour tout nombre donné $\beta < \Omega_0$,

$$Q_\beta = \sum_{\mathfrak{P}} q_\beta(\mathfrak{P}),$$

la sommation s'étendant à tous les ensembles parfaits de (0, 1).

On voit sans peine que les ensembles Q_β et $Q_{\beta'}$ seront sans points communs pour $\beta \neq \beta'$ et que tout ensemble Q_β contiendra au moins un point de tout ensemble parfait de (0, 1).

Or on démontre sans peine que tout ensemble situé dans (0, 1) et contenant au moins un point de tout ensemble parfait de (0, 1) a la mesure extérieure (lebesguienne) égale à l'unité. [Soit en effet $\mathcal{E}$ un tel ensemble : si la mesure extérieure de $\mathcal{E}$ était < 1, on pourrait enfermer (au sens étroit) $\mathcal{E}$ dans une infinité dénombrable d'intervalles dont la somme des longueurs est < 1 : les points de (0, 1) non intérieurs à aucun de ces intervalles formeraient donc un ensemble fermé de mesure positive, donc contenant un ensemble parfait. Il existerait donc dans (o, 1) un ensemble parfait sans point commun avec $\mathcal{E}$, contrairement à la propriété de $\mathcal{E}$.]

Désignons encore par Q_0 l'ensemble de tous les points de (o, 1) (s'il existe) n'entrant dans aucun des ensembles Q_β ($\beta < \Omega_0$) (on pourrait d'ailleurs démontrer sans peine que l'ensemble Q_0 est nul) : les ensembles

$$Q_0 + Q_1,\ Q_2,\ Q_3,\ \ldots,\ Q_\omega,\ \ldots,\ Q_\beta,\ \ldots\quad (\beta < \Omega_0)$$

donneront donc la décomposition cherchée de l'intervalle (o, 1) en une infinité de puissance du continu d'ensembles sans points communs deux à deux, dont chacun a la mesure extérieure égale à 1. On voit aussi sans peine que la mesure intérieure de chacun de ces ensembles sera nulle.

Nous signalerons encore en quelques mots une voie un peu différente pour arriver à une décomposition de l'intervalle dont nous venons de parler.

Soient, comme plus haut, Ω_0 le plus petit nombre transfini correspondant à la puissance du continu et

$$S_1,\ S_2,\ S_3,\ \ldots,\ S_\omega,\ \ldots,\ S_\alpha,\ \ldots\quad (\alpha < \Omega_0)$$

une suite transfinie du type Ω_0 formée de tous les ensembles parfaits contenus dans (o, 1).

Prenons maintenant un point s'_1 dans S_1, deux nouveaux points s''_1 et s''_2 dans S_2, trois nouveaux points s'''_1, s'''_2 et s'''_3 dans S_3, et ainsi de suite transfiniment (généralement prenons dans S_α les points $s_1^{(\alpha)}, s_2^{(\alpha)}, \ldots, s_\alpha^{(\alpha)}$). Pour tout nombre ordinal donné $\alpha < \Omega_0$, les points

$$s_\alpha^{(\alpha)},\ s_\alpha^{(\alpha+1)},\ s_\alpha^{(\alpha+2)},\ \ldots,\ s_\alpha^{(\alpha+\omega)},\ \ldots,\ s_\alpha^{(\alpha+\beta)},\ \ldots\quad (\beta < \Omega_0)$$

formeront, comme on le voit sans peine, un ensemble $\mathcal{E}_\alpha$, possédant des points communs avec tout ensemble parfait de (o, 1) (puisque tout ensemble parfait contient une infinité de puissance du continu de sous-ensembles parfaits).

Les ensembles $\mathcal{E}_\alpha$ ($\alpha < \Omega_0$) sont de mesure extérieure égale à 1, sans points communs deux à deux.

HISTORIA MATHEMATICA **25** (1998), 332–339
ARTICLE NO. HM972182

SOURCE

The Influence of P. A. Florensky on N. N. Luzin

Charles E. Ford

Department of Mathematics and Computer Science, Saint Louis University, Saint Louis, Missouri 63103

The study of the early history of the Moscow school of the theory of functions has been greatly enhanced by the discovery of correspondence between one of its founders, N. N. Luzin, and the priest, theologian, philosopher, and scientist P. A. Florensky. The correspondence reveals that Luzin experienced a profound spiritual crisis in 1905 when his materialist worldview collapsed. This crisis continued for three years and was finally resolved when Luzin had a decisive encounter with the religious philosophy of Florensky. After this, Luzin's interest in mathematics gradually revived until he was able, by 1909, to commit himself to a career in mathematics.

Les études concernant l'histoire des débuts de l'école moscovite de la théorie des functions ont été grandement éclairées par la découverte d'une correspondance entre l'un de ses fondateurs, N. N. Luzin, et P. A. Florenski, un prêtre, théologien, philosophe, et savant. Cette correspondance révèle que Luzin a traversé une crise spirituelle profonde à partir de 1905, lorsque sa perspective matérialiste du monde s'est effondrée. Cette crise a duré trois ans, et a trouvé sa résolution dans la rencontre décisive de Luzin avec la philosophie religieuse de Florenski. Par la suite, l'intérêt de Luzin pour les mathématiques s'est progressivement ravivé, jusqu'à ce qu'il soit en mesure de décider, en 1909, de consacrer sa vie aux mathématiques.

Изучение ранней истории московкой школы теории функций было значительно продвинуто обнаружением корреспонденции между одним из её основателей Н. Н. Лузиным и священником, теологом, философом, и учёным П. А. Флоренским. Корреспонденция обнаруживает, что Лузин переживал глубокий духовный кризис в 1905 году, когда его материалистические представления о мире рушились. Этот кризис продолжался в течение трёх лет и в конце концов был разрешён когда Лузин позкокомился с религиозной философией Флоренского, что имело для него решающее значение. После этого, интерес Лузина к математике постоянно возрождался до тех пор пока он был в состоянии к 1909 году посветить себе математике.

MSC 1991 subject classification: 01A30, 01A60, 01A80, 03-03.
Key Words: religion; philosophy; materialism; Luzin; Florensky; Moscow.

INTRODUCTION

Nikolai Nikolaevich Luzin (1883–1950) is a major figure in the history of 20th-century mathematics [8]. He is particularly important as a founder of the Moscow school of the theory of functions [2, 35; 8, 277]. The early history of the Moscow school has been greatly illuminated by the recent discovery that a close relationship had developed between Luzin and the philosopher, theologian, and scientist, Pavel Aleksandrovich Florensky (1882–1937) [1; 6].

This paper aims to highlight, through their correspondence [4], the profound influence of Florensky on Luzin as the latter struggled to find an appropriate worldview. Like many intellectuals of the early 20th century, Luzin embraced a materialist worldview, which collapsed around the time of his graduation from Moscow University in 1905. The ensuing crisis engulfed him for the next four years and was ultimately resolved when Luzin had a decisive encounter with the religious philosophy of Florensky. This enabled Luzin to launch his career as a mathematician.

Florensky entered Moscow University in 1900 and Luzin in 1901, both as students in mathematics and physics. Florensky left the University upon graduation in 1904 to enter the Moscow Theological Academy. Upon graduation in 1908, he was appointed directly to the faculty in the history of philosophy. In the spring of 1908 he wrote the treatise "On Spiritual Truth." This was expanded into what became his first doctoral dissertation in 1914. It was published under the title "The Pillar and Foundation of Truth" [5] and has become one of the most famous works of modern Russian theology.

THE CORRESPONDENCE

For the preservation of this correspondence we are indebted to the family of Florensky, especially to his courageous wife, Anna Mikhailovna. After his arrest she preserved much of his archive, including many unpublished manuscripts, papers from his student years, and correspondence, including the letters he had received from Luzin. The friendship between Florensky and Luzin had extended to include their wives.

Toward the end of his life, Luzin burned his diaries. After his death in 1950, his wife could easily have destroyed his remaining papers. Fortunately, she chose instead to give the letters her husband had received from Florensky to his widow, along with other related correspondence. Thus, at least parts of both sides of the correspondence have been preserved.

The publication of this correspondence, by Sergei Demidov and his colleagues, is a major event in the history of mathematics in Moscow. The correspondence has been augmented with excellent footnotes and significant quotations from related correspondence, as well as an informative introduction [3]. Another paper based on this correspondence traces Luzin's early mathematical and philosophical development.

Most of the correspondence covers the period from the time Florensky left Moscow for the Theological Academy in Sergiev Posad in 1904 until he moved back to Moscow in January 1921. Luzin frequently visited Sergiev Posad (renamed Zagorsk during the Soviet period and located about 30 miles northeast of Moscow) and often spent the summer there. He came to love the town and especially valued the opportunity to meet with Florensky. We can find echoes of their conversations in the correspondence. The correspondence shows the powerful influence Florensky had on turning Luzin toward an interest in the "general questions" of philosophy and science.

One feature of the correspondence that is lost in translation is its formality. As with several European languages, Russian has two forms of the second person singular pronoun "you." One form is familiar, the other formal. In this correspondence the latter is always used. This is rather surprising, given that the two had been students together and enjoyed a long and intimate relationship. In this connection, it may be noteworthy that the signatures at the end of the letters always include last names. After being ordained, Florensky nearly always signed himself as "priest." Perhaps the formality reflects the somewhat unequal relationship between them, a sign of Luzin's deference to the more mature Florensky. This deference is most pronounced in spiritual matters, in which Luzin repeatedly seeks the advice of Florensky. Whatever the case, the formality does not seem to be an obstruction to the closeness of the relationship, and, if anything, makes the (often religious) greetings in the correspondence seem the more intense.

SPIRITUAL CRISIS

After graduation, in the fall of 1905, Luzin left on his first commission to study abroad. Florensky wrote to him in the spring of 1906, by which time he was in Paris, in the midst of a deep spiritual crisis. Luzin responded to Florensky on 1 May 1906, in a letter [4, 135–138] which sheds light on the long period of indecision that he experienced before finally, in 1909, committing himself to mathematics. The letter confirms the account by Esther Phillips [7, 283] that Luzin considered studying medicine and attending lectures of a more philosophical nature, before finally settling on a career in mathematics in 1909. More than indecision, Luzin was experiencing an identity crisis that was tied up with his relationship with Florensky [6, 28] and that had begun a year earlier in the spring of 1905, shortly before he graduated from Moscow University.

Throughout this letter, Luzin addressed Florensky by the name Pyotr Afanasyevich, which has the same initials as Pavel Aleksandrovich, Florensky's real name. The reason for this is not known, but may have some religious significance, possibly connected with the names of Saints Peter and Athanasius. The editors note [4, 138–139, n. 1] that at this time Florensky was preparing a work entitled "Spiritual Renaming." Florensky was very interested in the question of names. The editors also remark [4, 138, first, unnumbered footnote] that the letter is addressed to "His Eminence P. A. Florensky" at the Theological Academy in Sergiev Posad, so there is no question that it was written to Florenksy. (Emphasis here and in succeeding passages is in the original.)

> I received your letter three weeks ago, and only now am I able (perhaps) to reply.... Life is too depressing for me, sometimes agonizingly depressing. I am left with nothing, no solid worldview; I am unable to find a solution to the "problem of life." My *self-image* is so frequently changing that life has become pure torment. That is why I was unable to reply to you immediately. You would not believe how precious your letter and your concern for my inner life were to me. It is horrible, horrible, infinitely horrible to feel yourself surrounded by egoism, nothing but unrelieved egoism.... There is such a lack of respect for people and even life itself.... If I were sure that in fact *there is not and cannot be* any absolute respect in the world for the soul of another, I would kill myself immediately. I cannot imagine living without that hope. Why? I do not know....

By now you will have understood how precious to me, against the background of the vast and empty night that oppresses the soul, the heart, and the brain to madness, how precious to me is your concern for my inner life.

I have found myself abroad, in Paris, I don't know how, seemingly by chance.

You found me a mere child at the University, knowing nothing. I don't know how it happened, but I cannot be satisfied any more with analytic functions and Taylor series.... To be precise, it happened about a year ago.... To see the misery of people, to see the torment of life, *to wend my way home from a mathematical meeting*, to wend my way through the Aleksandrovsky Garden [on the side of the Kremlin nearest the University], where, shivering in the cold, some women stand waiting in vain for dinner purchased with horror—this is an unbearable sight. It is unbearable, having seen this, to calmly study (in fact *to enjoy*) science. After that I could not study only mathematics, and I wanted to transfer to the medical school. I don't know what stopped me. All I know is that it had nothing to do with thoughts of a career. It seems to have been the thought: "Would this be the right decision?" Then came the terrible days for us all [the 1905 revolution], for me in particular.... I barely remember what happened to me. I could not work in science, and I seem to have begun to lose my mind as a result of the impossibility of living quietly and understanding where, where the *truth* is. Dmitrii Fedorovich (Professor Egorov), seeing me in such a state, sent me here to Paris. That is how I got here.... I have been here about 5 months, but have only recently begun to study. I lacked the self-awareness. [4, 135–136]

Thus, although the specific catalyst of the crisis was the sight of women reduced to prostitution, there are larger questions at issue. His letter continues by elaborating on this, revealing that his previous (materialist) worldview had collapsed.

Life is painfully depressing for me Pyotr Afanasyevich! The worldviews that I have known up to now (materialist worldviews) absolutely do not satisfy me. I may be wrong, but I believe there is some kind of vicious circle in all of them, some fatal reluctance to accept the contingency of matter, some reluctance, which I find absolutely incomprehensible, to sort out the fundamentals, the principles. I have only recently come to understand this. I used to *believe in* materialism, but I was not able to live according to it, and was miserable, infinitely miserable.

Yes, I now understand that "science," in essence, is metaphysical and based on nothing. The scientific quest for "differential resolvents for equations of degree six" is absolutely incomprehensible.... At the moment my scholarly interests are in *principles, symbolic logic, and set theory*. But I cannot live by science alone.... I have nothing, no worldview, and no education. I am absolutely ignorant of the philological sciences, history, philosophy.... It is painfully clear to me that it makes no sense to settle for a "scientific" education. To see all around that fatal lack of respect for the soul of another, its trampling underfoot, to see all this, and not to know the right way to deal with people, not to know how to stand up for them, *to sense* the mad folly of human relations and not to have or know the truth—Lord! what pain this is....

My dear Pyotr Afanasyevich! Forgive me, for God's sake, for such a personal letter, but I am so depressed that I can't see help anywhere.... Professor Egorov advised me to read Kant. But Kant's assumptions seem to me to be ultra-shaky, to say nothing of Poincaré's squalid philosophy.... I am willing to give up my personal life just to find out where to seek the truth. For to know what the truth is and not to go there is unthinkable and impossible.... Life depresses me perhaps because I don't understand what I need. What I am now thinking about doing is, upon returning to Moscow, to attend lectures in the Department of History and Philology. I am absolutely ignorant of these subjects, and I cannot live on just mathematics and natural science. This Department may perhaps enable me to make sense of human relations.... If I do not find a path to seek the truth ... then I will not go on living. [4, 136–137]

TOGETHER AGAIN

Luzin wrote again to Florensky for guidance on 14 March 1908. "I want to see you, and only you.... The longer I live, the more I feel that we must live differently ... " [4, 142]. Luzin described his current studies as follows. "I am studying mathematics a little. I don't want to. I am more looking through historical and philological journals and news of the theory of electrons" [4, 142]. He added as a postscript, "I cannot become a specialist: I have been poisoned by you. I thank you for that" [4, 143].

A month later, 11 April 1908, Luzin wrote an enthusiastic letter with a particularly strong religious greeting, indicating that it was probably written just after Easter: "Christ is Risen! Dear Pavel Aleksandrovich! It is a shame that it was not possible to greet you personally with these holy words on the Resurrection of Life itself" [4, 144].

Luzin was now spending all his time studying number theory. "It's a mysterious area that envelops me deeper and deeper" [4, 145]. He explained that his personal life had changed profoundly. "I have gotten married and am no longer alone" [4, 145]. He was as anxious as ever to meet with Florensky and to read his works, especially now because "my wife is also very interested and shares my commitment to the search for the profound truths of life" [4, 145].

Luzin spent the summer of 1908 with his parents at the Chernigov Skeete, a monastery near Sergiev Posad. In a long letter to his wife [4, 149–151, n. 2], he described a visit that he made to Florensky. By chance, Luzin's visit came at the very moment of Florensky's appointment to the faculty of the Moscow Theological Academy, to teach the history of philosophy. In the letter Luzin offers this evaluation of Florensky's mathematical work: "But as soon as he showed me his works in mathematics, my old opinion came back to me: that all his works are of no value in the area of mathematics. Suggestive hints, beautiful analogies—very attractive and promising, provocative, beckoning, but without any results" [4, 150].

This statement is somewhat surprising because Luzin knew that Florensky aimed not to produce results of interest to mathematicians, but rather to use the results of mathematics in his philosophical and theological writings. In fact, earlier in the letter Luzin described in great detail a discussion he had with Florensky about the "question about '*the number three*' ... why is God in *three persons*, not in two, not in four ..." [4, 149]. They had an extended discussion about this issue, including the drawing of diagrams and attempts at mathematical proofs. Luzin concluded "We were not able to do it after much thinking" [4, 150].

In any case, the crisis that Luzin had been experiencing continued. Indeed it seems to have reached its peak during the summer of 1908.

THE PILLAR AND FOUNDATION OF TRUTH

On 10 June 1908 Florensky defended his thesis "On Religious Truth." Florensky wanted to invite Luzin to attend the defense but was unable to reach him. That same month, Florensky's thesis was published under the title "The Pillar and Foundation of Truth." An expanded version with the same title appeared in 1914.

The book made a profound impact on Luzin. He wrote about it in a long letter to his wife on 29 June 1908 [4, 146–148, n. 3].

> Today I want to write at length from my personal experience—I don't know whether I'll be able. I had to endure a lot as a result of the work of Florensky: *"The Pillar and Foundation of Truth. . . ."* I read it all at once in a single day—skipping a lot, but the impression was overwhelming. As I read it, I was STUNNED the entire time by blows from a battering ram against a stronghold. . . . [4, 146]

As Luzin described it, Florensky had destroyed the foundation of contemporary intellectual life, which rejected religion in the name of reason, and he had brilliantly but logically defended religious concepts which were scorned or ignored by the contemporary intelligentsia. Luzin saw this as "a worldwide tragedy of life and reason."

> I don't hesitate to use such words, they are completely appropriate here. Yes precisely a world drama. For this is the stronghold by which the "essence" of our intelligentsia lives and moves. By which it *thinks* and *lives*. And not only ours. I repeat, the issue here is the human mind, its life and drama. . . .
>
> Our life, the life of the XIX and XX centuries, is full of confusion, full of contradictions. This was not true earlier, it was not true of the Greeks (hence their strength, born of unity and soundness). One need only to look at our intelligentsia [to see the confusion] . . . .
>
> Our culture is such, as you have said many times yourself while thinking out loud, that it is impossible to continue in it any further without harm to those of us who live in it.
>
> The work of Florensky is one solution to this impossible situation. [4, 146–147]

Luzin admitted that this solution lay "in sub-logical soil, and this means foreign to reason" [4, 147]. It could be understood in apocalyptic terms, "in the transformation of all flesh and all creation under the spirit of the judgment of God" [4, 147]. "How to respond to it," he asked. "By feeling or by reason? Are such matters amenable to reason?" Luzin believed that Florensky had created "a scandal for university philosophy."

> Because, in addition to discussing understanding through the senses ("Physics," "Natural Science") and understanding through the mind ("Mathematics," "Logic"), Florensky has given equal rights to another kind of understanding, which you never hear about at the university, namely "intuitive-mystical understanding" (Hindu). This is a great scandal. [4, 147]

Florensky had used "philosophical research" to give "a brilliant psychological analysis of the mind." Luzin regretted that this included so much scholarly material, otherwise it "would be understood by everyone, and cause everyone to turn pale" [4, 147]. Florensky had portrayed "the search of a mind suffering in a philosophical hell, a search for religion," and continued with

> a description of the *psychic* and *somatic* change at embarking on the Christian way and prospering in it. Then comes a theological defense of disputed points of doctrine: a defense of the doctrine of the Holy Spirit, a defense of the idea of that there is no contradiction between religion and the mind. Then a brilliant defense of *the doctrine of hell and its torments*, a logical defense, just brilliant. . . . Here there is a very good description of temptation by love—which perhaps Florensky has experienced himself. . . . [4, 147]

In summary he wrote:

> This work is so valuable because it deals with the most fundamental questions of life, not by taking anything on faith, but on the contrary, by showing the limits of the mind, and then going beyond, logically, intuitively, but based on reason. No work on religious questions has ever presented so little fantasy and so much logic, which religious people are usually criticized for lacking....
>
> The greatest of discoveries is when a pioneer builds a bridge between unconnected areas and passes over it.... If this work is completed, it will bring a wholeness to everything. [4, 147–148]

This powerful experience for Luzin came at the very time when he was contemplating suicide. Luzin explained this in a letter of 15 July 1908. He began by thanking Florensky "with all the gratitude of which I am able" for the conversations they had had that summer. "Two times I was very close to suicide—then I came here ... looking to talk with you, and both times I felt as if I had leaned on a 'pillar' and with this feeling of support I returned home. Excuse my openness and don't think of it as anything negative: I owe my interest in life to you ... "[4, 148]. Over the next year, Luzin would find his interest in mathematics gradually reviving.

AFTER THE REVOLUTION

Luzin was appointed to the Mathematics Faculty at Moscow University in 1917, just before the Revolution. Florensky continued to live in Sergiev Posad until the beginning of 1921.

In his final letter to Florensky, written 5 May 1919 (following the new calendar), Luzin turned anew to Florensky for spiritual guidance. It appears that the Revolution had precipitated a strong desire in Luzin for a renewed spiritual life. He turned to Florensky in "deep inner need." "I am overcome by a feeling of emptiness and isolation from a source of spiritual life" [4, 187]. On the outside nothing had changed, but inside "something has broken."

> I have to admit it: clearly, mistakes were made, requiring correction. I fear for the future, because for me it is now indefinite.
>
> It is difficult to say just why, but I need to stop and examine my inner life and this is inextricably connected with the need to draw closer to the way which you have followed and continue to follow.... That is why I ask you, dear Pavel Aleksandrovich, to help me.... I would like to have, if possible, all your published works. I *do* have many, but do not have your big works, the latest lectures on philosophy (it is said that they are already published) and above all, I do not have your thesis "The Pillar and Foundation of Truth." [4, 187–188]

This is the treatise that made such a profound impression on Luzin when he read it in preliminary form in June 1908. When he saw the completed version in 1914, he was rather critical of it. He now deeply regretted that he had not bothered to preserve a copy of this "basic and fundamental work." Knowing that the book was now very valuable and difficult to obtain, he turned to Florensky for help. He wanted not just to borrow a copy but to purchase one. "I do not want just *to read* it, it is necessary for me *to absorb* it." Luzin apologizes for imposing on Florensky, but has been overcome by "this overpowering spiritual need." "I need it *'now'* not *'later'* ... " [4, 188].

The correspondence ended much as it began, with Luzin again experiencing a strong desire to follow Florensky in matters of "the inner life." Florensky was always the more emotionally mature of the two and Luzin seems to have depended on this. This was also recognized by Luzin's wife, in two letters that she wrote to Florensky on her husband's behalf, which are included among the 39 letters published ([4, 184–186]). The editors have also included a footnote about two letters that she wrote to Florensky in 1914, when she and Luzin had temporarily separated. In response, Florensky noted

> Nikolai Nikolaevich is a very sweet and fine person; but in personal relationships he is not at all mature, especially in intuitively perceiving the hidden currents of life. You, as a woman and on the strength of that alone, would have a sense for the inner life. You will have to take the relationship in hand and create a family tone, simplicity. Instead, as I perceive it (excuse me, perhaps I am mistaken!), you have established the tone of an acquaintanceship rather than a family. [4, 175, n. 1]

In addition, the editors quote a letter she wrote to Florensky in 1915 [4, 183–184, n. 1] and one she wrote to his mother-in-law [4, 186, n. 1]. In the former, she pointed out how valuable Florensky's visits were for Luzin, and urged him to visit more often.

CONCLUSION

The influence of Florensky was decisive in turning Luzin toward religious philosophy. It was probably his friendship with Florensky that caused Luzin's materialist worldview to collapse in the first place, especially after Florensky gave up a promising career in mathematics to study theology. The Revolution caused Luzin to seek to reestablish and deepen his relationship with Florensky.

I express my sincere gratitude to an anonymous referee for generously providing the translation of the entire letter of 1 May 1906, and overall for helpful suggestions and comments.

REFERENCES

1. Sergei S. Demidov, On an Early History of the Moscow School of Theory of Functions, *Philosophia Mathematica* **3** (1988), 29–35.
2. Sergei S. Demidov, The Moscow School of the Theory of Functions, in *Golden Years of Moscow Mathematics*, ed. Smilka Zdravkovska and Peter L. Duren, History of Mathematics, vol. 6, Providence, RI: Amer. Math. Soc., 1993, 35–53.
3. Sergei S. Demidov, Aleksei N. Parshin, and Sergei M. Polovinkin, On the Correspondence of N. N. Luzin with P. A. Florensky, *Istoriko-Matematicheskiye Issledovaniya* **31** (1989), 116–124. [In Russian]
4. Sergei S. Demidov, Aleksei N. Parshin, Sergei M. Polovinkin, and Pavel V. Florensky, ed., The Correspondence of N. N. Luzin with P. A. Florensky, *Istoriko-Matematicheskiye Issledovaniya* **31** (1989), 125–191. [In Russian]
5. Pavel A. Florensky, *The Pillar and Foundation of Truth*, Moscow: Put, 1914 [in Russian]; facsimilie edition printed by Gregg International, Westmead, Farnborough, England, 1970.
6. Charles E. Ford, Dmitrii Egorov: Mathematics and Religion in Moscow, *The Mathematical Intelligencer* **13**(2) (1991), 24–30.
7. Esther R. Phillips, Nicolai Nicolaevich Luzin and the Moscow School of the Theory of Functions, *Historia Mathematica* **5** (1978), 275–305.

Vladimir Andreevich Steklov

Born – 9 Jan 1864 in Nizhny Novgorod (was Gorky from 1932–1990), Russia
Died – 30 May 1926 in Gaspra, Crimea, USSR

Vladimir Steklov's father was a clergyman who taught history and Hebrew at the Nizhni Novgorod seminary. His uncle on his mother's side of the family was a famous literary critic. Vladimir Steklov inherited many of these family talents and he himself had considerable literary and musical talents. Had he not made a career from mathematics he could well have made his profession as an opera singer.

Steklov entered the Alexander Institute in Nizhni Novgorod at the age of 10 years. He studied there for eight years graduating in 1884. He entered Moscow University but, after one year, he transferred to Kharkov University. There he became a student of Lyapunov and, after a highly successful university career, he graduated in 1887. Steklov remained at Kharkov University working towards becoming a university teacher. In 1891 he was appointed Lecturer in Mechanics and worked towards his Master's Degree.

For his Master's thesis Steklov worked on the equations of a solid body moving in an ideal non-viscous fluid. There were four cases to be considered in integrating the equations which arose from this problem, and two of these cases had been solved by Clebsch in 1871. Steklov solved the third case in his thesis, the final case being solved by Steklov's supervisor Lyapunov in 1893. In fact 1893 was the year that Steklov was awarded his Master's Degree. Then, in 1896, Steklov was appointed to an extraordinary professorship of mechanics and continued to work for his doctoral dissertation.

For his doctoral dissertation Steklov worked on problems that arose in potential theory, electrostatics and hydromechanics. He reduced problems of this type to boundary-value problems of Dirichlet type using rigorous mathematical analysis. Steklov was awarded his doctorate in 1902 and, in that year, his supervisor Lyapunov accepted an appointment at St Petersburg. Steklov was appointed to the Chair of Applied Mathematics at Kharkov University which Lyapunov had vacated.

Steklov had been an active member of the Kharkov Mathematical Society, serving the Society as secretary in 1891 and deputy chairman in 1899. In 1902, the year he was appointed to the Chair of Applied Mathematics, Steklov became chairman of the Society and held this post until 1906 when he moved to St Petersburg to take up the Chair of Mathematics there.

Steklov had strong political views at a time when political feelings in Russia ran high. He was strongly against the Tsarist Regime and sympathised with progressive students. In fact this made him a very popular teacher at St Petersburg and, coupled with his excellent lecturing skills, brought many students to study at the Department of Mathematics and Physics. Among the students that he taught at St Petersburg were Friedmann, Smirnov and Tamarkin

In 1910 Steklov had been elected to the Academy of Sciences. Then in 1919 he became vice-president of the Academy. In 1921 Steklov founded the Institute of Physics and Mathematics and served as its director until his death in 1926. In 1934 the Institute was split into two separate Institutes of Physics and of Mathematics and the Mathematics Institute was named the V A Steklov Mathematical Institute

Steklov made many important contributions to applied mathematics. In addition to the work for his master's thesis and his doctoral thesis referred to above, he reduced problems to boundary value problems of Dirichlet type where Laplace's equation must be solved on a surface. He wrote General *Theory of Fundamental Functions* in which he examined expansions of functions as series in an infinite system of orthogonal eigenfunctions. In fact the term "Fundamental Functions", which is due to Poincaré, means eigenfunctions in today's terminology. Steklov was not the first to examine series expansions in terms of infinite sets of orthogonal eigenfunctions, of course Fourier had examined a special case of this situation many years before. Steklov, however, produced many papers on this topic which led him to a general theory to replace the special cases examined by others. He studied a generalisation of Parseval's equality for Fourier series to his general setting showing this to be a fundamental property. In all his list of publications contains 154 items.

ТРУДЫ В. А. СТЕКЛОВА ПО МАТЕМАТИЧЕСКОЙ ФИЗИКЕ.

Н. М. Гюнтер.

(Речь, произнесённая в заседании Ленинградского физико-математического общества, посвященном памяти В. А. Стеклова, 9 октября 1926 г.)

Те годы прошлого столетия, в которые Владимир Андреевич Стеклов начинал свою деятельность, были годами поворота в математической физике.

После эпохи, когда Фурье*) [1] и Пуассон [2] создавали математическую теорию тепла, а Штурм и Лиувилль поочерёдно обогащали науку своими глубокими исследованиями [3], наступило относительное затишье.

В 1870 г. был подорван принцип minimum'a [4], а новые исследования, могущие его заменить при установлении теорем существования, не появлялись.

Разработанный в 1877 г. Нейманом [5] метод решения основных задач математической физики — задачи Дирихле и задачи, носящей ныне его имя, — нуждался в пополнениях и, будучи существенно связан с выпуклыми поверхностями, казалось, не мог быть распространённым на более общие.

Новое течение в математической физике началось с 1885 г. В этом году Шварцу [6] удалось дать в руки пионеров науки новое орудие, позволившее преодолеть те препятствия, перед которыми остановились старые мастера. Идея Шварца, которая была подхвачена прежде всего Пикаром, выяснившим в двух небольших заметках 1893 [7] и 1894 гг. [8] значение и силу нового метода, привела к мемуару Пуанкаре 1894 г. [9] в Rendiconti, а также, как естественному следствию его, к мемуару 1897 г. [10] о методе Неймана и фундаментальных функциях.

Этими мемуарами Пуанкаре как бы спешил поставить науку на новую дорогу; они полны новых творческих мыслей, но строго доказанного, что можно было бы счесть фундаментом будущего здания, в них почти нет; некоторые места в них похожи на набросок плана будущего исследования.

Ссылками на эти два мемуара пестрят первые работы Владимира Андреевича; их критика и попытки строгого установления высказанных в них утверждений составляют содержание первых его работ в области классических задач математической физики; поэтому, говоря об этих работах, нельзя не коснуться и содержания упомянутых мемуаров.

Остановимся, для примера, на задаче распределения температуры в однородном охлаждающемся теле. Эта температура V удовлетворяет прежде всего уравнению, устанавливающему для каждого элемента внутри тела изменение

*) Указания на литературу даны в примечаниях в конце статьи.

температуры в зависимости от течения тепла в теле. Это уравнение:

$$\frac{\partial V}{\partial t}=K\left\{\frac{\partial^2 V}{\partial x^2}+\frac{\partial^2 V}{\partial y^2}+\frac{\partial^2 V}{\partial z^2}\right\}=K\Delta V; \tag{1}$$

в нём K — коэффициент теплопроводности. Далее, температура V должна удовлетворять на границе тела некоторому так называемому предельному условию, характеризующему потерю тепла в окружающее тело пространство. Это условие на границе (S) тела:

$$\frac{\partial V}{\partial n}+hV=0; \tag{2}$$

здесь h — положительное число, зависящее от лучеиспускательной способности тела.

Функция V, кроме того, в начальный момент $t=0$ должна обращаться в некоторуı данную функцию — удовлетворять некоторому начальному условию; распределение температуры F именно в начальный момент считается известным.

Так как уравнения (1) и (2) линейны и однородны, сумма их решений есть их решение, так же как и произведение решения на постоянное.

Фурье и Пуассон обогатили науку идеей искать температуру V в виде суммы, слагаемые которой равны произведениям функции точки на показательную функцию времени и подобраны так, чтобы они удовлетворяли только условиям (1) и (2)

$$\sum Aue^{-\lambda t}. \tag{3}$$

Самую сумму надо подбирать так, чтобы она при $t=0$ обращалась в F.

Отыскивая эти функции точки u, легко установить, что они связаны уравнениями вида

$$\Delta u+\lambda u=0, \tag{4}$$

причём на границе соблюдено предельное усл вие (2).

Рассмотрим уравнение с неизвестной v, получаемое из (4) прибавлением некоторой функции f.

$$\Delta v+\lambda v+f=0, \tag{5}$$

подчиняя v тому же предельному условию.

Если искать решение уравнения (5) под видом ряда

$$v=v_0+v_1\lambda+v_2\lambda^2+\ldots, \tag{6}$$

расположенного по степеням λ, коэффициенты которого обозначены буквами v со значками, и подчинять функцию v требуемым условиям независимо от значения λ, то для нахождения коэффициентов v_k мы получим уравнения:

$$\Delta v_0+f=0, \quad \Delta v_k+v_{k-1}=0, \quad \frac{dv_k}{dn}+hv_k=0, \tag{7}$$

аналогичные уравнениям, встречавшимся у Шварца.

Не трудно убедиться, что если v мероморфная функция от λ и λ_k. её полюс, то при $\lambda=\lambda_k$ уравнению (5) можно удовлетворить не при всякой f; но зато можно составить функцию u_k, удовлетворяющую уравнению (4) и предельному условию (2); если, например, λ_k простой полюс, эта функция u_k будет вычетом v, соответствующим λ_k.

Таким образом, задача нахождения V сводится к трём:

a) к нахождению функций v_k, т. е. к составлению функции v;

b) к исследованию функции v, т. е. к нахождению её полюсов, называемых характеристическими числами, и функций u_k, называемых фундаментальными;

c) к подбору коэффициентов A таким образом, чтобы сумма ряда

$$\sum Au, \tag{8}$$

в который обращается искомый ряд (3) при $t=0$, была равна данной функции F.

При решении задачи (b) мы встречаем в мемуаре Пуанкаре замечательное доказательство мероморфности функции v; оценка полюсов этой функции выполнена при помощи леммы, которая даёт пример, сама по себе и по методу доказательств, удивительной интуиции.

Решение задачи (c), вследствие её трудности, не вызывает замечаний, хотя оно и достигнуто в мемуаре громоздким образом — непосредственным приложением теоремы Коши — и при тяжёлых ограничениях функции F; в первой своей работе [11] В. А., дав новый и изящный метод, тоже не смог освободиться от этих ограничений.

Но с рассуждениями, устанавливающими существование функций v_k, если исключить случай, когда $h=\infty$, говоря словами В. А., трудно согласиться, а это ставит под удар и решение задачи (b).

Нахождение функций v_k проведено при помощи особых функций, функции Грина и ей аналогичных, введённых тут же, причём не установлено, ограничены ли вблизи границы встречающиеся производные от этих функций, а, значит, также и от функций v_k, тогда как эти производные играют существенную роль в выкладках.

Иначе, впрочем, в 1894 г. и при выбранной широкой постановке задачи и не могло быть.

Метод «de balayage», данный Пуанкаре в 1889 г. [12], устанавливает существование решения задачи Дирихле для любой области, но ничего не говорит о существовании производных у решения вблизи границы.

Метод «арифметических средних», указанный за пятнадцать лет перед тем Нейманом, и о котором мы уже упоминали, основан исключительно на действиях с потенциалами двойного слоя. При наличии примера, принадлежащего А. М. Ляпунову и устанавливающего возможность отсутствия нормальной производной у таких потенциалов, он не давал возможности уверенно ими пользоваться, так как условия существования этих производных установлены не были.

При этом в рассуждениях Неймана, приводящих к основному его неравенству — принципу Неймана, — мы встречаем одно место [13], замечательное по остроумию, но нуждающееся в дополнениях, если дело идёт о произвольно выбранном задании на границе; это-то рассуждение, между прочим, совершенно неприменимо, когда область не выпуклая.

Установить существование решений задачи Дирихле и Неймана, обойдя эти два пункта метода Неймана, всё равно, что уничтожить ряд недочётов в мемуаре Пуанкаре.

В первой же работе по задаче Неймана, написанной в 1897 г. [14], В. А. становится на путь обхода обоих этих пунктов.

Во-первых, он представляет искомое решение в виде потенциала простого лоя.

Положим, что f — данное значение нормальной производной искомой функции, и знак $\frac{dV}{dn}$ обозначает значение нормальной производной в точке на границе (S) области.

Составим потенциалы простого слоя

$$V_1 = -\frac{1}{2\pi}\int f\frac{ds}{r}, \quad V_k = -\frac{1}{2\pi}\int \frac{dV_{k-1}}{dn}\frac{ds}{r}, \tag{9}$$

в первом из которых плотность получается делением f на -2π, и в каждом последующем таким же делением значения нормальной производной на границе предыдущего. Плотность потенциала простого слоя в решении В. А. равна сумме f с нормальными производными на границе от всех этих потенциалов (9):

$$f + \frac{dV_1}{dn} + \frac{dV_2}{dn} + \ldots + \frac{dV_k}{dn} + \ldots \tag{10}$$

Эту форму имеет решение и во всех последующих его работах.

Во-вторых, выводя неравенство, соответствующее принципу Неймана, он не прибегает к соображениям, аналогичным соображениям Неймана, хотя и ценою значительных ограничений поверхности (S).

Взяв точку m на границе (S), найдём верхний и, соответственно, нижний пределы радиуса сфер, касающихся границы в m и лежащих внутри (S), соответственно заключающих (S). Из этих радиусов, меняя положение точки m, берём наибольший и, соответственно, наименьший. Отношение σ найденных крайних радиусов, которое для сферы, очевидно, равно единице, В. А. называет уклонением (S) от сферы и доказывает сходимость своего процесса только для тех выпуклых поверхностей, в которых σ близко к единице.

В своём решении задачи о распределении электрического слоя по поверхности [15], данном в 1887 г., Робэн оперирует с нормальными производными на границе от потенциалов (9). На связь этой задачи Робэна с задачей Неймана указывает сам В. А. в заметке, помещённой тогда же в *CR* [16]; такое же указание, ничем, впрочем, не пояснённое, имеется и в конце работы Робэна.

Ссылаясь на потенциалы (9), мы будем называть их потенциалами Стеклова-Робэна.

Два года, последовавших за этой работой, несколько изменили обстановку. За это время появились исследования А. М. Ляпунова о задаче Дирихле [17], в которых, между прочим, устанавливаются достаточные условия существования нормальных производных у потенциала двойного слоя; этим ставится на твёрдую почву обращение с такими потенциалами.

В. А. снова возвращается к мемуару Пуанкаре 1897 года [10]. Этот мемуар посвящён, между прочим, — и это не главное содержание мемуара — методу Неймана.

В нём сделана попытка, пользуясь интегралами Шварца, распространить метод, применив некоторую подстановку, на не выпуклые поверхности; о границах предполагается, что они имеют определённые радиусы кривизны.

Пользование результатами Ляпунова позволяет В. А. построить доказательство таким образом [18], что он не только освобождается от тяжёлых ограничений, наложенных им самим на границу, но становятся излишними и ограничения, сделанные другими авторами; наибольшее достигается в предположениях о функции, заданной на границе: оказывается достаточным предположение об её непрерывности.

В основание решения задачи Дирихле и Неймана В. А. теперь кладёт решение задачи о распределении электричества, причём решает её, придерживаясь схемы, предложенной Робэном [15], но не придерживаясь его метода доказательства, строго Робэном не установленного. Ясно, что строгое установление сходимости процесса сводится к нахождению неравенства, заменяющего принцип Неймана. Такое неравенство устанавливается В. А. для всех поверхностей, допускающих преобразование Пуанкаре, состоящее в установлении однозначной зависимости между точками границы и точками сферы при помощи формул, имеющих ограниченными некоторые производные.

Положим, что поверхность допускает такое преобразование, и имеется потенциал простого слоя V, обладающий свойством, по которому интеграл по поверхности (S) от его внешней нормальной производной равен нулю:

$$\int \frac{dV_e}{dn}\, ds = 0. \tag{11}$$

Для всякого такого потенциала V отношение интегралов

$$J = \int \sum \left(\frac{\partial V}{\partial x}\right)^2 d\tau, \quad J' = \int \sum \left(\frac{\partial V}{\partial x}\right)^2 d\tau' \tag{12}$$

от его дифференциального параметра, взятых по области (D), ограниченной поверхностью (S), и по области внешней (D), заключается между двумя пределами, из которых нижний отличен от нуля; для сферы это утверждение, названное В. А. фундаментальной теоремой Пуанкаре, доказывается легко.

Приложим фундаментальную теорему к линейной функции с произвольными коэффициентами

$$\alpha V_k + \beta V_{k+1} \tag{13}$$

от двух смежных потенциалов (9) Стеклова-Робэна со значками $k+1$ и k. Эта функция, если интеграл от функции f, определяющей условия на границе, взятый по границе (S) равен нулю:

$$\int f\, ds = 0, \tag{14}$$

удовлетворяет условию (11) фундаментальной теоремы. Исследование полученной таким образом определённой квадратичной формы от указанных произвольных коэффициентов говорит, что отношение интегралов, взятых по всему пространству от дифференциальных параметров этих смежных потенциалов V_{k+1}, V_k меньше числа λ, меньшего единицы:

$$\frac{J_{k+1} + J'_{k+1}}{J_k + J'_k} < \lambda < 1. \tag{15}$$

Элементарные свойства потенциала простого слоя показывают тогда, что отношение интегралов от квадратов потенциалов V_k, взятых по поверхности, к интегралам по всему пространству от их дифференциальных параметров ограничено, откуда, на основании доказанного, уже ясно, что числители этих отношений меньше членов некоторой убывающей геометрической прогрессии:

$$\int V_k^2 ds < l(J_k + J'_k) < B\lambda^k; \tag{16}$$

после этого простая оценка V_k говорит, что если интеграл по поверхности от начальной функции f равен нулю, то потенциалы Стеклова-Робэна V_k абсолютно меньше членов убывающей геометрической прогрессии:

$$\text{если } \int f ds = 0, \text{ то } |V_k| < C\tau^k, \ 0 < \tau < 1. \tag{17}$$

Последний результат есть так называемый принцип Робэна.

Из приложения I ясно, каким образом принцип Робэна даёт возможность, применяя общеизвестные свойства потенциалов простого и двойного слоёв, получить как решение задачи Робэна, так и решение задач Неймана и Дирихле. Но более того, вследствие простой зависимости, указанной в таблице, между потенциалами Неймана и потенциалами, использованными при решении задачи Робэна, устанавливается принцип Неймана каждый раз, когда имеет место принцип Робэна.

Таким образом, решение основных задач, построенное В. А., не только даёт строгое доказательство приложимости процесса, но и выявляет, что многие ограничения, делаемые до тех пор, излишни, но только для поверхностей, допускающих указанное выше преобразование Пуанкаре.

Оставляя в стороне вопрос о тех явных условиях, которым подчинены поверхности при таком неявном ограничении, мы видим, что поверхность, во всяком случае, должна быть преобразуемой деформацией в сферу; следующий этап должен был освободить и от этого ограничения.

Это освобождение дано в 1901 г. польским учёным С. Зарембой [19], показавшим, что можно заменить применение преобразования Пуанкаре использованием его чрезвычайно замечательной леммы, о которой уже упомянуто. При этом с поверхности снимается значительная часть ограничений: остаётся только предположение, что она удовлетворяет трём известным условиям Ляпунова.

Можно с уверенностью утверждать, что толчком, вызвавшим работу Зарембы, был мемуар В. А. 1900 г. [20]; результаты этого мемуара, вместе с некоторыми другими, чрезвычайной важности, о которых будет речь ниже, собраны в его докторской диссертации [21].

Работа Зарембы немедленно вызвала появление в 1902 г. обширного мемуара [22], в котором В. А. доводит до конца заключения о задачах Робэна, Дирихле и Неймана, вытекающие из замечания Зарембы. Полученные результаты тут же прилагаются к восстановлению решений задач (a) и (b) мемуара из Rendiconti [9], на этот раз совершенно строгим образом и для поверхностей отмеченного выше чрезвычайно общего вида.

При этом также строго устанавливается существование функции Грина и некоторых других функций J и H, аналогичных ей и могущих служить для последовательного нахождения функций v_k в тепловой задаче при $h = 0$ и h положительном, подобно тому, как функция Грина решает эту задачу при $h = \infty$.

Эти функции встречаются и у Пуанкаре в Rendiconti [9]; попытка использовать их так, как это делал Пуанкаре, встречается в мемуаре В. А. 1897 г. [23]; но в этом месте они введены для решения задачи (c).

Кончая свой мемуар 1897 г. [10], Пуанкаре, по собственному признанию, не стараясь быть строгим, но желая дать идею о новых возможных приложениях метода фундаментальных функций к задачам стационарного характера, вводит новые функции, определяемые соответственно любой поверхности; это — потенциалы простого слоя, удовлетворяющие условию

$$\frac{dV_{ki}}{dn}-\frac{dV_{ke}}{dn}=-\lambda_k\left(\frac{dV_{ki}}{dn}+\frac{dV_{ke}}{dn}\right) \text{ на } (S), \tag{18}$$

при котором отношение разности внутренней и внешней нормальных производных к их сумме постоянно на (S) и равно числу — λ.

Эти потенциалы позволяют свести решение задач Неймана и Дирихле для любой поверхности к разложению по ним функций, определяющих предельные условия, подобно тому, как разложение начальной функции по фундаментальным в тепловой задаче служит, как выяснено, основанием для её решения.

Таким приёмом ранее решались задачи на сфере разложением по шаровым функциям; на эллипсоиде — при помощи введённых Ламе его произведений. Пуанкаре делает эти приёмы частными случаями одного общего метода.

Всё дело сводится, конечно, к доказательству существования нужных функций; этого у Пуанкаре, однако, нет.

Но за два года до появления мемуара Пуанкаре эта идея возбуждена В. А. В устном сообщении Харьковскому математическому обществу в 1895 г. В. А. вводит фундаментальные функции — потенциалы простого слоя, для которых отношение внутренней нормальной производной к значению потенциала постоянно на (S).

Установив существование этих потенциалов, он пользуется ими в последней главе докторской диссертации [21] для решения задач Неймана и Дирихле, как указано в приложении II, беря их в несколько более общем виде; они разобраны и в мемуаре 1902 г. [22].

Таким образом, приоритет в обобщении старых методов решения задач стационарного характера принадлежит В. А.

Мемуар Пуанкаре 1897 г. привёл Ле Руа к установлению в 1898 г. [24] особых функций, определяемых условием, по которому отношение разности между внутренней и внешней их нормальными производными к самой функции есть некоторая функция φ точки на (S):

$$\frac{dV_{ki}}{dn}-\frac{dV_{ke}}{dn}=\lambda_k\varphi V_k \text{ на } (S). \tag{19}$$

Только по недоразумению считалось некоторое время, что функции Пуанкаре получаются из функций Ле Руа при подобающем выборе φ, так что доказательства существования функций Пуанкаре, в действительности, до 1902 г. не было.

Это доказательство мы встречаем впервые опять-таки в том же мемуаре 1902 г. [22]. Вот оно в общих чертах.

Не трудно убедиться, что если V_k потенциалы Стеклова-Робэна, то функция V, разложенная по степеням параметра λ и имеющая коэффициентами эти потенциалы

$$V_1+V_2\lambda+V_3\lambda^2+\ldots, \tag{20}$$

даёт решение некоторого уравнения, аналогичного уравнению (5), лежащему в основе тепловой задачи:

(21) $$\frac{dV_i}{dn} - \frac{dV_e}{dn} = -2\lambda\frac{dV}{dn} - 2f \text{ на } (S);$$

для его получения из уравнения (5) надо лапласиан заменить разностью между внутренней и внешней нормальными производными от V, а функцию v значением удвоенной нормальной производной на границе.

Отсюда один шаг, требующий только повторения рассуждений из Rendiconti [9], для установления, что функция (20) V — функция мероморфная, числа λ в условии Пуанкаре (18) — её полюсы, а сами потенциалы Пуанкаре — вычеты, отвечающие этим полюсам.

Всё перечисленное до сих пор, однако, только подготовка к решению важнейшей задачи (c).

Переходя к ней, отметим прежде всего, что в каждой системе

(22) $$V_1, V_2, \ldots, V_k, \ldots$$

упомянутых до сих пор фундаментальных функций, соблюдено условие

(23) $$\int p V_k V_e d\omega = 0, \qquad k \neq l;$$

в этом условии интегрирование распространено или по области (D) в функциях, соответствующих тепловой задаче, или по поверхности (S) в функциях Пуанкаре, Ле Руа и Стеклова, а p — некоторая положительная функция. Этому условию удовлетворяют и функции Штурма-Лиувилля, применяемые в задаче об охлаждении бруса — частном случае тепловой задачи — и различные системы полиномов, определённых как решения уравнений 2-го порядка, зависящих от параметра, и как знаменатели подходящих различных непрерывных дробей. Это условие называется «условием ортогональности системы», а система (22) фундаментальных функций, при его соблюдении, «ортогональной».

Можно всегда считать, что интеграл, получаемый из интеграла условия ортогональности, при $k = l$ равен единице:

(24) $$\int p V_k^2 d\omega = 1.$$

При соблюдении последнего условия ортогональная система функций называется «нормальной».

При разложении некоторой функции в ряд

(25) $$A_1 V_1 + A_2 V_2 + \ldots + A_k V_k + \ldots,$$

расположенный по фундаментальным функциям (22), прежде всего, конечно, возникает вопрос о нахождении коэффициентов A_k этого ряда.

Вопрос этот тесно связан, являясь его частным случаем, с вопросом о наилучшем представлении данной функции f при помощи n первых фундаментальных функций (22) — суммой n первых членов ряда

(26) $$A_1 V_1 + A_2 V_2 + \ldots + A_n V_n;$$

под этим понимается, следуя пути, предложенному Чебышевым [25, 26], такой выбор коэффициентов A, при котором квадратичная погрешность

(27) $$\int [f - A_1 V_1 - A_2 V_2 - \ldots - A_n V_n]^2 \rho^2 d\omega$$

имела бы наименьшее значение.

При этом, под словами «квадратичная погрешность» понимается интеграл от произведения квадрата разности между функцией f и её приближением, на положительный множитель $\theta^2\, d\omega$.

Этот положительный множитель, равный произведению положительной функции p на $d\omega$

$$p\, d\omega = \theta^2\, d\omega, \tag{28}$$

есть выражение, получившееся при предельном переходе из выражения, характеризующего квадратичную погрешность каждого из данных n значений искомой функции в задаче вычисления её по этим n значениям.

Вычисление коэффициентов по этому правилу говорит, во-первых, что при соблюдении условий ортогональности и нормальности они вычисляются по тому же правилу, которое применимо при составлении ряда Фурье, и равны

$$A_k = \int pfV_k\, d\omega, \tag{29}$$

и что, во-вторых, если R_n разность между f и взятым приближением, (т. е. дополнительный член искомого бесконечного ряда), мы имеем для квадратичной погрешности следующее выражение:

$$\int pR_n^2\, d\omega = \int pf^2\, d\omega - A_1^2 - A_2^2 - \ldots - A_n^2. \tag{30}$$

Из последнего равенства ясно, что сходится ряд (25) или нет, квадратичная погрешность, равная интегралу

$$\int pR_n^2\, d\omega, \tag{31}$$

стоящему в левой части равенства (30), с увеличением n убывает.

Если ряд (25) сходится равномерно и сумма его равна f, то квадратичная погрешность (31) бесконечно мала.

Продолжаем предполагать, что сумма ряда (25), составленного соответственно некоторой функции f, равна f. Выберем из системы фундаментальных функций (22) некоторую совокупность так, чтобы при одной из функций (22), не входящих в эту совокупность, коэффициент в ряде не был равен нулю.

Мы получим новую систему, которая также ортогональна и нормальна.

Будем составлять квадратичные погрешности, соответствующие различным функциям φ, для этой новой системы; квадратичная погрешность всегда будет убывающей, но не всегда (когда, например, $\varphi = f$) будет бесконечно малой.

Итак, система фундаментальных функций (22) может быть ортогональной и нормальной, когда квадратичная погрешность не бесконечно мала.

Вопрос же о разложении произвольной функции или функции данного класса в ряд по функциям данной системы (22) может быть поставлен только тогда, когда квадратичная погрешность бесконечно мала для всех функций данного класса, т. е. когда для всех функций f данного класса интеграл от произведения квадрата f на p равен ряду из квадратов коэффициентов A:

$$\int pf^2\, d\omega = A_1^2 + A_2^2 + \ldots + A_n^2 + \ldots \tag{32}$$

Поверка соблюдения этого условия лежит в основании всех работ В. А. без исключения, начиная с первых работ 1896 г. в Сообщениях Харьковского математического общества о гармонических функциях Пуанкаре [11] и об охлаждении стержня [27], причём долгое время это равенство (32), доказываемое для различных систем фундаментальных функций, называлось просто «замечательным». В знаменитом мемуаре 1903 г. в Записках Академии [28] собраны все системы фундаментальных функций, для которых это равенство установлено В. А. Только в декабре 1910 г. в заметке в CR [29] В. А. назвал равенство (32) условием замкнутости, а системы функций (22), для которых это условие соблюдено при всякой интегрируемой ограниченной функции f, замкнутыми.

Из сказанного ясно, что замкнутость как бы, выражаясь не строго, гарантирует отсутствие пропуска в системе фундаментальных функций.

При доказательстве замкнутости только первый шаг — доказательство её для функций, имеющих некоторые производные, и удовлетворяющих, вообще говоря, граничным условиям, — существенно зависит от индивидуальных свойств системы.

Действительно, если, каково бы ни было ε, можно в соответствие данной функции f привести другую F, значения которой отличаются от значений f менее чем на ε, если даже и не во всей области (D), но в области, отличающейся от всей менее чем на ε, причём такую функцию F, что для неё условие замкнутости соблюдено, оно соблюдено и для функции f.

Если R_n и ρ_n дополнительные члены рядов, соответствующих функциям f и F.

$$f = A_1V_1 + \ldots + A_nV_n + R_n, \quad F = B_1V_1 + \ldots + B_nV_n + \rho_n, \tag{33}$$

легко получаем, замечая, что справедливы равенства

$$\int pR_n V_k \, d\omega = 0 \qquad k = 1, 2, \ldots, n, \tag{34}$$

что разность между квадратичными погрешностями, соответствующими f и F, равна интегралу, зависящему от произведения R_n на разность между f и F.

$$\int pR_n^2 \, d\omega = \int p\rho_n^2 \, d\omega + \int p(f - F) R_n \, d\omega. \tag{35}$$

Простое применение неравенства Шварца устанавливает тогда справедливость высказанного утверждения.

В первых работах В. А., в мемуарах 2-го [30] и 3-го [31] томов Annales de Toulouse мы встречаем при составлении функции F изящное применение теоремы Вейерштрасса.

Позднее, с заметки в ноябре 1910 г. в CR [32] В. А. с большей охотой пользуется особыми, введёнными им функциями, которым я позволил себе дать имя функций Стеклова.

Если f некоторая функция, останавливаясь для примера на случае функции одного переменного, то функция Стеклова Φ, соответствующая f, определяется равенством

$$\Phi(x) = \frac{1}{h} \int_x^{x+h} f(\xi) \, d(\xi). \tag{36}$$

В этом равенстве h настолько мало, что область $(x, x+h)$ в области задания функции f. Функция Φ непрерывная функция, если f только интегрируемая, и имеет одной производной больше, чем f, если f непрерывна. При этом, при непрерывности f, разность

$$f - \Phi \tag{37}$$

можно сделать, выбирая h достаточно малым, меньше наперёд заданного числа; подобное же заключение можно сделать об интегралах, зависящих от этой разности, если f только интегрируема.

Эта функция впервые появляется у В. А. в мемуаре 1903 г. о тригонометрических рядах [33]; но пользоваться ею как приближением функции В. А. начинает значительно позднее.

Как мы указали, первый шаг доказательства зависит от индивидуальных свойств системы.

Для системы полиномов равенство замкнутости (32) справедливо для всех полиномов, так как всякий полином равен конечной сумме полиномов.

Для системы функций Штурма-Лиувилля [31] доказательство достигается рассмотрением квадратичной погрешности производной — интеграла, под знаком которого стоит квадрат производной дополнительного члена R_n:

$$\int (R_n')^2\, dx; \tag{38}$$

в случае функции многих переменных роль квадратичной погрешности производной играет интеграл от дифференциального параметра дополнительного члена

$$\int \sum \left(\frac{\partial R_n}{\partial x}\right)^2 d\omega. \tag{39}$$

Во всех случаях простое установление ограниченности этой квадратичной погрешности производной непосредственно приводит к доказательству теоремы о замкнутости.

Эта погрешность производной, именно, один из интегралов Шварца в задаче, когда в исходном уравнении (5) функция f заменена через R_n, и теорема есть простое следствие утверждения, что в таком случае наименьший полюс решения имеет значок больший n.

Установление замкнутости системы даёт прежде всего важную теорему (А), полного доказательства которой для функций Штурма-Лиувилля нет у творцов их теории:

(А) Если ряд (25), расположенный по функциям замкнутой ортогональной системы и соответствующий непрерывной функции f, сходится равномерно, то сумма его равна f. Действительно, предел дополнительного члена R_n в этом случае непрерывная функция и из равенства нулю предела квадратичной погрешности непосредственно вытекает, что предел R_n равен нулю

$$R = \lim R_n = 0. \tag{40}$$

Мы встречаем безукоризненно строгое доказательство теоремы (A) во всех работах В. А., начиная с 1899 г.

Кроме этого, теорема о замкнутости приводит к такому следствию (В):

(В) Если φ функция с интегрируемым квадратом

$$\int \varphi^2 \, d\omega, \tag{41}$$

то справедливо равенство

$$\int\limits_{(D_0)} pf\varphi \, d\omega = A_1B_1 + A_2B_2 + \ldots, \quad B_k = \int\limits_{(D_0)} \varphi V_k \, d\omega, \tag{42}$$

в котором (D_0) любая часть (D).

Ряд (42) получается интегрированием произведения на $p\varphi$ ряда, соответствующего f; применение неравенства Шварца к его дополнительному члену даёт неравенство

$$\left(\int p\varphi R_n \, d\omega \right)^2 = \int p\varphi^2 \, d\omega \int pR_n^2 \, d\omega, \tag{43}$$

из которого непосредственно вытекает сходимость ряда (42).

Несмотря на его простоту, следствие (В) имеет большое значение в вопросе о разложении произвольной функции. Искусное его приложение составляет особенность работ В. А., начиная с заметок 1898 [34] и 1899 [35] гг. в CR.

В первых работах в Сообщениях Харьковского математического общества по разложению по функциям Штурма-Лиувилля [36] и гармоническим Пуанкаре [11] мы встречаем непосредственное приложение теоремы (А), для чего выполняется оценка членов ряда с целью доказательства его сходимости, впрочем, методом новым, отличным от метода Лиувилля.

Применение индивидуальных свойств системы ортогональных функций позволяет несколько раз преобразовывать интеграл, равный коэффициенту ряда, выделяя каждый раз множитель, зависящий от характеристического числа; но эти последовательные преобразования налагают тяжёлые условия на разложимую функцию, как показано в приложении III.

Привлечение теоремы (В) чрезвычайно облегчает дело; например, для разложения по гармоническим функциям Пуанкаре достаточно представить разлагаемую функцию f, использовав функцию Грина G в виде интеграла

$$f = -\int G \Delta f \, d\omega, \tag{44}$$

и воспользоваться некоторыми соотношениями

$$\left\{ \begin{aligned} &\int \Delta f V_k \, d\omega = \int f \Delta V_k \, d\omega = -\lambda_k \int f V_k \, d\omega; \\ &\int G_h V \, d\omega = -\frac{1}{\lambda_k} \int G \Delta V_k \, d\omega = \frac{V_k}{k^\lambda}, \end{aligned} \right. \tag{45}$$

характерными для рассматриваемых функций.

Для проведения доказательства достаточно допустить существование Δf, т. е. вторых производных у f.

В мемуаре 1902 г. [22] аналогичным образом использованы при решении тепловой задачи в случае, когда h не $= \infty$, те функции I и H, подобные функции Грина, о которых упомянуто выше.

При разложении по функциям Штурма-Лиувилля определяющее их уравнение

(46) $$\lambda p V = qV - V'', \quad q > 0,\ p > 0$$

позволяет выполнить операции, выполнявшиеся в пространственной задаче при помощи теоремы Грина. Для разложимости f достаточно существования второй производной.

Позднее, в 1910 г. [37] простое сс тношение, просто-напросто дающее значение интеграла от производной квадрата R_n:

(47) $$R^2_n(x) = R_n^2(\xi) + 2\int_\xi^x R_n R'_n\,dx,$$

позволило В. А. установить разложимость функции f, имеющей только первые производные; применение именно к последнему слагаемому в (47) неравенства Шварца и интегрирование по x приводят к некоторому неравенству

(48) $$R_n^2(\xi)\int_a^b p\,dx < \int_a^b pR_n^2\,dx + \frac{2}{a}\sqrt{\int_a^b pR_n^2\,dx}\sqrt{\int_a^b pR_n'^2\,dx \cdot \int_a^b p\,dx},$$

позволяющему, пользуясь заключением теоремы о замкнутости, непосредственно оценить дополнительный член R_n.

Ляпуновское распространение формулы интегрирования по частям на функции только интегрируемые, обобщив равенство (47), позволило [38, 39] дать разложение всех непрерывных функций, имеющих вид интеграла от интегрируемой ограниченной функции:

(49) $$\int_a^b \varphi\,dx.$$

Метод, применённый в этих работах, требует, чтобы коэффициенты p и q в уравнении, определяющем функции Штурма-Лиувилля (46), были положительны; других ограничений этих коэффициентов, кроме непрерывности, нет.

С начала своей деятельности до 1903 г. В. А. нераздельно господствовал в области разложения по фундаментальным функциям.

В 1903 г., однако, появился обширный мемуар Кнезера [40], который, применяя с небольшими изменениями старый Лиувиллевский метод асимптотических выражений, показал, что если коэффициенты исходного уравнения имеют четвёртые производные, разложение по функциям Штурма-Лиувилля допустимо при тех же условиях, как и в тригонометрические ряды.

Затем внезапно развился и вырос могучий соперник в теории интегральных уравнений; в двух последних из своих трёх знаменитых сообщений [41] Гильберт дал общий метод, охватывающий разнообразные виды фундаментальных функций, определённых по полюсам мероморфных решений интегральных уравнений; для разложимости по ним разлагаемая функция должна быть представляема в виде интеграла, аналогичного (44):

(50) $$\int K\varphi\,d\omega.$$

Этот интеграл соответствует некоторым образом интегральному уравнению, давшему рассматриваемые фундаментальные функции.

В. А. немедленно откликнулся на работу Гильберта обширным мемуаром 1905 г. [42]. В этом мемуаре, применяя свои методы, он устанавливает существование систем фундаментальных функций, соответствующих разного рода функциям Грина, ортогональных, нормальных и замкнутых, и даёт теоремы разложения, вполне аналогичные теоремам Гильберта, обнаруживая таким образом, что теоремы мемуара 1902 г.[22] являются только примерами этих теорем более общего характера.

Затем, в 1907 г. [43], призвав на помощь метод асимптотических выражений Лиувилля, он восстановляет результаты Кнезера с несколько меньшим ограничением коэффициентов исходного уравнения; в этом мемуаре имеются и аналогичные теоремы разложимости по некоторым системам полиномов.

Результаты, касающиеся функций Штурма-Лиувилля, повторены с некоторыми упрощениями в двух мемуарах 1910 г. [44, 45].

Возвращаясь к этим вопросам в 1913 г., В. А. снова отбрасывает метод асимптотических выражений; в обширном мемуаре [46] он обобщает задачу, рассматривая самый общий случай граничных условий и предполагая, что каждое из них зависит от значений функции и её производной на обоих концах промежутка, имея вид

$$(51) \qquad a_1^{(i)} V_k(a) + a_2^{(i)} V'_k(a) + a_3^{(i)} V_k(b) + a_4^{(i)} V'_k(b) = 0, \quad i = 1, 2.$$

Указав для коэффициентов $a_j^{(i)}$ этих граничных условий зависимости, необходимые для ортогональности фундаментальных функций, соответствующих им, В. А. устанавливает основные положения теории, откинув предположение, что коэффициент q положителен. Считаясь с последним обстоятельством, он приближается к своим методам применения теоремы (A), откинутым им с 1900 г.; пользуясь неравенством, вытекающим из (47), он устанавливает разложимость всякой функции, равной интегралу интегрируемой ограниченной функции, подчиняя её, однако, некоторым ограничениям на границах, когда из граничных условий (51) вытекает зависимость, не содержащая значений производных.

В 1902 г. в журнале Крелля помещён мемуар В. А., посвящённый разложению по полиномам Чебышева [47]. В этом мемуаре, после установления замкнутости систем полиномов Чебышева V_k, определённых в ограниченном промежутке условием:

$$(52) \qquad \int_a^b pPV_k\,dx = 0,$$

для всякого полинома P степени ниже k, если степень V_k равна k, рассмотрен вопрос о разложимости произвольной функции по полиномам Якоби, отвечающим функции

$$(53) \qquad p = C(x-a)^{\alpha-1}(b-x)^{\beta-1}, \qquad \alpha > 0, \ \beta > 0.$$

Эта разложимость установлена для всех возможных показателей α и β, для точек внутри области и для функций, имеющих вторые производные, причем доказана замечательная теорема, позволяющая в некоторых случаях расширить этот результат; ею устанавливается сходимость ряда во всем промежутке, если он сходится на одном из его концов.

В 1903 г. [28] разложимость внутри промежутка доказана без предположения о существовании вторых производных оригинальной оценкой дополнительного члена разложения.

В мемуаре 1907 г. [43] методом асимптотических выражений теорема распространена на функции ограниченной вариации.

Годы, начиная с 1914, посвящены обобщениям полученных результатов.

Три мемуара 1914 и 1921 гг. [48—50] посвящены различным обобщениям функции p, причём оказалось, что обобщение функции p заставляет налагать лишние условия на разлагаемую функцию.

Один из мемуаров 1921 г., написанный по поводу юбилея Чебышева [51], содержит доказательство разложимости без теории замкнутости, непосредственной оценкой дополнительного члена. Для разложимости функции оказывается достаточным существование интеграла

$$\int_a^b p(y) F^2(x, y)\, dy, \qquad F(x, y) = \frac{f(x) - f(y)}{x - y}; \tag{54}$$

условие это соблюдено, например, для полиномов Якоби, если f удовлетворяет условию Липшица.

Наконец, три работы посвящены распространению результатов на полиномы Лапласа — Эрмита — Чебышева и Чебышева — Лагерра, соответствующие бесконечно протяжённым промежуткам, что представляет своеобразные трудности. Результат этих мемуаров [52—54] перечислен в приложении IV.

С осени 1918 г. по весну 1920 г. В. А. читал избранной аудитории курс математической физики, сначала в университете, а затем, за невозможностью, в помещениях Академии.

Эти лекции послужили основанием для его обширного труда об основных задачах математической физики [55], первые два тома которого ему удалось издать при жизни.

Человек необыкновенной энергии, В. А. обнаруживал свою кипучую натуру не только в практической деятельности; таков он был и в своей научной работе.

Его творческая мысль работала непрерывно, и возникавшие у него новые идеи претворялись им в дело немедленно. Каждый год приносил несколько работ В. А.; за мемуаром, появившимся в ноябре, следовал в декабре другой, служивший уже развитием первого.

Эта спешность бывала причиной недочётов, незаметных, когда видишь целое; но она же двигала дело — так как новые мысли живут и рождают другие, только выпущенные в мир.

Его кипучей натуре и неустанной деятельности столь же обязаны мы, его современники, сколько будут обязаны наши потомки — пышный сад расцветёт из семян, разбросанных им в течение жизни.

Приложение I.

Схемы В. А. Стеклова решения задач Дирихле и Неймана:

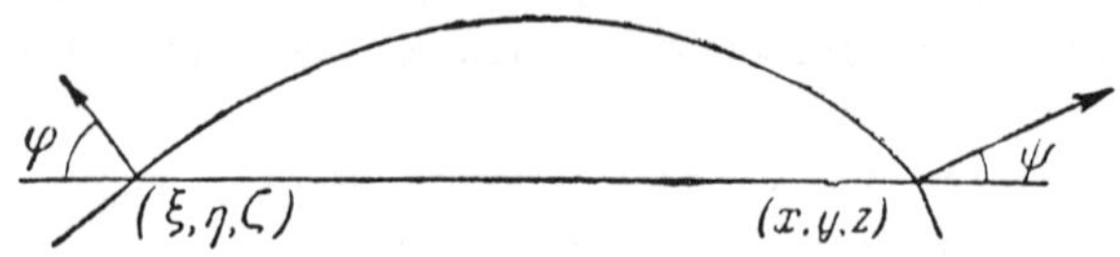

$$\rho_1=\frac{1}{2\pi}\int\rho_0\frac{\cos\psi}{r^2}d\sigma,\quad V_1=-\frac{1}{2\pi}\int\rho_0\frac{ds}{r},\quad V_{ki}=\overline{V}_k=V_{ke}\quad \rho_k=\frac{1}{2\pi}\int\rho_{k-1}\frac{\cos\psi}{r^2}ds,$$

$$V_k=-\frac{1}{2\pi}\int\frac{dV_{k-1}}{dn}\frac{ds}{r},\quad \overline{V}_k=\frac{1}{2\pi}\int\overline{V}_{k-1}\frac{\cos\varphi ds}{r^2},\quad \int\rho_k ds=\int\rho_0 ds,\quad \rho_k=\frac{dV_k}{dn}.$$

Принцип Робэна:

Если $\int\rho_0 ds=0$, то $|\rho_k|<N\tau^k$, $0<\tau<1$.

Следствие его:

Если $\int\rho_0 ds\neq 0$, то $|\rho_k-\rho_{k-1}|<N\tau^k$.

N_i. Если $V=\frac{1}{2\pi}\int\frac{\mu}{r}d\sigma$, $\mu=f+\rho_1+\rho_2+\ldots$, $\rho_0=f$, $\int f ds=0$, то $\frac{dV_i}{dn}=f$, R. $\overline{\rho}=\frac{1}{2\pi}\int\frac{\overline{\rho}\cos\psi}{r^2}ds$ на (S); $\rho=$ прд ρ_k, $k\to\infty$, N_e. Если $V=-\frac{1}{2\pi}\int\left(\mu+\frac{\rho}{2}\right)\frac{ds}{r}$, $\mu=f-\rho_1+\rho_2-\ldots$, $f=\rho_0$, то $\frac{dV_e}{dn}=f$.	простой слой

Потенциалы Неймана:

$$W_1=\frac{1}{2\pi}\int f\frac{\cos\varphi}{r^2}ds,\quad W_k=\frac{1}{2\pi}\int\overline{W}_{k-1}\frac{\cos\varphi}{r^2}ds.$$

Если

$$\frac{dW_{2i}}{dn}=\frac{dW_{2e}}{dn}=L_0,\quad V_2=-\frac{1}{2\pi}\int\frac{L_0}{r}ds,\quad V_k=-\frac{1}{2\pi}\int\frac{dV_{k-1}}{dn}\frac{ds}{r},$$

$$v_k=\frac{1}{2\pi}\int\overline{v}_{k-1}\frac{\cos\varphi}{r^2}ds,\quad v_2=V_2,$$

то

$$V_k=W_{k+1}-W_k,\quad v_k=W_{k+1}-W_{k-1}.$$

Принцип Неймана: $|\overline{W}_k-\overline{W}_{k-1}|<N_0\tau^k$, следствие принципа Робэна.

D_i. Если $W=\frac{1}{4\pi}\int\frac{\mu'\cos\varphi}{r^2}ds$, $\mu'=f-(\overline{W}_1-f)+(\overline{W}_2-\overline{W}_1)-\ldots$, то $W_e=f$. D_e. Если $W=\frac{C}{C_0}\int\frac{\rho ds}{r}-\frac{1}{4\pi}\int\frac{\mu'\cos\varphi}{r^2}ds$, $\mu'=f+(\overline{W}_1+f-2C)+(\overline{W}_2+\overline{W}_1-2C)+\ldots$, то $W_e=f$, $\lvert\overline{W}_k-C\rvert<\lambda\tau^k$, $C=\frac{\int\rho f ds}{\int\rho ds}$, $C_0=V_e$, $V=\int\frac{\rho ds}{r}$,	двойной слой

N_i. Если $W = -\frac{1}{2\pi}\int \frac{\mu'' \cos\varphi}{r^2} ds$,
$\mu'' = V_1 + V_3 + \overline{V}_5 + \ldots,\ \rho_0 = f$,
N_e. Если $W = -\frac{1}{2\pi}\int \frac{\rho}{r} ds - \frac{1}{2\pi}\int \frac{\mu'' \cos\varphi}{r^2} ds$,
то $\frac{dW_i}{dn} = f,\ \frac{dW_e}{dn} = f$.
} двойной слой

D_i. Если $\frac{dW_{1i}}{dn} = \frac{dW_{1e}}{dn} = L_0$, f удовлетворяет условию Липшица,
D_e. $V_1 = \frac{1}{2\pi}\int \frac{L_0}{r} ds,\ \ V_k = -\frac{1}{2\pi}\int \frac{dV_{k-1}}{dn}\frac{ds}{r},\ \ \rho_k = \frac{dV_k}{dn},\ \ \rho = \text{прд}\,\rho_k$,
$W = \frac{1}{2\pi}\int \frac{\mu^{(*)} ds}{r},\ \ \mu^{(*)} = \frac{\rho}{2} - L_0 + \rho_2 + \rho_4 + \ldots$, то $W_i = W_e = f$.
} простой слой

Приложение II.

Разложение по фундаментальным функциям.

1. Гармонические функции Пуанкаре [20].

$$\Delta V_k + \lambda_k V_k = 0 \text{ внутри } (D),\ V_k = 0 \text{ на } (S).$$

Если 1) f имеет производные 2-го порядка,
2) $f = 0$ на (S).

2. Функции Пуанкаре-Стеклова № 1 [42].

$$\Delta V_k + \lambda_k V_k = 0 \text{ внутри } (D),\ \ \frac{dV_k}{dn} = 0 \text{ на } (S).$$

Если 1) f имеет производные 2-го порядка,
2) $\frac{df}{dn} = 0$ на (S).

3. Функции Пуанкаре-Стеклова № 2 [22, 42].

$$\Delta V_k + \lambda_k V_k = 0 \text{ внутри } (D),\ \ \frac{dV_k}{dn} + hV_k = 0 \text{ на } (S).$$

Если 1) f имеет производные 2-го порядка,
2) $\frac{df_i}{dn} + hf = 0$ на (S).

4. Универсальные функции А. Корна [42].

$$\Delta V_k + \lambda_k V_k = 0 \text{ внутри } (D),\ \ \ \Delta V_k = 0 \text{ вне } (D).$$

V_k непрерывна с первыми производными во всём пространстве.

Если $f = \int \frac{\rho d\tau}{r}$, ρ интегрируемая и ограниченная.

5. Фундаментальные функции Стеклова [21].

$$\Delta V_k = 0 \text{ внутри } (D),\ \frac{dV_{ki}}{dn} = \lambda_k \varphi V_k \text{ на } (S),\ \varphi > 0.$$

$$\text{Если } \Phi = \sum_0^\infty A_k V_k,\ A_k = \int \varphi f V_k ds, \text{ то } \Phi = f \text{ на } (S).$$

$$\text{Если } \Phi = aV_0^2 + \sum_1^\infty \frac{B_k V_k}{\lambda_k},\ \ \ B_k = \int f V_k ds,$$

$$\text{то } \frac{d\Phi_i}{dn} = f,\ \ \ a = \int \varphi \Phi_i ds.$$

6. Фундаментальные функции Ле Руа [22].

$$\Delta V_k = 0 \text{ внутри } (D),\ \frac{dV_{ki}}{dn} - \frac{dV_{ke}}{dn} = \lambda_k \varphi V_k,\quad \varphi > 0.$$

Если $f = \int \frac{\mu ds}{r}$ и $\int \mu^2 ds$ имеет смысл.

7. Фундаментальные функции Пуанкаре [22]

$$\Delta V_k = 0 \text{ внутри } (D),\ \frac{dV_{ki}}{dn} - \frac{dV_{ke}}{dn} = -\lambda_k \left(\frac{dV_{ki}}{dn} + \frac{dV_{ke}}{dn}\right).$$

Решение задач Дирихле и Неймана.

Приложение III.

Разложение по фундаментальным функциям.

8. Функции Штурма-Лиувилля.

(Последовательное развитие метода)

$$V'' + [kp - q]\, V = 0,$$
$$V'(a) - hV(a) = 0,\quad V'(b) + HV(b) = 0,$$

$p > 0$ $q > 0$, и непрерывны.

a) 1896 г. [27] 1) f имеет четвёртые производные,

2) $f'(a) - hf(a) = 0,\quad f'(b) + Hf(b) = 0.$

3) $\psi'(a) - h\psi(a) = 0,\quad \psi'(b) + H\psi(b) = 0,\quad \psi = \frac{fq - f''}{p}$;

b) 1901 г. [31] 1) f имеет вторые производные,

2) $f'(a) - hf(a) = 0,\quad f'(b) + Hf(b) = 0$;

c) 1904 г. [28] $q = 0$ 1) f имеет одну производную, интегрируемую и ограниченную

$h = H = \infty$ 2) $f(a) = 0$, $f(b) = 0$;

d) 1907 г. [37] 1) f имеет одну производную, непрерывную и ограниченную;

e) 1910 г. [38] 1) $f = \int_a^x \varphi dx + C$, φ интегрируемая и ограниченная;

f) 1910 г. [39] 1) f удовлетворяет условию Липшица; $p > 0$, q какое угодно и непрерывны. Обобщение граничных условий;

g) 1913 г. [46] $f = \int_a^x \varphi dx + C$, φ ограниченная интегрируемая.

Если α) $V'(a) = \alpha V(a) + \beta V(b)$ $\alpha + \delta = 0$, без новых условий,
$V'(b) = \gamma V(a) + \delta V(b)$,

β) $\delta V(a) + \beta V(b) = 0$,
$\alpha V(a) + \lambda V'(a) + \gamma V(b) + \delta V'(b) = 0$,
$\beta^2 + \delta^2$ не $= 0$ при $\delta f(a) + \lambda f(b) = 0$,

γ) $V(a) = 0$, $V(b) = 0$ при $f(a) = 0$, $f(b) = 0$.

Метод асимптотических выражений:

A. 1907 г. [43] p и q имеют непрерывные третьи производные, f ограниченной вариации.

B. 1910 г. [45] p и q имеют вторые производные ограниченной вариации,
или 1) f непрерывная ограниченной вариации,
или 2) $|f(x+h) - f(x)| < \lambda h^\alpha$,
или 3) прд $\log h\, [f(x+h) - f(x)] = 0$ $h \to 0$.

Приложение IV.

Разложение по полиномам Чебышева.

$$p > 0 \int_a^b pPV_k dx = 0,$$

если V_k степени k и P произвольный полином степени $k-1$.

a) Полиномы Якоби

$$p = C(x-a)^{\alpha-1}(b-x)^{\beta-1}; \quad \alpha > 0, \; \beta > 0.$$

α) [47] f имеет производную первого порядка интегрируемую и ограниченную. Разложима в (a_1, b_1), $a < a_1 < b_1 < b$.

β) [51] (без теории замкнутости)

$$|f(x) - f(x')| < \lambda |x - x'|^\gamma, \quad 0 < \gamma \leqslant 1.$$

Разложима в (a, b).

γ) [43] (метод асимптотических выражений) f ограниченной вариации.

b) [48]$\cdot p = \int_a^x \varphi(x)\, dx + C$, $p > 0$, φ интегрируемая и ограниченная,

$$f = \int_a^x f' dx + C;$$

f' интегрируемая и ограниченная.

Разложима в (a, b)

c) [48]

$$p = (x-a)^{\alpha-1}(b-x)^{\beta-1} q(x), \quad q(x) = \int_a^x \varphi dx + C;$$

φ интегрируемая и ограниченная, $q > 0$

$$f = \int_a^x f' dx + C;$$

f' интегрируемая и ограниченная.

Разложима в (a_1, b_1), $a < a_1 < b_1 < b$

d) $p = (x-a)^{\alpha-1}(b-x)^{\beta-1}\Pi(x-x_j)^{2\alpha j}\cdot q$, $q > 0$ и интегрируема.

α) f удовлетворяет условию Коши,
разложима в (a_1, b_1), $a < a_1 < b_1 < b$.

β) f имеет вторые производные,
разложима в (a, b).

e) [51] (без замкнутости). Если $\int_a^b p(y) F^2(x, y)\, dy$ имеет смысл,

$$F(x, y) = \frac{f(x) - f(y)}{x - y};$$

разложима в (a_1, b_1), $a < a_1 < b_1 < b$.

f) [54] Полиномы Чебышева-Лагерра:

$$p = x^\beta e^{-x}.$$

α) Если $-1 < \beta < 0$, $f = \int_0^x \varphi\, dx + C$, φ интегрируемая, если $\int_0^\infty x^\beta e^{-x} f^2\, dx$, $\int_0^\infty x^{\beta+1} e^{-x} \varphi^2\, dx$ имеют смысл.

β) Если $-1 < \beta < m$, f имеет производные порядка $m+1$, если $\int_0^\infty x^{\beta+k} e^{-x} [f^{(k)}(x)]^2\, dx$, $k = 0, 1, 2 \ldots m+1$, имеют смысл.

Список литературы.

[1] Fourier, Théorie analytique de la chaleur, 1822.

[2] Poisson, Théorie mathématique de la chaleur, 1835.

[3] Journ. de Liouville, 1836, t. 1, pp. 106, 253, 269, 373; 1837, t. 2, pp. 16, 220, 418.

[4] Weierstrass. Über das sogenannte Dirichlet'sche Princip, 1870.

[5] C. Neumann. Untersuchungen über das logarithmische und Newtonische Potential, 1877.

[6] H. Schwarz, Über ein die Flächen kleinsten Flächeninhalts betreffendes Problem. Zweiter Teil. Integration der Differentialgleichung $\frac{\partial^2 u}{\partial x^2}+\frac{\partial^2 u}{\partial y^2}+pu=0$ unter vorgeschriebener Bedingungen, 1885.

[7] E. Picard, Sur les équations aux dérivées partielles qui se présentent dans la théorie de la vibration des membranes. CR, 1893.

[8] E. Picard, Sur les équations linéaires du second ordre renfermant un paramètre arbitraire, CR, 1894.

[9] H. Poincaré, Sur les équations de la physique mathématique. Rend. del'Acc. del Lincei, 1894.

[10] H. Poincaré, La méthode de Neumann et le problème de Dirichlet, 1897.

[11] В. А. Стеклов, О разложении данной функции в ряд по гармоническим функциям. Сообщения Харьковского математического общества, 1896.

[12] H. Poincaré, Sur les équations aux dérivées partielles de la physique mathématique. Amer. Journ., 1889, vol. 12.

[13] C. Neumann, Untersuchungen über das logarithmische und Newtonische Potential, 1877, Kap. 5, § 6, p. 183.

[14] В. А. Стеклов, К вопросу о существовании конечной и непрерывной внутри данной области функции координат, удовлетворяющей уравнению Лапласа, при заданном значении её нормальной производной на поверхности, ограничивающей область. Сообщ. Харьк. Мат. О-ва, 1897.

[15] Robin, Distribution de l'électricité sur une surface fermée convexe. CR, 1887, t. 104.

[16] W. Stekloff, Le problème de Neumann et le problème de la distribution de l'électricité, CR. 1897.

[17] A. Liapounoff, Sur quelques questions qui se rattachent au problème de Dirichlet. Journ. de Liouville, 1898.

[18] W. Stekloff, Sur la méthode de Neumann et le problème de Dirichlet. CR, 1897, t. 180.

[19] S. Zaremba, Sur la théorie de l'équations de Laplace et les méthodes de Neumann et de Robin. Bull. de l'Ac. des Sc. de Cracovie, 1901.

[20] W. Stekloff, Les méthodes générales pour résoudre les problèmes fondamentaux de la physique mathématique. Ann. de Toulouse, 1900, t. 2.

[21] В. А. Стеклов, Общая метода решения задач математической физики.

[22] W. Stekloff, Sur les problèmes fondamentaux de la physique mathématique. Ann. de l'Éc. Norm. Sup., 1902, t. 19.

[23] В. А. Стеклов, О дифференциальных уравнениях математической физики. Мат. Сб., 1897.

[24] Le Roy, Sur l'intégration des équations de la chaleur. Ann. de l'Éc. Norm. Sup., 1898.

[25] П. Л. Чебышев, О непрерывных дробях, 1855.

[26] П. Л. Чебышев, О разложении функции одной переменной, 1859.

[27] В. А. Стеклов, Задача об охлаждении неоднородного твёрдого стержня. Сообщ. Харьк. Мат. О-ва, 1896.

[28] W. Stekloff, Sur certaines égalités générales communes à plusieurs séries de fonctions souvent employées dans l'analyse. Зап. Имп. Ак. Наук по Физ.-Мат. Отд., VIII сер., 1904, т. XV.

[29] W. Stekloff, Sur la condition de fermeture des fonctions orthogonales, CR, 1910.

[30] W. Stekloff, Mémoire sur les fonctions harmoniques de M. Poincaré. Ann. de Toulouse, 1900, t. 2.

[31] W. Stekloff, Problème de refroidissement d'une barre hétérogène. Ann. de Toulouse, 1901, t. 3.

[32] W. Stekloff, Une application nouvelle de ma méthode de développement suivant les fonctions fondamentales. CR. 1910.

[33] W. Stekloff, Sur la théorie des séries trigonométriques. Bull. de l'Ac. des Sc. de Cracovie, 1903.

[34] W. Stekloff, Sur le développement d'une fonction donnée suivant les fonctions harmoniques. CR, 1893.

[35] W. Stekloff, Sur un probléme de la théorie analytique de la chaleur. CR, 1899.

[36] В. А. Стеклов, Задача об охлаждении неоднородного твёрдого стержня. Сообщ. Харьк. Мат. О-ва, 1896.

[37] W. Stekloff, Sur un probléme d'analyse, intimement lié avec le problème de refroidissement d'une barre hétérogène. CR, 1910.

[38] W. Stekloff, Sur le développement d'une fonction arbitraire en série de fonctions fondamentales. CR, 1910.

[39] W. Stekloff, Une application nouvelle de ma méthode de développement suivant les fonctions fondamentales. CR, 1910.

[40] Kneser, Untersuchungen über Darstellung willkülhricher Fonctionen in der mathematischer Physik. Math. Ann., 1903, Bd. 58.

[41] Gilbert, Grundlagen einer allgemeinen Theorie der linearen Integralgleichungen. Gött. Nachr., 1904.

[42] W. Stekloff, Théorie générale des fonctions fondamentales. Ann. de Toulouse, 1905, t. 6.

[43] W. Stekloff, Sur les expressions asymptotiques de certaines fonctions définies par les équations différenticlles du second ordre et leurs application au problème du développement d'une fonction arbitraire en séries, procédant suivant lesdites fonctions. Сообщ. Харьк. Мат. О-ва, 1907.

[44] W. Stekloff, Sur l'existence des fonctions fondamentales, correspondant à une équation différentielle du second ordre. Rend. del'Acc. dei Lincei, 1910.

[45] W. Stekloff, Solution générale du problème de développement d'une fonction arbitraire en séries suivant les fonctions fondamentales de Sturm — Liouville. Rend. del'Acc. dei Lincei, 1910.

[46] W. Stekloff, Sur certaines questions d'analyse qui se rattachent à plusieurs problèmes de la physique mathèmatique. Зап. Имп. Ак. Наук по Физ.-Мат. Отд., VIII сер., 1919, т. XXXI.

[47] W. Stekloff, Sur le développement d'une fonction donnée en séries procédant suivant les polynomes de Tchébychef. Journ. f. reine u. angew. Math., 1902.

[48] W. Stekloff, Sur une application de la théorie de fermeture au problème de développment des fonctions arbitraires en séries procédant suivant les polynomes de Tchébychef. Зап. Имп. Ак. Наук по Физ.-Матем. Отд., VIII сер., 1914, т. XXXIII.

[49] W. Stekloff, Sur le développement des fonctions continues en séries de polynomes de Tchébychef. Изв. Росс. Ак. Наук, 1921, т. XV.

[50] W. Stekloff, Une contribution nouvelle au problème du développement des fonctions arbitraires en séries de polynomes de Tchébychef. Изв. Росс. Ак. Наук, 1921.

[51] W. Stekloff, Une méthode de la solution du problème de développement des fonctions en séries de polynomes de Tchébychef indépendante de la théorie de fermeture. Изв. Росс. Ак. Наук, 1921.

[52] W. Stekloff, Théorème de fermeture pour les polynomes Laplace — Hermite — Tchébychef. Изв. Имп. Ак. Наук, 1916, т. X.

[53] W. Stekloff, Théorème de fermeture pour les polynomes de Tchébychef — Laguerre. Изв. Имп. Ак. Наук, 1916.

[54] W. Stekloff, Sur le développement des fonctions arbitraires en séries de polynomes de Tchébychef — Laguerre. Изв. Имп. Ак. Наук, 1916.

[55] В. А. Стеклов, Основные задачи математической физики, 1922, ч. I и 1923, ч. II.

Dimitrii Fedrovich Egorov

Born – 22 Dec 1869 in Moscow, Russia
Died – 10 Sept 1931 in Kazan, USSR

Dimitrii Egorov attended school in Moscow then entered Moscow University to study mathematics and physics, enrolling in 1887. The teacher who influence him most at this time was Bugaev. Egorov wrote his first paper in 1892 on numerical integrals and derivatives, it was clearly influenced by Bugaev's work.

Egorov taught at Moscow University from 1894, obtaining a doctorate in 1901. He spent a year abroad, then in 1903 he returned to become a professor at Moscow University.

Egorov worked on triply orthogonal systems and potential surfaces, making a major contribution to differential geometry. Some of Egorov's work was presented by Darboux in his famous four volume work *Leçons sur la théorie général des surfaces et les applications géométriques du calcul infinitésimal.*

Egorov also worked on integral equations and a theorem in the theory of functions of a real variable is named after him. Luzin was Egorov's first student and became a member of the school Egorov created in Moscow dealing with functions of a real variable.

In 1917 Egorov became secretary of the Moscow Mathematical Society. Then in 1921 he was elected vice-president, and becoming president the following year. In 1923 Egorov became director of the Institute for Mechanics and Mathematics at Moscow State University.

Egorov was a deeply religious man and when the Church was repressed after the revolution, Egorov defended them. In 1922–1923 there were mass execution of clergy and in 1928 the attack was renewed. Egorov was in a position of power in the Moscow Mathematical Society and he tried to shelter academics who had been dismissed from their posts. He tried to prevent the attempt to impose Marxist methodology on scientists.

In 1929 Egorov was dismissed as director of the Institute for Mechanics and Mathematics and given a public rebuke. Some time later he was arrested as a "religious sectarian" and put in prison. The Moscow Mathematical Society continued to support Egorov, refusing to expel him, and those who presented papers at the next meeting, including Kurosh, were to be expelled by an "Initiative group" took over the Society in November 1930. They expelled Egorov denouncing him as a "reactionary and a churchman".

Egorov went on a hunger strike in prison and eventually, by this time close to death, he was taken to the prison hospital in Kazan. It was reported that Egorov died at Chebotaryov's home.

Years Ago

Allen Shields*

Luzin and Egorov. Seventy-five years ago, Nikolaĭ Nikolaevič Luzin[1] published his famous theorem stating that a measurable function becomes continuous if one deletes a suitable set of arbitrarily small measure (Luzin [1912_1]). He based the proof on the theorem of his teacher Dmitriĭ Fedorovič Egorov, to the effect that if a sequence of measurable functions converges almost everywhere on a set of finite measure, then the convergence becomes uniform if we delete a suitable set of arbitrarily small measure (Egorov [1911]). (These authors stated the theorems for functions on an interval of the real axis, but the proofs work more generally.) Luzin [1912_2] showed that if a function f is measurable and finite almost everywhere on an interval, then there exists a continuous function F, differentiable almost everywhere on the interval, such that $F'(x) = f(x)$ almost everywhere.

Egorov was born in December 1859, became Privat Docent at Moscow University in 1893, and professor in 1903. He spent 1902–1903 abroad on leave of absence, paid by the university; the first summer he was in Berlin, the second summer in Göttingen, the fall and winter in Paris. He seems to have been one of the first foreign mathematicians to have embraced Lebesgue's theory, although during his life he worked also in number theory, partial differential equations, and differential geometry. Among his students were V. V. Golubev, N. N. Luzin, I. G. Petrovskiĭ, I. I. Privalov, and V. V. Stepanov (see Aleksandrov [1955], pp. 28, 43, and Kuznetsov [1971], p. 188). He was director of the Institute of Mathematics and Mechanics at Moscow University from 1921 until 1930.

Luzin was born in December 1883, and entered Moscow University in 1901. In the fall of 1905 the university was shut down for much of the semester because of the revolutionary situation in the country. Bari and Golubev [1951], p. 15, state that Luzin sometimes sheltered in his room people who were wanted by the police, and once a bomb was hidden under his bed. He continued to study with Egorov during this period. Some of Egorov's letters to Luzin from 1905–1914 have been published (see Egorov [1980]). These interesting letters contain much personal advice and support, in addition to mathematical ideas. We quote the letter of 4.VIII.1905 (old calendar).

> Respected Nikolaĭ Nikolaevič.
>
> Today I received a letter from V. A. Kostitsyn which caused me to worry very much about you. What is the matter? Why have you written me nothing, nothing all this time? Please believe me, I love you with all my soul. Why such despair? You must understand that human thought is not an empty game. It has accomplished much and will accomplish more; however, one must not expect the impossible from it and demand everything all at once. The problems of the world are not solved by the stroke of a pen and one must think through much and live through much before coming to a world outlook. Also, it changes with time; this is not surprising since life doesn't stand still. Don't disturb yourself with dismal conclusions from a gloomy philosophy.
>
> . . . Remember, you are a mathematician and the duty of a mathematician is to have a critical attitude toward everything! Maybe there are problems that we won't solve just yet by such a critical attitude but that is no reason to fall into ruinous mistakes.
>
> As regards a guide to life, people are not given reason alone; a sensitive conscience always reminds us that the task of life is to do good, to the limit of our abilities. In the end the world does contain love!
>
> Please forgive me that thus uninvited I have invaded your inner life; I hope you will answer me at once.
>
> Yours, D. Egorov

[1] We have chosen the spelling "Luzin," which is the correct transliteration of the Cyrillic into English, instead of "Lusin," which is the transliteration into German.

* **Column editor's address:** Department of Mathematics, University of Michigan, Ann Arbor, MI 48109 USA

Egorov arranged for Golubev and Luzin to go abroad, and they left for Paris in December 1905. Luzin was especially impressed with Poincaré, who seemed to be creating the mathematics during the lectures. Luzin was on a tight budget and took his meals at a Russian student cafeteria on rue St. Jacques; dinner cost 40 centimes. He returned to Moscow in 1906, passed the state examinations, and graduated. Egorov arranged for him to stay on at the university for "training to become a professor," in other words, as a graduate student. In 1910 he became Privat Docent, and was sent abroad on a paid leave of absence by the university, with a stipend of 1200 rubles for the year.

Luzin spent the fall in Göttingen and apparently spent most of his time in the library. We quote from Egorov's letter of 7.XII.1910.

> Thus far you seem to have been too shy to have a closer relationship with the Göttingen mathematicians. Is this really possible!? Then it was not worth traveling to Göttingen: you can read books in Moscow, and the lectures there won't contain much that's new for you. Everything comes down to having closer contacts; you will benefit much from them. Perhaps you haven't even been to the meetings of the math society. This is really unforgivable. So, make up for lost time as quickly as possible and go at least to Landau. He is very amiable, and if you have difficulties with the German language then fall back on French. And then as fast as possible you must perfect your spoken German: practice, talk to everyone you meet, with acquaintances, with strangers, at every opportunity! And then as soon as possible go to Klein and Hilbert.

Apparently Luzin took the advice to see Landau, for he published his first paper soon after, at Landau's insistence (Luzin [1911]). In it he gave an example of a power series with coefficients tending to zero that diverges at every boundary point of the unit disk in the complex plane. This initiated a series of investigations on the boundary convergence of power series. In 1912 W. Sierpiński gave an example of a power series converging at exactly one boundary point (see Landau [1919], Chap. 4). Later this was extended by Zygmund, using the methods of his teacher Rajchman: given an arbitrary closed subset E of the boundary, there is a power series converging at each point of E and diverging at each point of the complement (see Zygmund [1959], Chap. 9, Thm. 5.2). Then Herzog and Piranian [1949] showed that this remains true if E is a set of type F_σ. In the other direction the convergence set must be of type $F_{\sigma\delta}$, and it remains an open question whether every set of this type can occur.

Luzin stayed abroad until 1914, with Egorov fighting each year to have the Ministry of Higher Education extend the leave and raise the stipend to 2000 rubles. The ministry refused, saying they didn't have the funds to extend him at all, let alone increase the stipend. Then they agreed to extend his leave of absence, but without salary. Eventually Egorov was successful, however, and Luzin received the 2000 rubles (see Kuznetsov [1971], p. 199).

N. N. Luzin (1883–1950)

In 1916 Luzin defended his thesis for the "Magister" degree (at that time this was something like an American Ph.D., I think). The thesis, "Integral and trigonometric series," was judged so remarkable that he was directly awarded the degree of Doctor, something that had not happened in the preceding 60 years in Moscow University. The thesis has been reissued, with extensive commentaries (see Luzin [1951]). [On a personal note, I might mention that in 1952 I sent Antoni Zygmund a copy of this 1951 edition that I bought in a Russian bookstore in New York City. He wrote back thanking me and saying: "When still in Poland I had a copy of the original book, which I obtained from Luzin (of course it disappeared with all my library). The book had great influence on my work, and even now it is only partly out of date. A book of this type would be very valuable in America where even now the main interest in Fourier series flows mainly along a rather narrow and stagnant channel, filled with immobility, etc."]

Sierpiński was in Moscow at this time. We quote from Lavrent'ev [1974].

> The Polish mathematician W. Sierpiński happened to be in Moscow in 1915. Sierpiński was a German citizen and was evacuated to Moscow as a prisoner of war at the beginning of the war. As a result of the efforts of Egorov and Luzin he was allowed to live at liberty in Moscow and he was given what he needed for scientific work. He played an active part in founding Luzin's mathematical school, and he started his own when he returned to Warsaw at the end of the war.

Luzin was considered to be an outstanding lecturer. We quote from Aleksandrov [1955], p. 19.

> He had a most unusual ability to present mathematical results, compelling his listeners to take part in the very process of obtaining the results, transforming the lecture into a sort of laboratory of mathematical thought. The author . . . has never heard anyone else to compare with him in this respect.

D. F. Egorov (1869–1931), standing right; W. K. Sierpiński (1882–1969), standing left; N. N. Luzin (1883–1950), sitting.

Luzin had many students, including: P. S. Aleksandrov, N. K. Bari, A. Ya. Hinčin, L. V. Keldyš, A. N. Kolmogorov, M. A. Lavrent'ev, A. A. Lyapunov, L. A. Lyusternik, D. E. Menšov, P. S. Novikov, L. G. Šnirel'man, M. Ya. Suslin, and P. S. Uryson.

Fifty years ago the new journal *Uspehi Matem. Nauk*, in the first issue for 1937 (the second year of publication), printed the following unsigned lead article.

Get Rid of Luzinism in Scientific Circles!

The article in *Pravda* on Academician N. N. Luzin and the traditions of servility, their discussion in scientific institutions . . . and the examination of the case of N. N. Luzin by the Presidium of the Academy of Sciences USSR, placed before Soviet mathematicians, before all of Soviet science, a series of questions of the first importance.

M. Ya. Suslin (1894–1919)

P. S. Uryson (1898–1924)

What was clarified by the case of N. N. Luzin? First of all his double-dealing, combining enmity toward anything Soviet with insincere admiration for every manifestation of Soviet life, as soon as he had to openly express his opinion; where Luzin saw deficiencies and ought to have exposed them, he covered them instead with enthusiastic apologetics. He gave clearly dishonest laudatory opinions on weak works, deluding the institutions that had asked him. He wasn't ashamed to write an introduction to the ravings of a Fermatist, with the remark that the author was on the path of Fermat himself. He proposed granting a Doctorate, without defense of dissertation, to a mathematician who, according to Luzin, had discovered a "new" class of polynomials with remarkable properties, and what was most remarkable was that they reduced to the Legendre polynomials when multiplied by a constant! At the same time in his monographs and books he did not give credit to his own students, thereby "transferring" their results to himself. The condescending attitude of Luzin toward Soviet science combined with his servility toward representatives of foreign science sometimes surprised even these same foreign scientists.

There was a time when the reactionaries could speak relatively openly in scientific circles (recall the "Egorovists" in Moscow University), counting on the sympathy of a certain number of the scientists: in the first years after the October Revolution some scientists were worried that with the fall of the bourgeoisie, science would also die. The whole history of the Soviet Union shows the exact opposite: only in socialism is a true blooming of science possible. . . . That is why an enemy . . . now has to resort to duplicity, masquerading as a Soviet patriot. . . . That is why we need special vigilance. One must be able to expose enemies mercilessly, no matter behind what masks they may be hiding.

Thus we see that scientific circles still have not eliminated those habits, extraordinarily harmful in present conditions, which can be designated as "academic traditions." The facts in the Luzin case . . . were sufficiently well known; but they served only as the subject of "corridor gossip" and were never put to public discussion. The traditions of "academic decency" excluded a direct and firm public evaluation of inadequacies in the work of individual scholars, thereby obstructing the development of Bolshevik self-criticism.

The party and government are conducting a persistent struggle to raise the level of Soviet culture. . . . Liberal reviews of weak work, . . . liberalism in the awarding of degrees and titles, ambiguous reviews of bad books, . . . lead to the choking of scientific institutions. Objectively all of this is directed against the policies of the party and government . . . (although dictated not by a conscious desire to do harm but rather by cowardly obsequiousness, wishing to please people, or other equally "high" motives).

More responsibility is needed in recommendations, in reviews, and for the work of the collective. . . . If, for example, a poor textbook is approved and published, then along with the authors also the reviewers, editors, and the publishing house must answer.

Connected to these notorious academic traditions is the tradition of servility toward foreign science which came from the backwardness of prerevolutionary Russia. Our scientific press must occupy the same leading place as is occupied by Soviet science. But many of our scientists . . . publish in foreign organs. Some even publish in the "unified" fascist journals, which are boycotted by leading foreign scholars.

> The title of Soviet scholar is a high title and must be borne with honor. We must eradicate from scientific circles all remnants of the pitiful slavish past—then Soviet scholars . . . will be able to work still more fruitfully for the flourishing of Soviet science, of Soviet culture.

We mention three points of information in connection with the above article. First, a Fermatist is someone who claims to have proved Fermat's last theorem.

Second, Egorov died in exile in Kazan in September 1931, three months before his sixty-second birthday. He and Luzin are considered to be the founders of the Moscow school of pure mathematics. Egorov's work is discussed in Mat 15 [1932], Mat 30 [1948], and in Aleksandrov [1955]; Egorov [1970] is a collection of his papers in differential geometry. Kuznetsov [1971] is an appreciation of his life and work.

Third, Luzin was expelled from Moscow University but not from the Academy of Sciences. (Members of the academy receive a salary from the academy approximately equal to the salary of a professor, in addition to their salary from the university or institute where they work.) Subsequently Luzin worked in various institutes of the academy. Today his mathematical family tree is displayed in Moscow University. His work is discussed in several articles, including Mat 30 [1948], Mat 40 [1959], Menšov [1967], Keldyš [1974], and Dolženko [1985]. *Uspehi Mat. Nauk* (Vol. 40, No. 3, pp. 3–155) is devoted to articles about Luzin and his work, in connection with the centenary of his birth. For an English translation see *Russian Math. Surveys* (Vol. 40, No. 3 (1985), 1–180).

Incidentally, in reading Egorov's letters one sees that he was influenced by Hermann Weyl. In particular, Weyl [1909] introduced the concept of "essentially uniform" (*wesentlich gleichmässig*) convergence of a sequence. This means that the sequence converges almost everywhere (on a finite interval of the real axis), and to each $\epsilon > 0$ there corresponds a set of measure less than ϵ such that the convergence is uniform on the complement of this set. Weyl gave necessary and sufficient conditions for this to happen, but he missed the result, found by Egorov, that the second half of the condition is a consequence of the first half.

The article of Lavrent'ev [1974] is especially interesting for the human portrait it paints; one should read also Kolmogorov [1985]. After the present article was completed, we learned of Phillips [1978] which has much more detail on some of the topics covered here.

References

Abbreviations: MR = *Mathematical Reviews;* JRB = *Jahrbuch über die Fortschritte der Mathematik.*

P. S. Aleksandrov [1955], Mathematics in Moscow University in the first half of the XX century, *Istoriko-Matem. Issled.* 8, 9–54 (Russian). MR 17, 1.

Anonymous [1946], Nikolaĭ Nikolaevič Luzin, On his 60th birthday, *Uspehi Mat. Nauk* 1, No. 1 (1946), 226–228 (Russian).

N. K. Bari and V. V. Golubev, Biography of N. N. Luzin, in Luzin [1951], pp. 11–31 (Russian).

E. P. Dolženko and G. Ts. Tumarkin [1985], N. N. Luzin and the theory of boundary properties of analytic functions, *Uspehi Mat. Nauk* 40, No. 3 (243), 71–84, 239 (Russian). MR 86k:01033. English translation: *Russ. Math. Surveys* 40:3, 79–95.

D. F. Egorov [1911], Sur les suites des fonctions mesurables, *C.R. Acad. Sci. (Paris)* 152, 244–246. JRB 42, 423.

——— [1970], *Papers in differential geometry*, ed. S. P. Finikov, Moscow, "Nauka." MR 43, #1053.

——— [1980], Letters of D. F. Egorov to N. N. Luzin, ed. F. A. Medvedev and A. P. Yuškevič, *Istoriko-Matem. Issled.* 25, 335–361. MR 82k:01028.

F. Herzog and G. Piranian [1949], [1953], Sets of convergence of Taylor series I., II., *Duke Math. J.* 16, 529–534; ibid 20, 41–54. MR 11, 91; MR 14, 738.

L. V. Keldyš [1974], The ideas of N. N. Luzin in descriptive set theory, *Uspehi Mat. Nauk* 29, No. 5 (179), 183–196. MR 52, #7809. English translation: *Russian Math. Surveys* 29:5, 179–193.

A. N. Kolmogorov [1985], A pupil on his teacher, *Uspehi Mat. Nauk* 40, No. 3, 7–8 (Russian; Kolmogorov is interviewed by V. A. Uspenskiĭ). English translation: *Russ. Math. Surveys* 40:3, 5–7.

P. I. Kuznetsov [1971], Dmitriĭ Fedorovič Egorov, *Uspehi Mat. Nauk* 26, No. 5 (161), 168–206 (Russian). MR 52, #5311. English translation: *Russian Math. Surveys* 26:5, 125–164.

——— [1974], Nikolaĭ Nikolaevič Luzin, *Uspehi Mat. Nauk* 29, No. 5 (179), 197–210 (Russian). MR 52, #7610. English translation: *Russian Math. Surveys* 29:5, 195–208.

E. Landau [1919], Darstellung und Begründung einiger neuerer Ergebnisse der Funktionentheorie, JRB 46, 469; 2d Ed. [1929], JRB 55, 171–172.

M. A. Lavrent'ev [1974], Nikolaĭ Nikolaevič Luzin, *Uspehi Mat. Nauk* 29, No. 5 (179), 176–182 (Russian). MR 52, #7808. English translation: *Russian Math. Surveys* 29:5, 173–178.

N. N. Luzin [1911], Über eine Potenzreihe, *Rendiconti Circ. Mat. Palermo* 32, 386–390. JRB 42, 277.

——— [1912_1], Sur les propriétés des fonctions mesurables, *C.R. Acad. Sci. (Paris)* 154, 1688–1690. JRB 43, 484–485.

——— [1912_2], On the fundamental theorem of integral calculus, *Mat. Sbor.* 28, 266–294, 544 (Russian). JRB 43, 361.

——— [1951], Integral i trigonomet. ryad, edited with commentaries by N. K. Bari and D. E. Menšov, Gos. izdat. teh.-teor. lit., Moscow-Leningrad (Russian). MR 14, 2.

Mat 15 [1932], Matematika v SSSR za 15 let: 1917–1932, Moscow-Leningrad (Russian).

Mat 30 [1948], Matematika v SSSR za 30 let: 1917–1947, Gos. izdat. teh.-teor. lit., Moscow-Leningrad (Russian).

Mat 40 [1959], Matematika v SSSR za 40 let: 1917–1957, I, II, Gos. izdat. fiz.-mat. lit., Moscow-Leningrad (Russian).

E. R. Phillips [1978], Nicolai Nicolaevich Luzin and the Moscow school of the theory of functions, *Historia Mathematica* 5, 275–305. MR 81c:01027.

D. E. Menšov and P. L. Ul'yanov [1967], The metric theory of functions at Moscow University during the last fifty years, Vestnik Moskov. *Univ. Ser. I Mat. Meh.* 22, No. 5, 24–36 (Russian). MR 36, #2450.

H. Weyl [1909], Über die Konvergenz von Reihen, die nach Orthogonal-funktionen fortschreiten, *Math. Ann.* 67, 225–245. JRB 40, 310–311.

A. Zygmund [1959], *Trigonometric Series*, Cambridge University Press. MR 21, #6498.

Years Ago

Allen Shields*

Egorov and Luzin: Part 2

This column for vol. 9, no. 4 (1987) dealt with the mathematicians N. N. Luzin and D. F. Egorov. Since then additional material has come to my attention, some of it while I was visiting the Soviet Union for six weeks in May and June 1988.

We begin with Egorov. If one looks at vol. 35, no. 1 (1928) of the journal *Matematičeskiĭ Sbornik*, published by the Moscow Mathematical Society, one finds a list of all the members of the society at that time, with the date they joined the society. The president is Egorov (1893), and the vice president is Luzin (1910). (Egorov was born in 1869 and Luzin in 1883.) Curiously, Kolmogorov is not listed, although he had already published a number of important papers (he was born in 1903). Inside the front cover there is a note in French and in Russian to the effect that the journal will publish papers in Russian, French, German, English, and Italian. Papers in Russian must have a résumé in one of the other languages, and papers not in Russian must have a Russian résumé (which will be supplied by the editors upon request). The instructions about figures include the restriction that there be at most one figure for each four pages of text. Authors received 50 free reprints. Finally, all correspondence is to be sent to: Prof. D. F. Egorov, Moscow 69, Borisoglebskiĭ per., No. 8, Apartment 5. (Inside the back cover the addresses of the officers of the society are given; Luzin's address is: Moscow 2, Arbat, No. 25, Apartment 11, telephone 3-35-16.)

Three years later, inside the front cover of vol. 38, no. 1–2 (1931) of the journal one finds a different message from the editors:

> Starting with the next number the journal *Matematičeskiĭ Sbornik* will appear with the title *Sovetskiĭ Matematičeskiĭ Sbornik*.
>
> Until recently the Mathematical Society had retained its cliquish academic character. Prof. Egorov, a reactionary and a churchman, headed the society. He opposed the policies of Soviet power . . . under the guise of defending "academic traditions" and "pure apolitical science."
>
> The growth of the proletarian strata among the graduate students . . . and their struggle against the reactionary wing of the professors decisively turned the basic cadres of Soviet mathematicians . . . to active participation in social construction. . . .
>
> In the Mathematical Society an initiative group, including many prominent mathematicians, was formed to reorganize the society. The declaration of the initiative group was accepted by the Mathematical Society (it was published in the journal *Naučniĭ Rabotnik* no. 1, 1931). The Society expelled Egorov and other reactionaries, and replaced them mainly with graduate students. . . .
>
> . . . The editors of *Sovetskiĭ Matematičeskiĭ Sbornik* call upon Soviet mathematicians to close ranks around the journal and help to turn it into a fighting organ of Soviet mathematics.
>
> The editors invite the collaboration of foreign scholars sympathetic to the Soviet Union.

Inside the back cover is a notice that material for the journal should be sent to editor in chief Lazar Aronovič Lyusternik, or to the Secretary: Aleksandr Iosifovič Gel'fond. [Note: they seem to mean Aleksandr Osipovič Gel'fond; no A. I. Gel'fond is listed in Mat 40 [1959].]

The name "Sovetskiĭ Matem. Sbornik" was not in fact ever used by the journal.

As regards the initiative group [Lev], Note 12 gives the following reference: L. A. Lyusternik, L. G. Šnirel'man, and others, "Declaration of the Initiative Group for the Reorganization of the Mathematical Society," *Naučniĭ Rabotnik* 1930, no. 11/12, pp. 67–71. [The title is a translation of the Russian original. I have not seen this article, and if anyone could send me a copy I would be very grateful.]

In no. 3–4 of the same vol. 38 of the *Sbornik* one finds an appeal to Soviet mathematicians to support the journal, and to break with the old tradition of publishing their best works abroad. Because they want to have an international journal, articles in Russian will continue to have foreign résumés, and articles will be accepted in the four other languages indicated above. The editors in chief now are P. S. Aleksandrov and A. F. Bermant. Manuscripts are to be sent to Bermant. Authors now receive 100 free reprints.

In the same issue there is a report on the first all-

* **Column editor's address:** Department of Mathematics, University of Michigan, Ann Arbor, MI 48109-1003 USA

Ernest Kol'man

Russian conference on planning in mathematics, which met June 5–9, 1931. The report stresses the need for a close connection between theory and the practice of socialist construction, as called for by the Central Committee of the Party and by comrade Stalin. The necessary planning is said to be impossible in bourgeois countries. In bourgeois mathematics one observes the isolation of theory from practice. This tendency in the present period of the decay of capitalism is expressed in mathematics in formalism, which treats all mathematics as a chess game with meaningless symbols, and by subjective intuitionism.

There follows a long section presenting materials from the conference. First is a resolution based on a speech by E. Kol'man at the final session. It states that the attempt to fit the developing content of mathematics into an idealist formal scheme has led to the crisis in the foundations (Brouwer, Hilbert, and the French school of the theory of functions; Sierpiński and Luzin are mentioned as adherents of the French school). Further, although Soviet mathematics has attained one of the top places in world mathematics, it still does not have a form that expresses the socialist character of our Revolution. Until recently a basic obstacle had been the attitude of certain reactionary professors (Gyunter, Egorov, and others).

Other resolutions deal with the need for more institutes, for financial support, and for autonomy. Then there are reports on different areas of mathematics and their connection to applications.

Another reference, given to us by Prof. Šafarevič (Shafarevich), is to a journal called *Mathematical Sciences for Proletarian Cadres,* vol. 1, 1931. The article, entitled "Reorganization of the Moscow Mathematical Society," states that the Society expelled the reactionaries Egorov, Finikov, and Appel'rot. In addition, the pedagogical units were completely reorganized and have become a section in the Society, a tool in the struggle for Marxist pedagogy in mathematics. The Society elected a new presidium, including: Kol'man, Golubev, Gel'fond, Orlov, Burstin, Khinčin, Khotinski, Raikov, Frankl', Yanovskaya, Lyusternik, Lavrent'ev, and others.

Soon after being expelled from the Mathematical Society, Egorov lost his position at the University. He was arrested and intensively interrogated, apparently falling sick in prison, and then exiled to Kazan. Elsewhere in this issue of the *Intelligencer* there is a very interesting interview with Shafarevich by Smilka Zdravkovska. We quote: "As Chebotarev told me, Egorov died in Chebotarev's home, in his arms." (This was in September 1931.) Chebotarev was born in 1894, he was at Kazan University from 1928 until his death in 1947. There is more about Egorov and Luzin in the Shafarevich interview later in this issue.

Luzin's troubles in 1936 began innocently enough with a short article of his in the newspaper *Izvestiya* (June 27, p. 4) entitled "A pleasant disappointment." I have not seen this, but [Lev] gives an account, and a recent article in *Sovetskaya Kultura* has several quotes from it. Luzin describes attending the final exams in a secondary school in Moscow. He concluded by expressing complete satisfaction with the results of the exams, writing: "I could find no weak students, the students differed from one another only in that some answered quickly and some slowly, but all performed very well. On this occasion I found precisely that deep understanding of the laws of mathematics, of whose lack I have so often complained." He especially acknowledged the school director's personal contribution.

On July 3 *Pravda,* p. 3, carried an article by the school director entitled "An answer to Academician N. Luzin." He states that the students' performance was not outstanding, for example on the trigonometry exam out of 33 students 11 received the highest grade of 5, 12 received a grade of 4, and 10 received 3. The results for algebra and for geometry were similar. These results were normal, and did not at all justify Luzin's statement that he could not find any weak students. Luzin is severely criticized for overpraising the school, instead of providing comradely criticism.

The very next day *Pravda,* on p. 2, has an unsigned article entitled "On enemies in Soviet masks." It includes the following: "A careful examination of the activity of this academician in recent years shows that the deliberate overpraising of these students is far from accidental. It is but one link in a long chain, and is very instructive in showing the methods by which enemies mask themselves. We well know that N. Luzin is an anti-Soviet person. . . . Academician Luzin could have become an honest Soviet scholar, as did many from the older generation. But he didn't want this; he, Luzin, remained an enemy, counting on . . . the impenetrability of his mask. . . . It won't work, Gospodin Luzin!" ["Gospodin" means gentleman, or mister. After the Revolution it was replaced by "grazhdanin," which means citizen.]

Further articles in *Pravda* followed, on July 9, 10, 12 (p. 3 in each case). The article of July 9 was entitled

"Traditions of servility." It complains about the tradition of publishing abroad, of judging as valuable only those works that receive foreign approval:

> To take only mathematics, the majority of Soviet mathematicians (Aleksandrov, Kolmogorov, Khinčin, Bernšteĭn, and others) publish their works abroad and not here, in the USSR, in the Russian language. . . . It is known (see the articles in *Pravda* for July 2 and 3) that Academician Luzin sent his best works abroad, with the definite political intention of publishing in the USSR, as he himself cynically expressed it, "any rubbish. . . ." One must not suffer such a situation any longer. The Soviet Union is not Mexico, not some sort of Uruguay, but is a great socialist power. And the Russian language is the language of a mighty people, spoken by at least 150 million people. . . . The conditions have been created in our country for such a great flowering of science, as has not been known, and is not known, in any capitalist country.

As regards the reference to Uruguay, [Lev], Note 26, states that that country had recently broken diplomatic relations with the Soviet Union.

One of those who read the articles in *Pravda* on July 2 and 3 was the physicist Pyotr Leonidovič Kapitsa. He was outraged, and wrote a letter, dated July 6, 1936, to the Chairman of the Council of Ministers of the USSR, V. M. Molotov. Last year this letter was reprinted in the newspaper *Sovetskaya Kultura,* May 21, 1988, p. 6. We quote some excerpts:

> The article in *Pravda* on Luzin puzzled, astounded, and angered me and, as a Soviet scholar, I feel that I must tell you what I think about this.
>
> Luzin is accused of much. I don't know if these charges are correct but even if they are completely justified this does not change my negative attitude to the article.
>
> First let us begin with some accusations of a petty character. He did not publish his best works in the Soviet Union. Many of our scientists do this, and primarily for two reasons: 1) printing and paper are of poor quality here; 2) by international custom, priority is only given for works published in French, German, or English. And if Luzin published bad works in the Soviet Union, then this is the fault of the editors of the journals that accepted them.
>
> If he envied his students and therefore sometimes had inequitable relations with them, this phenomenon, unfortunately, occurs even among the strongest scholars.
>
> Thus there remains one, very serious, charge against Luzin, that he hid his anti-Soviet feelings behind flattery, though somehow this doesn't seem such a great offence. Here we have . . . a very important question of principle: how should one treat a scholar whose ethics do not correspond to the spiritual needs of the epoch?
>
> Newton, who gave mankind the law of gravity, was a religious maniac. Cardano, who found the formula for solving cubic equations and made a number of important discoveries in mechanics, was a wastrel and libertine. What would you have done with them if they had lived here in the Soviet Union?
>
> Or suppose that someone close to you fell ill. Would you call a brilliant doctor if his ethical and political convictions were contrary to your own? . . .
>
> I don't want to defend Luzin's moral qualities, but there is no doubt that he is a very strong mathematician, one of our four best mathematicians. Besides, he did more than anyone else to gather and to educate the pleiad of young Soviet mathematicians we now have.
>
> I consider that a country having strong scholars such as Luzin must first of all do everything so that their talents are used as fully as possible for mankind.
>
> People such as Luzin, who are ideologically not suitable to us, must, first, be given conditions in which they can continue their scientific work without having wide social influence. Second, everything possible must be done to reeducate them in the spirit of the epoch to make good Soviet citizens of them. . . .
>
> Was everything possible done to reeducate Luzin, and people like him in the Academy of Sciences? And can methods such as the article in *Pravda* attain this goal? . . .
>
> In general how did you go about reconstructing the Academy? You began by selecting party comrades for the Academy. This would be the best method if among the party members there were strong scholars. Leaving aside the social sciences, our party academicians are much weaker than the academicians of the older generations, and their influence is therefore small. . . . Our principal scientific capital continues to be the older generation. . . .
>
> Having in hand the economic achievements and political gains of our country, I don't understand why one cannot reeducate any academician, no matter whom, if only one proceeds in a considerate and individual manner. Recall, for example, the case of Pavlov. . . . [Editor's note: see below.]
>
> From these considerations . . . I see in the *Pravda* article only a hostile step towards our science and our Academy, since it does not reeducate our scholars and does not raise their prestige in the country.
>
> And Luzin's name is sufficiently well known in the West that such an article will not pass unnoticed. The article is weak and unconvincing, and can give rise to . . . ridicule.
>
> Seeing the harm that all this does to science in the Soviet Union, I felt that I had to write you about this.

Nikolaĭ Luzin

A day or two later a messenger arrived at Kapitsa's Institute with a package from Molotov for Kapitsa. Kapitsa was at his dacha, in the village of Zhukovka, and the messenger went there. Kapitsa opened the package and found inside his own letter. In the upper left corner of the first page, in a precise handwriting, were the words: "Not wanted, return to citizen Kapitsa. V. Molotov." Kapitsa was very angry at Molotov's answer, and in his archives there is the draft of a letter to Stalin about it. The letter was never sent, however.

After the *Pravda* articles meetings were held in universities and scientific institutes to discuss the articles. A commission was set up in the Academy to study the case. The commission's report is given in *Pravda*, Aug. 6, 1936, p. 3. Among other things Luzin is accused of publishing, without credit, results due to various students of his, including Suslin, Aleksandrov, Kolmogorov, Lavrent'ev, and Novikov (see point 5 in the report). Also, there is reference to Luzin's refusal in 1930

Luzin lost his position in the university but was not expelled from the Academy of Sciences.

to sign a letter of protest to French scientists for interfering in Soviet affairs. The French, and others in western Europe, were protesting the trial of the so-called Industrial Party. (See point 4 in the report, and see [Lev].) Luzin lost his position in the university but was not expelled from the Academy of Sciences.

Incidentally, the 1988 issue of *Sovetskaya Kultura* referred to above has two other letters by Kapitsa. One was written in 1937 to protest the arrest of the physicist Fok, and the third was written in 1980, to Andropov (then head of the KGB) to protest the treatment of the physicists Sakharov and Orlov. Kapitsa was in an unusual position when he wrote to Molotov about Luzin. For years Kapitsa had been at Cambridge University, working in the Cavendish Laboratory with Rutherford. The Soviet Union was interested in getting him back. Kol'man describes an occasion in the summer of 1931 when he and Bukharin were attending a conference in London. One day they went to Cambridge to see Kapitsa, who invited them home to dinner. After the meal Bukharin tried to convince him to return, and promised him "mountains of gold" if he would return (see [Kol], p. 176, in the chapter entitled "Return to science"). Kapitsa was hesitant, but his wife was against it. Kapitsa spent his summers at his dacha in the Soviet Union. He did this again in 1934, but when he wanted to return to England the Soviet authorities refused permission. He was given his own institute in Moscow, and since he knew they needed him he had a certain independence.

The 1980 letter of Kapitsa describes Lenin's relations with I. P. Pavlov. Pavlov was hostile to Soviet power after the Revolution. He took every opportunity to criticize the authorities, he wore his Tsarist medal, and he crossed himself whenever he passed a church. Lenin simply paid no attention to all this. For Lenin, Pavlov was a great scholar and Lenin did everything possible to provide him with good working conditions. In the early 1920s the food situation was catastrophic in Petrograd, but Lenin ordered that Pavlov's dogs should receive their full food rations. Kapitsa remembers Academician A. N. Krylov telling him of how he bumped into Pavlov one day and said to him: "Ivan Petrovič, may I ask a favor of you?" "Naturally," Pavlov replied. "Please take me as one of your dogs."

Finally we say a few words about Kol'man, who was mentioned above, and also in this column, vol. 10, no. 3, for his defense of the plant breeder and "geneticist" Lysenko against Kolmogorov. Kol'man (1892–1979) was born in Prague in a middle-class German Jewish family. He received his doctorate at the Charles University in Prague, was mobilized into the Austro-Hungarian army, and sent to the Russian front, where he was captured in 1916. He supported the Bolshevik revolution, joined the Communist Party, and stayed in the Soviet Union. He was connected for a time with the Institute of Red Professors, and with the Marx-Engels-Lenin Institute. After the Second World War he returned to Prague. In 1948, following the change in government in Czechoslovakia, he was arrested and deported to Moscow, where he spent three and a half years in Lubyanka Prison. After being released he took up science writing again.

In 1953 Kol'man was apparently the first (in the Soviet Union) to defend cybernetics, which had been much disparaged (see [Gr]). He returned to Prague where he took a critical attitude to the entry of Soviet troops in 1968. Eventually he emigrated to Sweden, where he published in memoirs (see [Kol]), in German and Russian, under the title *We should not have lived that way*. The memoirs have a list of all persons cited—there is no mention of Egorov or Luzin. There are three references to Kolmogorov—first that Kolmogorov and Aleksandrov read in manuscript a book of Kol'man in 1937, second that in 1974 they joined the chorus denouncing Solzhenitsyn when he was expelled from the Soviet Union, and third that Kolmogorov had first (October 1956) been negative towards cybernetics, and then in April 1957 changed his mind and supported it. One must conclude that the title of his book does not refer to any change of heart toward his role in the attacks on Egorov and Luzin.

Bibliography

[Kol] Arnošt (Ernest) Kol'man, We should not have lived that way, New York: Chalidze Publications (1982). [Russian].

[Gr] L. R. Graham, Science and philosophy in the Soviet Union, New York: A. Knopf (1972).

[Lev] Aleksey E. Levin, Anatomy of a public campaign: "Academician Luzin's case" in Soviet political history, preprint.

Mat 40 [1959] Matematika v SSSR za sorok let, 1917–1957, vol. I, II, *Gos. izdat. fiz.-mat. lit.*, Moscow-Leningrad (1959). (Russian).

Vladimir Ivanovich Smirnov

Born – 10 June 1887 in St Petersburg, Russia
Died – 11 Feb 1974 in Leningrad, USSR

Vladimir Smirnov was born in St Petersburg where he attended the 2nd Gymnasium, the oldest secondary school and there he won the gold medal for mathematics. From school he entered the Physics and Mathematics Faculty of St Petersburg University.

In 1910 Smirnov graduated from St Petersburg and remained at the University to study for the higher degrees which would allow him to become a university teacher. At the University a circle was formed in 1911 to study mathematical analysis and mechanics. Smirnov was a very active member of this circle, for example lecturing on the theory of algebraic equations, particularly the work of Goursat and Appell. In session 1911–1912 he gave nine lectures on Goursat's books.

Smirnov worked jointly with his friends from the 2nd Gymnasium. He published a joint paper with Friedmann in 1913 which was published in the Journal of the Russian Physico-Chemical Society (Physics Section). He wrote the first volume of his major five volume work *A Course in Higher Mathematics* jointly with Tamarkin.

From 1912 Smirnov taught at the St Petersburg Institute of Railway Engineering. He taught at Simferopol University in the southern Ukraine from 1919 to 1922, then he returned to St Petersburg (by now Leningrad). Smirnov was awarded his doctorate in 1936 and he became Head of the Institute of Mathematics and Mechanics. He became the Head of the Mathematics School at the University of Leningrad and was elected to the USSR Academy of Sciences.

In 1953 Smirnov organised the Leningrad Mathematical Seminar. To some extent this Seminar also filled the gap left when the Leningrad Mathematical Society disbanded due to political pressure in the late 1920s. Smirnov had been an active member of the Leningrad Mathematical Society through the 1920s and he was a strong believer in relaunching the Society. In 1959, mainly due to the efforts of Smirnov, it became possible to restart the Leningrad Mathematical Society and Smirnov was elected the honorary president of the Society.

Smirnov's mathematical activity was in both pure and applied mathematics. He wrote the five volume work *A Course in Higher Mathematics* referred to above, which was widely used in Russia. He worked on conjugate functions in multidimensional euclidean space and the theory of functions of a complex variable. With Sobolev he devised a method for obtaining solutions on the propagation of waves in elastic media with plane boundaries. Other applied mathematical work resulted in him developing methods for studying oscillations of elastic spheres.

THÉORIE DES FONCTIONS. — *Sur les polynomes orthogonaux à une variable complexe*. Note [1] de M. V. **Smirnoff**, présentée par M. J. Hadamard.

Soit Γ une courbe fermée et rectifiable de longueur l sur le plan de la variable complexe x. Les polynomes orthogonaux $P_n(x)$ ($n = 0, 1, 2, \ldots$) sur cette courbe satisfont aux conditions que voici [2] :

$$\frac{1}{l}\int_\Gamma P_n(x)\,\overline{P_m(x)}\,d\sigma = \varepsilon_{nm} \qquad (\varepsilon_{n,m} = 0, \text{ si } n \neq m;\ \varepsilon_{n,n} = 1),$$

où $d\sigma$ est la différentielle de l'arc. Le but de cette Note est de donner une méthode nouvelle pour obtenir un théorème assez général sur le développement suivant les polynomes $P_n(x)$.

---

(1) Séance du 28 novembre 1927.

(2) Szegö, *Math. Zeitschr.*, 9, 1921, p. 227.

Nous dirons que la fonction : $\omega(z)=\sum_{k=0}^{\infty} a_k z^k$, holomorphe dans le cercle-unité, appartient à la classe (H), si : 1° elle a presque partout des valeurs limites $\omega(e^{\theta i})$ sur les chemins non tangents; 2° $|\omega(e^{\theta i})|^2$ est sommable et 3° $\omega(z)$ est représentable par son intégrale de Cauchy. Il est facile de démontrer que : 1° la condition nécessaire et suffisante pour que $\omega(z)$ appartienne à la classe (H) est la convergence de la série $\sum_{k=0}^{\infty} |a_k|^2$; 2° cette condition étant remplie, on aura, en outre, l'équation de fermeture (formule de Parseval) :

$$\frac{1}{2\pi}\int_0^{2\pi} |\omega(e^{i\theta})|^2 d\theta = \sum_{k=0}^{\infty} |a_k|^2.$$

Soit $x=\varphi(z)$ la fonction qui fait la représentation conforme du cercle-unité sur le domaine limité par Γ. Alors les fonctions

$$R_n(z)=\sqrt{\frac{2\pi}{l}}\,P_n[\varphi(z)]\sqrt{\varphi'(z)}$$

appartiennent à la classe (H). En utilisant quelques résultats connus sur les valeurs limites des fonctions analytiques et sur les polynomes orthogonaux sur le cercle-unité ([1]), on peut démontrer que le tableau des coefficients de la série de Maclaurin pour $R_n(z)$ donne la transformation orthogonale complète. Pour démontrer ce fait on doit encore supposer que la fonction $\log|\varphi'(z)|$, harmonique dans le cercle-unité, est représentable par la formule de Poisson. Cela a lieu, par exemple, si le domaine limité par Γ est convexe ou a la forme de l'étoile. Probablement cette supposition est vraie dans le cas général du contour rectifiable.

Le fait mentionné conduit aux propositions suivantes :

I. Si la fonction $f(x)$, holomorphe à l'intérieur de Γ, a presque partout sur Γ les valeurs limites $f(\xi)$; si, en outre, $|f(\xi)|^2$ est sommable et les intégrales $\int_{\Gamma_r} |f(x)|^2 d\sigma$, où Γ_r sont les images des circonférences $|z|=r$,

([1]) Szegö, *Math. Annalen*, **84**, 1921, p. 233; *Math. Zeitschr.*, **6**, **9**, 1920, 1921, p. 176 et 175. — Riesz, *Math. Zeitschr.*, **18**, 1923, p. 94. — Privaloff, *Journal de l'École Polytechnique*, 2ᵉ série, **24**, 1924, p. 80, 84. — Fichtenholz, *Comptes rendus*, **184**, 1927, p. 1370 et 1528.

restent bornées quand $r \to 1$, on a alors l'équation de fermeture amsi que le théorème de développement en série de Fourier :

$$\frac{1}{l}\int_\Gamma |f(\xi)|^2 d\sigma = \sum_{k=0}^{\infty} |c_k|^2; \qquad f(x) = \sum_{k=0}^{\infty} c_k P_k(x) \qquad (x \text{ étant à l'intérieur de } \Gamma);$$

$$c_k = \frac{1}{l}\int_\Gamma f(\xi)\,\overline{P_k(\xi)}\,d\sigma.$$

II. Chaque série $\sum_{k=0}^{\infty} c_k P_k(x)$, pourvu que $\sum_{k=0}^{\infty} |c_k|^2$ soit convergente, est la série de Fourier d'une fonction admettant les propriétés mentionnées.

Si la courbe Γ est analytique, on peut facilement obtenir, moyennant les résultats de M. Szegö, le théorème sur le développement en série de Fourier pour la fonction possédant les valeurs limites intégrables sur Γ et s'exprimant par son intégrale de Cauchy. Pour la fonction qui n'est que holomorphe à l'intérieur de Γ, on peut également obtenir le développement suivant les polynomes $P_n(x)$, mais ce développement-ci diffère en général de celui de Fourier.

ANALYSE MATHÉMATIQUE. — *Sur les valeurs limites des fonctions analytiques.* Note de M. V. **Smirnoff**, présentée par M. Hadamard.

Soit $f(z)$ une fonction holomorphe dans le cercle K $(|z|<1)$. Nous dirons que $f(z)$ appartient à la classe H_δ $(\delta>0)$, si les intégrales

$$\int_0^{2\pi} |f(re^{i\vartheta})|^\delta \, d\vartheta$$

restent bornées pour $r\to 1$. On sait que, dans ce cas, $f(z)$ a presque partout les valeurs limites $f(e^{i\vartheta})$ sur les chemins non tangents et que $|g(e^{i\vartheta})|^\delta$ est sommable (¹). Notre but est de signaler deux théorèmes nouveaux sur les

(¹) F. Riesz, *Mathem. Zeitschrift*, 18, 1923, p. 94.

fonctions qui appartiennent à la classe mentionnée. Il nous semble qu'il est facile de tirer de ces théorèmes des conséquences importantes pour la théorie des fonctions.

Théorème I. — *Si $f(z)$ appartient à une classe $H_\delta(\delta > 0)$ et la fonction*

$$|f(e^{i\vartheta})|^\lambda \quad (\lambda > \delta)$$

est sommable, la fonction $f(z)$ appartient aussi à la classe H_λ.

Remarquons que la première condition du théorème est essentielle. Nous dirons que $f(z)$ est représentable par l'intégrale de Cauchy ou par l'intégrale de Cauchy-Stieltjès, si l'on a

$$f(z) = \frac{1}{2\pi i}\int_{(z'=e^{i\vartheta})} \frac{\varphi(z')}{z'-z}\,dz' \tag{1}$$

ou

$$f(z) = \frac{1}{2\pi i}\int_{(z'=e^{i\vartheta})} \frac{d\psi(z')}{z'-z}, \tag{2}$$

où $\varphi(z')$ est une fonction sommable et $\psi(z')$ une fonction à variation bornée.

Théorème II. — *Si $f(z)$ est représentable par l'intégrale de Cauchy ou de Cauchy-Stieltjès, elle appartient à toutes les classes H_δ pour $\delta < 1$.*

Faisons quelques remarques. On peut construire des fonctions qui sont représentables par l'intégrale de Cauchy de la forme (1), mais pour lesquelles la formule de Cauchy

$$f(z) = \frac{1}{2\pi i}\int_{z'=e^{\vartheta i}} \frac{f(e^{i\vartheta})}{z'-z}\,dz' \tag{3}$$

n'a pas lieu. L'exemple d'une telle fonction est

$$f(z) = \sum_{n=2}^{\infty} \frac{z^n}{\log n} \quad (^1).$$

Remarquons encore que toutes les classes H_δ pour $\delta > 1$ contiennent, outre les fonctions mentionnées dans le théorème II, quelques fonctions qui ne sont pas représentables par l'intégrale de Cauchy ou de Cauchy-Stieltjès. Par exemple,

$$f(z) = \frac{\log(1-z)}{1-z}.$$

Notons quelques conséquences immédiates des théorèmes précédents.

---

[1] Cet exemple m'a été communiqué par M. G. M. Fichtenholz.

1. Si une fonction est représentable par la formule de Cauchy (3), elle appartient à H_1.

2. Si $f(z)$ est représentable par l'intégrale de Cauchy, mais si la formule (3) n'a pas lieu, $f(e^{i\vartheta})$ n'est pas sommable.

3. Si une série trigonométrique, conjuguée de la série Fourier-Lebesgue, n'est pas une série Fourier-Lebesgue, elle est sommable par le procédé de Fejér vers une fonction qui n'est pas sommable.

4. On peut donner la représentation paramétrique pour toutes les fonctions de la classe H_δ (¹) :

$$f(z) = e^{\frac{1}{2\pi}\int_0^{2\pi} \log p(\vartheta) \frac{e^{i\vartheta}+z}{e^{i\vartheta}-z} d\vartheta} z^k \prod |z_s| \frac{1-\frac{z}{z_s}}{1-\bar{z}_s z} e^{\frac{1}{2\pi}\int_0^{2\pi} \frac{e^{i\vartheta}+z}{e^{i\vartheta}-z} d\psi(\vartheta)}.$$

Dans cette formule, $p(\vartheta)$ est une fonction quelconque positive pour $0 \leqq \vartheta \leqq 2\pi$, pour laquelle $[p(\vartheta)]^\delta$ et $\log p(\vartheta)$ sont sommables; $|z_s| < 1$ $(s = 1, 2, \ldots)$ et le produit $\Pi |z_s|$ est convergent, si ce produit est infini; $\psi(\vartheta)$ est une fonction quelconque décroissante pour $0 \leqq \vartheta \leqq 2\pi$ et $\psi'(\vartheta) - 0$ presque partout.

La démoustration des théorèmes I et II est très simple.

(¹) SZEGÖ, *Math. Annalen*, 84, 1921, p. 233.

PHYSIQUE MATHÉMATIQUE. — *Sur quelques problèmes de vibrations élastiques.* Note de MM. **V. Smirnoff** et **S. Soboleff**, présentée par M. Hadamard.

Dans une Note précédente [1] nous avons indiqué une méthode nouvelle pour la solution du problème plan des vibrations élastiques et nous avons appliqué cette méthode au cas des vibrations du demi-plan $y > 0$ sous l'action d'une source de vibrations longitudinales concentrée en un point $x = 0$, $y = f$ au moment $t = 0$. Maintenant nous allons considérer la source de vibrations transversales. Cette source est donnée par le potentiel

$$\psi + R[\Psi(\theta_1)], \tag{1}$$

où R est le symbole de la partie réelle et θ_1 est définie par l'équation

$$t - \theta_1 x + \sqrt{b^2 - \theta_1^2}\,(y - f) = 0 \tag{2}$$

qui fait correspondre aux rayons du faisceau conique

$$t^2 - b^2[x^2 + (y - f)^2] \geqq 0 \tag{3}$$

les points du plan θ_1 avec la coupure $(-b, +b)$. La moitié de ce faisceau coupe le plan $y = 0$ et les points d'intersection forment le domaine

$$t^2 - b^2(x^2 + f^2) \geqq 0. \tag{4}$$

De chaque point de ce domaine sortent deux rayons, l'un pour le potentiel réfléchi φ_1 des vibrations longitudinales et l'autre pour le potentiel ψ_1 des vibrations transversales. Les équations

$$t - \theta_2 x \quad \sqrt{a^2 - \theta_2^2}\,y - \sqrt{b^2 - \theta_2^2}\,f = 0, \qquad t - \theta_3 x - \sqrt{b^2 - \theta_3^2}\,(y + f) = 0 \tag{5}$$

font correspondre aux rayons de ces deux faisceaux réfléchis les demi-plans supérieurs des variables θ_2 et θ_3. Nous avons

$$\varphi_1 = R[\Phi_1(\theta_2)]; \qquad \psi_1 = R[\Psi_1(\theta_3)],$$

où

$$\Phi'_1(\theta) = \frac{4\theta(2\theta^2 - b^2)\sqrt{b^2 - \theta^2}}{F(\theta)}\Psi'(\theta);$$

$$\Psi'_1(\theta) = \frac{-(2\theta^2 - b^2)^2 + 4\theta^2\sqrt{a^2 - \theta^2}\sqrt{b^2 - \theta^2}}{F(\theta)}\Psi'(\theta);$$

$$F(\theta) = (2\theta^2 - b^2)^2 + 4\theta^2\sqrt{a^2 - \theta^2}\sqrt{b^2 - \theta^2}. \tag{6}$$

---

[1] *Comptes rendus*, **194**, 1932, p. 1437.

Aux points du domaine (4) du plan $y = 0$, les valeurs θ_1, θ_2 et θ_3 coïncident, et sur le segment de l'hyperbole $t^2 - b^2(x^2 + f^2) = 0$, dont les points correspondent aux valeurs réelles de θ qui satisfont aux conditions $b > |\theta| > a$, les rayons du potentiel φ_1 glissent sur le plan $y = 0$, les rayons réfléchis étant contenus dans ce plan ; et ce potentiel est différent de zéro hors du domaine (4). En effet, la première des équations (5) donne, après la séparation des parties réelle et imaginaire,

$$y = 0, \qquad t - x\theta^2 - \sqrt{b^2 - \theta_2^2}\, f = 0,$$

si θ_2 est réelle et $b > |\theta_2| > a$.

Le faisceau de rayons du potentiel ψ_1 forme un cône de sommet $\chi = 0$, $y = -f$, $t = 0$. Dans le cas présent nous devons prolonger le potentiel ψ_1 hors de ce cône. Soit U_α la partie du plan tangent à ce cône le long de la génératrice à laquelle correspond la valeur $\theta = \alpha$ entre le cône et le plan $y = 0$. Pour $-b < \alpha \leqq -a$ ou $a \leqq \alpha < b$, nous devons poser

$$\psi_1 = R[\Psi_1(\alpha)]$$

sur U_α. Remarquons que la partie réelle de $\Psi(\theta)$ est égale à zéro pour $-b < \theta < b$, mais en vertu des formules (6) les parties réelles de $\Phi_1'(\theta)$ et $\Psi_1'(\theta)$ sont différentes de zéro pour $-b < \theta < a$ et $a < \theta < b$.

La méthode exposée fournit la loi générale de la réflexion du plan dans l'espace (S) aux coordonnées (x, y, t) du faisceau de rayons d'un type spécial. Pour ce type le potentiel (longitudinal ou transversal) est la partie réelle d'une fonction de θ régulière sur le demi-plan supérieur où θ est la racine d'une équation de la forme

$$t - \theta x \pm \sqrt{c^2 - \theta^2}\, y = \chi(\theta) \qquad (c = a \text{ ou } b).$$

Il suit de ce fait que la méthode exposée plus haut donne la solution du problème des vibrations d'une couche limitée par des droites parallèles, ou des couches parallèles, aux constantes élastiques différentes, sous l'action de sources de vibrations du type signalé plus haut.

Si dans le cas du demi-plan $y > 0$ la source est située sur le contour, nous obtenons par une voie élémentaire la solution complète du problème traité par M. Lamb (¹).

En se servant de la méthode indiquée, on peut encore résoudre le pro-

(¹) *Philos. Trans. Roy. Soc. London*, série A, **203**, 1904, p. 1-42.

blème général de l'élasticité, avec les conditions initiales données, pour le cas du demi-plan ou d'une couche.

Dans le cas de l'espace où l'on a la symétrie axiale, la solution des problèmes peut être obtenue par la superposition des vibrations planes. Soient (ρ, z, ϑ) les coordonnées cylindriques et q, w les composantes du déplacement sur les axes ρ et z. Nous avons

$$q = \frac{\partial\varphi}{\partial\rho} - \frac{\partial\psi}{\partial z}; \qquad w = \frac{\partial\varphi}{\partial z} + \frac{\partial\psi}{\partial\rho} + \frac{1}{\rho}\psi.$$

Pour le potentiel φ, nous avons une expression de la forme

$$\varphi = \mathrm{R}\int_0^\pi \Phi(\theta_\lambda)\,d\lambda,$$

où θ_λ est définie par une équation de la forme

$$t - \theta_\lambda \rho \cos\lambda \pm \sqrt{a^2 - \theta_\lambda^2} = \chi(\theta_\lambda),$$

et pour ψ nous avons une expression de la forme

$$\psi = \mathrm{R}\int_0^\pi \Psi(\theta_\lambda)\cos\lambda\,d\lambda.$$

Sergi Natanovich Bernstein

Born – 5 March 1880 in Odessa, Ukraine
Died – 26 Oct 1968 in Moscow, USSR

Sergi Natanovich Bernstein was born in Odessa and he graduated from high school in 1898. Following this, he went to Paris where he studied at the Sorbonne and at Ecole d'Electrotechnique Supérieure. During his work in Paris, Bernstein spent the session 1902–1903 at Göttingen. Bernstein's doctoral dissertation submitted to the Sorbonne was a fine piece of work solving Hilbert's Nineteenth Problem. This problem, posed at the 1900 Congress, was on analytic solutions of elliptic differential equations. He received his doctorate from the Sorbonne in 1904.

Despite this excellent piece of work in his doctoral thesis, when Bernstein returned to Russia in 1905 he had to start his doctoral programme again since Russia did not recognise foreign qualifications for university posts. He studied for his Master's degree at Kharkov, continuing his way through Hilbert's Problems by solving the Twentieth on the analytic solution of Dirichlet's problem for a wide class of nonlinear elliptic equations.

In 1918 Bernstein was awarded his Master's degree and then, in 1913, he received his second doctorate, this time from Kharkov. He taught at Kharkov University for 25 years beginning in 1907.

From 1933 he lectured at Leningrad University and also lectured at the Polytechnic Institute. During this time he worked at the Mathematical Institute of the U.S.S.R. Academy of Sciences. In 1943 Bernstein moved to the University of Moscow and over the next seven years he worked on editing Chebyshev's complete works.

Bernstein worked on the theory of best approximation of functions. He greatly extended work begun by Chebyshev in 1854. In 1911 he introduced what are now called the Bernstein polynomials to give a constructive proof of Weierstrass' theorem (1885), namely that a continuous function on a finite subinterval of the real line can be uniformly approximated as closely as we wish by a polynomial.

At the International Congress at Cambridge in 1912, Bernstein talked about this work. He then continued to develop these ideas, solving problems in interpolation theory, methods of mechanical integration and, in 1914, introduced a new class of quasi-analytic functions.

Some of Bernstein's most important work was in the theory of probability. He attempted an axiomatisation of probability theory in 1917. He generalized Lyapunov's conditions for the central limit theorem, studied generalizations of the law of large numbers, and worked on Markov processes and stochastic processes. Bernstein also studied applications of probability, in particular to genetics.

Comptes Rendus (Doklady) de l'Académie des Sciences de l'URSS
1946. Volume LI, № 5

MATHÉMATIQUES

SUR LA MEILLEURE APPROXIMATION SUR TOUT L'AXE RÉEL DES FONCTIONS CONTINUES PAR DES FONCTIONS ENTIÈRES DE DEGRÉ FINI. I

Par SERGE BERNSTEIN, de l'Académie

1. Nous disons qu'une fonction entière

$$S_p(x) = \sum_{0}^{\infty} \frac{a_k x^k}{k!}$$

est de degré fini p ou bien inférieur ou égal à p, lorsque

$$\overline{\lim} \sqrt[k]{|a_k|} \leqslant p$$

J'ai démontré ailleurs([1]) que chaque fonction continue bornée sur tout l'axe réel admet une fonction $S_p(x)$ de degré donné, qui s'en écarte le moins possible parmi toutes les fonctions entières de même degré en ce sens qu'elle réalise le minimum de

$$\sup_{-\infty<x<\infty} |f(x) - S_p(x)| = A_p f(x).$$

Il résulte aussi de ce travail que la théorie de la meilleure approximation par des fonctions entières de degré fini présente un complément nécessaire de la théorie de la meilleure approximation par polynômes.

Cette Note et celle qui va suivre ont pour but d'ajouter quelques nouveaux résultats, qui en partie ont été exposés dans mon séminaire à l'Université de Moscou en novembre 1944.

Démontrons d'abord le théorème suivant.

Théorème I. *La fonction $f(x)$ étant bornée sur tout l'axe réel, la condition nécessaire et suffisante pour que $A_p f(x) \to 0$ avec $p \to \infty$ est que $f(x)$ soit uniformément continue $(-\infty < x < \infty)$.*

En effet, grâce au théorème fondamental([2]), relatif au maximum du module de la dérivée d'une fonction entière bornée de degré p, il résulte de $|S_p(x)| \leqslant L$ et de $|f(x) - S_p(x)| \leqslant A_p f(x)$ que

$$|f(x+h) - f(x)| \leqslant 2A_p f(x) + |S_p(x+h) - S_p(x)| \leqslant 2A_p f(x) + phL$$

Ainsi le maximum $\omega(\delta)$ de l'oscillation de la fonction $f(x)$ dans tout intervalle de longueur δ de l'axe réel satisfait à l'inégalité

$$\omega(\delta) \leqslant 2A_p f(x) + \varepsilon \tag{1}$$

où $\delta \leqslant \frac{\varepsilon}{Lp}$; par conséquent, si $A_p f(x) \to 0$ avec $p \to \infty$, il en résulte que $\omega(\delta)$ tend uniformément vers zéro avec $\delta \to 0$.

Admettons à présent qu'il soit donné que $\omega(\delta) \to 0$ avec $\delta \to 0$. Considérons un segment $(-\lambda, \lambda)$ et le polynôme d'approximation $R_n(x, \lambda)$

de degré n de $f(x)$ sur ce segment. Dans ces conditions $R_n(\lambda y, \lambda)$ servira de polynôme d'approximation de degré n sur le segment $(-1, +1)$ à la fonction $f(\lambda y)$. Or, puisque $|f(x)| \leqslant L$ pour $|x| \leqslant \lambda$, on aura $|R_n(\lambda y, \lambda)| \leqslant 2L$ pour $|y| \leqslant 1$. Nous en concluons, en utilisant les inégalités de W. Markoff, que

$$R_n(\lambda y, \lambda) = \sum_{k=0}^{n} a_{k,\lambda} y^k$$

où $|a_{k,\lambda}| \leqslant \frac{2Ln^k}{k!}$.

Nous aurons, par conséquent, en posant $p = n/\lambda$

$$R_n(x, \lambda) = \sum_{k=0}^{n} b_{k,\lambda} x^k$$

où $b_{k,\lambda} = \frac{a_{k,\lambda}}{\lambda^k}$ satisfont aux inégalités

$$|b_{k,\lambda}| \leqslant 2L\frac{p^k}{k!} \qquad (k = 0, 1, 2, \ldots) \tag{2}$$

Les polynômes $R_n(x, \lambda)$ appartiennent donc pour p fixe à la classe C_p de fonctions entières de degré p satisfaisant à l'inégalité (2).

En désignant par $E_n(f(x); \lambda)$ et par $A_p(f(x); \lambda)$, respectivement, la meilleure approximation de $f(x)$ sur le segment $(-\lambda, +\lambda)$ par polynômes de degré n et par des fonctions entières de la classe C_p, nous aurons ainsi

$$E_n(f(x); \lambda) \geqslant A_p(f(x); \lambda) \qquad (p = n/\lambda) \tag{3}$$

Or, λ croissant infiniment, $A_p(f(x); \lambda)$ ne peut que croître; donc $A_p(f(x); \lambda)$ tendra vers une limite $A_p(f(x); \infty)$ qui doit être identique à $A_p f(x)$, car la fonction $S_p(x)$ qui s'écarte le moins possible de $f(x)$ sur tout l'axe réel appartient également à la classe C_p, à cause du théorème fondamental relatif au maximum des dérivées successives de $S_p(x)$ rappelé précédemment [2]. Par conséquent, il résulte de (3) que

$$E_n(f(x); \lambda) \geqslant A_p f(x) - \varepsilon_n \tag{4}$$

où $\varepsilon_n \to 0$, lorsque $n = \lambda p \to \infty$.

Il reste à remarquer que, comme on sait, il existe une constante C telle que*

$$E_n(f(x); \lambda) = E_n(f(\lambda y); 1) < c\omega_1\left(\frac{1}{n}\right) = c\omega\left(\frac{\lambda}{n}\right)$$

où $\omega_1(\delta) = \omega(\lambda\delta)$ est le maximum de l'oscillation de $f(\lambda y)$, quand l'accroissement de y ne surpasse pas δ; on a, par conséquent, pour la même valeur de c

$$A_p f(x) < c\,\omega\left(\frac{1}{p}\right) \tag{5}$$

Le théorème est ainsi démontré. L'inégalité (4) est particulièrement importante; elle amène non seulement à l'inégalité (5), mais dans le cas, où la fonction est k fois dérivable et sa dérivée $f^{(k)}(x)$ d'ordre k admet sur tout l'axe réel le module de continuité $\omega_k(\delta)$, l'inégalité (4) conduit

* On peut poser, par example [3], $c = 2\pi$.

à l'inégalité

$$A_p f(x) < \frac{c_k}{p^k} \omega_k\left(\frac{1}{p}\right) \tag{7}$$

qui résulte de l'inégalité correspondante connue

$$E_n(f(x);\ \lambda) < \frac{c_k \lambda^k}{n^k} \omega_k\left(\frac{\lambda}{n}\right) \tag{6}$$

où c_k est une constante indépendante de $\omega_k(\delta)$. D'ailleurs, pour plusieurs classes de fonctions le signe d'égalité seul doit être conservé dans (4) à la limite pour $n=\infty$; ainsi, ceci a lieu, comme je l'ai montré dans l'article (1), lorsque $f(x)=|x|^s$, où $s>0$ est un nombre donné réel quelconque

$$\lim_{n=\infty} \lambda^s E_n(|x|^s;\ 1) = \lim_{\lambda=\infty} E_n(|x|^s;\ \lambda) = A_p|x|^s, \qquad (\lambda = n/p)$$

Cependant, en général, le signe d'inégalité dans (4) doit aussi être conservé. Mais il est facile de voir que l'inégalité (4) peut être renversée, si $p_1 < n/\lambda e$, et on a alors

$$E_n(f(x);\ \lambda) < A_{p_1} f(x) + \varepsilon_n$$

où $\varepsilon_n \to 0$, pour $n \to \infty$.

En effet, soit $S_{p_1}(x) = \sum_0^\infty \frac{a_k x^k}{k!}$ la fonction entière de degré p_1, qui s'écarte le moins de $f(x)$ sur l'axe réel; alors, en rejetant le reste $\rho_{n+1}(x) = \sum_{n+1}^\infty \frac{a_k x^k}{k!}$, nous obtiendrons un polynôme $P_n(x)$ de degré n, tel que pour $|x| \leqslant \lambda$

$$|S_{p_1}(x) - P_n(x)| = |\rho_{n+1}(x)| \leqslant \sum_{n+1}^\infty \frac{(\lambda p_1')^k}{k!} < \frac{1}{\sqrt{2\pi n}} \sum_{n+1}^\infty \left(\frac{\lambda p_1' e}{k}\right)^k$$

en supposant $p_1'/p_1 > 1$, où p_1' est aussi voisin de 1 que l'on veut, lorsque n est assez grand. Donc, $\rho_{n+1}(x) \to 0$, si $\lambda p_1 e/n = \theta < 1$.

Il importe aussi de rappeler que par un raisonnement, analogue à celui de la première partie de notre démonstration, j'ai montré il y a plus de 20 ans dans (2) (pp. 105—106) que la convergence de la série

$$\sum_{s=1}^\infty p_s^k A_{p_s} f(x)$$

où p_s est une suite de nombres infiniment croissants, est suffisante pour que $f(x)$ admette une dérivée $f^{(k)}(x)$ d'ordre k continue sur l'axe réel (de plus la dérivée d'ordre k de la différence $f(x) - S_{p_1}(x)$ est uniformément continue sur tout l'axe réel).

Ainsi, les relations qui existent entre la loi de décroissance de la meilleure approximation $A_p f(x)$ pour $p \to \infty$ et les propriétés différentielles de la fonction $f(x)$ bornée sur tout l'axe réel sont tout à fait analogues aux relations entre $E_n f(x)$ et les propriétés différentielles de $f(x)$ sur le segment $(-1, +1)$.

2. Démontrons encore un autre théorème qui montre que le problème de l'approximation par des fonctions de degré fini présente une généralisation naturelle de l'approximation des fonctions périodiques par des sommes trigonométriques d'ordre fini.

Théorème II. *Si $f(x)$ est périodique de période 2π, on a*

$$A_p f(x) = E^*_{p_0} f(x)$$

où $E^*_{p_0} f(x)$ représente la meilleure approximation de la fonction périodique par des sommes trigonométriques d'ordre p_0, où $p_0 = [p]$.

Démontrons d'abord que parmi les fonctions entières de degré non supérieur à p, qui s'écartent le moins possible de $f(x)$, il en existe au moins une qui est périodique de période 2π. En effet, la différence

$$f(x) - S_p(x + 2h\pi) = f(x + 2h\pi) - S_p(x + 2h\pi)$$

où h est un nombre entier quelconque, ne dépasse pas non plus $A_p f(x)$ en valeur absolue; par conséquent, la fonction

$$\sum\nolimits_{p,N}(x) = \frac{1}{2N+1} \sum_{h=-N}^{N} S_p(x + 2h\pi)$$

jouit de la même propriété. Or, les fonctions $\sum_{p,N}(x)$ de degré p tendent pour un choix convenable de la suite des nombres croissants N vers une fonction limite $\sum_p(x)$, qui est une fonction de degré inférieur ou égal à p dont l'écart de $f(x)$ est également inférieur ou égal à $A_p f(x)$; de plus, $\sum_p(x)$ doit être périodique de période 2π, car, $S_p(x)$ étant bornée ($|S_p(x)| \leqslant L$), la différence

$$\sum\nolimits_{p,N}(x + 2\pi) - \sum\nolimits_{p,N}(x) = \frac{1}{2N+1}[S_p(x + (2N+1)\pi) - S_p(x - 2N\pi)]$$

tend uniformément vers zéro pour $N \to \infty$; d'où il résulte que $\sum_p(x + 2\pi) - \sum_p(x) = 0$, quel que soit x $(-\infty < x < \infty)$. Ainsi

$$\sum\nolimits_p(x) = \sum_{k=0}^{\infty}(a_k \cos kx + b_k \sin kx) = \sum_{k=0}^{p_0}(a_k \cos kx + b_k \sin kx) + \delta(x) \quad (8)$$

$$(p_0 = [p])$$

et il est facile de voir que $\delta(x) = 0$. En effet, si l'on avait $a_k \gtrless 0$ pour une certaine valeur de $k > p$, on obtiendrait après avoir différentié n fois l'égalité (8) que

$$2L^2 p^{2n} \geqslant \frac{1}{\pi} \int_0^{2\pi} \left[\frac{d^n}{dx^n} \sum\nolimits_p(x)\right]^2 dx \geqslant k^{2n} a_k^2$$

ce qui est impossible pour n assez grand.

Manuscrit reçu
le 17. XII. 1945.

LITTÉRATURE CITÉE

[1] S. N. Bernstein, Bull. Acad. Sci. URSS, ser. math., 169 (1938). [2] S. Bernstein, Leçons sur les propriétés extrémales et la meilleure approximation des fonctions analytiques d'une variable réelle, Paris, 1926. [3] Н. И. Ахиезер, Лекции по теории апроксимации, Харьков, 1940.

ОБЗОРЫ И ПЕРЕВОДЫ.

ЦИКЛ СТАТЕЙ ПО УРАВНЕНИЯМ С ЧАСТНЫМИ ПРОИЗВОДНЫМИ ЭЛЛИПТИЧЕСКОГО ТИПА.

О ПЕРВОЙ КРАЕВОЙ ЗАДАЧЕ (ЗАДАЧЕ ДИРИХЛЕ) ДЛЯ УРАВНЕНИЙ ЭЛЛИПТИЧЕСКОГО ТИПА И О СВОЙСТВАХ ФУНКЦИИ, УДОВЛЕТВОРЯЮЩИХ ЭТИМ УРАВНЕНИЯМ.

(Обзор.)

С. Н. Бернштейн и И. Г. Петровский.

§ 1. Постановка задачи. Задача Дирихле для уравнения Лапласа.

Задача Дирихле, или первая краевая задача, состоит в следующем. Пусть на границе некоторой n-мерной области G задана непрерывная функция f. Требуется найти непрерывную в области G и на ее границе функцию u, которая внутри G удовлетворяет уравнению Лапласа

$$\frac{\partial^2 u}{\partial x_1^2}+\frac{\partial^2 u}{\partial x_2^2}+\ldots+\frac{\partial^2 u}{\partial x_n^2}=0,$$

а на границе G обращается в заданную функцию f. Всюду в этой статье мы будем рассматривать только конечные области G [1]).

Прежде при формулировке задачи Дирихле от функции u требовали также, чтобы она имела непрерывные или интегрируемые вторые производные. Это условие относительно вторых производных использовалось в доказательствах единственности решения задачи Дирихле, которые проводились при помощи формулы Грина. Но потом [2]) было дано новое доказательство единственности решения задачи Дирихле без использования каких-либо других свойств вторых производных функций u, кроме существования этих производных. Из этого доказательства следует, в частности, что всякое непрерывное в области G решение уравнения Лапласа обязательно имеет непрерывные производные всех по-

---

1) В случае бесконечной области G для обеспечения единственности решения задачи Дирихле надо потребовать некоторой правильности поведения функции на больших расстояниях от начала координат. Об этом см., например, Гурса, Курс математического анализа, т. III, ч. I, стр. 179, ГТТИ, 1933. См. также статьи Л. А. Люстерника и М. В. Келдыша в настоящем выпуске „Успехов математических наук".

2) См., например, работу И. И. Привалова [1]. [Цифры в квадратных скобках указывают номер работы соответствующего автора по списку литературы, приложенному в конце статьи.] Это доказательство воспроизведено в печатаемом в настоящем выпуске „Успехов математических наук" реферате И. Г. Петровского, стр. 107—114.

рядков даже в том случае, если стоящие в левой части вторые производные функции u понимать не в обычном смысле, а в смысле Шварца, т. е. как пределы при $h \to 0$ отношений

$$\frac{u(x_1,\ldots,x_{k-1},x_k+h,x_{k+1},\ldots,x_n)+u(x_1,\ldots,x_{k-1},x_k-h,x_{k+1},\ldots,x_n)-2u(x_1,\ldots,x_{k-1},x_k,x_{k+1},\ldots,x_n)}{h^2}$$

$$(k=1,\ldots,n).$$

Напротив, требование непрерывности самой функции u по совокупности $x_1,\ldots,x_n$ весьма существенно, как показывает следующий пример. Действительная часть функции

$$e^{-\frac{1}{(x+iy)^4}}$$

на всей плоскости (x, y) имеет вторые производные и удовлетворяет уравнениям Лапласа, но она неограниченна в окрестности начала координат. Поэтому всюду в дальнейшем, говоря о решении уравнения Лапласа или какого-нибудь другого эллиптического уравнения, мы будем считать это решение непрерывным внутри G, если даже нет особой оговорки об этом.

Первое решение задачи Дирихле было дано Пуассоном для того частного случая, когда областью G служит круг. Первую попытку доказать существование решения задачи Дирихле при широких предположениях относительно границы области G сделал Риман. Он рассуждал так. Поставим следующую вариационную задачу: из семейства функций v, которые непрерывны в области G и на ее границе, причем на границе G принимают значение заданной непрерывной функции f, а внутри G имеют непрерывные первые производные, выбрать ту функцию, которая дает минимум интегралу

$$I(v)=\int_G \sum_{k=1}^{n}\left(\frac{\partial v}{\partial x_k}\right)^2 dx_1\ldots dx_n.$$

Легко видеть, что уравнением Лагранжа-Эйлера для рассматриваемой вариационной задачи служит уравнение Лапласа. Поэтому если эта вариационная задача имеет решение, то оно и будет служить решением задачи Дирихле. Существование решения вариационной задачи Риман выводил из того, что множество значений интеграла $I(v)$ ограничено снизу, так как $I(v)\geqslant 0$, и потому имеет конечную нижнюю грань. Как потом показал Вейерштрасс, это заключение Римана было неправильным; Вейерштрасс привел примеры подобных вариационных задач, которые не имеют решения. Кроме того, Адамар (Hadamard) [1] привел пример такой непрерывной функции, заданной на границе круга, которую невозможно распространить непрерывным образом на внутренность круга так, чтобы для получившейся функции v интеграл $I(v)$ не был расходящимся. Но ошибка Римана была одной из тех творческих ошибок, которые способствуют развитию науки. Позже она была исправлена, и на идее Римана было дано несколько новых решений задачи Дирихле. Сюда относятся работы Гильберта (Hilbert) [1] [3]), Лебега (Lebesgue) [1], Люстерника [1] [4]) и др.

[3]) См. также Gesammelte Werke, т. II, стр. 10—37.

[4]) Переработанный перевод этой статьи см. в настоящем выпуске „Успехов математических наук", стр. 115—124.

Следующей была попытка Неймана (C. Neumann) [1][5]) доказать существование решения задачи Дирихле для выпуклых областей. Но доказательство самого Неймана вполне строго только для самых простых областей. В применении же метода Неймана к более сложным областям возникают трудности такого же порядка, как и в методе Римана. В последнее время Лебег [3, 4] точно выяснил эти трудности и внес нужные исправления в рассуждения Неймана. Развитием метода Неймана был метод интегральных уравнений[6]).

Шварц предложил „альтернирующий“ метод, позволяющий решать задачу Дирихле для области G, являющейся суммой таких областей G_1 и G_2, для каждой из которых мы умеем решать задачу Дирихле[7]).

На совершенно новой мысли был основан метод „выметания“ Пуанкаре. Этот метод был потом усовершенствован Перроном (Perron) [1][8]), Радо (Radó) и Риссом (F. Riesz) [1]. Метод Перрона позволяет решать задачу Дирихле в самых широких предположениях относительно границы рассматриваемой области. Чтобы охарактеризовать широту этого метода, введем понятие о „регулярной“ точке границы области. Мы будем точку P, принадлежащую границе области G, называть регулярной точкой, если, какова бы ни была заданная на границе G непрерывная функция f, построенная для нее по методу Перрона функция u, гармоническая внутри G и совпадающая с f на границе, непрерывна в точке P.

Для случая одномерной области задача Дирихле тривиальна: на каждом конечном интервале можно построить линейную функцию, которая на концах этого интервала принимает любые заданные значения. В этом случае обе граничные точки рассматриваемой области всегда регулярны.

У двумерной области будет регулярной всякая граничная точка P, для которой можно указать такую окрестность, в которой точку P нельзя окружить замкнутой линией, состоящей только из точек области G. С другой стороны, можно показать, что всякая изолированная точка границы области G нерегулярна.

У трехмерной области будут регулярными все те точки границы, до которых можно дотронуться извне острием конуса K, полученного вращением вокруг оси x_1 кривой

$$x_2 = f(x_1) = x_1^k,$$

где k — любое положительное число. Точнее это условие можно сформулировать так: в том пространстве (x_1, x_2, x_3), где расположена область G, можно так выбрать оси координат с началом в точке P, что все точки, лежащие внутри конуса K и имеющие положительные абсциссы x_1, не превосходящие некоторого положительного числа η, располагаются вне области G. С другой стороны, Лебегом [1] и независимо от него П. С. Урысоном [1] было показано, что нерегулярной будет всякая точка P границы области G, обладающая такой окрестностью $\mathfrak{U}_P$, что при соответствующем выборе координатных осей все точки этой окрест-

[5]) См. также Гурса, Курс математического анализа, т. III, ч. I, стр. 170—174, ГТТИ, 1933.

[6]) Гурса, Курс математического анализа, т. III, ч. II, стр. 167—178.

[7]) Гурса, Курс математического анализа, т. III, ч. I, стр. 177—179.

[8]) См. реферат И. Г. Петровского в настоящем выпуске „Успехов математических наук“, стр. 107—114.

ности, не принадлежащие области G, не выходят из конуса, образованного вращением вокруг оси x_1 кривой

$$x_2 = e^{-\frac{1}{x_1}}, \quad x_1 > 0.$$

То же самое будет, если эту кривую заменить кривой

$$x_2 = F(x_1) = e^{-|\ln x_1|^{1+\varepsilon}} = x_1^{|\ln x_1|^{\varepsilon}},$$

где ε — любое положительное число.

Для n-мерного пространства при $n > 2$ роль функции $f(x_1)$ играет функция

$$\frac{x_1}{|\ln x_1|^{\frac{1}{n-3}}}, \tag{1}$$

а роль функции $F(x_1)$ играет функция

$$\frac{x_1}{|\ln x_1|^{\frac{1}{n-3}+\varepsilon}} \text{ 9)}, \tag{1'}$$

где ε — любое положительное число. Уравнения соответствующих конусов получатся, если выражения (1) или (1′) приравнять $\sqrt{x_2^2 + \ldots + x_n^2}$. Заметим, что свойство точки P быть регулярной или нерегулярной есть локальное свойство, оно зависит только от строения границы области G вблизи этой точки. Необходимое и достаточное условие регулярности точки P было найдено Винером [10]).

К направлению Пуанкаре-Перрона примыкает недавно появившаяся работа Каратеодори (Carathéodory) [1] о решении задачи Дирихле.

Другим направлением в решении задачи Дирихле был метод конечных разностей. Без строгих обоснований этот метод применялся уже давно [11]). Но строго он был впервые проведен для двумерного случая Л. А. Люстерником [1]. Для случая любого числа измерений этот метод был проведен Курантом (Courant), Фридрихсом (Friedrichs) и Леви (Lewy) [1] [12]). Но эти авторы рассматривали только области с достаточно гладкой границей и притом показали лишь, что построенная ими внутри области G гармоническая функция принимает заданные на границе значения „в среднем". Используя один прием С. Н. Бернштейна [15], а также идею Пуанкаре о „барьере", можно показать [13]), что построенная методом конечных разностей гармоническая внутри области G функция принимает значения заданной на границе непрерывной функции во всех регулярных точках гран[иц]ы области. Этот метод, так же как и метод Куранта, Фридрихса и Ле[ви, п]ереносится на решение задачи Дирихле для общих линейных урав[нений э]ллиптического типа.

---

9) Соответствующие расчеты на основании критерия Винера были произведены Е. Д. Со[ло]менцевым.

10) С[м.] печатаемый в настоящем выпуске „Успехов математических наук" обзор М. В. Келдыша. [В] этом обзоре освещается также характер изменения решения задачи Дирихле при малых [изменениях] границы области.

11) См., например, J. le Roux [1].

12) См. перевод этой статьи в настоящем выпуске „Успехов математических наук", стр. 125—160.

13) См. печатаемую в настоящем выпуске „Успехов математических наук" статью И. Г. Петровского „Новое доказательство существования решения задачи Дирихле методом конечных разностей".

§ 2. Первая краевая задача для общих уравнений второго порядка эллиптического типа.

Линейным уравнением второго порядка эллиптического типа называется уравнение

$$\sum_{i,j=1}^{n} a_{ij}(x_1,\ldots,x_n)\frac{\partial^2 u}{\partial x_i \partial x_j}+\sum_{i=1}^{n} a_i\,(x_1,\ldots,\;x_n)\frac{\partial u}{\partial x_i}+a\,(x_1,\ldots,x_n)\,u=f\,(x_1,\ldots,x_n), \quad (2)$$

если для каждой точки $(x_1,\ldots,x_n)$ из рассматриваемой области квадратичная форма

$$\sum_{i,j=1}^{n} a_{ij}\,(x_1,\ldots,\;x_n)\,\alpha_i\alpha_j$$

строго дефинитна, т. е. не обращается в нуль ни при каких действительных значениях $\alpha_1,\ldots,\alpha_n$, для которых $\sum \alpha_k^2=1$.

На возможность решения первой краевой задачи для эллиптического уравнения (2) оказывает существенное влияние знак коэфициента $a\,(x_1,\ldots,x_n)$. Если этот коэфициент принимает положительное значение, то даже в случае постоянных коэфициентов уравнения (2) первая краевая задача для этого уравнения может не иметь решения, если область G достаточно велика. Так, например, уравнение

$$\frac{\partial^2 u}{\partial x^2}+\frac{\partial^2 u}{\partial y^2}+k^2 u=0 \quad (3)$$

имеет решение

$$u_0=\sin kx\sin ky,$$

которое обращается в нуль на границе квадрата Q со сторонами

$$x=0,\quad y=0,\quad x=\frac{\pi}{k},\quad y=\frac{\pi}{k}.$$

С другой стороны, легко показать, что если уравнение (3) имеет какое-нибудь решение, обращающееся в нуль на границе области, имеющей достаточно гладкую границу, и имеет там кусочно непрерывные нормальные производные то всякое другое решение u того же уравнения (3) должно на границе области G удовлетворять соотношению

$$\int\frac{\partial u_0}{\partial n}\,u\,ds=0, \quad (4)$$

где интегрирование производится по границе области G. Соотношение (4) получится, если произвести интегрирование по частям в левой части равенства

$$\iint\limits_G u_0\left(\frac{\partial^2 u}{\partial x^2}+\frac{\partial^2 u}{\partial y^2}+k^2 u\right)dx\,dy=0$$

таким образом, чтобы исчезли производные по x и y в интегралах по области G.

Из соотношения (4) следует, что первая краевая задача для уравнения (3), когда областью G служит квадрат Q, не может иметь решения, если заданная на краю функция не удовлетворяет этому соотношению. Вообще, при решении первой краевой задачи для эллиптического уравнения (2) случай, когда коэфициент a всюду $\leqslant 0$, существенно отличается от случая, когда этот коэфициент в некоторых местах положителен. В первом случае задача имеет единственное решение при любой непрерывной функции, заданной на границе области G, если

1) граница области G достаточно правильна;

2) коэфициенты a_{ij}, a_i и функция f непрерывны вместе с их производными достаточно высоких порядков.

Если же коэфициент a принимает в некоторых точках рассматриваемой области положительные значения, то для обеспечения существования и единственности решения достаточно потребовать еще, чтобы область G была достаточно малой. Как показал В. В. Немыцкий [1], даже для более общих уравнений (нелинейных) здесь важно, чтобы площадь области G была достаточно малой, диаметр же ее может быть как угодно большим. Другие условия, при которых существует единственное решение задачи Дирихле для линейных эллиптических уравнений, дали Пикар (Picard) [4], Параф (Paraf) [1] и С. Н. Бернштейн [10, 13, 20]. Ср. также Lichtenstein, Enzyklopädie, Bd. II, стр. 1300.

На эллиптические уравнения вида

$$\sum_i \frac{\partial}{\partial x_i}\left(a_{ij}\frac{\partial u}{\partial x_j}\right)+au=f, \tag{5}$$

которые называются самосопряженными, можно распространить метод Пуанкаре-Перрона, если вторые производные от коэфициентов a_{ij} удовлетворяют условиям Гельдера с положительным показателем, а функции a и f имеют непрерывные первые производные [14]). Тогда класс областей, для которых решается первая краевая задача, расширяется до таких же пределов, как в случае уравнения Лапласа: все те граничные точки, которые были регулярны при решении первой краевой задачи для уравнения Лапласа, остаются регулярными и при решении этой задачи для общего уравнения (5) обобщенным методом Пуанкаре-Перрона. При этом, конечно, в том случае, когда коэфициент a принимает положительные значения, рассматриваемую область надо брать достаточно малой. Теоремы существования для общего линейного эллиптического уравнения (2) при более общих условиях относительно коэфициентов доказаны Шаудером (Schauder) [2].

В последние годы появился целый ряд работ Лихтенштейна, Брело, Жиро и др., посвященных решению первой краевой задачи для эллиптических уравнений вида (2) с коэфициентами, имеющими в некоторых точках, линиях или поверхностях разрывы.

§ 3. Первая краевая задача для нелинейных уравнений.

Первая краевая задача (или задача Дирихле) для нелинейных уравнений была разрешена Пикаром [1, 2, 3] для уравнения

$$\frac{\partial^2 z}{\partial x^2}+\frac{\partial^2 z}{\partial y^2}=f(x, y, z).$$

Метод Пикара заключался в соединении метода последовательных приближений с соответствующим образом видоизмененным альтернирующим методом Шварца. Идея Пикара могла бы быть использована и для несколько более широких классов уравнений, в частности, и тогда, когда число независимых пере-

[14]) См. Pueschel [1].

менных $n \geqslant 3$. Другой общий метод для решения задачи Дирихле, *параметрический* метод, представляющий соединение метода последовательных приближений с аналитическим продолжением по некоторому параметру, был предложен и развит С. Н. Бернштейном [15]).

Метод С. Н. Бернштейна тесно связан с доказанной им теоремой об аналитическом характере решений эллиптических уравнений. Он рассматривает, в первую очередь, уравнения эллиптического типа

$$F(r, s, t, p, q, z, x, y) = 0, \quad 4F'_r F'_t - F'^2_s > 0, \tag{6}$$

где F есть аналитическая функция в данной области $S(x, y)$ при всех вещественных конечных значениях r, s, t, p, q, z, и подробно развивает свой метод для случая, когда имеет место условие

$$F'_r F'_z \leqslant 0, \tag{7}$$

благодаря которому задача Дирихле для уравнения (6) не может иметь более одного решения (если неравенство (7) соблюдается тождественно). Кроме того, в работах [10] и [13] все доказательства ведутся в предположении, что контур C (внутри S), для которого ставится задача Дирихле, есть окружность радиуса R. Но так как при конформном отображении на круг области, ограниченной аналитическим контуром, уравнение (6) сохраняет свои общие свойства, то формулировки общих теорем остаются в силе для любых аналитических контуров. Напротив, при рассмотрении частных видов уравнений (6) соответствующее преобразование переменных (x, y) может нарушить некоторые свойства функции F, и здесь геометрические свойства контура могут иногда играть существенную роль; этому вопросу уделено значительное внимание в работе [17].

В основе параметрического метода С. Н. Бернштейна лежит следующая фундаментальная лемма [16]):

Если z_0 есть аналитическое решение аналитического уравнения эллиптического типа

$$F(r, s, t, p, q, z, x, y, \alpha) = 0 \quad F'_r F'_z \leqslant 0, \tag{8}$$

---

[15]) Руководящая идея указана уже в работе [2]; основные теоремы доказаны в работах [10] и [13]. Различные дополнения и приложения имеются, в особенности, в работах [16], [17], [19] и [20]. Работа [17] воспроизведена в настоящем выпуске „Успехов математических наук", стр. 32—74.

[16]) Доказательство дано в работах [10], § 29, стр. 139, и [20], § 14, стр. 634.

Даваемая здесь формулировка учитывает, что благодаря методу вспомогательных функций, о котором речь будет впереди, при помощи верхней границы для модулей производных первых двух порядков от решений возможно установить верхнюю границу для производных любого данного порядка.

В работе [17] С. Н. Бернштейн в формулировке фундаментальной леммы освобождается от требования $F'_r F'_z \leqslant 0$, заменяя его предположением, что рассматриваемое решение z_0 есть простое решение уравнения (8), т. е. такое, что регулярное решение u, при $\alpha = \alpha_0$, $z = z_0$, так называемого вариационного (линейного) уравнения

$$F'_r \frac{\partial^2 u}{\partial x^2} + F'_s \frac{\partial^2 u}{\partial x \, \partial y} + F'_t \frac{\partial^2 u}{\partial y^2} + F'_p \frac{\partial u}{\partial x} + F'_q \frac{\partial u}{\partial y} + F'_z u = 0$$

(которое является усеченным уравнением для всех уравнений, из которых определяются последовательные производные от z по α при $\alpha = \alpha_0$) должно быть тождественно равно нулю, если оно обращается в нуль на рассматриваемом контуре C.

соответствующее вещественному значению параметра $\alpha=\alpha_0$, имеющее ограниченные производные первых двух порядков в замкнутой области $\overline{G}$, ограниченной аналитическим контуром C, на котором $z_0=\varphi(s)$ обращается в данную аналитическую функцию дуги, то существует такое положительное число ε_0, что при всех значениях (вещественных или комплексных) параметра α, удовлетворяющих неравенству $|\alpha-\alpha_0|\leqslant\varepsilon_0$, уравнение (8) *также допускает аналитическое решение, совпадающее с z_0 на контуре C и имеющее ограниченные производные первых двух порядков во всей замкнутой области $\overline{G}$.*

Таким образом ε_0 представляет собой нижнюю границу радиуса сходимости по степеням $\alpha-\alpha_0$ аналитической вблизи α_0 функции $z(x, y, \alpha)$, причем ε_0 зависит только от верхней границы P_0 модулей частных производных z_0 по x, y первых двух порядков. Поэтому, применяя ту же лемму к решению $z_1=z(x, y, \alpha_1)$, где $\alpha_1=\alpha_0+\varepsilon_0$, у которого производные первых двух порядков будут ограничены некоторым числом P_1, мы найдем, что решение уравнения (8) существует также и для α в промежутке $(\alpha_1, \alpha_1+\varepsilon_1)$, где $\varepsilon_1>0$; однако P_1 вообще может быть больше P_0, а следовательно, $\varepsilon_1<\varepsilon_0$. Для того чтобы аналитическое продолжение функции $z(x, y, \alpha)$ было возможно от α_0 до β_0, причем функция $z(x, y, \alpha)$ оставалась бы аналитической по α на всем замкнутом отрезке (α_0, β_0), необходимо и достаточно, чтобы на этом отрезке радиус сходимости имел положительную нижнюю границу. Отсюда нетрудно было вывести следующую общую теорему [17]):

Теорема А. *Для того чтобы первая краевая задача для данного уравнения* (6) *допускала регулярное* [18]) *решение, обращающееся в некоторую определенную аналитическую функцию дуги $\varphi(s)$ на данном аналитическом контуре C, необходимо и достаточно, чтобы в данное уравнение* (6) *мог быть введен параметр α так, чтобы оно обратилось в уравнение* (8), *для которого*: 1) *при всех α $(0\leqslant\alpha\leqslant1)$ из предположения существования регулярного аналитического решения, соответствующего тем же краевым заданиям, вытекала равномерная ограниченность его модуля, как и модулей его частных производных по x, y первых двух порядков*; 2) *при $\alpha=1$ уравнение* (8) *обращалось в данное уравнение* (6); 3) *при $\alpha=0$ оно имело очевидное регулярное аналитическое решение.*

Если уравнение (6) квазилинейно, т. е. имеет вид

$$A(x, y, p, q)r+2B(x, y, p, q)s+C(x, y, p, q)t=D(x, y, z, p, q)$$

$$(AC-B^2>0, \quad AD'_z\geqslant0), \tag{9}$$

то с помощью метода вспомогательных функций ограниченность вторых производных выводится из ограниченности первых производных (см. примечание {4} к печатаемой в настоящем выпуске „Успехов математических наук", стр. 32—74, работе [17]), поэтому в теореме А в условии 1) фигурируют лишь первые производные.

---

[17]) [13], § 20, стр. 131 и § 22, стр. 135.

[18]) Регулярным решением задачи Дирихле называется решение, которое вместе со своими производными первых двух порядков ограничено в рассматриваемой замкнутой области, т. е. решение, удовлетворяющее условию фундаментальной леммы.

В работе [17] теорема А распространена на случай, когда знак F_z' произволен, но известно, что соответствующее решение уравнения (8) простое (см. сноску [16])) при всех α в промежутке $0 \leqslant \alpha \leqslant 1$.

Таким образом для уравнения (9), в котором A, B, C могут также содержать z, С. Н. Берештейн получает следующую теорему:

Теорема В. *Если (при любом знаке D_z') задача Дирихле ни при каком задании $f(s)$ на данном контуре C не имеет более одного регулярного решения, то задача Дирихле имеет регулярное решение при любой аналитической функции $\varphi(s)$ на этом контуре тогда и только тогда, когда:* 1) *она имеет регулярное решение при некотором определенном задании $\varphi_0(s)$;* 2) $|z|$, $|p|$, $|q|$ *могут быть* a priori *ограничены.*

В частном случае, когда $D=0$, условие 1) соблюдается благодаря существованию решения $z=0$; кроме того, $|z| \leqslant \max|\varphi(s)|$, так как кривизна поверхности z отрицательна; верхняя граница p^2+q^2 по той же причине может быть указана, если контур C выпуклый, так как касательная плоскость в любой точке поверхности пересекает ее край (имеющий проекцией C) по крайней мере в трех точках. Таким образом в этом случае задача Дирихле имеет решение внутри любого выпуклого контура [19]).

В работах [10], стр. 163, и [13], стр. 133, дается теорема об аналитическом продолжении решения, о котором речь будет впереди; из этой теоремы вытекает, что *если условие* 2) *теоремы* В *соблюдено для окружности любого радиуса R с центром в начале, то условие* 1) *автоматически выполнено на любом контуре,* так как рассматриваемое уравнение допускает тогда решения, регулярные в любой данной области.

Не останавливаясь на приложениях этих теорем к квазилинейным уравнениям (9), данным главным образом в работах [16] и [17], из которых последняя помещена в этом выпуске „Успехов математических наук", ограничимся указанием приложений теоремы А к уравнению (6) в работе [16].

Для уравнения

$$rt - s^2 = k^2 > 0, \tag{10}$$

где k — постоянная, задача Дирихле имеет два различных решения при любых аналитических заданиях на всякой окружности C.

Напротив, через замкнутую аналитическую кривую S, имеющую проекцией окружность C, можно провести две поверхности постоянной гауссовой кривизны (одну выпуклую книзу, другую — кверху), удовлетворяющие уравнению

$$\frac{rt - s^2}{(1+p^2+q^2)^2} - k^2 \tag{10^{bis}}$$

только при достаточно малых [20]) значениях k. Каждой кривой S соответствуют два значения $k_0 \leqslant k_1$ таких, что задача имеет два регулярных решения при

---

[19]) См. примечание {2} к помещенной в этом выпуске „Успехов математических наук", стр. 32—74, работе [17].

[20]) Пользуюсь случаем, чтобы внести две небольшие поправки. В конце статьи [16] указывается, что поверхность нулевой кривизны, получающаяся в пределе, когда $k \to 0$, может быть нерегулярной (т. е. не развертывающейся) и принадлежать таким образом к классу нелинейчатых поверхностей Лебега, налагающихся на плоскость. Однако пример, который там указан, приводит к двум цилиндрическим поверхностям, а не к по-

$0<k<k_0$, одно решение при $k_0<k<k_1$, ни одного решения при $k>k_1$ (при $k=k_0$, $k=k_1$ одно из решений иррегулярно).

Теперь нам необходимо остановиться вкратце на методе доказательств С. Н Бернштейна. Метод этот существенно опирается на доказанную им теорему об аналитическом характере решений эллиптических уравнений, а техническое оформление связано с теми же нормальными рядами, посредством которых упомянутая теорема была доказана [21]). Таким образом предварительные неравенства, относящиеся к линейным уравнениям, необходимые для доказательства, выражаются при помощи линейных функционалов, называемых нормализованными модулями [22]). Не останавливаясь на довольно сложном способе построения этих функционалов, будем их обозначать без указания характеризующих их параметров с помощью фигурных скобок.

Тогда для линейного уравнения

$$Ar+2Bs+Ct+2Dp+2Eq+Fz=M \quad (AC-B^2>k>0,\ AF\leqslant 0) \quad (11)$$

можно указать неравенства, устанавливающие верхние грани для функционала $\{z\}$ и его производных первых двух порядков при помощи того же функционала от правой части M.

А именно: *если z есть решение линейного уравнения, имеющее непрерывные производные первых двух порядков внутри круга C радиуса R и обращаю-*

верхностям Лебега. Другая поправка относится к доказательству утверждения, что два решения уравнения (10) z и u, совпадающие на некотором контуре C (и регулярные внутри), совпадают, если $\frac{\partial^2 u}{\partial x^2}$ и $\frac{\partial^2 z}{\partial x^2}$ имеют одинаковые знаки, т. е. выпуклость их направлена в одну и ту же сторону. Обозначая $\delta=z-u$, я замечаю на стр. 247 указанной работы [16], что δ удовлетворяет обоим линейным уравнениям

$$\frac{\partial^2 z}{\partial y^2}\,\frac{\partial^2\delta}{\partial x^2}+\frac{\partial^2 u}{\partial x^2}\,\frac{\partial^2\delta}{\partial y^2}-\left(\frac{\partial^2 z}{\partial x\partial y}+\frac{\partial^2 u}{\partial x\partial y}\right)\frac{\partial^2\delta}{\partial x\partial y}=0,$$

$$\frac{\partial^2 u}{\partial y^2}\,\frac{\partial^2\delta}{\partial x^2}+\frac{\partial^2 z}{\partial x^2}\,\frac{\partial^2\delta}{\partial y^2}-\left(\frac{\partial^2 z}{\partial x\partial y}+\frac{\partial^2 u}{\partial x\partial y}\right)\frac{\partial^2\delta}{\partial x\partial y}=0.$$

Но высказанное далее утверждение, что одно по крайней мере из этих уравнений — эллиптического типа, ошибочно. Вместо этого следует сказать, что уравнение, получающееся от сложения их, будет всегда эллиптическим, так как из неравенств $AC-B^2>0$, $A_1C_1-B_1^2>0$ при $AA_1>0$ вытекает

$$(A+A_1)(C+C_1)-(B+B_1)^2=$$

$$=(A+A_1)=\left(\frac{AC-B^2}{A}+\frac{A_1C_1-B_1^2}{A_1}\right)+\left(B\sqrt{\frac{A_1}{A}}-B_1\sqrt{\frac{A}{A_1}}\right)^2>0.$$

Таким образом заключение, что $\delta=0$ внутри контура C, если оно равно нулю на самом контуре, вполне справедливо. *С. Б.*

[21]) См. следующий параграф.

[22]) Если функция $f(x)$ переменной x на отрезке OR разлагается в нормальный ряд на части его OA, где она имеет норму $[f(x)]_{Aa}$ внутри контура Γ_{OAa} (см. ниже сноску [30])), а на остальной части отрезка максимум $|f(x)|$ равен M, то наибольшее из этих двух чисел обозначается через $[f(x)]_{Aa}^{OR}$ и называется нормализованным внутри контура Γ_{OAa} модулем $f(x)$ на отрезке OR. Нормализованный тригонометрический модуль для функции двух переменных $f(x,y)=f(\rho\cos\theta,\ \rho\sin\theta)$ определяется аналогично.

щееся в нуль на его окружности, далее, если все коэфициенты уравнения (11) *являются аналитическими функциями внутри C и, кроме того, если производные первых семи порядкев от A, B, C и производные первых двух порядков от D, E, F ограничены по модулю в замкнутом круге некоторым числом $P > 0$, то имеют место неравенства*

$$\begin{gathered}\{z\} < \lambda\{M\}, \quad \left\{\frac{\partial z}{\partial x}\right\} < \lambda\{M\}, \quad \left\{\frac{\partial z}{\partial y}\right\} < \lambda\{M\}, \\ \left\{\frac{\partial^2 z}{\partial x^2}\right\} < \lambda\{M\}, \left\{\frac{\partial^2 z}{\partial x \partial y}\right\} < \lambda\{M\}, \left\{\frac{\partial^2 z}{\partial y^2}\right\} < \lambda\{M\},\end{gathered} \tag{12}$$

где множитель λ зависит только от R и P (после того, как способ построения функционала $\{\ \}$ зафиксирован).

После установления неравенств (12) указанная выше фундаментальная лемма доказывается посредством соответствующего применения метода мажорант, однако, лишь при условии ограниченности производных до девятого порядка.

Но метод вспомогательных функций позволяет последовательно установить верхние границы для модулей частных производных всех порядков $p \leqslant n$ для любого аналитического решения z уравнения (6) в замкнутом круге C, на границе которого z обращается в функцию, диференцируемую $n+2$ раза, после того как известны верхние границы модулей его производных первых двух порядков. В случае же квазилинейного уравнения

$$A(x, y, z, p, q)r + 2B(x, y, z, p, q)s + C(x, y, z, p, q)t = D(x, y, z, p, q)$$

достаточно, чтобы были известны верхние границы производных первого порядка (см. примечание {4} к помещенной в этом выпуске „Успехов математических наук“, стр. 32—74, статье [17]).

Случай неаналитических уравнений и неаналитических краевых заданий не был подробно исследован С. Н. Бернштейном[23]). Прием, который он предлагает, заключается в применении теоремы Вейерштрасса, т. е. в рассмотрении данного неаналитического уравнения как предела аналитических уравнений, и то же самое для краевых заданий. Таким образом результаты С. Н. Бернштейна распространяются на достаточно гладкие функции. Кроме того, в заметке [15] С. Н. Бернштейн указывает прием для равномерного ограничения производных регулярных решений во всякой фиксированной области S в н у т р и контура, из которого вытекает существование решения задачи Дирихле для любой непрерывной краевой функции, если таковое существует для любой аналитической функции.

Заканчивая обзор этих исследований С. Н. Бернштейна, обратим внимание на заключительную главу его работы [13], где, подводя итоги своей теории решения задачи Дирихле, он указывает, что те же принципы могут быть применены к решению других краевых задач, в частности, задачи Неймана, когда на контуре задана производная по нормали; для этого нужно доказать, что и в этих случаях справедлива теорема А. Позднейшие исследования Жиро, Шаудера, Лерэ и других отчасти осуществили эту программу; при этом прин-

---

[23]) [10], стр. 111 и 136; на стр. 163 С. Н. Бернштейн пишет: „Роль аналитических функций в нашем доказательстве была, таким образом, по существу служебная, и вполне вероятно, что при дальнейшем развитии анализа можно будет более простым путем притти к тем же выводам“.

ципиально новым была замена метода аналитического продолжения по параметру α более общими методами непрерывных изменений при формулировке и доказательстве теорем, аналогичных теореме А. Сущность этих методов достаточно полно освещена в статье В. В. Немыцкого „Метод неподвижных точек в анализе", помещенной в первом выпуске „Успехов математических наук".

Следует упомянуть, что немного позже С. Н. Бернштейна Корн (Korn) [1] и Мюнц (Müntz) [1] доказали теорему, являющуюся частным случаем указанной выше фундаментальной леммы, а именно при предположении, что уравнение (8) имеет вид

$$\frac{\partial^2 z}{\partial x^2}+\frac{\partial^2 z}{\partial y^2}=\alpha f\left(\frac{\partial^2 z}{\partial x^2},\ \frac{\partial^2 z}{\partial x\,\partial y},\ldots,\ y\right)$$

(α предполагается достаточно малым), т. е. z_0 есть решение уравнения Лапласа. (Ввиду такого ограничения эту теорему нельзя использовать для аналитического продолжения.) При доказательстве авторы выводили неравенства, аналогичные неравенствам (12), но только соответствующие уравнению Пуассона. Функционалы, которыми они пользуются, связаны с условием Гельдера. Позднейшие авторы, которые доказывали теоремы, аналогичные фундаментальной лемме С. Н. Бернштейна, пользовались тем же функционалом Гельдера[24]) и приспособили к нему метод вспомогательных функций С. Н. Бернштейна.

Таким образом были получены некоторые новые результаты, относящиеся к случаям, когда единственность решения не обеспечена и рассматриваемая область ограничена несколькими контурами.

В частности, наиболее полные результаты имеются в работе Шаудера и Лерэ (Leray) [1][25]), где доказано существование решения однородного эллиптического квазилинейного уравнения

$$A(x,\ y,\ z,\ p,\ q)r+2B(x,\ y,\ z,\ p,\ q)s+C(x,\ y,\ z,\ p,\ q)t=0$$

без предположения о единственности этого решения. Более существенно распространение в этой же работе теоремы А на случай любого числа n переменных, так как до настоящего времени аппарат (нормы и нормализованные модули), использованный С. Н. Бернштейном для случая двух переменных, не был применен при любом $n>2$, и хотя, без сомнения, неравенства, тождественные неравенствам (12), распространяются на любое число n переменных (при $n=3$ тригонометрические ряды заменяются шаровыми функциями), но техника доказательства требовала бы значительных изменений. Приведем формулировку Шаудера и Лерэ.

Пусть требуется найти решение $z(x_1,\ldots,x_n)$ уравнения

$$F\left(x_1,\ldots,x_n,\ z,\ \frac{\partial z}{\partial x_1},\ldots,\frac{\partial z}{\partial x_n},\ \frac{\partial^2 z}{\partial x_1^2},\ \frac{\partial^2 z}{\partial x_1\partial x_2},\ldots,\ \frac{\partial^2 z}{\partial x_n^2},\ k\right)=0 \qquad (13)$$

внутри некоторой области G, обращающееся в 0 на границе этой области. Допустим, что при $k=0$ поставленная для уравнения (13) задача имеет решение,

---

[24]) См. Schauder [2], стр. 259.

[25]) Упоминая, что этот результат является обобщением соответствующего результата С. Н. Бернштейна, авторы упускают из виду, что в случае единственности (даже при вхождении z в коэфициенты A, B, C) их результат является следствием более общей теоремы Бернштейна (§ 11 гл. IV работы [17]). Заметим кстати, что результаты Шаудера в работе [2] полностью покрываются работой [17] С. Н. Бернштейна.

вторые производные которого удовлетворяют условию Гельдера, и что вблизи этого решения уравнение (13) эллиптично. Станем изменять k. Тогда поставленная краевая задача может перестать иметь решение при некотором значении $k=k_1$ только в том случае, если произойдет одно из следующих двух обстоятельств:

а) Появится решение, обращающееся в 0 на краю G, такое, что вблизи него уравнение (13) не эллиптично.

б) Значениям параметра, заключенным между 0 и k_1, соответствуют такие решения поставленной задачи, что, хотя у каждого из них вторые производные удовлетворяют условию Гельдера, однако не существует одного и того же условия Гельдера, т. е. условия Гельдера с одинаковыми показателями и коэфициентами, которому удовлетворяли бы вторые производные всех этих решений.

Следует отметить также работы Жиро (Giraud) [1, 2], тоже рассматривающего случай любого числа переменных, а в работе [3] изучающего также и другие краевые задачи, и работу Реллиха (Rellich) [1], рассматривающего уравнения Монжа-Ампера.

§ 4. Об аналитическом характере решений эллиптических уравнений.

Для обыкновенного диференциального уравнения

$$\frac{d^n y}{dx^n}=f\left(x,\ y,\ \frac{dy}{dx},\ \ldots,\ \frac{d^{n-1}y}{dx^{n-1}}\right)$$

хорошо известно, что если функция f аналитична по всем своим аргументам, то и все его решения $y(x)$ аналитичны по x [26]). Для уравнений с частными производными так бывает далеко не всегда. Например, решениями волнового уравнения

$$\frac{\partial^2 u}{\partial t^2}-a^2\frac{\partial^2 u}{\partial x^2}=0$$

будут все $u(t,\ x)$, даваемые формулой

$$u(t,\ x)=\varphi(x+at)+\psi(x-at),$$

где φ и ψ — любые непрерывные вместе с их производными до второго порядка функции [27]). С другой стороны, все непрерывные решения уравнения Лапласа

$$\frac{\partial^2 u}{\partial x^2}+\frac{\partial^2 u}{\partial y^2}=0$$

аналитичны по x и y.

Кроме уравнения Лапласа, существует много других уравнений с частными производными, все достаточно гладкие (т. е. имеющие непрерывные производные

---

[26]) Точнее было бы говорить здесь о голоморфности вместо аналитичности, но мы здесь, как и всюду в дальнейшем, будем пользоваться общепринятым в литературе термином „аналитичность", понимая под ним голоморфность.

[27]) В работе С. Н. Бернштейна [2] (глава V) доказано (стр. 60), что решение $u(x,\ y)$ уравнения гиперболического типа $\dfrac{\partial^2 u}{\partial x \partial y}=f\left(x,\ y,\ u,\ \dfrac{\partial u}{\partial x},\ \dfrac{\partial u}{\partial y}\right)$, где f есть аналитическая функция, ни при каком значении y не будет аналитической функцией переменной x, если $u(x,\ 0)$ не аналитична.

достаточно высоких порядков), решения которых аналитичны. Пикар давно уже [28]) доказал, например, что все непрерывные вместе со всеми их производными до второго порядка решения уравнений

$$\frac{\partial^2 u}{\partial x^2} + \frac{\partial^2 u}{\partial y^2} = e^u$$

и

$$a\frac{\partial^2 u}{\partial x^2} + 2b\frac{\partial^2 u}{\partial x \partial y} + c\frac{\partial^2 u}{\partial y^2} + d\frac{\partial u}{\partial x} + e\frac{\partial u}{\partial y} + fu = F,$$

где a, b, c, d, e, f, F — аналитические функции от x и y и $ac > b^2$ (условие эллиптичности), аналитичны по x и y.

То обстоятельство, что существуют диференциальные уравнения с частными производными, все достаточно гладкие решения которых аналитичны, Гильберт считал одним из наиболее замечательных свойств аналитических функций. В 1901 г. на Международном конгрессе математиков в Париже он поставил среди других следующую задачу: доказать, что все достаточно гладкие решения диференциального уравнения

$$\Phi(x, y, u, p, q, r, s, t) = 0, \qquad (14)$$

у которого левая часть аналитична по всем ее аргументам и

$$4\Phi_r \Phi_t > \Phi_s^2$$

(условие эллиптичности), аналитичны по x и y [29]).

Как сообщает С. Н. Бернштейн в статье [6] (в 1905 г.), когда он спросил Гильберта, на чем было основано утверждение последнего, что все решения регулярных задач вариационного исчисления должны быть аналитическими Гильберт ответил, что он считает, что решения всех проблем естествознания выражаются аналитическими функциями. Уточняя эту мысль Гильберта С. Н. Бернштейн формулировал следующий общий математический принцип, который устанавливает тесную связь между философским детерминизмом и идеей аналитического продолжения: если (достаточно гладкая) функция $f(x_1, \ldots, x_n)$, определенная в любой области $S(x_1, \ldots, x_n)$ и подчиняющаяся аналитическому диференциальному уравнению, может быть единственным образом продолжена за пределы S относительно одной из своих переменных x_i, то f должна быть аналитической функцией этой переменной.

В предшествующей работе [2] (от 1904 г.) С. Н. Бернштейн доказал указанную выше теорему Гильберта для уравнения (14) при условии существования и конечности производных третьего порядка [30]). Кроме того, в той же работе

---

[28]) См. Picard [1].

[29]) При этом, как это он вообще часто делал, Гильберт связывал рассматриваемое диференциальное уравнение с некоторой вариационной задачей: он ограничивался предположением, что это диференциальное уравнение является уравнением Лагранжа-Эйлера для некоторой регулярной вариационной задачи. Но это ограничение оказалось несущественным.

[30]) Доказательство, исходным пунктом которого является способ последовательных приближений Пикара, потребовало введения нового типа рядов

$$F(x) = \sum_{p=0}^{\infty} \sum_{q=0}^{\infty} A_{pq} x^p (R - x)^q.$$

было отмечено, что все решения аналитического уравнения $\frac{\partial^2 u}{\partial x^2}=f\left(\frac{\partial u}{\partial x}, u, x, y\right)$ аналитичны относительно x, а в заметке „Sur les équations du type parabolique" [Comptes rendus, т. 140 (1905), стр. 137—139] С. Н. Бернштейн доказал также, в согласии с формулированным выше принципом, что все решения линейного уравнения параболического типа

$$\frac{\partial^2 u}{\partial x^2}=a\frac{\partial u}{\partial y}+b\frac{\partial u}{\partial x}+cu+d,$$

где a, b, c, d — аналитические функции по x, y, должны всегда быть аналитическими функциями относительно x. Существование и непрерывность третьих производных у решений уравнения (14) в более или менее общих предположениях устанавливали С. Н. Бернштейн [31]), Лихтенштейн (Lichtenstein) [2], Радо [1, 2], Гопф [1] и др.

Недавно, отправляясь от одной идеи Пикара, Р. Курант [32]) и независимо от него С. А. Христианович дали новое простое доказательство теоремы Гиль-

---

Существенную роль в нем играет функционал

$$[F(x)]_{Rr}=\inf\sum_{p=0}^{\infty}\sum_{q=0}^{\infty}|A_{pq}|R^{p+q}\frac{p^p q^q}{(p+q)^{p+q}}\left(\frac{R+r}{R-r}\right)^q,$$

называемый (низшей) *нормой* $F(x)$ в комплексной области Γ_{ORr} переменной x, образованной касательными из точки R к окружности радиуса $r<R$ с центром в O и большей дугой этой окружности между точками касания.

Если $F(x)$ регулярна в замкнутой области Γ_{ORr}, то норма $[F(x)]_{Rr}$ конечна; обратно, если $[F(x)]_{Rr}$ конечна, то $F(x)$ регулярна внутри Γ_{ORr} и конечна и непрерывна на контуре Γ_{ORr}. Точно так же, если

$$F(x, y)=F(\rho\cos\theta, \rho\sin\theta)=\sum_{n=0}^{\infty}[A_n(\rho)\cos n\theta+B_n(\rho)\sin n\theta],$$

где

$$A_n(\rho)=\rho^n\sum_{p=0}^{\infty}\sum_{q=0}^{\infty}a_{pq}^{(n)}\rho^p(R^2-\rho^2)^q,\qquad B_n(\rho)=\rho^n\sum_{p=0}^{\infty}\sum_{q=0}^{\infty}b_{pq}^{(n)}\rho^p(R^2-\rho^2)^q,$$

то

$$[F(x, y)]_{Rr}=\sum_{n=0}^{\infty}\{[A_n(\rho)]_{Rr}+[B_n(\rho)]_{Rr}\}$$

называется (низшей) нормой $F(x, y)$ внутри Γ_{ORr}, и конечность ее является признаком того, что $F(x, y)$ аналитична внутри круга C радиуса R по обеим переменным x, y и разлагается в абсолютно и равномерно сходящийся ряд Фурье также и на его окружности. Ход доказательства был совершенно тождественным с доказательством, имеющимся в работе [22], напечатанной в этом выпуске „Успехов математических наук", стр. 82—99, с той лишь разницей, что для решения u уравнения Пуассона

$$\frac{\partial^2 u}{\partial x^2}+\frac{\partial^2 u}{\partial y^2}=f(x, y),$$

обращающегося в нуль на окружности C радиуса R, в неравенствах, подобных неравенствам (11) работы [22], фигурируют нормы. Кроме того, благодаря приведенному выше свойству норм, последняя часть рассуждения работы [22] становится излишней, так как из конечности нормы $[u]_{Rr}$ сразу вытекает ее аналитичность по обеим переменным.

[31]) [10], стр. 132, [17], стр. 484.

[32]) Hilbert und Courant, Methoden der mathematischen Physik, Bd. II, Kap. 5, Anhang.

берта-Бернштейна. На этой же идее были основаны работы Жеврея (Gevrey) и Жиро. Они доказали аналитичность всех непрерывных решений линейного эллиптического уравнения, у которого все коэфициенты и правая часть аналитичны.

Жеврей в Comptes rendus Парижской Академии за 1913 и 1914 гг. сообщал, что ему удалось не только доказать теорему Гильберта-Бернштейна для нелинейных эллиптических уравнений с двумя независимыми переменными, но и значительно ее обобщить даже для уравнений высших порядков и систем таких уравнений. Так, например, ему принадлежит следующая формулировка. Будем называть действительную функцию $\varphi(x_1, \ldots, x_n)$, имеющую на некоторой n-мерной области производные всех порядков, функцией класса α_k по x_k $(k = 1, \ldots, n)$ если для любой системы $k_1, \ldots, k_n$

$$\left|\frac{\partial^{k_1+\cdots+k_n}\varphi}{\partial x_1^{k_1}\cdots\partial x_n^{k_n}}\right| \leqslant M\frac{k_1!^{\alpha_1}\cdots k_n!^{\alpha_n}}{R_1^{k_1}\cdots R_n^{k_n}}, \qquad \alpha_1>0, \ldots, \alpha_n>0,$$

где M и $R_1, \ldots, R_n$ — некоторые постоянные. Если α есть наибольшее из всех чисел $\alpha_1, \ldots, \alpha_n$, то мы будем называть функцию $\varphi(x_1, \ldots, x_n)$ функцией класса α по совокупности $x_1, \ldots, x_n$. Функции первого класса аналитичны. Пусть теперь дано диференциальное уравнение (14). Будем предполагать, что функция Φ или класса $\alpha \geqslant 1$ по x и непрерывна по y, или класса $\beta \geqslant 1$ по y и непрерывна по x, или класса $\alpha \geqslant 1$ по x и класса $\beta \geqslant 1$ по y, по всем же остальным аргументам u, p, q, r, s, t эта функция класса $\nu \geqslant 1$. Кроме того, будем предполагать, что уравнение (14) эллиптично вблизи некоторого решения $u(x, y)$, имеющего непрерывные производные до третьего порядка включительно. Тогда функция $u(x, y)$ — такого же класса относительно x и y, как и функция Φ относительно x и y. В 1918 г. Жевреем [33]) было дано подробное изложение результатов, относящихся только к уравнению (14) [34]).

Еще в 1908 г. Е. Е. Леви (E. Levi) [1] [35]) дал доказательство аналитичности всех решений линейных эллиптических систем с частными производными по двум независимым переменным и аналитическими коэфициентами. При этом предполагалось только, что эти решения непрерывны вместе с их производными до тех порядков, какие входят в рассматриваемые уравнения. Определение эллиптичности для систем будет дано позже.

В 1931 г. Э. Гопф [2] распространил теорему Гильберта-Бернштейна на нелинейные эллиптические уравнения второго порядка с любым числом независимых переменных.

На новой, очень простой и естественной мысли основано доказательство аналитичности решений уравнения (14), данное в 1927—1928 гг. Г. Леви (H. Lewy) [1, 2] [36]). Чтобы доказать аналитичность функции $u(x, y)$, он про-

---

[33]) Ann. Éc. Norm. (3), т. 35 (1918), стр. 129—190.

[34]) В 1926 г. Т. Радо указал на имеющиеся здесь неточности. С этими указаниями Жеврей согласился и обещал [Bulletin de Sciences math., т. 50 (1926), стр. 113] сделать нужные исправления в ближайшей работе, но до сих пор эти исправления не опубликованы. Из всех других уравнений, аналитичность решения которых обещал доказать Жеврей, до сих пор опубликованы только его работы, относящиеся к уравнению (14).

[35]) Перевод этой статьи помещен в настоящем выпуске „Успехов математических наук", стр. 249—292.

[36]) Перевод первой из этих статей помещен в настоящем выпуске „Успехов математических наук", стр. 100—106.

должает ее в пространство (x, x^*, y, y^*) таким образом, чтобы удовлетворялись уравнения Коши-Римана по x и x^*, y и y^*. Это — задача Коши. Вся трудность в том, что она ставится для системы Коши-Римана, которая не является гиперболической (определение гиперболичности системы дано, например, в статье И. Г. Петровского „О задаче Коши в области неаналитических функций" в III выпуске „Успехов математических наук"). Для произвольной функции $u(x, y)$, хотя бы имеющей как угодно большое число непрерывных производных, эта задача, конечно, не имеет решения; иначе получилось бы, что эта функция аналитична. Но Г. Леви удалось свести задачу продолжения функции $u(x, y)$ на пространство (x, x^*, y, y^*) к задаче Коши для некоторой гиперболической системы, пользуясь тем, что функция $u(x, y)$ удовлетворяет эллиптическому уравнению (14). Метод Г. Леви распространяется на эллиптические уравнения как угодно высоких порядков, но только с частными производными по двум независимым переменным. При большем числе независимых переменных задачу продолжения рассматриваемых функций на комплексное пространство не удается свести к задаче Коши для некоторой гиперболической системы диференциальных уравнений. В применении к уравнению (14) метод Г. Леви позволяет доказать аналитичность всех его решений в предположении, что эти решения имеют непрерывные частные производные до четвертого порядка.

Для систем уравнений с частными производными по любому числу независимых переменных И. Г. Петровский [1, 2] доказал следующую теорему. Пусть дана система

$$F_i(x_0, x_1, \ldots, x_n; u_1, \ldots, u_N; \ldots) = 0, \quad i = 1, \ldots, N; \tag{15}$$

входящие в левые части частные производные от функций u_j $(j = 1, \ldots, N)$ не выше n_j-го порядка не выписаны явно. Мы предполагаем, что в некоторой комплексной по x_0, всем u_j и их производным и действительной по $x_1, \ldots, x_n$ области все эти левые части аналитичны по x_0, всем u_j и их производным и имеют непрерывные производные достаточно высоких порядков по $x_1, \ldots, x_n$ Допустим еще, что в рассматриваемой области при всяких действительных $\alpha_0, \alpha_1, \ldots, \alpha_n$ сумма квадратов которых равна 1, детерминант матрицы

$$\left\| \sum_{(k)} \frac{\partial F_i}{\partial \left\{ \dfrac{\partial^{k_0 + \cdots + k_n} u_j}{\partial x_0^{k_0} \ldots \partial x_n^{k_n}} \right\}} \alpha_0^{k_0} \ldots \alpha_n^{k_n} \right\| \tag{16}$$

не обращается в 0. Здесь $\sum\limits_{(k)}$ означает суммирование по всем целым неотрицательным $k_0, k_1, \ldots, k_n$, сумма которых равна n_j. Условие, чтобы детерминант матрицы (16) не обращался в 0, когда $\sum \alpha_s^2 = 1$, мы будем называть условием эллиптичности системы (15).

При выполнении этих условий все решения системы (15), имеющие непрерывные производные по $x_0, x_1, \ldots, x_n$ достаточно высоких порядков, аналитичны по x_0. Мы говорим здесь для краткости, что решение „аналитично по x_0" или „имеет непрерывные производные до порядка δ", понимая под этим, что все функции, составляющие это решение, аналитичны по x_0 или имеют непрерывные производные до порядка δ. Рассматриваемое решение, аналитичность которого по x_0 надо доказать, состоит из функций u_j $(j = 1, \ldots N)$, принимающих, вообще говоря, комплексные значения. Эти функции предполагаются заданными на некоторой $(n+1)$-мерной области действительных значений $x_1, \ldots, x_n$ и действительной части x_0 при постоянной мнимой части x_0. Говоря, что эти функции удовлетворяют системе (15), мы понимаем входящие в F_i производные от них по x_0 как производные по действительной части x_0.

При доказательстве этой теоремы, подобно тому, как это делал Г. Леви, И. Г. Петровский продолжает заданное решение на комплексное пространство $x_0^*, x_0, x_1, \ldots, x_n$. Но это продолжение выполняется методом, отличным от метода Г. Леви, без сведения к задаче Коши для некоторой гиперболической системы.

Из сформулированной только что теоремы следует, в частности, что если левые части F_i уравнений эллиптической системы (15) аналитичны по всем их аргументам, то все решения этой системы, непрерывные вместе с их производными достаточно высоких порядков, аналитичны по всем линейным комбинациям от $x_0, x_1, \ldots, x_n$ с постоянными действительными коэфициентами и, следовательно, аналитичны по всей совокупности x_k $(k=0, 1, \ldots, n)$.

С другой стороны, можно показать, что у системы (15) есть решения, имеющие непрерывные производные как угодно высоких порядков и тем не менее не состоящие из аналитических по всем x_k функций u_j, при следующих условиях:

1) Существует такая система действительных $\alpha_0, \alpha_1, \ldots, \alpha_n$, не состоящая из одних нулей, которая обращает детерминант матрицы (16) в нуль.

2) Этот детерминант не равен нулю тождественно по $\alpha_0, \alpha_1, \ldots, \alpha_n$.

3) F_i суть линейные функции с постоянными коэфициентами от u_j $(j=1, \ldots, N)$ и их производных, члены же, свободные от u_j и их производных, голоморфны по $x_0, x_1, \ldots, x_n$.

Как показывают примеры, условие, чтобы детерминант матрицы (16) не был равен нулю тождественно при всех α_k, здесь существенно. Условие 3), вероятно, несущественно, но это доказано пока только для некоторых частных случаев.

Таким образом из этих работ И. Г. Петровского вытекает доказательство принципа С. Н. Бернштейна для линейных систем с постоянными коэфициентами, у которых характеристический детерминат не равен тождественно нулю. Для линейных же систем с переменными коэфициентами, а также и для нелинейных систем до сих пор остается недоказанной возможность продолжения решения за действительную характеристическую поверхность не единственным образом.

§ 5. Некоторые другие свойства решений эллиптических уравнений.

Решения эллиптических уравнений обладают многими свойствами, похожими на свойства гармонических функций. Отметим некоторые из этих свойств.

1. Если однородное эллиптическое уравнение (2) с достаточно гладкими коэфициентами имеет в некоторой ограниченной области G одно (и, следовательно, только одно) решение первой краевой задачи для всякой непрерывной функции, заданной на границе, то для таких решений имеет место первая теорема Харнака: если последовательность таких функций равномерно сходится на границе G, то она также равномерно сходится и внутри G, и притом к функции, удовлетворяющей тому же уравнению (2).

2. Аналог второй теоремы Харнака: Если ряд

$$u_1(x_1, \ldots, x_n) + u_2(x_1, \ldots, x_n) + \ldots,$$

составленный из неотрицательных на некоторой конечной области G решений однородного уравнения (2), сходится хотя бы в одной точке этой области, то он также равномерно сходится во всякой области G^*, лежащей вместе со своей границей внутри G, и притом предельная функция удовлетворяет уравнению (2) [37]).

3. Если $a \equiv 0$, то всякое непрерывное решение однородного уравнения (2) принимает наибольшее и наименьшее значения на границе G.

4. Решения самосопряженного однородного эллиптического уравнения (5) обладают свойством среднего арифметического, если их рассматривать в римановом пространстве с соответствующим образом подобранной метрикой. Для таких уравнений можно строить функции, аналогичные $\ln \rho$, потенциалу простого или двойного слоя для обычного уравнения Лапласа с двумя независимыми переменными [38]). Аналогичные $\ln \rho$, так называемые „фундаментальные" решения, построены Е. Е. Леви [1] [39]) также для некоторых эллиптических систем.

5. Теорема Лиувилля для аналитических функций также распространяется на эллиптические уравнения второго порядка [40]).

6. Если все функции u_i, удовлетворяющие некоторой однородной линейной эллиптической системе с n независимыми переменными и аналитическими коэфициентами, одновременно обращаются в нуль на некоторой $(n-1)$-мерной аналитической поверхности вместе со всеми их производными до $(n_i - 1)$-го порядка, то они тождественно равны нулю во всей той области, где они удовлетворяют рассматриваемой системе.

Это утверждение получается, например, как следствие теоремы Гольмгрена (Holmgren) [1] о единственности решения задачи Коши для линейных систем с аналитическими коэфициентами, у которых нет действительных характеристик.

7. Обобщение теоремы Шварца для гармонических функций: Если $u(x, y)$ есть регулярное [41]) внутри области S, ограниченной аналитическим контуром C, решение аналитического уравнения эллиптического типа

$$F(x, y, u, \ldots, r, s, t) = 0,$$

---

[37]) Lichtenstein [1].

[38]) Feller [1]. (Перевод этой статьи печатается в настоящем выпуске „Успехов математических наук", стр. 232—248.)

[39]) Перевод этой статьи помещен в настоящем выпуске „Успехов математических наук", стр. 249—292. См. также Herglotz [1, 2].

[40]) Бернштейн [18]. (Перевод этой статьи печатается в настоящем выпуске „Успехов математических наук", стр. 75—81.) Уравнение $A(x, y, \ldots, r) r + 2B(x, y, \ldots, r) s + C(x, y, \ldots, r) t = 0$, где $AC - B^2 > 0$, не имеет иных регулярных решений, ограниченных на всей плоскости по абсолютному значению: $|u| < h$, кроме постоянной. Однако теорема Лиувилля в узком смысле не всегда верна: так, требование, чтобы $u \geqslant 0$ на всей плоскости, не для всех указанных уравнений влечет, что u постоянная: например, уравнение $2(1 + 2y^2) r + 4ys + t = 0$ имеет положительное на всей плоскости решение $u = e^{x - y^2}$.

[41]) Т. е. имеющее ограниченные производные первых двух порядков в замкнутой области S.

обращающееся на C в аналитическую функцию $\varphi(s)$ дуги, то оно может быть аналитически продолжено за пределы C[42]).

8. Если все коэфициенты a_{ij} эллиптического уравнения (2) в некоторой конечной области G имеют ограниченные производные до p-го порядка, а коэфициенты a_i, a и f имеют в этой области ограниченные производные до $(p-1)$-го порядка, то все непрерывные решения этого уравнения имеют во всякой области G^*, лежащей вместе со своей границей внутри G, ограниченные производные до p-го порядка.

9. Допустим, что оператор $L[M, f(P)]$ обладает следующими двумя свойствами:

1) $L[M, f_1(P)+f_2(P)] = L[M, f_1(P)] + L[M, f_2(P)]$,

$L[M, af(P)] = aL[M, f(P)]$ $(a = \text{const})$,

2) $L[M, f(P)] \leqslant 0$, если $f(P)$ имеет в точке M неотрицательный максимум.

Тогда для всех достаточно гладких функций $f(P)$

$$L[M, f(P)] = -a(M)f(M) + \sum_{i=1}^{n} a_i(M)\frac{\partial f(M)}{\partial x_i} + \sum_{i=1}^{n}\sum_{j=1}^{n} a_{ij}(M)\frac{\partial^2 f(M)}{\partial x_i\,\partial x_j},$$

где $\sum_{i=1}^{n}\sum_{j=1}^{n} a_{ij}\alpha_i\alpha_j \geqslant 0$ и $a(M) \geqslant 0$[43]).

Список литературы.

Айзенштат Н. Д.

[1] Об одном типе аддитивного оператора. Ученые записки МГУ, выпуск пятнадцатый. Математика (ГОНТИ, 1939), 95—103.

Бернштейн С. Н.

[1] Sur la nature analytique des solutions de certaines équations aux dérivées partielles du second ordre. Comptes rendus Ac. Sci. Paris, 137 (1903), 778—781.

[2] Sur la nature analytique des solutions des équations aux dérivées partielles du second ordre. Math. Annalen, 59 (1904), 20—76.

[3] Sur certaines équations aux dérivées partielles du second ordre. Comptes rendus Ac. Sci. Paris, 139 (1904), 627—628.

[4] Sur les équations aux dérivées partielles du type elliptique. Comptes rendus Ac. Sci. Paris, 140 (1905), 1440—1442.

[5] Sur les surfaces minima. Comptes rendus Ac. Sci. Paris, 141 (1905), 558—559.

[6] Sur la déformation des surfaces. Math. Annalen, 60 (1905), 434—436.

[7] Sur les singularités des solutions des équations aux dérivées partielles du type elliptique. Comptes rendus Ac. Sci. Paris, 142 (1906), 564—565.

[8] Sur la généralisation du problème de Dirichlet. (Première partie.) Math. Annalen, 62 (1906), 253—271.

[9] Méthode générale pour la résolution du problème de Dirichlet. Comptes rendus Ac. Sci. Paris, 144 (1907), 1025—1027.

[42]) Полное доказательство в работе С. Н. Бернштейна [20]. В работах [7] и [10] исследован вопрос об особенностях функции $u(x, y)$ вне контура C в зависимости от особенностей функции $\varphi(s)$ в комплексной области в случае линейного уравнения. В частности, если все коэфициенты уравнения являются целыми функциями, то $u(x, y)$, как и гармоническая функция, будет целой функцией тогда и только тогда, когда $\varphi(s)$ будет целой функцией.

[43]) Айзенштат [1].

[10] Исследование и интегрирование диференциальных уравнений с частными производными второго порядка эллиптического типа. Сообщения Харьковского математического общества, вторая серия, XI (1908—1909), 1—164.

[11] Sur le principe de Dirichlet et le développement des fonctions harmoniques en séries de polynomes. Comptes rendus Ac. Sci. Paris, 148 (1909), 1306—1308.

[12] Conditions nécessaires et suffisantes pour la possibilité du problème de Dirichlet. Comptes rendus Ac. Sci. Paris, 150 (1910), 514—515.

[13] Sur la généralisation du problème de Dirichlet. (Deuxième partie.) Math. Annalen, 69 (1910), 82—136.

]14] Sur les équations du calcul des variations. Comptes rendus Ac. Sci. Paris, 151 (1910), 195—198.

[15] Sur une généralisation des théorèmes de Liouville et de M. Picard. Comptes rendus Ac. Sci. Paris, 151 (1910), 636—639.

[16] Sur les surfaces définies au moyen de leur courbure moyenne ou totale. Ann. de l'Éc. Norm. (3), 27 (1910), 233—256.

[17] Sur les équations du calcul des variations. Ann. de l'Éc. Norm. (3), 29 (1912), 431—485. Русский перевод см. в настоящем выпуске „Успехов математических наук", стр. 32—74.

[18] Sur un théorème de géométrie et son application aux équations aux dérivées partielles du type elliptique. Сообщения Харьковского математического общества XV (1915), 38—45. Русский перевод см. в настоящем выпуске „Успехов математических наук", стр. 75—81.

[19] Sur l'intégration des équations aux dérivées partielles du type elliptique. (Réponse à une critique.) Math. Annalen, 95 (1925), 585—594.

[20] Sur l'intégration des équations aux dérivées partielles du type elliptique. II. Note. Math. Annalen, 96 (1927), 633—647.

[21] Sur la nature analytique des solutions des équations différentielles aux dérivées partielles du type elliptique. Math. Zeitschrift, 25 (1926), 505—513.

[22] Démonstration du théorème de M. Hilbert sur la nature analytique des solutions des équations du type elliptique sans l'emploi des séries normales. Math. Zeitschrift, 28 (1928), 330—348. Русский перевод см. в настоящем выпуске „Успехов математических наук", стр. 82—99.

Carathéodory C.

[1] On Dirichlet's problem. Amer. Journ. Math., 59 (1937), 709—731.

Courant R., K. Friedrichs und H. Lewy.

[1] Über die partiellen Differenzengleichungen der mathematischen Physik. Math. Annalen, 100 (1928), 32—74. Русский перевод см. в настоящем выпуске „Успехов математических наук", стр. 125—160.

Feller W.

[1] Über die Lösungen der linearen partiellen Differentialgleichungen zweiter Ordnung vom elliptischen Typus. Math. Annalen, 102 (1930), 633—649. Русский перевод см. в настоящем выпуске „Успехов математических наук", 232—248.

Gevrey M.

[1] Sur la nature analytique des solutions des équations aux dérivées partielles. Premier mémoire. Ann. de l'Éc. Norm. (3), 35 (1918), 129 — 190.

Giraud G.

[1] Sur le problème de Dirichlet généralisé. Ann. de l'Éc. Norm. (3), 46 (1929), 131—245.

[2] Sur différentes questions relatives aux équations du type elliptique. Ann. de l'Éc. Norm. (3), 47 (1930), 197—266.

[3] Sur certains problèmes non linéaires de Neumann et sur certains problèmes non linéaires mixtes. Ann. de l'Éc. Norm., 49 (1932), 1—104.

Hadamard J.

[1] Sur le principe de Dirichlet. Bull. Soc. Math. France, 34 (1906), 135—138.

Herglotz G.

[1] Über die Integration linearer, partieller Differentialgleichungen mit konstanten Koeffizienten. I. Berichte Sächs. Akad., Math.-phys. Klasse, 78 (1926), 93—126, 287—318.

[2] Über die Integration linearer, partieller Differentialgleichungen mit konstanten Koeffizienten. III. Berichte Sächs. Akad., Math. phys. Klasse, 80 (1928), 69—114.

Hilbert D.

[1] Über das Dirichletsche Prinzip. Math. Annalen, 59 (1904), 161—186.

Holmgren E.

[1] Über Systeme von linearen partiellen Differentialgleichungen. Öfversigt af Kongl. Svenska Vetenskaps-Akademiens Förhandlingar, 58 (1901), 91—103.

Hopf E.

[1] Zum analytischen Charakter der Lösungen regulärer zweidimensionaler Variationsprobleme. Math. Zeitschrift, 30 (1929), 404—413.

[2] Über den funktionalen, insbesondere den analytischen Charakter der Lösungen elliptischer Differentialgleichungen zweiter Ordnung. Math. Zeitschrift, 34 (1931), 194—233.

Korn A.

[1] Über Minimalflächen, deren Randkurven wenig von ebenen Kurven abweichen. Abhandlungen Preuss. Akad., 1909.

Lebesgue A.

[1] Sur le problème de Dirichlet, Rendiconti del Circolo Matematico di Palermo, 24 (1907), 371—402.

[2] Sur le cas d'impossibilité du problème de Dirichlet ordinaire. Société mathématique de France, Comptes rendus des séances de l'année 1913, 17.

[3] Sur la méthode de Carl Neumann. Journ. Math. pures et appl. (IX), 16 (1937), 205—217.

[4] Sur la méthode de Carl Neumann. II. Journ. Math. pures et appl. (IX), 16 (1937), 421—423.

Leray J.

[1] Discussion du problème de Dirichlet. Comptes rendus Ac. Sci. Paris, 205 (1937), 268 — 270.

[2] Sur la résolution du problème de Dirichlet. Comptes rendus Ac. Sci. Paris, 205 (1937), 784—786.

[3] Majoration des dérivées secondes des solutions d'un problème de Dirichlet. Journ Math. pures et appl. (9), 17 (1938), 89 — 104.

Levi E. E.

[1] Sulle equazioni lineari totalmente ellitiche alle derivate parziali. Rendiconti del Circolo Matematico di Palermo, 24 (1907), 275—317. Русский перевод см. в настоящем выпуске „Успехов математических наук“, стр. 249—292.

Lewy H.

[1] Über den analytischen Charakter der Lösungen elliptischer Differentialgleichungenl. Göttinger Nachrichten (1927), 178—186. Русский перевод см. в настоящем выпуске „Успехов математических наук“, стр. 100—106.

[2] Neuer Beweis des analytischen Charakters der Lösungen elliptischer Differentialgleichungen. Math. Annalen, 101 (1929), 609—619, и 107 (1934), 804.

Lichtenstein L.

[1] Beiträge zur Theorie der linearen partiellen Differentialgleichungen zweiter Ordnung vom elliptischen Typus. Unendliche Folgen positiver Lösungen. Rendiconti del Circolo Matematico di Palermo, 33 (1912), 201—211, и 34 (1912), 278—279.

[2] Über den analytischen Charakter der Lösungen regulärer zweidimensionaler Variationsprobleme. Bulletin international de l'Académie des sciences de Cracovie, Classe des sciences mathématiques et naturelles, Série A, 1912, 915—941.

[3] Randwertaufgaben der Theorie der linearen partiellen Differentialgleichungen (zweiter Ordnung vom elliptischen Typus. I, Journ. reine u. angew. Math. 142 (1912), 1—40.

Люстерник Л. А.

[1] Über einige Anwendungen der direkten Methoden in Variationsrechnung. Математический сборник, 33 (1926), 173—200. Переработанный перевод см. в настоящем выпуске „Успехов математических наук", стр. 115—124.

[2] Bemerkung zur Lösung des Dirichlet'schen Problems. Математический сборник, 33 (1926), 363—364.

Müntz Ch.

[1] Zum Randwertproblem der partiellen Differentialgleichung der Minimalflächen. Journ. reine u. angew. Math., 139 (1911), 52—79.

Немыцкий В. В.

[1] Теоремы существования и единственности для нелинейных интегральных уравнений. Математический сборник, 41 (1934), 438—452. Французский перевод там же, 421—438.

Neumann C.

[1] Untersuchungen über das logarithmische und Newton'sche Potential Leipzig, 1877.

Paraf A.

[1] Sur le problème de Dirichlet et son extension au cas de l'équation linéaire générale du second ordre. Ann. Fac. Sc. Toulouse (1), 6 (1892), 1—75.

Perron O.

[1] Eine neue Behandlung der ersten Randwertaufgabe für $\Delta u = 0$. Math. Zeitschrift, 18 (1923), 42—54.

Петровский И. Г.

[1] О системах дифференциальных уравнений, все решения которых аналитичны. Доклады Академии наук СССР, XVII (1937), 339—342.

[2] Sur l'analyticité des solutions des systèmes d'équations différentielles. Математический сборник, 5 (47) (1939), 3—68.

Picard E.

[1] Mémoire sur la théorie des équations aux dérivées partielles et la méthode des approximations successives. Journ. Math. pures et appl. (4), 6 (1890), 145—210 и 231—232.

[2] Sur la détermination des intégrales d'une équation aux dérivées partielles par ses valeurs sur un contour fermé. Journ. Math. pures et appl. (5), 2 (1896) 295—304.

[3] Sur la détermination des intégrales de certaines équations linéaires du second ordre par leurs valeurs sur un contour fermé. Journ. Math. pures et appl. (5), 6 (1900), 129—140.

[4] Sur les équations linéaires aux dérivées partielles et la généralisation du problème de Dirichlet. Acta Mathematica, 25 (1902), 121—137.

Привалов И. И.

[1] Sur les fonctions harmoniques. Математический сборник, 32 (1925), 464—469.

Pueschel W.

[1] Die erste Randwertaufgabe der allgemeinen selbstadjungierten elliptischen Differentialgleichung zweiter Ordnung im Raum für beliebige Gebiete. Math. Zeitschrift, 34 (1932), 535—553.

Radó T.

[1] Über den analytischen Charakter der Minimalflächen. Math. Zeitschrift, 24 (1926), 321—327.

[2] The problem of the least area and the problem of Plateau. Math. Zeitschrift, 32 (1930), 763—796.

Radó T, u. F. Riesz.

[1] Über die erste Randwertaufgabe für $\Delta u = 0$. Math. Zeitschrift, 22 (1925), 41—44.

Rellich F.

[1] Zur ersten Randwertaufgabe bei Monge-Ampère Differentialgleichungen vom elliptischen Typus. Math. Annalen, 107 (1932), 505—513 и 804.

Roux J., le

[1] Sur le problème de Dirichlet. Journ. Math. pures et appl. (6), 10 (1914), 189—230.

Schauder J.

[1] Über den Zusammenhang zwischen der Eindeutigkeit und Lösbarkeit partieller Differentialgleichungen zweiter Ordnung. Math. Annalen, 106 (1932), 661—721.

[2] Über lineare elliptische Differentialgleichungen zweiter Ordnung. Math. Zeitschrift 38 (1934), 257—282.

Schauder J. et J. Leray.

[1] Topologie et équations fonctionnelles. Journ. de l'Éc. Norm. (3), 51 (1934), 45—78.

Урысон П. С.

[1] Zur ersten Randwertaufgabe der Potentialtheorie. Ein Fall der Unlösbarkeit. Math Zeitschrift, 23 (1925), 155—158.

Pavel Samuilovich Urysohn

Born – 3 Feb 1898 in Odessa, Ukraine
Died – 17 Aug 1924 in Batz-sur-Mer, France

Pavel Urysohn is also known as Pavel Uryson. His father was a financier in Odessa, the town in which Pavel Samuilovich was born. He came from a family descended from the sixteenth century Rabbi M Jaffe. It was a well-off family and Urysohn received his secondary education in Moscow at a private school there.

In 1915 Urysohn entered the University of Moscow to study physics and in fact he published his first paper in this year. Being interested in physics at this time, his first paper was on a physics topic, and it was on Coolidge tube radiation. However his interest in physics soon took second place for after attending lectures by Luzin and Egorov at the University of Moscow. From then on he began to concentrate on mathematics.

Urysohn graduated in 1919 and continued his studies there working towards his doctorate. Luzin persuaded Urysohn to stay on in order to study for a doctorate during 1919–1921. At this stage Urysohn was interested in analysis, in particular integral equations, and this was the topic of his habilitation. He was awarded his habilitation in June 1921 and, following this, became an assistant professor at the University of Moscow.

Urysohn soon turned to topology. He was asked two questions by Egorov and it was these which occupied him during the summer of 1921. The first question that Egorov posed was to find a general intrinsic topological definition of a curve which when restricted to the plane became Cantor's notion of a continuum which is nowhere dense in the plane. The second of Egorov's question was a similar one but applied to surfaces, again asking for an intrinsic topological definition.

Now these were difficult questions that had been around for some time. It was not that Egorov had come up with new questions, rather he was giving the bright young mathematician Urysohn two really difficult problems in the hope that he might come up with new ideas. Egorov was not to be disappointed, for Urysohn attacked the questions with great determination. He did not sit still waiting for inspiration to strike, rather he tried one idea after another to see if it would give him the topological definition of dimension that he was looking for.

A holiday with other young Moscow mathematicians to the village of Burkov, on the banks of the river Kalyazmy near to the town of Bolshev, did not stop him trying to find the "right" definition of dimension. Quite the opposite, it was a good chance for him to think in congenial surroundings, and one morning near the end of August he woke up with an idea in his mind which he felt, even before working through the details, was right. Immediately he told his friend Aleksandrov about his inspiration.

Of course there was a lot of hard work after the moment of inspiration. During the following year Urysohn worked through the consequences building a whole new area of dimension theory in topology. It was an exciting time for the topologists in Moscow for Urysohn lectured on the topology of continua and often his latest results were presented in the course shortly after he had proved them. He published a series of short notes on this topic during 1922. The complete theory was presented in an article which Lebesgue accepted for publication in the *Comptes rendus* of the Academy of Sciences in Paris. This gave Urysohn an international platform for his ideas which immediately attracted the interest of mathematicians such as Hilbert.

Urysohn published a full version of his dimension theory in *Fundamenta mathematica.* He wrote a major paper in two parts in 1923 but they did not appear in print until 1925 and 1926. Sadly Urysohn had died before even the first part was published. In fact Urysohn set out to do far more in this paper than to answer the two questions that Egorov had posed to him.

Urysohn visited Göttingen in 1923. His reports to the Mathematical Society of Göttingen interested Hilbert and while in Göttingen he learnt of Brouwer's contributions to the area made in the paper of 1913 to which we referred above. Urysohn spotted an error in Brouwer's paper regarding a definition of dimension while he was studying it in Göttingen and easily constructed a counter-example.

In the summer of 1924 Urysohn set off again with Aleksandrov on a European trip through Germany, Holland and France. Again the two mathematicians visited Hilbert and, by 7 May, they must have left since Hilbert wrote to Urysohn on that day telling him his paper with Aleksandrov was accepted for publication in *Mathematische Annalen.* This letter also thanks Urysohn for caviar he had given Hilbert, and expresses the hope that Urysohn will visit again the following summer.

They then met Hausdorff who was impressed with Urysohn's results. He also wrote a letter to Urysohn which was dated 11 August 1924. The letter discusses Urysohn's metrization theorem and his construction of a universal separable metric space. The construction of a universal metric space, containing an isometric image of any metric space, was one of Urysohn's last results. Like Hilbert, Hausdorff expressed the hope that Urysohn would visit again the following summer.

After this visit the two mathematicians continued their holiday to Brittany where they rented a cottage. Urgsohn drowned in rough sees while on one of their regular swims off the coast.

Urysohn was not only an "inseparable friend" to Aleksandrov but the two collaborated on important publications such as *Zur Theorie der topologischen Räume published in Mathematische Annalen* in 1924. Urysohn's main contributions, in addition to the theory of dimension discussed above, are the introduction and investigation of a class of normal surfaces, metrization theorems, and an important existence theorem concerning mapping an arbitrary normed space into a Hilbert space with countable basis. Urysohn's name is familiar to anyone who has taken a course in general topology, he is remembered particularly for "Urysohn's lemma": *Given two disjoint closed sets in a normal (T_4) topological space, there exists a continuous real-valued function that is* 0 *on one of the sets and* 1 *on the other.*

After Urysohn's death Brouwer and Aleksandrov made sure that the mathematics he left was properly dealt with.

Zum Metrisationsproblem*).

Von

Paul Urysohn †.

1. Das *Metrisationsproblem*, d. h. das Problem, die Bedingungen aufzustellen, unter denen ein topologischer Raum[1]) einem metrischen Raume[2]) homöomorph ist, hat in der letzten Zeit sowohl eine allgemeine, aber ziemlich komplizierte Lösung, wie auch eine völlig befriedigende Lösung in einigen wichtigen Spezialfällen gefunden[3]). In dieser Arbeit will ich zeigen, wie man das Metrisationsproblem in sehr einfacher Weise für alle *Räume mit* II. *Abzählbarkeitsaxiom*[4]) lösen kann: es tritt hier nämlich die

*) Der Inhalt dieses Aufsatzes wurde von Paul Urysohn im April 1924 der Moskauer Mathematischen Gesellschaft und am 8. Juli 1924 der Göttinger Mathematischen Gesellschaft vorgetragen. Von vorliegendem Text wurde nur der erste Paragraph von ihm selbst redigiert, das übrige nach seinem Tode vom Unterzeichneten, dem auf Grund mündlicher Mitteilungen sowohl der ganze Plan der Arbeit wie sämtliche Einzelheiten der Beweisführung genau bekannt waren.

Paul Alexandroff.

[1]) Hausdorff, Grundzüge der Mengenlehre, Leipzig 1914 (im folgenden einfach „Hausdorff" zitiert), S. 213.

[2]) Hausdorff, S. 211.

[3]) Siehe P. Alexandroff et P. Urysohn, *Une condition nécessaire et suffisante* ..., Comptes Rendus Paris **177**, S. 1274; P. Urysohn, Über die Metrisation der kompakten topologischen Räume, Math. Annalen 92, S. 275—293; P. Alexandroff, Über die Metrisation der im Kleinen kompakten topologischen Räume, ebenda S. 294 u. f. In diesen Arbeiten wird man auch weitere Literaturangaben finden. Man vgl. auch die Arbeit P. Urysohn, Über die Mächtigkeit der zusammenhängenden Mengen, Math. Annalen **94**, S. 262—295, insbesondere Fußnote [7]) und § 28.

[4]) Hausdorff, S. 263; vgl. übrigens „Über die Mächtigkeit der zusammenhängenden Mengen", Fußnote [8]). Man kann das II. Abzählbarkeitsaxiom (wie man leicht einsieht) auch folgendermaßen formulieren: *Es existiert im Raume ein abzählbares System von Gebieten*[7]), *so daß jedes Gebiet die Vereinigungsmenge gewisser Gebiete dieses Systems ist.*

Normalität[5]) des Raumes als notwendige und hinreichende Bedingung der Metrisierbarkeit auf. Ich erinnere daran, daß ein topologischer Raum *normal* heißt, wenn in ihm folgende Eigenschaft (das Tietzesche[6]) III. Trennungsaxiom") erfüllt ist: wenn A und B zwei elementenfremde abgeschlossene Mengen sind, so sind sie durch Gebiete[7]) trennbar, d. h. es können zwei zueinander fremde Gebiete G_A und G_B gefunden werden, die diese Mengen bzw. enthalten: $G_A \supset A$, $G_B \supset B$.

2. Wir wollen also den folgenden Satz beweisen:

Hauptsatz. *Jeder dem* II. *Abzählbarkeitsaxiom genügende normale topologische Raum ist metrisierbar* (*d. h. einem metrischen Raume homöomorph*).

Da ein metrischer Raum, wie man leicht einsieht, immer normal ist[8]), so ist die Bedingung der Normalität eine nicht nur hinreichende, sondern auch notwendige Bedingung für die Metrisierbarkeit der topologischen Räume mit dem II. Abzählbarkeitsaxiom.

Da endlich in metrischen Räumen das II. Abzählbarkeitsaxiom mit der Forderung der Existenz einer abzählbaren, überall dichten Teilmenge äquivalent ist, so kann man den soeben ausgesprochenen Satz auch folgendermaßen formulieren:

Damit ein topologischer Raum E einem metrischen Raume mit abzählbarer überall dichter Teilmenge homöomorph ist, ist notwendig und hinreichend, daß in E die beiden Voraussetzungen der Normalität und des II. *Abzählbarkeitsaxioms erfüllt seien*[9]).

Wir stützen uns beim Beweise auf den folgenden, in der Arbeit „Über die Mächtigkeit der zusammenhängenden Mengen", Anhang III bewiesenen

Hilfssatz. Es seien A und B zwei elementenfremde abgeschlossene, im normalen Raume E liegende Mengen. Man kann dann in E eine stetige Funktion $f(x)$ erklären, so daß

$$\begin{aligned} f(x) &= 0 \quad \text{auf } A, \\ f(x) &= 1 \quad \text{auf } B, \\ 0 \leqq f(x) &\leqq 1 \quad \text{in allen übrigen Punkten von } E. \end{aligned}$$

---

[5]) P. Alexandroff und P. Urysohn, Zur Theorie der topologischen Räume, Math. Annalen **92**, S. 263, § 4.

[6]) Tietze, Beiträge zur allgemeinen Topologie I, Math. Annalen **88**, S. 290.

[7]) Gebiet = offene Menge. Vgl. Hausdorff, S. 215.

[8]) Siehe Tietze, l. c. [6]), S. 310. Vgl. auch P. Urysohn, l. c. [4]), § 3.

[9]) Dagegen ist (wie man durch Beispiele zeigen kann) das Vorhandensein einer abzählbaren überall dichten Teilmenge in einem normalen topologischen Raume für die Metrisierbarkeit dieses Raumes keineswegs hinreichend; vgl. darüber P. Alexandroff, l. c. [3]), § 3.

Um die Lektüre des vorliegenden Aufsatzes von der soeben zitierten Arbeit unabhängig zu machen, geben wir hier den Beweis des Hilfssatzes kurz wieder.

Es sei $G_1 = E - B$ das zur Menge B komplementäre Gebiet; es ist $A \subset G_1$; A und B sind elementenfremd, also durch Gebiete $G_0 \supset A$, $G^* \supset B$ trennbar. Aus $G_0 \cdot G^* = 0$ folgt sofort $\bar{G}_0 \cdot G^* = 0$ (wobei $\bar{M}$ immer die abgeschlossene Hülle $\bar{M} = M + M'$ der Menge M bedeutet), d. h. $A \subset G_0 \subset \bar{G}_0 \subset G_1$.

Es seien jetzt alle Gebiete $G_{\frac{m}{2^n}}$ ($n \geqq 0$ feste ganze Zahl, $0 \leqq m \leqq 2^n$) bereits in der Weise definiert, daß dabei aus $\frac{m_1}{2^n} < \frac{m_2}{2^n}$ die Inklusion $\bar{G}_{\frac{m_1}{2^n}} \subset G_{\frac{m_2}{2^n}}$ folgt (für den Fall $n = 0$ ist uns dies ja gelungen). $\bar{G}_{\frac{m}{2^n}}$ und $E - G_{\frac{m+1}{2^n}}$ sind abgeschlossene elementenfremde Mengen. Man kann sie also durch Gebiete $G_{\frac{2m+1}{2^{n+1}}} \supset \bar{G}_{\frac{m}{2^n}}$ und $G^\dagger \supset E - G_{\frac{m+1}{2^n}}$ trennen, und da auch $\bar{G}_{\frac{2m+1}{2^{n+1}}} \cdot G^\dagger = 0$ ist, so ist

$$\bar{G}_{\frac{m}{2^n}} \subset G_{\frac{2m+1}{2^{n+1}}} \subset \bar{G}_{\frac{2m+1}{2^{n+1}}} \subset G_{\frac{m+1}{2^n}}.$$

In dieser Weise fortfahrend erhält man ein System G_{r_i} $\left(\text{wo } r_i = \frac{m}{2^n},\ 0 \leqq r_i \leqq 1,\ n = 1, 2, \ldots \text{ ist}\right)$ von Gebieten, die allen zwischen 0 und 1 gelegenen dyadischen Brüchen entsprechen und die so beschaffen sind, daß

1. $A \subset G_0$,
2. aus $r_i < r_j$ die Inklusion $\bar{G}_{r_i} \subset G_{r_j}$ folgt.

Man kann dieses System $\{G_{r_i}\}$ zu einem System $\{G_t\}$, wo t *alle* Werte zwischen 0 und 1 annimmt, ergänzen, indem man

$$G_t = \sum_{r_k < t} G_{r_k}$$

setzt. Aus $t_1 < t_2$ folgt dann $\bar{G}_{t_1} \subset G_{t_2}$ (da, falls $t_1 < r_i < r_j < t_2$ ist, ist $G_{t_1} \subset G_{r_i}$, also $\bar{G}_{t_1} \subset \bar{G}_{r_i} \subset G_{r_j} \subset G_{t_2}$). Wir setzen endlich $G_t = 0$ (0 ist hier das Zeichen für die leere Menge) für $t < 0$ und $G_t = E$ für $t > 1$; dann ist die Bedingung

$$\text{aus } t_1 < t_2 \text{ folgt } \bar{G}_{t_1} \subset G_{t_2}$$

für beliebige reelle t_1 und t_2 erfüllt. Wenn nun x irgendein Punkt von E ist, so folgt aus dieser Bedingung, daß die Indizes t derjenigen Gebiete G_t, die den Punkt x *nicht erhalten*, eine *Halbgerade* $-\infty < t \leqq \tau_x$, oder $-\infty < t < \tau_x$ bilden; und wir wollen $f(x) = \tau_x$ setzen.

Für einen zu G_0 (also insbesondere zu A) gehörenden Punkt x ist dann $f(x) = 0$, außerhalb G_1 (d. h. auf B) ist $f(x) = 1$, in den übrigen Raumpunkten ist $f(x)$ nicht negativ und $\leqq 1$. Die Funktion $f(x)$ ist stetig. Es seien in der Tat ein Punkt $a \subset E$ und ein $\varepsilon > 0$ beliebig gegeben. Offenbar ist $a \subset G_{\tau_a + \varepsilon} - \bar{G}_{\tau_a - \varepsilon}$; es existiert also eine Umgebung U_a von a, die in demselben *Gebiete* $G_{\tau_a + \varepsilon} - \bar{G}_{\tau_a - \varepsilon}$ liegt; für alle Punkte $x \subset U_a$ ist also $\tau_a - \varepsilon < \tau_x < \tau_a + \varepsilon$, d. h. $|f(x) - f(a)| < \varepsilon$. Der Hilfssatz ist hiermit bewiesen.

3. Es sei nun

$$(1) \qquad \{U_1, U_2, \ldots, U_n, \ldots\}$$

ein *abzählbares* Umgebungssystem, das den gegebenen Raum E als topologischen Raum definiert[10]).

Es seien

$$(2) \qquad \pi_1, \pi_2, \ldots, \pi_n, \ldots$$

alle Paare von Umgebungen, $\pi_n = (U_i, U_k)$, von der Beschaffenheit $\overline{U}_i \subset U_k$. Es sei endlich, für beliebiges ganzes $n > 0$, $f_n(x)$ die dem Paare $\pi_n = (U_i, U_k)$ entsprechende, dem Hilfssatze gemäß vorhandene, in allen Punkten des Raumes E erklärte, stetige Funktion mit den Eigenschaften:

$$f_n(x) = 0 \quad \text{auf } \overline{U}_i,$$
$$f_n(x) = 1 \quad \text{auf } E - U_k,$$
$$0 \leqq f_n(x) \leqq 1 \quad \text{in allen Punkten von } E.$$

Wir setzen für zwei beliebige Punkte x, y des Raumes E

$$(3) \qquad \varrho(x, y) = \sum_{n=1}^{\infty} \frac{1}{2^n} |f_n(x) - f_n(y)|.$$

Offenbar ist

$$\varrho(x, x) = 0, \qquad \varrho(x, y) = \varrho(y, x), \qquad \varrho(x, y) + \varrho(y, z) \geqq \varrho(x, z).$$

Wir beweisen noch, daß für zwei *verschiedene* Punkte $x \neq y$ unmöglich $\varrho(x, y) = 0$ sein kann. In der Tat, nach dem Hausdorffschen Axiom (D)[10a]) gibt es zwei Umgebungen U_k und U_j aus (1), so daß $U_k \supset x$, $U_j \supset y$, $U_k \cdot U_j = 0$ ist. Nun sind die beiden abgeschlossenen Mengen x und $E - U_k$ durch Gebiete $G_1 \supset x$, $G_2 \supset E - U_k$ trennbar, so daß $G_1 \cdot G_2 = \overline{G}_1 \cdot G_2 = 0$ und folglich $x \subset G_1 \subset U_k$ ist. Da G_1 ein Gebiet ist, so existiert eine Umgebung U_i des Punktes x, die in G_1 enthalten ist, und also *a fortiori* $x \subset U_i \subset \overline{U}_i \subset \overline{G}_1 \subset U_k$.[11]) Das Umgebungspaar (U_i, U_k) ist also ein gewisses π_n der Folge (2). Da $x \subset \overline{U}_i$, so ist $f_n(x) = 0$; da aber $y \subset E - U_k$, so ist $f_n(y) = 1$, also ist $|f_n(x) - f_n(y)| = 1$ und $\varrho(x, y) \geqq \frac{1}{2^n}$.

Indem wir die durch (3) definierte Funktion $\varrho(x, y)$ als Entfernung der beiden Punkte x und y betrachten, erhalten wir also einen metrischen

---

[10]) Vgl. hierüber P. Urysohn, l. c. [4]), Fußnote [8]).

[10a]) Hausdorff, S. 213.

[11]) Wir haben also bewiesen, daß *in einem normalen Raume jede Umgebung U_x eines beliebigen Punktes x die abgeschlossene Hülle $\overline{V}_x$ einer gewissen Umgebung V_x desselben Punktes x enthält*, d. h. daß ein normaler Raum immer *regulär* ist. Vgl. P. Alexandroff u. P. Urysohn, Zur Theorie der topologischen Räume, a. a. O. [5]), sowie P. Urysohn, l. c. [4]), §§ 2, 3.

Raum M, dessen Punktvorrat derselbe, wie der des topologischen Raumes E ist; es bleibt uns also nur übrig zu beweisen, daß auch die Limesbildungen in den beiden Räumen dieselben sind, d. h. daß E und M homöomorph sind. Dazu brauchen wir nur folgende Tatsachen I und II zu bestätigen:

I. Für einen beliebigen Punkt $a \subset E$ und ein beliebiges $\varepsilon > 0$ gibt es eine Umgebung U_n des Punktes a, die in der, auf Grund der soeben gegebenen Entfernungsdefinition erklärten, Sphäre[12]) $S(a, \varepsilon)$ enthalten ist.

II. Zu jeder Umgebung U_i des beliebigen Punktes $a \subset E$ gibt es eine gewisse, in dieser Umgebung enthaltene Sphäre $S(a, \varepsilon)$.

Beweis von I. Die Funktion $\varphi(x) = \varrho(a, x)$ ist (zufolge der gleichmäßigen Konvergenz der Reihe (3)) stetig; es gibt also für jedes ε eine Umgebung U_n des Punktes $x = a$, so daß

$$|\varphi(x) - \varphi(a)| = \varphi(x) = \varrho(a, x) < \varepsilon$$

für alle $x \subset U_n$, d. h. daß $U_n \subset S(a, \varepsilon)$.

Beweis von II. Es sei U_k eine beliebige Umgebung des Punktes a. Vorausgesetzt, es gebe kein ε von der Beschaffenheit, daß $S(a, \varepsilon) \subset U_k$. Es existiert dann für jedes m ein Punkt x_m, der den Bedingungen $\varrho(a, x_m) < \frac{1}{2^m}$, $x_m \subset E - U_k$ genügt. Es existiert nun[11]) eine Umgebung U_i des Punktes a, so daß $\overline{U}_i \subset U_k$ ist. Es sei π_N das Paar (U_i, U_k); da, für beliebiges m, $x_m \subset E - U_k$ und $a \subset \overline{U}_i$ ist, so ist $f_N(a) = 0$, $f_N(x_m) = 1$, also $\varrho(a, x_m) \geqq \frac{1}{2^N}$ ist, was für $m \geqq N$ zu einem Widerspruch mit der Definition des Punktes x_m führt.

Unser Hauptsatz ist hiermit bewiesen.

4. Korollar. *Das* II. *Abzählbarkeitsaxiom ist eine notwendige und hinreichende Bedingung, damit ein kompakter topologischer Raum einem metrischen Raume homöomorph sei*[13]).

Die Notwendigkeit der soeben ausgesprochenen Bedingung ist schon bei Hausdorff, S. 274, X und S. 273 enthalten; um einzusehen, daß diese Bedingung auch hinreichend ist, brauchen wir jetzt nur die Normalität der kompakten Räume mit II. Abzählbarkeitsaxiom zu beweisen, was sehr einfach und zwar wie folgt geschieht.

Es seien F_1 und F_2 zwei elementenfremde abgeschlossene Mengen, die

[12]) Die Sphäre $S(a, \varepsilon)$ ist in einem metrischen Raume definitionsgemäß die Menge aller Punkte, deren Entfernung vom „Mittelpunkt" a weniger als ε („der Radius") ist.

[13]) Der erste Beweis dieses Satzes bildet den Inhalt der Arbeit P. Urysohn. Über die Metrisation der kompakten topologischen Räume, Math. Annalen **92**, S. 294 f.

in einem kompakten, dem II. Abzählbarkeitsaxiome genügenden topologischen Raume E liegen. Es sei

$$(4) \qquad \{U_1, U_2, \ldots, U_n, \ldots\}$$

das abzählbare Umgebungssystem des Raumes E. Wir greifen irgendeinen, zuerst fest zu denkenden, Punkt x aus der Menge F_1; nach dem Hausdorffschen Axiome (D) gibt es zu jedem Punktepaare (x, y), wo x der schon ausgewählte Punkt der Menge F_1 und y ein beliebiger Punkt der zweiten Menge F_2 ist, ein Umgebungspaar $U_{j_{(xy)}}$, $U_{k_{(xy)}}$ von der Beschaffenheit $U_{j_{(xy)}} \supset x$, $U_{k_{(xy)}} \supset y$, $U_{j_{(xy)}} \cdot U_{k_{(xy)}} = 0$. Indem y alle Punkte der Menge F_2 durchläuft (während x fest bleibt), erhalten wir eine gewisse Teilmenge der Umgebungsmenge (4), die aus sämtlichen $U_{k_{(xy)}}$ besteht und die Menge F_2 überdeckt; nach dem (wegen der Kompaktheit des Raumes E gültigen) Heine-Borelschen Theorem kann man also aus den $U_{k_{(xy)}}$ eine endliche Menge, es sei $\{U_{k_1}, U_{k_2}, \ldots, U_{k_\lambda}\}$, so auswählen, daß $G_{x_1 F_2} = U_{k_1} + U_{k_2} + \ldots + U_{k_\lambda}$ die Menge F_2 enthält. Die unter den $U_{k_{(xy)}}$ in dieser Weise ausgewählten Umgebungen $U_{k_1}, U_{k_2}, \ldots, U_{k_\lambda}$ entsprechende $U_{j_{(xy)}}$ seien $U_{j_1}, U_{j_2}, \ldots, U_{j_\lambda}$. Die Durchschnittsmenge $U_{j_1} \cdot U_{j_2} \cdots U_{j_\lambda}$ ist ein den Punkt x enthaltendes Gebiet G_x, es existiert also eine bestimmte (z. B. mit möglichst kleinem Index versehene) Umgebung U_{i_x} (aus dem System 4) des Punktes x, die in G_x enthalten ist. Es besteht die Relation

$$\begin{aligned} U_{i_x} \cdot G_{x, F_2} \subset G_x \cdot G_{x, F_2} &= U_{j_1} \cdot U_{j_2} \cdots U_{j_\lambda} \cdot (U_{k_1} + U_{k_2} + \ldots + U_{k_\lambda}) \\ &\subset U_{j_1} \cdot U_{k_1} + U_{j_2} \cdot U_{k_2} + \ldots + U_{j_\lambda} \cdot U_{k_\lambda} = 0. \end{aligned}$$

Wir bestimmen jetzt zu jedem Punkte $x \subset F_1$ die soeben erklärte Umgebung U_{i_x} und das Gebiet G_{x, F_2}. Eine Anwendung des Borelschen Satzes erlaubt wieder endlich viele U_{i_x}, es seien $U_{i_1}, U_{i_2}, \ldots, U_{i_\mu}$, derart auszuwählen, daß $G_1 = U_{i_1} + U_{i_2} + \ldots + U_{i_\mu} \supset F_1$. Die diesen, unter den U_{i_x} ausgewählten Umgebungen $U_{i_1}, U_{i_2}, \ldots, U_{i_\mu}$ entsprechende Gebiete G_{x, F_2} seien $G_{1, F_2}, G_{2, F_2}, \ldots, G_{\mu, F_2}$. Der Durchschnitt $G_{1, F_2} \cdot G_{2, F_2} \cdot \ldots \cdot G_{\mu, F_2}$ ist wieder ein die Menge F_2 enthaltendes Gebiet G_2. Da

$$\begin{aligned} G_1 \cdot G_2 &= (U_{i_1} + U_{i_2} + \ldots + U_{i_\mu}) \cdot G_{1, F_2} \cdot G_{2, F_2} \cdot \ldots \cdot G_{\mu, F_2} \\ &\subset U_{i_1} \cdot G_{1, F_2} + U_{i_2} \cdot G_{2, F} + \ldots + U_{i_\mu} \cdot G_{\mu, F_2} = 0 \end{aligned}$$

ist, so ist die Normalität von E und infolgedessen der Metrisationssatz für kompakte Räume bewiesen[14]).

[14]) Der hier dargestellte Beweis der Normalität der kompakten, dem II. Abzählbarkeitsaxiom genügenden Räume enthält offenbar den Beweis der Normalität einer viel breiteren Klasse von Räumen, nämlich der sogenannten *bikompakten* Räume. Vgl. hierzu P. Alexandroff und P. Urysohn, Zur Theorie der topologischen Räume [5]), wo auch auf andere Arbeiten derselben Verfasser hingewiesen wird.

Es sei am Ende auf folgendes Problem hingewiesen. In der Fußnote [11]) ist bereits bemerkt worden, daß ein jeder normale Raum auch regulär ist. Die Umkehrung der letzten Behauptung ist dagegen falsch. Nun fragt es sich, ob nicht diese Umkehrung doch richtig bleibt, falls man sich nur auf topologische Räume mit dem II. Abzählbarkeitsaxiom beschränkt, d. h. *ob nicht schon die Regularität allein für die Metrisierbarkeit der das* II. *Abzählbarkeitsaxiom erfüllenden Räume genügt.*

(Eingegangen am 28. 9. 1924.)

Über nulldimensionale Punktmengen.

Von

Paul Alexandroff in Moskau und Paul Urysohn † *).

1. Unter einer *nulldimensionalen Punktmenge* verstehen wir im folgenden einen nulldimensionalen [1]) metrisierbaren [2]) separablen [3]) topologischen Raum. Die Benennung „nulldimensionale Punktmenge" scheint für diese Räume besonders angebracht zu sein, weil jeder derartige Raum einer Teilmenge der Menge aller (reellen) Irrationalzahlen homöomorph ist [4]).

Unter einer (*absolut*) *abgeschlossenen* nulldimensionalen Punktmenge verstehen wir einen kompakten nulldimensionalen metrisierbaren Raum.

*) Die Resultate der vorliegenden Arbeit stammen im wesentlichen vom Frühjahr 1924. Der vorliegende Text ist aber erst im April 1926 vom Unterzeichneten endgültig redigiert worden. Paul Alexandroff.

[1]) Vgl. wegen Terminologie folgende Arbeiten von P. Urysohn:

P. Urysohn, Über die Mächtigkeit der zusammenhängenden Mengen (Math. Annalen **94**, S. 262–296, insbesondere § 8 (S. 270).

P. Urysohn, Mémoire sur les multiplicités Cantoriennes, I Partie: „La dimension" (Fund. Math. **7**, S. 30–137, und **8**, S. 225–360, insbesondere **7**, S. 49–79), II Partie: „Les lignes Cantoriennes" (Verhandelingen der Koninklijke Akademie van Wetenschappen te Amsterdam, I Sectie, Deel XIII, 1927).

P. Alexandroff, Darstellung der Grundzüge der Urysohnschen Dimensionstheorie (Math. Annalen **98**).

Im folgenden werden diese Arbeiten bzw. als „Zusammenhängende Mengen", „Mémoire I", „Mémoire II", „Darstellung" zitiert.

[2]) Ein topologischer Raum heißt *metrisierbar*, wenn er einem metrischen Raume homöomorph ist.

[3]) Ein metrisierbarer Raum heißt *separabel*, falls in ihm eine abzählbare Teilmenge dicht ist („dicht" im folgenden stets im Sinne „überall dicht" gemeint).

[4]) In der Tat, erstens ist jeder metrisierbare separable Raum einer Teilmenge eines kompakten metrischen Raumes homöomorph (P. Urysohn, Der Hilbertsche Raum usw., Math. Annalen **92**, S. 302), zweitens ist jede nulldimensionale Teilmenge eines kompakten metrischen Raumes einer Teilmenge der Menge aller Irrationalzahlen homöomorph (Mémoire **1**, ch. I, § 16; Fund. Math. **7**, S. 76, 77); vgl. auch die daselbst zitierte Arbeit von Sierpiński (Fund. Math. **2**, S. 89), in welcher derselbe Satz bis auf unwesentliche Einschränkungen bewiesen ist. Übrigens wird der Leser im folgenden eine kurze Wiedergabe des Beweises des soeben erwähnten Satzes finden.

(Bekanntlich[5]) folgt die Separabilität aus dem gleichzeitigen Bestehen der Kompaktheit und Metrisierbarkeit.) Die (absolut) abgeschlossenen nulldimensionalen Punktmengen sind also mit den beschränkten abgeschlossenen nirgends dichten linearen Mengen topologisch identisch.

Wir führen noch ein für allemal folgende Bezeichnung ein:

Es sei R ein beliebiger topologischer Raum und $\mathfrak{B}$ ein System von in R gelegenen Gebieten (= offenen Mengen). Das Gebietssystem $\mathfrak{B}$ soll eine *Basis* des Raumes R heißen, falls *jedes* in R gelegene Gebiet als Vereinigungsmenge gewisser Gebiete des Systems $\mathfrak{B}$ darstellbar ist[6]).

Letztere Erklärung gestattet folgenden Satz bequem auszusprechen:

I. *Zu jeder unendlichen nulldimensionalen Punktmenge M gibt es* (*wenn man M als topologischen Raum betrachtet*) *eine abzählbare Basis* $\mathfrak{B}$, *die folgenden Bedingungen genügt*:

1. *Von je zweien nicht zueinander fremden, sonst beliebigen Gebieten der Basis $\mathfrak{B}$ ist das eine in dem andern enthalten.*

2. *Es gibt keine unendliche Folge wachsender, zu $\mathfrak{B}$ gehörender Gebiete* (d. h. jede wachsende Folge

$$G_1 \subset G_2 \subset \ldots, \qquad G_n \neq G_{n+1}$$

von Gebieten aus $\mathfrak{B}$ bricht im Endlichen ab).

Beweis. Es sei M ein nulldimensionaler separabler metrischer Raum. Damit ein Gebietsystem $\mathfrak{B}$ eine Basis des metrischen Raumes M sei, ist, wie leicht ersichtlich, notwendig und hinreichend, daß man für jedes positive ε und jeden Punkt x des Raumes M ein, den Punkt x enthaltendes, zu $\mathfrak{B}$ gehörendes Gebiet von einem Durchmesser $< \varepsilon$ finden kann.

Andererseits ist M ein *nulldimensionaler* Raum, woraus leicht folgt[7]), daß man für jedes $\varepsilon > 0$ M in endlich oder abzählbar viele zueinander fremde Gebiete vom Durchmesser $< \varepsilon$ zerlegen kann. Durch sukzessive Anwendung dieser Bemerkung erhält man die Zerlegungen

$$
\begin{array}{lll}
 & M & = \sum\limits_{(i_1)} M_{i_1}, \\
(\text{für jedes } i_1) & M_{i_1} & = \sum\limits_{(i_2)} M_{i_1, i_2}, \\
 & \multicolumn{2}{l}{\ldots\ldots\ldots\ldots\ldots} \\
 & M_{i_1, i_2, \ldots, i_k} & = \sum\limits_{(i_{k+1})} M_{i_1, i_2, \ldots, i_k, i_{k+1}}, \\
 & \multicolumn{2}{l}{\ldots\ldots\ldots\ldots\ldots,}
\end{array}
$$

---

[5]) Hausdorff, Grundzüge der Mengenlehre (erste Auflage), S. 274. Leipzig, Veit, 1914. Im folgenden wird das soeben erwähnte Buch kurz als „Hausdorff" zitiert.

[6]) Falls man jedes Gebiet des Systems $\mathfrak{B}$ als Umgebung jedes seiner Punkte betrachtet, so stimmt der Begriff einer Basis mit demjenigen eines den Raum R definierenden Umgebungssystems überein. Den Beweis der letzteren Behauptung dürfen wir wohl dem Leser überlassen.

[7]) S. z. B. Mémoire **1**, ch. I, § 17 (Fund. Math. **7**, S. 77).

wobei alle Gebiete $M_{i_1, i_2, \ldots i_k}$ den Bedingungen

$$\delta(M_{i_1, i_2, \ldots, i_k}) < \frac{1}{k}, \qquad M_{i_1, i_2, \ldots, i_k, i'_{k+1}} \cdot M_{i_1, i_2, \ldots, i_k, i''_{k+1}} = 0$$

genügen. Da jeder Punkt x des Raumes M in einem $M_{i_1, i_2, \ldots, i_k}$ mit beliebig großem k enthalten ist, so genügt das System $\mathfrak{B}$ aller Gebiete $M_{i_1, i_2, \ldots, i_k}$ dem oben aufgestellten Kriterium einer Basis. Da weiter bei einer wachsenden Folge von Gebieten $M_{i_1, i_2, \ldots, i_k}$ die Indizesanzahl k notwendig abnehmen muß, so bricht jede derartige Folge im Endlichen ab.

Endlich sind zwei Gebiete des Systems $\mathfrak{B}$

$$M_{i_1, i_2, \ldots, i_k} \quad \text{und} \quad M_{j_1, j_2, \ldots, j_h} \qquad (h \leqq k)$$

stets zueinander fremd, es sei denn, daß

$$j_1 = i_1, \quad j_2 = i_2, \quad \ldots, \quad j_h = i_k$$

ist, in welchem Falle $M_{j_1, j_2, \ldots, j_h}$ (das mit $M_{i_1, i_2, \ldots, i_h}$, $h \leqq k$ identisch ist) $M_{i_1, i_2, \ldots, i_k}$ enthält.

Das Gebietsystem $\mathfrak{B}$ ist also eine allen Bedingungen des Satzes I genügende Basis.

2. Aus dem Satze I folgt leicht die bereits am Anfang des § 1 erwähnte Tatsache, nämlich, daß jeder nulldimensionale separable metrisierbare Raum (also, den bekannten Metrisationssätzen gemäß, jeder reguläre, dem II. Abzählbarkeitsaxiom genügende nulldimensionale topologische Raum) sich eineindeutig und beiderseits stetig auf eine Menge von Irrationalzahlen abbilden läßt: jedem Punkt x des Raumes M entspricht, in der Tat, eindeutig eine Folge natürlicher Zahlen

$$i_1, i_2, \ldots, i_k, \ldots$$

(die durch die Bedingung $x \subset M_{i_1, i_2, \ldots, i_k}$ ausgezeichnet wird), also eine bestimmte Irrationalzahl

$$t(x) = \cfrac{1}{i_1 + \cfrac{1}{i_2 + \ddots + \cfrac{1}{i_k + \ddots}}}$$

Die auf diese Weise erhaltenen Irrationalzahlen $t(x)$ bilden, wenn x den ganzen Raum M durchläuft, eine Menge T, von der man leicht beweist, daß sie (mittels der Beziehung $x \sim t(x)$) eineindeutig und beiderseits stetig auf M abgebildet ist[8]).

[8]) Ausführlicher a. a. O. [7]).

3. Aus dem Satze I folgt weiter der Satz

II. *Damit eine nulldimensionale Punktmenge M (absolut) abgeschlossen sei, ist notwendig und hinreichend, daß sie sich unmöglich in unendlich viele zueinander fremde Teilgebiete spalten läßt.*

Die Bedingung ist notwendig. Falls in der Tat

$$M = \sum_{n=1}^{\infty} M_n, \qquad M_i \cdot M_k = 0 \qquad \text{(für } i \neq k\text{)}$$

ist, wobei jedes M_n eine (in M) offene Menge ist und x_n irgendeinen in M_n enthaltenen Punkt bedeutet, so sieht man ohne weiteres ein, daß die unendliche Menge

$$x_1, x_2, \ldots, x_n, \ldots$$

im Raum M keinen Häufungswert besitzt, also M nicht kompakt ist[9]).

Die Bedingung ist hinreichend. Es läßt sich noch mehr behaupten: es gilt in der Tat der Satz

II*. *Es sei ε eine beliebige positive Zahl und M ein nicht kompakter separabler metrischer Raum. Dann läßt sich M in abzählbar viele zueinander fremde Teilgebiete*

$$M_1, M_2, \ldots, M_n, \ldots$$

spalten, von denen jedes einen Durchmesser $< \varepsilon$ hat.

Es sei D eine divergente (= keinen Häufungspunkt besitzende) Teilmenge von M und $\mathfrak{B}$ eine den Bedingungen des Satzes I genügende Basis des Raumes M. Für jeden Punkt $x \subset M$ läßt sich alsdann eine, höchstens einen Punkt der Menge D enthaltende, „Umgebung“ $U(x)$ (in unserem Falle ein Gebiet des Systems $\mathfrak{B}$) wählen, von der außerdem vorausgesetzt werden darf, daß

$$\delta(U(x)) < \varepsilon$$

ist. Es sei $\mathfrak{B}^*$ das auf diese Weise gewonnene Teilsystem des Gebietsystems $\mathfrak{B}$. Zufolge der Bedingung 2. des Satzes I, der das System $\mathfrak{B}$ (also a fortiori $\mathfrak{B}^*$) genügt, läßt sich für jedes in $\mathfrak{B}^*$ vorkommende Gebiet ein größtes (d. h in keinem weiteren Gebiet des Systems $\mathfrak{B}^*$ enthaltenes) Gebiet desselben Systems angeben. Es seien nun

$$M_1, M_2, \ldots, M_n, \ldots \tag{1}$$

diese „größten“ Gebiete (von denen keine zwei identische in (1) aufgenommen werden sollen).

Offenbar füllt die Vereinigungsmenge aller Gebiete (1) den ganzen Raum M aus; die Gebiete (1) sind außerdem paarweise zueinander fremd

[9]) Die Notwendigkeit unserer Bedingnng gilt (wie aus dem obigen Beweis ersichtlich) auch für Räume positiver Dimension.

und sämtlich vom Durchmesser $< \varepsilon$ (da sie ja alle zu $\mathfrak{B}^*$ gehören); endlich ist die Folge (1) sicher eine unendliche, weil jedes M_n höchstens einen Punkt der unendlichen Menge D enthält (die, als Teilmenge von M, in der Vereinigungsmenge aller M_n enthalten ist).

Der Satz II*, also auch der Satz II, ist hiermit bewiesen.

4. Wir nennen eine Punktmenge *nirgends abgeschlossen*, falls sie, als topologischer Raum betrachtet, in keinem ihrer Punkte kompakt ist[10]).

Über solche Punktmengen läßt sich folgender Satz aussprechen:

III. *Damit eine nulldimensionale Punktmenge nirgends abgeschlossen sei, ist notwendig und hinreichend, daß sie sich auf eine überall dichte Teilmenge des Linearkontinuums eineindeutig und stetig abbilden läßt.*

Das Hinreichen der Bedingung bedarf keines Beweises (eine nulldimensionale überall dichte Teilmenge des Linearkontinuums kann ja, als topologischer Raum betrachtet, in keinem Punkte kompakt sein).

Die Notwendigkeit läßt sich wie folgt zeigen.

Es sei M eine nirgends abgeschlossene Punktmenge. Zufolge dem Satze II* gibt es eine Zerlegung

$$M = \sum_{i_1=1}^{\infty} M_{i_1} \tag{2}$$

des Raumes M in unendlich viele zueinander fremde Gebiete $M_1, M_2, \ldots, M_{i_1}, \ldots$, von denen jedes einen Durchmesser < 1 hat.

Wir setzen jetzt voraus, die Gebiete $M_{i_1, i_2, \ldots, i_k}$ wären bereits konstruiert. Da M nirgends abgeschlossen ist, können wir den Satz II* auf jede Punktmenge $M_{i_1, i_2, \ldots, i_k}$ anwenden und sie in unendlich viele zueinander fremde Gebiete

$$M_{i_1, i_2, \ldots, i_k, 1}, \quad M_{i_1, i_2, \ldots, i_k, 2, \ldots}, \quad M_{i_1, i_2, \ldots, i_k, i_{k+1}, \ldots}$$

unter der Geltung der Bedingung

$$\delta(M_{i_1, i_2, \ldots, i_k, i_{k+1}}) < \frac{1}{k+1} \tag{3}$$

zerlegen.

Auf diese Weise werden für alle Werte der natürlichen Zahlen $k, i_1, i_2, \ldots, i_k$ die Gebiete $M_{i_1, i_2, \ldots, i_k}$ in vollständiger Induktion definiert.

Jedem Punkt x wird mithin eindeutig eine Folge natürlicher Zahlen

$$i_1, i_2, \ldots, i_k, \ldots, \tag{4}$$

---

[10]) Dabei heißt ein topologischer Raum R *im Punkte* ξ kompakt (P. Alexandroff, Über die Metrisation der im Kleinen kompakten topologischen Räume, Math. Annalen 92, S. 294), falls es eine Umgebung des Punktes ξ gibt, deren abgeschlossene Hülle, als Relativraum betrachtet, kompakt ist.

also eine Irrationalzahl

(5) $$t(x) = i_1 + \cfrac{1}{i_2 + \cdots + \cfrac{1}{i_k + \cdots}}$$

mittels der Gleichung

$$x = \prod_{k=1}^{\infty} M_{i_1, i_2, \ldots, i_k}$$

(in der x als gegeben, $i_1, i_2, \ldots, i_k, \ldots$ als unbekannt zu denken sind) zugeordnet. Die Menge der auf diese Weise definierten Irrationalzahlen $t(x)$ (wobei x alle Punkte von M durchläuft) soll T heißen. Man beweist, wie üblich in solchen Fällen, daß M und T nicht nur eineindeutig, sondern auch beiderseits stetig einander entsprechen[11]).

Es bleibt nur noch übrig zu zeigen, daß T auf der positiven Zahlengerade überall dicht liegt.

Es sei $\varepsilon > 0$ und irgendeine positive Irrationalzahl

$$t_0 = i_1^0 + \cfrac{1}{i_2^0 + \cdots + \cfrac{1}{i_k^0 + \cdots}}$$

gegeben. Wir wollen zeigen, daß es Punkte x des Raumes M gibt, für die

(5) $$|t_0 - t(x)| < \varepsilon$$

ist. Wir wählen dazu die natürliche Zahl k so groß, daß für jedes

$$t = i_1 + \cfrac{1}{i_2 + \cdots + \cfrac{1}{i_k + \cdots}}$$

die Ungleichung

$$|t_0 - t| < \varepsilon$$

aus den Identitäten

$$i_1 = i_1^0, \; i_2 = i_2^0, \; \ldots, \; i_k = i_k^0$$

folgt. Dann ist für jeden Punkt x der Menge $M_{i_1^0, i_2^0, \ldots, i_k^0}$ die Ungleichung (5) erfüllt, w. z. b. w.

5. Wir wenden uns jetzt zu einem wichtigen Spezialfall des Satzes III und setzen voraus, daß die Punktmenge M, die als nulldimensionaler metrisierbarer separabler Raum erklärt war, außerdem die Eigenschaft der *Vollständigkeit* besitzt. Letztere Eigenschaft (deren Begriff von Fréchet herrührt) lautet dabei wie folgt:

[11]) Siehe a. a. O. [7]).

Ein *metrisierbarer* topologischer Raum heißt *vollständig*, falls er sich durch eine passend gewählte Metrik in einen, *dem Cauchyschen Konvergenzkriterium genügenden* metrischen Raum verwandeln läßt.

Der auf diese Weise metrisierte Raum besitzt dann (als *metrischer* Raum betrachtet) die wesentliche Eigenschaft der *metrischen Unerweiterbarkeit* (Hausdorff[12])), d. h. des Abgeschlossenseins in jedem umfassenderen metrischen Raume.

Es sei nun M ein *vollständiger* metrisierbarer separabler nulldimensionaler Raum, der in keinem seiner Punkte kompakt ist, wir sagen kürzer: eine *vollständige, nulldimensionale, nirgends abgeschlossene Punktmenge*, die wir uns mit einer, die Vollständigkeit zutage bringenden Metrik versehen denken werden.

Da jede Menge $M_{i_1, i_2, \ldots, i_k}$ (wir denken uns das Verfahren des vorigen Paragraphen auf M angewandt) eine abgeschlossene[13]) und $\delta(M_{i_1, i_2, \ldots, i_k}) < \frac{1}{k}$ ist, so bestimmt (nach einem bekannten Hausdorffschen Satze[14])) jede Folge natürlicher Zahlen

$$i_1, i_2, \ldots, i_k, \ldots$$

einen und nur einen Punkt

$$x_{i_1, i_2, \ldots, i_k, \ldots} = \prod_{k=1}^{\infty} M_{i_1, i_2, \ldots, i_k}$$

des Raumes M, und zwar ist dabei

$$t(x_{i_1, i_2, \ldots, i_k, \ldots}) = i_1 + \cfrac{1}{i_2 + \cdot_{\cdot_{\cdot + \cfrac{1}{i_k + \cdot_{\cdot_{\cdot}}}}}}$$

Das heißt aber nichts anderes, als daß *jede* positive Irrationalzahl t in T enthalten ist, mit anderen Worten, daß in unserem Falle M mit der Menge aller positiven (also auch mit der Menge aller reellen) Irrationalzahlen homöomorph ist.

Wir gelangen mithin zum folgenden Satze:

IV. *Jede nulldimensionale vollständige nirgends abgeschlossene Menge ist mit der Menge aller Irrationalzahlen des Linearkontinuums homöomorph.*

---

[12]) A. a. O. S. 315f.

[13]) Die Komplementärmenge $M - M_{i_1, i_2, \ldots, i_k}$ ist ja (als Vereinigungsmenge aller von $M_{i_1, i_2, \ldots, i_k}$ verschiedener *Gebiete* $M_{j_1, j_2, \ldots, j_k}$) ein Gebiet, folglich ist $M_{i_1, i_2, \ldots, i_k} = M - (M - M_{i_1, i_2, \ldots, i_k})$ abgeschlossen.

[14]) Hausdorff, S. 318 („Jede absteigende Folge nicht leerer abgeschlossener Mengen, deren Durchmesser nach Null konvergiert, hat einen und nur einen gemeinsamen Punkt“).

6. Letzteres Ergebnis läßt sich auch anders aussprechen. Dazu brauchen wir einige Vorbemerkungen.

Wir nennen einen topologischen Raum eine *absolute G_δ-Menge*, wenn er einer in einem kompakten metrischen Raume gelegenen G_δ-Menge homöomorph ist[15]).

Nun ist jede absolute G_δ-Menge (und nur eine solche Menge) einem vollständigen metrischen Raume homöomorph[16]), was denselben Satz IV wie folgt ausdrücken läßt:

IVbis. *Jede nirgends abgeschlossene nulldimensionale absolute G_δ-Menge ist mit der Menge aller Irrationalzahlen homöomorph*[16a]).

7. Wir wollen jetzt einen einfachen direkten Beweis des Satzes IVbis geben.

Es sei M eine nulldimensionale nirgends abgeschlossene absolute G_δ-Menge, wir sagen kürzer: eine nulldimensionale *irreduzible* G_δ-Menge. Auf Grund des Satzes III und der topologischen Invarianz der G_δ-Mengen können wir uns M als eine auf der Einheitsstrecke liegende und daselbst überall dichte, aus Irrationalzahlen bestehende G_δ-Menge denken. Es sei dementsprechend

$$M = \prod_{n=1}^{\infty} G_n, \tag{6}$$

---

[15]) Zufolge dem unter [4]) in erster Linie zitierten Urysohnschen Satze und der topologischen Invarianz der in vollständigen Räumen gelegenen G_δ-Mengen ist die Klasse der absoluten G_δ-Mengen mit derjenigen der *in vollständigen separablen* Räumen gelegenen G_δ-Mengen topologisch identisch.

[16]) P. Alexandroff, Les ensembles de la 1re classe et les espaces abstraits, Comptes Rendus **178** (1924), S. 185. Die soeben ausgesprochene Behauptung ist in dieser Arbeit (zum erstenmal) auf dem Wege bewiesen, daß ein *topologisches* Kriterium für vollsändige separable Räume einerseits und für in solchen Räumen gelegene G_δ-Mengen anderseits gegeben wird, wobei es sich zeigt, daß das betreffende Kriterium in beiden Fällen *dasselbe* ist. Hausdorff hat danach (Fund. Math. **6**) direkt bewiesen, daß die in vollständigen (*sogar nicht separablen*) metrischen Räumen gelegenen G_δ-Mengen mit vollständigen Räumen topologisch identisch sind. (Bei diesem Beweise wird aber keine topologische Charakterisierung der betreffenden Räume gegeben, sondern direkt gezeigt, daß man jede in Betracht kommende Menge so „ummetrisieren" kann, daß sie sich in einen vollständigen metrischen Raum verwandelt.) — Anmerkung bei der Korrektur (5. II. 1927). In der letzten Zeit ist es Herrn Wedenissoff gelungen, die in meiner soeben zitierten Comptes Rendus-Note gegebene Charakterisierung mit Hilfe des Satzes und eines Satzes von Sierpiński (Fund. Math. 6, S. 106) auf beliebige (nicht separable) vollständige Räume zu erweitern. Wie mir Herr Hausdorff mitgeteilt hat, hat auch er das gleiche Resultat bekommen.

[16a]) Im wesentlichen ist dieser Satz (für a priori als linear vorausgesetzte Mengen) in einem Theorem von Brouwer erhalten (Proceed. Kon. Ak. Amsterdam **20** (April 1917), S. 1193).

wobei $G_n \supset G_{n+1}$ und jedes G_n ein lineares Gebiet (also Vereinigungsmenge von endlich oder abzählbar vielen zueinander fremden offenen Intervallen) ist.

Wir betrachten für einen Augenblick irgendein lineares Gebiet G (das ganz unabhängig von den Gebieten G_n zu denken ist). Es seien

$$(7) \qquad (a_1, b_1), (a_2, b_2), \ldots, (a_k, b_k), \ldots,$$

die (endlich- oder abzählbar-viele) Intervalle, in die G zerfällt. Wir wollen eine allgemeine Operation $\varphi(G)$ definieren, die dem Gebiete G ein aus *unendlich* vielen zueinander fremden Intervallen gebildetes Teilgebiet, und zwar auf folgende Weise, entsprechen läßt. Wir wählen innerhalb eines jeden Intervalles (a_k, b_k) eine *feste* abzählbare Menge *rationaler* Punkte

$$\ldots, c_{k,-i}, c_{k,-(i-1)}, \ldots, c_{k,-1}, c_{k,0}, c_{k,1}, \ldots, c_{k,i-1}, c_{k,i}, \ldots,$$

die den Bedingungen

1. $c_{k,i} < c_{k,j}$ (für $i < j$, i, j beliebige *ganze* Zahlen),
2. $\lim\limits_{i \to -\infty} c_{k,i} = a_k$, $\lim\limits_{i \to \infty} c_{k,i} = b_k$,
3. $|c_{k,i} - c_{k,i+1}| < \frac{b_k - a_k}{2}$

genügen, sonst aber beliebig sind.

Das Gebiet $\Gamma = \varphi(G)$ ist dann definitionsgemäß die Vereinigungsmenge aller offener Intervalle $(c_{k,i}, c_{k,i+1})$ (für alle in Betracht kommenden k und *alle* ganzzahligen i $(-\infty < i < \infty)$).

8. Wir setzen jetzt der Reihe nach:

$$\begin{cases} G_1^* = G_1; & \Gamma_1 = \varphi(G_1^*), \\ \ldots\ldots\ldots & \ldots\ldots\ldots \\ G_n^* = \Gamma_n \cdot G_{n+1}; & \Gamma_{n+1} = \varphi(G_n^*). \end{cases}$$

Wir bezeichnen weiter durch Δ_{i_1} $(i_1 = 1, 2, \ldots$ in inf.) alle (abzählbar vielen) Komponenten von Γ_1, und allgemein durch

$$\Delta_{i_1, i_2, \ldots, i_k, 1}, \Delta_{i_1, i_2, \ldots, i_k, 2}, \ldots, \Delta_{i_1, i_2, \ldots, i_{k+1}}, \ldots$$

alle (ebenfalls *abzählbar viele*) *in* $\Delta_{i_1, i_2, \ldots, i_k}$ *enthaltenen Komponenten* von Γ_{k+1}.

Da die Länge jedes Intervalles $\Delta_{i_1, i_2, \ldots, i_k}$ kleiner als $\frac{1}{2^k}$ ist und jedes $\Delta_{i_1, i_2, \ldots, i_k, i_{k+1}}$ samt seiner beiden Endpunkte in $\Delta_{i_1, i_2, \ldots, i_k}$ enthalten ist, so entspricht jeder positiven Irrationalzahl

$$t = i_1 + \cfrac{1}{i_2 + \cdots + \cfrac{1}{i_k + \cdots}}$$

ein einziger Punkt

$$x_{i_1, i_2, \ldots, i_k, \ldots} = \prod_{k=1}^{\infty} \Delta_{i_1, i_2, \ldots, i_k},$$

der offenbar zu

$$\prod_{k=1}^{\infty} \Gamma_k = M \tag{7}$$

gehört[17]).

Auf diese Weise wird wieder eine eineindeutige Beziehung zwischen M und der Menge aller positiven Irrationalzahlen hergestellt, von der man leicht beweist, daß sie auch eine beiderseits stetige ist.

8. Neben den G_δ-Mengen bilden bekanntlich die F_σ-Mengen die wichtigste Mengenklasse. Unserem bisherigen Sprachgebrauch gemäß wollen wir als *absolute F_σ-Menge* jeden topologischen Raum, der einer F_σ-Menge eines kompakten metrischen Raumes homöomorph ist, bezeichnen.

Wir beschränken uns im folgenden auf nulldimensionale F_σ-Mengen. Es sei dabei folgender grundlegender Satz erwähnt:

Damit eine absolute F_σ-Menge nulldimensional sei, ist notwendig und hinreichend, daß sie kein Teilkontinuum, oder (was in unserem Falle auf dasselbe auskommt) *keine, mehr als einen Punkt enthaltende zusammenhängende Teilmenge enthält*[18]).

Unter den nulldimensionalen absoluten F_σ-Mengen zeichnen sich diejenigen besonders aus, die nicht nur nirgends abgeschlossen, sondern gleichzeitig auch *nirgends abzählbar* sind. Solche F_σ-Mengen wollen wir *irreduzibel* nennen. *Eine absolute nirgends abgeschlossene F_σ-Menge M* soll also irreduzibel heißen, falls jedes Relativgebiet von M unabzählbar ist.

Bemerkung. Für den Fall der G_δ-Mengen haben wir die letztere Bedingung nicht besonders erwähnt, weil jede nirgends abgeschlossene absolute G_δ-Menge auch nirgends abzählbar ist.

9. Es gilt nun folgender Satz:

V. *Alle nulldimensionalen irreduziblen F_σ-Mengen sind untereinander homöomorph.*

---

[17]) Zum Beweise der Identität (7) braucht nur bemerkt zu werden, daß einerseits jedes Γ_k in G_k, also $\prod_{k=1}^{\infty} \Gamma_k$ in $\prod_{k=1}^{\infty} G_k = M$ enthalten ist, anderseits aber (da M aus Irrationalzahlen besteht) zufolge der Definition der Operation $\varphi(G)$, M in jedem Γ_k, also in $\prod_{k=1}^{\infty} G_k$, enthalten ist.

[18]) Siehe wegen des Beweises Mémoire **1**, ch. VI, § 8 (Fund. Math. **8**, S. 341). Bei Menger (Monatshefte f. Math. u. Phys. **33**, S. 147) wird der Satz für kompakte Räume bewiesen, woraus er für F_σ unmittelbar folgt mittels des Satzes, daß die Summe abzählbar vieler in sich kompakter nulldimensionaler Mengen wieder nulldimensional ist.

Zum Beweise des Satzes V genügt es zu zeigen, daß jede irreduzible nulldimensionale F_σ-Menge M einer und derselben Menge Ψ, die wir die *Cantorsche F_σ-Menge* nennen wollen, homöomorph ist.

Die Menge Ψ wird folgendermaßen konstruiert.

Es sei U irgendeine geradlinige Strecke. Wir bezeichnen durch $P(U)$ die durch Anwendung des bekannten Cantorschen Dreiteilungsverfahrens entstehende perfekte Teilmenge von U.

Es sei weiter Π eine beiderseits (= in positiver und negativer Richtung) sich ins Unendliche erstreckende perfekte nirgends dichte Menge reeller Zahlen. Wir bezeichnen durch

$$U_1, U_2, \ldots, U_n, \ldots$$

die zu Π komplementären Intervalle, und setzen definitionsgemäß

$$\Psi(\Pi) = \Pi + \sum_{n=1}^{\infty} P(U_n).$$

Indem wir *ein für allemal* für jedes ganzzahlige k $(-\infty < k < \infty)$ durch Δ_k die Strecke $k \leqq x \leqq k+1$ bezeichnen, setzen wir weiter

$$P_0 = \sum_{-\infty}^{+\infty} P(\Delta_k),$$

dann allgemein:

$$P_{n+1} = \Psi(P_n) \qquad (\text{für jedes } n = 0, 1, 2, \ldots \text{ in inf.})$$

und schließlich

$$\Psi = \sum_{n=1}^{\infty} P_n. \tag{8}$$

10. Beim Beweise des Satzes, daß jede nulldimensionale irreduzible F_σ-Menge der soeben konstruierten Menge Ψ homöomorph ist, dürfen wir uns auf Grund des Satzes II auf den Fall beschränken, wo M eine auf der Zahlengerade überall dicht liegende Menge ist. Der wesentliche Punkt des Beweises ist der folgende

Hilfssatz. Es sei M eine auf der ganzen Zahlengerade überall dichte irreduzible F_σ-Menge. M läßt sich dann als Vereinigungsmenge lauter *beiderseits unbegrenzter* perfekter Mengen Π_n

$$M = \sum_{n=1}^{\infty} \Pi_n$$

darstellen, und zwar unter Geltung folgender Nebenbedingungen:

1. Für jedes n ist Π_n in Π_{n+1} enthalten.

2. Es seien

$$U_1^n = (a_1^n, b_1^n), \quad U_2^n = (a_2^n, b_2^n), \quad \ldots, \quad U_k^n = (a_k^n, b_k^n), \quad \ldots$$

die zu Π_n komplementären Intervalle. Falls wir

$$(9) \qquad V_k^n = \Pi^{n+1} \cdot U_k^n$$

setzen, so ist a_k^n din untere, b_k^n die obere Grenze von V_k^n.

Um nun diesen Hilfssatz zu beweisen, bemerken wir zuerst, daß M a priori als Vereinigungsmenge von abgeschlossenen Mengen gegeben ist:

$$(10) \qquad M = \sum_{n=1}^{\infty} F_n;$$

man sieht leicht ein, daß man stets voraussetzen darf, jede der Mengen F_n wäre beiderseits unbegrenzt.

Um jetzt zu erreichen, daß jede F_n perfekt sei, umgeben wir jeden isolierten Punkt ξ_m^n von F_n mit einem „isolierenden" Intervalle (α_m^n, β_m^n), und sorgen dabei dafür, daß zwei verschiedenen isolierten Punkten $\xi_{m'}^n$ und $\xi_{m''}^n$ entsprechende Intervalle $(\alpha_{m'}^n, \beta_{m'}^n)$ und $(\alpha_{m''}^n, \beta_{m''}^n)$ zueinander fremd bleiben. Indem wir weiter für jedes m und k eine zwischen $\xi_m^n - \frac{\xi_m^n - \alpha_m^n}{2^k}$ und $\xi_m^n - \frac{\xi_m^n - \alpha_m^n}{2^{k+1}}$ gelegene perfekte Teilmenge $\Xi_{m,k}^n$ der Menge M auszeichnen und

$$\Xi_m^n = \xi_m^n + \sum_{k=1}^{\infty} \Xi_{m,k}^n$$

setzen, erhalten wir eine den Punkt ξ_m^n enthaltende, in (α_m^n, β_m^n) gelegene perfekte Teilmenge von M. Wenn wir nun F_n durch $F_n + \sum_{(m)} \Xi_m^n$ ersetzen, verwandelt sich F_n in eine perfekte Menge.

11. Wir sind also so weit, um voraussetzen zu dürfen, daß jede in (10) vorkommende Menge F_n eine *perfekte, beiderseits unbegrenzte Menge ist.* An dieser Voraussetzung halten wir jetzt fest.

Wir definieren alsdann

$$\Pi_1 = F_1.$$

Es sei Π_n bereits konstruiert. Wir bezeichnen durch

$$U_1^n = (a_1^n, b_1^n), \quad U_2^n = (a_2^n, b_2^n), \quad \ldots, \quad U_k^n = (a_k^n, b_k^n), \quad \ldots$$

die zu Π_n komplementären Intervalle und betrachten die Menge $W_k^n = U_k^n \cdot F_{n+1}$ (die ja mit $U_k^n \cdot (F_{n+1} + \Pi_n)$ identisch ist).

Falls die untere bzw. obere Grenze von W_k^n mit a_k^n bzw. b_k^n zusammenfällt, setzen wir $T_k^n = 0$; falls dies aber nicht zutrifft, bezeichnen wir durch T_k^n eine (mittels der Methoden des § 10 leicht zu konstruierende) auf der Strecke $[a_k^n, b_k^n]$ gelegene und die Punkte a_k^n, b_k^n enthaltende perfekte Teilmenge von M.

Wir setzen nun endlich

$$\Pi_{n+1} = \Pi_n + F_{n+1} + \sum_{k=1}^{\infty} T_k^n.$$

Auf diese Weise werden die perfekten beiderseits unbegrenzten Mengen Π_n für jedes n definiert; offenbar ist

$$\sum_{n=1}^{\infty} \Pi_n = M, \tag{11}$$

und man sieht leicht ein, daß alle Bedingungen des Hilfssatzes erfüllt sind.

12. Jetzt sind wir aber auch mit dem Beweise des Satzes V beinahe fertig: wir brauchen uns nämlich nur an die Definitionsgleichung (8) der Menge Ψ zu erinnern. Die Anfangsmengen P_1 und Π_1 sind als beiderseits unbeschränkte, perfekte, nirgendsdichte lineare Mengen einander ähnlich (im Sinne der Theorie der geordneten Mengen). Diese Ähnlichkeitsbeziehung läßt sich aber unserer Konstruktion gemäß unmittelbar von P_n bzw. Π_n auf P_{n+1} bzw. Π_{n+1} erweitern, was schließlich die Ähnlichkeit und also um so mehr die Homöomorphie der Vereinigungsmengen Ψ und M ergibt, w. z. b. w.

Bemerkung. Betrachten wir für einen Augenblick die *beschränkte* irreduzible F_σ-Menge Ψ_0, die sich von Ψ nur dadurch unterscheidet, daß man die Operation $\Psi(\Pi)$ für eine beliebige *beschränkte* perfekte Menge Π unter alleiniger Berücksichtigung ihrer, im Endlichen gelegenen, komplementären Intervalle definiert und in der Konstruktion des § 9 die Cantorsche perfekte Menge $P(\Delta_0)$ zum Ausgangspunkt wählt, so sind Ψ_0 und Ψ nach dem Satze V homöomorph (was ja a priori kaum als selbstverständlich betrachtet werden kann [19]), so daß man auch Ψ_0 als den topologischen Typus aller irreduziblen F_σ-Mengen betrachten kann.

13. Die irreduziblen G_δ- bzw. F_σ-Mengen bilden nicht nur je einen einzigen topologischen Typus, und zwar denjenigen der Menge J aller Irrationalzahlen, bzw. den der Menge Ψ, sie besitzen außerdem eine gewisse *Homogenitätseigenschaft*, die, genau formuliert, darin besteht, daß man jede der uns hier interessierenden Mengen eineindeutig und stetig auf sich selbst derart abbilden kann, daß dabei von zwei beliebigen Punkten der eine in den anderen übergeht [20]): in der Tat trifft diese Eigenschaft sowohl für die Menge J als für Ψ zu (im Falle dieser Mengen läßt sich ja die betreffende Abbildung sogar ohne Verletzung der Ordnungsrelationen, die in unseren Mengen, als in linearen Mengen, herrschen, herstellen), woraus

[19]) Man bemerke nur, daß Ψ unbeschränkt nach beiden Richtungen ist, Ψ_0 dagegen sowohl ihre untere, als obere Grenze *enthält*.

[20]) Auch die Homogenitätseigenschaft der linearen G_δ-Mengen war zuerst von Brouwer (a. a. O. [16a])) bewiesen.

(da es sich ja um eine topologisch invariante Eigenschaft handelt) ihre allgemeine Gültigkeit ohne weiteres folgt. Dieses Ergebnis gestattet folgende Verallgemeinerung der beiden Sätze IVbis bzw. V.

VI. *Falls A und B zwei nulldimensionale irreduzible F_σ- oder G_δ-Mengen sind, weiter a ein beliebiger Punkt von A und b ein beliebiger Punkt von B ist, so gibt es eine eineindeutige und beiderseits stetige Beziehung zwischen A und B, die die beiden Punkte a und b einander entsprechen läßt.*

14. Wir bemerken noch (und werden bald diese Bemerkung auch benutzen), daß der *Satz VI auch für beschränkte perfekte nirgendsdichte lineare Mengen* („irreduzible abgeschlossene Mengen") gilt[21]). Um letztere Behauptung zu beweisen, genügt es eine bestimmte perfekte Menge, z. B. die Cantorsche Menge Π als Typus zu wählen, einen bestimmten Punkt ξ_0 der Menge Π, z. B. den Punkt

$$\xi_0 = 0{,}000\ldots \quad \text{(im triadischen Zahlensystem)}$$

auszuzeichnen, und dann zeigen, daß, wie auch die perfekte Menge P und einer ihrer Punkte x_0 gegeben sein mag, eine eineindeutige und stetige Beziehung zwischen Π und P vorhanden ist, die die beiden Punkte ξ_0 und x einander entsprechen läßt.

Die Menge Π ist bekanntlich die Menge aller derjenigen Punkte der abgeschlossenen Einheitsstrecke, deren triadische Entwickelung ohne Benutzung der Ziffer 1 sich darstellen läßt; jeder Punkt ξ von Π hat hiermit eine nur aus 0 und 2 bestehende triadische Entwickelung

$$\xi = 0, i_1, i_2 \ldots i_k \ldots \qquad (i_k = 0, 2)$$

und umgekehrt gehört jeder eine derartige Entwickelung besitzende Punkt zu Π.

15. Mit der Menge P kann man nun eine Reihe von Zerlegungen in $2, 2^2, \ldots, 2^k, \ldots$ zueinander fremde abgeschlossene Teilmengen $P_{i_1}, P_{i_1 i_2}, \ldots, P_{i_1 i_2 \ldots i_k}, \ldots$ vornehmen, wobei jedesmal die Menge $P_{i_1 i_2 \ldots i_h}$ in zwei Mengen $P_{i_1 i_2 \ldots i_h i_{h+1}}$ zerlegt wird, und die Durchmesser der Mengen $P_{i_1 i_2 \ldots i_k}$ mit $\frac{1}{k}$ nach Null konvergieren. Der Symmetrie halber bezeichnen wir die beiden Werte von jedem i_k auch durch 0 und 2, und *sorgen dabei nur dafür, daß jedesmal diejenige der Mengen $P_{i_1}, P_{i_1 i_2}, \ldots, P_{i_1 i_2 \ldots i_k}, \ldots$, die den a priori gegebenen Punkt x_0 enthält, alle ihre Indizes gleich 0 hat.*

---

[21]) Wenn diese an sich ganz elementare Eigenschaft der perfekten Mengen nicht ohne weiteres ins Auge springt, so ist das ausschließlich darauf zurückzuführen, daß im Falle der perfekten Mengen die betreffende stetige Beziehung eine *die Ordnungsrelationen störende* ist.

Jedem Punkte x von P entspricht auf diese Weise eine unendliche Folge

$$i_1, i_2, \ldots, i_k, \ldots \qquad (i_k = 0{,}2),$$

also ein bestimmter Punkt ξ der Menge Π. Auf diese Weise entsteht eine eineindeutige Beziehung zwischen den beiden Mengen Π und P, die die Punkte ξ_0 und x_0 einander zuordnet, und die, wie leicht ersichtlich, eine stetige ist.

Aus dem soeben Bewiesenen folgt insbesondere die topologische Homogenität der perfekten nirgends dichten Mengen:

Man kann mittels einer eineindeutigen stetigen Transformation einer perfekten nirgendsdichten Menge in sich selbst jeden ihrer Punkte in jeden anderen überführen.

16. Aus den Resultaten des vorigen Paragraphen folgt weiter unmittelbar, daß *alle Mengen, die aus einer beschränkten perfekten nirgendsdichten linearen Menge durch Weglassung eines einzigen Punktes entstehen, untereinander homöomorph sind.*

Insbesondere sind alle diese Mengen einer (beliebigen) *unbeschränkten* perfekten Menge homöomorph, da ja letztere mittels Hinzufügung des unendlich fernen Punktes (topologisch) in eine beschränkte perfekte Menge übergeht. Nun sind aber diese Mengen topologisch nichts anderes, als die nulldimensionalen im Kleinen kompakten separablen metrisierbaren Räume ohne isolierte Punkte: in der Tat läßt sich jeder derartige Raum mittels Hinzufügung eines einzigen Punktes zu einem kompakten metrisierbaren nulldimensionalen Raume, der ebenfalls keine isolierten Punkte besitzt, also zu einer „beschränkten perfekten nirgendsdichten linearen Menge" topologisch ergänzen [22]).

Wenn man anderseits einen beliebigen nulldimensionalen, im Kleinen kompakten separablen metrisierbaren Raum direkt auf das Linearkontinuum abbildet (was ja zufolge seiner Nulldimensionalität stets möglich ist), so wird er offenbar zu einer linearen F_ϱ-Menge [23]). Eine solche F_ϱ-Menge wollen wir *irreduzibel* nennen, wenn sie nirgends abzählbar ist (also einfach keine isolierte Punkte enthält). Man könnte natürlich auch wieder über *absolute* F_ϱ-Mengen sprechen, da wir aber im nulldimensionalen Falle wieder nur dieselben Mengen erhalten, gehen wir darauf nicht weiter ein.

Die nulldimensionalen irreduziblen F_ϱ-Mengen sind also mit den nulldimensionalen im Kleinen kompakten Räumen topologisch identisch. Man kann nun leicht direkt zeigen, daß jede nulldimensionale irreduzible F_ϱ-

---

[22]) Siehe die unter [10]) zitierte Arbeit.

[23]) Eine in irgendeinem Raume gelegene Menge heißt nach Sierpiński eine F_ϱ-Menge, falls sie sich als Differenz zweier abgeschlossener Mengen darstellen kann.

Menge sich durch Hinzufügung eines Punktes zu einer perfekten Menge ergänzen läßt. Es sei in der Tat

$$M = P - F$$

eine irreduzible F_ϱ-Menge. Dann kann P als eine beschränkte perfekte, F als eine abgeschlossene Menge vorausgesetzt werden, und es ist

$$M = \sum_{(n)} P \cdot U_n,$$

wo $U_n = (a_n, b_n)$ die komplementären Intervalle zu F sind.

Wenn man nun alle (a_n, b_n) kongruent (in der Ebene) überträgt, so daß alle a_n in einen und denselben Punkt ξ zusammenfallen, und darauf auch die zweiten Endpunkte b_n sämtlich mit ξ identifiziert (was alles in der Ebene geschehen kann), so bleibt es nur noch übrig, der Menge M (die während dieser ganzen Operation nur eineindeutige und stetige Transformationen erlitten hat) den Punkt ξ hinzuzufügen, um eine perfekte diskontinuierliche (also nulldimensionale) ebene Menge $M + \xi$ zu erhalten. $M + \xi$ ist alsdann einer nirgendsdichten linearen Menge homöomorph.

Wir können mithin zu den früher bewiesenen Sätzen noch folgende hinzufügen:

VII. *Alle nulldimensionalen irreduziblen F_ϱ-Mengen sind untereinander homöomorph; insbesondere sind diese Mengen sämtlich einer beliebigen unbeschränkten perfekten nirgendsdichten linearen Menge homöomorph.*

17. Nach der Baire-Lebesgueschen Klassifikation der Borelschen Mengen gehören die abgeschlossenen Mengen zur Klasse 0, die F_σ- und G_δ-Mengen dagegen (soweit sie nicht abgeschlossen sind) zur Klasse 1 (insbesondere rechnen wir auch die F_ϱ-Mengen zur Klasse 1). Wenn wir den früher eingeführten Sinn der Bezeichnung irreduzibel für jede dieser Mengenklasse beibehalten (insbesondere soll eine abgeschlossene Menge irreduzibel heißen, falls sie „nirgends endlich", d. h. *perfekt* ist), können wir folgendermaßen die Resultate der §§ 7—16 zusammenfassen:

Unter den irreduziblen nulldimensionalen Mengen von den Klassen 0 *und* 1 *gibt es nur folgende vier topologischen Typen*:

1. *die beschränkten perfekten nirgendsdichten linearen Mengen* („Fall F"),
2. *die unbeschränkten* „ „ „ „ („Fall F_ϱ"),
3. *die Menge* Ψ („Fall F_σ"),
4. *die Menge* J *aller Irrationalzahlen* („Fall G_δ").

Alle diese Mengen sind dabei topologisch homogen

Für Borelsche Mengen höherer Klassen läßt sich der Begriff der irreduziblen Mengen noch konsequenter einführen („eine Menge von der Klasse α

heißt irreduzibel, falls sie in keiner Umgebung keiner ihrer Punkte zu einer niedrigeren Klasse gehört"); der Begriff einer Menge, die im *absoluten* Sinne von der gegebenen Klasse ist, bleibt dabei genau derselbe, wie wir ihn in den §§ 6 und 8 für den Spezialfall der G_δ- und F_σ-Mengen erläutert haben. Man ist dabei gewissermaßen gezwungen, auch das Problem der Aufzählung (oder wenigstens der Mächtigkeitsbestimmung der Menge) aller topologischen Typen der *irreduziblen* nulldimensionalen Mengen einer gegebenen Klasse aufzustellen. Analoge Probleme könnte man auch für sogenannte A-Mengen untersuchen.

Anhang. Über Mengen erster und zweiter Kategorie.

In diesem Anhange werden wir keinerlei Voraussetzungen über die Dimension der im folgenden vorkommenden Mengen machen; ebenso brauchen wir nicht die Separabilität des Raumes vorauszusetzen, so daß unsere Betrachtungen für *beliebige* vollständige metrisierbare Räume gelten.

Eine in einem derartigen Raume R gelegene Menge M heißt bekanntlich[24]) von der *I. Kategorie* (in bezug auf R), falls sie sich als Vereinigungsmenge höchstens abzählbar vieler in R nirgendsdichter Teilmengen darstellen läßt.

Im Anschlusse an einen, von Herrn Lusin herrührenden Sprachgebrauch (der von der ursprünglichen Terminologie Baires abweicht) sagen wir, daß *eine Menge M von der II. Kategorie* (in bezug auf den Raum R) *ist, falls ihre Komplementärmenge $R - M$ daselbst von der I. Kategorie ist.*

Es gilt nun folgender Satz:

Eine in einem vollständigen metrisierbaren Raume R gelegene Menge M ist dann und nur dann von der II. Kategorie (*Lusinschen Sinne*), *wenn sie eine in R überalldichte G_δ-Menge enthält.*

Die Bedingung ist notwendig. Es sei in der Tat M von der zweiten, also $R - M$ von der ersten Kategorie. Dann ist

$$R - M = \sum_{k=1}^{\infty} N_k,$$

wobei jede N_k in R nirgendsdicht ist (unter den N_k dürfen auch leere Mengen vorkommen).

$\overline{N}_k$ ist aber dann auch nirgendsdicht, so daß

$$\Phi = \sum_{k=1}^{\infty} \overline{N}_k$$

---

[24]) Baire, Leçons sur les fonctions discontinues. Allgemein zuerst bei Hausdorff, S. 328.

eine F_σ-Menge von der ersten Kategorie ist, woraus folgt, daß die in M enthaltene Menge $R - \Phi$ in R überalldicht ist. $R - \Phi$ ist aber eine G_δ-Menge.

Die Bedingung ist hinreichend. Es sei in der Tat in M eine im ganzen Raume dichte G_δ-Menge H enthalten. Dann ist

$$H = \prod_{n=1}^{\infty} G_n,$$

wobei jedes G_n ein (in R dichtes) Gebiet ist. Die abgeschlossenen Mengen $F_n = R - G_n$ sind sämtlich in R nirgendsdicht, also ist $\Phi = \sum_{n=1}^{\infty} F_n$ von der ersten Kategorie. Φ ist aber mit $R - H$ identisch, woraus die Inklusion

$$\Phi \supset R - M$$

folgt; letztere Menge ist also a fortiori von der ersten Kategorie, d. h. M ist von der zweiten Kategorie, w. z. b. w.

Man erkennt sofort, daß der soeben bewiesene Satz trotz seiner Selbstverständigkeit doch von Nutzen sein dürfte, da er stillschweigend oft mitbewiesen wird.

Manche Sätze aus der reellen Funktionentheorie, und auch aus der Topologie, lassen sich übrigens direkt auf unseren Satz zurückführen. Um nur ein topologisches Beispiel zu geben, erwähnen wir folgende, aus unserem Satze unmittelbar sich ergebende Tatsachen.

I. Falls auf einer n-dimensionalen (absolut) abgeschlossenen (oder auch nur vollständigen) Punktmenge M die Menge der Punkte x, in denen $\dim_x M \leqq k$ $(0 \leqq k \leqq n)$ ist, überalldicht ist, so ist sie von der zweiten Kategorie.

II. Die Menge der Endpunkte einer allgemeinen Kurve C ist dann und nur dann überalldicht (in C), wenn die Menge aller übrigen Punkte von C daselbst von der ersten Kategorie ist[25]).

Beide Sätze folgen einfach aus der bekannten Tatsache, daß die Punkte, in denen $\dim_x M \leqq k$ bzw. $\operatorname{ind}_x C \leqq k$ ist, eine G_δ-Menge (rel M bzw. C) bilden[26]).

---

[25]) Siehe wegen Terminologie die unter [1]) zitierten Urysohnschen dimensionstheoretischen Arbeiten, insbesondere die Einleitung zum Mém. I und auch das erste Kapitel des Mém. II.

[26]) Vgl. wegen der Sätze I und II Mém. **1**, ch. IV, § 15 (Fund. Math. **8**, S. 281), Mém. **2**, ch. I, § 10 (Verh. Akad. Amsterdam **13**, Nr. 4), wo sie auf der hier gegebenen Grundlage bewiesen sind. Vgl. auch den Beweis des Satzes II dieses Anhanges, den Menger in Math. Annalen **95**, S. 285 (Satz VI) gegeben hat.

(Eingegangen am 8. 5. 1926.)

Nikolai Grigorievich Chebotaryov

Born – 15 June 1894 in Kamenets–Podolsk, Ukraine
Died – 2 July 1947 in Moscow, USSR

Nikolai Chebotaryov was barely 15 years old when he began his "love affair" with Mathematics. Nikolai's mother was a strong influence on his education. Yet strangely enough, her lack of knowledge on the subject of Mathematics seemed to be one of the main reasons Nikolai went in that direction. His education was disrupted by ill health and he spent the winters of 1910–1911 in Italy recovering from pneumonia, with his mother.

In 1912, Nikolai entered the University in Kiev where he became a student of Grave. He obtained his degree by 1916 and was awarded his master's degree in the year 1918. During this period of time, he earned his keep through teaching in schools and giving private tuition.

Nikolai moved to Odessa to help out his parents in 1921. Sadly, his father passed away the following year. But in that very same year, Nikolai proved the theorem that he was best known for, the density theorem. It was first conjured by Frobenius in a paper written in 1880 and published in 1896. The density theorem generalized Dirichlet's theorem on primes in an arithmetical progression giving a method used by Artin in 1927 in his reciprocity law, a result that was considered the best of class field theory.

By 1924, Nikolai had a permanent job in Moscow, but he was appointed to the post that Egorov had been dismissed from due to political reasons. The situation was rather unhappy and Nikolai resigned after a few months. He returned to Odessa and found a poorly paid secretarial job. In 1925, he managed to attend a meeting of the German Mathematical Society in Danzig, where he met Emmy Noether, Hensel and Hasse. He also met Schur in Berlin and visited Göttingen after the meeting.

With a thesis based on his results on the density theory in 1922, Nikolai received his doctorate from the Ukrainian Academy of Science. In 1928, he became a professor at Kazan University, having been offered posts both at Kazan and Leningrad. Throughout the rest of his life, he never left Kazan.

Nikolai worked on the algebra of polynomials, in particular, examining the distributions of the zeros. He also studied Galois theory and wrote a two-volume textbook *Basic Galois Theory* in 1934 and 1937. He was the moving force in having the first translation of Galois' work into Russian in 1936. Other important works of Nikolai included Lie groups, he wrote the first Russian text *Theory of Lie Groups* in 1940. He also made important contributions on the inverse Galois problem and the theory of resolvants.

Nikolai Grigorievich Chebotaryov passed away on 2 July 1947 in Moscow, USSR at the age of 53.

Известия Российской Академии Наук. 1923.

(Bulletin de l'Académie des Sciences de Russie).

Определение плотности совокупности простых чисел, принадлежащих к заданному классу подстановок.

Н. Чеботарева.

Представлено академиком Я. В. Успенским в заседании Отделения Физико-Математических Наук 7 марта 1923 года.

I.

Задача, решение которой является целью настоящей работы, была поставлена Frobenius'ом («Über Beziehungen u. s. w.», Sitzungsber. der Berl. Akad., 1896, S. 689). Она состоит в следующем. Дано неприводимое нормальное уравнение n-ой степени

$$(1) \qquad f(x) = 0.$$

Обозначим область, полученную от присоединения к области рациональных чисел его корня, через $\Omega(x)$, а через $\mathfrak{p}$ простой идеал внутри $\Omega(x)$, взаимно простой с дискриминантом уравнения (1). Тогда имеют место сравнения:

$$(2) \qquad x_1^p \equiv x_{\alpha_1},\ x_2^p \equiv x_{\alpha_2}, \ldots\ x_n^p \equiv x_{\alpha_n} \pmod{\mathfrak{p}},$$

если через $x_1, x_2, \ldots x_n$ обозначить сопряженные корни уравнения (1), а

$$S = \begin{pmatrix} 1 & 2 & 3 \cdots n \\ \alpha_1 & \alpha_2 & \alpha_3 \cdots \alpha_n \end{pmatrix}$$

является подстановкой над $1, 2, 3, \ldots n$, которая, как известно, входит в группу G уравнения (1) (см. Dedekind, Zur Theorie der Ideale, Gött. Nachr., 1894; также Frobenius, loc. cit.). Тогда будем говорить, что простой идеал $\mathfrak{p}$ *принадлежит* к подстановке S, а рациональное простое число p, кратное $\mathfrak{p}$, *принадлежит* к классу подстановок TST^{-1}, где T пробегает все подстановки группы G.

Рассмотрим совокупность всех простых чисел, принадлежащих к этому классу подстановок. Выражение

$$\lim_{s=1} \frac{\sum\limits_{p} p^{-s}}{\lg \frac{1}{s-1}}, \tag{3}$$

где сумма распространяется на все простые числа этой совокупности, носит название *плотности* совокупности. Наша задача заключается в определении величины плотности этой совокупности.

Понятие плотности впервые встречается у Dirichlet (Werke, Bd. I, S. 307, 313; 1837; также его Vorles. üb. Zahlentheorie), который доказал, что простые числа *равномерно* распределяются по классам сравнений любого модуля k, взаимно простым с k, т. е. что плотность каждой такой совокупности равна $\frac{1}{\varphi(k)}$. Исследования Kummer'а связывают этот результат с определением плотности совокупности простых чисел, принадлежащих к каждой из подстановок в области каких-нибудь корней из единицы (см. Kummer, «Über die Divisoren u. s w.», Journ. f. Math., B-de 30; 35, также § II моей работы). Kronecker (Monatsber. Berl. Akad., 1880, S. 156: «Über Irreducibilität von Gleichungen») нашел величину плотности совокупности простых чисел, принадлежащих к тождественной подстановке в произвольной области.

Вопросом о простых числах, принадлежащих к любому классу подстановок в произвольной области, занялся Frobenius. Скелет его результата был ему известен еще в 1882 году, и в 101 томе Journ. f. Math. он публикует групповую часть этой проблемы.

Однако, свой результат в полном объеме он опубликовал лишь после того, как Dedekind (loc. cit.) доказал обратную теорему: подстановка, к которой принадлежит какой-нибудь простой идеал, входит в группу области. В своем основном мемуаре Frobenius (loc. cit.), применяя результат Kronecker'а последовательно к различным делителям основной области, определяет плотность совокупности простых чисел, принадлежащих к *отделу* данной подстановки, т. е. к совокупности классов, образованных всеми ее первообразными степенями. Frobenius'у однако не удалось определить плотностей совокупностей простых чисел, принадлежащих к отдельным классам подстановок, о чем он говорит в своей работе в следующих выражениях:

«Indem man hier für G der Reihe nach allen cyklischen Untergruppen von H setzt, erhält man eine Reihe von Gleichungen, die aber nicht ausreichen, um schliessen zu können, dass

$$(16) \qquad \sum p_\lambda^{-1-w} = \frac{h_\lambda}{h} \lg\left(\frac{1}{w}\right) + \mathfrak{P}_\lambda(w)$$

ist.»

И далее:

«Wenn es gelänge die Formel (16) zu beweisen, so würde sich für die Dichtigkeit der Primzahlen p_λ, die der λ^{ten} Classe von Substitutionen entsprechen, der einfache Ausdruck:

$$(18) \qquad D_\lambda = \frac{h_\lambda}{h} = \frac{1}{v_\lambda}$$

ergeben, es würde also der Satz gelten:

V. Jeder Classe von Substitutionen der Gruppe H entsprechen unzählig viele rationale Primzahlen. Ihre Dichtigkeit ist der Anzahl der verschiedenen Substitutionen der Classe proportional.

Oder:

Die Dichtigkeit der Primzahlen, die einer Classe von Substitutionen entsprechen, ist der Dichtigkeit der Classe gleich».

Этому общему результату Frobenius'а не было суждено сыграть в науке такой большой роли, какую сыграл результат Hilbert'а, который был опубликован почти одновременно с Frobenius'овским и может быть рассматриваем, как его частный случай, выраженный в несколько видоизмененной форме. Это теорема Hilbert'а о существовании в любой нормальной области, заключающей в себе l-ые корни из единицы, бесчисленного множества простых идеалов с наперед заданной вычетностью (см. Hilbert, Zahlbericht, S. 424—428; также главу VI настоящей работы). Основываясь на этой теореме, Furtwängler (Algemeiner Existenzbeweis u. s. w. Math. Ann., Bd. 63) доказывает для любой области существование соответствующей ей *области классов* (Klassenkörper). Эта же теорема играет большую роль при выводе общего закона взаимности (см. Furtwängler, Allgemeine Reziprozitätsgesetze, Math. Ann., Bde. 72, 74).

Дальнейшая литература по этому вопросу невелика. В Arch. d. Math. u. Phys. Bd. 6 (3) (1904) помещены три небольшие статьи M. Bauer'а, в которых он дает непосредственные приложения Frobenius'овского результата к констатированию тождественности или независимости двух областей, а также к выводу характеристического свойства областей деления круга. Эти результаты являются частными случаями теоремы, доказываемой в главе V настоящей работы. Кроме этих статей, в Прот. Харьк. Мат. Общ. за 1915 год помещена статья Б. Н. Делоне[1], который, пользуясь Frobenius'овским результатом и законом взаимности Eisenstein'а, доказал (правда, не для всех случаев) теорему Kronecker'а-Weber'а о том, что все абелевы области суть области деления круга.

Моя работа построена по следующему плану:

В § I я в существенных чертах воспроизвожу исследование Frobenius'а, доделывая все выкладки до конца. В ходе рассуждений я значительно уклоняюсь от Frobenius'а, главным образом потому, что я ограничиваюсь рассмотрением неприводимых уравнений, что дает возможность избежать применения трудных и мало известных теорем из теории групп.

В § II я обобщаю найденную Kummer'ом (loc. cit.) связь между прогрессиями и областями деления круга. Вместо прогрессии, я ввожу в рассмотрение более общее понятие *комплекса*.

В § III я обобщаю теорему Dirichlet о прогрессиях. Именно, я доказываю существование в любой области бесчисленного множества простых идеалов, нормы которых лежат в наперед заданных допустимых комплексах. Подобные обобщения делались и другими авторами. Так, Weber в своей статье: «Über Zahlgruppen u. s. w.», (Math. Ann., Bd. 49) доказывает даже более общую теорему, но высказывает ее не вполне отчетливо (именно, расплывчато формулировано понятие, соответствующее моим *допустимым комплексам*) и, кроме того, он предполагает существование области классов (Klassenkörper), чего я избегаю, несколько суживая обьем результата и вводя понятие комплексов, допустимых в *узком* и в *широком смысле*. При доказательстве приходится обобщать результат Dirichlet-Minkowsk'ого (связь между кратными интегралами и рядами Dirichlet). Hilbert в своей статье: «Relativquadratischer Zahlkörper» (Math. Ann., Bd. 51, Sätze 30, 31, S. 53 ff.) делает это же самое обобщение и применяет его к теореме, подобной моей, но только имея в виду комплексы более частного типа.

В § IV я определяю плотность совокупности простых чисел, принадлежащих к данному классу подстановок. Для этого я присоединяю к нашей области некоторую область деления круга. Тогда, если определяющее эту последнюю уравнение остается по модулю p неприводимым, то теорема 12

---

[1] См. также B. Delaunay, Zur Bestimmung algebraischer Zahlkörper u. s. w., Journal für die reine und ang. Math., Bd. 152, S. 120.

показывает, что известные уравнения из распространенной области имеют **или** не имеют рациональные корни в зависимости от того, к какому из классов отдела принадлежит p. Пользуясь результатом § III, мы отсюда докажем существование в каждом из классов отдела бесчисленного множества принадлежащих к нему простых чисел. Для того же, чтобы определить величину их плотности, я присоединяю не одну область деления круга а k, причем k безгранично увеличиваю. Этот прием дает возможность определить искомую плотность. Однако, оценить величину остаточного члена мне не удалось.

§ V посвящен выводу критерия родственности областей. Он содержит теорему, частные случаи которой рассмотрены в статьях M. Bauer'a и Б. Н. Делоне (loc. cit.).

Наконец, в § VI я показываю, что теорема Hilbert'a (loc. cit.), притом в более общей формулировке, может быть легко выведена из результата § V.

§ I. Исследование Frobenius'a.

Возьмем неприводимое нормальное уравнение

$$f(x) = 0 \tag{1a}$$

и обозначим его корни через $x_1, x_2, \ldots x_n$. Тогда справедлива

Теорема 1. Имеют место сравнения

$$x_1^p \equiv x_{\alpha_1},\ x_2^p \equiv x_{\alpha_2}, \ldots\ x_n^p \equiv x_{\alpha_n} \pmod{\mathfrak{p}}, \tag{2a}$$

где p простое число, не входящее в дискриминант уравнения (1), $\mathfrak{p}$ его простой идеальный множитель в $\Omega(x)$, а

$$S = \begin{pmatrix} 1 & 2 & 3 \ldots n \\ \alpha_1 & \alpha_2 & \alpha_3 \cdots \alpha_n \end{pmatrix}$$

подстановка над числами $1, 2, 3, \ldots n$.

Доказательство см. Frobenius, S. 691.

Определение 1. В случае, когда имеют место сравнения (2a), будем говорить, что простой идеал $\mathfrak{p}$ *принадлежит* к подстановке S.

Теорема 2. Подстановка S входит в группу Galois G уравнения (1a).

Доказательство см. Fr., S. 696.

Теорема 3. Если $\mathfrak{p}$ принадлежит к S, то простой идеал $\mathfrak{p}/T$, в который $\mathfrak{p}$ переходит посредством подстановки к T из G, принадлежит к $T^{-1}ST$.

Доказательство см. Fr., S. 700.

Определение 2. Совокупность подстановок $T^{-1}ST$, где T пробегает всю группу G, будем называть *классом* подстановки S.

Так как p является произведением всех различных простых идеалов, сопряженных с $\mathfrak{p}$, то теорема 3 позволяет ввести следующее новое определение:

Определение 3. В случае, если имеют место сравнения (2a), простое число p принадлежит к классу подстановки S.

Теорема 4. Если z величина из $\Omega(x)$, принадлежащая к группе H, а

$$F(z) = 0 \tag{4}$$

неприводимое уравнение, которому она удовлетворяет, то сравнение

$$F(z) \equiv 0 \pmod{p} \tag{5}$$

имеет рациональные корни тогда и только тогда, если хоть одна из подстановок S класса, к которому принадлежит p, входит в H, и число этих рациональных корней равно числу входящих в H подстановок в ряду $T_\nu S T_\nu^{-1}$, если T_ν пробегает все значения $T_1 = 1$, $T_2, T_3 \ldots$ в разложении

$$G = H + T_2H + T_3H + \ldots$$

Доказательство см. Fr., § 5, *S.* 701.

Определение 4. Под символом $P(s-1)$ будем подразумевать функцию от s, остающуюся конечной при $s=1$.

Теорема 5. (Kronecker'a). Имеет место формула

$$\sum_p \nu_p p^{-s} = \lg \frac{1}{s-1} + P(s-1), \tag{6}$$

где сумма распространяется на все простые числа p, а ν_p обозначает число рациональных корней сравнения (5).

Доказательство см. Landau, Verteilung der Primzahlen in den Idealklassen (Math. Ann., Bd. 63, S. 150).

Определение 5. Под *отделом* (Abteilung) мы будем понимать совокупность подстановок TS^iT^{-1}, где T пробегает все подстановки группы G, а i—все числа, не превышающие и взаимно простые с порядком f подстановки S.

Определение 6. Совокупность простых чисел, принадлежащих ко всем классам отдела, будем называть совокупностью, принадлежащей к отделу.

Определение 7. Под *плотностью* совокупности простых чисел мы будем разуметь значение выражения

$$\lim_{s=1} \frac{\sum\limits_{p} p^{-s}}{\lg \frac{1}{s-1}},$$

где сумма распространяется на все простые числа этой совокупности.

Пусть

$$S_\lambda = (x_{11}, x_{12}, \ldots x_{1f_\lambda})(x_{21}, x_{22}, \ldots x_{2f_\lambda}) \ldots (x_{e_\lambda 1}, x_{e_\lambda 2}, \ldots x_{e_\lambda f_\lambda}) \tag{7}$$
$$(e_\lambda f_\lambda = n)$$

будет подстановка порядка f_λ, входящая в G. Определим плотность совокупности простых чисел, принадлежащих к отделу S_λ.

Введем следующие обозначения:

f_λ — для порядка подстановки S_λ;

n_λ — для числа различных подстановок, входящих в класс S_λ;

k_λ — для числа различных классов, входящих в отдел S_λ;

H_λ — для группы подстановок, входящих в G и перестановочных с S_λ;

$\overline{H}_\lambda$ — для группы таких подстановок T, входящих в G, что подстановки $TS_\lambda T^{-1}$ являются степенями S_λ.

Тогда мы сможем придти к следующей

Главной теореме. Плотность совокупности простых чисел, принадлежащих к отделу S_λ, равна $k_\lambda . \frac{n_\lambda}{n}$.

Для *доказательства* изберем индуктивный путь, доказав теорему: A) для тождественной подстановки, B) для случая, когда порядок S_λ есть простое число и C) для общего случая в предположении, что теорема доказана для тех степеней S_δ подстановки S_λ, порядки которых f_δ суть *настоящие* (echte) делители числа f_λ.

A) Применим формулу (6) к уравнению (1). Так как оно нормально, то сравнение

$$(8) \qquad f(x) \equiv 0 \pmod{p}$$

имеет рациональные корни тогда и только тогда, если имеют место следующие сравнения:

$$(9) \qquad x_1^p \equiv x_1,\ x_2^p \equiv x_2, \ldots\ x_n^p \equiv x_n \pmod{\mathfrak{p}},$$

а тогда все корни сравнения (8) рациональны, т. е. $\nu_p = n$. Формула (6) принимает вид:

$$(10) \qquad \sum_{p_0} p_0^{-s} = \frac{1}{n} \lg \frac{1}{s-1} + P(s-1),$$

и таким образом плотность нашей совокупности равна $\frac{1}{n}$.

B) Рассмотрим теперь $\frac{f_\lambda}{q_1}$ ую степень подстановки S_λ:

$$S_1 = S_\lambda^{\frac{f_\lambda}{q_1}},$$

где q_1 какой-нибудь простой делитель f_λ. Образуем величину ξ_1, принадлежащую к группе $(S_1)^i$ (символ $(S_1)^i$ обозначает циклическую группу, состоящую из степеней S_1) и неприводимое уравнение

$$(11) \qquad \Phi(\xi_1) = 0,$$

которому она удовлетворяет. Сравнение

$$\Phi(\xi_1) \equiv 0 \pmod{p}$$

имеет, в силу теоремы 4, рациональные корни тогда и только тогда, если подстановка класса, к которому принадлежит простой модуль p, входит хоть в одну из групп $T_\nu (S_1)^i T_\nu^{-1}$, и их число равно числу таких групп. Это может иметь место в двух случаях: 1) p принадлежит к тождественной подстановке; все $\frac{n}{q_1}$ корней рациональны; это простые числа уже рассмотренного нами типа p_0; 2) или p принадлежит к одной из первообразных подстановок группы $(S_1)^i$, т. е. к одному из классов нашего отдела $T_\nu (S_1)^i T_\nu^{-1}$. Из теоремы 4 следует, что искомое число равно индексу $(\overline{H}_1, (S_1)^i)$. Но так как

$$(\overline{H}_1, (S_1)^i) = (\overline{H}_1, H_1)(H_1, (S_1)^i),$$

то дело сводится к выражению обоих множителей через n, n_1, q_1 и k_1.

С другой стороны

$$(H_1, (S_1)^i) = (G, (S_1)^i) \colon (G, H_1),$$

а

$$(G, (S_1)^i) = \frac{n}{q_1}.$$

Если мы теперь введем на время обозначение $(G, H_1) = a$, то из разложения

$$G = H_1 + T_2 H_1 + \cdots + T_a \cdot H_1$$

следует, что для того, чтобы две подстановки из G преобразовали S_1 в одну и ту же подстановку, необходимо и достаточно, чтобы они входили в одну и ту же сопряженную систему. Это показывает, что число различных подстановок типа $T_\nu S_1 T_\nu^{-1}$ равно a, т. е. $a = n$. Поэтому

$$(H_1, (S_1)^i) = \frac{n}{q_i\, n_1}.$$

Чтобы определить величину $(\overline{H}_1, H_1)$, заметим, что все $\varphi(q_1)$ первообразных степеней подстановки S_1 образуют k_1 классов отдела, а поэтому в каждый из этих классов входит по $\frac{\varphi(q_1)}{k_1}$ степеней S_1. Поэтому из разложения

$$\overline{H}_1 = H_1 + T_2 H_1 + \cdots$$

следует, что индекс $(\overline{H}_1, H_1)$ равен $\frac{\varphi(q_1)}{k_1}$. Сопоставляя все полученные результаты, мы видим, что искомое число рациональных корней сравнения $\Phi(\xi_1) \equiv 0 \pmod p$ равно $\frac{n \cdot \varphi(q_1)}{q_1 \cdot n_1 \cdot k_1}$.

Формула (6) принимает вид

$$\frac{n}{q_1} \sum_{p_0} p_0^{-s} + \frac{n(q_1 - 1)}{q_1 \cdot n_1 \cdot k_1} \sum_{p_1} p_1^{-s} + P(s-1) = \lg \frac{1}{s-1} + P(s-1),$$

если мы через p_1 будем обозначать простые числа, принадлежащие к нашему отделу. Пользуясь формулой (10), мы приходим к следующему результату:

$$(12) \quad \sum p_{11}^{-s} + \sum p_{12}^{-s} + \cdots + \sum p_{1k_1}^{-s} = k_1 \frac{n_1}{n} \lg \frac{1}{s-1} + P(s-1),$$

где мы совокупность (p_1) разбили на совокупности $(p_{11}), (p_{12}), \ldots (p_{1k_1})$, принадлежащие к различным классам нашего отдела.

C) Теперь мы в состоянии доказать, что плотность совокупности простых чисел, принадлежащих к отделу подстановки S_λ, равна $\frac{k_\lambda n_\lambda}{n}$, если мы будем наше утверждение считать доказанным для тех степеней S_λ, порядок которых f_δ есть настоящий (echter) делитель f_λ $(f_\delta < f_\lambda)$.

Пусть ξ_λ будет величина из $\Omega(x)$, принадлежащая к группе $(S_\lambda)^i$ и пусть

$$\Phi(\xi_\lambda) = 0 \tag{13}$$

будет неприводимое уравнение, которому она удовлетворяет. Степень этого уравнения равна $\frac{n}{f_\lambda}$. Чтобы сравнение

$$\Phi(\xi_\lambda) \equiv 0 \pmod p \tag{14}$$

имело рациональные корни, необходимо и достаточно, чтобы p принадлежало к одному из отделов, образованных степенями подстановки S_λ. Рассмотрим простые числа типа p_δ, принадлежащие к отделу S_δ. Порядок S_δ пусть равен f_δ (пока мы не исключаем случая $f_\delta = f_\lambda$). Число рациональных корней сравнения (14) в этом случае равно числу таких подстановок T_ν в разложении

$$G = (S_\lambda)^i + T_2\,(S_\lambda)^i + \cdots + T_{\frac{n}{f_\lambda}}\cdot(S_\lambda)^i, \tag{15}$$

которые преобразуют подстановку S_δ в одну из ее степеней. Но так как группа $(S_\lambda)^i$ в виду соотношения

$$S_\lambda \cdot S_\delta = S_\lambda \cdot S_\lambda{}^i = S_\lambda{}^i \cdot S_\lambda = S_\delta \cdot S_\lambda$$

является делителем H_δ, то искомое число равно индексу

$$(\overline{H}_\delta, (S_\lambda)^i) = (\overline{H}_\delta, H_\delta)\,(H_\delta, (S_\lambda)^i).$$

Далее мы можем, точно также, как в B), доказать, что $(\overline{H}_\delta, H_\delta)$ равно $\frac{\varphi(f_\delta)}{k_\delta}$. Что касается до индекса $(H_\delta, (S_\lambda)^i)$ то он равен

$$(H_\delta, (S_\lambda)^i) = (H_\delta, (S_\delta)^i) : ((S_\lambda)^i, (S_\delta)^i) = (H_\delta, (S_\delta)^i) : \frac{f_\lambda}{f_\delta}.$$

Наконец прием, примененный нами уже в B), дает для индекса $(H_\delta, (S_\delta)^i)$ значение $\frac{n}{f_\delta \cdot n_\delta}$, откуда следует, что число рациональных корней сравнения (14) в нашем случае равно

$$\frac{n \cdot \varphi(f_\delta)}{f_\lambda \cdot n_\delta \cdot k_\delta}.$$

Формула (6) для уравнения (13) принимает поэтому следующий вид:

$$\sum_{f_\delta | f_\lambda} \frac{n\varphi(f_\delta)}{f_\lambda \cdot n_\delta \cdot k_\delta} \sum_{p_\delta} p_\delta^{-s} = \lg\frac{1}{s-1} + p(s-1). \tag{16}$$

Но для каждого $f_\delta < f_\lambda$ имеет место, в силу нашего предположения, формула

$$\sum_{p_\delta} p_\delta^{-s} = k_\delta \frac{n_\delta}{n} \lg\frac{1}{s-1} + p(s-1) \quad (f_\delta < f_\lambda). \tag{17}$$

Подставляя эту формулу для всевозможных p_δ в (16), мы получим:

$$\sum_{\substack{f_\delta | f_\lambda \\ f_\delta < f_\lambda}} \frac{n\varphi(f_\delta)}{f_\lambda \cdot n_\delta \cdot k_\delta} \cdot k_\delta \cdot \frac{n_\delta}{n} \cdot \lg\frac{1}{s-1} + \frac{n\varphi(f_\lambda)}{f_\lambda \cdot n_\lambda \cdot k_\lambda} \sum_{p_\lambda} p_\lambda^{-s} = \tag{18}$$

$$= \lg\frac{1}{s-1} + p(s-1),$$

что может быть преобразовано так:

$$\sum_{p_\lambda} p_\lambda^{-s} = k_\lambda \frac{n_\lambda}{n} \cdot \frac{\left\{f_\lambda - \sum\limits_{\substack{f_\delta | f_\lambda \\ f_\delta < f_\lambda}} \varphi(f_\delta)\right\}}{\varphi(f_\lambda)} \cdot \lg\frac{1}{s-1} + p(s-1), \tag{19}$$

откуда мы, в силу известной формулы

$$\sum_{f_\delta | f_\lambda} \varphi(f_\delta) = f_\lambda,$$

получаем:

$$\sum_{p_\lambda} p_\lambda^{-s} = k_\lambda \cdot \frac{n_\lambda}{n} \cdot \lg\frac{1}{s-1} + P(s-1), \tag{20}$$

и мы приходим к искомому результату.

Этот результат принадлежит Frobenius'у. Однако, ему не удалось доказать, что плотность совокупности простых чисел, принадлежащих к каждому из классов отдела, равна $\frac{n_\lambda}{n}$. Решение этой задачи и является целью настоящего сочинения.

§ II. Области деления круга.

Существует целая категория областей, для которых упомянутая задача решается при помощи известных ранее методов. Это области деления круга.

Не ставя вопроса во всей общности, я рассмотрю случай, который играет важную роль для дальнейшего.

Пусть l будет простое число типа $ef+1$, где f заданное число, не подчиненное никаким дальнейшим ограничениям. Рассмотрим уравнение

$$x^l - 1 = 0. \tag{21}$$

Далее, пусть ρ будет один из его первообразных корней. Каждый из e-членных Gauss'овых периодов

$$\begin{aligned} \eta_0 &= \rho + \rho_f + \rho_{2f} + \cdots + \rho_{(e-1)f}, \\ \eta_1 &= \rho_1 + \rho_{f+1} + \rho_{2f+1} + \cdots + \rho_{(e-1)f+1}, \\ &\cdots\cdots\cdots\cdots\cdots\cdots \\ \eta_{f-1} &= \rho_{f-1} + \rho_{2f-1} + \rho_{3f-1} + \cdots + \rho_{ef-1}, \end{aligned} \tag{22}$$

где под ρ_k мы будем разуметь ρ^{g^k}, а g какой-нибудь первообразный корень сравнения $x^{ef} \equiv 1 \pmod{l}$, удовлетворяет неприводимому уравнению f-ой степени:

$$\varphi_e(y) = 0. \tag{23}$$

Возьмем теперь какое-нибудь простое число p. Возникает вопрос, с какой из сопряженных величин η сравнима степень $\eta_\nu{}^p$ по модулю p. Пусть $p \equiv g^\alpha \pmod{l}$. Тогда имеет место следующее сравнение:

$$\begin{aligned} \eta_\nu{}^p &= (\rho_\nu + \rho_{f+\nu} + \cdots + \rho_{(e-1)f+\nu})^p \equiv \\ &\equiv \rho_\nu{}^p + \rho^p{}_{f+\nu} + \cdots + \rho^p{}_{(e-1)f+\nu} \equiv \rho_\nu^{g^\alpha} + \rho_{f+\nu}^{g^\alpha} + \cdots + \cdots + \rho_{(e-1)f+\nu}^{g^\alpha} \equiv \\ &\equiv \rho_{\nu+\alpha} + \rho_{f+\nu+\alpha} + \cdots + \rho_{(e-1)f+\nu+\alpha} \equiv \eta_{\nu+\alpha} \pmod{p}, \end{aligned} \tag{24}$$

где значек $\nu+\alpha$ необходимо привести к его наименьшему вычету по модулю f. Следовательно, корни сравнения

$$\varphi_e(y) \equiv 0 \pmod{p} \tag{25}$$

претерпевают при возвышении их в p-ую степень подстановку $(\nu, \nu+\alpha)$ тогда и только тогда, если p является членом одной из прогрессий

$$lx + g^\alpha,\ lx + g^{\alpha+f}, \ldots\ lx + g^{\alpha+(e-1)f}.$$

Подстановку $(\nu, \nu+\alpha)$ можно рассматривать, как α-ую степень подстановки $S = (\nu, \nu+1) = (0, 1, 2, \ldots f-1)$. Введем следующее

Определение 8. Совокупность прогрессий

(26) $$lx + g^{\alpha},\ lx + g^{\alpha+f}, \dots\ lx + g^{\alpha+(e-1)f}$$

назовем *комплексом* индекса α. Про все числа, представляемые одною из прогрессий (26), будем говорить, что они *лежат* в комплексе индекса α.

Тогда результат настоящей главы мы можем выразить в виде следующего предложения:

Главная теорема. Дано неприводимое уравнение (23), которому удовлетворяет система Gauss'овых периодов (22). Тогда возможно таким образом приурочить первообразной подстановке U из его группы Galois первообразный корень сравнения $x^{ef} \equiv 1 \pmod{l}$, чтобы каждое простое число, принадлежащее к подстановке U^{α}, лежало в комплексе индекса α.

Следует подчеркнуть, что теорема Dirichlet (loc. cit.) позволяет нам таким образом определить плотность совокупности простых чисел, принадлежащих к каждой заданной подстановке.

§ III. О распределении норм и простых чисел по различным комплексам.

В настоящей главе я намерен доказать теорему, являющуюся в известном смысле обобщением теоремы Dirichlet.

Рассмотрим неприводимое уравнение

(27) $$f(x) = 0,$$

которое мы здесь не будем предполагать непременно нормальным. Введем еще в рассмотрение произвольное нечетное число $L = l_1^{\lambda_1} l_2^{\lambda_2} \cdots l_k^{\lambda_k}$ и систему таких чисел $f_1, f_2, \dots f_k$, что каждое f_i является делителем

$$\varphi\left(l_i^{\lambda_i}\right) \qquad (i=1, 2, 3, \dots k).$$

Обозначив первообразный корень сравнения

(28) $$x^{\varphi\left(l_i^{\lambda_i}\right)} \equiv 1 \pmod{l_i^{\lambda_i}} \qquad (i=1, 2, 3, \dots k)$$

через g_i, мы можем представить все взаимно простые с l_i классы по модулю $l_i^{\lambda_i}$ в виде системы

$$1,\ g_i,\ g_i^2, \dots\ g_i^{\varphi\left(l_i^{\lambda_i}\right)-1} \qquad (i=1, 2, 3, \dots k).$$

Распределим эти классы по комплексам (*частичным*), называя каждую систему классов

$$g_i^{\alpha_i},\quad g_i^{\alpha_i+f_i},\ \ldots\ g_i^{\alpha_i+f_i\left(\frac{\varphi\left(l_i^{\lambda_i}\right)}{f_i}-1\right)} \qquad (i=1,\ 2,\ 3,\ldots k)$$

частичным комплексом индекса α_i. Индекс α_i будем считать приведенным по модулю f_i.

Классы сравнений по модулю L могут быть распределены следующим образом по *полным комплексам*:

Определпние 9. Дан взаимно простой с L класс A сравнений по модулю L. Если $A\equiv a_i \left(\mathrm{mod}\ l_i^{\lambda_i}\right)$, а a_i лежит в комплексе индекса α_i $(i=1,\ 2,\ldots k)$, то будем говорить, что A лежит в *полном комплексе* индекса $(\alpha_1,\ \alpha_2,\ldots \alpha_k)$.

Если мы будем умножать числа из комплекса индекса $(\alpha_1,\ \alpha_2,\ldots \alpha_k)$ на числа из комплекса индекса $(\beta_1,\ \beta_2,\ldots \beta_k)$, то будем получать числа, лежащие в комплексе индекса $(\alpha_1+\beta_1,\ \alpha_2+\beta_2,\ldots \alpha_k+\beta_k)$. Стало быть, комплексы образуют абелеву группу порядка $f_1,\ f_2\ldots f_k$. В ней роль единицы играет *нулевой комплекс* $(0,\ 0,\ldots 0)$.

Вернемся опять к области $\Omega(x)$, образованной при помощи одного из корней уравнения (27). Пусть

$$\omega_1,\ \omega_2,\ldots \omega_n$$

будут элементы его фундаментального базиса. Тогда каждое целое число из $\Omega(x)$ может быть представлено так:

$$\mu=c_1\omega_1+c_2\omega_2+\ldots+c_n\omega_n, \tag{29}$$

где $c_1,\ c_2,\ldots c_n$ суть целые рациональные числа, которые называют *координатами* числа μ.

Если возможно подобрать координаты $c_1,\ c_2,\ldots c_n$ для числа μ таким образом, чтобы норма $N(\mu)$ лежала в комплексе $(\alpha_1,\ \alpha_2,\ldots \alpha_k)$, то назовем этот комплекс *допустимым* (*в узком смысле*), говоря в то же время про число μ, что оно *лежит* в комплексе индекса $(\alpha_1,\ \alpha_2,\ldots \alpha_k)$.

Для допустимых комплексов можно доказать следующую теорему:

Теорема 6. Все допустимые комплексы образуют группу, которая является делителем группы всех комплексов.

Доказательство. Если $N(\mu)\equiv a$ и $N(\nu)\equiv b \pmod L$, то имеет место также сравнение $N(\mu\nu)\equiv ab \pmod L$. Если теперь a лежит в комплексе

$(\alpha_1, \alpha_2, \ldots \alpha_k)$, а b в комплексе $(\beta_1, \beta_2, \ldots \beta_k)$, то ab лежит в комплексе $(\alpha_1 + \beta_1, \alpha_2 + \beta_2, \ldots \alpha_k + \beta_k)$. Далее, нулевой комплекс допустим, так как сравнение

$$N(\mu) \equiv 1 \pmod{L}$$

имеет очевидное решение $\mu = 1$.

Возьмем теперь определенный допустимый комплекс и рассмотрим совокупность всех чисел μ, лежащих в этом комплексе. Назовем эту совокупность совокупностью *решений комплекса.*

Чтобы ввести понятие *числа* решений комплекса, необходимо предварительно условиться в некоторых определениях:

Определение 10. Два целых числа μ и μ' из $\Omega(x)$ будем называть *равными по модулю* L тогда и только тогда, если их координаты c_i и c_i' удовлетворяют сравнениям

$$c_i \equiv c_i' \pmod{L} \qquad (i=1, 2, \ldots n).$$

Из определения фундаментального базиса следует, что эти сравнения необходимы и достаточны для сравнения $\mu \equiv \mu' \pmod{L}$. Это утверждение справедливо и тогда, если мы, вместо фундаментального базиса, положим в основу базис какого-нибудь идеала, взаимно простого с L.

Определение 11. Если для координат c_i числа μ удовлетворяются неравенства

$$0 \leqq c_i < L \qquad (i=1, 2, \ldots n), \tag{30}$$

то число μ мы будем называть *приведенным* по модулю L.

Для каждого целого числа μ из $\Omega(x)$ можно подобрать равное ему по модулю L приведенное число μ', и притом только одно.

Очевидно, что каждый допустимый комплекс имеет лишь конечное число приведенных решений. Чтобы определить их число для каждого допустимого комплекса, докажем следующую теорему:

Теорема 7. Каждый допустимый комплекс имеет одно и то же число приведенных решений.

Доказательство. Пусть нам задан допустимый комплекс $(\alpha_1, \alpha_1, \ldots \alpha_k)$, и пусть μ будет одно из его решений. Если числа

$$\eta_1, \eta_2, \ldots \eta_v \tag{31}$$

обозначают полную систему решений нулевого комплекса, то числа

(32) $$\mu\eta_1, \mu\eta_2, \ldots \mu\eta_v$$

после их приведения по модулю L представят полную систему решений комплекса $(\alpha_1, \alpha_2, \ldots \alpha_k)$. В самом деле:

1) все числа (32) лежат в комплексе $(\alpha_1, \alpha_2, \ldots \alpha_k)$;

2) все они различны по модулю L;

3) каждое число, лежащее в комплексе $(\alpha_1, \alpha_2, \ldots \alpha_k)$, должно быть сравнимо по модулю L с одним из чисел (32).

1) Следует из правила умножения комплексов.

2) Если бы имело место, например, $\mu\eta_i \equiv \mu\eta_j \pmod{L}$, то, умножая сравнение на целое алгебраическое число $\frac{N(\mu)}{\mu}$ и затем деля на взаимно простое с L целое рациональное число $N(\mu)$, мы получим $\eta_i \equiv \eta_j \pmod{L}$, что стоит в противоречии с определением системы (31).

3) Пусть μ' будет число, лежащее в комплексе $(\alpha_1, \alpha_2, \ldots \alpha_k)$. Тогда число $\mu' . \frac{N(\mu)}{\mu} \cdot [N(\mu)]^{\varphi L - 1}$ лежит в нулевом комплексе, так как его норма равна

$$N(\mu') \cdot [N(\mu)]^{n-1} \cdot [N(\mu)]^{n\varphi L - n} \equiv N(\mu') \cdot [N(\mu)]^{-1} \pmod{L}.$$

Поэтому это число должно быть сравнимо с одним из чисел системы (31). Пусть

$$\mu' \cdot \frac{N(\mu)}{\mu} \cdot [N(\mu)]^{\varphi L - 1} \equiv \eta_i \pmod{L}.$$

Умножая это сравнение на μ, мы придем к сравнению

$$\mu' \equiv \mu\eta_i \pmod{L},$$

что оправдывает утверждение 3). Таким образом, число решений каждого допустимого комплекса равно v, т. е. оно одинаково для всех комплексов, и т. д.

Это число легко определить. В самом деле, обозначим через F число всех допустимых комплексов. Все приведенные решения всех допустимых комплексов равномерно распределяются по отдельным допустимым комплексам. Но число всех приведенных решений всех допустимых комплексов равно числу взаимно простых с L классов по модулю L внутри $\Omega(x)$, т. е. равно

(33) $$\bar{\varphi}(L) = N(L)\left(1 - \frac{1}{N(l_1)}\right)\left(1 - \frac{1}{N(l_2)}\right) \cdots \left(1 - \frac{1}{N(l_p)}\right),$$

где l_1, l_2, ... l_p представляют собой совокупность всех простых идеальных делителей числа L внутри $\Omega(x)$. Таким образом, число решений каждого допустимого комплекса равно

$$v = \frac{\overline{\varphi}(L)}{F}. \tag{34}$$

То же число решений мы получим, если, вместо фундаментального базиса, положим в основу базис какого-нибудь идеала, взаимно простого с L.

Вспомним теперь, что все целые и делящиеся на идеал $\mathfrak{m}$, взаимно простой с L, числа из $\Omega(x)$ могут быть представлены в форме

$$c_1\omega_1 + c_2\omega_2 + \cdots + c_n\omega_n, \tag{35}$$

где $(\omega_1, \omega_2, \ldots \omega_n)$ базис идеала $\mathfrak{m}$, а $c_1, c_2, \ldots c_n$ пробегают все целые рациональные значения. С другой стороны, все целые и делящиеся на $\mathfrak{m}$ числа из $\Omega(x)$, которые, кроме того, лежат в допустимом комплексе $(\alpha_1, \alpha_2, \ldots \alpha_k)$, могут быть представлены в форме одного из следующих выражений:

$$\omega_i^{(\alpha_1, \cdots \alpha_k)} + L\,(c_1\omega_1 + c_2\omega_2 + \cdots + c_n\omega_n) \qquad (i=1, 2, \ldots v) \tag{36}$$

где $\omega_i^{(\alpha_1, \cdots \alpha\ )}$ $(i = 1,\ 2, \ldots v)$ обозначает систему приведенных решений комплекса $(\alpha_1,\ \alpha_2, \ldots \alpha_k)$, в то время как остальные обозначения имеют то же значение, что и для (35).

Рассмотрим теперь бесконечный ряд

$$\sum_{\mathfrak{a}}{}' \frac{1}{N(\mathfrak{a})^s},$$

где сумма распространяется на все делящиеся на $\mathfrak{m}$ и взаимно простые с L главные идеалы. Этот ряд вблизи $s = 1$ может быть представлен так:

$$\sum_{\mathfrak{a}}{}' \frac{1}{N(\mathfrak{a})^s} = \frac{\overline{\varphi}(L)}{L^n} \cdot \frac{g}{N(\mathfrak{m})} \cdot \frac{1}{s-1} + P(s-1), \tag{37}$$

где g имеет следующее независимое от s значение:

$$g = \frac{2^{\nu}\pi^{n-\nu} R}{w.\sqrt{\pm\Delta}}. \tag{38}$$

В этом выражении ν обозначает сумму числа вещественных корней и числа пар сопряженных комплексных корней уравнения (27), R регулятор области $\Omega(x)$, w число корней из единицы внутри $\Omega(x)$, Δ дискриминант области $\Omega(x)$. $P(s-1)$, как обычно, обозначает функцию, остающуюся вблизи $s=1$ конечной.

Проследим путь получения этой формулы, сообразуясь с обозначениями, принятыми у Weber'a, Lehrb. d. Alg. Bd. II, XXI Abschnitt, S. 693—735. В ней коэффициент при $\frac{1}{s-1}$ равен $\frac{1}{w}$-ой предела выражения $\frac{T}{t}$ при $t=\infty$, где T означает число систем целых значений $c_1, c_2, \ldots c_n$, подчиненных неравенствам

$$(39) \qquad |N(c_1\omega_1+c_2\omega_2+\cdots+c_n\omega_n)| \leqq t,$$

$$(40) \qquad \begin{cases} 0 \leqq \xi_1(c_1\omega_1+c_2\omega_2+\cdots+c_n\omega_n) < 1, \\ 0 \leqq \xi_2(c_1\omega_1+c_2\omega_2+\cdots+c_n\omega_n) < 1, \\ \cdots\cdots\cdots\cdots\cdots\cdots\cdots\cdots \\ 0 \leqq \xi_{\nu-1}(c_1\omega_1+c_2\omega_2+\cdots+c_n\omega_n) < 1, \end{cases}$$

где функции $\xi(\mu)$ суть так называемые показатели числа

$$\mu = c_1\omega_1+c_2\omega_2+\cdots+c_n\omega_n$$

и определяются при помощи системы следующих ν уравнений:

$$(41) \qquad \begin{aligned} &\xi_1\cdot l_{11}+\xi_2\cdot l_{21}+\cdots+\xi_{\nu-1}\cdot l_{\nu-1,1}+\frac{\delta_1}{n}\lg|N(\mu)| = \lambda_1(\mu), \\ &\xi_1\cdot l_{12}+\xi_2\cdot l_{22}+\cdots+\xi_{\nu-1}\cdot l_{\nu-1,2}+\frac{\delta_2}{n}\lg|N(\mu)| = \lambda_2(\mu), \\ &\cdots\cdots\cdots\cdots\cdots\cdots\cdots\cdots\cdots\cdots \\ &\xi_1\cdot l_{1,\nu}+\xi_2\cdot l_{2,\nu}+\cdots+\xi_{\nu-1} l_{\nu-1,\nu}+\frac{\delta_\nu}{n}\lg|N(\mu)| = \lambda_\nu(\mu), \end{aligned}$$

из которых только $\nu-1$ независимы. В них δ_i обозначает единицу, если $\Omega(x_i)$, т. е. i-ая сопряженная с $\Omega(x)$ область, вещественна, и 2, если она комплексна. Символы $\lambda_i(\mu)$ выражают величины $\delta_i \lg|\mu_i|$, где μ_i — i-ая сопряженная с μ величина. Наконец, $l_{1i}, l_{2i}, \ldots l_{\nu-1,i}$ система логарифмов от основных единиц области $\Omega(x_i)$.

Из формул (41) легко заключить, что функции $\xi(\mu)$ однородны и нулевого измерения относительно $c_1, c_2, \ldots c_n$, функция же $N(\mu)$ однородна и n-го измерения. Поэтому, если в неравенствах (39) и (40) мы сделаем подстановку

$$(42) \qquad c_1=\frac{z_1}{\tau},\ c_2=\frac{z_2}{\tau},\ldots\ c_n=\frac{z_n}{\tau},\ t=\tau^{-n},$$

то они примут вид:

$$(43)\qquad |N(z_1\omega_1 + z_2\omega_2 + \cdots + z_n\omega_n)| \leqq 1,$$

$$(44)\qquad \begin{cases} 0 \leqq \xi_1\,(z_1\omega_1 + z_2\omega_2 + \cdots + z_n\omega_n) < 1, \\ 0 \leqq \xi_2\,(z_1\omega_1 + z_2\omega_2 + \cdots + z_n\omega_n) < 1, \\ \dots\dots\dots\dots\dots\dots\dots\dots \\ 0 \leqq \xi_{\nu-1}\,(z_1\omega_1 + z_2\omega_2 + \cdots + z_n\omega_n) < 1, \end{cases}$$

и нахождение $\lim\limits_{t=\infty} \frac{T}{t}$ приведется к нахождению интеграла

$$\iint \cdots \int dz_1 \cdot dz_2 \cdots dz_n,$$

распространенного на n-мерное пространство, ограниченное неравенствами (43) и (44), который, как известно, равен

$$\frac{2^\nu \pi^{n-\nu} R}{N(\mathfrak{m})\sqrt{\pm\Delta}}.$$

Обратимся теперь к ряду

$$(45)\qquad \sum_{\mathfrak{a}}^{(\alpha_1, \ldots \alpha_k)} \frac{1}{N(\mathfrak{a})^s},$$

в котором сумма распространена на все главные идеалы, делящиеся на $\mathfrak{m}$ и, кроме того, лежащие в допустимом комплексе $(\alpha_1, \alpha_2, \ldots \alpha_k)$. Числа, соответствующие этим главным идеалам, и только эти числа, могут быть представлены при помощи v формул (36). Разобьем сумму (45) на v частичных сумм, относя каждый главный идеал $\mathfrak{a}$ к той из формул (36), при помощи которой выражается соответствующее ему приведенное число (приведенное в смысле см. Weber, § 193, S. 708). Конечно, будем все время иметь в виду, что ассоциированные числа лежат в одном и том же комплексе. Поэтому коэффициент при $\frac{1}{s-1}$ в выражении каждой (i-ой) частичной суммы равен

$$(46)\qquad \frac{1}{w} \cdot \lim_{t=\infty} \frac{T_i}{t},$$

где T_i число систем $c_1, c_2, \ldots c_n$, удовлетворяющих неравенствам

$$(47)\qquad \left| N\left(\omega_i^{(\alpha_1, \alpha_2, \ldots \alpha_k)} + Lc_1\omega_1 + Lc_2\omega_2 + \cdots + Lc_n\omega_n\right)\right| \leqq t,$$

$$(48)\qquad \begin{cases} 0 \leqq \xi_1\left(\omega_i^{(\alpha_1, \ldots \ldots \alpha_k)} + Lc_1\omega_1 + Lc_2\omega_2 + \cdots + Lc_n\omega_n\right) < 1, \\ 0 \leqq \xi_2\left(\omega_i^{(\alpha_1, \ldots \alpha_k)} + Lc_1\omega_1 + Lc_2\omega_2 + \cdots + Lc_n\omega_n\right) < 1, \\ \cdots\cdots\cdots\cdots\cdots\cdots\cdots\cdots\cdots \\ 0 \leqq \xi_{\nu-1}\left(\omega_i^{(\alpha_1, \ldots \alpha_k)} + Lc_1\omega_1 + Lc_2\omega_2 + \cdots + Lc_n\omega_n\right) < 1. \end{cases}$$

Вводя подстановку

$$(49)\qquad c_1 = \frac{z_1}{\tau},\ c_2 = \frac{z_2}{\tau}, \cdots\ c_n = \frac{z_n}{\tau},\ t = L^n \cdot \tau^{-n},$$

мы преобразуем неравенства (47) и (48) так:

$$(50)\qquad \left| N\left(\frac{\omega_i^{(\alpha_1, \ldots \alpha_k)}}{L} \cdot \tau + z_1\omega_1 + z_2\omega_2 + \cdots + z_n\omega_n\right)\right| \leqq 1,$$

$$(51)\qquad \begin{cases} 0 \leqq \xi_1\left(\dfrac{\omega_i^{(\alpha_1, \ldots \alpha_k)}}{L} \cdot \tau + z_1\omega_1 + z_2\omega_2 + \cdots + z_n\omega_n\right) < 1, \\ 0 \leqq \xi_2\left(\dfrac{\omega_i^{(\alpha_1, \ldots \alpha_k)}}{L} \cdot \tau + z_1\omega_1 + z_2\omega_2 + \cdots + z_n\omega_n\right) < 1, \\ \cdots\cdots\cdots\cdots\cdots\cdots\cdots\cdots\cdots \\ 0 \leqq \xi_{\nu-1}\left(\dfrac{\omega_i^{(\alpha_1, \ldots \alpha_k)}}{L} \cdot \tau + z_1\omega_1 + z_2\omega_2 + \cdots + z_n\omega_n\right) < 1. \end{cases}$$

Как мы видим, неравенства (50) и (51) отличаются от неравенств (43) и (44) только членами

$$\frac{\omega_i^{(\alpha_1, \ldots \alpha_k)}}{L} \cdot \tau,$$

которые при $t = \infty$, т. е. при $\tau = 0$, обращаются в нуль. Теорема 30 в цитированной статье Hilbert'a (Math. Ann., Bd. 51, S. 53) доказывает, что эти члены не влияют ни на величину интеграла

$$\iint \cdots \int dz_1\, dz_2 \ldots dz_n,$$

ни на оценку остаточного члена. С другой стороны, нетрудно убедиться (см. подстановку (49)), что величина (46) равна

$$\frac{1}{w \cdot L^n} \int\int \cdots \int dz_1\, dz_2 \ldots dz_n,$$

где интеграл имеет прежнее значение. Поэтому каждая из частичных сумм имеет вблизи $s = 1$ выражение

$$\frac{1}{w \cdot L^n} \cdot \frac{2^\nu \cdot \pi^{n-\nu} \cdot R}{N(\mathfrak{m}) \sqrt{\pm \Delta}} \cdot \frac{1}{s-1} + P(s-1).$$

Чтобы получить выражение для всей суммы (45), мы должны сложить v подобных сумм. Вспоминая, что

$$v = \frac{\overline{\varphi}(L)}{F},$$

мы получим:

$$(52) \qquad \sum_{\mathfrak{a}}^{(\alpha_1, \ldots \alpha_k)} \frac{1}{N(\mathfrak{a})^s} = \frac{\overline{\varphi}(L)}{F} \cdot \frac{2^\nu \cdot \pi^{n-\nu} \cdot R}{L^n \cdot w \cdot N(\mathfrak{m}) \sqrt{\pm \Delta}} \cdot \frac{1}{s-1} + P(s-1).$$

Как мы видим, коэффициент при $\frac{1}{s-1}$ в каждой из этих сумм в F раз меньше, чем в сумме (37).

Выберем теперь в каждом из идеальных классов области $\Omega(x)$ по одному идеалу $\mathfrak{m}$, взаимно простому с L. Если $\mathfrak{m}$ для какого-нибудь класса лежит в допустимом комплексе, то все взаимно простые с L идеалы этого класса должны лежать в допустимых комплексах; если же существуют классы, идеалы которых лежат в недопустимых комплексах, то обозначим через $\overline{K}$ группу тех классов, идеалы которых будут лежать в допустимых комплексах; группу же всех классов будем обозначать через K. Пусть порядки этих групп будут соответственно $\overline{h}$ и h.

Введем теперь новое определение:

Определение 12. Комплекс, в котором лежит какой-нибудь идеал области $\Omega(x)$, будем называть *допустимым в широком смысле.*

Рассуждения, подобные примененным нами при доказательстве теоремы 6, убеждают нас в том, что все допустимые в широком смысле комплексы образуют абелеву группу, для которой группа допустимых в узком смысле комплексов является делителем. Обозначим первую группу через $\overline{A}$, вторую через A. Тогда имеет место

Теорема 8. Hölder'овские дополнительные группы $\left(\frac{K}{\overline{K}}\right)$ и $\left(\frac{\bar{A}}{A}\right)$ однозначно (holoëdrisch) изоморфны.

Доказательство. Разложим группу K на сопряженные системы по подгруппе $\overline{K}$:

$$K = \overline{K} + k_2\overline{K} + k_3\overline{K} + \cdots + k_{\mathfrak{y}}\overline{K}. \tag{53}$$

Так как классы $k_2, k_3, \ldots k_{\mathfrak{y}}$ не входят в $\overline{K}$, то идеалы каждого из них лежат в комплексах, не входящих в A. Пусть $a_2, a_3, \ldots a_{\mathfrak{y}}$, будут комплексы, в которых лежат какие-нибудь идеалы соответственно из $k_2, k_3, \ldots k_{\mathfrak{y}}$. Рассмотрим систему

$$A,\ a_2A,\ a_3A, \ldots\ a_{\mathfrak{y}}A. \tag{54}$$

Каждая из систем a_iA исчерпывает все комплексы, в которых лежат какие бы то ни было взаимно простые с L идеалы из k_i. Действительно, пусть идеал $\mathfrak{m}_i$ из k_i лежит в a_i. В каком комплексе лежит какой-нибудь другой идеал $\mathfrak{n}_i$ из k_i? В силу известной теоремы (см., напр., Weber, S. 593, Satz 3) можно найти такой взаимно простой с L идеал $\mathfrak{r}$, чтобы идеал $\mathfrak{m}_i\mathfrak{r}$ был главным. Тогда и идеал $\mathfrak{n}_i\mathfrak{r}$ будет главным. Пусть $\mathfrak{m}_i\mathfrak{r}\sim\mu$ лежит в комплексе b, а $\mathfrak{n}_i\mathfrak{r}\sim\nu$ в комплексе c. Оба комплекса, согласно определению A, должны входить в A. Но $\mathfrak{n}_i\sim\mathfrak{m}_i\frac{\nu}{\mu}$, а потому $\mathfrak{n}_i$ лежит в комплексе a_icb^{-1}, который входит в систему a_iA.

Каждая из систем a_iA исчерпывает все комплексы, в которых лежат какие бы то ни было взаимно простые с L идеалы из $k_i\overline{K}$. Действительно, пусть идеал $\mathfrak{n}_i$ лежит в классе $k_i\bar{k}$, где $\bar{k}$ какой-нибудь класс из $\overline{K}$. Выберем в классе $\bar{k}^{-1}$ идеал $\mathfrak{r}$, лежащий в комплексе b, входящем в A. Идеалы $\mathfrak{m}_i$ и $\mathfrak{n}_i\mathfrak{r}$ оба лежат в классе k_i, т. е. эквивалентны. Предыдущее рассуждение позволяет нам заключить, что идеал $\mathfrak{n}_i\mathfrak{r}$ лежит в комплексе c, входящем в систему a_iA. Значит, идеал $\mathfrak{n}_i$ лежит в комплексе $c.b^{-1}$, который тоже входит в систему a_iA.

С другой стороны, ни одна из систем (54) не может иметь с какой-нибудь другой общих элементов. В самом деле, если бы, например, a_iA и a_jA имели общие элементы, то тогда комплекс $a_ia_j^{-1}$, а с ним и система $a_iA.(a_jA)^{-1}$, входила бы в A, а потому все идеалы из классов системы $k_i\overline{K}(k_j\overline{K})^{-1}$ должны были бы лежать в комплексах из A, т. е. входить в классы группы $\overline{K}$, что противоречит разложению (53).

Таким образом, мы доказали, 1) что системы (54) образуют группу $\overline{A}$, 2) что системы (53) и (54) однозначно изоморфны. Это доказывает, что группы

$$\frac{K}{\overline{\overline{K}}} \text{ и } \frac{\overline{A}}{A}$$

однозначно изоморфны.

Следствие. Порядок группы $\overline{A}$ равен $\overline{F} = F.\mathfrak{y}$.

Возьмем теперь какой-нибудь допустимый в широком смысле комплекс $(-\alpha_1, -\alpha_2, \ldots -\alpha_k)$. Идеалы, лежащие в нем, входят в $\overline{h}$ идеальных классов, образующих одну из систем (53). Выберем в каждом из этих классов по одному взаимно простому с L идеалу $\mathfrak{m}_i$ $(i = 1, 2, \ldots \overline{h})$, и пусть каждый идеал $\mathfrak{m}_i$ лежит в комплексе $(\beta_{1i}, \beta_{2i}, \ldots \beta_{ki})$. Тогда комплексы $(-\alpha_1 - \beta_{1i}, -\alpha_2 - \beta_{2i}, \ldots - \alpha_k - \beta_{ki})$ $(i = 1, 2, \ldots \overline{h})$, а с ними и комплексы $(\alpha_1 + \beta_{1i}, \alpha_2 + \beta_{2i}, \ldots \alpha_k + \beta_{ki})$, будут лежать в A. Поэтому мы можем применить к ним формулу (52), выбирая в роли базиса $(\omega_1, \omega_2, \ldots \omega_n)$ базис идеала $\mathfrak{m}_i$. Таким образом, мы получим:

$$(55)\quad \sum_{\mathfrak{m}_i \mathfrak{a}}^{(\alpha_1+\beta_{1i}, \ldots \alpha_k+\beta_{ki})} \frac{1}{N(\mathfrak{m}_i)^s N(\mathfrak{a})^s} = \frac{1}{F}\cdot\frac{\overline{\varphi}(L)}{L^n}\cdot\frac{2^\nu\cdot\pi^{n-\nu}\cdot R}{w\cdot\sqrt{\pm\Delta}}\cdot\frac{1}{s-1} + P(s-1) \qquad (i = 1, 2, \cdots \overline{h}),$$

где $\mathfrak{a}$ пробегает все идеалы, входящие в противоположный с $\mathfrak{m}_i$ класс и лежащие в комплексе $(\alpha_1, \alpha_2, \ldots \alpha_k)$. Умножим формулу (55) на

$$N(\mathfrak{m}_i)^s = N(\mathfrak{m}_i) + (s-1)\,P(s-1)$$

и просуммируем по i. Получим:

$$(56)\qquad \sum_{\mathfrak{a}}^{(\alpha_1, \alpha_2, \ldots \alpha_k)} \frac{1}{N(\mathfrak{a})^s} = \frac{\overline{h}}{F}\cdot\frac{\overline{\varphi}(L)}{L^n}\cdot\frac{2^\nu\cdot\pi^{n-\nu}\cdot R}{w\cdot\sqrt{\pm\Delta}}\cdot\frac{1}{s-1} + P(s-1),$$

где теперь $\mathfrak{a}$ пробегает идеалы всех классов, противоположных с классами всех $\mathfrak{m}_i$. Но это будут вообще все классы, в которые только могут входить идеалы, лежащие в комплексе $(\alpha_1, \alpha_2, \ldots \alpha_k)$.

Преобразуем формулу (56), положив в ней

$$\frac{\overline{h}}{F} = \frac{\overline{h}\cdot\mathfrak{y}}{F\cdot\mathfrak{y}} = \frac{h}{\overline{F}}.$$

— 228 —

Сравнивая полученную формулу с известной формулой

$$(57) \qquad \sum_{\mathfrak{a}}{}' \frac{1}{N(\mathfrak{a})^s} = h \cdot \frac{\overline{\varphi}(L)}{L^n} \cdot \frac{2^\nu \cdot \pi^{n-\nu} \cdot R}{w \cdot \sqrt{\pm \Delta}} \frac{1}{s-1} + P(s-1).$$

где $\mathfrak{a}$ пробегает все взаимно простые с L идеалы, мы получим:

$$(58) \qquad \sum_{\mathfrak{a}}^{(\alpha_1, \ldots \alpha_k)} \frac{1}{N(\mathfrak{a})^s} = \frac{1}{\overline{\overline{F}}} \sum_{\mathfrak{a}}{}' \frac{1}{N(\mathfrak{a})^s} + P(s-1).$$

Эта формула справедлива для всех допустимых в широком смысле комплексов $(\alpha_1, \alpha_2, \ldots \alpha_k)$.

Для группы $\bar{A}$ возможно найти $\overline{F}$ различных систем *характеров*. Выберем одну из таких систем и введем для каждого характера этой системы обозначение $\chi(\alpha_1, \alpha_2, \ldots \alpha_k)$. Между характерами будет иметь место соотношение

$$(59) \qquad \chi(\alpha_1 + \beta_1, \ldots \alpha_k + \beta_k) = \chi(\alpha_1, \ldots \alpha_k) . \chi(\beta_1, \ldots \beta_k).$$

Далее, каждому взаимно простому с L идеалу $\mathfrak{a}$ можно тоже приурочить характер, беря характер комплекса, в котором лежит идеал $\mathfrak{a}$. Для этих характеров введем обозначение $\chi(\mathfrak{a})$. Для них будет иметь место формула

$$(60) \qquad \chi(\mathfrak{a}) \cdot \chi(\mathfrak{b}) = \chi(\mathfrak{a}\mathfrak{b}),$$

а потому также

$$(61) \qquad \frac{\chi(\mathfrak{a})}{N(\mathfrak{a})^s} \cdot \frac{\chi(\mathfrak{b})}{N(\mathfrak{b})^s} = \frac{\chi(\mathfrak{a}\mathfrak{b})}{N(\mathfrak{a}\mathfrak{b})^s}.$$

Умножим теперь каждую из формул (58) на $\chi(\alpha_1, \alpha_2, \ldots \alpha_k)$ и сложим их. Если мы выбрали систему *главных* характеров, то придем опять к формуле (57). Если же выбранная система характеров не главная, то, в силу известной формулы

$$(62) \qquad \sum_{(\alpha_1, \alpha_2, \ldots \alpha_k)} \chi(\alpha_1, \alpha_2, \ldots \alpha_k) = 0,$$

мы придем к формуле

$$\sum_{\mathfrak{a}} \frac{\chi(\alpha_1, \alpha_2, \ldots \alpha_k)}{N(\mathfrak{a})^s} = P(s-1),$$

ш, пользуясь вторым обозначением характеров:

(63) $$\sum_{\mathfrak{a}} \frac{\chi(\mathfrak{a})}{N(\mathfrak{a})^s} = P(s-1).$$

Эту формулу можно, в силу соотношения (60), преобразовать так:

(64) $$\prod_{\mathfrak{p}} \frac{1}{1-\frac{\chi(\mathfrak{p})}{N(\mathfrak{p})^s}} = P(s-1),$$

где произведение распространяется на все простые идеалы $\mathfrak{p}$ области $\Omega(x)$, отличные от множителей L (ср., напр., Weber, Lehrb. d. Alg. Bd. II, S. 725). Прологарифмируем формулу (64) и возьмем от логарифма вещественную часть (для обозначения вещественных частей от выражений введем символ $R[\ldots]$). При этом могут встретиться два случая: 1) Левая часть (64) имеет при $s=1$ предел, отличный от нуля. Тогда вещественная часть ее логарифма будет тоже типа $P(s-1)$. 2) Левая часть (64) стремится при $s=1$ к нулю. Тогда вещественная часть ее логарифма беспредельно растет по абсолютной величине, оставаясь все время отрицательной. Этот результат запишем следующим образом:

$$R\left[\sum_{\mathfrak{p}} \frac{\chi(\mathfrak{p})}{N(\mathfrak{p})^s} + \frac{1}{2}\sum_{\mathfrak{p}} \frac{\chi(\mathfrak{p})^2}{N(\mathfrak{p})^{2s}} + \frac{1}{3}\sum_{\mathfrak{p}} \frac{\chi(\mathfrak{p})^3}{N(\mathfrak{p})^{3s}} + \cdots\right] \leqq P(s-1).$$

Нетрудно показать (см. Lejeune-Dirichlet, Vorles. üb. Zahlenth, 4-te Aufl., S. 358), что сумма всех сумм, за исключением первой, при $s=1$ стремится к конечной величине. Кроме того, сумма тех членов в первой сумме, которые соответствуют идеалам порядков $\geqq 2$, тоже остается при $s=1$ конечной, так как для таких идеалов $N(\mathfrak{p}) \geqq p^2$, а число идеалов, имеющих одну и ту же норму, не превышает числа n. Перенося все упомянутые члены внутрь символа $P(s-1)$, мы получим:

(65) $$R\left[\sum_{p} \frac{\nu_p\, \chi(p)}{p^s}\right] \leqq P(s-1),$$

где ν_p означает число идеалов 1-го порядка с нормой p, а $\chi(p)$ мы пишем вместо $\chi(\mathfrak{p})$.

Теперь вернемся к обозначению $\chi(\alpha_1, \alpha_2, \ldots \alpha_k)$ и соединим идеалы с одними и теми же характерами. Тогда формула (65) преобразуется так:

(66) $$R\left[\sum_{(\alpha_1,\ldots\alpha_k)} \chi(\alpha_1, \ldots a_k) \sum_{p_{(\alpha_1,\ldots\alpha_k)}} \frac{\nu_p}{p^s_{(\alpha_1,\ldots\alpha_k)}}\right] \leqq P(s-1).$$

Подобным же образом может быть преобразована и формула (57):

$$(67)\qquad \sum_{(\alpha_1,\dots\alpha_k)} \sum_{p_{(\alpha_1,\dots\alpha_k)}} \frac{\nu_p}{p^s_{(\alpha_1,\dots\alpha_k)}} = \lg\frac{1}{s-1} + P(s-1).$$

Складывая формулу (67) со всеми формулами (66), полученными для $\bar{F}-1$ возможных систем не главных характеров, мы, в силу того, что сумма

$$\sum_{\chi} \chi(\alpha_1, \alpha_2, \dots \alpha_k),$$

распространенная на все характеры одного и того же комплекса, равна $\bar{F}$, если $(\alpha_1, \alpha_2, \dots \alpha_k)$ нулевой комплекс, и равна нулю для всех других комплексов, получим:

$$(68)\qquad \bar{F} \sum_{p_{(0,\dots 0)}} \frac{\nu_p}{p^s_{(0,\dots 0)}} \leqq \lg\frac{1}{s-1} + P(s-1).$$

В формуле (68) можно, идя другим путем, констатировать *знак равенства*. Но так как доказательство для общего случая довольно громоздко, то мы проделаем его только для тех случаев, котврые представятся нам необходимыми для дальнейшего.

Этим мы будем заниматься в следующем параграфе. Здесь же отметим, что если в формуле (68) имеет место знак равенства, то он будет иметь место и во всех формулах (66). Действительно, если бы хоть в одной из них имел место знак $<$, то после сложения всех формул (66) и (67) формула (68) получилась бы тоже со знаком $<$.

В случае знака равенства легко получить:

$$(69)\qquad \sum_{p_{(\alpha_1,\dots\alpha_k)}} \frac{\nu_p}{p^s_{(\alpha_1,\dots\alpha_k)}} = \frac{1}{\bar{F}} \lg\frac{1}{s-1} + P(s-1),$$

где сумма распространяется на все простые числа, лежащие в допустимом в широком смысле комплексе $(\alpha_1, \alpha_2, \dots \alpha_k)$ (см. напр., Dirichlet, loc. cit., S. 357), и мы приходим к следующей

Главной теореме. Если в формуле (68) имеет место *знак равенства*, то совокупность простых чисел первого порядка *равномерно* распределяется по всем допустимым в широком смысле комплексам.

Известия Российской Академии Наук. 1923.

(Bulletin de l'Académie des Sciences de Russie).

Определение плотности совокупности простых чисел, принадлежащих к заданному классу подстановок [1].

Н. Чеботарева.

(Представлено академиком Я. В. Успенским в заседании Отделения Физико-Математических Наук 7 марта 1923 г.).

II.

§ IV. О плотностях совокупностей простых чисел, принадлежащих к отдельным классам подстановок.

Даны неприводимое нормальное уравнение n-ой степени

$$f(x) = 0 \tag{70}$$

и подстановка f-го порядка

$$S = (x_{11}, x_{12}, \dots x_{1f})(x_{21}, x_{22}, \dots x_{2f}) \dots (x_{e1}, x_{e2}, \dots x_{ef}) \; (ef = n), \tag{71}$$

входящая в его группу G. Выберем k простых чисел $l_1, l_2, \dots l_k$ вида $fx + 1$, взаимно простых с дискриминантом D уравнения (70) (k произвольное число, которое мы в дальнейшем будем безгранично увеличивать), и образуем при помощи корней уравнений

$$x^{l_1} - 1 = 0, \; x^{l_2} - 1 = 0, \dots \; x^{l_k} - 1 = 0 \tag{72}$$

k циклических уравнений, каждое степени f (например, при помощи Gauss'овых периодов). Дискриминанты этих уравнений суть степени простых чисел $l_1, l_2, \dots l_k$, а потому области, образованные при помощи их корней, взаимно просты между собой и с $\Omega(x)$ (т. е. не имеют, кроме рацио-

[1] См. стр. 205—230.

нальных чисел, общих элементов). Выберем затем из каждого из наших уравнений по одному корню η_ν ($\nu = 1, 2, \ldots k$) и по подстановке U_ν, *производящей* группу каждого из наших уравнений. Далее рассмотрим область $\Omega(x, \eta_1, \eta_2, \ldots \eta_k)$. Ее группа однозначно (holoëdrisch) изоморфна с произведением $G.\ (U_1)^{i_1}.(U_2)^{i_2}\ldots(U_k)^{i_k}$. Не вводя для подстановок области $\Omega(x, \eta_1, \eta_2, \ldots \eta_k)$ новых обозначений, мы будем для них пользоваться символами $S.\ U_1^{\alpha_1}.U_2^{\alpha_2}\ldots U_k^{\alpha_k}$, где подстановка S пробегает группу G, а каждое α_ν ($\nu = 1, 2, \ldots k$) принимает значения $0, 1, 2, \ldots f-1$. Область $\Omega(x, \eta_1, \eta_2, \ldots \eta_k)$ тоже нормальна, и ее порядок равен $n.f^k$.

Рассмотрим теперь совокупность простых чисел, принадлежащих к отделу подстановки S в $\Omega(x)$. Исследуем, как распределяются простые числа этой совокупности по комплексам, образованным областью $\Omega(\eta_1, \eta_2, \ldots \eta_k)$. Число этих комплексов равно f^k, и каждый комплекс $(\alpha_1, \alpha_2, \ldots \alpha_k)$ соответствует подстановке $U_1^{\alpha_1}.U_2^{\alpha_2}\ldots U_k^{\alpha_k}$ в том смысле, что простое число принадлежит к этой подстановке тогда и только тогда, если оно лежит в комплексе $(\alpha_1, \alpha_2, \ldots \alpha_k)$ (см. § II настоящего сочинения, Главная теорема).

В основу предстоящего исследования положим следующую классификацию комплексов:

Определение 13. 1) Те комплексы $(\alpha_1, \alpha_2, \ldots \alpha_k)$, в которых общий наибольший делитель чисел $\alpha_1, \alpha_2, \ldots \alpha_k, f$ равен единице (другими словами, подстановка $U_1^{\alpha_1}.U_2^{\alpha_2}\ldots U_k^{\alpha_k}$ f-го порядка), назовем *первообразными.*

2) Числа $\alpha_1, \alpha_2, \ldots \alpha_k, f$ имеют общий наибольший делитель d $\left(\text{подстановка } U_1^{\alpha_1}.U_2^{\alpha_2}\ldots U_k^{\alpha_k} \text{ порядка } \frac{f}{d}\right)$, причем $1 < d < f$. В этом случае будем называть комплексы $(\alpha_1, \alpha_2 \ldots \alpha_k)$ *особенными.*

3) *Нулевой* комплекс $(0, 0, \ldots 0)$.

Чтобы определить число различных первообразных комплексов, заметим, 1) что число всех комплексов *порядка* d равно d^k, 2) что каждый особенный комплекс $(\alpha_1, \alpha_2, \ldots \alpha_k)$ порядка f можно, деля его индексы на d, преобразовать в первообразный комплекс порядка $\delta = \frac{f}{d}$. В силу всего этого, имеет место формула

$$\sum_{\delta | f} \psi(\delta) = f^k, \tag{73}$$

где мы через $\psi(f)$ обозначаем искомое число комплексов, а сумма в левой части распространяется на все делители δ числа f. Так как эта формула справедлива для всех значений f, то, пользуясь обычным Dedekind'овским приемом обращения, мы получим для $\psi(f)$ следующее выражение:

$$(74)\quad \psi(f) = \sum_{d\,|f} \mu(d)\left(\frac{f}{d}\right)^k = f^k\left(1 - \frac{1}{q_1{}^k}\right)\left(1 - \frac{1}{q_2{}^k}\right)\ldots\left(1 - \frac{1}{q_a{}^k}\right),$$

где $q_1, q_2, \ldots q_a$ обозначают систему всех простых делителей числа f.

Если мы возьмем k настолько большим, чтобы k-ая степень наименьшего из всех простых чисел $q_1, q_2, \ldots q_a$ (назовем его Q) была больше, чем каждый из биномиальных коэффициентов $C_a^{\ 1}, C_a^{\ 2}, \ldots C_a^{\ a-1}$, то легко убедиться в справедливости следующего неравенства:

$$(75)\qquad \psi(f) > f^k\left(1 - \frac{a}{Q^k}\right).$$

Это неравенство играет в дальнейшем весьма важную роль.

Убедимся в равномерности распределения совокупности простых чисел, принадлежащих к отделу подстановки S, по комплексам. Для этого изберем индуктивный путь. Именно, сначала убедимся в этом для совокупности простых чисел, принадлежащих к тождественнной подстановке в $\Omega(x)$. Их общая плотность равна $\frac{1}{n}\cdot$ Введем следующее

Определение 14. Систему комплексов

$$(r\alpha_1, r\alpha_2, \ldots r\alpha_k),$$

где $(\alpha_1, \alpha_2, \ldots \alpha_k)$ какой-нибудь первообразный комплекс, а r пробегает значения $0, 1, 2, \ldots f - 1$, назовем *лучем* $(\alpha_1, \alpha_2, \ldots \alpha_k)$.

Если же r пробегает только взаимно простые с f значения из ряда $0, 1, 2, \ldots f - 1$, то полученную систему комплексов будем называть *первообразной частью* луча $(\alpha_1, \alpha_2, \ldots \alpha_k)$.

Если число (или идеал) лежит в одном из комплексов луча $(\alpha_1, \alpha_2, \ldots \alpha_k)$, то будем говорить, что оно *лежит в луче* $(\alpha_1, \alpha_2 \ldots \alpha_k)$.

Из этого определения вытекают следующие очевидные теоремы:

Теорема 9. Первообразный комплекс не может входить более, чем в один луч.

Теорема 10. Число всех первообразных частей лучей равно

$$\frac{\psi(f)}{\varphi(f)} = f^{k-1} \prod_{i=1}^{a} \frac{\left(1-\frac{1}{q_i^k}\right)}{\left(1-\frac{1}{q_i}\right)}. \tag{76}$$

Теорема 11. Чтобы простое число p одновременно принадлежало к тождественной подстановке в $\Omega(x)$ и лежало в первообразной части луча $(\alpha_1, \alpha_2, \ldots \alpha_k)$, необходимо и достаточно, чтобы оно в области $\Omega(x, \eta_1, \eta_2, \ldots \eta_k)$ принадлежало к отделу подстановки $1 \,.\, U_1^{\alpha_1} U_2^{\alpha_2} \ldots U_k^{\alpha_k}$.

Существование же подобных простых чисел непосредственно следует из результата Frobenius'а. С другой стороны, мы можем рассматривать эти простые числа, как нормы идеалов в $\Omega(x)$, а поэтому по крайней мере один из комплексов, лежащих в первообразной части луча $(\alpha_1, \alpha_2, \ldots \alpha_k)$, допустим в широком смысле. Стало быть, все эти комплексы допустимы, так как все комплексы луча можно рассматривать, как степени какого-нибудь одного первообразного. Поэтому здесь роль $\overline{F}$ играет f^k.

Чтобы констатировать для нашего случая знак равенства в формуле (68), заметим, что плотность совокупности простых чисел, принадлежащих к тождественной подстановке в $\Omega(x)$ и лежащих во всех f^k комплексах, равна $\frac{1}{n}\cdot$ С другой стороны, совокупность простых чисел, принадлежащих к тождественной подстановке в $\Omega(x)$ и одновременно лежащих в нулевом комплексе, имеет плотность

$$\frac{1}{n \cdot f^k},$$

так как эти и только эти простые числа принадлежат к тождественной подстановке в $\Omega(x, \eta_1, \eta_2, \ldots \eta_k)$, а порядок группы этой области равен

$$nf^k.$$

Обе плотности относятся, как $f^k : 1$, т. е. $\overline{F} : 1$, что доказывает, что в формуле (68) должен быть знак равенства. Таким образом, условия, достаточные для равномерности распределения наших простых чисел по комплексам, удовлетворены.

Докажем теперь равномерность распределения по комплексам простых чисел, принадлежащих к отделу подстановки (71) f-го порядка, предполагая, что равномерность доказана для тех ее степеней, порядок которых меньше f. Тогда будет достаточно доказать равномерность распределения простых чисел p, удовлетворяющих тому условию, что неприводимое уравнение

$$F(z) = 0, \tag{77}$$

корень которого принадлежит к циклической группе $(S)^i$, рассматриваемое, как сравнение по модулю p, имеет рациональные корни. Чтобы провести это доказательство, нужно сперва убедиться в том, что все комплексы для уравнения (77) допустимы. Это следует уже из того, что они допустимы для меньшей совокупности простых чисел, принадлежащих в $\Omega(x)$ к тождественной подстановке. Далее мы должны доказать, что в формуле (68) для нашего случая имеет место знак равенства. Для этого образуем неприводимое уравнение

$$\Phi(\zeta) = 0, \tag{78}$$

корень которого ζ принадлежит к группе $(S)^i$ в $\Omega(x, \eta_1, \ldots \eta_k)$. Величину ζ можно рационально выразить через $z, \eta_1, \eta_2, \ldots \eta_k$, и каждой сопряженной с z величине будет соответствовать f^k сопряженных с ζ величин, которые мы получим, если в выражении ζ через $z, \eta_1, \eta_2, \ldots \eta_k$ будем над величинами η_1, η_2, η_k производить всевозможные подстановки типа соответственно $U_1^{\alpha_1}, U_2^{\alpha_2}, \ldots U_k^{\alpha_k}$ $(\alpha_\nu = 0, 1, \ldots f-1)$. Если поэтому простое число p лежит в нулевом комплексе, то каждому рациональному корню сравнения

$$F(z) \equiv 0 \pmod p \tag{79}$$

должны соответствовать f^k различных корней сравнения

$$\Phi(\zeta) \equiv 0 \pmod p. \tag{80}$$

В этом случае величины $\eta_1, \eta_2, \ldots \eta_k$ сравнимы с рациональными числами, а корни сравнения (80) все различны между собой, если только p не входит в дискриминант уравнения (78). С другой стороны, из групповых соображений следует, что z может быть рационально выражено через ζ, а поэтому каждому рациональному корню сравнения (80) соответствует рациональный корень сравнения (79). Если, поэтому, мы применим формулу (6) к уравнению (77) и к уравнению (78), то значения ν_p для простых чисел, лежащих в нулевом комплексе, во втором случае в f^k раз больше, чем в первом, а для других простых чисел все ν_p во втором случае равны нулю. Это доказывает, что в нашем случае в формуле (68) имеет место знак равенства. Таким образом, все условия для равномерности распределения рассматриваемых простых чисел выполнены, и мы заключаем отсюда, что плотность совокупности простых чисел, принадлежащих в $\Omega(x)$ к отделу подстановки S_λ f_λ-го порядка и одновременно лежащих в одном из комплексов, равна

$$\frac{1}{f_\lambda^k}\cdot k_\lambda \frac{n_\lambda}{n},$$

а плотность совокупности простых чисел, принадлежащих в $\Omega(x)$ к этому же отделу и лежащих во всех первообразных комплексах, равна

$$\frac{\psi(f_\lambda)}{f_\lambda^k}\cdot k_\lambda\cdot\frac{n_\lambda}{n}.$$

Теперь исследуем распределение простых чисел нашего отдела по различным его классам. Для этого образуем неприводимое уравнение

$$\Phi(\xi_\lambda) = 0, \tag{81}$$

корень которого ξ_λ принадлежит к циклической группе $\left(S_\lambda\cdot U_1^{\alpha_1}\cdot U_2^{\alpha_2}\cdots U_k^{\alpha_k}\right)^t$ в $\Omega(x, \eta_1, \eta_2, \ldots \eta_k)$, где $(\alpha_1, \alpha_2, \ldots \alpha_k)$ какой-нибудь первообразный комплекс. Тогда мы сможем доказать следующую теорему:

Теорема 12. Из трех утверждений:

1°. Простое число p принадлежит в $\Omega(x)$ к классу подстановки S_λ;

2°. Простое число p лежит в одном из комплексов

$$(r'_\mu\,\alpha_1, r'_\mu\,\alpha_2, \ldots r'_\mu\,\alpha_k) \quad \left(\mu = 1, 2, \ldots t = \frac{\varphi(f_\lambda)}{k_\lambda}\right).$$

Здесь числа $r'_1, r'_2, \ldots r'_t$ суть корни соответственно сравнений

$$r_\mu\cdot x \equiv 1 \pmod{f_\lambda} \quad (\mu = 1, 2, \ldots t), \tag{82}$$

а числа $r_1, r_2, \ldots r_t$ взяты таким образом, чтобы подстановки

$$S_\lambda^{r_1}, S_\lambda^{r_2}, \ldots S_\lambda^{r_t}$$

принадлежали к классу подстановки S_λ (все r_μ взаимно просты с f_λ, так как порядок подстановки $S_\lambda^{r_\mu}$ должен быть равен f_λ).

3°. Сравнение

$$\Phi(\xi_\lambda) \equiv 0 \pmod{p} \tag{83}$$

имеет рациональные корни,—
каждое является следствием двух остальных.

A) 3° выводится из 1° и 2° следующим образом. Корни уравнения (81) рационально выражаются через $x, \eta_1, \eta_2, \ldots \eta_k$. Пусть

$$\xi_\lambda = \varphi(x, \eta_1, \eta_2, \ldots \eta_k).$$

Возвысим это выражение в p-ую степень и станем рассматривать результат по модулю $\mathfrak{p}$, где $\mathfrak{p}$ один из простых идеальных делителей p в $\Omega(x)$. Тогда мы увидим, что x в нем претерпевает подстановку S_λ, в то время как $\eta_1, \eta_2, \ldots \eta_k$, в силу 2°, соответственно подстановки

$$U_1^{r'\mu^{\alpha_1}},\ U_2^{r'\mu^{\alpha_2}}, \ldots U_k^{r'\mu^{\alpha k}}$$

(последние можно рассматривать по модулю p, так как $\Omega(\eta_i)$ суть области деления круга). Рассмотрим теперь тот корень уравнения (81), который принадлежит к группе $\left(S_\lambda^{r_\mu}\, U_1^{\alpha_1}\ U_2^{\alpha_2} \ldots U_k^{\alpha_k}\right)^i$. Из теоремы Schönemann'a следует, что этот корень сравним с собственной p-ой степенью по модулю $\mathfrak{p}$ и, значит, подавно по модулю некоторого простого идеала в $\Omega(x, \eta_1, \eta_2, \ldots \eta_k)$, а поэтому, в силу обобщенной теоремы Fermat'a (см. Weber, S. 618, 3. и 4.), этот корень сравним по этому модулю с рациональным числом, которое должно быть корнем сравнения (83).

B) Предположим теперь, что удовлетворено условие 3°. Пусть p в $\Omega(x)$ принадлежит к классу какой-нибудь подстановки $\overline{S}$, а в $\Omega(\eta_1, \eta_2 \ldots \eta_k)$ к подстановке $U_1^{\beta_1} . U_2^{\beta_2} \ldots U_k^{\beta k}$ (другими словами, p лежит в комплексе $(\beta_1, \beta_2, \ldots \beta_k)$. Тогда, как мы убедились, в $\Omega(x, \eta_1, \eta_2, \ldots \eta_k)$ оно принадлежит к классу подстановки $\overline{S} . U_1^{\beta_1} . U_2^{\beta_2} \ldots U_k^{\beta k}$. Но Frobenius доказал (loc. cit., S. 701), что для существования рациональнаго корня сравнения (83) необходимо и достаточно, чтобы хоть одна из подстановок

$$T\overline{S}T^{-1} . U_1^{\beta_1} . U_2^{\beta_2} \ldots U_k^{\beta_k}$$

входила в группу, к которой принадлежит ξ_λ, т. е. в

$$\left(S_\lambda . U_1^{\alpha_1} . U_2^{\alpha_2} \ldots U_k^{\alpha k}\right)^i.$$

Пусть, таким образом,

$$T\overline{S}T^{-1} . U_1^{\beta_1} . U_2^{\beta_2} \ldots U_k^{\beta k} = S_\lambda^R . U_1^{R\alpha_1} . U_2^{R\alpha_2} \ldots U_k^{R\alpha_k}.$$

где R — одно из чисел ряда $0, 1, 2, \ldots f_\lambda - 1$. Но так как в S_λ и $\bar{S}$ входят только величины x, а в U_ν только η_ν и все величины $x, \eta_1, \eta_2, \ldots \eta_k$ рационально независимы, то должно иметь место:

$$(84) \qquad T\bar{S}T^{-1} = S_\lambda^R, \; U_1^{\beta_1} = U_1^{R\alpha_1}, \; U_2^{\beta_2} = U_2^{R\alpha_2}, \ldots U_k^{\beta_k} = U_k^{R\alpha_k}.$$

Из этого следует, что p лежит в луче $(\alpha_1, \alpha_2, \ldots \alpha_k)$.

C) Если мы еще предположим, что выполнено условие 1°, т. е. что $\bar{S} = S_\lambda$, то тогда в группе G должна находиться такая подстановка T, что $TS_\lambda T^{-1} = S_\lambda^R$. Это показывает, что R является одним из чисел ряда $r_1, r_2, \ldots r_t$. Но ряд $r_1, r_2, \ldots r_t$ может отличаться от ряда $r_1', r_2', \ldots r_t'$ только порядком своих членов. В самом деле, возводя соотношение

$$T_i \, S_\lambda \, T_i^{-1} = S_\lambda^{r_i}$$

в r_i'-ую степень, мы получим

$$T_i \, S_\lambda^{r'_i} \, T_i^{-1} = S_\lambda,$$

или

$$T_i^{-1} S_\lambda \, T_i = S_\lambda^{r'_i},$$

что показывает, что r_i' находится в ряду $r_1, r_2, \ldots r_t$. Таким образом, утверждение 2° выполнено.

D) Из 3° и 2° следует 1°. Действительно, утверждение 2° характеризуют следующие сравнения:

$$\beta_i \equiv r_\mu \cdot \alpha_i \pmod{f_\lambda} \qquad (i=1, 2, \ldots k),$$

которые вместе с сравнениями

$$\beta_i \equiv R\alpha_i \pmod{f_\lambda} \qquad (i=1, 2, \ldots k) \quad \text{(см. B)}$$

дают:

$$(R - r_\mu)\, \alpha_i \equiv 0 \pmod{f_\lambda} \qquad (i=1, 2, \ldots k).$$

Образуем с помощью этих сравнений следующее:

$$(R - r_\mu)\,(X_1 \alpha_1 + X_2 \alpha_2 + \cdots + X_k \alpha_k) \equiv 0 \pmod{f_\lambda}.$$

Подберем $X_1, X_2, \ldots X_k$ так, чтобы $X_1\alpha_1 + X_2\alpha_2 + \cdots + X_k\alpha_k$ было взаимно просто с f_λ, что возможно, в силу первообразности комплекса $(\alpha_1, \alpha_2, \ldots \alpha_k)$. Тогда необходимо, чтобы имело место

$$R \equiv r_\mu \pmod{f_\lambda},$$

откуда $\bar{S} = T^{-1} S_\lambda^{r_\mu} T$, т. е. $\bar{S}$ принадлежит к классу подстановки S_λ. Итак, теорема доказана во всех своих частях.

Теперь докажем равномерность распределения по допустимым комплексам таких простых чисел p, что сравнение (83) имеет рациональные корни. Из предыдущего (см. В) доказательства теоремы (12) ясно, что для этого уравнения могут быть допустимы только f_λ комплексов луча $(\alpha_1, \alpha_2, \ldots \alpha_k)$. С другой стороны, они действительно допустимы, так как группа этих комплексов циклическая, а совокупность первообразных комплексов из луча $(\alpha_1, \alpha_2, \ldots \alpha_k)$ соответствует отделу $S_\lambda . U_1^{\alpha_1} . U_2^{\alpha_2} \ldots U_k^{\alpha_k}$ в $\Omega(x, \eta_1, \eta_2, \ldots \eta_k)$. Далее, в формуле (68) имеет место знак равенства. В самом деле, если p лежит в нулевом комплексе, то левая часть сравнения (83) распадается на линейные множители, а потому $\nu_p = n \cdot f_\lambda^{k-1}$ (ведь, уравнение (81) $n \cdot f_\lambda^{k-1}$-ой степени). С другой стороны, эти и только эти простые числа принадлежат к тождественной подстановке в $\Omega(x, \eta_1, \eta_2, \ldots \eta_k)$, а поэтому плотность их совокупности равна

$$\frac{1}{n \cdot f_\lambda^k}.$$

Таким образом, обе плотности стоят в отношении $f_\lambda : 1$. Но это число играет здесь роль $\bar{F}$. Таким образом, равномерность распределения наших простых чисел по комплексам может считаться доказанной.

Рассмотрим теперь те из наших простых чисел, которые лежат в первообразной части луча $(\alpha_1, \alpha_2, \ldots \alpha_k)$. Если какое-нибудь из них принадлежит к классу S_λ в $\Omega(x)$, то, в силу D), оно лежит в одном из комплексов

$$(r_\mu\ \alpha_i) \quad \begin{pmatrix} \mu = 1, 2, \ldots t = \frac{\varphi(f_\lambda)}{k_\lambda} \\ i = 1, 2, \ldots k \end{pmatrix}$$

(для краткости мы, вместо $(\alpha_1, \alpha_2, \ldots \alpha_k)$, вводим обозначе́ние (α_i)). Нетрудно видеть, что ряд $r_1, r_2, \ldots r_t$ образует группу относительно умножения. Действительно, пусть, например,

— 240 —

$$S_\lambda^{r_i} = T_i S_\lambda T_i^{-1}, \quad S_\lambda^{r_j} = T_j S_\lambda T_j^{-1};$$

тогда

$$S_\lambda^{r_i r_j} = \left(S_\lambda^{r_i}\right)^{r_j} = \left(T_i S_\lambda T_i^{-1}\right)^{r_j} = T_i S_\lambda^{r_j} T_i^{-1} = T_i T_j S_\lambda T_j^{-1} T_i^{-1} =$$
$$= \left(T_i T_j\right) S_\lambda \left(T_i T_j\right)^{-1}.$$

Эта группа является делителем группы всех взаимно простых с f_λ классов по модулю f_λ. Обозначив обе группы соответственно через $\mathfrak{r}$ и $\mathfrak{R}$, мы сможем разбить R на сопряженные системы по r:

$$\mathfrak{R} = \mathfrak{r} + R_2\mathfrak{r} + R_3\mathfrak{r} + R_{k_\lambda}\mathfrak{r}. \tag{85}$$

Таким образом, классы подстановок S_λ, $S_\lambda^{R_2}, \ldots S_\lambda^{R_{k_\lambda}}$ исчерпывают весь отдел S_λ. Приурочим к этим классам значки соответственно 1, 2, 3, . . . k_λ. Вообще около каждого простого числа будем ставить два значка: первый будет обозначать класс, к которому оно принадлежит, второй — комплекс, в котором оно лежит. Для простых чисел типов

$$p_{\nu, \alpha_i}, \quad p_{\nu, r_2 \alpha_i}, \; \ldots \; p_{\nu, r_i \alpha_i}$$

введем общий символ $p_{\nu, \mathfrak{r}\alpha_i}$.

Каждое простое число, принадлежащее к отделу

$$S_\lambda . U_1^{\alpha_1} . U_2^{\alpha_2} \ldots U_k^{\alpha_k} \text{ в } \Omega(x, \eta_1, \eta_2, \ldots \eta_k),$$

должно принадлежать к одному из следующих типов:

$$p_{1, r\alpha_i}, \quad p_{2, R_2 r\alpha_i}, \ldots \; p_{k_\lambda, R_{k_\lambda} r\alpha_i}.$$

Равномерность распределения этих простых чисел по комплексам может быть выражена следующей формулой:

$$\sum_p p_{1, \mathfrak{r}\alpha_i}^{-s} = \sum_p p_{2, R_2 \mathfrak{r}\alpha_i}^{-s} + P(s-1) = \ldots = \sum_p p_{k_\lambda, R_{k_\lambda}\mathfrak{r}\alpha_i}^{-s} +$$

(86)

$$+ P(s-1) = \frac{1}{k_\lambda} \sum_{\mu=1}^{k_\lambda} \sum_p p_{\mu, R_\mu \mathfrak{r}\alpha_i}^{-s} + P(s-1).$$

Возьмем в этой формуле в роли (α_i) каждый из комплексов (α_i), $(R_2\,\alpha_i), \ldots$ $(R_{k_\lambda}\,\alpha_i)$ и сложим все полученные формулы. При этом заметим, 1) что все сопряженные системы (85) после их умножения на R_μ могут только изменить порядок своего расположения, 2) что при этом ни одна из систем не может остаться не измененной. Поэтому сумма, получаемая в правой части новой формулы, распространяется на все такие простые числа, которые принадлежат к отделу S_λ в $\Omega(x)$ и вместе с тем лежат в $k_\lambda . t = \varphi(f_\lambda)$ комплексах $(R_\mu \mathfrak{r} \alpha_i)$, образующих первообразную часть луча (α_i). Но так как простые числа, принадлежащие к отделу S_λ в $\Omega(x)$, равномерно распределены по всем $f_\lambda^{\,k}$ комплексам, то общая плотность рассматриваемой совокупности простых чисел равна

$$\frac{\varphi(f_\lambda)}{f_\lambda^{\,k}} \cdot k_\lambda \cdot \frac{n_\lambda}{n},$$

и мы, таким образом, приходим к формуле:

$$\sum_{\mu=1}^{k_\lambda} \sum_p p_1^{-s},\ R_\mu\,\alpha_i = \frac{\varphi(f_\lambda)}{f_\lambda^{\,k}} \cdot \frac{n_\lambda}{n} \cdot \lg\frac{1}{s-1} + P(s-1), \tag{87}$$

где вся сумма в левой части распространяется на все простые числа, принадлежащие к классу S_λ в $\Omega(x)$ и одновременно лежащие в первообразной части луча (α_i).

Образуем теперь формулы, подобные (87), для всех $\dfrac{\psi(f_\lambda)}{\varphi(f_\lambda)}$ первообразных частей лучей (см. теорему 10), и сложим их. Этим мы еще можем не исчерпать всех простых чисел, принадлежащих к классу S_λ в $\Omega(x)$, так как еще могут существовать принадлежащие к классу S_λ в $\Omega(x)$ простые числа, лежащие вместе с тем в нулевом или в особенных комплексах. Таким образом, выражение в левой части получаемого уравнения

$$\leqq \sum_p p_1^{-s} + P(s-1),$$

откуда вытекает неравенство:

$$\sum_p p_1^{-s} \geqq \frac{\psi(f_\lambda)}{f_\lambda^{\,k}} \cdot \frac{n_\lambda}{n} \cdot \lg\frac{1}{s-1} + P(s-1). \tag{88}$$

Возьмем k настолько большим, чтобы удовлетворялось неравенство (75). Тогда вместо (88) мы можем написать:

$$(89)\qquad \sum_p p_1^{-s} > \left(1-\frac{a}{Q^k}\right)\cdot\frac{n_\lambda}{n}\cdot\lg\frac{1}{s-1} + P(s-1).$$

Эта формула дает нам возможность получить искомый результат. Сперва докажем, что

$$(90)\qquad \lim_{s=1} inf \frac{\sum_p p_1^{-s}}{\lg\frac{1}{s-1}} \geqq \frac{n_\lambda}{n}.$$

В самом деле, если бы имело место

$$\lim_{s=1} inf \frac{\sum_p p_1^{-s}}{\lg\frac{1}{s-1}} = \frac{n_\lambda}{n} - \alpha,$$

где $\alpha > 0$, то взяв

$$k > \frac{\lg n_\lambda - \lg n + \lg a - \lg\alpha}{\lg Q},$$

мы бы имели:

$$(91)\qquad \frac{n_\lambda}{n} - \alpha < \frac{n_\lambda}{n}\left(1-\frac{a}{Q^k}\right).$$

С другой стороны, деля неравенство (88) на $\lg\frac{1}{s-1}$ и беря от обеих частей $\lim\limits_{s=1} inf$, мы получим:

$$(92)\qquad \frac{n_\lambda}{n} - \alpha \geqq \left(1-\frac{a}{Q^k}\right)\frac{n_\lambda}{n}.$$

Несовместимость (91) и (92) доказывает справедливость неравенства (90).

Неравенства, подобные (90), имеют место для каждого из классов нашего отдела. Чтобы *оценить* величину

$$\lim_{s=1} \sup \frac{\sum_p p_1^{-s}}{\lg\frac{1}{s-1}},$$

перепишем формулу (20) таким образом:

$$(93)\quad \frac{\sum_p p_1^{-s}}{\lg\frac{1}{s-1}} = k_\lambda\cdot\frac{n_\lambda}{n} - \left\{\frac{\sum_p p_2^{-s}}{\lg\frac{1}{s-1}} + \frac{\sum_p p_3^{-s}}{\lg\frac{1}{s-1}} + \cdots + \frac{\sum_p p_{k_\lambda}^{-s}}{\lg\frac{1}{s-1}}\right\} + \frac{P(s-1)}{\lg\frac{1}{s-1}}.$$

Взяв от обеих частей $\lim\limits_{s=1}\sup$, мы, в силу очевидного неравенства

$$\lim\inf(\alpha+\beta+\cdots)\geqq\lim\inf\alpha+\lim\inf\beta+\cdots$$

и неравенства (90), примененного ко 2-му, 3-му,... k_λ-му классу нашего отдела, получим:

$$\lim_{s=1}\sup\frac{\sum\limits_{p}p_1^{-s}}{\lg\frac{1}{s-1}}\leqq k_\lambda\frac{n_\lambda}{n}-\Big(\overbrace{\frac{n_\lambda}{n}+\frac{n_\lambda}{n}+\cdots+\frac{n_\lambda}{n}}^{k_\lambda-1\text{ раз}}\Big)=\frac{n_\lambda}{n}. \tag{94}$$

Сопоставим неравенства (90) и (94):

$$\frac{n_\lambda}{n}\leqq\lim_{s=1}\inf\frac{\sum\limits_{p}p_1^{-s}}{\lg\frac{1}{s-1}}\leqq\lim_{s=1}\sup\frac{\sum\limits_{p}p_1^{-s}}{\lg\frac{1}{s-1}}\leqq\frac{n_\lambda}{n}. \tag{95}$$

Эти соотношения могут быть справедливы только тогда, если в них везде будет иметь место знак равенства. Стало быть:

$$\lim_{s=1}\inf\frac{\sum\limits_{p}p_1^{-s}}{\lg\frac{1}{s-1}}=\lim_{s=1}\sup\frac{\sum\limits_{p}p_1^{-s}}{\lg\frac{1}{s-1}}=\lim_{s=1}\frac{\sum\limits_{p}p_1^{-s}}{\lg\frac{1}{s-1}}=\frac{n_\lambda}{n}, \tag{96}$$

и искомая плотность найдена.

§ V. Критерий родственности областей.

В этой главе я исследую связь между *степенью родства* нескольких алгебраических областей и разложением определяющих их уравнений, рассматриваемых, как сравнения по простым модулям. До сих пор были разобраны два следующих частных случая этой задачи:

1) если одна область является делителем другой, то принадлежность простого числа к известному классу подстановок во второй области влечет за собой принадлежность его к вполне определенному классу в первой. В частности, если простое число принадлежит во второй области к тождественной подстановке, то в первой оно тоже

принадлежит к тождественной подстановке. Имеет место также обратная теорема, пользуясь которой, Б. Н. Делоне (loc. cit.) доказал частный случай известной теоремы Kronecker'a-Weber'a о том, что всякая Абелева область есть область деления круга.

2) две области взаимно просты. Тогда, взяв по произволу из обеих областей по классу подстановок, можно найти бесчисленное множество простых чисел, принадлежащих к классам этих подстановок в соответственных областях. Этот случай рассмотрен M. Bauer'ом (loc. cit.).

Рассмотрим самый общий случай. Возьмем две произвольных области ω_1 и ω_2 и составим их *нормы*, т. е. наименьшие заключающие их нормальные области, Ω_1 и Ω_2. Образуем их наименьшую общую область Ω, которая тоже будет нормальна, и пусть ее группа будет G, а группы, которым принадлежат области, соответственно G_1 и G_2. Известно, что G_1 и G_2 являются нормальными делителями G. Их Hölder'овские дополнения

$$H_1 = \frac{G}{G_1} \text{ и } H_2 = \frac{G}{G_2}$$

изоморфны с группами Galois внутри областей соответственно Ω_1 и Ω_2.

Обозначим пересечение областей Ω_1 и Ω_2 через $\overline{\Omega}$. Группа K, к которой оно принадлежит внутри Ω, является наименьшей группой, содержащей группы G_1 и G_2, и потому, в силу нормальности последних, может быть выражена, как их символическое произведение:

$$K = G_1 \cdot G_2.$$

Внутри каждой из областей Ω_1 и Ω_2 оно принадлежит соответственно к группам

$$K_1 = \frac{K}{G_1} \text{ и } K_2 = \frac{K}{G_2}.$$

Индексы (G, K), (H_1, K_1), (H_2, K_2) будут, очевидно, равны одному и тому же числу j.

Разложим теперь группу G на сопряженные системы по K, группу H_1 по K_1, группу H_2 по K_2:

$$G = K + K U_1 + K U_2 + \cdots + K U_{j-1}, \tag{97}$$

$$H_1 = K_1 + K_1 V_1 + K_1 V_2 + \cdots + K_1 V_{j-1}, \tag{98}$$

$$H_2 = K_2 + K_2 W_1 + K_2 W_2 + \cdots + K_2 W_{j-1}. \tag{99}$$

Подстановки V_i и W_i выберем так, чтобы они соответствовали системам соответственно $G_1\ U_i$ и $G_2\ U_i$ (вспомним определение Hölder'овского дополнения).

Ясно, что если простое число p принадлежит в Ω к классу подстановки, входящей в систему KU_i, то в Ω_1 и Ω_2 оно будет принадлежать к классам подстановок, входящих в системы соответственно $K_1 V_i$ и $K_2 W_i$. Докажем, что если две подстановки T_1 и T_2 внутри Ω_1 и Ω_2 будут входить в системы соответственно $K_1 V_i$ и $K_2 W_i$, то можно найти соответствующую им подстановку внутри Ω, входящую в систему KU_i. Достаточно доказать, это для систем K_1, K_2, K. Чтобы найти подстановку из K, соответствующую подстановкам T_1 из K_1 и T_2 из K_2, достаточно найти подстановку, соответствующую T_1 из K_1 и 1 из K_2, а затем найти подстановку, соответствующую 1 из K_1 и T_2 из K_2. Найдем первое. Для этого достаточно показать, что внутри Ω существует подстановка, оставляющая величины из Ω_2 на месте (т. е. входящая в G_2) и производящая над величинами из Ω_1 любую подстановку группы $K_1 = \frac{K}{G_1}$. Допустим противное, т. е. что подстановки из G_2 производят над величинами из Ω_1 не всевозможные подстановки группы K_1, а только некоторые, которые должны образовать группу $\overline{K}_1$, являющуюся делителем группы K_1.

Рассмотрим величины, принадлежащие внутри Ω к группе $\overline{K}_1$. С одной стороны они входят в Ω_1; с другой стороны, не изменяясь в Ω от подстановок группы G_2, они входят в Ω_2; значит, они входят в их пересечение $\overline{\Omega}$. Это же невозможно в том случае, если группа $\overline{K}_1$ является *настоящим* делителем группы K_1. Все это позволяет считать нашу теорему доказанной.

Если Ω_2 целиком входит в Ω_1, то $\Omega_1 = \Omega$, $\Omega_2 = \overline{\Omega}$, $K = G_2$, $K_2 = 1$. Здесь для каждой подстановки в G_1 мы, в силу разложений (97), (98), (99), получим вполне определенную подстановку в G_2. В частности, если мы в G_1 выбрали тождественную подстановку, то в G_2 соответствующая подстановка должна быть тоже тождественной.

Если Ω_2 не входит в Ω_1, то G_2 является *настоящим* делителем K, K_2 содержит не тождественные элементы, а потому существуют простые числа, принадлежащие в Ω_1 к тождественной подстановке, а в Ω_2 к подстановке, отличной от тождественной. Это теорема, противоположная предыдущей; следовательно, справедлива и обратная.

Наконец, если Ω_1 и Ω_2 взаимно просты, т. е. если $\overline{\Omega}$ есть область рациональных чисел, то $K = G$, $K_1 = G_1$, $K_2 = G_2$ $(j = 1)$. В этом случае в обеих областях можно выбрать подстановки совершенно независимо.

Если мы имеем дело с бо̀льшим количеством областей, то исследование придется вести следующим образом: сначала надо исследовать, подобно предыдущему, какие-нибудь две из заданных областей; затем взять их общую область вместе с третьей; зàтем область, заключающую все три области, вместе с четвертой, и т. д. В этом случае более симметричного закона подыскать нельзя. Чтобы убедиться в этом, достаточно рассмотреть области $\Omega(\sqrt{2})$, $\Omega(\sqrt{3})$, $\Omega(\sqrt{6})$. Любые две из этих областей взаимно просты, и потому подстановки в них независимы; когда же мы выбрали в обеих областях по подстановке, то этим мы вполне определяем подстановку в третьей области, так как третья область входит делителем в наименьшую область, заключающую в себе обе остальные области.

Для разрешения в каждом отдельном случае вопроса, существуют ли простые числа, одновременно принадлежащие к заданным классам подстановок в нескольких различных областях, достаточно выяснить, существует ли в группе области, образованной из этих областей, подстановка, которая внутри каждой из заданных областей производит соответствующую заданную подстановку. Вспомним, что каждому простому числу соответствует класс подстановок, а каждому простому идеалу одна определенная подстановка (ср. Frobenius, loc. cit., конец). Этот принцип более общ, так как здесь не играет никакой роли нормальность областей.

§ VI. Теорема Hilbert'а о существовании простых идеалов с заданной вычетностью.

Принцип, высказанный в конце предыдущей главы, позволяет доказать, притом в несколько расширенной формулировке, теорему Hilbert'а о существовании бесчисленного множества простых идеалов с заданной вычетностью (см. Hilbert, Zahlbericht, S. 426, Satz 152).

Дана нормальная область Ω, содержащая в себе l-ые корни из единицы. Пусть α будет целое алгебраическое число из Ω, а $\mathfrak{p}$ простой идеал, не входящий в α. На основании обобщенной теоремы Fermat'a

$$\alpha^{p^f-1} \equiv 1 \pmod{\mathfrak{p}}, \tag{100}$$

где f *порядок* идеала $\mathfrak{p}$. Но так как Ω содержит в себе l-ый корень из единицы $\zeta = e^{\frac{2\pi \cdot i}{l}}$, то внутри области $\omega(\zeta)$ порядок простого идеального делителя p является делителем f, а потому должно иметь место сравнение

$$p^f \equiv 1 \pmod{l},$$

т. е. $p^f - 1$ должно делиться на l.

Из формулы (100) нетрудно получить:

$$\alpha^{\frac{p^f-1}{l}} \equiv \zeta^c \pmod{\mathfrak{p}}, \tag{101}$$

где c одно из чисел $0, 1, 2, \ldots l-1$. Условимся в этом случае в обозначении $\zeta^c = \left\{\frac{\alpha}{\mathfrak{p}}\right\}$ (см. Zahlber, S. 365). Теорема, которую я собираюсь доказать, заключается в следующем:

Даны t целых чисел из Ω:

$$\alpha_1, \alpha_2, \ldots \alpha_t,$$

притом так, что произведение

$$\alpha_1^{m_1} . \alpha_2^{m_2} \ldots \alpha_t^{m_t}$$

может только тогда быть l-ой степенью числа из Ω, если каждое из чисел $m_1, m_2, \ldots m_t$ делится на l. Зададим по произволу t l-ых корней из единицы: $\zeta^{c_1}, \zeta^{c_2}, \ldots \zeta^{c_t}$. Тогда в Ω существует бесчисленное множество таких простых идеалов $\mathfrak{p}$, для которых одновременно имеют место равенства

$$\left\{\frac{\alpha_1}{\mathfrak{p}}\right\} = \zeta^{c_1}, \left\{\frac{\alpha_2}{\mathfrak{p}}\right\} = \zeta^{c_2}, \ldots \left\{\frac{\alpha_t}{\mathfrak{p}}\right\} = \zeta^{c_t}. \tag{102}$$

Ограничимся рассмотрением простых идеалов первого порядка, т. е. таких, для которых $f=1$; иначе говоря, эти идеалы будут в Ω принадлежать к тождественной подстановке.

Присоединим к Ω величины

$$\beta_1 = \sqrt[l]{\alpha_1}, \; \beta_2 = \sqrt[l]{\alpha_2}, \ldots \; \beta_t = \sqrt[l]{\alpha_t}.$$

Считая Ω областью рациональности, мы получим абелеву область, группа которой состоит из подстановок, переводящих величины β_i в $\zeta_i \beta_i$, где ζ_i какой-нибудь корень из единицы. Докажем, что порядок этой области по отношению к Ω равен l^t. Допустим противное, т. е. что ее порядок ниже.

Тогда, если мы станем присоединять к Ω последовательно корни уравнений

$$(103)\qquad z^l - \alpha_1 = 0, \quad z^l - \alpha_2 = 0, \ldots \quad z^l - \alpha_t = 0,$$

то какое-нибудь из этих уравнений должно будет сделаться приводимым в области, образованной из области Ω, посредством присоединения к ней корней предыдущих уравнений. Но в силу нормальности каждого из этих уравнений внутри Ω уравнение должно будет распасться на множители одной и той же степени, которая, благодаря тому, что l простое число, должна быть первой. Пусть первое из таких уравнений будет $z^l - \alpha_\nu = 0$. Тогда мы получим:

$$(104)\qquad \beta_\nu = \varphi(\beta_1, \beta_2, \ldots \beta_{\nu-1}).$$

С другой стороны, область $\Omega(\beta_1, \beta_2, \ldots \beta_{\nu-1})$ должна быть порядка $l^{\nu-1}$ относительно Ω. Обозначим через S_i операцию над $\beta_1, \beta_2, \ldots \beta_{\nu-1}$, заключающуюся в умножении β_i на ζ и оставляющую величины $\beta_1, \ldots \beta_{i-1}$, $\beta_{i+1}, \ldots \beta_{\nu-1}$ без перемены. Группа Galois области $\Omega(\beta_1, \beta_2, \ldots \beta_{\nu-1})$ состоит из подстановок типа $S_1^{\xi_1} S_2^{\xi_2} \ldots S_{\nu-1}^{\xi_{\nu-1}}$. Для того, чтобы эта группа была порядка $l^{\nu-1}$, необходимо, чтобы показатели $\xi_1, \xi_2, \ldots \xi_{\nu-1}$ пробегали независимо друг от друга все значения $0, 1, 2, \ldots l-1$. В частности, все подстановки S_i $(i = 1, 2, \ldots \nu-1)$ должны входить в группу области. Таким образом, величина $\varphi(\beta_1, \ldots \beta_{i-1}, \zeta\beta_i, \beta_{i+1}, \ldots \beta_{\nu-1})$ будет сопряженной с β_ν, т. е. она должна быть равна $\zeta^c . \beta_\nu$, где c какое-нибудь из чисел $0, 1, 2, \ldots l-1$. Соотношение (104) можно переписать следующим образом:

$$\beta_\nu = A_0 + A_1\beta_{\nu-1} + A_2\beta_{\nu-1}^2 + \cdots A_{l-1}\beta_{\nu-1}^{l-1},$$

где $A_0, A_1, A_2, \ldots A_{l-1}$ элементы области $\Omega(\beta_1, \beta_2, \ldots \beta_{\nu-2})$. Тогда

$$\zeta^c\beta_\nu = A_0 + A_1\zeta\beta_{\nu-1} + A_2\zeta^2\beta_{\nu-1}^2 + \cdots + A_{l-1}\cdot\zeta^{l-1}\cdot\beta_{\nu-1}^{l-1}.$$

Далее

$$\zeta^{2c}\beta_\nu = A_0 + A_1\zeta^2\beta_{\nu-1} + A_2\cdot\zeta^4\beta_{\nu-1}^2 + \cdots + A_{l-1}\cdot\zeta^{2(l-1)}\cdot\beta_{\nu-1}^{l-1},$$

. . . .

$$\zeta^{(l-1)c}\beta_\nu = A_0 + A_1\zeta^{l-1}\beta_{\nu-1} + A_2\zeta^{2(l-1)}\cdot\beta_{\nu-1}^2 + \cdots + \\ + A_{l-1}\cdot\zeta^{(l-1)^2}\cdot\beta_{\nu-1}^{l-1}.$$

Из этих равенств легко заключить, что все A, кроме A_c, равны нулю. Продолжая рассуждение подобным же образом применительно к

$$\beta_{\nu-2},\ \beta_{\nu-3}, \ldots \beta_2,\ \beta_1,$$

мы увидим, что наше соотношение должно иметь вид

$$\beta_\nu = A\beta_1^{c_1} \cdot \beta_2^{c_2} \cdots \beta_{\nu-1}^{c_{\nu-1}},$$

где A элемент области Ω. Возведем это соотношение в l-ую степень. Получим:

$$\alpha_\nu = A^l \cdot \alpha_1^{c_1} \cdot \alpha_2^{c_2} \cdots \alpha_{\nu-1}^{c_{\nu-1}},$$

Возможность такого соотношения исключена при формулировке теоремы. Поэтому область $\Omega(\beta_1, \beta_2, \ldots \beta_t)$ должна быть l-го порядка относительно Ω, и ее группа должна состоять из подстановок типа

$$S_1^{\xi_1} \cdot S_2^{\xi_2} \cdots S_t^{\xi_t},$$

где каждая подстановка S_i изменяет β_i в $\zeta\beta_i$ и оставляет величины

$$\beta_1, \ldots \beta_{i-1},\ \beta_{i+1} \ldots \beta_t$$

без перемены, а показатели $\xi_1, \xi_2, \ldots \xi_t$ пробегают независимо друг от друга все значения $0, 1, 2, \ldots l-1$. Эти же подстановки должны входить в группу *нормы* области $\Omega(\beta_1, \beta_2, \ldots \beta_t)$, т. е. наименьшей заключающей ее нормальной области, а потому, в силу результата главы IV, существует бесчисленное множество простых чисел p, принадлежащих к классу подстановки

$$S = S_1^{c_1} S_2^{c_2} \cdots S_t^{c_t}.$$

Эти простые числа внутри Ω разлагаются на идеальные множители первого порядка, так как подстановка S оставляет величины из Ω без перемены. Внутри же нормы области $\Omega(\beta_1, \beta_2, \ldots \beta_t)$ каждое такое простое число p разобьется на простые идеалы таким образом, что хоть один из них, например, π, будет принадлежать к подстановке S. Применяя характеризующее принадлежность к S сравнение к величинам $\beta_1, \beta_2, \ldots \beta_t$, мы получим ряд сравнений

$$(105) \qquad \beta_1^p \equiv \zeta^{c_1}\beta_1,\ \ \beta_2^p \equiv \zeta^{c_2}\beta_2, \ldots\ \beta_t^p \equiv \zeta^{c_t}\beta_t \pmod{\pi}.$$

Разделив эти сравнения соответственно на $\beta_1, \beta_2, \ldots \beta_t$, мы будем иметь:

$$(106) \qquad \beta_1^{p-1} \equiv \zeta^{c_1}, \ \beta_2^{p-1} \equiv \zeta^{c_2}, \ldots \ \beta_t^{p-1} \equiv \zeta^{c_t} \pmod{\pi}.$$

Но так как, в силу того, что $f = 1$, число $p - 1$ делится на l, мы можем переписать сравнения (106) так:

$$(107) \qquad \alpha_1^{\frac{p-1}{l}} \equiv \zeta^{c_1}, \ \alpha_2^{\frac{p-1}{l}} \equiv \zeta^{c_2}, \ldots \ \alpha_t^{\frac{p-1}{l}} \equiv \zeta^{c_t} \pmod{\pi}.$$

В сравнениях (107) фигурируют величины из Ω, а потому эти сравнения будут справедливы и для простого внутри Ω модуля $\mathfrak{p}$, в который π входит множителем.

Далее, из сравнения

$$(108) \qquad \alpha^{\frac{p-1}{l}} \equiv \left\{\frac{\alpha}{\mathfrak{p}}\right\} \pmod{\mathfrak{p}}$$

и сравнений (107) мы получаем:

$$(109) \qquad \left\{\frac{\alpha_1}{\mathfrak{p}}\right\} \equiv \zeta^{c_1}, \ \left\{\frac{\alpha_2}{\mathfrak{p}}\right\} \equiv \zeta^{c_2}, \ldots \left\{\frac{\alpha_t}{\mathfrak{p}}\right\} \equiv \zeta^{c_t} \pmod{\mathfrak{p}}.$$

Эти сравнения должны быть заменены равенствами, так как различные корни из единицы не могут быть сравнимы по модулю $\mathfrak{p}$, не входящему в дискриминант уравнения $\zeta^{l-1} + \zeta^{l-2} + \cdots + \zeta + 1 = 0$. Значит, идеалы $\mathfrak{p}$ удовлетворяют поставленным требованиям, т. е. теорема доказана.

Ivan Matveevich Vinogradov

Born – 14 Sept 1891 in Milolyub, Velikie Luki, Pskov province, Russia
Died – 20 March 1983 in Moscow, Russia

Ivan Matveevich Vinogradov studied at St Petersburg, beginning his studies in 1910. Two of his teachers there, A. A. Markov and Ya V. Uspenskii, both had interests in probability and number theory and Vinogradov's interest in number theory stems from this period. He graduated with his first degree in 1914 and continued his studies, supervised by Uspenskii.

His master's degree was completed in 1915 and he worked on quadratic residues. He generalised results of Voronoy on the Dirichlet divisor problem.

Vinogradov taught at the State University of Perm from 1918 to 1920. The State University of Perm had been founded in 1916. It was called Molotov University for a time, and is now the Gorky State University. He returned to St Petersburg to two posts, one at the polytechnic and the other at the university. He gave a course on number theory at the university which was to be the basis for his famous text on the subject. He was promoted to professor at the university in 1925, becoming head of the probability and number theory section.

From around 1930 he became heavily involved with mathematics administration on a national level but his research work was amazingly unaffected by the heavy workload. He moved to Moscow to become the first director of the Steklov Institute in 1934, a post he held until his death. As an indication of his research activity during this period, it is worth noting that he published around 12 papers in each of the years 1934 to 1938.

The importance of trigonometric sums in the theory of numbers was first shown by Weyl in 1916. In the 1920s the work of Hardy and Littlewood developed Weyl's methods to attack other problems in analytic number theory. However it was Vinogradov who, in a series of papers in the 1930s, brought the method to its full potential.

His methods reached their height in *Some theorems concerning the theory of prime numbers* written in 1937 which provides a partial solution to the Goldbach conjecture. In it Vinogradov proved that every sufficiently large odd integer can be expressed as the sum of three odd primes.

Vinogradov made many other contributions, for example to the theory of distribution of power residues, non-residues, indices and primitive roots. He often returned to the topics of his first research paper on the error term in an asymptotic formula discovered by Gauss.

Vinogradov's influence outside the Soviet Union was soon apparent. Even in Landau's three volume work on number theory, published in 1927, prominence is given to Vinogradov's methods. However he seldom travelled outside the Soviet Union although he did visit St Andrews in 1958 as the leader of the Soviet delegation to the International Mathematical Union. He then went on to the International Congress at Edinburgh.

Always a fit man, and proud of his physical fitness, he remained healthy and active into his early 90s.

Representation of an Odd Number as the Sum of Three Primes

Doklady Akademii Nauk SSSR, **15** no. 6–7 (1937) 291–294.

Simple examples[1] illustrating the application of my method to the theory of primes were given in 1934.

In this paper I shall apply this method to estimate the sum

$$\sum_{p \leqq N} e^{2\pi i \alpha p}.$$

Using this estimate and a new theorem on the distribution of primes in an arithmetical progression[2] (the common difference increasing slowly with the increasing number of terms), I shall derive an asymptotic expression for the number of representations of an odd number $N > 0$ in the form

$$N = p_1 + p_2 + p_3.$$

It directly follows that any sufficiently large odd number can be represented in the form of a sum of three primes. This is the complete solution of the Goldbach problem for odd numbers.

The estimates derived in this paper can be improved further.

Notations. $N > 0$ is a sufficiently large odd number; $n = \ln N$; h, h_1, h_2 are arbitrary large constants > 3; $\tau = Nn^{-3h}$; $\tau_1 = Nn^{-h}$; θ is a real number; $|\theta| \leqq 1$; $A \ll B$ or $A = O(B)$ denotes that the ratio $|A|/B$ does not exceed a certain constant; (d) stands for a sequence of divisors $d \leqq N$ of the product H of all primes $\leqq \sqrt{N}$; (d_0) denotes that part of this sequence which contains all d with an even number of prime divisors, while (d_1) denotes that part which contains all d with an odd number of prime divisors. The sequence (d) can also be divided into two subsequences (d') and (d'') as follows: the first contains the numbers d satisfying the condition that all prime divisors are less than n^{3h} and the second contains all the remaining numbers d.

Accordingly, the sequences (d_0) and (d_1) are divided into subsequences (d_0'), (d_0'') and (d_1'), (d_1''), respectively.

Lemma. Let (x) and (y) denote two increasing sequences of positive integers:

$$1 < U_0 < U_1 \leqq N_1 \leqq N; \quad m = \text{an integer} > 0;$$

[1] On some new problems in the theory of numbers; Dokl. akad. nauk SSSR (1934, vol. 3, no. 1) 1–6. Some theorems in analytic number theory; Dokl. akad. nauk SSSR (1934, vol. 4, no. 4) 185–187.

[2] Walfisz, A.: Zur additiven Zahlentheorie II; Math. Z. **40** (1935) 592–607.

$$\alpha = \frac{a}{q} + \frac{\theta}{q\tau}; \quad (a, q) = 1; \quad 0 < q \leqq \tau; \quad m = m_1 \delta; \quad q = q_1 \delta;$$

$$\delta = (m, q), \quad T = \sum_x \sum_y e^{2\pi i \alpha m x y},$$

where x runs through all the members of (x) satisfying the condition

$$U_0 < x \leqq U_1$$

and y, for a given x, runs through all the members of (y) satisfying the condition

$$0 < y \leqq \frac{N_1}{x}.$$

Then

$$T \ll N_1 n \sqrt{\frac{n}{U_0} + \frac{U_1}{N_1} + \frac{q_1 n}{N_1} + \frac{1}{q_1} + \frac{m_1}{\tau}}.$$

Theorem 1. Let

$$\alpha = \frac{a}{q} + \frac{\theta}{q\tau}; \quad (a, q) = 1; \quad n^{3h} \leqq q \leqq \tau.$$

Then

$$S = \sum_{p \leqq N} e^{2\pi i \alpha p} \ll N n^{2-h}.$$

Proof. We have

$$S = \sum_{(d)} \mu(d) S_d + O(\sqrt{N}); \qquad S_d = \sum_{m=1}^{N/d} e^{2\pi i \alpha m d}. \tag{1}$$

Hence we find

$$\begin{gathered} S = \sum_{d > \tau_1} \mu(d) S_d + O(N n^{-h+1}) = T_0 - T_1 + O(N n^{-h+1}); \\ T_0 = \sum_{(d_0)} S_d; \qquad T_1 = \sum_{(d_1)} S_d, \end{gathered} \tag{2}$$

where d runs through those values which are $> \tau_1$. We shall only estimate T_0, because T_1 can be estimated in the same manner. By changing the order of summation, we obtain

$$T_0 = \sum_m T(m); \qquad T(m) = \sum_d e^{2\pi i \alpha m d}, \tag{3}$$

where m runs through

$$m = 1, \ldots, [n^h],$$

and d, for each given m, runs through the numbers (d_0) satisfying the condition:

$$\tau_1 < d \leqq \frac{N}{m}.$$

Furthermore, we find that

$$T(m) = T''(m) + O\left(\frac{N}{m} n^{-h}\right), \tag{4}$$

where $T''(m)$ contains only those terms in the sum $T(m)$ that correspond to the values of d in the sequence (d_0''). Indeed, the part $T'(m)$ of the sum $T(m)$ which corresponds to the values of d in the sequence (d_0') does not exceed the number of those terms in the sequence (d') which are not greater than N/m. But the order of this number is far less than

$$\frac{N}{m} n^{-h}.$$

If d belongs to (d_0'') and k is the number of prime divisors of d, then

$$k < n.$$

Therefore

$$T''(m) = \sum_{k<n} T_k(m), \tag{5}$$

where $T_k(m)$ contains those terms in $T''(m)$ which contain exactly k prime divisors greater than n^{3h}. Moreover,

$$T_k(m) = \frac{1}{k} T_{k_0}(m) + O\left(\frac{N}{mk} u^{-3h}\right), \tag{6}$$

where

$$T_{k_0}(m) = \sum_u \sum_v e^{2\pi i \alpha m u v}.$$

Here u runs through the primes $\geqq n^{3h}$ belonging to (d), while v runs, for a given u, through the numbers satisfying the condition

$$\frac{\tau_1}{u} < v \leqq \frac{N}{mu}.$$

Applying the lemma to the sum T_{k_0}, we obtain

$$T_{k_0}(m) \ll N \frac{n^{(3/2)-(3h/2)}}{\sqrt{m}}.$$

Thus, from (6), (5), (4), (3) and (2) we obtain Theorem 1.

Theorem 2. The number I_N of the representations of N in the form

$$N = p_1 + p_2 + p_3$$

is given by the expression

$$I_N = RS + O(N^2 n^{-c}),$$

where c is an arbitrarily large constant >3 and

$$S=\sum_{q=1}^{\infty}\frac{\mu(q)}{q_1^3}\sum_{\substack{0\leqq a<q\\(a,q)=1}}e^{2\pi i\frac{a}{q}N};\quad R=\frac{N^2}{2n^3}(1+\lambda);\quad \lim_{N\to\infty}\lambda=0;\quad q_1=\varphi(q).$$

Proof. a) We have

$$I_N=\int_0^1 S_\alpha^3 e^{-2\pi i\alpha N}\cdot d\alpha;\qquad S_\alpha=\sum_{p\leqq N}e^{2\pi i\alpha p}.$$

Divide the range of integration into intervals of two classes as follows:
Class I contains all intervals satisfying the conditions

$$\alpha=\frac{a}{q}+z;\qquad (a,q)=1;\qquad 0<q\leqq n^{3h};\qquad -\frac{1}{\tau}\leqq z\leqq\frac{1}{\tau}.$$

Class II contains all other remaining intervals for which

$$\alpha=\frac{a}{q}+z;\qquad (a,q)=1;\qquad n^{3h}<q\leqq\tau;\qquad |z|\leqq\frac{1}{q\tau}.$$

Accordingly, we have

$$I_N=I_{N_1}+I_{N_2}. \tag{7}$$

b) Applying Theorem 1 we find that

$$I_{N_2}\ll Nn^{2-h}\int_0^1|S_\alpha|^2\,d\alpha\ll Nn^{2-h}\int_0^1\sum_{p\leqq N}\sum_{p_1\leqq N}e^{2\pi i\alpha(p-p_1)}d\alpha$$
$$\ll Nn^{2-h}\frac{N}{n}\ll N^2n^{1-h}.$$

c) It is not difficult to calculate I_{N_1}. It is calculated by the procedure that was used in Waring's problem, but here we have to use a new theorem on the distribution of primes in an arithmetical progression.

If α belongs to an interval of the first class, then we find

$$S_\alpha=\frac{\mu(q)}{q_1}V(z)+O(Nn^{-h_1});\qquad V(z)=\int_2^N\frac{e^{2\pi izx}}{\ln x}dx,$$

Hence that part of I_{N_1} which corresponds to a given fraction a/q is represented as

$$R\frac{\mu(q)}{q_1^3}e^{-2\pi i\frac{a}{q}N}+O(Nn^{-h_2});\qquad R=\int_{-1/\tau}^{1/\tau}[V(z)]^3e^{-2\pi izN}dz, \tag{8}$$

where R can be expressed in the form

$$R=\frac{N^2}{2n^3}(1+\lambda);\qquad \lim_{N\to\infty}\lambda=0.$$

Hence, without any difficulty we find that

$$I_{N_1}=RS+O(N^2n^{-h_3}),$$

and by virtue of (7) and (8) the theorem follows immediately.

Estimates of Certain Simple Trigonometric Sums with Prime Numbers

Izvestiya Akademii Nauk SSSR, Ser. Mat., 3 no. 4 (1939) 371–398.

The most important estimates reported in my recent papers[1] for trigonometric sums with primes that can be evaluated by my methods developed in the years 1937–38 are the best possible results. In some fairly important particular cases, however, the accuracy of my result can be improved further. This paper is devoted to a reconsideration of one such particular case. Here I shall estimate sums of the type

$$\sum_{N-A<p\leqq N} e^{2\pi i \frac{a}{q} p^n}; \qquad (a,q)=1\,. \tag{1}$$

It should be noted that for small q ($q<e^u$, $u=r^\delta$, where $r=\ln N$ and δ is a proper fraction slightly greater than 1/2) and A a quantity of an order close to N, quite exact estimates of the sum (1) follow directly from the theorems on the distribution of primes in arithmetic progressions (Siegel, Esterman and others).

My method gives estimates for (1) close to the limiting value under sufficiently general condition; the main difficulty lies in applying my method to the case where q is small. Nevertheless, as I have already shown[2], my method together with Brun's method gives a possibility of treating the case of small q with success.

At the end of the paper I shall consider a sum more general than (1) for the case of $n=1$. Thus, I shall derive a quite general theorem on the distribution of remainders for the division of primes by a given large number (distribution of primes with respect to a given modulus) with high accuracy.

Notations. The symbolic inequalities $A \ll B$ or $B \gg A$ denote that the ratio $|A|/|B|$ does not exceed a certain constant. The letter p stands for a prime, θ for a number satisfying the condition $|\theta| \leqq 1$. Introduce the symbols:

$$\{x\} = x - [x]; \qquad (x) = \min(\{x\}, 1-\{x\})\,.$$

Lemma 1 (Brun). Let ε, c_0, c be any positive constants <1 and ϱ an arbitrarily small positive constant. Let $N>2$, $r=\ln N$

$$0<q\leqq e^{r^c}; \qquad 0<q_1<e^{r^c}; \qquad (q_1,q)=1; \qquad 0\leqq l<q; \qquad (q,l)=1,$$

---

[1] Some general lemmas and their application to the estimation of trigonometrical sums, Math. sbornik (n.s.) **3(45)** (1938) 435–471. Estimation of certain sums containing primes, Izv. akad. nauk SSSR (ser. mat.) **2** (1938) 399–416.

[2] A new improvement of the method of estimation of trigonometrical sums with primes [in English], Dokl. akad. nauk **22,** no. 2 (1939) 59.

$$Ne^{-r^{c_0}} \ll A < N; \qquad p_0 = e^{r^{1-\varepsilon}},$$

and let D be the product of all primes $\leqq p_0$ which are not factors of qq_1.

Then the number T of numbers $\equiv 1 \pmod q$ coprime to D lying in the interval

$$N - A < qx + l \leqq N,$$

satisfies the inequality

$$T \ll \frac{A(qq_1)^{\varrho}}{r^{1-\varepsilon} q}.$$

Proof.[1] It suffices to take $N > C$, where C is a sufficiently large constant >2. Let δ denote the number of prime divisors $p_1, \ldots, p_\delta$ of D.

$$m = 2[3\ln r - 1],$$

$\Omega(d)$ denotes the number of different prime divisors of d. Then we have

$$T < \sum_{\substack{d \backslash D \\ \Omega(d) \leqq m}} \mu(d) \left(\frac{A}{qd} + \theta_d \right).$$

Hence we find

$$T < \sum_{\substack{d \backslash D \\ \Omega(d) \leqq m}} \mu(d) \frac{A}{qd} + \sum_{h=0}^{m} \binom{\sigma}{h}.$$

But

$$\sum_{d \backslash D} \mu(d) \frac{A}{qd} = \frac{A}{q} \frac{\prod\limits_{p \leqq p_0} \left(1 - \frac{1}{p}\right)}{\prod\limits_{p \backslash qq_1} \left(1 - \frac{1}{p}\right)} \ll \frac{A(qq_1)^{\varrho}}{r^{1-\varepsilon} q}.$$

Now we have

$$\left| \sum_{\substack{d \backslash D \\ m < \Omega(d) \leqq \sigma}} \mu(d) \frac{A}{qd} \right| < \frac{A}{q} \sum_{\substack{d \backslash D \\ m < \Omega(d) \leqq \sigma}} \frac{1}{d} < \frac{A}{q} \sum_{n>m}^{\sigma} S_n,$$

where S_n denotes the n-th elementary symmetric function of

$$\frac{1}{p_1}, \ldots, \frac{1}{p_\sigma}.$$

It is readily seen that

$$\frac{A}{q} \sum_{n>m}^{\sigma} S_n < \frac{A}{q} \sum_{n>m}^{\sigma} \frac{S_1^n}{n!} < \frac{A}{q} \sum_{n>m}^{\sigma} \left(\frac{eS_1}{n}\right)^n \ll \frac{A}{q} \sum_{n>m}^{\sigma} \left(\frac{4\ln r}{m}\right)^n \ll \frac{A}{q} \left(\frac{4}{5}\right)^{6\ln r} \ll \frac{A}{rq}.$$

[1] Brun, V.: Videnskapsselskapets Skrifter I; Mat.-naturv. Kl. **3**. Titchmarsh, E.C.: A divisor problem, Rend. circ. mat. Palermo 54 (1930) 414–524 (Correction) *ibid.* **57** (1932) 478–9. Landau, E.: Vorlesungen über Zahlentheorie, vol. 1, II, ch. 2 (Leipzig, 1927).

Finally,

$$\sum_{h=0}^{m}\binom{\sigma}{h} < \sum_{h=0}^{m}\sigma^h < p_\sigma^m < \frac{A}{rq}\,\frac{rq}{A}\,e^{6r^{1-\epsilon}\ln r} < \frac{A}{rq},$$

which proves our lemma.

Lemma 2. Let η be an arbitrarily small positive constant, n a positive constant, q an integer >0, and let x run through the reduced system of residues modulo q. Then for $(a, q) = 1$, the number of solutions of the congruence

$$x^n \equiv a \pmod q$$

is

$$\ll q^\eta.$$

Proof. Let

$$q = p_1^{\alpha_1} \dots p_k^{\alpha_k}$$

be the canonical expansion of q. The congruence

$$x^n \equiv a \pmod q$$

is equivalent to the system of congruences

$$x^n \equiv a \pmod{p_1^{\alpha_1}},$$

$$\dots\dots\dots\dots$$

$$x^n \equiv a \pmod{p_k^{\alpha_k}}.$$

Each of these congruences has at most n solutions, except possibly one which may have $\leqq 2n$ solutions (the case of $p_1 = 2$). Hence, from

$$n^k \ll q^\eta$$

the lemma follows directly.

Lemma 3. Let ϱ be an arbitrarily small positive constant and n a positive constant,

$$(a, q) = 1; \qquad q > 1;$$

Let $\varphi(x)$ and $\psi(y)$ denote the functions which take non-negative values for the values of x and y under consideration and let

$$\sum_{\substack{0 \leqq x < q \\ (x,q)=1}} (\varphi(x))^2 \leqq X; \qquad \sum_{\substack{0 \leqq y < q \\ (y,q)=1}} (\psi(y))^2 \leqq Y.$$

Then the sum

$$S = \sum_{\substack{0 \leqq x < q \\ (x,q)=1}} \sum_{\substack{0 \leqq y < q \\ (y,q)=1}} \varphi(x)\,\psi(y)\,e^{2\pi i \frac{a}{q} x^n y^n}.$$

obeys the inequality

$$S^2 \ll XYq^{1+\varrho}.$$

Proof. We have

$$S^2 \ll X \sum_x \sum_y \sum_{y_1} \psi(y)\,\psi(y_1)\, e^{2\pi i \frac{a}{q} x^n (y^n - y_1^n)}$$

$$\ll X q^{\frac{\varrho}{3}} \sum_{n=0}^{q-1} \sum_y \sum_{y_1} \psi(y)\,\psi(y_1)\, e^{2\pi i \frac{a}{q} n (y^n - y_1^n)}.$$

First summing over u, we obtain a sum equal to either q or 0, depending on whether $y^n \equiv y_1^n \pmod q$ or not. Hence the lemma follows.

Lemma 4. Let ϱ be an arbitrarily small positive constant, n a positive constant

$$(a, q) = 1; \qquad q > 1$$

and let z run through a reduced system of residues modulo q. Then the sum

$$T = \sum_z e^{2\pi i \frac{a}{q} z^n}$$

obeys the inequality

$$T \ll q^{1/2 + \varrho}.$$

Proof. Evidently

$$T = \frac{1}{\Phi(q)} S; \qquad S = \sum_{\substack{0 \leqq x < q \\ (x,q)=1}} \sum_{\substack{0 \leqq y < q \\ (y,q)=1}} e^{2\pi i \frac{a}{q} x^n y^n},$$

where $\Phi(q)$ is the Euler totient function. According to Lemma 3, we have

$$S \ll q^{3/2 + \varrho/2}.$$

Hence, since

$$\Phi(q) \gg q^{1 - \varrho/2},$$

Lemma 4 follows immediately.

Note. The same estimate will also hold true for a more general sum

$$T_1 = \sum_{0 \leqq z < q} \chi(z)\, e^{2\pi i \frac{a}{q} z^n},$$

where χ is a certain character modulo q. Indeed,

$$T_1 = \frac{1}{\Phi(q)} S_1; \qquad S_1 = \sum_{0 \leqq x < q} \sum_{0 \leqq y < q} \chi(x)\,\chi(y)\, e^{2\pi i \frac{a}{q} x^n y^n}.$$

The sum S_1 is estimated in a similar way to the sum S in Lemma 3. Consequently, we always have

$$T_1 \ll q^{\frac{1}{2} + \varrho}.$$

Theorem 1. Let ε be a positive constant $\leqq 1/3$, η an arbitrarily small positive constant $<\varepsilon$, $N>2$, $r=\ln N$

$$Ne^{-r^{1-2\varepsilon}} \ll A \leqq N, \qquad (a,q)=1, \qquad 0<q\leqq e^{r^\varepsilon},$$

n a positive integer,

$$S=\sum_{N-A<p\leqq N} e^{2\pi i \frac{a}{q} p^n}.$$

Then

$$S \ll \frac{A}{r^{1-4\varepsilon} q^{\frac{1}{2}-\eta}}.$$

Proof. 1°. It suffices to assume that $N \geqq N_0$, where N_0 is a sufficiently large number >2. Let

$$\eta = 3\varrho; \qquad p_0 = e^{r^{1-\varepsilon}}$$

and D denote the product of all primes $\leqq p_0$ which do not divide q. Now let d run through all the divisors of D and P through the numbers coprime to Dq. Then

$$\sum_{N-A<P\leqq N} e^{2\pi i \frac{a}{q} P^n} = \sum_d \mu(d) S_d; \qquad S_d = \sum_{\substack{\frac{N-A}{d}<m\leqq\frac{N}{d} \\ (m,q)=1}} e^{2\pi i \frac{a}{q} d^n m^n}. \tag{1}$$

2°. From the left-hand side, omit all those terms where P is divisible by the square of an integer >1. The error arising from such an elimination is

$$\ll \sum_{p_0<P\leqq\sqrt{N}} \left(\frac{A}{P^2}+1\right) \ll \frac{A}{p_0} + \sqrt{N} \ll \frac{A}{r\sqrt{q}}.$$

3°. Now let us estimate the error Δ that arises if those terms which correspond to

$$d > N^{3/4}.$$

are omitted from the right-hand side. Let such a d have k prime divisors. Then, we find

$$p_0^k \geqq N^{3/4}; \qquad k \geqq \tfrac{3}{4} r^\varepsilon.$$

But

$$\Delta \leqq \sum_{N^{3/4}<d\leqq N} \sum_{\frac{N-A}{d}<m\leqq\frac{N}{d}} 1 \leqq \sum_{0<m\leqq N^{1/4}} \sum_{\frac{N-A}{m}<d\leqq\frac{N}{m}} 1.$$

Now from a well-known formula, it follows that

$$\tau\left[\frac{N-A}{m}+1\right]+\ldots+\tau\left[\frac{N}{m}\right] \ll \frac{A}{m} r.$$

Therefore

$$2^{\frac{3}{4}r^\varepsilon} \sum_{\frac{N-A}{m} < d \leq \frac{N}{m}} 1 \ll \frac{A}{m} r; \qquad \sum_{\frac{N-A}{m} < d \leq \frac{N}{m}} 1 \ll \frac{A}{m} r \cdot 2^{-\frac{3}{4}r^\varepsilon},$$

$$\Delta \ll Ar^2 \cdot 2^{-\frac{3}{4}r^\varepsilon} \ll \frac{A}{r\sqrt{q}}.$$

4°. By the results of 2° and 3°, Eq. (1) takes the form

$$\sum_{N-A<Q\leq N} e^{2\pi i \frac{a}{q} Q^n} = \sum_{0<d\leq N^{3/4}} \mu(d) S_d + O\left(\frac{A}{r\sqrt{q}}\right),$$

$$S_d = \sum_{\substack{\frac{N-A}{d} < m \leq \frac{N}{d} \\ (m,q)=1}} e^{2\pi i \frac{a}{q} d^n m^n}. \tag{2}$$

Here Q in the left-hand side runs through only those numbers which are not divisible by the square of an integer >1 and coprime to Dq.

5°. Let us now estimate one of the sums S_d in (2). We have

$$S_d = \left[\frac{A}{qd}\right] T + O(q),$$

where T does not depend on d, and by Lemma 4,

$$T \ll q^{\frac{1}{2}+\varrho}.$$

Therefore

$$\sum_{0<d\leq N^{3/4}} \mu(d) S_d = T \sum_{0<d\leq N^{3/4}} \mu(d) \left[\frac{A}{qd}\right] + O(N^{3/4} q).$$

Arguing as in 3°, we find

$$\left| T \sum_{N^{3/4}<d\leq \frac{A}{q}} \mu(d) \left[\frac{A}{qd}\right] \right| < |T| \sum_{N^{3/4}<d\leq \frac{A}{q}} \sum_{0<m\leq \frac{A}{qd}} 1 \ll \frac{A}{r\sqrt{q}}.$$

Consequently,

$$\sum_{0<d\leq N^{3/4}} \mu(d) S_d \ll q^{\frac{1}{2}+\varrho} Z + O\left(\frac{A}{r\sqrt{q}}\right); \qquad Z = \sum_{0<d\leq \frac{A}{q}} \mu(d) \left[\frac{A}{qd}\right].$$

Obviously, Z is the number of those numbers which are coprime to D and are not greater than A/q. Therefore, by Lemma 1 (with 1 in place of q, q in place of q_1, A/q in place of N and A, and $\ln A/q$ in place of r), we have

$$Z \ll \frac{Aq^\varrho}{qr^{1-\varepsilon}}.$$

Therefore Eq. (2) can be rewritten as

$$\sum_{N-A<Q\leqq N} e^{2\pi i \frac{a}{q} Q^n} \ll \frac{A}{r^{1-\varepsilon} q^{\frac{1}{2}-2\varrho}}. \tag{3}$$

6°. Consider the left-hand side of (3). Let k be the number of prime divisors of some Q. Then we have

$$p_0^k \leqq N; \qquad k \leqq r^{\varepsilon}.$$

Consequently, Eq. (3) can be rewritten as

$$\sum_{0<k\leqq r^{\varepsilon}} \Omega_k \ll \frac{A}{r^{1-\varepsilon} q^{\frac{1}{2}-2\varrho}}, \tag{4}$$

where Ω_k is the sum of the terms on the left-hand side of (3) which correspond to the values of Q having exactly k prime divisors.

7°. Now we shall estimate the value of a term Ω_k, assuming that $k > 1$. First we shall estimate the sum

$$T_k = \sum_u \sum_v e^{2\pi i \frac{a}{q} u^n v^n},$$

where u runs through such primes as satisfy the condition

$$p_0 < u \leqq N p_0^{-k+1}$$

and v, for a given u, runs through Q's in the interval

$$\frac{N-A}{u} < v \leqq \frac{N}{u},$$

which are not divisible by the square of an integer >1 and have exactly $(k-1)$ prime factors $>p_0$ and $\leqq \sqrt{N}$.

We shall divide the interval

$$p_0 < u \leqq N p_0^{-k+1}$$

into $\ll r$ subintervals of the type

$$U < u \leqq U_0; \qquad 2U \leqq U_0 \leqq 4U$$

and then assuming that

$$H_0 = e^{(\ln U)^{1-\varepsilon}},$$

we shall divide each subinterval into $\ll H_0$ still smaller subintervals of the type

$$u_0 < u \leqq u'; \qquad U H_0^{-1} < u' - u_0 \leqq 2 U H_0^{-1}.$$

Let

$$W = \sum_{u_0 < u \leqq u'} \sum_{\frac{N-A}{u} < v \leqq \frac{N}{u}} e^{2\pi i \frac{a}{q} u^n v^n}.$$

Then we have

$$\frac{N-A}{u}=\frac{N-A}{u_0}+O\left(\frac{N}{U}H_0^{-1}\right); \qquad \frac{N}{u}=\frac{N}{u_0}+O\left(\frac{N}{U}H_0^{-1}\right).$$

Therefore the sum W will differ from the sum

$$W_1=\sum_{u_0\leqq u\leqq u'}\ \sum_{\frac{N-A}{u_0}<v\leqq\frac{N}{u_0}} e^{2\pi i\frac{a}{q}u^n v^n}$$

by a quantity of the order of

$$\frac{N}{U}H_0^{-1}UH_0^{-1}=NH_0^{-2}.$$

The sum of all such errors for all W into which T_k is divided will be

$$\ll N\sum_U H_0^{-1}\ll Ae^{r^{1-2\varepsilon}-r^{(1-\varepsilon)^2}}r\ll\frac{A}{rq}.$$

8°. Let l be a number in the sequence $0, 1, \ldots, (q-1)$ such that $(l, q)=1$. Applying Lemma 1, we shall estimate the number of u's of the form $(qx+l)$ in the interval

$$u_0<u\leqq u'.$$

Here we shall put $q_1=1$. In place of N, now we have to take u'. Consequently, in place of r we have to take $\ln u'$. For ε we may take the same ε as in the theorem that is being proved. For A we take $u'-u_0$ which is

$$\gg u'e^{-(\ln u')^{1-\varepsilon}}.$$

Then the conditions of the lemma are satisfied and we have

$$B\ll\frac{u'-u_0}{r^{(1-\varepsilon)^2}q^{1-\varrho}}\ll M; \qquad M=\frac{UH_0^{-1}}{r^{(1-\varepsilon)^2}q^{1-\varrho}}.$$

9°. Now, applying Lemma 1, we shall also estimate the number B_1 of numbers $v\equiv l \pmod q$ which are coprime to D and lie in the interval

$$\frac{N-A}{u_0}<v\leqq\frac{N}{u_0}.$$

Here instead of the N used in Lemma 1, we have to take N/u_0, and instead of r we have to take $\ln N/u_0$, and finally in place of A we have to take A/u_0. Evidently, then

$$r^{1-\varepsilon}\leqq\ln\frac{N}{u_0}\leqq r-r^{1-\varepsilon}.$$

Therefore we obtain

$$B_1\ll M_1; \qquad M_1=\frac{A}{Ur^{(1-\varepsilon)^2}q^{1-\varrho}}.$$

10°. We can now express W_1 in the form

$$W_1=\sum_l\sum_{l_1}e^{2\pi i\frac{a}{q}l^n l_1^n},$$

where l and l_1 run, independently of each other, through those numbers from $0, 1, \ldots, (q-1)$ which are coprime to q. Moreover, l takes each value $\ll M$ times and l_1 takes each value $\ll M_1$ times. Therefore, according to Lemma 3, we have

$$W_1 \ll MM_1 q^{\frac{3}{2}+\varrho} \ll \frac{AH_0^{-1}}{r^{2-4\varepsilon+2\varepsilon^2} q^{\frac{1}{2}-3\varrho}}.$$

At the same time, from 7°, we find that

$$T_k \ll \frac{A}{r^{1-4\varepsilon+2\varepsilon^2} q^{\frac{1}{2}-3\varrho}}.$$

11°. Evidently, every term

$$e^{2\pi i \frac{a}{q} Q^n}$$

of the sum Ω_k is also contained in the sum T_k as well, and each term occurs exactly k times. Moreover, the sum T_k may also contain terms of the type

$$e^{2\pi i \frac{a}{q} u^{2n} v_1^n}.$$

But the number of such terms is

$$\ll \sum_{p_0 < u \leq \sqrt{N}} \left(\frac{A}{u^2} + 1\right) \ll \frac{A}{p_0}.$$

Therefore, for $k > 1$,

$$\Omega_k = \frac{1}{k} T_k + O\left(\frac{1}{k}\frac{A}{p_0}\right) \ll \frac{A}{kr^{1-4\varepsilon+2\varepsilon^2} q^{\frac{1}{2}-3\varrho}}.$$

Hence from (4) our theorem follows immediately.

Note. Looking for asymptotic formulae in additive problems with prime numbers (I mean the modified method of Hardy and Littlewood used in my papers and the problems which have already been studied) reduces to solving two problems, viz. (1) estimation of an integral on large arcs (small q) and (2) estimation of an integral on small arcs (large q).

The methods used to solve the first problem, based on the theory of L-series, which are equally suitable both for mixed problems (the arguments of all terms are not necessarily primes) as well as for pure problems (arguments of all terms are primes), have been developed in full, just before the publication of my estimates for trigonometric sums with primes. These are the method based on Siegel's lemma and Esterman's method, which gives a much weaker order for the remainder term, not based on Siegel's lemma. Both the methods are based on certain results of Titchmarch and Page.

There is no need to use my estimates in mixed problems. My estimates are however essential in solving all pure problems.

Here it should be mentioned that to a considerable degree, my method can replace the theory of L-series in solving the first problem. For instance, using Theorem 1, we can easily derive my asymptotic expression for the number of representations of an integer N in the form

$$N = p_1^n + \ldots + p_s^n; \qquad n > 1; \qquad s \geqq s_0$$

with the help of the usual formula for $\pi(x; q, l)$ proved only for the case $q < r^\varepsilon (r = \ln N)$, where ε is an arbitrarily small positive constant. If only Theorem 1 and Page's results are used, the remainder terms obtained in all the problems are almost as accurate as those derived with the help of the Esterman method. Thus, in my formula the remainder term is of the order of

$$N^{s\nu - 1} r^{-s+4+\varepsilon}; \qquad \nu = \frac{1}{n},$$

i.e. it is of the same order as that which is obtained if the Esterman method is used correctly.

Lemma 5. Let N be an integer $> N_0$, where N_0 is a sufficiently large constant >1, $r = \ln N$, D denote the product of all primes $\leqq \sqrt{N}$, h a positive number $\leqq 1/6$, $\lambda = 1 + h$. Then all the divisors of D which are less than or equal to N can be divided into

$$< r^{\ln r/\ln \lambda}$$

classes satisfying the condition so that for each of these classes there exists a positive integer τ and 3τ numbers $\gamma_1, \ldots, \gamma_\tau; G_1, \ldots, G_\tau; l_1, \ldots, l_\tau$ having the following properties:

$$G_1 \geqq G_2 \geqq \ldots \geqq G_\tau; \qquad \gamma_s^\lambda = G_s; \qquad l_s \geqq 0;$$

Every d belonging to a given class can be represented in the form of a product

$$d = g_1 \ldots g_\tau,$$

where g_s is the product of l_s primes p satisfying the condition

$$\gamma_s \leqq p^{l_s} < G_s,$$

so that

$$\gamma_s \leqq g_s < G_s.$$

Proof. 1°. We shall divide all the prime divisors of D into groups as follows. Let τ denote the greatest integer satisfying the condition

$$2^{\lambda^{\tau-1}} \leqq \sqrt{N}.$$

Then we obtain τ groups of primes, if we include into the s-th group $(s = 1, 2, \ldots, \tau)$ the numbers p satisfying the condition:

$$2^{\lambda^{s-1}} \leqq p < 2^{\lambda^s}.$$

2°. By the definition of τ, it follows that

$$\tau < \frac{\ln r}{\ln \lambda}.$$

3°. For the number ν of prime factors of some $d \leqq N$, the upper bound is

$$\nu \leqq r - 1 .$$

4°. We shall divide the values of d under consideration into classes as follows: the values of d with different number of factors belonging to the same group shall be put into the same class. By virtue of 2° and 3°, the number of all such classes is

$$< r^{\ln r/\ln \lambda} .$$

5°. Let f_s denote the product of prime factors of the s-th group which occur in the number d of a chosen class, and let l_s denote the number of such factors ($f_s = 1$, $l_s = 0$, if there are no such factors). Thus we have

$$d = f_1 \ldots f_\tau .$$

Putting

$$\varphi_s = 2^{\lambda^{s-1} l_s}; \qquad F_s = 2^{\lambda^s l_s},$$

we obtain

$$\varphi_s \leqq f_s < F_s .$$

The following numbers

$$\varphi_1, \ldots, \varphi_\tau,$$
$$f_1, \ldots, f_\tau,$$
$$F_1, \ldots, F_\tau,$$

on being rearranged in such a way the F_s form a non-increasing sequence, shall be denoted by the symbols:

$$\gamma_1, \ldots, \gamma_\tau,$$
$$g_1, \ldots, g_\tau,$$
$$G_1, \ldots, G_\tau .$$

Accordingly, we shall change the numbers of the groups of p's as well. Thus, the proof of the lemma is complete.

Lemma 6. Let (u) and (v) be two sequences of positive integers coprime to q, $N \geqq N_0$, where N_0 is a sufficiently large constant >1, $r = \ln N$ and

$$1 < U_1 < U_2 \leqq N; \qquad 1 \leqq A \leqq N,$$
$$(a, q) = 1; \qquad 1 \leqq q \leqq N,$$

n is an integer $\geqq 1$,

$$S = \sum_u \varphi(u) \sum_v e^{2\pi i \frac{a}{q} u^n v^n},$$

where u runs through the numbers of the sequence (u) satisfying the condition

$$U_1 < u \leqq U_2$$

and v runs through, for a given u, the numbers of the sequence (v) satisfying the condition

$$\frac{N-A}{u} < v \leqq \frac{N}{u}.$$

Now let $\varphi(u)$ denote a function which for all u in the interval

$$U_1 < u \leqq U_2$$

take non-negative values, and which, for any u_1 and u_2 satisfying the condition

$$U_1 \leqq u_1 < u_2 \leqq U_2,$$

obeys the inequality

$$\sum_{u_1 < u \leqq u_2} (\varphi(u))^s \ll u_2 F_s.$$

Finally, let ϱ be an arbitrarily small positive constant. Then

$$S \ll Aq^{\varrho - \frac{1}{2}} \sqrt{\left(1 + \frac{q}{U_1}\right)\left(1 + \frac{U_2 q}{A}\right)\left(1 + \frac{U_2}{A}\right)}\, rF_2^{1/2} + \left(\frac{Nq}{U_1} + U_2\right) rF_1.$$

Proof. 1°. Divide the interval

$$U_1 < u \leqq U_2$$

into $\ll r$ subintervals of the type:

$$U < u U_0; \qquad U_0 \leqq 2U.$$

2°. Consider that part S_0 of the sum S which corresponds to one such subinterval. After dividing this subinterval into $\ll (U/q + 1)$ still smaller subintervals of the type

$$u_0 < u \leqq u'; \qquad u' - u_0 \leqq q,$$

we obtain

$$S_0 = \sum\nolimits_0 S_1; \qquad S_1 = \sum_{u_0 < u \leqq u'} \varphi(u) \sum_{\frac{N-A}{u} < v \leqq \frac{N}{u}} e^{2\pi i \frac{a}{q} u^n v^n}.$$

The sum S_1 is approximately equal to

$$S_2 = \sum_{u_0 < u \leqq u'} \varphi(u) \sum_{\frac{N-A}{u_0} < v < \frac{N}{u_0}} e^{2\pi i \frac{a}{q} u^n v^n}.$$

Since

$$\frac{N-A}{u} = \frac{N-A}{u_0} + O\left(\frac{Nq}{U^2}\right); \qquad \frac{N}{u} = \frac{N}{u_0} + O\left(\frac{Nq}{U^2}\right),$$

we obtain

$$S_1 - S_2 \ll \left(\frac{Nq^{3/2}}{U^2} + 1\right) \sum_{u_0 < u \leqq u'} \varphi(u).$$

Therefore

$$\sum\nolimits_0 S_1 - \sum\nolimits_0 S_2 \ll \left(\frac{Nq}{U^2} + 1\right) \sum_{U_1 < u \leqq U_0} \varphi(u) \ll \left(\frac{Nq}{U} + U\right) F_1 .$$

Now we have

$$\left(\sum\nolimits_0 S_2\right)^2 \ll \left(\frac{U}{q} + 1\right) \sum\nolimits_0 |S_2|^2 .$$

But each sum S_2 can be divided into

$$\ll \frac{A}{Uq} + 1$$

subsums of the type

$$S_3 = \sum_{u_0 < u \leqq u'} \varphi(u) \sum_{v_0 < v \leqq v'} e^{2\pi i \frac{a}{q} u^n v^n}; \qquad v' - v_0 \leqq q ,$$

where

$$S_2^2 \ll \left(\frac{A}{Uq} + 1\right) \sum |S_3|^2 .$$

Summation is taken over all subsums S_3. Applying Lemma 3 to each subsum S_3, we obtain

$$S_3^2 \ll q^{1+\varrho} \sum_{u_0 < u \leqq u'} (\varphi(u))^2 \sum_{v_0 < v \leqq v} 1 ,$$

$$S_2^2 \ll q^{1+\varrho} \left(\frac{A}{Uq} + 1\right) \sum_{u_0 < u \leqq u'} (\varphi(u))^2 \left(\frac{A}{U} + 1\right).$$

At the same time, we have

$$\left(\sum\nolimits_0 S_2\right)^2 \ll \left(\frac{U}{q} + 1\right) U F_2 \left(\frac{A}{Uq} + 1\right)\left(\frac{A}{U} + 1\right) q^{1+\varrho}$$

$$= A^2 q^{\varrho - 1} \left(1 + \frac{q}{U}\right)\left(1 + \frac{Uq}{A}\right)\left(1 + \frac{U}{A}\right) F_2 ,$$

$$S_0 \ll A q^{\varrho/2 - 1/2} \sqrt{\left(1 + \frac{q}{U}\right)\left(1 + \frac{Uq}{A}\right)\left(1 + \frac{U}{A}\right)}\, F_2^{1/2} + \left(\frac{Nq}{U} + U\right) F_1 .$$

Hence, Lemma 6 follows immediately.

Theorem 2. Let ε, η and h be arbitrarily small positive constants <1, $N > 2$, $r = \ln N$ $(a, q) = 1$, $e^{r^\varepsilon} < q$, $N^{2/3+h} q^{3/2} \leqq A \leqq N$, and n a positive integer:

$$S = \sum_{N - A < p \leqq N} e^{2\pi i \frac{a}{q} p^n} .$$

Then

$$S \ll A q^{\eta - 1/2} .$$

Proof.[1] 1°. Let $\eta = 2\varrho$. Then we have

$$S = \sum_d \sum_m \mu(d)\, e^{2\pi i \frac{a}{q} d^n m^n} + O(\sqrt{N}),$$

where d runs through the divisors of the product of all primes at most $\sqrt{N}$ that are not contained in q, while m runs through the natural numbers that are coprime to q, and the summation is taken over the range

$$N - A < dm \leqq N.$$

Then we find

$$\sum_d \sum_m \mu(d)\, e^{2\pi i d^n m^n} = T_0 - T_1,$$

where T_0 is the sum of terms which correspond to the values of d with even number of prime divisors, while T_1 is the sum of the terms which correspond to the values of d with an odd number of prime divisors. Now consider one of the sums T_0 or T_1, and denote it by T_2.

2°. Divide the numbers $d \leqq N$ into classes satisfying the conditions stated in Lemma 5 (the notations being as in this lemma).

3°. Divide the numbers m into $\ll r$ classes. Put all numbers m in a certain interval of the type

$$M < m \leqq M_0; \qquad M_0 \leqq 2M.$$

into one class.

4°. Denote by T_3 that part of the sum T_2, which corresponds to the chosen classes of d and m.

5°. Let

$$MG_1 \ldots G_\tau > N^{1/3},$$

where β is the first term in the sequence $1, 2, \ldots, \tau$ that satisfies the condition

$$MG_1 \ldots G_\beta > N^{1/3}.$$

6°. Let $\beta > 1$. Then, putting

$$u = mg_1 \ldots g_\beta; \qquad v = g_{\beta+1} \ldots g_\tau,$$

we obtain

$$MG_1 \ldots G_{\beta-1} \leqq N^{1/3}; \qquad G_\beta \leqq G_{\beta-1}; \qquad MG_1 \ldots G_\beta \leqq N^{2/3}.$$

Therefore for the case under consideration

$$T_3 = \sum_u \varphi(u) \sum_v e^{2\pi i \frac{a}{q} u^n v^n},$$

[1] An improved estimate for a trigonometrical sum containing primes; Izv. akad. nauk SSSR (ser. mat.) **2** (1938) 1–24.

where u runs through integers satisfying the condition

$$N^{1/3\lambda} < u \leqq 2N^{2/3},$$

while v, for a given u, runs through integers satisfying the condition

$$\frac{N-A}{u} < v \leqq \frac{N}{u}.$$

Here $\varphi(u)$ does not exceed the number $\tau(u)$ of divisors of u. Consequently, the numbers F_1 and F_2 stated in Lemma 6 will satisfy the conditions:

$$F_1 \ll r; \qquad F_2 \ll r^3.$$

Applying Lemma 6, we obtain

$$T_3 \ll Aq^{\varrho-1/2}\sqrt{\left(1+\frac{q}{N^{1/3\lambda}}\right)\left(1+\frac{N^{2/3}q}{A}\right)}\, r^{5/2} + (N^{1-1/3\lambda}q + N^{2/3})\, r^2$$
$$\ll Ar^{5/2}q^{\varrho-1/2}.$$

7°. Let $\beta = 1$. For the sake of brevity, assuming that $l_1 = \chi$ and denoting by χ_1 the least integer that satisfies the condition

$$MG^{\chi_1/\chi} > N^{1/3},$$

we shall represent each d in the form:

$$d = u_0 v,$$

where u_0 is the product of χ_1 prime factors of g_1, while v is the product of the remaining $(\chi-\chi_1)$ factors of g_1 and all $g_2, \ldots, g_\tau$. For $\chi_1 > 1$, we have

$$T_3 = \frac{1}{\binom{\chi}{\chi_1}} T_3'; \qquad T_3' = \sum_m \sum_{u_0} \sum_v e^{2\pi i \frac{a}{q} m^n u_0^n v^n},$$

where u_0 and v run through, independently of each other, the products of χ_1 and $(\chi-\chi_1)$ prime factors of g_1; here T_3' may contain only the terms with

$$(u_0, v) = 1.$$

Thus, we find

$$T_3' = \sum_\delta \mu(\delta) W_\delta; \qquad W_\delta = \sum_m \sum_{u_1} \sum_{v_1} e^{2\pi i \frac{a}{q}\delta^{2n} m^n u_1^n v_1^n},$$

where δ runs through all possible divisors of g_1, while u_1 and v_1, for a given δ, run through the quotients of division of u_0 and v by δ.
Putting $mu_1 = u$, we obtain

$$W_\delta = \sum_u \varphi(u) \sum_{v_1} e^{2\pi i \frac{a}{q}\delta^{2n} u^n v_1^n};$$

$$F_1 \ll r; \qquad F_2 \ll r^3.$$

The number u in the sum W_δ runs through the values satisfying the condition

$$\frac{N^{1/3\lambda}}{\delta} < u \leqq 2\frac{N^{2/3}}{\delta},$$

while v_1, for a given u, runs through the values satisfying the condition:

$$\frac{N-A}{\delta^2 u} < v_1 \leqq \frac{N}{\delta^2 u}.$$

8°. Evidently, $\delta \leqq \sqrt{N}$. In addition, let

$$\delta \leqq q.$$

Now we can apply Lemma 6. Thus, we have

$$W_\delta \ll \frac{A}{\delta^2} q^{\varrho-1/2} \sqrt{\left(1+\frac{q\delta}{N^{1/3\lambda}}\right)\left(1+\frac{N^{2/3} q\delta}{A}\right)\left(1+\frac{N^{2/3}\delta}{A}\right)}\, r^{5/2}$$

$$+\left(\frac{N^{1-1/3\lambda}q}{\delta}+\frac{N^{2/3}}{\delta}\right) r^2 \leqq \frac{A}{\delta} q^{\varrho-1/2} r^{5/2}.$$

9°. If, however, $\delta > q$, then W_δ does not exceed the number of systems of values of m, u_0/δ, v_1 satisfying the condition

$$\frac{N-A}{\delta^2} < m\frac{u_0}{\delta} v_1 \leqq \frac{N}{\delta^2}$$

and, consequently, by the well known property of $\tau(a)$, we have

$$W_\delta \ll \left(\frac{A}{\delta^2}+\frac{\sqrt{N}}{\delta}\right) r^2,$$

which is contained as a particular case in the inequality proved in 8°.

10°. Therefore, since $\delta \leqq \sqrt{N}$ and by virtue of 7°, we have

$$T_3 \ll T_3' \ll A r^{7/2} q^{\varrho-1/2}.$$

Similarly, we obtain the same estimate even when $\chi_1 = 1$.

11°. Now consider the case, where

$$M \geqq N^{1/3}.$$

Here $d < N^{2/3}$, $A > dq$. The sum

$$S_d = \sum_{\frac{N-A}{d} < m \leqq \frac{N}{a}}$$

can be divided into

$$\ll \frac{A}{dq}$$

subsums, each of which, according to Lemma 4, is

$$\ll q^{\varrho+1/2}.$$

Therefore

$$S_d \ll \frac{A}{d} q^{\varrho - 1/2}; \qquad T_3 \ll Arq^{\varrho - 1/2}.$$

12°. Finally, if

$$MG_1 \ldots G_\tau < N^{1/3},$$

then evidently we have

$$T_3 \ll N^{1/3} r \ll Aq^{-1/2}.$$

13°. From the estimates derived for T_3 (in 6°, 10°, 11°, 12°), Lemma 5 and 1° and 4°, we find that our lemma is true.

More General Results For the Case of $n = 1$. In view of the special importance of the case $n = 1$, we shall demonstrate a more general theorem than Theorem 2.

Lemma 7.[1] Let $N \geqq N_0$, where N_0 is a sufficiently large constant >1, $r = \ln N$,

$$W \geqq 1; \qquad 1 \leqq V \leqq N; \qquad k - \text{integer} > 0,$$

$$\lambda = \frac{a}{q} + \frac{\theta}{q^2}; \qquad (a, q) = 1; \qquad 0 < q \leqq N,$$

t runs through (all or some) integers in a certain interval of length $\ll V$

$$S = \sum \min \left(W, \frac{1}{2(\lambda k t)} \right).$$

Then

$$S \ll \left(\frac{V}{q} + 1 \right) (kW + qr).$$

Lemma 8. Let (u) and (v) be two sequences of positive integers, $N \geq N_0$, where N_0 is a sufficiently large constant >1, $r = \ln N$;

$$1 \leqq U_1 < U_2 \leqq N, \qquad U_2 \leqq A \leqq N; \qquad k - \text{integer} > 0,$$

$$\lambda = \frac{a}{q} + \frac{\theta}{q^2}; \qquad (a, q) = 1; \qquad 0 < q \leqq N,$$

$$S = \sum_u \varphi(u) \sum e^{2\pi i \lambda k u v},$$

where u runs through the members of the sequence (u) satisfying the condition:

$$U_1 < u \leqq U_2,$$

while v runs, for a given u, through the members of the sequence (v) satisfying the condition:

$$\frac{N - A}{u} < v \leqq \frac{N}{u}.$$

[1] Estimates for trigonometrical sums Izv. akad. nauk SSSR (ser. mat.) **2** (1938) 505–524.

Here $\varphi(u)$ is a function which takes, for all u in the interval:

$$U_1 < u \leqq U_2$$

non-negative values; and for any u_1 and u_2 satisfying the condition

$$U_1 \leqq u_1 < u_2 \leqq U_2,$$

we always have

$$\sum_{u_1 < v \leqq u_2} (\varphi(u))^2 \ll u_2 F_2.$$

Hence

$$S \ll A\sqrt{F_2}\sqrt{\frac{k}{q} + \frac{kU_2}{A} + \frac{N}{U_1 A} + \frac{Nq}{A^2}}\, r^{3/2}.$$

Proof. 1°. It suffices to consider only the case

$$A \geqq \max\left(U_2, \frac{N}{U_1}, \sqrt{Nq}\right).$$

Divide the interval

$$U_1 < u \leqq U_2$$

into $\ll r$ subintervals of the type

$$U < u \leqq U_0; \qquad U_0 \leqq 2U.$$

2°. Consider that part S_0 of the sum S which corresponds to one such subinterval. Now dividing this subinterval into

$$\ll \frac{N}{A}$$

new subintervals of the type

$$u_1 < u \leqq u_2; \qquad u_2 - u_1 \ll \frac{AU}{N},$$

we find that

$$S_0 = \sum\nolimits_1 S_1; \qquad S_1 = \sum_{u_1 < u \leqq u_2} \varphi(u) \sum_{\frac{N-A}{u} < v \leqq \frac{N}{u}} e^{2\pi i\lambda kuv},$$

$$S_0^2 \ll \frac{N}{A}\sum\nolimits_1 |S_1|^2; \qquad S_1^2 \ll \sum_{u_1 < u \leqq u_2} (\varphi(u))^2 R,$$

$$R = \sum_{u_1 < u \leqq u_2} \sum_{\frac{N-A}{u} < v \leqq \frac{N}{u}} \sum_{\frac{N-A}{u} < v_1 \leqq \frac{N}{u}} e^{2\pi i\lambda ku(v-v_1)},$$

where u runs through the integers in the above interval.

To estimate R, chenge the order of summation. Evidently, v and v_1 may take only those values which satisfy the condition

$$\frac{N-A}{u_2} < v \leqq \frac{N}{u_1}; \qquad \frac{N-A}{u_2} < v_1 \leqq \frac{N}{u_1},$$

and u, for the chosen v and v_1, runs through the values in the interval

$$u' < u \leqq u''; \qquad u' = \max\left(u_1, \frac{N-A}{v}, \frac{N-A}{v_1}\right),$$
$$u'' = \min\left(u_2, \frac{N}{v}, \frac{N}{v_1}\right).$$

Therefore, since $u'' - u' \leqq u_2 - u_1 \ll AU/N$, we obtain

$$R \ll \sum_{v} \sum_{v_1} \min\left(\frac{AU}{N}, \frac{1}{2(\lambda k(v - v_1))}\right),$$

Observing that v, for a given v_1, runs through the numbers in the interval of length

$$\frac{N}{u_1} - \frac{N-A}{u_2} = \frac{N(u_2 - u_1) + Au_1}{u_1 u_2} \ll \frac{A}{U},$$

by Lemma 7, we find that

$$R \ll R_0; \qquad R_0 = \frac{A}{U}\left(\frac{A}{Uq} + 1\right)\left(k\frac{AU}{N} + qr\right).$$

At the same time, we have

$$S_1^2 \ll \sum_{u_1 < u \leqq u_2} (\varphi(u))^2 R_0; \qquad S_0^2 \ll \frac{N}{A} R_0 \sum_{U < u \leqq U_0} (\varphi(u))^2$$
$$\ll \frac{NU}{A} F_2 R_0 \ll A^2 F_2 \left(\frac{1}{Uq} + \frac{1}{A}\right)\left(kU + \frac{Nqr}{A}\right),$$
$$S_1 \ll A\sqrt{F_2}\sqrt{\frac{k}{q} + \frac{kU}{A} + \frac{Nr}{UA} + \frac{Nqr}{A^2}}.$$

Hence Lemma 8 follows immediately.

Lemma 9[1]. Let k be an integer >0

$$N > 2; \qquad r = \ln N; \qquad 1 < d_0 < A \leqq N,$$
$$\lambda = \frac{a}{q} + \frac{\theta}{q^2}; \qquad (a, q) = 1; \qquad 0 < q \leqq N.$$

Then

$$\sum_{0 < d \leqq d_0} \min\left(\frac{A}{d}, \frac{1}{2(\lambda k d)}\right) \ll \left(d_0 + q + \frac{k^2 A}{q}\right) r.$$

Theorem 3. Let h be an arbitrarily small positive constant $\leqq 1/6$, $N > 2$, $r = \ln N$, k an integer > 0,

$$\alpha = \frac{a}{q} + \frac{\theta}{q^2}; \qquad (a, q) = 1; \qquad 0 < q \leqq N, \qquad 1 \leqq A \leqq N,$$
$$S = \sum_{N - A < p \leqq N} e^{2\pi i \alpha k p}.$$

[1] An improved estimate for a trigonometrical sum containing primes; Izv. akad. nauk SSSR (ser. mat.) **2** (1938) 1–24.

Then

$$(\lambda = 1 + h)$$

$$S \ll A r^{\frac{\ln r}{\ln \lambda} + 6} \sqrt{\frac{kN^{\frac{2+h}{3}}}{A} + \frac{Nq}{A^2} + \frac{k}{q} + \frac{k^4}{q^2}}.$$

Proof. 1°. We have

$$S = \sum_d \sum_m \mu(d) \, e^{2\pi i \alpha k d m} + O(\sqrt{N}),$$

where d runs through the divisors of the product of all primes at most $\sqrt{N}$ and m runs through natural numbers. Summation is taken over the range

$$N - A < dm \leqq N.$$

Now we find

$$\sum_d \sum_m \mu(d) \, e^{2\pi i \alpha k d m} = T_0 - T_1,$$

where T_0 is the sum of the terms corresponding to the values of d with even number of prime divisors, while T_1 is the sum of the terms corresponding to the values of d with odd number of prime divisors. Now consider one of the sums T_0 or T_1 and denote it by T_2.

2°. Divide the numbers $d \leqq N$ into classes satisfying the conditions stated in Lemma 5, all notations being as in Lemma 5.

3°. Also divide the numbers m into $\ll r$ classes. Put all the numbers contained in a certain interval of the type

$$M < m \leqq M_0; \qquad M_0 \leqq 2M$$

into the same class.

4°. Denote by T_3 that part of the sum T_2 which corresponds to the chosen classes of the values of d and m.

5°. First let

$$MG_1 \ldots G_\tau > N^{1/3},$$

here β is the first of the members in the sequence $1, 2, \ldots, \tau$ that satisfies the condition:

$$MG_1 \ldots G_\beta > N^{1/3}.$$

6°. Let $\beta > 1$. Then putting

$$u = mg_1 \ldots g_\beta; \qquad v = g_{\beta+1} \ldots g_\tau,$$

we obtain

$$MG_1 \ldots G_{\beta-1} \leqq N^{1/3}; \qquad G_\beta \leqq G_{\beta-1}; \qquad MG_1 \ldots G_\beta \leqq N^{2/3}.$$

Therefore for our case

$$T_3 = \sum_u \varphi(u) \sum_v e^{2\pi i\alpha kuv},$$

where u runs through the integers satisfying the condition

$$N^{1/3\lambda} < u \leqq 2N^{2/3},$$

while v, for a given u, runs through the integers satisfying the condition

$$\frac{N-A}{u} < v \leqq \frac{N}{u}.$$

Here $\varphi(u)$ does not exceed the number $\tau(u)$ of divisors of u. Consequently, the number F_2 stated in Lemma 8, will satisfy the condition

$$F_2 \ll r^3.$$

Applying Lemma 8, we obtain

$$T_3 \ll A\sqrt{r^3}\sqrt{\frac{k}{q} + \frac{kN^{2/3}}{A} + \frac{N^{1-1/3\lambda}}{A} + \frac{Nq}{A^2}}\, r^{3/2}.$$

7°. Let $\beta = 1$. For the sake of brevity, assuming that $l_1 = \chi$ and denoting by χ_1 the least integer satisfying the condition

$$MG_1^{\chi_1/\chi} > N^{1/3},$$

we shall represent each d in the form

$$d = u_0 v,$$

where u_0 is the product of χ_1 prime factors of g_1, while v is the product of the remaining $(\chi - \chi_1)$ factors of g_1 and all $g_2, \ldots, g_\tau$.

For $\chi_1 > 1$, we have

$$T_3 = \frac{1}{\binom{\chi}{\chi_1}} T_3'; \qquad T_3' = \sum_m \sum_{u_0} \sum_v e^{2\pi i\alpha k m u_0 v},$$

where u_0 and v run through, now independently of each other, the product of χ_1 and $(\chi - \chi_1)$ different prime divisors of g_1; only those terms for which $(u_0, v) = 1$ are contained in T_3'.

Just as before, we conclude that

$$N^{1/3\lambda} < u \leqq 2N^{2/3}.$$

Now we find

$$T_3' = \sum_\delta \mu(\delta) W_\delta; \qquad W_\delta = \sum_m \sum_{u_1} \sum_{v_1} e^{2\pi i\alpha k\delta^2 m u_1 v_1},$$

where δ runs through all possible divisors of g_1, whereas u_1 and v_1, for a given δ, run through the quotients of division of u_0 and v by δ.

Putting $mu_1 = u$, we obtain

$$W_\delta = \sum_u \varphi(u) e^{2\pi i\alpha\delta^2 uv_1}; \qquad F_2 \leqq r^3.$$

In the sum W_δ, u runs through the values satisfying the condition:

$$\frac{N^{1/3\lambda}}{\delta} < u \leqq \frac{2N^{2/3}}{\delta},$$

while v_1, for a given u, runs through the values satisfying the condition:

$$\frac{N-A}{\delta^2 u} < v_1 \leqq \frac{N}{\delta^2 u}.$$

8°. Evidently, we have $\delta \leqq \sqrt{N}$. Furthermore, let

$$\delta \leqq \delta_0; \qquad \delta_0 = \min(N^{1/3\lambda}, AN^{-2/3}).$$

Now we can apply Lemma 8. Thus, we have

$$W_\delta \ll \frac{A}{\delta^2}\sqrt{r^3}\sqrt{\frac{k}{q} + \frac{kN^{2/3}\delta}{A} + \frac{N^{1-1/3\lambda}\delta}{A} + \frac{Nq\delta^2}{A^2}} \cdot r^{3/2}.$$

9°. If, however, $\delta > \delta_0$, then W_δ does not exceed the number of systems of the values of m, u_0/δ, v_1 satisfying the condition

$$\frac{N-A}{\delta^2} < m\frac{u_0}{\delta}v_1 \leqq \frac{N}{\delta^2}.$$

Consequently, by the well known property of the function $\tau(a)$, we have

$$W_\delta \ll \left(\frac{A}{\delta^2} + \frac{\sqrt{N}}{\delta}\right) r^2,$$

which is a particular case of the inequality proved in 8°.

10°. By virtue of the statements proved above, on summing over all δ, we find

$$T_3 \ll A\sqrt{\frac{k}{q} + \frac{kN^{2/3}}{A} + \frac{N^{1-1/3\lambda}}{A} + \frac{Nq}{A^2}} \cdot r^5.$$

Similarly, we obtain the same estimate even when $\chi_1 = 1$.

11°. Now consider the case where

$$M > N^{1/3}.$$

Here we have $d < N^{2/3}$ and

$$T_3 = \sum S_d; \qquad S_d = \sum_m e^{2\pi i\alpha kdm},$$

where d and m run through the numbers in the chosen classes. Here m in the sum S_d runs through the values satisfying the condition

$$M < m \leqq M_0; \qquad \frac{N-A}{d} < m \leqq \frac{N}{d}.$$

For $A \geqq N^{2/3}$, we have

$$S_d \ll \min\left(\frac{A}{d}, \frac{1}{2(\alpha kd)}\right)$$

therefore, by Lemma 9,

$$T_3 \ll \left(N^{2/3} + q + \frac{k^2 A}{q}\right) r.$$

12°. Finally, for the case where

$$MG_1 \ldots G_\tau \leqq N^{1/3},$$

evidently we have

$$T_3 \ll N^{1/3} r.$$

13°. Collecting all the results proved in 1°, 2°, 4°, 6°, 11° and 12°, we obtain

$$S \ll A r^{\frac{\ln r}{\ln \lambda}+6} \sqrt{\frac{k}{q} + \frac{kN^{2/3}}{A} + \frac{N^{1-1/3\lambda}}{A} + \frac{Nq}{A^2} + \frac{k^4}{q^2}}.$$

Hence Theorem 3 follows directly.

Lemma 10[1]**.** Let ε be an arbitrarily small positive constant,

$$\lambda = \frac{a}{q} + \frac{\theta}{q^2}; \qquad (a, q) = 1; \qquad q > 0.$$

Let k and g run, independently of each other, through the values

$$1, 2, \ldots, K$$
$$1, 2, \ldots, G.$$

Then the number of values of the symbol (λkg) which do not exceed q^{-1} is

$$\ll (KGq^{-1} + 1)(KG)^\varepsilon.$$

Lemma 11. Let h be an arbitrarily small positive constant <1, $N > N_0$, where N_0 is a sufficiently large constant >2,

$$r = \ln N: \qquad \max(N^{2/3}, N^{1/2} q^{1/2}) \leqq A \leqq N,$$

$$\lambda = \frac{a}{q} + \frac{\theta}{q^2}; \qquad (a, q) = 1; \qquad 0 < q \leqq N.$$

Let k run through $1, \ldots, K$, where the number g satisfying the condition

$$\lambda k = \frac{c}{g} + \frac{\theta_1}{gq}; \qquad (c, g) = 1; \qquad 0 < g \leqq q; \qquad |\theta_1| < 1.$$

is put into correspondance with each k. Then, putting

$$D_k = \min\left(\frac{1}{k}\sqrt{\frac{N^{2+h/3}}{A} + \frac{Ng}{A^2} + \frac{1}{g}},\ \frac{1}{k}\sqrt{\frac{kN^{2+h/3}}{A} + \frac{Nq}{A^2} + \frac{k^2}{q}}\right),$$

$$S = \sum_{0<k\leqq q} D_k,$$

[1] An analytic proof of the theorem on the distribution of the fractional parts of the values of an integral polynomial; Izv. akad. nauk SSSR (ser. mat.) (1927, no. 7–8) 567–578, Lemma 5.

we have

$$S \ll r^2 \sqrt{\frac{N^{2/3+h}}{A} + \frac{N^{1+h}q^{1-h}}{A^2} + \frac{1}{q^{1-h}}}.$$

Proof. 1°. First consider the case where

$$q > N^{5/6}, \qquad g > N^{1/6}.$$

Evidently, we have

$$D_k \ll \frac{1}{k}\sqrt{\frac{Nq}{A^2}}.$$

Consequently, that part of the sum S which corresponds to such a case is

$$\ll r\sqrt{\frac{Nq}{A^2}}.$$

2°. Now consider the cases where

$$q \leqq N^{5/6}; \qquad q > N^{5/6}, \qquad g \leqq N^{1/6}.$$

First we shall estimate that part of the sum S which includes the terms satisfying the condition

$$kg \geqq q.$$

For this purpose, we shall divide all values of k, and then in an exactly similar manner, all the values of g into groups as follows:

$$\begin{array}{ll} k = 1, & g = 1, \\ k = 2, 3, & g = 2, 3, \\ k = 4, 5, 6, 7, & g = 4, 5, 6, 7, \\ \cdots\cdots\cdots & \\ k = 2^s, 2^s+1, \ldots, q, & g = 2^s, 2^s+1, \ldots, q. \end{array}$$

Let $\chi, \chi+1, \ldots, \chi+\chi'; \gamma, \gamma+1, \ldots, \gamma+\gamma'$ be two such groups of k's and g's. The numbers k and g in these groups may satisfy the inequality $kg \geqq q$ if and only if

$$4\chi\gamma \geqq q.$$

Assume that this condition is fulfilled. Then, we have

$$\lambda k = \frac{c}{g} + \frac{\theta}{gq}; \qquad \lambda kg = c + \frac{\theta}{q}$$

and therefore, by Lemma 10, the sum of the terms of S satisfying the condition that k and g belong to the groups chosen, is equal to

$$\ll \chi^{\varepsilon}\gamma^{1+\varepsilon}q^{-1}\sqrt{\frac{N^{(2+h)/3}}{A} + \frac{N\gamma}{A^2} + \frac{1}{\gamma}}$$

$$\ll q^{2\varepsilon}\sqrt{\frac{N^{(2+h)/3}}{A} + \frac{N\gamma}{A^2} + \frac{1}{q}} \ll \sqrt{\frac{N^{2/3+h}}{A} + \frac{N^{1+h}q^{1-h}}{A^2} + \frac{1}{q^{1-h}}}.$$

Summing over all possible pairs of groups, we obtain a value

$$\ll r^2 \sqrt{\frac{N^{2/3+h}}{A} + \frac{N^{1+h} q^{1-h}}{A^2} + \frac{1}{q^{1-h}}}.$$

3°. Finally, we shall estimate that part of the sum S which corresponds to the case where

$$kg < q.$$

Comparing the two different expressions for λk, we obtain

$$\frac{c}{g} + \frac{\theta_1}{gq} = \frac{ak}{q} + \frac{\theta k}{q^2}; \qquad |akg - cq| < 2.$$

Consequently, either the difference $akg - cq$ is zero or $v = \pm 1$. The first case is impossible since $(a, q) = 1$, $kg < q$. Therefore

$$akg - cq = v.$$

Denoting by z_0 the least positive solution of the congruence

$$az \equiv v \pmod{q},$$

we obtain $kg = z_0$. Since the number of solutions in positive integral k and g of this equation is $\ll q^{\varepsilon}$, the number of terms in our sum satisfying the condition $kg < h$ will also be of the same order. Each term is

$$\ll \frac{1}{k} \cdot \sqrt{\frac{k N^{(2+h)/3}}{A} + \frac{Nq}{A^2} + \frac{k^2}{q}}.$$

Therefore, that part of the sum S which corresponds to our case is

$$\ll r \sqrt{\frac{N^{2/3+h}}{A} + \frac{N^{1+h} q^{1-h}}{A^2} + \frac{1}{q^{1-h}}}.$$

The lemma follows immediately from 1°, 2° and 3°.

Theorem 4. Let ε and h be arbitrarily small positive constants $\leqq 1/6$, $N > N_0$, where N_0 is a sufficiently large constant > 2, $r = \ln N$,

$$\lambda = \frac{a}{q} + \frac{\theta}{q^2}; \qquad (a, q) = 1; \qquad 1 \leqq q \leqq N, \qquad 1 \leqq A \leqq N; \qquad 0 < \delta \leqq 1.$$

Let T denote the number of all values of the fractional part

$$\{\lambda p\}; \qquad N - A < p \leqq N$$

and T_1 the number of values of the same fractional part satisfying the condition

$$0 \leqq \{\lambda p\} \leqq \delta.$$

Then

$$T_1 = \delta T + O(A\Delta),$$

$$\Delta = e^{r^{\varepsilon}} \sqrt{\frac{N^{2/3+h}}{A} + \frac{N^{1+h} q^{1-h}}{A^2} + \frac{1}{q^{1-h}}}.$$

Proof. 1°. According to Lemma 14 in Chap. I of my book "A New Method in Analytic Number Theory" [1] (assume $\Delta < 0.5$, otherwise Theorem 4 is trivial), there exists a periodic function $\psi(x)$ of period 1 with the following properties.
Let α and β be real numbers satisfying the condition

$$0 \leqq \beta - \alpha \leqq 1 - 2\Delta .$$

Then

1. $\psi(x) = 1$ in the interval $\alpha \leqq x \leqq \beta$.
2. $0 \leqq \psi(x) \leqq 1$ in the intervals $\alpha - \Delta \leqq x \leqq \alpha$, $\beta \leqq x \leqq \beta + \Delta$.
3. $\psi(x) = 0$ in the interval $\beta + \Delta \leqq x \leqq 1 + \alpha - \Delta$.
4. $\psi(x)$ can be expanded as a Fourier series of the type

$$\psi(x) = \beta - \alpha + \Delta + \sum_{k=1}^{\infty} (a_k \cos 2\pi kx + b_k \sin 2\pi kx),$$

where

$$a_k \ll \frac{1}{k}, \quad b_k \ll \frac{1}{k}, \qquad \text{if} \quad k < \frac{1}{\Delta},$$

$$a_k \ll \frac{1}{\Delta k^2}, \ b_k \ll \frac{1}{\Delta k^2}, \quad \text{if} \quad k \geqq \frac{1}{\Delta}.$$

2°. Using 1° and Theorem 3, we find

$$\sum_{N-A<p\leqq N} \psi(\lambda p) = (\beta - \alpha + \Delta) T + AR,$$

$$\frac{1}{\Delta k^2} \sqrt{\frac{k^4}{q^2}} = \frac{1}{\Delta \sqrt{q}} \frac{1}{k} \sqrt{\frac{k^2}{q}} \ll \frac{1}{k} \sqrt{\frac{k^2}{q}},$$

$$R \ll r^{\frac{\ln r}{\ln \lambda}+6} \sum_{0<k\leqq q} D_k + \sum_{k>q} \frac{1}{\Delta k^2} \ll \Delta + \frac{1}{\Delta q} \ll \Delta .$$

Hence, arguing in the same way as in my book (Chap. VIII), we find that Theorem 4 is true.

Theorem 5. (Distribution of primes with respect to a given modulus). Let η be an arbitrarily small positive constant <1, $N > 2$,

$$1 \leqq A \leqq N; \qquad 1 \leqq Q \leqq N,$$
$$0 \leqq \delta \leqq 1 .$$

Replace all primes p in the interval

$$N - A < p \leqq N$$

by their remainders P on division by Q. Let T denote the number of all these remainders (i.e. the number of all primes in the interval $N - A < p \leqq N$) and T_1 the number of those primes which lie in the interval

$$0 \leqq P \leqq \delta Q .$$

---

[1] Trudy mat. inst. Steklov **10** (1937).

Then

$$T_1 = \delta T + O(A\Delta),$$

$$\Delta = N^{\eta}\sqrt{\frac{N^{2/3}}{A} + \frac{NQ}{A^2} + \frac{1}{Q}}.$$

Proof. In Theorem 4, let

$$\lambda = \frac{1}{Q}.$$

Then, expressing λ as

$$\lambda = \frac{1}{q} + \frac{\theta}{q^2}; \qquad q = [Q],$$

we arrive at Theorem 5 as a direct corollary of Theorem 4.

Particular case. The law of distribution of primes with respect to a given modulus Q which can be derived from Theorem 5 is of the simplest form when

$$A \geqq N^{2/3}Q.$$

Here we have

$$T_1 = \delta T + O\left(N^{\eta}\frac{A}{\sqrt{Q}}\right).$$

Example 1. Let $A = N^{5/6}$, $Q = N^{1/6}$, $\delta = 1/2$. On replacing all primes in the interval

$$N - N^{5/6} < p \leqq N$$

according to our theorem (particular case) by their remainders on division by $N^{1/6}$, and denoting by T the number of all remainders, by T_1 the number of all those remainders which are $\leqq \frac{1}{2}N^{1/6}$ we obtain

$$T_1 = \tfrac{1}{2}T + O(N^{5/6+\varepsilon-1/12}).$$

Example 2. Let $A = N$, $Q = N^{1/3}$, $\delta = 1/2$. On replacing all primes $p \leqq N$ by their remainders on division by $N^{1/3}$, and denoting by T the number of all remainders and by T_1 the number of remainders which are $\leqq N^{1/3}$, we obtain

$$T_1 = \tfrac{1}{2}T + O(N^{1+\eta-1/6}).$$

Note. Using my method, we can prove a theorem similar to Theorem 5 even in the case where, in place of all primes in the interval

$$N - A < p \leqq N$$

we take those primes which possess certain arithmetical properties, say, for instance, those which are the quadratic residues or non-residues to some modulus, or residues or non-residues of power $n > 2$, etc.

Pavel Sergeevich Aleksandrov

Born – 7 May 1896 in Bogorodsk, Russia
Died – 16 Nov 1982 in Moscow, USSR

Pavel Aleksandrov's father was a medical graduate from Moscow University who had decided not to follow an academic career but instead had chosen to use his skills in helping people. So he worked as a general practitioner in Yaroslavskii. Later he worked in more senior positions in a hospital in Bogorodskii (also called Noginsk), which is where he was when Pavel Sergeevich was born.

When Pavel Sergeevich was one year old his father moved to Smolensk State hospital, where he was to earn the reputation of being a very fine surgeon, and the family lived from this time in Smolensk. The city of Smolensk is on the Dnieper River 420 km west of Moscow. Pavel Sergeevich's early education was from his mother, who applied all her considerable talents to bringing up and educating her children. It was from her that Aleksandrov learnt French and also German. His home was one that was always filled with music as his brothers and sisters all great talent in that area.

The fine start that his mother gave him meant that he always excelled at the grammar school in Smolensk that he attended. His mathematics teacher Alexsander Romanovich Eiges soon realized that his pupil had a remarkable talent for the subject. In 1913 Aleksandrov graduated from the grammar school being dux of the school and winning the gold medal. Eiges was the role model who he was aspiring to match at this stage, for Eiges had done more than teach Aleksandrov mathematics, he had also influenced his tastes in literature and the arts.

Aleksandrov entered Moscow University in 1913 and immediately he was helped by Stepanov. Stepanov was an important influence on Aleksandrov at this time and suggested that Aleksandrov join Egorov's seminar even in the first year of his studies in Moscow. In Aleksandrov's second year of study he came in contact with Luzin who had just returned to Moscow.

Aleksandrov proved his first important result in 1915, namely that every non-denumerable Borel set contains a perfect subset. It was not only the result that was important for set theory, but also the methods that Aleksandrov used that turned out to be one of the most useful methods in descriptive set theory.

Aleksandrov then went to Novgorod–Severskii and became a theatre producer. He then went to Chernikov where, in addition to theatrical work, he lectured on Russian and foreign languages, becoming friends with poets, artists and musicians. After a short term in jail in 1919 at the time of the Russian revolution, Aleksandrov returned to Moscow in 1920. Luzin and Egorov had built up an impressive research group at the University of Moscow which the students called 'Luzitania' and they, together with Privalov and Stepanov, were very welcoming to Aleksandrov on his return.

It was not an immediate return to Moscow for Aleksandrov, however, for he spent 1920–1921 back home in Smolensk where he taught at the University. During this time he worked on his research, going to Moscow about once every month to keep in touch with the mathematicians there and to prepare himself for his examinations. At around this time Aleksandrov became friendly with Urysohn, who was a member of "Luzitania", and the friendship would soon develop into a major mathematical collaboration.

After taking his examinations in 1921, Aleksandrov was appointed as a lecturer at Moscow University and lectured on a variety of topics including functions of a real variable, topology and Galois theory. In July 1922 Aleksandrov and Urysohn went to spend the summer at Bolshev, near to Moscow, where they began to study concepts in topology. Aleksandrov and Urysohn now began to push the theory forward with work on countably compact spaces producing results of fundamental importance. The notion of a compact space and a locally compact space is due to them.

In the summers of 1923 and 1924 Aleksandrov and Urysohn visited Göttingen and impressed Emmy Noether, Courant and Hilbert with their results. The mathematicians in Göttingen were particularly impressed with their results on when a topological space is metrisable. In the summer of 1924 they also visited Hausdorff in Bonn and he was fascinated to hear the major new directions that the two were taking in topology. Aleksandrov and Urysohn then visited Brouwer in Holland and Paris in August 1924 before having a holiday in the fishing village of Bourg de Batz in Brittany. On the morning of 17 August Urysohn began to write a new paper but tragically he drowned while swimming in the Atlantic later that day. Aleksandrov determined that no ideas of his great friend and collaborator should be lost and he spent part of 1925 and 1926 in Holland working with Brouwer on preparing Urysohn's paper for publication.

The atmosphere in Göttingen had proved very helpful to Aleksandrov, particularly after the death of Urysohn, and he went there every summer from 1925 until 1932. He became a close friend with Hopf and the two held a topological seminar in Göttingen. Of course Aleksandrov also taught in Moscow University and from 1924 he organised a topology seminar there. At Göttingen, Aleksandrov also lectured and participated in Emmy Noether's seminar. In fact Aleksandrov always included Emmy Noether and Hilbert among his teachers, as well as Brouwer in Amsterdam and Luzin and Egorov in Moscow.

The year 1929 marks not only the beginning of the friendship with Kolmogorov but also the appointment of Aleksandrov as Professor of Mathematics at Moscow University. In 1935 Aleksandrov went to Yalta with Kolmogorov, then finished the work on his Topology book in the nearby Crimea and the book was published in that year.

Aleksandrov wrote about 300 scientific works in his long career. As early as 1924 he introduced the concept of a locally finite covering which he used as a basis for his criteria for the metrisability of topological spaces. He laid the foundations of homology theory in a series of fundamental papers between 1925 and 1929. His methods allowed arguments of combinatorial and algebraic topology to be applied to point set topology and brought together these areas. Aleksandrov's work on homology moved forward with his homological theory of dimension around 1928–1930.

Aleksandrov was the first to use the phrase "kernel of a homomorphism" and around 1940–1941 he discovered the ingredients of an exact sequence. He worked on the theory of continuous mappings of topological spaces. In 1954 he organised a seminar on this last topic aimed at first year students at Moscow University and in this he showed one of the aspects of his career which was of major importance to him, namely the education of students.

Many honours were given to Aleksandrov for his outstanding contribution to mathematics. He was president of the Moscow Mathematical Society from 1932 to 1964, vice president of the International Congress of Mathematicians from 1958 to 1962, a corresponding member of the Soviet Academy of Sciences from 1929 and a full member from 1953. Many other societies elected Aleksandrov to membership including the Göttingen Academy of Sciences, the Austrian Academy of Sciences, the Leopoldina Academy in Halle, the Polish Academy of Sciences, the National Academy of Sciences of the United States, the London Mathematical Society, the American Philosophical Society, and the Dutch Mathematical Society.

He edited several mathematical journals, in particular the famous Soviet Journal *Uspekhi Matematicheskikh Nauk*, and he received many Soviet awards, including the Stalin Prize in 1943 and five Orders of Lenin. Also in memory of Aleksandrov's contributions to topology at Moscow University and his work with the Moscow Mathematical Society, there is an annual topological symposium *Aleksandrov Proceedings* held every May.

Sur la notion de dimension des ensembles fermés;

PAR PAUL ALEXANDROFF
(Moscou).

(Extrait d'une lettre à M. Fréchet.)

1. Les notions fondamentales dont nous aurons besoin sont celles de *cycles* et *d'homologie;* elles sont introduites par Poincaré dans ses Mémoires classiques sur l'*Analysis situs* (*Journ. Éc. Pol.*, 1895, et *Rend. Palermo*, 1899)([1]). L'idée intuitive de cycle est donnée dans le cas le plus simple par une courbe polygonale fermée ou par une surface polyédrale fermée; il s'agit dans la notion générale de cycle des généralisations de cette idée intuitive pour le cas d'un nombre quelconque de dimensions; *il est toutefois essentiel de remarquer que la notion de cycle n-dimensionnel ne présuppose aucune notion générale de dimension :* elle appartient entièrement à l'*Analysis situs* combinatoire, où la dimension apparaît tout simplement comme le nombre de sommets de certaines figures que l'on appelle *tétraédroïdes* ou *simplexes* et qui servent de « briques » ou de « cellules » pour édifier les figures dont l'étude forme l'objet de l'*Analysis situs* classique. Les définitions qui suivent élucideront ces considérations un peu trop vagues.

2. Soient donnés, dans l'espace euclidien à n dimensions que nous

([1]) Le lecteur désirant faire une étude de la Topologie combinatoire consultera, outre les Mémoires de Poincaré, surtout les Traités américains de M. Veblen (*Analysis situs*) et de M. Lefschetz (*Topology*); on trouvera dans le livre de M. Lefschetz une bibliographie bien complète. *Voir* aussi le petit livre de l'auteur *Einfachste Grundbegriffe der Topologie*, qui vient de paraître chez M. Springer (Berlin, 1932).

désignerons toujours par E^n, un certain nombre $k+1$ de points différents,

(1) $$a_0, a_1, \ldots, a_k.$$

En concentrant dans ces points des masses arbitraires (non négatives),

$$m_0, m_1, \ldots, m_k,$$

on obtient un certain point $p(m_0, m_1, \ldots, m_k)$ comme centre de gravité de ces masses; en variant de toutes les façons possibles les masses

$$m_0, m_1, \ldots, m_k,$$

placés dans les points (1), on obtient un certain ensemble de points, qui est appelé le *simplexe* (ou le *tétraédroïde*) *aux sommets* $a_0, a_1, \ldots, a_k$; on démontre aisément que c'est un ensemble de points fermé et convexe et que de plus c'est *le plus petit parmi tous les ensembles fermes et convexes contenant les points donnés* (1).

Cette dernière propriété peut servir de définition d'un simplexe :

Le simplexe aux sommets (1) *est le plus petit ensemble fermé et convexe contenant les points* (1). *Le nombre de sommets du simplexe moins un s'appelle la dimension du simplexe.*

C'est ainsi que le simplexe aux sommets (1) est un simplexe à k dimensions.

5. Un simplexe aux sommets $a_0, a_1, a_2, \ldots, a_k$ sera désigné par

(2) $$|a_0, a_1, a_2, \ldots, a_k|,$$

quelquefois aussi par $|x^k|$.

Si les sommets (1) n'appartiennent à aucun hyperplan à $k-1$ dimensions, le simplexe est dit *non dégénéré*, autrement *il dégénère*.

Un simplexe non dégénéré à une dimension est évidemment un segment de droite, tandis qu'un simplexe (non dégénéré) à deux dimensions est un triangle, celui à trois dimensions est un tétraèdre, etc.; la dénomination « tétraédroïde » (employée d'ailleurs par M. Hadamard) se trouve donc bien justifiée.

Dans ce qui suit, la « dégénération éventuelle » d'un simplexe n'affectera en rien nos raisonnements.

4. VOICI LE POINT ESSENTIEL. — *Un simplexe est entièrement défini par ses sommets*, de façon qu'on se trouve, en *Analysis situs* combinatoire, devant la possibilité d'*identifier un simplexe donné avec le système de ses sommets*. Cette remarque est d'une importance tout à fait capitale pour tout ce qui suit. Elle nous suggère, en effet, la définition fondamentale suivante :

Définition I. — Soit donc un ensemble M d'éléments quelconques appelés *sommets*. Nous supposons de plus que certains sous-ensembles finis de M sont choisis *a priori* et appelés les *simplexes*; si $k+1$ est le nombre de sommets formant un simplexe, le nombre k est dit *sa dimension* (1). L'ensemble M s'appelle dans les conditions un champ de sommets.

Le cas le plus important est celui où M est un espace (D) « espace distancié » au sens de M. Fréchet.

Un simplexe à k dimensions de l'espace M est par définition un ensemble *quelconque* de $k+1$ points de cet espace. Si le diamètre de cet ensemble fini est inférieur au nombre positif ε (c'est-à-dire la distance maximale entre deux sommets quelconques du simplexe est inférieure à ε), le simplexe est dit un ε-simplexe. Un simplexe étant un ensemble fini de points de M, on appelle tout sous-ensemble de cet ensemble une *face* du simplexe donné; un simplexe à n dimensions a évidemment

$$\frac{n(n-1)(n-2)\dots(n-r+1)}{1.2\dots r}$$

faces à r dimensions. Les plus importantes parmi les faces (d'un simplexe à n dimensions) sont celles à $n-1$ dimensions.

5. *Définition II.* — Donner à un simplexe

$$|x^n| = |a_0, a_1, \dots, a_n|$$

une *orientation* veut dire *choisir un certain ordre* dans lequel ses sommets sont considérés. Un *simplexe orienté* à n dimensions n'est donc autre chose qu'un système de $n+1$ points [de l'espace (D) donné]

(1) Il est supposé que tout sous-ensemble de l'ensemble fini, qui est un simplexe, est lui aussi un simplexe.

considéré avec un certain ordre de succession de ces points. On définit, cependant, comme identiques deux ordres provenant l'un de l'autre par une permutation *paire; chaque simplexe ne possède donc que deux orientations;* l'une d'entre elles est choisie comme l'orientation *positive* du simplexe en question. Si

$$|x^n| = |a_0, a_1, \ldots, a_n|$$

est le simplexe donné, on désigne par x^n et $-x^n$ les deux simplexes orientés auquel $|x^n|$ donne naissance et l'on écrit par exemple $x^n = (a_0, a_1, \ldots, a^n)$.

6. Cela posé, on parvient à la notion fondamentale de l'*Analysis situs* combinatoire, celle du *complexe*. Cette notion suppose que nous possédons, outre le champ des sommets que nous avons désigné par M, un *champ des coefficients*, c'est-à-dire un ensemble quelconque de grandeurs qui seront attribuées aux simplexes (du champ de sommets choisi) comme coefficients. Les plus importants champs de coefficients sont :

1° L'ensemble J_0 de tous les nombres entiers;

2° L'ensemble J_m des *résidus* (des nombres entiers) d'après le *module m* ($m = 2, 3, 4, \ldots$);

3° L'ensemble R des nombres rationnels.

Définition fondamentale. — Un système fini de simplexes orientés (du champ des sommets donné), chacun considéré avec un certain coefficient (tiré du champ des coefficients choisi), s'appelle un *complexe algébrique* (des deux champs donnés), ou simplement un *complexe*.

Remarque importante. — Nous ne considérons que des complexes algébriques *homogènes*, c'est-à-dire formés de simplexes ayant tous le même nombre de dimensions.

Évidemment on peut dire aussi :

Un complexe algébrique à n dimensions est une forme linéaire

$$C^n = \sum_{(i)} t^i x_i^n,$$

dont les variables x^n sont les simplexes orientés (du champ des sommets donné) et dont les coefficients sont les éléments du champ des coefficients choisi.

7. *Frontière d'un simplexe et d'un complexe.* — C'est là une autre notion fondamentale, également due à Poincaré. Considérons d'abord un triangle orienté, c'est-à-dire le simplexe orienté $x^2 = (a_0, a_1, a_2)$, l'orientation étant donnée par cet ordre de succession des sommets a_0, a_1, a_2. Il est alors conforme à l'usage ordinaire, de considérer

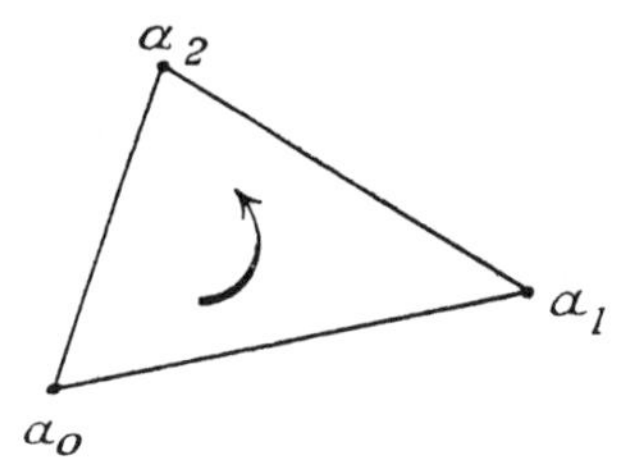

comme frontière de ce *triangle orienté* le système de ses côtés orientés comme il suit :

$$(a_0, a_1); \quad (a_1, a_2); \quad (a_2, a_0).$$

En remarquant que $(a_2, a_0) = -(a_0, a_2)$, on peut dire que la frontière du triangle orienté

$$x^2 = (a_0, a_1, a_2)$$

est le complexe algébrique formé de simplexes orientés (a_1, a_2), (a_0, a_2) et (a_0, a_1) considérés respectivement avec les coefficients $+1$, -1, $+1$. On a donc, en désignant la frontière de x_2 par $F_r(x^2)$,

$$\mathrm{Fr}(x^2) = (a_1, a_2) - (a_0, a_2) + (a_0, a_1).$$

C'est cette même formule qui, écrite pour $n+1$ sommets, nous donne la frontière d'un simplexe orienté

$$x^n = (a_0, a_1, \ldots, a_n) :$$

on entend par sa frontière le *complexe algébrique* à $n-1$ dimensions suivant :

$$\mathrm{Fr}(x^n) = \sum (-1)^i (a_0, a_1, \ldots, a_{i-1}, a_{i+1}, \ldots, a_n). \tag{3}$$

Le lecteur qui examinerait le cas d'un simplexe à trois dimensions (c'est-à-dire d'un tétraèdre ordinaire) s'apercevra combien la définition donnée par la formule (3) est élémentaire et naturelle.

Remarque importante. — Si le champ de coefficients choisi est J_m, on entend sous -1 le résidu de ce nombre par rapport au module m de sorte que la formule (3) est parfaitement applicable dans ce cas. Dans le cas particulier où le module choisi est $m=2$, on a, par rapport à ce module, $+1=-1$, de sorte que dans la « Topologie module 2 » la notion d'orientation se trouve éliminée; c'est pourquoi tous les théorèmes et toutes les démonstrations deviennent extrêmement simples quand on les considère « module 2 ».

La frontière d'un simplexe orienté étant définie, la définition de *la frontière d'un complexe algébrique arbitraire* ne présente plus aucune difficulté : la frontière d'un complexe algébrique est par définition la somme algébrique des frontières de ses simplexes (considérés avec les coefficients qui leur sont attribués). En d'autres mots : si le complexe algébrique donné est

$$C^n=\sum_{(i)} t^i x_i^n,$$

on définit comme sa frontière le complexe algébrique à $n-1$ dimensions suivant :

$$\mathrm{Fr}(C^n)=\Sigma t^i \mathrm{Fr}(x_i^n), \tag{4}$$

où, bien entendu, $\mathrm{Fr}(x_i^n)$ est donnée par la formule (3).

8. Il arrive souvent que la $\mathrm{Fr}(C^n)$ calculée d'après la formule (4) est identiquement nulle [c'est-à-dire que la forme linéaire obtenue en substituant pour les $\mathrm{Fr}(x_i^n)$ leurs valeurs données par (2) a tous ses cofficients nuls]. Dans ce cas, le complexe algébrique C^n s'appelle *un cycle* (*à n dimensions*) (¹).

Tout cela s'applique à un champ de sommets et un champ de coefficients quelconque (si le champ de coefficients choisi est par

(¹) On prouve sans peine que la frontière d'un complexe algébrique arbitraire est toujours un cycle.

exemple J_m, une forme linéaire identiquement nulle est, bien entendu, une forme dont les coefficients sont égaux à zéros mod m, ou, ce qui revient au même, des entiers divisibles par m).

Remarque — Si le *champ de sommets donné* est formé par les points de l'espace euclidien E^n, on peut substituer à un simplexe quelconque $(a_0, a_1, \ldots, a_k)$ le simplexe « géométrique » défini par les sommets $a_0, a_1, \ldots, a_k$ (comme nous l'avons fait au début de cette exposition); on parlera alors de *complexes algébriques polyédraux* (donc en particulier des *cycles polyédraux*.

9. Notre champ de sommets sera maintenant un espace (D) compact que nous désignons par F. Le cas le plus important est d'ailleurs celui où F est un ensemble fermé et borné situé dans un espace enclidien E^n quelconque.

Les simplexes de F étant des sous-ensembles finis de F, la notion de *complexes algébriques de* F (par rapport à un champ de coefficients choisi) est définie sans ambiguïté. Si tous les simplexes d'un complexe donné $C^n = \sum_i t^i x_i^n$ sont des ε-simplexes de F (*voir* n° 4), C^n est dit un ε-complexe de F. On peut donc en particulier parler des ε-*cycles de* F.

Définition III. — Une suite

$$Z^n = (z_1^n, z_2^n, \ldots, z_k^n, \ldots), \tag{5}$$

où z^n est un ε_k-cycle de F avec $\lim \varepsilon_k = 0$, s'appelle *un vrai cycle à n dimensions de* F.

Les cas particuliers suivants de la notion du vrai cycle sont extrêmement importants :

1° Tous les cycles z_k^n sont des *cycles ordinaires* (c'est-à-dire des cycles par rapport à l'ensemble J_0 de tous les nombres entiers, choisi comme champs de coefficients); (5) est dit alors un vrai cycle *ordinaire*.

2° Tous les cycles z_k^n sont cycles « module m », c'est-à-dire cycles considérés par rapport à J_m choisi comme champ de coefficients; dans ce cas, (5) est dit *vrai cycle module m*, de F.

3° Le cycle z_k^n est un cycle mod m_k, *m_k dépendant*, en général, *de k*. Dans ce cas (5) est appelé *vrai cycle de* F *d'après un module variable*.

10. *Remarque très importante*. — (5) étant défini comme un vrai cycle de F, il se peut qu'il est en même temps un vrai cycle d'un sous-ensemble F′ de F : en effet, si F′ contient les sommets de tous les z_k^n appartenant à la suite (5), le cycle (5) est évidemment un vrai cycle d'un tel ensemble F′.

Tout sous-ensemble F′ *tel que* (5) *peut être considéré comme vrai cycle de* F′ *s'appelle un ensemble associé au vrai cycle* (5).

Soit donné un vrai cycle (5) de F; supposons qu'on peut trouver pour tout ε'_k-complexe C_k^{n+1} de F ayant pour sa frontière le cycle z_k^n et cela d'une facon que ε'_k tende vers zéro quand k augmente indéfiniment. Dans ce cas nous dirons que le vrai cycle (5) est *homologue à zéro* dans F,

$$Z^n \sim 0 \quad (\text{dans F}).$$

Bien entendu, le complexe C_k^{n+1} doit être pris par rapport au même champ des coefficients que le cycle z_k^n; donc C_k^{n+1} est un complexe module m, si Z^n est un vrai cycle module m; dans le cas du module variable, C_k^{n+1} est un complexe d'après le module m_k (par rapport auquel le cycle z_k^n est envisagé), ce module variant avec k. *Dans le cas « ordinaire », quand Z^n est un cycle « aux coefficients entiers », il est cependant préférable de considérer, dans la définition d'homologie, des complexes C^{n+1} aux coefficients rationnels;* ou de supposer (ce qui revient au même) qu'un cycle C_k^{n+1} (aux coefficients entiers) admet pour sa frontière non nécessairement le cycle z_k^n, mais un cycle de la forme $t_k z_k^n$, où t_k est un nombre entier convenablement choisi (en d'autres mots, c'est *l'homologie « avec division »* au sens de Poincaré qui doit être envisagée).

11. *Définition IV*. — Un vrai cycle de F s'appelle *essentiel*, s'il existe pour lui au moins un ensemble associé dans lequel il n'est pas homologue à zéro.

Soient par exemple F l'ensemble de tous les points du triangle $|a_0 a_1 a_2|$ de notre figure, F′ l'ensemble de tous les points de la frontière de F (au sens élémentaire). Si nous désignons par z_1 la frontière du triangle *orienté* $(a_0 a_1 a_2)$, telle qu'elle était définie au n° **7**, et par

$$z_2, z_3, \ldots$$

des subdivisions successives du cycle z_1 (c'est-à-dire du polygone orienté $a_0 a_1 a_2 a_0$), la suite

$$z_1, z_2, \ldots \tag{6}$$

forme évidemment un cycle de F, qui est ~ 0 dans F, *mais n'est pas* ~ 0 dans F′; (6) est donc un cycle essentiel de F.

12. Les définitions précédentes permettent de faire une étude approfondie des espaces (D) compacts au point de vue de l'*Analysis situs* classique; elles permettent en particulier d'édifier une théorie générale de la dimension, dans laquelle la plus étroite connection avec les méthodes d'*Analysis situs* combinatoire est respectée de manière que cette théorie apparaît comme une simple application desdites méthodes au domaine général des espaces abstraits.

Voici la définition sur laquelle repose cette théorie :

Définition de dimension. — On appelle la dimension de F *le plus grand nombre r tel qu'il existe dans* F *un vrai cercle r* — 1 *dimensionnel et essentiel qui cependant est homologue à zéro dans* F.

13. On voit de suite que cette définition embrasse en réalité une infinité de définitions différentes : on obtient, en effet, des définitions différentes selon qu'on entend sous *vrai cycle* un vrai cycle ordinaire, un vrai cycle module m ($m = 2, 3, \ldots$) ou bien un vrai cycle d'après un module variable. On obtient par conséquent :

1° La dimension basée sur la considération des cycles ordinaires; nous l'appellerons (pour des raisons qui deviendront claires dans un moment) *dimension sans torsion* et la désignerons par $\Delta_0(F)$;

2° La dimension module m, $\Delta_m(F)$;

3° La dimension à module variable, $\Delta_\omega(F)$.

Les fonctions d'ensemble ainsi obtenues sont d'ailleurs différentes : M. Pontrjagin [1] a en effet réussi à construire pour tout m un ensemble

[1] *Comptes rendus*, t. **190**, 1930, p. 1102, et **191**, 1930, p. 475.

fermé situé dans l'espace à 4 dimensions pour lequel $\Delta_m(F) = 2$ tandis que pour tout nombre m' premier avec m,

$$\Delta'_m(F) = 1;$$

il existe aussi des ensembles F pour lesquels $\Delta_0(F)$ est différent de $\Delta_\omega(F)$ et de tous les $\Delta_m(F)$. *Pour tous les polyèdres à un nombre quelconque de dimensions toutes nos définitions donnent cependant le même résultat, d'ailleurs conforme avec le nombre de dimensions du* polyèdre au sens élémentaire.

On a enfin le suivant :

Premier théorème fondamental. — *Pour tout ensemble fermé situé dans un espace euclidien à un nombre arbitraire de dimensions, la dimension au sens de Poincaré-Brouwer-Urysohn-Menger coïncide avec la dimension à module variable.*

14. Les fonctions d'ensemble que nous venons d'introduire comme dimensions étant, d'après le résultat cité de M. Pontrjagin, différentes, c'est seulement par *une suite infinie* de nombres entiers

$$(7) \qquad \Delta_\omega(F), \quad \Delta_0(F), \quad \Delta_2(F), \quad \ldots, \quad \Delta_m(F), \quad \ldots,$$

qu'on peut espérer de caractériser la dimension d'un ensemble fermé; c'est d'ailleurs une question non résolue, si les dimensions différentes que nous avons introduites sont les seules à envisager; cette question ne peut être traitée, d'ailleurs, que par la voie axiomatique.

Quoi qu'il en soit, nous dirons pour le moment actuel que deux ensembles F_1 et F_2 appartiennent au même *type dimensionnel*, si chacune des fonctions (7) prend pour F_1 la même valeur que pour F_2. Il est alors facile à voir que *l'ensemble de types de dimensions différentes a la puissance du continu*. Comme il peut bien arriver qu'on a, par exemple,

$$\Delta_2(F_1) = 1, \quad \Delta_3(F_1) = 2 \quad \text{et en même temps} \quad \Delta_3(F_2) = 1, \quad \Delta_2(F_2) = 2,$$

on voit qu'il est impossible d'ordonner tous les types de dimension « d'après leur grandeur » : on aura toujours des types de dimensions *incomparables*,

Cela conduit d'une manière inévitable au problème suivant :

Quelles sont les relations entre les types de dimension que nous venons d'introduire et ceux de M. Fréchet?

Les types de dimensions introduits, il y a un quart de siècle, par M. Fréchet, présentent maintenant, en effet, un intérêt tout à fait nouveau, puisqu'on voit que dans toute théorie qui voudra tenir compte de divers caractères présentés par la structure dimensionnelle des ensembles, on aura nécessairement affaire aux types de dimensions incomparables.

15. Chacune des définitions de dimension que nous avons proposées conduit à une « théorie de dimension ». Toutes ces théories se développent d'ailleurs d'une manière parallèle; on a, par exemple, dans chacune d'elles les théorèmes suivants :

I. Un ensemble fermé de dimension n ne peut pas être représenté comme somme d'un nombre fini ou d'une infinité dénombrable d'ensembles fermés dont la dimension est inférieure à n.

II. Il existe pour tout ensemble fermé F un nombre positif $\varepsilon = \varepsilon(F)$ tel qu'il est impossible de transformer F, par une déformation continue déplaçant chaque point de F de moins de ε, en un ensemble dont la dimension est inférieure à celle de F lui-même.

III. Tout ensemble fermé de dimension n contient une *variété cantorienne* à n dimensions. (On entend, par variété cantorienne à n dimensions, un continu n-dimensionnel qui reste connexe après la suppression d'un ensemble fermé quelconque de ses points, sous la condition que cet ensemble a une dimension $n - 2$ au plus.)

Remarque. — Dans chacun de ces théorèmes, le mot « dimension » peut être compris dans le sens d'une quelconque de définitions que nous avons données.

16. On croyait, il y a quelques ans, que les théorèmes précédents expriment des propriétés particulières à la dimension au sens de Poincaré-Brouwer-Urysohn-Menger. On voit maintenant que ces propriétés appartiennent toutes à une infinité de fonctions d'ensemble, fonctions, qui méritent d'être appelées « dimensions ».

17. Il y a une hypothèse en Topologie qu'il est naturel de nommer *hypothèse du Produit.* Rappelons d'abord qu'on appelle « produit » des deux espaces X et Y (dans lesquels les voisinages sont définis), l'espace $W = X \times Y$ dont les points w sont par définition les couples de points (x, y) appartenant respectivement à X et à Y; quant aux voisinages, on les définit dans W de la facon suivante : un voisinage arbitraire d'un point $w_0 = (x_0, y_0)$ est constitué de tous les points $w = (x, y)$, x et y appartenant aux voisinages arbitraires $V(x)$ et $V(y_0)$ des points x_0 et y_0 dans leurs espaces respectifs. C'est là une notion qui remonte, en réalité, jusqu'à Descartes : puisque la façon dont le plan est envisagé en géométrie analytique est précisément celle de le considérer comme produit de deux droites. Il est à remarquer que, X et Y étant deux espaces (D), il en est de même de leur produit $X \times Y$: il suffit de définir comme distance $\rho(w_1, w_2)$ entre deux points $w_1 = (x_1, y_1)$ et $w_2 = (x_2, y_2)$ de $W = X \times Y$ le nombre positif $\sqrt{[\rho(x_1, x_2)]^2 + [\rho(y_1, y_2)]^2}$, $\rho(x_1, x_2)$ et $\rho(y_1, y_2)$ désignant les distances entre x_1 et x_2 entre y_1 et y_2 dans leurs espaces respectifs; cette définition de distance est bien entendu conforme avec la définition de voisinage précédemment adoptée. On démontre facilement que le produit de deux espaces compacts est lui aussi compact.

Cela posé, « l'hypothèse du Produit » est celle-ci :

La dimension de $X \times Y$ *est égale à la somme des dimensions de* X *et de* Y.

Or, cette hypothèse se trouve vérifiée si l'on entend par dimension la dimension dite sans torsion [c'est-à-dire $\Delta_0(F)$] *ou bien la dimension module m, m étant un nombre premier.*

L'hypothèse du Produit ne se trouve pas réalisée pour la dimension au sens de Brouwer-Urysohn : les exemples de M. Pontrjagin montrent en effet qu'il existe dans l'espace euclidien à quatre dimensions deux ensembles fermés et bornés possédant tous les deux la dimension (au sens de Brouwer-Urysohn) égale à 2, tandis que leur produit a la dimension 3.

18. Je passe maintenant au théorème qui est peut-être le plus important dans toute cette théorie.

Soit J^n l'intérieur d'une sphère S^{n-1} dans l'espace E^n. Considérons un ensemble fermé F quelconque situé dans E^n et ayant des points communs avec J^n; considérons enfin un *cycle polyédral* z (*voir* n° **8**, *Remarque*) situé dans J^n et n'ayant avec F aucun point commun. Nous dirons que z est *enlacé avec* F *dans* J^n, si, quel que soit le complexe C situé dans J et ayant le cycle z pour sa frontière, les simplexes constituant C ont nécessairement des points communs avec F.

Cette définition s'applique aussi bien dans le cas « module m » que dans le cas des cycles et de complexes dits « ordinaires ». On parlera donc des cycles enlacés « module m », ou bien des *cycles simplement enlacés*.

Remarque très importante. — Soit z un cycle simplement enlacé avec F dans J^n; aucun complexe C situé dans J^n et étranger à F ne possède le cycle z pour sa frontière, *le complexe* C *étant supposé*, bien entendu, *ordinaire*, c'est-à-dire *ayant des coefficients entiers*. Mais il peut fort bien arriver qu'il existe dans J^n un complexe C aux coefficients rationnels étranger à F et ayant z pour sa frontière; ou, ce qui est la même chose, il est possible qu'un certain complexe C aux coefficients entiers situé dans J^n et étranger à F a pour frontière, non le cycle z, mais le cycle tz, où t est un entier positif convenablement choisi. En d'autres mots encore : il se peut que z, n'étant pas ~ 0 dans la partie de J^n étrangère à F, est, dans cette même région, ~ 0 *avec division* (au sens de Poincaré). Dans ce cas, évidemment, ladite région de l'espace présente ce qu'on appelle une *torsion* (correspondant au nombre de dimensions du cycle z envisagé).

On est donc conduit à introduire, à côté de l'enlacement simple, une autre notion d'enlacement, qui soit *plus forte* et que nous appellerons *enlacement sans torsion :* nous dirons notamment que le cycle z (supposé toujours situé dans J^n et étranger à F) est, dans J^n, *enlacé avec* F *sans torsion*, si, quel que soit l'entier positif t, tout complexe C situé dans J^n et ayant tz pour sa frontière, possède nécessairement des points communs avec F.

19. On a alors le résultat suivant :

SECOND THEORÈME FONDAMENTAL. — *Soit* F *un ensemble fermé situé dans* E^n. *Pour que la dimension brouwerienne de* F *soit égale à* r, *il faut et il suffit qu'il existe un cycle (ordinaire) à* $n-r-1$ *dimensions simplement enlacé avec* F *dans l'intérieur* J^n *d'une sphère convenablement choisis; et qu'en même temps, aucun cycle à* $n-r-2$ *dimensions au plus ne soit enlacé avec* F.

Si, dans ce même énoncé, on considère l'enlacement sans torsion au lieu d'enlacement simple, on obtient une condition nécessaire pour que la dimension sans torsion soit égale à r.

La dimension module m *est enfin caractérisée par la condition obtenue en supposant que les cycles en question sont enlacés avec* F *(module* m*).*

On s'apercoit facilement que, si un cycle à zéro dimensions est enlacé avec F dans J^n (dans n'importe quel sens), c'est que F décompose le domaine connexe J^n en plusieurs domaines sans points communs. Il en résulte :

Corollaire I. — *Pour que l'ensemble* F *(situé dans* E^n*) possède la dimension* $n-1$ *(dans n'importe quel sens), il faut et il suffit que cet ensemble décompose (coupe) un certain domaine connexe de cet* E^n.

Corollaire II. — *Un ensemble* F, *situé dans* R^n *et possédant la dimension* $n-1$ *dans le sens d'une quelconque de nos définitions, possède la même dimension dans tous les autres sens envisagés ici.*

On prouve aisément, d'autre part, qu'un résultat analogue au corollaire II subsiste encore pour les ensembles à zéro dimensions et pour les sous-ensembles n-dimensionnels du E^n [quant à ce dernier cas, on démontre que la condition nécessaire et suffisante pour que l'ensemble F, situé dans E^n, ait la dimension n (dans n'importe quel sens), est qu'il contient des points intérieurs]. Il s'ensuit :

Corollaire III. — *Pour tous les ensembles fermés de l'espace tridimensionnel, toutes les définitions précédentes de dimension sont équivalentes entre elles.*

Le second théorème fondamental nous explique la vraie raison pour l'impossibilité d'avoir *une seule* « parfaite » définition de dimension; cette raison consiste dans l'existence des coefficients de torsion, qui semblent prédestinés à causer des ennuis aux topologues ([1]); si le monde était créé sans ces inopportuns coefficients, la dimension brouwerienne serait identique avec celle dite « sans torsion » et toutes les considérations, impliquant des modules différents, seraient disparues aussi; on ne pourrait donc écrire, dans ce cas imaginaire, qu'un seul livre sur la théorie de la dimension, tandis que maintenant, le danger est grand qu'il y en aura une infinité !

20. Mais la dimension brouwerienne possède toujours des propriétés qui lui sont particulières et dont certaines sont extrêmement intuitives; cette définition de la dimension attira donc toujours, et même de préférence, l'attention des géomètres. L'une des plus importantes parmi lesdites propriétés est celle-ci :

Pour que la dimension de F soit au plus égale à r, il faut et il suffit qu'on puisse déformer continûment l'ensemble F en un polyèdre à r dimensions au plus, et cela de facon que tout point de F soit déplacé, au cours de ladite déformation, aussi peu que l'on veut.

Une autre propriété importante de la dimension brouwerienne exige une remarque préliminaire.

Soit donnée une représentation continue f d'un ensemble fermé F sur un cube Q à un nombre quelconque de dimensions; soit F' l'ensemble de tous les points de F auxquels correspondent, en vertu de la représentation f, des points de la frontière de Q. Considérons une *modification continue de* f, c'est-à-dire une famille de représentations continues f_t de F sur Q dépendant d'une facon continue d'un paramètre t, $0 \leqq t \leqq 1$, avec $f_0 = f$. Parmi toutes les modifications continues de f, nous ne considérons comme *permises* que celles qui laissent f invariable dans tous les points de F' [c'est-à-dire qui vérifient, pour toute valeur du paramètre t et pour tout point x de F', la condition $f_t(x) = f(x)$]. Nous dirons que f est une *représentation essentielle* (de F sur Q), si, quelle que soit la représentation f_1 obtenue par une modification *permise* de f, le cube Q tout entier est couvert par l'image de F : $f_1\,\mathrm{F} = \mathrm{Q}$; dans le cas contraire f est dite *non essen-*

([1]) On se souvient, en effet, que Poincaré n'a découvert les coefficients de torsion que dans son second Mémoire et qu'il y a dû reprendre par conséquent les matières déjà traitées dans son premier Mémoire.

tielle (dans ce cas, il est évidemment possible de parvenir par une modification permise de f à une représentation f_1 de f, qui fait correspondre à chaque point de F un point *de la frontière de* Q). On a alors le théorème suivant :

Pour que la dimension de F (au sens de M. Brouwer) *soit au moins égale à r, il faut et il suffit qu'il soit possible de représenter* F *d'une facon essentielle sur un cube r-dimensionnel.*

21. On peut enfin modifier, dans le cas de la dimension brouwerienne, le second théorème fondamental de facon à le rendre indépendant de toute notion de l'*Analysis situs* combinatoire.

Soit de nouveau J^n l'intérieur d'une sphère $n-1$ dimensionnelle de l'espace E^n; considérons dans E^n un ensemble fermé F quelconque et un polyèdre P situé dans J^n et étranger à F. Nous désignons comme *permise* toute déformation continue *simultanée* de F et de P vérifiant les conditions suivantes :

1° Elle laisse invariables tous les points de F situés à l'extérieur de la sphère S^{n-1} ou sur cette sphère même;

2° Au cours de cette déformation, F et P restent toujours sans points communs.

Cela posé, nous dirons que F et P sont *enchaînés* dans J^n, s'il est impossible de réduire, par une déformation permise de F et P, le polyèdre P en un seul point.

On a alors le théorème suivant :

Pour que la dimension de l'ensemble fermé F *situé dans* E^n *soit égale à r, il faut et il suffit qu'il existe un polyèdre à* $n-r-1$ *dimensions enchaîné avec* F *dans l'intérieur* J^n *d'une certaine sphère et qu'en même temps aucun polyèdre à* $n-r-2$ *dimensions au plus ne soit enchaîné avec* F.

Le lecteur trouvera des démonstrations détaillées des résultats ci-dessus mentionnés (et de quelques autres) dans mon Mémoire *Dimensions theorie* (*Mathematische Annalen*, t. 106, 1932, p. 161-238). On y trouvera aussi plusieurs problèmes se rapportant à ce sujet et qui ne sont pas résolus encore.

Dmitrii Evgenevich Menshov

Born – 18 April 1892 in Moscow, Russia
Died – 25 Nov 1988

Dmitrii Menshov's father was a medical doctor who worked in the New Ekaterininskii Hospital and the Lazarevskii Institute of Oriental Languages. His mother was an important early influence in Menshov's education. She was a highly educated woman who, as well as educating her son, also sometimes gave French lessons to other children.

In 1904, at the age of 12 years, Menshov began his secondary schooling. He attended the gymnasium section of the Lazarevskii Institute of Oriental Languages where his father acted as school doctor. Influenced by his mother's tuition in foreign languages, Menshov's first love at school was indeed for languages. He went on to study French, German, English, Latin, and Armenian at school. However, Menshov had an outstanding mathematics teacher and, like many children who are influenced by an outstanding teacher, Menshov began to show a strong interest in mathematics from about the age of 13. He also was strongly attracted to physics so, with wide interests across many subjects, he graduated from the school in 1911 with the gold medal for outstanding achievement.

After leaving school, Menshov sat the entrance examination for the Moscow Engineering College and began his studies there in the autumn of 1911. However he only studied there for six months before deciding to leave and work on his own on learning advanced mathematics. Then in the autumn of 1912 Menshov entered the Department of Physics and Mathematics at Moscow University. There he attended lectures by Egorov, Lakhtin, Andreev and he took his first course on functions of a real variable given by Byushgens. Perhaps the most significant event for Menshov, however, was that Luzin returned from Göttingen to Moscow in the autumn of 1914 and began to lecture on functions of a real variable.

Menshov attended Luzin's lecture course, and when Luzin posed the open problem of whether the Denjoy integral and the Borel integral were equivalent, he was able to solve the problem. The Denjoy integral is the more general of the two and Menshov showed that this was the case. He showed Luzin his solution to the problem that Luzin had just posed and before the end of 1914 the two had begun a firm mathematical friendship. Menshov's discovery, made while still an undergraduate, became his first publication. It appeared as the paper *The relationship between the definitions of the Denjoy and Borel integrals* in 1916.

Luzin quickly established a School of Mathematics at Moscow University and Menshov became one of his first research students along with P.S. Aleksandrov, M. Ya Suslin, and A. Ya Khinchin. Menshov's first degree was awarded in 1916 for the thesis that he wrote on *The Riemann theory of trigonometric series* that was examined by Egorov and Luzin. However, only three weeks after he graduated, Menshov discovered one of his most fundamental results on the uniqueness problem for trigonometric series.

By the end of 1918 Menshov had been awarded his Master's degree and he went to Ivanovo north-east of Moscow, which at that time was the temporary capital of the revolutionary government, but he soon moved to Nizhnii–Novgorod where he was appointed as a professor at the University. He taught at Nizhnii–Novgorod during 1919 and early 1920 but he returned to Ivanovo in May 1920 where he was appointed as a professor at the Ivanovo Pedagogic Institute. In addition to this appointment he also taught at the Polytechnic Institute at Ivanovo from January 1921. At this time Luzin and other members of his research school were in Ivanovo so Menshov was certainly in the mainstream of the exciting mathematics that was being developed.

In the autumn of 1922 Menshov returned to Moscow and began teaching at the University. He also taught for a few years at the Moscow Institute of Forest Technology. He never submitted a thesis for a doctorate but, despite this, he was awarded the doctorate in 1935. Together with the award of the doctorate came Menshov's appointment to a professorship at Moscow University.

In 1933 a new chair of Analysis and Theory of Functions was created at Moscow University and Lavrentev was appointed. In 1938 the Faculty of Mechanics and Mathematics at Moscow University founded two chairs, the chair of the Theory of Functions and the chair of Functional Analysis. Privalov held this first chair up to 1941 but then, on Privalov's early death in that year, Menshov was appointed to the chair of the Theory of Functions. Lusternik held the chair of Functional Analysis from 1938. In 1943 these two chairs were combined and the Department of Theory of Functions and Functional Analysis was created with Menshov as its head. Menshov also worked a the Steklov Mathematical Institute of the Academy of Sciences of the USSR from 1934 to 1941 and then again from 1947.

For his work on the representation of functions by trigonometric series, Menshov was awarded a State Prize in 1951. He was then elected a Corresponding Member of the Academy of Sciences of the USSR in 1953. In 1958 Menshov attended the International Congress of Mathematicians in Edinburgh and he was invited to address the Congress with his paper *On the convergence of trigonometric series.*

Доклады Академии Наук СССР
1949. Том LXVII, № 5

МАТЕМАТИКА

Д. МЕНЬШОВ

О РЯДАХ ФУРЬЕ НЕПРЕРЫВНЫХ И СУММИРУЕМЫХ ФУНКЦИЙ

(Представлено академиком Н. Н. Лузиным 14 VI 1949)

Если задана какая-нибудь функция $f(x)$, непрерывная на сегменте $[-\pi, \pi]$, то, как известно, ее можно изменить на множество сколь угодно малой меры таким образом, чтобы для полученной новой функции ряд Фурье сходился равномерно на $[-\pi, \pi]$ ([1]). В этой теореме множество E, на котором происходит изменение функции $f(x)$, зависит вообще от вида этой функции.

Несколько видоизменяя доказательство только что формулированной теоремы, можно, однако, получить более общее утверждение, а именно, можно доказать следующую теорему.

Теорема 1. *Для любой положительной неубывающей функции $\rho(\delta)$ положительного аргумента δ, удовлетворяющей условию $\lim_{\delta\to+0} \rho(\delta)=0$, и для любого положительного числа ε существует измеримое множество E, обладающее свойствами:*

a) $\operatorname{mes} E < \varepsilon, \quad E \subset [-\pi, \pi]$;

b) *любую функцию $f(x)$, непрерывную на сегменте $[-\pi, \pi]$, модуль непрерывности которой на этом сегменте не превосходит $\rho(\delta)$, можно изменить на множестве E таким образом, чтобы для полученной новой функции ряд Фурье сходился равномерно на $[-\pi, \pi]$.*

Из теоремы 1 следует, что в теореме, формулированной в начале заметки, множество E можно считать зависящим только от числа ε и от модуля непрерывности функции $f(x)$, но не от структуры этой функции.

Возникает вопрос, нельзя ли множество E, на котором происходит изменение функции $f(x)$, считать зависящим только от числа ε, но не зависящим от модуля непрерывности функции.

Можно доказать, что дальнейшее усиление теоремы 1 невозможно; а именно, можно доказать следующую теорему.

Теорема 2. *Для любого совершенного множества P положительной меры, лежащего на $[-\pi, \pi]$, и для любой точки плотности x_0 этого множества можно определить функцию $f(x)$, непрерывную на $[-\pi, \pi]$ и обладающую следующим свойством:*

Какова бы ни была функция $\psi(x)$, ограниченная и измеримая на $[-\pi, \pi]$ и совпадающая с $f(x)$ на множестве P, ряд Фурье функции $\psi(x)$ расходится в точке x_0.

Так как дополнение к множеству P относительно сегмента $[-\pi, \pi]$ может иметь сколь угодно малую меру, то из теоремы 2 следует, что теорему 1 нельзя усилить.

Возьмем теперь произвольное совершенное нигде не плотное множество P, расположенное на сегменте $[-\pi, \pi]$.

Возникает вопрос, можно ли любую функцию $f(x)$, непрерывную на $[-\pi, \pi]$, изменить вне множества P таким образом, чтобы для полученной новой непрерывной функции $\psi(x)$ ряд Фурье сходился почти всюду.

Этот вопрос остается до сих пор не решенным, даже если мы ограничимся требованием, чтобы $\psi(x)$ была измеримой функцией с суммируемым квадратом на $[-\pi, \pi]$.

В самом деле, Г. П. Толстов доказал, что существует на сегменте $[-\pi, \pi]$ совершенное нигде не плотное множество P_0 положительной меры, обладающее следующим свойством. Каковы бы ни были измеримые функции $f(x)$ и $\psi(x)$ с суммируемым квадратом на $[-\pi, \pi]$, совпадающие на множестве P_0, почти всюду на этом множестве ряды Фурье функций $f(x)$ и $\psi(x)$ или одновременно сходятся или одновременно расходятся [2].

Таким образом, если на некотором множестве E положительной меры, расположенном на P_0, ряд Фурье от непрерывной функции $f(x)$ расходится, то при любом изменении функции $f(x)$ вне множества P_0 ряд Фурье полученной новой функции $\psi(x)$ будет расходиться почти всюду на том же множестве E при условии, что $\psi(x)$ есть измеримая функция с суммируемым квадратом на $[-\pi, \pi]$. Отсюда следует, что поставленный выше вопрос равносилен вопросу о сходимости почти всюду рядов Фурье от непрерывных функций.

Однако можно доказать следующую теорему.

Теорема 3. *Какова бы ни была функция $f(x)$, суммируемая на сегменте $[-\pi, \pi]$, и каково бы ни было совершенное нигде не плотное множество P, расположенное на том же сегменте, всегда можно изменить функцию $f(x)$ вне множества P таким образом, чтобы для полученной новой суммируемой функции $\psi(x)$ ряд Фурье сходился почти всюду на $[-\pi, \pi]$.*

Доказательство теоремы 3 опирается на следующие леммы.

Лемма А. *Возьмем произвольное действительное число c_0, произвольные положительные числа δ и δ', удовлетворяющие условию $\delta' \leqslant \delta$, и произвольные натуральные числа $p > 2$ и $\nu > 8$.*

Положим

$$c_s = c_0 + \delta\nu s, \qquad a_s = c_s - \delta',$$

$$b_s = c_s + 2\delta, \qquad a'_s = c_s - 2\delta,$$

где $s = 1, 2, \ldots, p\nu$.

Тогда

$$\left| \sum_{s=1}^{p\nu} \int_{a_s}^{c_s} \frac{\sin n(t-x)}{t-x}\, dt \right| < L \frac{\delta'}{\delta}$$

для любых натуральных чисел n и для любых точек x множества $\sum_{s=p+1}^{p(\nu-1)} [b_{s-1}, a'_s]$, где L есть абсолютная постоянная.

Лемма В. *Пусть заданы: сегмент $[c, d]$, действительное число γ, положительное число ε, натуральное число $\nu > 8$ и замкнутое нигде не плотное множество $\Pi \subset [c, d]$.*

Тогда можно определить функцию $\psi(x)$ *и измеримое множество* E, *которые обладают следующими свойствами*:

a) $\psi(x)$ *постоянна по отрезкам на сегменте* $[c, d]$ *;

b) $\int_c^d |\psi(t)|\, dt \leqslant 2|\gamma|(d-c)$;

c) $\left|\int_{c'}^{d'} \psi(t)\, dt\right| < \varepsilon$

для любого сегмента $[c', d'] \subset [c, d]$;

d) $\left|\int_c^d \psi(t) \frac{\sin n(t-x)}{t-x}\, dt\right| \leqslant B\nu|\gamma|$

для любого натурального числа n *и для любой точки* $x \in E$, *где* B — *абсолютная постоянная*;

e) $\psi(x) = \gamma$

для любой точки $x \in \Pi$;

f) $\operatorname{mes} E > (d-c)\left(1 - \frac{8}{\nu}\right), \quad E \subset [c, d]$.

Поступило
7 VI 1949

ЦИТИРОВАННАЯ ЛИТЕРАТУРА

[1] Д. Меньшов, ДАН, **32**, 245 (1941); Матем. сб., **11** (53), 67 (1942). [2] Г. Толстов, Матем. сб., **10** (52), 249 (1942).

* Функцию $\psi(x)$ мы будем называть постоянной по отрезкам на сегменте $[c, d]$, если этот сегмент можно разбить на конечное число не перекрывающихся сегментов, внутри каждого из которых $\psi(x)$ постоянна.

Moscow University Mathematics Bulletin
Vol. 47, No. 4, pp. 4-7, 1992

Vestnik Moskovskogo Universiteta. Matematika

DMITRI EVGENIEVICH MENSHOV*

(To the Centennial Anniversary of Menshov's Birth)

E.P. Dolzhenko and P.L. Ul'yanov

On April 18, 1992 it was 100 years since the birth of Dmitri Evgenievich Menshov (1892–1988), prominent Russian mathematician, one of the first representatives of the Moscow Mathematical School, winner of the State Prize, holder of five orders of the Soviet Union, one of the oldest professors of Moscow University, Corresponding Member of the USSR Academy of Sciences.

Dmitri Evgenievich was born on April 18(6), 1892 in the family of a physician. His father, Evgeni Titovich Menshov (1852–1904) was the son of an official from the town of Tula. Having graduated from the medical faculty of Moscow University, E.T. Menshov worked as a physician at the Moscow Ekaterininskaya Hospital and at Lazarev Institute of Oriental Languages. Dmitri Evgenievich's mother, Alexandra Nikolaevna (1856–1918), descended from the ancient noble lineage of Tatishchevs. Her ancestor was awarded the title of count for heroism in the Patriotic War of 1812. Alexandra Nikolaevna brought up her children, Dmitri and Maria, in the family. When her husband died, she began giving French lessons.

Young Dmitri Menshov inherited from his mother not only the knowledge of French, but also the interest to foreign languages in general. While being a student of the high-school department of Lazarev Institute of Oriental Languages (1904–1911), he eagerly studied Latin, German, and Armenian; English was the subject of his independent studies. D.E. Menshov displayed great abilities for physics and mathematics already in his school days. In his remembrances of that time Dmitri Evgenievich always spoke with particular warmth about his teachers of mathematics V.N. Sedashev and L. Sevast'yanov and about his teacher of German K.Ch. Sorgenfrei.

Dmitri Evgenievich finished high school in 1911 with a gold medal. Feeling drawn to scientific work, but having no idea where he could actually do it, Dmitri Evgenievich entered Moscow Engineering School after having passed serious competitive examinations (more serious, D.E. Menshov remarked afterward, than those at the Physico-Mathematical Department of Moscow University). Half a year later, however, Dmitri Evgenievich became convinced that his future speciality was absolutely at variance with his vocation. So, he left this School and started to study higher mathematics independently, earning his living by giving private lessons. It took Dmitri Evgenievich only half a year to master mathematics within the scope of the first and partially of the second university course. In the fall of 1912 Dmitri Evgenievich entered the Physico-Mathematical Department of Moscow University. There he attended the lectures of such remarkable professors as D.F. Egorov, L.K. Lakhtin, K.A. Andreev. At the lectures of S.S. Bushgens D.E. Menshov became acquainted with the theory of functions of a real variable; then he attended the same course of lectures read by N.N. Luzin, then a young privat-docent. In 1914 N.N. Luzin returned from his scientific mission to Göttingen and Paris and in the fall of the same year began reading a special course in the theory of functions. His lectures aroused great interest of the students not only because the topic was novel, but also because the lecturer managed to formulate naturally a wealth of unsolved problems. At one of those lectures N.N. Luzin posed the question about the relationship between two generalizations of Lebesgue's integral: Borel's integral and Denjoy's integral, and announced that he was going to open a seminar devoted to this subject. D.E. Menshov, then a student of the third course, decided to take part in the work of this seminar and began preparing for it. A month later he proved that Borel's integral was narrower than Denjoy's integral. When N.N. Luzin learned about it, he invited D.E. Menshov for a talk. When N.N. Luzin saw that Menshov's proof was correct, he canceled the planned seminar and, instead of it, announced a seminar on monogenic functions.

On December 14 of the same year Dmitri Evgenievich reported his result at the session of Moscow Mathematical Society (he was elected a member of this Society later, in 1919). The same result was also the subject of the first Menshov's publication (in *Matematicheskii Sbornik*, 1916). After that the interest of mathematicians to Borel's integral began to decline.

* Spelt D. Menchoff in French references.

Moscow University Mathematics Bulletin *Vol. 47, No. 4*

It should be pointed out that D.F. Egorov contributed to the development of the theory of functions at Moscow University. In particular, it was by his recommendation that N.N. Luzin was commissioned abroad. N.N. Luzin played a decisive part in laying the basis of the Moscow school of the theory of functions. Luzin impersonated a tremendous mathematical gift together with an exceptional ability to recruit the most talented young men to scientific investigations. P.S. Alexandrov, D.E. Menshov, M.Ya. Suslin, A.Ya. Khinchin were among the first Luzin's disciples.

The school of N.N. Luzin grew rapidly owing to the influx of talented youth. Later on N.N. Luzin's friends, V.V. Golubev, I.I. Privalov, and V.V. Stepanov, joined this group. The Moscow Mathematical School was thus formed.

In 1916 Dmitri Evgenievich Menshov graduated from Moscow University and was left there for preparing to the professorial rank. His graduation thesis "Riemann's Theory of Trigonometric Series" was written under the scientific guidance of D.F. Egorov and N.N. Luzin. Three weeks after the graduation from the University D.E. Menshov constructed the so-called trigonometric zero-series. This result is of such a fundamental importance, that it should be discussed in more detail.

Any one who wants to learn the theory of trigonometric series starts with the Fourier series

$$(a_0/2) + \sum_{n=1}^{\infty} a_n \cos nt + b_n \sin nt \tag{1}$$

of continuous 2π-periodic functions $f(t)$. The coefficients a_n and b_n of this series are expressed in terms of $f(t)$ by simple formulas:

$$\begin{aligned} a_n &= (1/\pi)\int_0^{2\pi} f(t)\cos nt\,dt, \\ b_n &= (1/\pi)\int_0^{2\pi} f(t)\sin nt\,dt \quad (n = 0, 1, \ldots). \end{aligned} \tag{2}$$

There is a lot of criteria of uniform convergence for the Fourier series. It is easy to see that if a certain trigonometric series of form (1) converges to a function $f(t)$, then its coefficients are expressed by formulas (2), i.e., in this case series (1) is Fourier's series of this function.

As to the uniqueness of expansion of an *arbitrary* function into a trigonometric series, even in the late nineteenth century it was known (G. Cantor, W. Young) that none measurable and everywhere finite 2π-periodic function can be a sum of two different trigonometric series each of which converges to it everywhere, with the exception of only a certain denumerable set of points. Stated otherwise, if trigonometric series (1) converges to zero at each t from $[0, 2\pi]\backslash E$, where E is no more than a denumerable set, then all the coefficients of this series are equal to zero. Insofar as in the metric theory of functions sets of measure zero can often be neglected (for instance, a change of the summable function in the set of measure zero has no effect on the Fourier coefficients of this function), many outstanding mathematicians in the early twentieth century tried to prove that in the theorem of uniqueness just formulated it will be sufficient to require that the set E should have Lebesgue's measure zero. For this very reason the trigonometric zero-series constructed by D.E. Menshov, i.e., the series converging almost everywhere to zero, not all the coefficients of which are equal to zero, was extremely surprising to everybody. Further development of this topic has shown that the problem of characterizing sets $E \subset [0, 2\pi]$ which possess the property that from the convergence of the series to zero on $[0, 2\pi]\backslash E$ there follows the equality of all its coefficients to zero (such sets E are called U-sets) proved to be extremely difficult. It remains unsolved up till now, even for perfect sets E.

Dmitri Evgenievich passed his examinations for master's degree at Moscow University ahead of schedule, less than in two years. In 1918 his mother died. At that time hunger reigned in Moscow, the Civil War was going on in Russia. At the end of 1918 Dmitri Evgenievich moved to the town of Ivanovo, but soon he left Ivanovo for Nizhni Novgorod and began working at the university there as a professor. In the spring of 1920 he returned to Ivanovo and taught at the Pedagogical Institute there; since January 1921 D.E. Menshov taught at the Polytechnical Institute as well. At that time N.N. Luzin and his disciples A.Ya. Khinchin, V.S. Fedorov, and others worked in Ivanovo.

In the fall of 1922 Dmitri Evgenievich returned to Moscow and began teaching at the Forestry Engineering Institute (till 1925). At the same time he worked at Moscow University: first as a non-staff teacher, then

Moscow University Mathematics Bulletin *Vol. 47, No. 4*

as a docent (1927), and since 1935 to the last day of his life as a professor. In 1935 for great achievements in the theory of functions D.E. Menshov was granted the academic degree of Ph.D. without dissertation. From 1941 to 1979 he held the chair of the theory of functions and functional analysis, having succeeded I.I. Privalov (1891–1941).

Dmitri Evgenievich also worked at the Institute of Mathematics and Mechanics at Moscow State University (1923–1934), at V.I. Lenin Moscow Pedagogical Institute as a professor (1929–1935), and at V.A. Steklov Mathematical Institute of the USSR Academy of Sciences (1934–1941, 1947–1988).

The main investigations of D.E. Menshov relate to the theory of trigonometric series, to the theory of orthogonal series, to the theory of conformal mappings of plane domains, and to the theory of monogenic functions. The results obtained by him in each of these fields of the theory of functions determined the trends of their development for many years to come. This stemmed from the approach of Dmitri Evgenievich to mathematics: he always tackled the most difficult problems of fundamental importance and finally succeeded in solving the chosen problem, sometimes after many years of perseveration.

We have already spoken above about the first outstanding contribution made by Dmitri Evgenievich to the theory of trigonometric series: about the zero-series. It deals with the uniqueness of representation of the measurable function by the trigonometric series. Later on D.E. Menshov returned more than once to the problem of representation of the measurable function as a sum of the trigonometric series, posed by N.N. Luzin.

In 1915 N.N. Luzin demonstrated that for any measurable and almost everywhere finite 2π-periodic function there exists a trigonometric series, almost everywhere summable to it by the methods of Abel–Poisson and Riemann. I.I. Privalov extended this Luzin's theorem to Cesaro's methods (C, α) at all $\alpha > 1$. These results were, naturally, far from the solution of the above-mentioned problem of N.N. Luzin. In 1937 Dmitri Evgenievich found a new approach to solving this problem. The theorem he published in 1940 asserts that any measurable and almost everywhere finite 2π-periodic function can be represented by a trigonometric series converging to it almost everywhere (certainly, such a representation is not unique in view of the existence of zero-series). For proving this theorem Dmitri Evgenievich resorted to his another brilliant result in the theory of convergence of trigonometric series: any measurable and almost everywhere finite function $f(t)$ $(t \in [0, 2\pi])$ can be modified in such a manner on a certain set U with a measure smaller than any preset number $\sigma > 0$ that the obtained continuous function will have a Fourier series uniformly converging to it.

Thus, the problem of N.N. Luzin for the case of *almost everywhere finite* functions was solved completely by D.E. Menshov by 1940. He demonstrated also that for an *arbitrary* function $f(t)$, measurable on $[0, 2\pi]$ there exists a trigonometric series with coefficients tending to zero which converges to $f(t)$ on $[0, 2\pi]$ *in terms of measure* (1950). Later, in 1971, D.E. Menshov considered a very delicate problem of representation of the measurable, almost everywhere finite 2π-periodic function $f(t)$ by the trigonometric series summable to it by Cesaro's method (C, α) of a preset negative nonintegral order α, and proved that the function $f(t)$ can be modified on the set $E \subset [0, 2\pi]$ whose measure is smaller than the number σ specified in advance, in such a manner that in the obtained function $g(t)$ a certain subsequence of Cesaro's α-means of its Fourier series will converge uniformly to $g(t)$ on $[0, 2\pi]$.

Important and in some cases fundamental results were obtained by Dmitri Evgenievich also in the study of universal trigonometric series (relative to different kinds of convergence), limits of the uncertainty of partial sums of divergent (in particular, trigonometric) series, different methods of summation of numerical, trigonometric, and other functional series.

Since 1920 to the end of his life D.E. Menshov periodically dealt with the theory of orthogonal series. He obtained fundamental results in this field, particularly, related to the convergence and summability of such series. A detailed review of these results is given in the paper written by P.L. Ul'yanov and published in this issue.

In 1923–1935 Dmitri Evgenievich worked actively at the fundamentals of the theory of conformal mappings and monogenic functions. In particular, owing to his works, the theory of conformal mappings of plane domains, geometrical by its nature, has become forever a chapter in the theory of analytic functions. A review of these results can be found in the paper written by E.P. Dolzhenko, published in the present issue.

The second field of the interests of Dmitri Evgenievich, after the theory of functions, was quantum mechanics, though he, punctilious to his own works, never prepared a single paper on this subject for the press. He was a constant and active participant of the seminar on theoretical physics at the Physical Institute

Moscow University
Mathematics Bulletin *Vol. 47, No. 4*

of the USSR Academy of Sciences and of the seminar on the quantum theory of field at the Department of Physics of Moscow University.

Overall, D.E. Menshov published about 100 scientific papers in Russian and foreign periodicals. Many of them proved to be the source of inspiration and new ideas for numerous specialists in the theory of functions in different countries of the world.

To the end of his days Dmitri Evgenievich carried out pedagogical work. At Moscow University, where he taught since 1922, he read practically all the general mathematical courses and different special courses, conducted annually special seminars on the theory of functions of real variable: from seminars for junior students to the permanently functioning All-Moscow seminar for scientific workers. These seminars were noted for a combination of respectful attitude to the person making a report and to the participants of the seminar, on the one hand, and high requirements to the quality of the report, striving to get to all the "technicalities", with thorough verification of all the calculations on the blackboard, on the other hand. Such a style attracted both venerable mathematicians and capable young students to the seminars.

Dmitri Evgenievich devoted much time to individual work with his pupils, readily responded to any request for consultation. This used to take all his free evenings and weekends.

Dmitri Evgenievich trained more than 35 doctors and candidates of science, G.P. Tolstov, S.B. Stechkin, L.A. Brudno, V.G. Chelidze, N.P. Kuptsov, I.I. Volkov, Yu.B. Germeier, A.A. Talalyan, E.P. Dolzhenko, V.A. Skvortsov among them.

All who were lucky to know Dmitri Evgenievich, feel deep respect and love to him, are conquered forever by his unique personality. In the last decades of his life D.E. Menshov in his appearance was a tall, graceful, thin, cultured old man with a small beard and grayish light hair, speaking in a pleasant rumbling deep voice. He was a kind, responsive, sincere man. A man of perfect honesty, non-compromizing in this respect, a man of tremendous mathematical talent, an excellent pedagogue and lecturer, a real scientist and patriot—such was Dmitri Evgenievich Menshov.

Aleksandr Osipovich Gelfond

Born – 24 Oct 1906 in St Petersburg, Russia
Died – 7 Nov 1968 in Moscow, Russia

Aleksandr Gelfond taught mathematics at the Moscow Technological College (1929–1930) and then from 1931 at Moscow State University where he held chairs of analysis, theory of numbers and the history of mathematics.

In addition to his important work in the number theory of transcendental numbers (that is, numbers that cannot be expressed as the root or the solution of an algebraic equation with rational coefficients), Gelfond made significant contributions to the theory of interpolation and the approximation of functions of a complex variable.

In 1934 Gelfond proved a special case of his conjecture namely that a^b is transcendental if a is algebraic ($a \neq 0, 1$) and b is an irrational algebraic number. This result is now known as Gelfond's theorem and solved Problem 7 of the list of Hilbert 23 problems. It was solved independently by Schneider. In 1966 A. Baker proved Gelfond's Conjecture in general.

Gelfond's major contributions to transcendental numbers are in *Transtendentnye i algebraicheskie chisla* (1952) and to approximation and interpolation theories are in *Ischislenic konechnykh raznostey* (1952).

ИЗВЕСТИЯ АКАДЕМИИ НАУК СССР. 1934
BULLETIN DE L'ACADEMIE DES SCIENCES DE L'URSS
Classe des sciences mathématiques et naturelles
Отделение математических и естественных наук

SUR LE SEPTIÈME PROBLÈME DE HILBERT

Par A. GELFOND

(*Presenté par* I. Vinogradov, *Membre de l'Académie*)

Au Congrès international des Mathématiciens, tenu à Paris en 1900, D. Hilbert, parmi ses 23 problèmes, en posa un concernant la nature arithmétique des nombres ω^η, où ω est un nombre algébrique et η un nombre algébrique irrationnel.

Le présent travail donne l'exposé complet d'une solution générale du problème, dont la marche se trouve déjà résumée dans ma note sous le même titre, parue aux «Comptes Rendus de l'Académie des Sciences de l'URSS» (1934, vol. II, № 1, p. 1). La même note renferme un bref aperçu de la situation à ce jour du problème posé par Hilbert.

Soient $\alpha \neq 0,1$ et $\beta \neq 0,1$ des nombres algébriques. Admettons que $\eta = \frac{\ln \alpha}{\ln \beta}$ soit un nombre algébrique irrationnel. Admettons de plus que les nombres α, β et η appartiennent, respectivement, à des équations de degrés p_1, p_2 et p_3 à coefficients rationnels entiers.

Désignons par $\alpha_1, \alpha_2, \ldots, \alpha_{p_1-1}$ des nombres conjugués à α, par $\beta_1, \beta_2, \ldots, \beta_{p_2-1}$ et par $\eta_1, \eta_2, \ldots, \eta_{p_3-1}$ des nombres conjugués à β et η. Soit $c > 0$ un nombre entier rationnel tel que les nombres $c\alpha$, $c\beta$ et $c\eta$ soient des nombres entiers algébriques.

Introduisons une série de constantes qui nous seront nécessaires par la suite.

Définissons d'abord des constantes $\gamma, \lambda_1, \lambda_2, \lambda_3, \lambda_4, \lambda_5$ par les égalités suivantes:

$$\begin{gathered}\gamma = e^{2\pi} \operatorname{Max}[3, |\alpha|, |\alpha_1|, \ldots, |\alpha_{p_1-1}|, \\ |\beta|, |\beta_1|, \ldots, |\beta_{p_2-1}|, |\eta|, |\eta_1|, \ldots, |\eta_{p_3-1}|].\end{gathered} \tag{1}$$

$$\lambda_1 = 2 \ln 3 \cdot p_1 \cdot p_2 \cdot p_3. \tag{2}$$

$$\lambda_2 = (2 \ln \gamma + 4 \ln c) p_1 \cdot p_2 \cdot p_3. \tag{3}$$

$$\lambda_3 = 2p_1 \cdot p_2 \cdot p_3. \tag{4}$$

$$\lambda_4 = \ln[2 \ln \gamma]. \tag{5}$$

$$\lambda_5 = 2 \ln \gamma. \tag{6}$$

Il est évident que les constantes définies de la sorte dépendent exclusivement des propriétés des nombres algébriques α, β et η.

Introduisons ensuite dans nos opérations des constantes

$$q_1,\ q_2,\ q_3,\ q_4,\ \ldots,\ q_8,$$

qui seront toutes des nombres entiers positifs définis par les conditions que pour chaque nombre entier positif $q \geqq q_i$, $i = 1, 2, \ldots, 8$ doit être remplie une certaine inégalité. Ces inégalités se présentent comme suit:

Pour $q \geqq q_1$,

$$\ln q > \ln(\gamma + \gamma^2) + 2 \ln c; \tag{7}$$

pour $q \geqq q_2$,

$$\frac{\ln q}{\ln \ln q} > 4(3 \ln 3 + \lambda_4 + \lambda_5); \tag{8}$$

pour $q \geqq q_3$,

$$\frac{\ln q}{\ln \ln q} > 4(\lambda_1 + \lambda_2 + \lambda_3); \tag{9}$$

pour $q \geqq q_4$,

$$\ln \ln q > 6(2 \ln 6 + \lambda_5); \tag{10}$$

pour $q \geqq q_5$,

$$\ln \ln q > 12(\lambda_1 + \lambda_2 + \lambda_3); \tag{11}$$

pour $q \geqq q_6$,

$$q^{7/3} < \frac{1}{2} \frac{q^{5/2}}{\ln \ln q} - (2 \ln 6 + \lambda_5) \frac{q^{5/2}}{\ln q \ln \ln q}; \tag{12}$$

pour $q \geqq q_7$,

$$q^{7/3} > 5q^2 \ln(5q^2) + \ln 2; \tag{13}$$

pour $q \geqq q_8$,

$$q^{7/3} > 5(\lambda_1 + \lambda_2 + \lambda_3) q^2 \ln q. \tag{14}$$

Comme les constantes précédemment introduites λ_i, $i = 1, 2, \ldots, 5$, dépendaient seulement de la nature des nombres α, β et η, de même les constantes q_k, $k = 1, 2, \ldots, 8$, ne dépendent rien que des propriétés algébriques de ces mêmes nombres.

Posons également

$$r_1 = \left[\frac{q_0^2}{\ln q_0 \ln \ln q_0}\right], \quad r_2 = [\ln^2 \ln q_0] \tag{15}$$

et

$$q_0 = \mathbf{1} + \mathrm{Max}[3^{3^4},\ q_1,\ q_2,\ q_3,\ q_4,\ q_5,\ q_6,\ q_7,\ q_8]. \tag{16}$$

Par cela nous fixons ferme les nombres q_0, r_1, r_2.

Nous considérerons dans la suite des nombres s et t qui seront toujours non négatifs, entiers et variables.

Examinons maintenant la fonction

$$f(x)=\sum_{k=0}^{k=q_0}\sum_{l=0}^{l=q_0} C_{k,l}\,\alpha^{kx}\,\beta^{lx}, \tag{17}$$

où tous les nombres $C_{k,l}$; $k=0, 1, \ldots, q_0$; $l=0, 1, \ldots, q_0$, doivent être entiers et rationnels.

Pour de telles fonctions nous allons démontrer trois lemmes qui nous seront nécessaires par le suite.

Lemme I. *Quels que soient les nombres entiers rationnels* $C_{k,l}$ *entrant dans la fonction* $f(x)$ *[égalité (17)], si* $C_{k,l}\leqq 3^{q_0^2}$ *et si nous admettons que les trois nombres* α, β et $\eta=\frac{\ln\alpha}{\ln\beta}$ *soient algébriques en même temps, alors ou bien est remplie l'inégalité*

$$|f^{(s)}(t)|>e^{-(\lambda_1 q_0^2+\lambda_2 q_0 t+\lambda_3 s\ln q_0)}, \tag{18}$$

où les nombres q_0, λ_1, λ_2, λ_3 *sont définis par les égalites* (16), (2), (3), (4), *ou bien*

$$f^{(s)}(t)=0. \tag{19}$$

Démonstration. Considérons l'expression $\left[\frac{d^{(s)}f(x)}{dx^s}\right]_{x=t}$:

$$\begin{aligned} f^{(s)}(t) &= \sum_{k=0}^{k=q_0}\sum_{l=0}^{l=q_0} C_{k,l}(k\ln\alpha+l\ln\beta)^s\,\alpha^{kt}\,\beta^{lt}= \\ &=\ln^s\beta\sum_{k=0}^{k=q_0}\sum_{l=0}^{l=q_0} C_{k,l}(k\eta+l)^s\,\alpha^{kt}\,\beta^{lt}. \end{aligned} \tag{20}$$

Puisque les nombres α, β et η sont algébriques et les nombres $c\alpha$, $c\beta$ et $c\eta$ entiers algébriques, il en résulte que pour des $C_{k,l}$, s et t fixés le nombre

$$f^{(s)}(t)\ln^{-s}\beta=\sum_{k=0}^{k=q_0}\sum_{l=0}^{l=q_0} C_{k,l}(k\eta+l)^s\,\alpha^{kt}\,\beta^{lt} \tag{21}$$

doit être algébrique et le nombre

$$c^{s+2q_0t}f^{(s)}(t)\ln^{-s}\beta$$

doit être entier algébrique et appartenir à une équation irréductible de dégré m ne dépassant pas $p_1p_2p_3$.

Les nombres qui lui sont conjugués d'après cette équation s'obtiendront à l'aide de substitution dans (21) de tous les nombres conjugués à α, β et η ou de quelques-uns d'entre eux.

Il n'est pas difficile d'evaluer aussi bien le nombre

$$c^{s+2q_0t} f^{(s)}(t) \ln^{-s} \beta$$

que ses conjugués. On obtient immédiatement l'evaluation:

$$\left| c^{s+2q_0t} \overline{[f^{(s)}(t) \ln^{-s} \beta]} \right| \leqq (q_0+1)^2\, 3^{q_0^2}\, q_0^s (\gamma+1)^s \gamma^{2q_0t} c^{s+2q_0t} \leqq \\ \leqq e^{2\ln 3 q_0^2} \cdot e^{s\ln q_0+s\ln(1+\gamma)+s\ln c} \cdot e^{(2\ln\gamma+2\ln c)q_0t}, \tag{22}$$

où γ est défini par l'égalité (1).

Puisque la norme d'un nombre algébrique différent de zéro n'est pas moindre que 1 en valeur absolue il s'ensuit immédiatement de l'évaluation (22)

$$\left| c^{s+2q_0t} f^{(s)}(t) \ln^{-s} \beta \right| > e^{-p_1p_2p_3[2\ln 3 q_0^2+s\ln q_0+s\ln(1+\gamma)+s\ln c+(2\ln\gamma+2\ln c)q_0t]} \tag{23}$$

ou, en dernier lieu, puisque $q_0 > q_1$ (inégalité (7)),

$$\left| f^{(s)}(t) \right| > e^{-(\lambda_1 q_0^2+\lambda_2 q_0 t+\lambda_3 s\ln q_0)} \tag{24}$$

où λ_1, λ_2, λ_3 sont définis par les inégalités (2), (3), (4).

Lemme II. *A la condition de* $|C_{k,l}| \leqq 3^{q_0^2}$ *les coefficients entiers et dans leur ensemble différents de zéro* $C_{k,l}$ *de la fonction* $f(x)$ *peuvent toujours être choisis de facon à ce que soient simultanément remplies* r_1, r_2 *conditions, où*

$$r_1 = \left[\frac{q_0^2}{\ln q_0 \ln\ln q_0} \right], \quad r_2 = [\ln^2 \ln q_0],$$

et

$$\left| f^{(s)}(t) \right| < e^{-\frac{1}{6}\frac{q_0^2 \ln q_0}{\ln\ln q_0}}, \quad 0 \leqq s \leqq r_1 - 1, \quad 0 \leqq t \leqq r_2 - 1. \tag{25}$$

Démonstration. Dans l'ex ression de $f^{(s)}(t)$

$$f^{(s)}(t) = \sum_{k=0}^{q_0} \sum_{l=0}^{q_0} C_{k,l} (k \ln \alpha + l \ln \beta)^s \alpha^{kt} \alpha^{lt} \tag{26}$$

il entre $(q_0+1)^2$ coefficients qui parcourent toutes les valeurs entières entre zéro et $3^{q_0^2}$. Pour satisfaire aux conditions (25) profitons du principe bien connu de Dirichlet. Envisageons un espace à $2r_1r_2$ dimensions et déposons suivant les axes de cet espace les nombres

$$Rf^{(s)}(t) \quad \text{et} \quad If^{(s)}(t) \quad 0 \leqq s \leqq r_1 - 1, \quad 0 \leqq t \leqq r_2 - 1. \tag{27}$$

Sur chacun des axes nous obtiendrons $(3^{q_0^2}+1)^{(q_0+1)^2}$ points. Les points seront disposés le long de chaque axe sur un intervalle dont la longueur n'excédera pas

$$2 \max_{\substack{0 \leq s \leq r_1 - 1 \\ 0 \leq t \leq r_2 - 1}} |f^{(s)}(t)| < (q_0+1)^2\, 3^{q_0^2}\, q_0^{r_1} (2\gamma)^{r_1} \gamma^{2q_0 r_2} < e^{(2\ln 3 + \lambda_4 + \lambda_5)q_0^2}, \quad (28)$$

où λ_4 et λ_5 sont définis par les égalités (5) et (6).

Divisons chacun de ces intervalles en

$$\left[e^{\frac{\ln 3}{2} \frac{q_0^2 \ln q_0}{\ln \ln q_0}} \right] \quad (29)$$

parties égales. Faisant passer dans notre espace des plans parallèles aux plans de coordonnés par les points de division, nous y obtenons des cubes au nombre de

$$\left[e^{\frac{\ln 3}{2} \frac{q^2 \ln q_0}{\ln \ln q_0}} \right]^{2r_1 r_2} < e^{\ln 3\, q_0^4}. \quad (30)$$

Or, la partie de l'espace occupée par ces cubes renferme

$$(3^{q_0^2}+1)^{(q_0+1)^2}$$

de nous susdits points. Par conséquent au moins un de ces cubes devra renfermer au moins deux points différents, dont chacun correspond à l'une de nos $(3^{q_0^2}+1)^{(q_0+1)^2}$ fonctions $f(x)$. Donc, les coordonnées du point qui correspond à la différence de ces deux fonctions devront toutes satisfaire aux inégalités

$$\begin{gathered} |Rf^{(s)}(t)| < e^{(2\ln 3 + \lambda_4 + \lambda_5) q_0^2 - \frac{\ln 3}{2} \frac{q_0^2 \ln q_0}{\ln \ln q_0} + \ln 3} \\ |If^{(s)}(t)| < e^{(2\ln 3 + \lambda_4 + \lambda_5) q_0^2 - \frac{\ln 3}{2} \frac{q_0^2 \ln q_0}{\ln \ln q_0} + \ln 3}, \\ 0 \leq s \leq r_1 - 1, \quad 0 \leq t \leq r_2 - 1. \end{gathered} \quad (31)$$

De là s'ensuivent les inégalités

$$\begin{gathered} |f^{(s)}(t)| < e^{(3\ln 3 + \lambda_4 + \lambda_5) q_0^2 - \frac{\ln 3}{2} \frac{q_0^2 \ln q_0}{\ln \ln q_0}} < e^{-\frac{1}{4} \frac{q_0^2 \ln q_0}{\ln \ln q_0}} \\ 0 \leq s \leq r_1 - 1, \quad 0 \leq t \leq r_2 - 1, \end{gathered} \quad (32)$$

puisque $q_0 > q_2$ (inégalité (8)).

Ainsi donc, la différence de nos deux fonctions représente une fonction satisfaisent aux exigences du lemme.

Lemme III. *Si* $| C_{k,e} | \leqq 3^{q_0}$, *alors* $f(x) = \sum_{k=0}^{k=q_0} \sum_{l=0}^{l=q_0} C_{k,e}\, \alpha^{kx} \beta^{lx}$ *satisfait, pour n'importe quelle valeur complexe.de* x *à l'inegalité*

$$| f(x) | < e^{2 \ln 3\, q_0^2 + \lambda_5 q_0 | x |}, \tag{33}$$

où A_5 *est défini par l'egalité* (6).

La démonstration de ce lemme est évidente.

Théorème. *Si* $\alpha \neq 0, 1$ et $\beta \neq 0, 1$ *sont des nombres algébriques, le nombre* $\eta = \frac{\ln \alpha}{\ln \beta}$ *sera ou bien rationnel on bien transcendant.*

Démonstration. Supposons tous les trois nombres α, β et η algebriques et η, de plus, irrationnel. Admettons encore qu'ils satisfassent à toutes les conditions imposées aux nombres α, β et η au commencement de notre mémoire. Alors, suivant le lemme II, on peut trouver une fonction

$$f(x) = \sum_{k=0}^{k=q_0} \sum_{l=0}^{l=q_0} C_{k,l}\, \alpha^{kx} \beta^{lx}, \quad | C_{k,l} | \leqq 3^{q_0}, \tag{34}$$

dont les coefficients $C_{k,l}$ seront des nombres entiers, rationnels et non égaux à zéro dans leur ensemble et pour laquelle seront remplies les inégalités

$$| f^{(s)}(t) | < e^{-\frac{1}{4} q_0^2 \frac{\ln q_0}{\ln \ln q_0}}, 0 \leqq s \leqq r_1 - 1; \quad 0 \leqq t \leqq r_2 - 1. \tag{35}$$

Pour cette fonction doit être valable, avec des entiers quelconques $s \geqq 0$ et $t \geqq 0$, le lemme I, c'est-à-dire que l'on a ou bien

$$| f^{(s)}(t) | > e^{-(\lambda_1 q_0^2 + \lambda_2 q_0 t + \lambda_3 s \ln q_0)} \tag{36}$$

ou bien $f^{(s)}(t) = 0$.

Juxtaposant les inégalités (35) et (36) nous obtenons, puisque $q_0 > q_3$, en vertu de l'inégalité (9), pour notre fonction $f(x)$

$$f^{(s)}(t) = 0 \quad 0 \leqq s \leqq r_1 - 1, \quad 0 \leqq t \leqq r_2 - 1. \tag{37}$$

Donc, la fonction $f(x)\, [x(x-1)\ldots(x - r_2 + 1)]^{-r_1}$ doit être entière. De là s'ensuit, conformément au principe du maximum et au lemme III, l'inégalité

$$| f(x) [x(x-1)\ldots(x - r_2 + 1)]^{-r_1} | < e^{(2 \ln 3 + \lambda_6) q_0^2 - r_1 r_2 \ln \frac{q_0}{2}} \tag{38}$$

pour $| x | < q_0$.

De (38) il dérive immédiatemant que pour $|x| \leqq q_0^{2/3}$ le module de $f(x)$ satisfait à l'inégalité

$$| f(x) | < e^{(2 \ln 3 + \lambda_5) q_0^2 - r_1 r_2 \ln \frac{q_0}{2} + \frac{2}{3} r_1 r_2 \ln 2q_0} \tag{39}$$

ou, puisque $q_0 > q_4$, en vertu de l'inégalité (10)

$$| f(x) | < e^{-\frac{1}{6} q_0^2 \ln \ln q_0}, \ | x | \leqq {}_0q^{2/3} \tag{40}$$

Mais comme $f(x)$ est une fonction analytique de x nous obtiendrons en en représentant les dérivées par une intégrale de Cauchy, que

$$f^{(s)}(t) = \frac{s!}{2\pi i} \int \frac{f(x)}{(x-t)^s} dx, \qquad 0 \leqq t \leqq [\sqrt{q_0}] \tag{41}$$

Cela nous permet d'évaluer le module de $f^{(s)}(t)$ pour un t quelconque $0 \leqq t \leqq [\sqrt{q_0}]$. En effet, l'on a

$$| f^{(s)}(t) | < e^{\frac{q_0^2}{\ln \ln q_0}} e^{-\frac{1}{6} q_0^2 \ln \ln q_0} < e^{-\frac{1}{12} q_0^2 \ln \ln q_0}, \qquad 0 \leqq s \leqq r_1, \ \ 0 \leqq t \leqq [\sqrt{q_0}]. \tag{42}$$

Juxtaposant les inégalités (42) et (18) nous obtiendrons, puisque $q_0 > q_5$, et partant est remplie l'inégalité (11), en vertu du lemme I, que

$$f^{(s)}(t) = 0, \quad 0 \leqq s \leqq r_1, \quad 0 \leqq t \leqq [\sqrt{q_0}]. \tag{43}$$

Mais alors, la fonction $f(x)[x(x-1)\ldots(x-[\sqrt{q_0}])]^{-r_1}$ doit être entière. Utilisant derechef le principe du maximum d'une fonction analytique, nous aurons l'inégalité

$$| f(x)[x(x-1)\ldots(x-[\sqrt{q_0}])]^{-r_1} | < e^{(2 \ln 3 + \lambda_5) q_0^2 - r_1 \sqrt{q_0} \ln \frac{q_0}{2}}, \tag{44}$$

vraie pour $| x | < q_0$. Il suit de la que chaque x, $| x | = 2$, se vérifie l'inégalité

$$| (x) < e^{(2 \ln 3 + \lambda_5) q_0^2 - r_1 \sqrt{q_0} \ln \frac{q_0}{2} + r_1 (\sqrt{q_0} + 1) \ln 2 \sqrt{q_0}} \tag{45}$$

Mais, puisque $q_0 > q_6$ et que, pour cette raison, l'inégalité (13) est remplie, nous avons

$$f(x) | < e^{-2 q_0^{7/3}}, \ | x | = 2. \tag{46}$$

Ecrivant l'éxpression de $f^{(s)}(0)$ à l'aide de l'integrale de Cauchy,

$$f^{(s)}(0)=\frac{s!}{2\pi i}\int\frac{f(x)\,dx}{x^{s+1}}, \tag{47}$$

nous pouvons évaluer $|x|=2$ la valeur du module de $f^{(s)}(0)$ pour un s quelconque, $0\leqq s\leqq 5q_0^2$.

En effet, vu l'inégalité (46), nous avons

$$|f^{(s)}(0)|< e^{5q_0^2\ln 5q_0^2-2q_0^{7/3}},\quad 0\leqq s\leqq 5q_0^2, \tag{48}$$

ou, puisque $q_0 < q_7$ et que l'inégalilé (13) est remplie,

$$|f^{(s)}(0)|\leqq e^{-q_0^{7/3}},\quad 0\leqq s\leqq 5q_0^2. \tag{49}$$

En comparant les inégalités (49) et (18) nous pouvons établir que

$$f^{(s)}(0)=0,\quad 0\leqq s\leqq 5q_0^2, \tag{50}$$

puisque $q_0 > q_8$ et, par conséquent, est remplie l'inégalité (14).

Or, le système des égalités (50) peut être regardé comme un système d'équations linéaires homogènes à $(q_0+1)^2$ inconnues $C_{k,l}$,

$$f^{(s)}(0)=\sum_{k=0}^{k=q_0}\sum_{l=0}^{l=q_0}C_{k,l}(k\ln\alpha+l\ln\beta)^s=0,\quad 0\leqq s\leqq(q_0+1)^2-1, \tag{51}$$

puisque $5q_0^2>(q_0+1)^2$.

Le déterminant de ce système est évidemment un déterminant de Vandermonde. Il doit être différent de zéro, car la relation

$$k_1\ln\alpha+l\ln\beta=k_2\ln\alpha+l_2\ln\beta;\quad (k_1-k_2)^2+(l_1-l_2)^2>0 \tag{52}$$

est impossible en vertu de l'irrationalité du nombre $\eta=\frac{\ln\alpha}{\ln\beta}$.

Par conséquent, tous les $C_{k,l}$, $0\leqq k\leqq q_0$, $0\leqq l\leqq q_0$, doivent être égaux à zéro.

Nous aboutissons à une contradiction, parce que les coefficients de la fonction $f(x)$, de par le mode son choix, ne devaient pas, dans leur ensemble être égaux à zéro et le théorème est démontré.

Signalons-en, pour terminer, une conséquence. Un logarithme quelconque de nombre algébrique, pris suivant une base algébrique, doit être un nombre transcendant ou rationnel.

А. О. ГЕЛЬФОНД. О СЕДЬМОЙ ПРОБЛЕМЕ ГИЛЬБЕРТА

РЕЗЮМЕ

Настоящая работа представляет собой более полное изложение моей заметки помещенной под тем же заглавием в Докладах Академии наук СССР (1934 т. II, № 1, стр, 1). Гильбертом была поставлена проблема трансцендентности чисел вида ω^{η}, где ω и η два алгебраические числа, а η, кроме того, иррационально. Современное, до моей работы, состояние этой проблемы было приведено в выше указанной заметке.

Пусть $\alpha \neq 0,1$ и $\beta \neq 0,1$ будут алгебраические числа. Допустим, что $\eta = \frac{\ln \alpha}{\ln \beta}$ будет алгебраическим и иррациональным числом. Пусть, далее, числа α, β и η будут корнями уравнений (неприводимых) степеней p_1, p_2, p_3 с целыми рациональными коэффициентами. Обозначим через $\alpha_1, \alpha_2 \ldots \alpha_{p_1-1}$, $\beta_1, \beta_2, \ldots, \beta_{p_2-1}$ и $\eta_1, \eta_2 \ldots \eta_{p_3-1}$, числа сопряженные, соответственно, с α, β и η. Пусть c будет целое положительное число, и притом такое, что $c\alpha$, $c\beta$ и $c\eta$ будут целыми алгебраическими числами.

Введем теперь ряд констант, которые будут нужны в дальнейшем. Определим константы γ, $\lambda_1, \ldots \lambda_5$ равенствами:

$$\left.\begin{aligned}&\gamma = e^{2\pi}\,\mathrm{Max}\,[3, |\alpha|, \ldots, |\alpha_{p_1-1}|, |\beta|, |\beta_1|, \ldots, |\beta_{p_2-1}|, |\eta|, |\eta_1|, \ldots |\eta_{p_3-1}|];\\ &\lambda_1 = 2\ln 3\, p_1 p_2 p_3; \quad \lambda_2 = (2\ln\gamma + 4\ln c)\, p_1 p_2 p_3; \quad \lambda_3 = 2 p_1 p_2 p_3;\\ &\lambda_4 = \ln[2\ln\gamma]; \quad \lambda_5 = 2\ln\gamma.\end{aligned}\right\} \quad (1)$$

Вводим в рассмотрение целые, положительные числа $q_1, q_2, \ldots, q_8$, которые будут определяться тем, что для всякого $q > q_i$ должно выполняться соответствующее η-ое неравенство:

$$\left.\begin{gathered}\text{при } q \geqq q_1, \quad \ln q > \ln(\gamma + \gamma^2) + 2\ln c; \quad \text{при } q > q_2,\\ \frac{\ln q}{\ln\ln q} > 4(3\ln 3 + \lambda_4 + \lambda_5); \quad \text{при } q > q_3,\\ \frac{\ln q}{\ln\ln q} > 4(\lambda_1 + \lambda_2 + \lambda_3); \quad \text{при } q \geqq q_3, \ln\ln q > 6(2\ln 6 + \lambda_5);\\ \text{при } q \geqq q_5, \ln\ln q > 12(\lambda_1 + \lambda_2 + \gamma_3); \quad \text{при } q \leqq q_6,\\ 2q^{7/3} < \frac{1}{2}\frac{q^{5/2}}{\ln\ln q} - (2\ln 6 + \lambda_5)\frac{q^{5/2}}{\ln q \ln\ln q}; \quad \text{при } q \geqq q_7,\\ q^{7/3} > 5q^2 \ln(5q^2)\ln 2; \quad \text{при } q \geqq q_8,\\ q^{7/3} > 5(\lambda_1 + \lambda_2 + \lambda_3) q^2 \ln q.\end{gathered}\right\} \quad (2)$$

Очевидно, что как константы $\gamma, \lambda_1, \ldots, \lambda_5$, так и константы $q_1, q_2, \ldots q_8$ зависят только от алгебраических чисел α, β, η.

Положим так же

$$\left.\begin{aligned} q_0 &= 1 + \mathrm{Max}\,[3^{3^4}, q_1, \ldots, q_8], \\ r_1 &= \left[\frac{q_0^2}{\ln q_0 \ln\ln q_0}\right], \quad r_2 = [\ln^2 \ln q_0]. \end{aligned}\right\} \tag{3}$$

Числа s и t будут, в дальнейшем, всегда целыми неотрицательными и переменными.

Рассмотрим теперь функцию

$$f(x) = \sum_{k=0}^{k=q_0} \sum_{l=0}^{l=q_0} C_{k,l}\, \alpha^{kx} \beta^{lx} \tag{4}$$

где все числа $C_{k,l}$; $k=0, 1, \ldots, g_0$; $l=0, 1, \ldots, q_0$, будут целыми и рациональными.

Относительно подобных функций можно доказать три необходимых, в дальнейшем, леммы.

Лемма I. *Каковы бы ни были целые числа* $C_{k,l}$, $|C_{k\,l}| \leqq 3^{q^2_0}$ *функция* (4) *при предположении алгебраичности чисел* α, β и η, *обладает тем свойством, что или выполнено неравенство*

$$|f^{(s)}(t)| > e^{-(\lambda_1 q_0^2 + \lambda_2 q_0 t + \lambda_3 s \ln q_0)}, \tag{5}$$

или $f^s(t) = 0$.

Доказательство этой леммы основано только на том факте, что

$$f^{(s)}(t) = \ln^{-s}\beta \sum_{k=q0}^{k=q_0} \sum_{l=0}^{l=q_0} C_{k,l} (k\eta + l)^s \alpha^{kt} \beta^{lt},$$

является, на основании предположения об алгебраичности α, β и η, алгебраическим, с точностью до множителя, числом. Но норма этого алгебраического числа, которое мы делаем целым умножением на $c^{s+2q_0 t}$, должна быть или больше 1 или оно должно быть нулем. Простые подсчеты дают (5) неравенство с константами λ_1, λ_2, λ_3, определенными равенством (1), при использовании условия $q_0 > q_1$.

Лемма II. *При условии* $|C_{k,l}| \leqq 3^{q_0^2}$ *целые, в совокупности не равные нулю коэффициенты функции* $f(x)$, *могут быть выбраны так, что одновременно будут выполняться неравенства, в числе* $r_1 r_2$,

$$|f^{(s)}(t)| < e^{-\frac{1}{6}\frac{q_0^2 \ln\ln q_0}{\ln\ln q_0}}, \quad 0 \leqq s \leqq r_1 - 1, \quad 0 \leqq t \leqq r_2 - 1. \tag{6}$$

Доказательство этой леммы строится с помощью принципа Дирихле в пространстве $2r_1 r_2 \simeq 2 q_2{}^0 \dfrac{\ln\ln q_0}{\ln q_0}$ измерений, на осях которого откладываются числа $Rf^{(s)}(t)$ и $If^{(s)}(t)$. Число точек в этом пространстве с этими координатами равно $\sim 3^{q_0^4}$, а число измерений его $\sim 2\,q_0^{\,2}\,\frac{\ln\ln q_0}{\ln q_0}$. Все точки умещаются в гиперкубе со стороной $\sim B_0\,e^{d_0 q_0^2}$, где B_0 и d_0 малые константы. Из этих соображений и следует, непосредственно, лемма II.

Лемма III. *Если* $|C_{k,l}| \leqq 3^{q_0^2}$ *то функция* $f(x) = \sum_{k=0}^{k=q_0} \sum_{l=0}^{l=q_0} C_{k,l}\,\alpha^{kx}\,\beta^{lx}$ *при любом комплексном удовлетворяет неравенству*:

$$|f(x)| < e^{2\ln 3\, q_0^2 + \lambda_5 q_0 |x|} \tag{7}$$

где λ_5 *определена равенствами* (1).

Доказательство получается непосредственной оценкой $f(x)$.

Теперь может быть доказана теорема:

Если $\alpha \neq 0{,}1$ и $\beta \neq 0{,}1$ *будут алгебраическими числами, то число* $\eta = \frac{\ln\alpha}{\ln\beta}$ *или рационально или трансцендентно.*

Выбираем, с помощью леммы II, функцию $f(x)$. Сравнивая неравенства (6) с неравенством (5), которое выполнено, так как мы делаем предположение об иррациональности и алгебраичности η, мы приходим к равенствам $f^{(s)}(t) = 0$, $0 \leqq s \leqq r_1 - 1$, $0 \leqq t \leqq r_2 - 1$, так как $q_0 > q_3$. Это дает нам, что функция $f(x)\,[x(x-1)\ldots(x-r_2+1)]^{-r_1}$ будет целой. С помощью принципа максимума мы выводим отсюда неравенство, верное для всякого x, $|x| < q_0^{2/3}$,

$$|f(x)| < e^{-\frac{1}{6} q_0^3 \ln\ln q^0}, \tag{8}$$

из которого, в свою очередь при помощи интеграла Коши следует верное для $t \leqq [\sqrt{q_0}]$ неравенство:

$$|f^{(s)}(t)| < e^{-\frac{1}{12} q_0^2 \ln\ln q_0}, \quad 0 \leqq s \leqq r_1 - 1, \quad 0 \leqq t \leqq r_2 - 1. \tag{9}$$

Сравнивая (5) и (9) неравенства мы приходим. так-как $q_0 > q_5$, к условиям $f^{(s)}(t) = 0$, $0 \leqq s \leqq r_1 - 1$, $0 \leqq t \leqq r_2 - 1$. Из существования значительно большего, чем ранее это было установлено, нулей $f(x)$, с помощью аналогичных рассуждений, может быть получено для $0 \leqq s \leqq 5\,q_0^{\,2}$, $t = 0$, неравенство

$$|f^{(s)}(0)| < e^{-q^{7/3}}, \tag{10}$$

Сравнивая опять неравенства (5) и (10), мы получим, что $f^{(s)}(0) = 0$, $0 \geqq s \leqq 5q_0^{\,2}$, так как $q_0 > q_8$.

Итак, мы имеем относительно $C_{k,e}$ ($k = 0, 1, \ldots q_0$, $l = 0, 1, \ldots q_0$) $(q_0 + 1)^2$ уравнений, очевидно однородных,

$$f^{(s)}(0) = \sum_{k=0}^{k=q_0} \sum_{l=0}^{l=q_0} C_{k,l}\,(k\ln\alpha + l\ln\beta)^s = 0, \quad 0 \leqq s \leqq (q_0 + 1)^2 - 1. \tag{11}$$

Детерминантом этой системы будет Вандермондов детерминант. Он отличен от нуля, так как равенство

$$k_1 \ln\alpha + l_1 \ln\beta = k_2 \ln\alpha + l_2 \ln\beta, \quad (k_1 - k_2)^2 + (l_1 - l_2)^2 \neq 0 \tag{12}$$

невозможно в силу иррацональности, по предположению числа, $\eta = \frac{\ln\alpha}{\ln\beta}$.

Значит все $C_{k,l} = 0$ и мы пришли к противоречию, так как по способу их выбора это невозможно. Этим наша теорема доказана.

Из нее, между прочим, непосредственно следует, что логарифм алгебраического числа, взятый при алгебраическом основании, трансцендентен или рационален.

Aleksandr Yakovlevich Khinchin

Born – 19 July 1894 in Kondrovo, Kaluzhskaya guberniya, Russia
Died – 18 Nov 1959 in Moscow, USSR

Aleksandr Khinchin's father was an engineer. Khinchin attended the technical high school in Moscow where he became fascinated by mathematics. However mathematics was certainly not his only interest when he was at secondary school for he also had a passionate love of poetry and of the theatre. He completed his secondary education in 1911 and entered the Faculty of Physics and Mathematics of Moscow University in that year.

Khinchin worked with Luzin and others at university in Moscow. He was an outstanding student being particularly interested in the metric theory of functions and before he graduated in 1916 he had already written his first paper on a generalization of the Denjoy integral. This first paper began a series of publications by Khinchin on properties of functions which are retained after deleting a set of density zero at a given point. He summarised his contributions to this area with the paper *Recherches sur la structure des fonctions measurables in Fundamanta mathematica* in 1927.

After graduating in 1916, Khinchin remained at Moscow University undertaking research for his dissertation that would allow him to become a university teacher. After a couple of years he began teaching in a number of different colleges both in Moscow and Ivanovo. The town of Ivanovo, east of Moscow, was a centre for the textile industry and it plays a surprisingly important part in the development of Russian mathematics with several of the major figures teaching in the town.

Around 1922 Khinchin took up new mathematical interests when he began to study the theory of numbers and probability theory. In the following year he strengthened results of Hardy and Littlewood with his introduction of the iterated logarithm published in *Mathematische Zeitschrift*. With these ideas he also strengthened the law of large numbers due to Borel.

In 1927 Khinchin was appointed as a professor at Moscow University and, in the same year, he published *Basic laws of probability theory*. Between 1932 and 1934 he laid the foundations for the theory of stationary random processes culminating in a major paper in *Mathematische Annalen* in 1934. Khinchin left Moscow in 1935 to spend two years at Saratov University but returned to Moscow University in 1937 to continue his role of building the school of probability theory there in partnership with Kolmogorov and others, including in particular their student Gnedenko. From the 1940s his work changed direction again and this time he became interested in the theory of statistical mechanics. In the last few years of his life his interests turned to developing Shannon's ideas on information theory.

Khinchin first published the book *Continued Fractions* in 1936 with a second edition being published in 1949. The book consists of three chapters, the first two of which present the classical theory of continued fractions. The third chapter, the longest and most important, contains an account of Khinchin's own contributions to the topic of the metrical theory of Diophantine approximations. Another contribution by Khinchin to number theory is the short book *Three pearls of number theory* which appeared in an English translation in 1952.

The book *Eight lectures on mathematical analysis* by Khinchin ran to several editions. It was first published in 1943 and the eight lectures it contains are: Continuum; Limits; Functions; Series; Derivative; Integral; Series expansions of functions; and Differential equations. The book was designed to be used to supplement a standard course on the calculus and gives a careful treatment of some of the basic notions of mathematical analysis.

Khinchin published *Mathematical Principles of Statistical Mechanics* in 1943. It showed how to make classical statistical mechanics into a mathematically rigorous subject, developing a consistent presentation of the topic. In 1951 he extended the work of this 1943 book when he published *Mathematical foundations of quantum statistics.* This new publication on the topic appeared in a German translation in 1956 and then in an English translation in 1960. The book was written in such a way as to be useful both to mathematicians who wanted to become better acquainted with some applications of analysis to physics, and also to physicists who wanted to understand more about the mathematical foundations for their subject.

Khinchin's book *Mathematical Foundations of Information Theory*, translated into English from the original Russian in 1957, is important. It consists of English translations of two articles: *The entropy concept in probability theory and On the basic theorems of information theory* which were both published earlier in Russian. The second of these articles provides a refinement of Shannon's concepts of the capacity of a noisy channel and the entropy of a source. Khinchin generalised some of Shannon's results in this book which was written in an elementary style yet gave a comprehensive account with full details of all the results.

Among the many honours that Khinchin received for his work was election to the Soviet Academy of Sciences in 1939 and the award of a State Prize for scientific achievements in the following year.

CHAPTER I

VAN DER WAERDEN'S THEOREM ON ARITHMETIC PROGRESSIONS

§ 1

In the summer of 1928 I spent several weeks in Göttingen. As usual, many foreign scholars had arrived there for the summer semester. I got to know many of them, and actually made friends with some. At the time of my arrival, the topic of the day was the brilliant result of a young Hollander, van der Waerden,* who at that time was still a youthful beginner, but is now a well-known scholar. This result had just been obtained here in Göttingen, and in fact, only a few days before my arrival. Nearly all mathematicians whom I met told me about it with enthusiasm.

The problem had the following history. One of the mathematicians there (I forget his name**) had come upon the following problem in the course of his scientific work: Imagine the set of all natural numbers to be divided in any manner whatsoever into two parts (e. g., into even and odd numbers, or into prime and composite numbers, or in any other way). Can one then assert that arithmetic progressions of arbitrary length can be found in at least one of these parts? (By the length of an arithmetic progression I mean here, and in what follows, simply the number of its terms.) All to whom this question was put regarded the problem at first sight as quite simple; its solution in the affirmative appeared to be almost self-evident. The first attempts to solve it, however, led to nought. And as the mathematicians of Göttingen and their foreign guests were by tradition in constant association with one another, this problem, provoking in its resistance, soon became the object of general mathematical interest. Everyone took it up, from the venerable scholar to the young student. After several weeks of strenuous exertions, the problem finally yielded to the attack of a young man who had come to Göttingen to study, the Hollander, van der Waerden. I made his acquaintance, and learned the solution of the problem from him personally. It was elementary, but not simple by any means. The problem turned out to be deep, the appearance of simplicity was deceptive.

*B. L. van der Waerden, *Beweis einer Baudetschen Vermutung*, Nieuw Arch. Wiskunde 15, 212-216 (1927). (Trans.)

**Most probably Baudet; cf. the preceding footnote. (Trans.)

12

Quite recently, M. A. Lukomskaya (of Minsk) discovered and sent me a considerably simpler and more transparent proof of van der Waerden's theorem, which, with her kind permission, I am going to show you in what follows.

§ 2

Actually, van der Waerden proved somewhat more than what was required. In the first place, he assumes that the natural numbers are divided, not into two, but into arbitrarily many, say k, classes (sets). In the second place, it turns out that it is not necessary to decompose the entire sequence of natural numbers in order to guarantee the existence of an arithmetic progression of prescribed (arbitrarily large) length l in at least one of these classes; a certain segment of it suffices for this purpose. The length, $n(k, l)$, of this segment is a function of the numbers k and l. Of course it doesn't matter where we take this segment, so long as there are $n(k, l)$ successive natural numbers.

Accordingly, van der Waerden's theorem can be formulated as follows:

Let k and l be two arbitrary natural numbers. Then there exists a natural number $n(k, l)$ such that, if an arbitrary segment, of length $n(k, l)$, of the sequence of natural numbers is divided in any manner into k classes (some of which may be empty), then an arithmetic progression of length l appears in at least one of these classes.

This theorem is true trivially for $l=2$. To see this, it suffices to set $n(k, 2)=k+1$; for if $k+1$ numbers are divided into k classes, then certainly at least one of these classes contains more than one number, and an arbitrary pair of numbers forms an arithmetic progression of length 2, which proves the theorem. We shall prove the theorem by induction on l. Consequently, we shall assume throughout the following that the theorem has already been verified for some number $l \geqq 2$ and for arbitrary values of k, and shall show that it retains its validity for the number $l+1$ (and naturally also for all values of k).

§ 3

According to our assumption, then, for every natural number k there is a natural number $n(k, l)$ such that, if an arbitrary segment, of length $n(k, l)$, of the natural numbers, is divided in any manner into k classes, there exists in at least one of these classes an arithmetic progression of length l. We must then prove that, for every natural number k, an $n(k, l+1)$ also exists. We solve this problem by actually constructing the number $n(k, l+1)$. To this end we set

$$q_0 = 1, \quad n_0 = n(k, l)$$

and then define the numbers $q_1, q_2, \ldots, n_1, n_2, \ldots$ successively as follows: If q_{s-1} and n_{s-1} have already been defined for some $s > 0$, we put

$$(1) \qquad q_s = 2n_{s-1}q_{s-1}, \quad n_s = n(k^{q_s}, l) \quad (s = 1, 2, \ldots).$$

The numbers n_s, q_s are obviously defined hereby for an arbitrary $s \geq 0$. We now assert that for $n(k, l+1)$ we may take the number q_k. We have to show then that if a segment, of length q_k, of the sequence of natural numbers is divided in any manner into k classes, then there is an arithmetic progression of length $l+1$ in at least one of these classes. The remainder of the chapter is devoted to this proof.

In the sequel we set $l+1 = l'$ for brevity.

§ 4

Suppose then that the segment Δ, of length q_k, of the sequence of natural numbers is divided in an arbitrary way into k classes. We say that two numbers a and b of Δ are of the same type, if a and b belong to the same class, and we then write $a \approx b$. Two equally long subsegments of Δ, $\delta = (a, a+1, \ldots, a+r)$ and $\delta' = (a', a'+1, \ldots, a'+r)$, are said to be of the same type, if $a \approx a'$, $a+1 \approx a'+1$, $\ldots$, $a+r \approx a'+r$, and we then write $\delta \approx \delta'$. The number of different possible types for the numbers of the segment Δ is obviously equal to k. For segments of the form $(a, a+1)$ (i. e., for segments of length 2) the number of possible types is k^2; and in general, for segments of length m, it is k^m. (Of course not all these types need actually appear in the segment Δ.)

Since $q_k = 2n_{k-1}q_{k-1}$ (see (1)), the segment Δ can be regarded as a sequence of $2n_{k-1}$ subsegments of length q_{k-1}. Such subsegments, as we have just seen, can have $k^{q_{k-1}}$ different types. The left half of the segment Δ now contains n_{k-1} such subsegments, where $n_{k-1} = n(k^{q_{k-1}}, l)$ according to (1). Because of the meaning of the number $n(k^{q_{k-1}}, l)$, we can assert* that the left half of the segment Δ contains an arithmetic progression of l of these subsegments of

*Work with the initial numbers of the n_{k-1} subsegments. (Trans.)

the same type,

$$\Delta_1, \Delta_2, \ldots, \Delta_l$$

of length q_{k-1}; here we say for brevity that equally long segments Δ_i form an arithmetic progression, if their initial numbers form such a progression. We call the difference between the initial numbers of two neighboring segments of the progression $\Delta_1, \Delta_2, \ldots, \Delta_l$ the difference d_1 of this progression. Naturally the difference between the second (or third, fourth, etc.) numbers of two such neighboring segments is likewise equal to d_1.

To this progression of segments we now add the succeeding, $(l+1)$-st, term $\Delta_{l'}$ (we recall that $l'=l+1$) which may already project beyond the boundary of the left half of the segment Δ, but which in any case still belongs entirely to the segment Δ. The segments $\Delta_1, \Delta_2, \ldots, \Delta_l, \Delta_{l'}$ then form an arithmetic progression, of length $l'=l+1$ and difference d_1, of segments of length q_{k-1}, where $\Delta_1, \Delta_2, \ldots, \Delta_l$ are of the same type. We know nothing about the type of the last segment $\Delta_{l'}$.

This completes the first step of our construction. It would be well if you thought it through once more before we continued.

§5

We now proceed to the second step. We take an arbitrary one of the first l terms of the progression of segments just constructed. Let this term be Δ_{i_1}, so that $1 \leqq i_1 \leqq l$; Δ_{i_1} is a segment of length q_{k-1}. We treat it the same way as we treated the segment Δ. Since $q_{k-1} = 2n_{k-2}q_{k-2}$, the left half of the segment Δ_{i_1} can be regarded as a sequence of n_{k-2} subsegments of length q_{k-2}. For subsegments of this length there are $k^{q_{k-2}}$ types possible, and on the other hand $n_{k-2} = n(k^{q_{k-2}}, l)$ because of (1). Therefore the left half of Δ_{i_1} must contain a progression of l of these subsegments of the same type, $\Delta_{i_1 i_2}$ $(1 \leqq i_2 \leqq l)$, of length q_{k-2}. Let d_2 be the difference of this progression (i.e., the distance between the initial numbers of two neighboring segments). To this progression of segments we add the $(l+1)$-st term $\Delta_{i_1 l'}$, about whose type, of course, we know nothing. The segment $\Delta_{i_1 l'}$ does not have to belong to the left half of the segment Δ_{i_1} any more, but must obviously belong to the segment $\Delta_{i_1'}$.

We now carry over our construction, which we have executed up to now in only one of the segments Δ_{i_1}, congruently to all the other segments Δ_{i_1} $(1 \leqq i_1 \leqq l')$. We thus obtain a set of segments $\Delta_{i_1 i_2}$ $(1 \leqq i_1 \leqq l', 1 \leqq i_2 \leqq l')$ with two indices. It is clear that two arbitrary segments of this set with indices not exceeding l are of the same type:

$$\Delta_{i_1 i_2} \approx \Delta_{i_1' i_2'} \qquad (1 \leqq i_1, i_2, i_1', i_2' \leqq l)$$

You no doubt see now that this process can be continued. We carry it out k times. The results of our construction after the first step were segments of length q_{k-1}, after the second step, segments of length q_{k-2}, etc. After the k-th step, therefore, the results of the construction are segments of length $q_0 = 1$, i. e., simply numbers of our original segment Δ. Nevertheless we denote them as before by

$$\Delta_{i_1 i_2 \ldots i_k} \quad (1 \leqq i_1, i_2, \ldots, i_k \leqq l').$$

For $1 \leqq s \leqq k$ and $1 \leqq i_1, \ldots, i_s, i_1', \ldots, i_s' \leqq l$ we have

$$\Delta_{i_1 i_2 \ldots i_s} \approx \Delta_{i_1' i_2' \ldots i_s'} \tag{2}$$

We now make two remarks which are important for what follows.

1) In (2), if $s<k$ and if $i_{s+1}, i_{s+2}, \ldots, i_k$ are arbitrary indices taken from the sequence $1, 2, \ldots, l, l'$, then the number $\Delta_{i_1 i_2 \ldots i_s i_{s+1} \ldots i_k}$ appears in the same position in the segment $\Delta_{i_1 \ldots i_s}$ as the number $\Delta_{i_1' i_2' \ldots i_s' i_{s+1} \ldots i_k}$ does in the segment $\Delta_{i_1' \ldots i_s'}$. Since these two segments are of the same type because of (2), it follows that

$$\Delta_{i_1 i_2 \ldots i_s i_{s+1} \ldots i_k} \approx \Delta_{i_1' i_2' \ldots i_s' i_{s+1} \ldots i_k} \tag{3}$$

if $1 \leqq i_1, \ldots, i_s, i_1', \ldots, i_s' \leqq l$ and $1 \leqq i_{s+1}, i_{s+2}, \ldots, i_k \leqq l' (1 \leqq s \leqq k)$.

2) For $s \leqq k$ and $i_s' = i_s + 1$, $\Delta_{i_1 \ldots i_{s-1} i_s}$ and $\Delta_{i_1 \ldots i_{s-1} i_s'}$ are obviously neighboring segments in the s-th step of our construction. Therefore for arbitrary indices $i_{s+1}, \ldots, i_k$, the numbers $\Delta_{i_1 \ldots i_{s-1} i_s i_{s+1} \ldots i_k}$ and $\Delta_{i_1 \ldots i_{s-1} i_s' i_{s+1} \ldots i_k}$ appear in the same position in two such neighboring segments, so that (with $i_s' = i_{s+1}$)

16

$$(4)\qquad \Delta_{i_1\ldots i_{s-1}\, i_s'\, i_{s+1}\ldots i_k} - \Delta_{i_1\ldots i_{s-1}\, i_s\, i_{s+1}\ldots i_k} = d_s.$$

§6

Now we are near our goal. We consider the following $k+1$ numbers of the segment Δ:

$$(5)\qquad \begin{cases} a_0 = \Delta_{l'l'l'\ldots l'} \\ a_1 = \Delta_{1\,l'l'\ldots l'} \\ a_2 = \Delta_{1\,1\,l'\ldots l'} \\ \cdot\ \cdot\ \cdot\ \cdot\ \cdot\ \cdot\ \cdot\ \cdot \\ a_k = \Delta_{1\,1\,1\,\cdots\,1}\,. \end{cases}$$

Since the segment Δ has been divided into k classes, and we have $k+1$ numbers in (5), there are two of these numbers which belong to the same class. Let these be the numbers a_r and a_s $(r<s)$, so that

$$(6)\qquad \Delta_{\underbrace{1\ldots1}_{r}\underbrace{l'\ldots l'}_{k-r}} \approx \Delta_{\underbrace{1\ldots1}_{s}\underbrace{l'\ldots l'}_{k-s}}\,.$$

We consider the $l+1$ numbers

$$(7)\qquad c_i=\Delta_{\underbrace{1\ldots1}_{r}\underbrace{i\ldots i}_{s-r}\underbrace{l'\ldots l'}_{k-s}} \qquad (1\leqq i\leqq l')\,:$$

The first l numbers of this group (i. e., those with $i<l'$) belong to the same class because of (3). The last $(i=l')$, however, is of the same type as the first because of (6). Consequently all $l+1$ numbers in (7) are of the same type, and to prove our assertion we have only to show that these numbers form an arithmetic progression, i. e., that the difference $c_{i+1}-c_i$ $(1\leqq i\leqq l)$ does not depend on i.

We set $i+1=i'$ for brevity. Further let

$$c_{i,m}=\Delta_{\underbrace{1\ldots1}_{r}\underbrace{i'\ldots i'}_{m}\underbrace{i\ldots i}_{s-r-m}\underbrace{l'\ldots l'}_{k-s}} \qquad (0\leqq m\leqq s-r),$$

so that $c_{i,0} = c_i$, $c_{i,s-r} = c_{i+1}$, and hence

$$c_{i+1} - c_i = \sum_{m=1}^{s-r} (c_{i,m} - c_{i,m-1}).$$

Because of (4) we have

$$c_{i,m} - c_{i,m-1} = \Delta_{\underbrace{1\ldots1}_{r}\underbrace{i'\ldots i'}_{m}\underbrace{i\ldots i}_{s-r-m}\underbrace{l'\ldots l'}_{k-s}} - \Delta_{\underbrace{1\ldots1}_{r}\underbrace{i'\ldots i'}_{m-1}\underbrace{i\ldots i}_{s-r-m+1}\underbrace{l'\ldots l'}_{k-s}} = d_{r+m}.$$

Thus the difference

$$c_{i+1} - c_i = d_{r+1} + d_{r+2} + \ldots + d_s,$$

and is indeed independent of i, which completes the proof of our assertion.

You see how complicated a completely elementary construction can sometimes be. And yet this is not an extreme case: in the next chapter you will encounter just as elementary a construction which is considerably more complicated. Besides, it is not out of the question that van der Waerden's theorem admits of an even simpler proof, and all research in this direction can only be welcomed.

CHAPTER II

THE LANDAU-SCHNIRELMANN HYPOTHESIS AND MANN'S THEOREM

§1

You have perhaps heard of the remarkable theorem of Lagrange, that *every natural number is the sum of at most four squares.* In other words, every natural number is either itself the square of another number, or else the sum of two, or else of three, or else of four such squares. For the purpose at hand it is desirable to understand the content of this theorem in a somewhat different form. Let us write down the sequence of all perfect squares, beginning with zero:

$$(S) \qquad 0, 1, 4, 9, 16, 25, \dots .$$

This is a certain sequence of whole numbers. We denote it by S, and imagine four completely identical copies of it, S_1, S_2, S_3, S_4, to be written down. Now we choose an arbitrary number a_1^2 from S_1, an arbitrary number a_2^2 from S_2, an arbitrary number a_3^2 from S_3, and an arbitrary number a_4^2 from S_4, and add these numbers together. The resulting sum

$$(*) \qquad n = a_1^2 + a_2^2 + a_3^2 + a_4^2$$

can be

1) zero (if $a_1 = a_2 = a_3 = a_4 = 0$);

2) the square of a natural number (if in some representation (*) of the number n three of the numbers a_1, a_2, a_3, a_4 are zero and the fourth is not zero);

3) the sum of two squares of natural numbers (if in some representation (*) of the number n two of the numbers a_1, a_2, a_3, a_4 are zero and the other two are not zero);

4) the sum of three squares of natural numbers (if in some representation (*) of the number n one of the numbers a_1, a_2, a_3, a_4 is equal to zero and the remaining three are not zero);

5) the sum of four squares of natural numbers (if in some representation of the number n all four numbers are different from zero).

Thus the resulting number n is either zero or else a natural number which can be represented as the sum of at most four squares, and it is clear that conversely every natural number can be obtained by the process which we have described.

Now let us arrange all natural numbers n which can be obtained by means of our process (i.e., by the addition of four numbers taken respectively from the sequences S_1, S_2, S_3, S_4), in order of magnitude, in the sequence

$$(A) \qquad 0, n_1, n_2, n_3, \ldots$$

(where $0<n_1<n_2<n_3<\ldots$, so that if there are equal numbers among those constructed, only one of them appears in (A)). The theorem of Lagrange now asserts simply that the sequence (A) contains all the natural numbers, i. e., that $n_1=1$, $n_2=2$, $n_3=3$, etc.

We shall now generalize our process. Let there be given k monotonically increasing sequences of integers which begin with zero:

$$\begin{array}{ll} (A^{(1)}) & 0, a_1^{(1)}, a_2^{(1)}, \ldots, a_m^{(1)}, \ldots, \\ (A^{(2)}) & 0, a_1^{(2)}, a_2^{(2)}, \ldots, a_m^{(2)}, \ldots, \\ \cdots\cdots & \cdot\;\cdot\;\cdot\;\cdot\;\cdot\;\cdot\;\cdot\;\cdot\;\cdot\;\cdot \\ (A^{(k)}) & 0, a_1^{(k)}, a_2^{(k)}, \ldots, a_m^{(k)}, \ldots . \end{array}$$

We choose arbitrarily a single number from each sequence $A^{(i)}$ $(1\leq i\leq k)$ and add these k numbers together. The totality of all numbers constructed in this manner, if we order them according to magnitude, yields a new sequence

$$(A) \qquad 0, n_1, n_2, \ldots, n_m, \ldots$$

of the same type, which we shall call the *sum* of the given sequences $A^{(1)}, A^{(2)}, \ldots, A^{(k)}$:

$$A=A^{(1)}+A^{(2)}+\ldots+A^{(k)}=\sum_{i=1}^{k} A^{(i)} .$$

20

The content of Lagrange's theorem is that the sum $S+S+S+S$ contains the entire sequence of natural numbers.

Perhaps you have heard of the famous theorem of Fermat, that *the sum* $S+S$ *contains all prime numbers which leave a remainder of* 1 *when divided by* 4 (i. e., the numbers 5, 13, 17, 29, ...). Perhaps you also know that the famous Soviet scholar Ivan Matveyevitch Vinogradov proved the following remarkable theorem, on which many of the greatest mathematicians of the preceding two centuries had worked without success:

If we denote by P *the sequence*

$$(P) \qquad 0, 2, 3, 5, 7, 11, 13, 17, \ldots$$

consisting of zero and all prime numbers, then the sum $P+P+P$ *contains all sufficiently large odd numbers.*

I have cited all these examples here for only one very modest purpose: to familiarize you with the concept of the sum of sequences of numbers and to show how some classical theorems of number theory can be formulated simply and conveniently with the aid of this concept.

§2

As you have undoubtedly observed, in all the examples mentioned we are concerned with showing that the sum of a certain number of sequences represents a sequence which contains either completely or almost completely this or that class of numbers (e. g., all the natural numbers, all sufficiently large odd numbers, and others of the same sort). In all other similar problems the purpose of the investigation is to prove that the sum of the given sequences of numbers represents a set of numbers which is in some sense "dense" in the sequence of natural numbers. It is often the case that this set contains the entire sequence of natural numbers (as we saw in our first example). The theorem of Lagrange says that the sum of the four sequences S contains the whole sequence of natural numbers. Now it is customary to call the sequence A a *basis* (of the sequence of natural numbers) *of order* k if the sum of k identical sequences A contains all the natural numbers. The theorem of Lagrange then states that the sequence S of perfect squares is a basis of order four. It was shown later that the sequence of perfect cubes forms a basis of order nine. A little reflection shows that every basis of order k is also a basis of order $k+1$.

In all these and in many other examples the "density" of the sum which is to be established is determined by particular properties of the sequences that are added together, i. e., by the special arithmetical nature of the numbers which go to make up these sequences (these numbers being either perfect squares, or primes, or others of a similar nature). Sixteen years ago the distinguished Soviet scholar Lev Genrichovitch Schnirelmann first raised the question: To what extent is the density of the sum of several sequences determined solely by the density of the summands, irrespective of their arithmetical nature. This problem turned out to be not only deep and interesting, but also useful for the treatment of some classical problems. During the intervening fifteen years it received the attention of many outstanding scholars, and it has given rise to a rich literature.

Before we can state problems in this field precisely and write the word "density" without quotation marks, it is evident that we must first agree on what number (or on what numbers) to use to measure the "density" of our sequences with (just as in physics the words "warm" and "cold" do not acquire a precise scientific meaning until we have learned to measure temperature).

A very convenient measure of the "density" of a sequence of numbers, which is now used for all scientific problems of the kind we are considering, was proposed by L. G. Schnirelmann. Let

$$(A) \qquad 0,\ a_1,\ a_2,\ \ldots,\ a_n,\ \ldots$$

be a sequence of numbers, where, as usual, all the a_n are natural numbers and $a_n < a_{n+1}$ $(n=1, 2, \ldots)$. We denote by $A(n)$ the number of natural numbers in the sequence (A) which do not exceed n (zero is not counted), so that $0 \leqq A(n) \leqq n$. Then the inequality

$$0 \leqq \frac{A(n)}{n} \leqq 1$$

holds. The fraction $A(n)/n$, which for different n naturally has different values, can obviously be interpreted as a kind of average density of the sequence (A) in the segment from 1 to n of the sequence of natural numbers. Following the suggestion of Schnirelmann, the *greatest lower bound* of all values of this fraction is called the *density* of the sequence (A) (in the entire sequence of natural numbers). We shall denote this density by $d(A)$.

In order to become familiar at once with the elementary properties of this concept, I recommend that you convince yourself of the validity of the following theorems:

1. If $a_1 > 1$ (i. e., the sequence (A) does not contain unity), then $d(A) = 0$.

2. If $a_n = 1 + r(n-1)$ (i. e., the sequence (A), beginning with a_1, is an arithmetic progression with initial term 1 and difference r), then $d(A) = 1/r$.

3. The density of every geometric progression is equal to zero.

4. The density of the sequence of perfect squares is equal to zero.

5. For the sequence (A) to contain the entire sequence of natural numbers ($a_n = n$, $n = 1, 2, \ldots$), it is necessary and sufficient that $d(A) = 1$.

6. If $d(A) = 0$ and A contains the number 1, and if $\epsilon > 0$ is arbitrary, then there exists a sufficiently large number m such that $A(m) < \epsilon m$.

If you have proved all this, you are familiar enough with the concept of density to be able to use it. Now I want to acquaint you with the proof of the following remarkable, albeit very simple, lemma of Schnirelmann:

$$d(A+B) \geq d(A) + d(B) - d(A)\, d(B). \tag{1}$$

The meaning of this inequality is clear: the density of the sum of two arbitrary sequences of numbers is not smaller than the sum of their densities diminished by the product of these densities. This "Schnirelmann inequality" represents the first tool, still crude to be sure, for estimating the density of a sum from the densities of the summands. Here is its proof. We denote by $A(n)$ the number of natural numbers which appear in the sequence A and do not exceed n, and by $B(n)$ the analogous number for the sequence B. For brevity we set $d(A) = \alpha$, $d(B) = \beta$, $A + B = C$, $d(C) = \gamma$. The segment $(1, n)$ of the sequence of natural numbers contains $A(n)$ numbers of the sequence A, each of which also appears in the sequence C. Let a_k and a_{k+1} be two consecutive numbers of this group. Between them there are $a_{k+1}-a_k-1=l$ numbers which do not belong to A. These are the numbers

$$a_k+1,\ a_k+2,\ \ldots,\ a_k+l=a_{k+1}-1.$$

Some of them appear in C, e.g., all numbers of the form a_k+r, where r occurs in B (which we abbreviate as follows: $r\,\varepsilon B$). There are as many numbers of this last kind, however, as there are numbers of B in the segment $(1,l)$, that is, $B(l)$ of them. Consequently every segment of length l included between two consecutive numbers of the sequence A contains at least $B(l)$ numbers which belong to C. It follows that the number, $C(n)$, of numbers of the segment $(1,n)$ appearing in C is at least

$$A(n)+\Sigma\, B(l)$$

where the summation is extended over all segments which are free of the numbers appearing in A. According to the definition of density, however, $B(l)\geqq\beta l$, so that

$$C(n)\geqq A(n)+\beta\Sigma l=A(n)+\beta\{n-A(n)\},$$

because Σl is the sum of the lengths of all the segments which are free of the numbers appearing in A, which is simply the number $n-A(n)$ of numbers of the segment $(1,n)$ which do not occur in A. But $A(n)\geqq\alpha n$, and hence

$$C(n)\geqq A(n)(1-\beta)+\beta n\geqq\alpha n(1-\beta)+\beta n,$$

which yields

$$C(n)/n\geqq\alpha+\beta-\alpha\beta.$$

Since this inequality holds for an arbitrary natural number n, we have

$$\gamma-d(C)\geqq\alpha+\beta-\alpha\beta, \qquad \text{Q. E. D.}$$

Schnirelmann's inequality (1) can be written in the equivalent form

$$1-d(A+B)\leqq\{1-d(A)\}\{1-d(B)\},$$

and in this form can easily be generalized to the case of an arbitrary number of summands:

$$1-d(A_1+A_2+\ldots+A_k)\leqq\prod_{i=1}^{k}\{1-d(A_i)\}.$$

It is proved by a simple induction; you should have no trouble in carrying it out yourself. If we write the last inequality in the form

$$d(A_1+A_2+\ldots+A_k)\geqq 1-\prod_{i=1}^{k}\{1-d(A_i)\}, \tag{2}$$

it again enables one to estimate the density of a sum from the densities of the summands. L. G. Schnirelmann derived a series of very remarkable results from his elementary inequality, and obtained above all the following important theorem:

Every sequence of positive density is a basis of the sequence of natural numbers.

In other words, if $\alpha=d(A)>0$, then the sum of a sufficiently large number of sequences A contains the entire sequence of natural numbers. The proof of this theorem is so simple that I should like to tell you about it, even though this will divert us a bit from our immediate problem.

Let us denote for brevity by A_k the sum of k sequences, each of which coincides with A. Then by virtue of inequality (2),

$$d(A_k)\geqq 1-(1-\alpha)^k.$$

Since $\alpha>0$, we have, for sufficiently large k,

$$d(A_k)>\tfrac{1}{2}. \tag{3}$$

Now one can easily show that the sequence A_{2k} contains the whole sequence of natural numbers. This is a simple consequence of the following general proposition.

LEMMA. *If* $A(n)+B(n)>n-1$, *then* n *occurs in* $A+B$.

Indeed, if n appears in A or in B, everything is proved. We may therefore assume that n occurs in neither A nor B. Then $A(n)=A(n-1)$ and $B(n)=B(n-1)$, and consequently

$$A(n-1)+B(n-1)>n-1.$$

Now let $a_1, a_2, \ldots, a_r$ and $b_1, b_2, \ldots, b_s$ be the numbers of the segment $(1, n-1)$ which appear in A and B, respectively, so that $r=A(n-1)$, $s=B(n-1)$. Then all the numbers

$$a_1, a_2, \ldots, a_r,$$
$$n-b_1, n-b_2, \ldots, n-b_s$$

belong to the segment $(1, n-1)$. There are $r+s=A(n-1)+B(n-1)$ of these numbers, which is more than $n-1$. Hence one of the numbers in the upper row equals one of the numbers in the lower row. Let $a_i=n-b_k$. Then $n=a_i+b_k$, i. e., n appears in $A+B$.

Returning now to our objective, we have, on the basis of (3), for an arbitrary n:

$$A_k(n)>\tfrac{1}{2}n>\tfrac{1}{2}(n-1)$$

and therefore

$$A_k(n)+A_k(n)>n-1.$$

According to the lemma just proved, it follows that n appears in $A_k+A_k=A_{2k}$. But n is an arbitrary natural number, and hence our theorem is proved.

This simple theorem led to a series of important applications in the papers of L. G. Schnirelmann. For example, he was the first to prove that the sequence P consisting of unity and all the prime numbers is a basis of the sequence of natural numbers. The sequence P, it is true, has density zero, as Euler had already shown, so that the theorem which we just proved is not directly applicable to it. But Schnirelmann was able to prove that $P+P$ has positive density. Hence $P+P$ forms a basis, and therefore P indeed also. From this it is easy to infer that an arbitrary natural number, with the exception of 1, can, for sufficiently large k, be represented as the sum of at most k primes. For that time (1930) this result was fundamental and evoked the greatest interest in the scientific world. At present, thanks to the remarkable work of I. M. Vinogradov, we know considerably more in this direction, as I already related to you at the beginning of this chapter.

§3

In the preceding it was my purpose to introduce you in the shortest way possible to the problems of this singular and fascinating branch of number theory, whose study began with L. G. Schnirelmann's remarkable work. The immediate goal of the present chapter, however, is a specific problem in this field, and I now proceed to its formulation.

In the fall of 1931, upon his return from a foreign tour, L. G. Schnirelmann reported to us his conversations with Landau in Göttingen, and related among other things that in the course of these conversations they had discovered the following interesting fact: In all the concrete examples that they were able to devise, it was possible to replace the inequality

$$d(A+B) \geqq d(A)+d(B)-d(A)d(B),$$

which we derived in §2, by the sharper (and simpler) inequality

$$d(A+B) \geqq d(A)+d(B). \tag{4}$$

That is, the density of the sum always turned out to be at least as large as the sum of the densities of the summands (under the assumption, of course, that $d(A)+d(B) \leqq 1$). They therefore naturally assumed that inequality (4) was the expression of a universal law, but the first attempts to prove this conjecture were unsuccessful. It soon became evident that if their conjecture was correct, the road to its proof would be quite difficult. We wish to note at this point that if the hypothetical inequality (4) does represent a universal law, then this law can be generalized immediately by induction to the case of an arbitrary number of summands; i.e., under the assumption that

$$\sum_{i=1}^{k} d(A_i) \leqq 1$$

we have

$$d\Big(\sum_{i=1}^{k} A_i\Big) \geqq \sum_{i=1}^{k} d(A_i). \tag{5}$$

This problem could not help but attract the attention of scholars, because of the simplicity and elegance of the general hypothetical law (4) on the one hand, and on the other because of the sharp contrast between the elementary character of the problem and the difficulty of its solution which became apparent already after the first attacks. I myself was fascinated by it at the time, and neglected all my other researches on its account. Early in 1932, after several months of hard work, I succeeded in proving inequality (4) for the most important special case, $d(A)=d(B)$ (this case must be considered as the most important because in the majority of concrete problems all the summands are the same). At the same time I also proved the general inequality (5) under the assumption that $d(A_1)=d(A_2)=\dots=d(A_k)$ (it is easy to see that this result cannot be derived from the preceding one simply by induction, but requires a special proof). The method which I used was completely elementary, but very complicated. I was later able to simplify the proof somewhat.

Be that as it may, it was but a special case. For a long time it seemed to me that a none too subtle improvement of my method should lead to a full solution of the problem, but all my efforts in this direction proved fruitless.

In the meantime the publication of my work had attracted the attention of a wide circle of scholars in all countries to the Landau-Schnirelmann hypothesis. Many insignificant results were obtained, and a whole literature sprang up. Some authors carried over the problem from the domain of natural numbers to other fields. In short, the problem became "fashionable". Learned societies offered prizes for its solution. My friends in England wrote me in 1935 that a good half of the English mathematicians had postponed their usual work in order to try to solve this problem. Landau, in his tract devoted to the latest advances in additive number theory, wrote that he "should like to urge this problem on the reader". But it proved to be obstinate, and withstood the efforts of the most able scholars for a whole series of years. It was not until 1942 that the young American mathematician Mann finally disposed of it: he found a complete proof of inequality (4) (and hence also of inequality (5)). His method is wholly elementary and is related to my work in form, although it is based on an entirely different idea. The proof is long and very complicated, and I could not bring myself to present it to you here. A year later, however, in 1943, Artin and Scherk published a new proof of the same theorem, which rests on an altogether different idea. It is considerably shorter and more transparent, though still quite elementary. This is the proof that I should like to tell you about; I have written this chapter on its account, and it forms the content of all the succeeding sections.

§4

Suppose then that A and B are two sequences. We set $A+B=C$. Let $A(n)$, $d(A)$, etc. have their usual meaning. We recall that all our sequences begin with zero, but that only the natural numbers appearing in these sequences are considered when calculating $A(n)$, $B(n)$, $C(n)$. We have to prove that

$$d(C) \geqq d(A)+d(B) \tag{6}$$

provided that $d(A)+d(B) \leqq 1$. For brevity we set $d(A)=\alpha$, $d(B)=\beta$ in what follows.

FUNDAMENTAL LEMMA. *If n is an arbitrary natural number, there exists an integer m $(1 \leqq m \leqq n)$ such that*

$$C(n)-C(n-m) \geqq (\alpha+\beta)\, m.$$

In other words, there exists a "remainder" $(n-m+1, n)$ of the segment $(1, n)$, in which the average density of the sequence C is at least $\alpha+\beta$.

We are now faced with two problems: first, to prove the fundamental lemma, and second, to show that inequality (6) follows from the fundamental lemma. The second of these problems is incomparably easier than the first, and we shall therefore begin with the second problem.

Suppose then that the fundamental lemma has already been proved. This means that in a certain "remainder" $(n-m+1, n)$ of the segment $(1, n)$ the average density of the sequence C is at least $\alpha+\beta$. By the fundamental lemma, however, the segment $(1, n-m)$ again has a certain "remainder" $(n-m-m'+1, n-m)$ in which the average density of C is at least $\alpha+\beta$. It is clear that by continuing this process, the segment $(1,n)$ is eventually divided into a finite number of subsegments, in each of which the average density of C is at least $\alpha+\beta$. Therefore the average density of C is also at least $\alpha+\beta$ in the whole segment $(1,n)$. Since n was arbitrary, however, we have

$$d(C)\geqq\alpha+\beta, \qquad \text{Q.E.D.}$$

Thus the problem is now reduced to proving the fundamental lemma. We now turn to this proof, which is long and complicated.

§5
NORMAL SEQUENCES

In all that follows we shall regard the number n as fixed, and all sequences which we investigate will consist of the number 0 and certain numbers of the segment $(1,n)$. We agree to call such a sequence N *normal*, if it possesses the following property: If the arbitrary numbers f and f' of the segment $(1,n)$ do not appear in N, then neither does the number $f+f'-n$ appear in N (where the case $f=f'$ is not excluded).

If the number n belongs to the sequence C, then

$$C(n)-C(n-1)=1\geqq(\alpha+\beta)\cdot 1,$$

so that the fundamental lemma is trivially correct $(m=1)$. Consequently we shall assume in the sequel–I beg you to keep this in mind–that n does not occur in C.

To begin with, the fundamental lemma is easy to prove in case the sequence C is normal. Indeed, let us denote by m the smallest positive number which does not appear in C ($m\leqq n$ because n, by assumption, does not occur in C). Let s be an arbitrary integer lying between $n-m$ and n; $n-m<s<n$. Then $0<s+m-n<m$. I say that $s\varepsilon C$. For if this were not the case, the number $s+m-n$, because of the normality of C, would not appear in C. But we have just seen that this number is smaller than m, whereas m, by definition, is the *smallest* positive integer which does not occur in C.

Hence all numbers s of the segment $n-m<s<n$ appear in C, and therefore

$$C(n)-C(n-m)=m-1.$$

On the other hand, by the lemma on p. 24, since m does not occur in $C=A+B$ we have $A(m)+B(m)\leqq m-1$. Consequently

$$C(n)-C(n-m)\geqq A(m)+B(m)\geqq(\alpha+\beta)m, \tag{7}$$

which again proves the validity of the fundamental lemma.

§6
CANONICAL EXTENSIONS

We now turn our attention to the case where the sequence $C=A+B$ is not normal. In this case we shall add to the set B, according to a very definite scheme, numbers which it does not contain, and thereby pass from B to an extended set B_1. The set $A+B_1=C_1$ evidently will then be a certain extension of the set C. As I said before, this extension of the sets B and C (the set A remains unaltered) will be defined precisely and unambiguously; it is possible if and only if the set C is not normal. We shall call this extension a *canonical extension* of the sets B and C. Some important properties of canonical extensions will be derived, with whose aid the proof of the fundamental lemma will be completed.

I now come to the definition of the canonical extension of the sets B and C. If C is not normal, there exist two numbers c and c' in the segment $(0,n)$, such that

$$c\notin C,\ c'\notin C,\ c+c'-n\,\varepsilon C.$$

Since $C=A+B$, it follows that

$$c+c'-n=a+b \qquad (a\varepsilon A,\ b\varepsilon B). \tag{8}$$

Let β_0 be the smallest number of the set B which can play the role of the number b in equation (8). In other words, β_0 is the smallest integer $b\varepsilon B$ which satisfies equation (8) for suitably chosen numbers $c\notin C$, $c'\notin C$, $a\varepsilon A$ of the segment $(0,n)$. This number β_0 will be called the *basis* of our extension.

Thus the equation

$$c+c'-n=a+\beta_0 \tag{9}$$

necessarily has solutions in the numbers c, c', a satisfying the conditions

$$c \notin C, \; c' \notin C, \; a \in A,$$

where all three numbers belong to the segment $(0, n)$. We write all numbers c and c' which satisfy equation (9) and the enumerated conditions, to form a set C^*. Clearly the sets C and C^* do not have a single element in common. We call their union* (i. e., the totality of all numbers which occur either in C or in C^*)

$$C \cup C^* = C_1$$

the *canonical extension* of the set C.

Let us now examine the expression $\beta_0 + n - c$. If c here is allowed to run through all the numbers of the set C^* just constructed, the values of this expression form a certain set B^*. According to equation (9), every such number $\beta_0 + n - c$ $(c \in C^*)$ can be written in the form $c' - a$, where $c' \in C^*$, $a \in A$.

Let b^* be an arbitrary number occurring in B^*. Since it is of the form $\beta_0 + n - c$, it is $\geqq \beta_0 \geqq 0$; and since it is also of the form $c' - a$ $(c' \in C^*, \; a \in A)$, it is $\leqq c' \leqq n$. Hence all numbers of the set B^* belong to the segment $(0, n)$. Moreover, if $b^* \in B^*$, then $b^* \notin B$, because otherwise it would follow from $b^* = c' - a$ that $c' = a + b^* \in A + B = C$, which is false.

Accordingly, the set B^* is embedded in the segment $(0, n)$ and has no elements in common with the set B. We put

$$B \cup B^* = B_1$$

and call the set B_1 a canonical extension of the set B.

Let us show that

$$A + B_1 = C_1.$$

First, let $a \in A$, $b_1 \in B_1$. We shall prove that $a + b_1 \in C_1$. From $b_1 \in B_1$ it follows that either $b_1 \in B$ or $b_1 \in B^*$. If $b_1 \in B$, then $a + b_1 \in A + B = C \subset C_1$. If $b_1 \in B^*$, however, then $a + b_1$ either occurs in C, and hence also in C_1, or $a + b_1 \notin C$. But in this case (since b_1, as an element of the set B^*, is of the form $\beta_0 + n - c'$, $c' \notin C$) we obtain

$$c = a + b_1 = a + \beta_0 + n - c' \notin C.$$

Therefore

*Here and in the sequel we use the symbol $\cup$ to denote the union of sets, since we are using the symbol + in another sense.

32

$$c+c'-n=a+\beta_0 \varepsilon A+B=C,$$

where $c \notin C$ and $c' \notin C$. But then according to the definition of the set C^*,

$$c=a+b_1 \varepsilon C^* \subset C_1, \qquad \text{Q.E.D.}$$

Thus we have shown that $A+B_1 \subset C_1$.

To prove the inverse relation, let us assume that $c \varepsilon C_1$, which means that either $c \varepsilon C$ or $c \varepsilon C^*$. If $c \varepsilon C$, then $c=a+b$, $a \varepsilon A$, $b \varepsilon B \subset B_1$. If, however, $c \varepsilon C^*$, then, for a certain $a \varepsilon A$, the number $b^*=c-a$, as we know, occurs in B^*. We have $c=a+b^* \varepsilon A+B^* \subset A+B_1$. Therefore $C_1 \subset A+B_1$. We also proved above that $A+B_1 \subset C_1$. Consequently $C_1=A+B_1$.

Now recall that according to our assumption, $n \notin C$. It is easy to see–and this is important–that the number n does not appear in the extension C_1. For if we had $n \varepsilon C^*$, we could, by the definition of C^*, put $c'=n$ in equation (9), which would yield $c=a+\beta_0 \varepsilon A+B=C$, whereas $c \notin C$ according to (9).

If the extended sequence C_1 is not yet normal, then, because of $A+B_1=C_1$ and $n \notin C_1$, the sets A, B_1, and C_1 form a triple with all the properties of the triple A, B, C that are necessary for a new canonical extension. We take a new basis β_1 of this extension, define the complementary sets B_1^*, C_1^* as before, put

$$B_1 \cup B_1^* = B_2, \quad C_1 \cup C_1^* = C_2,$$

and are able to assert once more that $A+B_2=C_2$ and $n \notin C_2$. It is evident that this process can be continued until one of the extensions C_h proves to be normal. Obviously this case must certainly take place, because in every extension we add new numbers to the sets B_μ and C_μ without overstepping the bounds of the segment $(0,n)$.

In this way we obtain the finite sequences of sets

$$B=B_0 \subset B_1 \subset \ldots \subset B_h,$$
$$C=C_0 \subset C_1 \subset \ldots \subset C_h,$$

where every $B_{\mu+1}$ (respectively $C_{\mu+1}$) contains numbers which do not appear in B_μ (C_μ) and which go to make up the set B_μ^* (C_μ^*), so that

$$B_{\mu+1}=B_\mu \cup B^*_\mu, \quad C_{\mu+1}=C_\mu \cup C^*_\mu \qquad (0 \leqq \mu \leqq h-1).$$

We denote by β_μ the basis of the extension which carries (B_μ, C_μ) into $(B_{\mu+1}, C_{\mu+1})$. We have

$$A+B_\mu=C_\mu, \quad n \notin C_\mu \qquad (0 \leqq \mu \leqq h).$$

Finally, the set C_h is normal, whereas the sets C_μ $(0 \leqq \mu \leqq h-1)$ are not.

§7
PROPERTIES OF THE CANONICAL EXTENSIONS

We shall now formulate and prove in the form of three lemmas those properties of the canonical extensions which are needed later. Only Lemma 3 will have further application; Lemmas 1 and 2 are required solely for the proof of Lemma 3.

LEMMA 1. $\beta_\mu > \beta_{\mu-1}$ $(1 \leqq \mu \leqq h-1)$; *i. e., the bases of successive canonical extensions form a monotonically increasing sequence.*

In fact, since $\beta_\mu \in B_\mu = B_{\mu-1} \cup B^*_{\mu-1}$, either $\beta_\mu \in B^*_{\mu-1}$ or $\beta_\mu \in B_{\mu-1}$. If $\beta_\mu \in B^*_{\mu-1}$, then β_μ is of the form

$$\beta_\mu = \beta_{\mu-1}+n-c,$$

where $c \in C^*_{\mu-1} \subset C_\mu$ and therefore $c<n$, so that $\beta_\mu > \beta_{\mu-1}$, and Lemma 1 is proved. If $\beta_\mu \in B_{\mu-1}$, however, then by the definition of the number β_μ there exist integers $a \in A$, $c \notin C_\mu$, $c' \notin C_\mu$ such that

$$c+c'-n=a+\beta_\mu \in C_\mu.$$

But for $\beta_\mu \in B_{\mu-1}$, we have

$$c+c'-n=a+\beta_\mu \in A+B_{\mu-1}=C_{\mu-1}, \tag{10}$$

where $c \notin C_{\mu-1}$, $c' \notin C_{\mu-1}$. Hence, because of the minimal property of $\beta_{\mu-1}$, $\beta_\mu \geqq \beta_{\mu-1}$. If $\beta_\mu = \beta_{\mu-1}$, it would follow from (10) and the definition of the set $C^*_{\mu-1}$ that

$$c \in C^*_{\mu-1} \subset C_\mu, \qquad c' \in C^*_{\mu-1} \subset C_\mu.$$

Both are false, however, and therefore $\beta_\mu > \beta_{\mu-1}$.

In the sequel we shall denote by m the smallest positive integer which does not appear in C_h.

LEMMA 2. *If* $c \varepsilon C^*_\mu$ $(0 \leq \mu \leq h-1)$ *and* $n-m<c<n$, *then* $c>n-m+\beta_\mu$. *That is, all numbers c of the set* C^*_μ *which lie in the interval* $n-m<c<n$ *are embedded in that part of this segment which is characterized by the inequalities* $n-m+\beta_\mu<c<n$.

We have to show that

$$c+m-n>\beta_\mu.$$

It follows from $n-m<c<n$ that

$$0<m+c-n<m.$$

Therefore, by the definition of the number m,

$$m+c-n \ \varepsilon \ C_h.$$

Now

$$C_h = C_\mu \cup C^*_\mu \cup C^*_{\mu+1} \cup \ldots \cup C^*_{h-1}.$$

We consider two cases.

1) If $m+c-n \ \varepsilon \ C_\mu$, then

$$m+c-n=a+b_\mu, \qquad a \ \varepsilon \ A, \ b_\mu \ \varepsilon \ B_\mu.$$

But $m \notin C_\mu$ and $c \notin C_\mu$ (the latter because $c \ \varepsilon \ C^*_\mu$). Therefore because of the minimal property of β_μ we must have $b_\mu \geqq \beta_\mu$. If $b_\mu=\beta_\mu$ it would follow from the definition of the set C^*_μ that $m \ \varepsilon \ C^*_\mu$, which is false because $C^*_\mu \subset C_{\mu+1} \subset C_h$ and $m \notin C_h$. Consequently $b_\mu>\beta_\mu$, so that

$$m+c-n=a+b_\mu \geqq b_\mu>\beta_\mu,$$

and Lemma 2 is proved.

2) If $c'=m+c-n \ \varepsilon \ C^*_\nu$ $(\mu \leq \nu \leq h-1)$, then, by the definition of the set C^*_ν, c' satisfies an equation of the form (9),

$$c'-a=\beta_\nu+n-c'',$$

where $a \ \varepsilon \ A$, $c'' \varepsilon \ C^*_\nu$. Hence $c' \geqq c'-a>\beta_\nu \geqq \beta_\mu$ (where the last inequality is given by Lemma 1), and Lemma 2 is again proved.

LEMMA 3. *We have*

$$C_\mu^*(n)-C_\mu^*(n-m)=B_\mu^*(m-1) \qquad (0\leq\mu\leq h-1).$$

That is, the number of integers $c\in C_\mu^*$ *in the segment* $n-m<c<n$ *is exactly the same as the number of integers* $b\in B_\mu^*$ *in the segment* $0<b<m$ *(of the same length).*

Let us examine the relation

$$(11) \qquad b=\beta_\mu+n-c.$$

By the very definition of the sets B_μ^* and C_μ^*, $c\in C_\mu^*$ implies $b\in B_\mu^*$, and conversely. If, in addition, $n-m+\beta_\mu<c<n$, then $\beta_\mu<b<m$, and conversely. Hence

$$C_\mu^*(n)-C_\mu^*(n-m+\beta_\mu)=B_\mu^*(m-1)-B_\mu^*(\beta_\mu).$$

By Lemma 2, $C_\mu^*(n-m+\beta_\mu)=C_\mu^*(n-m)$. On the other hand, every $b\in B_\mu^*$ can be expressed in the form (11), where $c<n$; b therefore exceeds β_μ, and consequently $B_\mu^*(\beta_\mu)=0$. It follows that

$$C_\mu^*(n)-C_\mu^*(n-m)=B_\mu^*(m-1), \qquad \text{Q.E.D.}$$

§ 8

PROOF OF THE FUNDAMENTAL LEMMA

It is very easy now to prove the fundamental lemma by proceeding from the results in §5 and appealing to Lemma 3 which was just proved.

If we apply the result of §5 in the form of inequality (7) to the sequences A, B_h, and C_h (which is permissible because of the normality of C_h), we find that

$$(12) \qquad C_h(n)-C_h(n-m)\geq A(m)+B_h(m),$$

where m is the smallest positive integer which does not occur in C_h. Obviously $m\notin A$ and $m\notin B_h$, so that we may write $A(m-1)$ and $B_h(m-1)$ instead of $A(m)$ and $B_h(m)$, respectively.

We have

$$C_h=C\cup C^*\cup C_1^*\cup\ldots\cup C_{h-1}^*,$$
$$B_h=B\cup B^*\cup B_1^*\cup\ldots\cup B_{h-1}^*,$$

where the sets appearing in any one of these two unions are mutually exclusive, so that

$$C_h(n)-C_h(n-m)=C(n)-C(n-m)+\sum_{\mu=0}^{h-1}\{C_\mu^*(n)-C_\mu^*(n-m)\},$$

$$B_h(m)=B_h(m-1)=B(m-1)+\sum_{\mu=0}^{h-1}B_\mu^*(m-1);$$

we have of course put $C_0^*=C^*$, $B_0^*=B^*$. On account of (12) it follows that

$$C(n)-C(n-m)+\sum_{\mu=0}^{h-1}\{C_\mu^*(n)-C_\mu^*(n-m)\}$$

$$\geqq A(m)+B(m-1)+\sum_{\mu=0}^{h-1}B_\mu^*(m-1).$$

By Lemma 3, however,

$$C_\mu^*(n)-C_\mu^*(n-m)=B_\mu^*(m-1) \qquad (0\leqq\mu\leqq h-1),$$

so that the preceding inequality becomes

$$C(n)-C(n-m)\geqq A(m)+B(m-1)=A(m)+B(m)\geqq(\alpha+\beta)\,m,$$

which proves the fundamental lemma.

As we saw in § 4, this also completes the proof of Mann's theorem which solves the fundamental metric problem of additive number theory.

Doesn't Artin and Scherk's construction have the stamp of a magnificent masterpiece? I find the outstanding combination of structural finesse and the extremely elementary form of the method especially attractive.

CHAPTER III

AN ELEMENTARY SOLUTION OF WARING'S PROBLEM

§1

You will recall the theorem of Lagrange, which was discussed at the beginning of the preceding chapter. It says that every natural number can be expressed as the sum of at most four squares. I also showed you that this theorem could be stated in entirely different terms: If four sequences, each identical with

(A_2) $$0, 1^2, 2^2, \ldots, k^2, \ldots,$$

are added together, the resulting sequence contains all the natural numbers. Or even more briefly, the sequence (A_2) is a basis (of the sequence of natural numbers) of order four. I also mentioned that, as had been shown later, the sequence of cubes

(A_3) $$0, 1^3, 2^3, \ldots, k^3, \ldots$$

was a basis of order nine. All these facts lead in a natural manner to the hypothesis that, for an arbitrary natural number n, the sequence

(A_n) $$0, 1^n, 2^n, \ldots, k^n, \ldots$$

is a basis (whose order of course depends on n). This conjecture was also actually propounded by Waring as early as the eighteenth century. The problem proved to be very difficult, however, and it was not until the beginning of the present century that the universal validity of Waring's hypothesis was demonstrated, by Hilbert (1909). Hilbert's proof is not only ponderous in its formal aspect and based on complicated analytical theories (multiple integrals), but also lacks transparency in conceptual respects. The eminent French mathematician Poincaré wrote in his survey of Hilbert's creative scientific work, that once the basic motivations behind this proof were understood, arithmetical results of great importance would probably flow forth as from a cornucopia. In a certain sense he was right. Ten to fifteen years later, new proofs of Hilbert's theorem were furnished by Hardy and Littlewood in England and by I. M. Vinogradov in the USSR. These proofs were again analytic and formally unwieldy, but differed favorably from Hilbert's proof in their clarity of method and their conceptual simplicity, which left nothing to be desired. In fact, because of this, both methods became mighty scources of new arithmetical theorems.

But when our science is concerned with such a completely elementary problem as Waring's problem, it invariably attempts to find a solution which requires no concepts or methods transcending the the limits of elementary arithmetic. The search for such an elementary proof of Waring's hypothesis is the third problem which I should like to tell you about. Such a fully elementary proof of Hilbert's theorem was first obtained in 1942, by the young Soviet scholar Y. V. Linnik.

You are already accustomed to the fact that "elementary" does not mean "simple". The elementary solution of Waring's problem discovered by Linnik is, as you will see, not very simple either, and it will take considerable effort on your part to understand and digest it. I shall endeavor to make this task as easy as possible for you through my exposition. But you must remember that in mathematics (as probably in any other science) the assimilation of anything really valuable and significant involves trying labor.

The ideas of Schnirelmann which I described to you in the beginning of the second chapter play an essential role in Linnik's proof. You will recall (I mentioned it at that time) how Schnirelmann proved his famous theorem that the sequence P consisting of zero, unity, and all the primes, is a basis of the sequence of natural numbers: He showed that the sequence $P+P$ has a positive density. This immediately yields the assertion, however, because, according to the general theorem of Schnirelmann which we proved on pp.24-25, every sequence of positive density is a basis of the sequence of natural numbers. The same method also lies at the basis of the proof of Hilbert's theorem discovered by Linnik. It all boils down to the proof that the sum of a sufficiently large number of sequences (A_n) is a sequence of positive density. As soon as this is accomplished, we can, by virtue of the same general theorem of Schnirelmann, regard Hilbert's theorem as proved.

§2

THE FUNDAMENTAL LEMMA

If we add together k sequences, identical with A_n, according to the rule in Chapter II, we evidently obtain a sequence $A_n^{(k)}$ which contains zero and all those natural numbers which can be expressed as a sum of at most k summands of the form x^m, where x is an arbitrary natural number. In other words, the number m belongs to the sequence $A_n^{(k)}$, if the equation

(1) $$x_1^n + x_2^n + \dots + x_k^n = m$$

is solvable in nonnegative integers x_i $(1 \leq i \leq k)$. As we saw in §1, the problem is to show that, for sufficiently large k, the sequence $A_n^{(k)}$ has a positive density.

For preassigned k and m, equation (1) in general can be solved in several different ways. In the sequel we shall denote by $r_k(m)$ the number of these ways, i. e., the number of systems of nonnegative integers $x_1, x_2, \dots, x_k$ which satisfy equation (1). It is clear that the number m occurs in $A_n^{(k)}$ if and only if $r_k(m) > 0$.

In the following, we shall assume the number n to be given and fixed, and shall therefore call all numbers which depend only upon n, constants. Such constants will be denoted by the letter c or $c(n)$, where such a constant c may have different values in different parts of our discussion, provided merely that these values are constants. Perhaps you are rather unused to such "freedom" of notation, but you will soon become familiar with it. It has proved to be very convenient, and appears more and more frequently in modern research.

FUNDAMENTAL LEMMA. *There exist a natural number $k = k(n)$, depending only on n, and a constant c, such that, for an arbitrary natural number N,*

(2) $$r_k(m) < c N^{(k/n)-1} \quad (1 \leq m \leq N).$$

Once more, as in the preceding chapter, we are faced with two problems: first, to prove the fundamental lemma, and second, to draw from the fundamental lemma the conclusion that we need, viz., that the sequence $A_n^{(k)}$ has a positive density. This time again the second problem is considerably easier than the first, and we shall therefore begin with the second problem.

It follows immediately from the definition of the number $r_k(m)$, that the sum

$$r_k(0)+r_k(1)+\ldots+r_k(N)=R_k(N)$$

represents the number of systems $(x_1, x_2, \ldots, x_k)$ of k nonnegative integers for which

$$x_1^n+x_2^n+\ldots+x_k^n \leqq N. \tag{3}$$

Every group of numbers for which

$$0 \leqq x_i \leqq (N/k)^{1/n} \qquad (1 \leqq i \leqq k),$$

obviously satisfies this condition. To satisfy these inequalities, every x_i can evidently be chosen in more than $(N/k)^{1/n}$ different ways $(x_i=0, 1, \ldots, [(N/k)^{1/n}])$.* After an arbitrary choice of this sort, the numbers $x_1, x_2, \ldots, x_k$ may be combined, and so we have more than $(N/k)^{k/n}$ different possibilities for choosing the complete system of integers x_i $(1 \leqq i \leqq k)$ so as to satisfy condition (3). This shows that

$$R_k(N) \geqq (N/k)^{k/n}. \tag{4}$$

We assume that the fundamental lemma has been shown to be correct, and that inequality (2) is satisfied for an arbitrary N. We now have to verify that inequality (2) is consistent with inequality (4) which we proved, only if the sequence $A_n^{(k)}$ has a positive density. The idea behind the following deduction is very simple: In the sum $R_k(N)$, only those summands $r_k(m)$ are different from zero, for which m occurs in $A_n^{(k)}$. If $A_n^{(k)}$ had density zero, then for large N the number of such summands would be relatively small; because of (2), however, every summand cannot be very large. Their sum $R_k(N)$, therefore, would also be relatively small, whereas according to (4) it must be rather large.

* $[a]$ denotes the largest integer $\leqq a$.

It remains to carry out the calculations. Suppose that $d(A_n^{(k)})=0$. Then, for an arbitrary small $\epsilon>0$ and a suitably chosen N,

$$A_n^{(k)}(N)<\epsilon N.$$

Here the number N may be assumed to be arbitrarily large, because $A_n^{(k)}$ (for an arbitrary k) contains the integer 1 (bear in mind Problem 6 on p. 22, which you solved). Applying the estimate (2) we get

$$R_k(N)=\sum_{m=0}^{n} r_k(m)=r_k(0)+\sum_{m=1}^{N} r_k(m)<1+cN^{(k/n)-1}A_n^{(k)}(N)<1+c\epsilon N^{(k/n)},$$

and hence, for sufficiently large N,

$$R_k(N)<2c\epsilon N^{k/n}.$$

For sufficiently small ϵ,

$$2c\epsilon<(1/k)^{k/n},$$

so that

$$R_k(N)<(N/k)^{k/n},$$

which contradicts (4). Therefore we must have

$$d(A_n^{(k)})>0.$$

But, as we already know, this proves Hilbert's theorem.

You see how simply it all comes out. But we still have to prove the fundamental lemma, and to do this we shall have to travel a long and difficult road, as in the preceding chapter.

§3
LEMMAS CONCERNING LINEAR EQUATIONS

We shall have to go far back. It will therefore be well for you to forget completely for a while the problem which has been posed. I shall call your attention to it when we return to it later.

Right now, however, we have to find some estimates for the number of solutions of systems of linear equations. The lemmas of this paragraph, moreover, are perhaps also of intrinsic interest, independent of the problem for whose solution they are required here.

LEMMA 1. *In the equation*

(5) $$a_1 z_1 + a_2 z_2 = m,$$

let a_1, a_2, m *be integers with* $|a_2| \leqq |a_1| \leqq A$, *and let* a_1 *and* a_2 *be relatively prime. Then the number of solutions of equation* (5) *satisfying the inequalities* $|z_1| \leqq A$, $|z_2| \leqq A$, *does not exceed* $3A/|a_1|$.

Proof: We may assume that $a_1 > 0$, because otherwise we have merely to replace z_1 by $-z_1$ in every solution.

Let $\{z_1, z_2\}$ and $\{z_1', z_2'\}$ be two different solutions of equation (5). Then from

$$a_1 z_1 + a_2 z_2 = m,$$
$$a_1 z_1' + a_2 z_2' = m$$

we get

$$a_2 (z_2' - z_2) = a_1 (z_1 - z_1')$$

by subtraction. Accordingly the left-hand side of this equation must be divisible by a_1. But* $(a_1, a_2) = 1$, and consequently $z_2' - z_2$ must be divisible by a_1. Now $z_2' \neq z_2$, and therefore $|z_2' - z_2|$, as a multiple of a_1, is not smaller than a_1. Thus, for two distinct solutions $\{z_1, z_2\}$ and $\{z_1', z_2'\}$ of equation (5), we must have $|z_2' - z_2| \geqq a_1$.

In every solution $\{z_1, z_2\}$ of equation (5), let us agree to call z_1 the first member and z_2 the second. It is obvious that the number of solutions of equation (5) which satisfy the conditions $|z_1| \leqq A$, $|z_2| \leqq A$, is not more than the number t of second members which occur in the interval $\langle -A, A\rangle$. Since we have proved that two such second members are at least the distance a_1 apart, the difference between the largest and smallest second members occurring in the interval $\langle -A, A\rangle$ is at least $a_1(t-1)$. On the other hand, this difference does not exceed $2A$, so that

$$a_1(t-1) \leqq 2A,$$
$$(t-1) \leqq 2A/a_1,$$
$$t \leqq (2A/a_1) + 1 \leqq 3A/a_1$$

(because, by assumption, $a_1 \leqq A$, and therefore $1 \leqq A/a_1$). This proves Lemma 1.

LEMMA 2. *In the equation*

(6) $$a_1 z_1 + a_2 z_2 + \dots + a_l z_l = m,$$

* (a_1, a_2) denotes the greatest common divisor of the integers a_1 and a_2.

*let the a_i and m be integers satisfying the conditions**

$$|a_i| \leq A \quad (1 \leq i \leq l), \qquad (a_1, a_2, \ldots, a_l) = 1.$$

Then the number of solutions of equation (6) satisfying the inequalities $|z_i| \leq A$ $(1 \leq i \leq l)$, does not exceed

$$c(l)\, A^{l-1}/H,$$

where H is the largest of the numbers $|a_1|, |a_2|, \ldots, |a_l|$, and $c(l)$ is a constant depending only on l.

Proof: If $l=2$, Lemma 2 obviously becomes Lemma 1 (with $c(2)=3$). Accordingly Lemma 2 is already verified for $l=2$. We shall therefore assume that $l \geq 3$ and that the truth of Lemma 2 has already been established for the case of $l-1$ unknowns. Since the numbering is unimportant, we may assume that $|a_l|$ is the largest of the numbers $|a_1|, |a_2|, \ldots, |a_l|$, i.e., $H = |a_l|$.

There are two cases to consider.

1) $a_1 = a_2 = \ldots = a_{l-1} = 0$. Since $(a_1, a_2, \ldots, a_l) = 1$, we have $|a_l| = H = 1$, so that the given equation is of the form $\pm z_l = m$. In this equation each of the unknowns $z_1, z_2, \ldots, z_{l-1}$ can obviously assume an arbitrary integral value in the interval $\langle -A, A \rangle$, and hence at most $2A+1 \leq 3A$ values all told. As for z_l, however, it can assume at most one value. Consequently the number of solutions of the given equation satisfying the inequalities $|z_i| \leq A$ $(1 \leq i \leq l)$, does not exceed

$$(3A)^{l-1} = c(l)\, A^{l-1} = c(l)\, A^{l-1}/H,$$

which proves Lemma 2 for this case.

2) If at least one of the numbers $a_1, a_2, \ldots, a_{l-1}$ is different from zero, then

$$(a_1, a_2, \ldots, a_{l-1}) = \delta$$

exists. Let us denote by H' the largest of the numbers

$$|a_i|/\delta \qquad (1 \leq i \leq l-1).$$

Suppose now that the numbers $z_1, z_2, \ldots, z_l$ satisfy the given equation (6) and the inequalities $|z_i| \leq A$ $(1 \leq i \leq l)$. We set

$$(a_1/\delta)\, z_1 + (a_2/\delta) z_2 + \ldots + (a_{l-1}/\delta)\, z_{l-1} = m', \tag{7}$$

*$(a_1, a_2, \ldots, a_l)$ denotes the greatest common divisor of the integers in parentheses.

and hence

$$a_1z_1+a_2z_2+\dots+a_{l-1}z_{l-1}=\delta m'.$$

Then obviously

$$(8)\qquad \delta m'+a_lz_l=m$$

and

$$|\delta m'|\leq \sum_{i=1}^{l-1}|a_i|\,|z_i|\leq l\,\delta\,H'A,$$

which implies that

$$|m'|\leq lH'A.$$

Thus, if the numbers $z_1, z_2, \dots, z_l$ satisfy equation (6) and the inequalities $|z_i|\leq A$ $(1\leq i\leq l)$, then the integer m' exists, which, with these numbers, satisfies equations (7) and (8), where $|m'|\leq lH'A$. But in equation (8) clearly $\delta\leq|a_l|$ and $(\delta, a_l)=1$ (otherwise we should have $(a_1, a_2, \dots, a_{l-1}, a_l)>1$). Hence, by Lemma 1, the number of solutions of equation (8) (in the unknowns m', z_l), for which $|m'|\leq lH'A$, $|z_l|\leq A<lH'A$, does not exceed $3lH'A/|a_l|$. For the same m', equation (7), according to Lemma 2 for equations in $l-1$ unknowns, has at most $c(l)A^{l-2}/H'$ solutions in integral z_i with $|z_i|\leq A$.

It is evident, from what has been said, that the number of solutions $\{z_1, z_2, \dots, z_l\}$ of equation (6) which satisfy the inequalities $|z_i|\leq A$ $(1\leq i\leq l)$, does not exceed

$$(3lH'A/|a_l|)\,c(l)A^{l-2}/H'=c(l)A^{l-1}/|a_l|$$
$$=c(l)A^{l-1}/H,$$

which completes the proof of Lemma 2.*

We shall now investigate the totality of equations of the form

$$(9)\qquad a_1z_1+a_2z_2+\dots+a_lz_l=0,$$

where $|a_i|\leq A$ $(1\leq i\leq l)$ and, as always, all a_i are integers. Let B be a positive number whose relation to the number A is described by the inequalities $1\leq A\leq B\leq c(l)A^{l-1}$, and let $l>2$. We now want to estimate

*You have probably noticed that in the last chain of equations the symbol $c(l)$ occurred in different places with different meanings. On p. 39 I prepared you for such a use of this symbol.

the sum of the numbers of solutions z_i, $|z_i| \leqq B$ $(1 \leqq i \leqq l)$ of all the equations (9) of this family.

1° First let us make a separate examination of equation (9) for $a_1 = a_2 = \ldots = a_l = 0$ (it is a member of our family) and estimate the number of its solutions which satisfy the inequalities $|z_i| \leqq B$ $(1 \leqq i \leqq l)$. Our equation is obviously satisfied by an arbitrary system of numbers $z_1, z_2, \ldots, z_l$, and we have merely to calculate how many such systems exist which satisfy the inequalities $|z_1| \leqq B$, $|z_2| \leqq B$, $\ldots$, $|z_l| \leqq B$. Since the interval $\langle -B, +B \rangle$ contains at most $2B+1$ integers, each z_i can assume at most $2B+1$ different values. Consequently the number of systems $\{z_1, z_2, \ldots, z_l\}$ of the type in which we are interested does not exceed $(2B+1)^l \leqq (3B)^l = c(l)B^l$. By our hypothesis, however, $B \leqq c(l)A^{l-1}$, so that $c(l)B^l = c(l)B^{l-1}B \leqq c(l)(AB)^{l-1}$. Hence, for the case where $a_1 = a_2 = \ldots = a_l = 0$, equation (9) has at most $c(l)(AB)^{l-1}$ solutions of the type we are interested in.

2° Even if only one of the coefficients a_i is different from zero, the greatest common divisor of these coefficients, $(a_1, a_2, \ldots, a_l) = \delta$, exists. Suppose first that $\delta = 1$, and let H be the largest of the numbers $|a_i|$ $(i = 1, 2, \ldots, l)$. Clearly H is one of the integers in the interval $\langle 1, A \rangle$. Hence, H is either between A and $A/2$, or between $A/2$ and $A/4$, or between $A/4$ and $A/8$, etc. It is therefore possible to find an integer $m \geqq 0$ such that

$$(10) \qquad A/2^{m+1} < H \leqq A/2^m.$$

According to Lemma 2, for an equation of the form (9) in which $\delta = 1$ and H satisfies the inequalities (10), the number of solutions z_i, $|z_i| \leqq B$, does not exceed

$$c(l)B^{l-1}/H \leqq c(l)B^{l-1}/(A/2^{m+1}) = c(l)B^{l-1}2^m/A.$$

On the other hand, it follows from (10) that

$$(11) \qquad |a_i| \leqq A/2^m \quad (1 \leqq i \leqq l).$$

Consequently the number of equations of type (9) for which the inequalities (10) are satisfied is at most equal to the number of equations of the same type which satisfy the conditions (11), i.e., at most

$$(2(A/2^m)+1)^l \leqq (3A/2^m)^l = c(l)A^l 2^{-ml}.$$

46

Thus the sum of the numbers of solutions $|z_i| \leq B$ of all such equations of type (9) for which $\delta = 1$ and $A2^{-(m+1)} < H \leq A2^{-m}$, doesn't exceed

$$(c(l)B^{l-1}2^m/A) \cdot c(l)A^l 2^{-ml} = c(l)(AB)^{l-1}2^{-(l-1)m}.$$

Summing this estimate over all $m \geq 0$, we reach the following conclusion: The sum of the numbers of solutions $|z_i| \leq B$ of all equations (9) for which $|a_i| \leq A$ $(1 \leq i \leq l)$ and $\delta = 1$ is at most

$$c(l)(AB)^{l-1}.$$

3° It remains for us to figure out the numbers of solutions of the required type for equations with $\delta > 1$. In this case equation (9) is evidently synonymous with the equation

$$(a_1/\delta)z_1 + (a_2/\delta)z_2 + \ldots + (a_l/\delta)z_l = 0,$$

where only

$$(a_1/\delta, a_2/\delta, \ldots, a_l/\delta) = 1$$

and the number A has to be replaced by the number A/δ. As we saw in 2°, the sum of the numbers of solutions $|z_i| \leq B$ of all such equations, for a given, fixed δ, does not exceed*

$$c(l)(A\delta^{-1} \cdot B)^{l-1} = c(l)(AB)^{l-1}\delta^{-(l-1)}.$$

Clearly now we have merely to sum this expression over all the possible values of δ $(1 \leq \delta \leq A)$.

Thus we find that the sum of the numbers of required solutions of all equations of the form (9), where $|a_i| \leq A$ $(1 \leq i \leq l)$ and not all a_i are equal to zero, does not exceed the value

$$c(l)(AB)^{l-1} \sum_{\delta=1}^{A} \delta^{-(l-1)} < c(l)(AB)^{l-1} \cdot \frac{l-1}{l-2} = c(l)(AB)^{l-1}.$$

[To obtain the first relation we employ the inequality

$$\sum_{n=1}^{A} (1/n^{q+1}) < (q+1)/q,$$

*Since instead of A we now have to take the smaller number A/δ, it is conceivable that the assumed condition $B \leq c(l)A^{l-1}$ is violated. You can verify, however, without any trouble, that we made no use of this assumption in Case 2°, and that the result in 2° therefore does not depend on it.

which is valid for an arbitrary natural number q and for an arbitrary $A\geq 1$ (we denote by q the number $l-2$, which is positive because we assumed that $l>2$). Here is a simple proof: For $n\geq 1$ we have

$$\begin{aligned} n^{-q}-(n+1)^{-q} &= \{(n+1)^q-n^q\}/n^q(n+1)^q \\ &= (n^q+qn^{q-1}+\ldots+1-n^q)/n^q(n+1)^q \\ &\geq qn^{q-1}/n^q(n+1)^q > q/(n+1)^{q+1}, \end{aligned}$$

and hence

$$(n+1)^{-(q+1)}<q^{-1}\{n^{-q}-(n+1)^{-q}\}.$$

By substituting successively $n=1,2,\ldots,A-1$ in this inequality and adding all the resulting inequalities together we find that

$$\sum_{n=2}^{A} n^{-(q+1)}<q^{-1}(1-A^{-q})<1/q,$$

which implies that

$$\sum_{n=1}^{A} n^{-(q+1)}<1+(1/q)=(q+1)/q, \qquad \text{Q.E.D.]}$$

Comparing this with the result in 1°, where we obtained an estimate for the case $a_1=a_2=\ldots=a_l=0$, we reach the following conclusion:

LEMMA 3. *Let $l>2$ and $1\leq A\leq B\leq c(l)A^{l-1}$. Then the sum of the numbers of solutions $|z_i|\leq B$ ($1\leq i\leq l$) of all equations of the form*

$$a_1z_1+a_2z_2+\ldots+a_lz_l=0, \tag{9}$$

where $|a_i|\leq A$ ($1\leq i\leq l$), does not exceed

$$c(l)(AB)^{l-1}.$$

§4
TWO MORE LEMMAS

Before proceeding to prove the fundamental lemma, we have to derive two more lemmas of a special type. They are both very simple, in idea as well as in form, and yet their assimilation might cause you some difficulty because they are concerned with the enumeration of all possible combinations, whose construction is rather involved. The difficulty with such an abstract combinatorial problem is that it is hard to put it in mathematical symbols: one has to express more in words than in signs. This is of course a difficulty of presentation, however, and not of the subject itself, and I shall take pains to outline all questions that arise, and their solution, as concretely as possible.

We shall denote by A a finite complex (i. e., collection) of numbers, not all of which are necessarily distinct. If the number a occurs λ times in the complex A, we shall say that its multiplicity is λ. Let $a_1, a_2, \ldots, a_r$ be the distinct numbers which appear in A, and let $\lambda_1, \lambda_2, \ldots, \lambda_r$ be their respective multiplicities (because the complex A contains all together $\sum_{i=1}^{r}\lambda_i$ numbers). Let B be another complex of the same type, which consists of the distinct numbers b_1, $b_2, \ldots, b_s$ with the respective multiplicities $\mu_1, \mu_2, \ldots, \mu_s$.

Let us investigate the equation

$$x+y=c, \tag{12}$$

where c is a given number and x and y are unknowns. We are interested in such solutions $\{x, y\}$ of this equation in which x is one of the numbers of the complex A (abbreviated $x \in A$) and y is one of the numbers of the complex B ($y \in B$). If the numbers $x=a_i$ and $y=b_k$ satisfy equation (12), this yields $\lambda_i\mu_k$ solutions of the required kind, because any one of the λ_i "specimens" of the number a_i, which occur in the complex A, can be combined with an arbitrary one of the μ_k specimens of the number b_k appearing in the complex B. But we have* $\lambda_i\mu_k \leqq \frac{1}{2}(\lambda_i^2+\mu_k^2)$. Therefore the number of such solutions of quation (12), where $x=a_i$, $y=b_k$, is not greater than $\frac{1}{2}(\lambda_i^2+\mu_k^2)$. It follows that the number of all solutions $x \in A$, $y \in B$ of equation (12) is not more than the sum $\Sigma\frac{1}{2}(\lambda_i^2+\mu_k^2)$. Here the summation is over all pairs of indices $\{i, k\}$ for which $a_i+b_k=c$. Our sum is enlarged if we sum λ_i^2 over all i and μ_k^2 over all k (because every b_k can be combined with at most one a_i.) It finally follows, therefore, that the number of solutions $x \in A$, $y \in B$ of equation (12) does not exceed the number

$$\tfrac{1}{2}\left(\sum_{i=1}^{r}\lambda_i^2+\sum_{k=1}^{s}\mu_k^2\right).$$

*"The geometric mean is not greater than the arithmetic mean". Here is the simplest proof:

$$0 \leqq (\lambda_i-\mu_k)^2=\lambda_i^2+\mu_k^2-2\lambda_i\mu_k, \text{ and hence } 2\lambda_i\mu_k \leqq \lambda_i^2+\mu_k^2.$$

On the other hand, let us consider the equation

(13) $$x-y=0$$

and calculate the number of its solutions $x\in A$, $y\in A$. Clearly every such solution is of the form $x=y=a_i$ $(1\leq i\leq r)$. For a given i we obtain λ_i^2 solutions, because the numbers x and y can coincide, independently of one another, with any one of the λ_i specimens of the number a_i appearing in A. Accordingly the total number of solutions $x\in A$, $y\in A$ of equation (13) is equal to $\sum_{i=1}^{r}\lambda_i^2$. In exactly the same way we find, of course, that the number of solutions $x\in B$, $y\in B$ of the same equation is equal to $\sum_{k=1}^{s}\mu_k^2$. If we compare these results with the one found above, we reach the following conclusion:

LEMMA 4. *The number of solutions of the equation*

$$x+y=c,\quad x\in A,\ y\in B$$

does not exceed half the sum of the numbers of solutions of the equations

$$x-y=0,\quad x\in A,\ y\in A$$

and

$$x-y=0,\quad x\in B,\ y\in B.$$

For the special case in which the complexes A and B coincide we obtain the following

COROLLARY. *The number of solutions of the equation*

$$x+y=c,\quad x\in A,\ y\in A$$

does not exceed the number of solutions of the equation

$$x-y=0,\quad x\in A,\ y\in A.$$

Now let k and s be two arbitrary natural numbers. We put $k\cdot 2^s=l$, and investigate the equation

$$x_1+x_2+\dots+x_l=c.$$

Let $A_1, A_2, \dots, A_l$ be finite complexes of numbers. Suppose that the complex A_i $(1\leq i\leq l)$ consists of the distinct numbers $a_{i1}, a_{i2}, \dots$

with the respective multiplicities $\lambda_{i1}, \lambda_{i2}, \ldots$. We are interested in the number of solutions of the equation

(14) $$x_1+x_2+\ldots+x_l=c, \quad x_i \in A_i \ (1 \leq i \leq l).$$

If we set

$$x_1+x_2+\ldots+x_{l/2}=x, \quad x_{(l/2)+1}+\ldots+x_l=y$$

($l/2$ is of course an integer), then the given equation can be written in the form

$$x+y=c,$$

and Lemma 4, which we have just proved, can be applied to it. We have only to find out to which complexes the numbers x and y belong. Since $x_i \in A_i$ $(1 \leq i \leq l)$, x can be an arbitrary number of the form $z_1+z_2+\ldots+z_{l/2}$, where $z_i \in A_i$ $(1 \leq i \leq l/2)$. Similarly y can be an arbitrary number of the same form, where, however, $z_i \in A_{(l/2)+i}$ $(1 \leq i \leq l/2)$.

Hence, by Lemma 4, the number of solutions of equation (14) does not exceed half the sum of the numbers of solutions of the equation

(15) $$x-y=0$$

under the following two hypotheses:

1) $$x=z_1+z_2+\ldots+z_{l/2}, \quad y=z_1'+z_2'+\ldots+z_{l/2}',$$

where

(16) $$z_i \in A_i, \quad z_i' \in A_i \quad (1 \leq i \leq l/2);$$

2) x and y have the same form, but

(17) $$z_i \in A_{(l/2)+i}, \quad z_i' \in A_{(l/2)+i} \ (1 \leq i \leq l/2).$$

In both cases equation (15) may be rewritten in the form

(18) $$(z_1-z_1')+(z_2-z_2')+\ldots+(z_{l/2}-z_{l/2}')=0.$$

We conclude therefore that the number of solutions of equation (14) does not exceed half the sum of the numbers of solutions of equation

(18) under the hypotheses (16) and (17), i.e., it does not exceed half the sum of the numbers of solutions of the equations

$$\text{(18a)} \quad \sum_{i=1}^{l/2}(z_i - z_i') = 0, \quad z_i \varepsilon A_i, \qquad z_i' \varepsilon A_i \qquad (1 \leqq i \leqq l/2)$$

and

$$\text{(18 b)} \quad \sum_{i=1}^{l/2}(z_i - z_i') = 0, \quad z_i \varepsilon A_{(l/2)+i}, \quad z_i' \varepsilon A_{(l/2)+i} \quad (1 \leqq i \leqq l/2).$$

Equation (18) has $l/2$ summands on the left-hand side, i.e., half as many as the original equation (14).

We set

$$\sum_{i=1}^{l/4}(z_i - z_i') = x, \qquad \sum_{i=(l/4)+1}^{l/2}(z_i - z_i') = y,$$

and thereby bring equation (18) into the form

$$x + y = 0.$$

To this we can apply Lemma 4 anew. It is evident that, just as we arrived at equation (18) from equation (14), we now get from equation (18) to the equation

$$\text{(19)} \qquad \sum_{i=1}^{l/4}(u_i + u_i' - u_i'' - u_i''') = 0,$$

where we have to consider the sum of the numbers of solutions of this equation under the following (now four) hypotheses:

$$\left.\begin{array}{l} 1)\ u_i,\ u_i',\ u_i'',\ u_i''' \varepsilon A_i, \\ 2)\ u_i,\ u_i',\ u_i'',\ u_i''' \varepsilon A_{(l/4)+i}, \\ 3)\ u_i,\ u_i',\ u_i'',\ u_i''' \varepsilon A_{(l/2)+i}, \\ 4)\ u_i,\ u_i',\ u_i'',\ u_i''' \varepsilon A_{(3l/4)+i}. \end{array}\right\} \quad (1 \leqq i \leqq l/4)$$

Since $l = k \cdot 2^s$, we can repeat this process s times. We evidently end up then with the equation

$$\text{(20)} \qquad \sum_{i=1}^{k}\{y_i^{(1)} + y_i^{(2)} + \ldots + y_i^{(2^{s-1})} - y_i^{(2^{s-1}+1)} - \ldots - y_i^{(2^s)}\} = 0,$$

where we have to consider the sum of the numbers of solutions of this equation under 2^s different hypotheses, viz.:

$$\left.\begin{array}{l} 1)\ y_1^{(j)} \in A_1, y_2^{(j)} \in A_2, \ldots, y_k^{(j)} \in A_k, \\ 2)\ y_1^{(j)} \in A_{k+1}, y_2^{(j)} \in A_{k+2}, \ldots, y_k^{(j)} \in A_{2k}, \\ \ldots\ldots\ldots\ldots\ldots\ldots \\ 2^s)\ y_1^{(j)} \in A_{k2^s-k+1}, \ldots, y_k^{(j)} \in A_{k2^s}. \end{array}\right\} \quad (1 \leqq j \leqq 2^s)$$

If we put

$$y^{(j)} = y_1^{(j)} + y_2^{(j)} + \ldots + y_k^{(j)} \qquad (1 \leqq j \leqq 2^s),$$

then equation (20) takes on the simple form

$$(21) \qquad y^{(1)} + y^{(2)} + \ldots + y^{(2^{s-1})} - y^{(2^{s-1}+1)} - \ldots - y^{(2^s)} = 0.$$

Here we are concerned with the sum of the numbers of solutions of equation (21) under the following 2^s hypotheses, which differ from one another in the value of the parameter w $(0 \leqq w \leqq 2^s - 1)$:

$$y^{(j)} = y_1^{(j)} + y_2^{(j)} + \ldots + y_k^{(j)},$$

where

$$y_1^{(j)} \in A_{wk+1}, \ y_2^{(j)} \in A_{wk+2}, \ \ldots, \ y_k^{(j)} \in A_{(w+1)k} \quad (j = 1, 2, \ldots, 2^s).$$

Thus we can express the final result of our deduction in the form of the following proposition:

LEMMA 5. *If $l = k \cdot 2^s$, the number of solutions of the equation*

$$(14) \qquad x_1 + x_2 + \ldots + x_l = c, \qquad x_i \in A_i \quad (1 \leqq i \leqq l)$$

does not exceed the sum of the numbers of solutions of the equation

$$(21) \qquad y^{(1)} + y^{(2)} + \ldots + y^{(2^{s-1})} - y^{(2^{s-1}+1)} - \ldots - y^{(2^s)} = 0,$$

$$\left.\begin{array}{c} y^{(j)} = y_1^{(j)} + y_2^{(j)} + \ldots + y_k^{(j)}, \\ y_1^{(j)} \in A_{wk+1}, \ y_2^{(j)} \in A_{wk+2}, \ \ldots, \ y_k^{(j)} \in A_{(w+1)k} \end{array}\right\} \quad (j = 1, 2, \ldots, 2^s)$$

under the hypotheses $w = 0, 1, \ldots, 2^s - 1$.

Notice the connection between Lemma 4 and Lemma 5 for $k = s = 1$, $l = 2$.

This winds up our preliminaries, and we are ready now to begin the direct assault on the fundamental lemma.

§5
PROOF OF THE FUNDAMENTAL LEMMA

We are going to prove the fundamental lemma by the method of induction on n. It is often the case in inductive proofs, that a strengthening of the proposition to be proved, considerably facilitates its proof by the given method (and sometimes is actually what makes the proof feasible in the first place). The reason for this is easy to understand. In inductive proofs, the proposition is assumed to be correct for the number $n-1$, and is proved for the number n. Hence, the stronger the proposition, the more that is given to us by the case $n-1$; of course, so much the more has to be proved for the number n, but in many problems the first consideration turns out to be more important than the second.

And so it is, in fact, in the present case. Of immediate interest for us is the number of solutions of the equation $x_1^n+x_2^n+\dots+x_k^n=m$ $(1\leq m\leq N)$ (where, according to the very meaning of the problem, $0\leq x_i\leq m^{1/n}\leq N^{1/n}$). But x^n is the simplest special case of an n-th degree polynomial

$$f(x)=a_0x^n+a_1x^{n-1}+\dots+a_{n-1}x+a_n,$$

and it will be to our advantage to replace the given equation (1) by the more general equation

$$f(x_1)+f(x_2)+\dots+f(x_k)=m, \tag{22}$$

where the unknowns are subjected to the weaker conditions $|x_i|\leq N^{1/n}$ $(1\leq i\leq k)$. The proof of our proposition for equation (22) will give us more than we really need; but, as you will see, it is just this strengthening of our proposition which creates the possibility of an induction. And so, for $m\leq N$, let us denote by $r_k(m)$ the number of solutions of equation (22) which satisfy the conditions $|x_i|\leq N^{1/n}$ $(1\leq i\leq k)$. Of course we are still free to dispose arbitrarily of the coefficients of the polynomial $f(x)$ in the interest of the induction to be performed (provided only that the imposed conditions are satisfied in the case $f(x)=x^n$). We are going to prove the following proposition:

Let the coefficients of the polynomial $f(x)$ satisfy the inequalities

$$|a_i|\leq c(n)N^{i/n} \qquad (0\leq i\leq n). \tag{23}$$

Then, for a suitably chosen $k=k(n)$,

$$r_k(m)<c(n)\,N^{(k/n)-1}\quad (1\leqq m\leqq N).$$

Since the inequalities (23) are obviously satisfied in the case $f(x)=x^n$ for $c(n)=1$, this theorem is indeed a sharpening of our fundamental lemma.

Let us first consider the case $n=1$, $f(x)=a_0x+a_1$. We set $k(1)=2$, so that equation (22) acquires the form

$$a_0(x_1+x_2)=m-2a_1.$$

We are interested in solutions of this equation which satisfy the requirements $|x_1|\leqq N$, $|x_2|\leqq N$. Thus at most $2N+1\leqq 3N$ values are possible for x_1. But at most one x_2 corresponds to every x_1, so that

$$r_2(m)\leqq 3N,$$

which completes the proof of our proposition for $n=1$ $(k=2)$.

Now let $n>1$, and suppose that our assertion has already been verified for the exponent $n-1$. Put $k(n-1)=k'$ and choose

$$k=k(n)=2n\cdot 2^{[4\log_2 k']},$$

where the exponent means the greatest integer not exceeding $4\log_2 k'$. In the sequel we shall set $[4\log_2 k']-1=s$, for brevity, so that

$$k=2n\cdot 2^{s+1}. \tag{24}$$

To estimate the number, $r_k(m)$, of solutions of equation (22), we first apply Lemma 4 to it, setting

$$x=\sum_{i=1}^{\frac{1}{2}k}f(x_i),\qquad y=\sum_{i=\frac{1}{2}k+1}^{k}f(x_i).$$

The complex A (and the complex B which coincides with it in this case) consists of all sums of the form

$$\sum_{i=1}^{\frac{1}{2}k}f(x_i),\ \text{where}\ |x_i|\leqq N^{1/n}\ (1\leqq i\leqq \tfrac{1}{2}k).$$

By the Corollary of Lemma 4, $r_k(m)$ does not exceed the number of solutions of the equation $x-y=0$, where $x\in A$, $y\in A$, i.e.,

$$x=\sum_{i=1}^{k/2} f(x_i),\quad y=\sum_{i=1}^{k/2} f(y_i),$$

$$|x_i|\leqq N^{1/n},\quad |y_i|\leqq N^{1/n}\quad (1\leqq i\leqq k/2).$$

In other words, $r_k(m)$ does not exceed the number of solutions of the equation

$$\sum_{i=1}^{k/2}\{f(x_i)-f(y_i)\}=0, \tag{25}$$

where $|x_i|\leqq N^{1/n}$, $|y_i|\leqq N^{1/n}$ $(1\leqq i\leqq k/2)$. We now set $x_i-y_i=h_i$ $(1\leqq i\leqq k/2)$ and replace the system of unknowns $\{x_i, y_i\}$ by the system $\{y_i, h_i\}$; here we allow y_i and h_i $(1\leqq i\leqq k/2)$ to assume all possible integral values in the interval $<-2N^{1/n}, +2N^{1/n}>$, which can only increase the number of solutions of our equation. This means that every summand $f(x_i)-f(y_i)$ in equation (25) is replaced by the expression

$$f(y_i+h_i)-f(y_i)=\sum_{v=0}^{n-1} a_v\{(y_i+h_i)^{n-v}-y_i^{n-v}\}$$
$$=\sum_{v=0}^{n-1} a_v\sum_{t=1}^{n-v}\binom{n-v}{t}h_i^t y_i^{n-v-t}.$$

If we change the variable t of summation by putting

$$v+t=u,$$

so that

$$n-v-t=n-u,\qquad t=u-v,$$

we obtain

$$f(y_i+h_i)-f(y_i)=h_i\sum_{v=0}^{n-1} a_v\sum_{u=v+1}^{n}\binom{n-v}{u-v}h_i^{u-v-1}y_i^{n-u}$$
$$=h_i\sum_{u=1}^{n} y_i^{n-u}\sum_{v=0}^{u-1} a_v\binom{n-v}{u-v}h_i^{u-v-1}$$
$$=h_i\sum_{u=1}^{n} a_{i,u}y_i^{n-u}=h_i\phi_i(y_i),$$

where

$$\phi_i(y)=\sum_{u=1}^{n} a_{i,u}\,y^{n-u}$$

is a polynomial of degree $n-1$ with coefficients

$$a_{i,u}=\sum_{v=0}^{u-1} a_v \binom{n-v}{u-v} h_i^{u-v-1} \quad (1 \leqq i \leqq k/2)$$

which depend on the numbers h_i.

Thus, in the new variables $\{y_i, h_i\}$, equation (25) assumes the form

$$h_1\phi_1(y_1)+h_2\phi_2(y_2)+\ldots+h_{\frac{1}{2}k}\phi_{\frac{1}{2}k}(y_{\frac{1}{2}k})=0. \tag{26}$$

In this equation the numbers h_i and y_i may take on arbitrary integral values in the interval $<-2N^{1/n}, +2N^{1/n}>$, where we must bear in mind that the coefficients of the polynomials $\phi_i(y)$ (of degree $n-1$) depend on the numbers h.

Mark well that we have proved the following so far: *The number $r_k(m)$ which we are estimating, does not exceed the sum of the numbers of solutions in integers y_i, $|y_i| \leqq 2N^{1/n}$ $(1 \leqq i \leqq k/2)$, of all the equations* (26) *which can be obtained from all possible values of the numbers h_i, $|h_i| \leqq 2N^{1/n}$ $(1 \leqq i \leqq k/2)$.*

§ 6

CONTINUATION

We are now going to examine one of the equations (26), i.e., we shall regard the numbers h_i $(1 \leqq i \leqq k/2)$ for a while as fixed. Let us apply Lemma 5 to this equation; the numbers $h_i\phi_i(y_i)$ play the role of the unknowns x_i, the number $\frac{1}{2}k = 2n\cdot 2^s$ plays the role of the number l, and we set $2n = k_0$ for brevity. Recall once more that the numbers h_i appear in equation (26) not only explicitly but also through the coefficients of the polynomials $\phi_i(y)$. The complex A_i to which the numbers $x_i = h_i\phi_i(y_i)$ must belong consists, in the present case, of all numbers of the form $h_i\phi_i(y_i)$, where the numbers h_i have given, fixed values and the numbers y_i run through the interval $<-2N^{1/n}$, $+2N^{1/n}>$.

According to Lemma 5, the number of solutions of equation (26) satisfying the requirements just described, does not exceed the sum of the numbers of solutions of the equation

$$y^{(1)}+y^{(2)}+\ldots+y^{(2^{s-1})}-y^{(2^{s-1}+1)}-\ldots-y^{(2^s)}=0 \tag{21}$$

under the following 2^s hypotheses which correspond to the values of the parameter $w=0, 1, \ldots, 2^s-1$:

$$\left.\begin{array}{l} y^{(j)}=y_1^{(j)}+y_2^{(j)}+\ldots+y_{k_0}^{(j)}, \\ y_i^{(j)}\in A_{wk_0+i} \quad (1\leqq i\leqq k_0) \end{array}\right\} \qquad (1\leqq j\leqq 2^s),$$

where, remember, A_r $(1\leqq r\leqq 2^s)$ is the complex of numbers of the form $h_r\phi_r(y_r)$ with prescribed h_r and arbitrary y_r, $|y_r|\leqq 2N^{1/n}$.

For the case $w=0$ (which we choose merely as an example), equation (21) in expanded form looks as follows:

$$\begin{aligned} &\{y_1^{(1)}+y_2^{(1)}+\ldots+y_{k_0}^{(1)}\} \\ &+\{y_1^{(2)}+y_2^{(2)}+\ldots+y_{k_0}^{(2)}\} \\ &+\cdots\cdots\cdots \\ &+\{y_1^{(2^{s-1})}+y_2^{(2^{s-1})}+\ldots+y_{k_0}^{(2^{s-1})}\} \\ &-\{y_1^{(2^{s-1}+1)}+y_2^{(2^{s-1}+1)}+\ldots+y_{k_0}^{(2^{s-1}+1)}\} \\ &-\cdots\cdots\cdots \\ &-\{y_1^{(2^s)}+y_2^{(2^s)}+\ldots+y_{k_0}^{(2^s)}\}=0, \end{aligned}$$

or, rearranging the summands,

$$\begin{aligned} &\{y_1^{(1)}+y_1^{(2)}+\ldots+y_1^{(2^{s-1})}-y_1^{(2^{s-1}+1)}-\ldots-y_1^{(2^s)}\} \\ &+\{y_2^{(1)}+y_2^{(2)}+\ldots+y_2^{(2^{s-1})}-y_2^{(2^{s-1}+1)}-\ldots-y_2^{(2^s)}\} \\ &+\cdots\cdots\cdots \\ &+\{y_{k_0}^{(1)}+y_{k_0}^{(2)}+\ldots+y_{k_0}^{(2^{s-1})}-y_{k_0}^{(2^{s-1}+1)}-\ldots-y_{k_0}^{(2^s)}\}=0; \end{aligned}$$

every one of the numbers $y_i^{(j)}$ is a number of the form $h_i\phi_i(v_i^{(j)})$, where $|v_i^{(j)}|\leqq 2N^{1/n}$. Hence the last equation can be rewritten in the form

$$\begin{aligned} &h_1\{\phi_1(v_1^{(1)})+\phi_1(v_1^{(2)})+\ldots+\phi_1(v_1^{(2^{s-1})})-\phi_1(v_1^{(2^{s-1}+1)})-\ldots-\phi_1(v_1^{(2^s)})\} \\ &+h_2\{\phi_2(v_2^{(1)})+\ldots-\phi_2(v_2^{(2^s)})\}+\ldots+h_{k_0}\{\phi_{k_0}(v_{k_0}^{(1)})+\ldots-\phi_{k_0}(v_{k_0}^{(2^s)})\}=0. \end{aligned}$$

By putting, for brevity,

$$\phi_i(v_i^{(1)})+\phi_i(v_i^{(2)})+\ldots+\phi_i(v_i^{(2^{s-1})})-\phi_i(v_i^{(2^{s-1}+1)})-\ldots-\phi_i(v_i^{(2^s)})=z_i,$$
$$(1\leqq i\leqq k_0)$$

this equation can be written quite shortly as follows:

(27) $$h_1z_1+h_2z_2+\ldots+h_{k_0}z_{k_0}=0.$$

All together we have 2^s equations of this sort, and their totality can be written down in the compact form

$$\sum_{i=1}^{k_0} h_{wk_0+i}\,z_{wk_0+i}=0 \qquad (0\leqq w\leqq 2^s-1).$$

For the present, however, we shall confine our investigation to equation (27), which may of course be regarded as typical. To estimate the number of solutions of this equation which interest us, we must first see within what limits the quantity $\phi_i(v_i^{(j)})$ can vary. To this end we recall that (p. 55)

$$\phi_i(y)=\sum_{u=1}^{n} a_{i,u}\,y^{n-u},$$

where

$$a_{i,u}=\sum_{v=0}^{u-1} a_v\binom{n-v}{u-v}h_i^{u-v-1} \qquad (1\leqq i\leqq k/2).$$

Hence it follows from our hypotheses $|a_v|<c(n)N^{v/n}$ and $|h_i|\leqq 2N^{\frac{1}{n}}$, that

$$|a_{i,u}|<\sum_{v=0}^{u-1}c(n)N^{v/n}\binom{n-v}{u-v}c(n)N^{(u-v-1)/n}=c(n)N^{(u-1)/n}\sum_{v=0}^{u-1}\binom{n-v}{u-v};$$

i. e., in view of $u\leqq n$,

(28) $$|a_{i,u}|<c(n)N^{(u-1)/n}.$$

On the other hand, because of $|v_i^{(j)}|\leqq 2N^{1/n}$, we have $|v_i^{(j)}|^{n-u}\leqq c(n)\cdot N^{(n-u)/n}$ and consequently

$$|a_{i,u}|\cdot|v_i^{(j)}|^{n-u}\leqq c(n)N^{(u-1)/n}N^{(n-u)/n}=c(n)N^{(n-1)/n}.$$

The same estimate (with another $c(n)$) holds for the whole $\phi_i(v_i^{(j)})$, since the number of terms of this polynomial is equal to n. Accordingly

$$|\phi_i(v_i^{(j)})|<c(n)N^{(n-1)/n} \quad (1\leqq i\leqq k_0, 1\leqq j\leqq 2^s).$$

But every z_i is the sum of $2^s = c(n)$ summands of the form $\pm\phi_i(v_i^{(j)})$, and therefore

$$|z_i| < c(n) N^{(n-1)/n} \qquad (1 \leq i \leq k_0)$$

(with another $c(n)$, naturally). This means that in equation (27) every z_i can assume only the values lying in the interval $<-c(n)N^{(n-1)/n}, +c(n)N^{(n-1)/n}>$.

Let $\bar{m}$ be one of these numbers. The equation $z_i = \bar{m}$ can be satisfied in general not only in one but in several ways, because the definition of the number z_i (p. 58) is such that one and the same value of z_i can very well result from different choices of the numbers $v_i^{(j)}$ $(1 \leq j \leq 2^s)$. We now have to estimate the number of solutions of the relation $z_i = \bar{m}$, i. e., of the equation

$$(29) \qquad \phi_i(v_i^{(1)}) + \dots + \phi_i(v_i^{(2^{s-1})}) - \phi_i(v_i^{(2^{s-1}+1)}) - \dots - \phi_i(v_i^{(2^s)}) = \bar{m}.$$

For this purpose we shall finally have to apply the long-promised induction. We proceed as follows.

First we rewrite equation (29) in the form

$$\phi_i(v_i^{(1)}) + \phi_i(v_i^{(2)}) + \dots + \phi_i(v_i^{(k')})$$
$$= \bar{m} - \phi_i(v_i^{(k'+1)}) - \dots + \phi_i(v_i^{(2^{s-1}+1)}) + \dots + \phi_i(v_i^{(2^s)}).$$

This is possible because for $k' = k(n-1) > 1$ (and we have seen that already $k(1) = 2$) we have

$$2^{s-1} = 2^{[4 \log_2 k'] - 2} > k'.$$

(In detail: $k' \geq 2$, $\log_2 k' \geq 1$, $3 \log_2 k' \geq 3$, $[4 \log_2 k'] - 2 > 4 \log_2 k' - 3 \geq \log_2 k'$, $2^{s-1} = 2^{[4 \log_2 k'] - 2} \geq k'$.)

If we denote the right-hand side of the last equation by m', we get

$$(30) \qquad \phi_i(v_i^{(1)}) + \dots + \phi_i(v_i^{(k')}) = m'.$$

Let us choose some particular values for the numbers

$$v_i^{(j)} \qquad (k'+1 \leq j \leq 2^s)$$

(in the interval $<-2N^{1/n}, +2N^{1/n}>$, naturally); then m' also acquires a definite value. To equation (30) we now apply the theorem to be proved, since $\phi_i(y)$ is a polynomial of degree $n-1$. We have to verify that all the necessary hypotheses are fulfilled. We have

$$\phi_i(y) = \sum_{u=1}^{n} a_{i,u} y^{n-u},$$

where, according to (28),

$$|a_{i,u}| < c(n) N^{\frac{u-1}{n}} = c(n)(N^{\frac{n-1}{n}})^{\frac{u-1}{n-1}}, \tag{31}$$

and, as is easily seen,

$$|m'| < c(n) N^{\frac{n-1}{n}}$$

(because $\bar{m}$ and all $\phi_i(y_i^{(j)})$ satisfy this inequality).

In virtue of the last inequality, the role of N can be assumed by the number $c(n)N^{(n-1)/n}$; then the conditions (31), which the coefficients of the polynomial $\phi_i(y)$ satisfy, are precisely the conditions (23) with n replaced by $n-1$. Thus all the hypotheses are indeed fulfilled, and we can assert that the number of solutions of equation (30), for which $|v_i^{(j)}| \leq 2N^{1/n} = 2(N^{(n-1)/n})^{1/(n-1)}$, does not exceed the number

$$c(n)(N^{\frac{n-1}{n}})^{\frac{k'}{n-1} - 1} = c(n) N^{\frac{k'-n+1}{n}}. \tag{32}$$

This estimate is obtained for the fixed values $v_i^{(k'+1)}, \ldots, v_i^{(2^s)}$. Clearly we have at most

$$(4N^{1/n}+1)^{2^s-k'} < c(n) N^{\frac{2^s-k'}{n}} \tag{33}$$

such systems of values. The total number of solutions of the required type, of equation (29), therefore does not exceed the product of the right-hand sides of (32) and (33), i. e., it is at most

$$c(n) N^{\frac{2^s-n+1}{n}}. \tag{34}$$

We now return to equation (27). We saw before (p. 59) that every z_i can assume only the values lying in the interval $\langle -c(n)N^{(n-1)/n}, +c(n)N^{(n-1)/n}\rangle$. Now we see that the "multiplicity" of each of these values (i. e., the number of ways of choosing the $y_i^{(j)}$ so as to satisfy the equation) does not exceed the number (34).

This result makes it possible to reduce the whole problem to an estimation of the numbers of solutions of linear equations. For at the end of §5 we reduced the estimation of $r_k(m)$ to the estimation of the numbers of solutions of equations of the form (26). But as we proved by an application of Lemma 5, the number of solutions of equation (26), for which $|y_i| \leq 2N^{1/n}$, is at most equal to the sum of the numbers of solutions of 2^s equations of type (27), i. e., already linear equations. In this connection we obtained limits within which the unknowns z_i are allowed to vary. A certain new difficulty (the price we have had to pay for the transition to linear equations) is that the new unknowns z_i have to be considered with certain multiplicities (for which we have also determined limits).

Finally we must not forget that all these calculations are made under the assumption that the numbers h_i are chosen and fixed. Therefore we still have to multiply the result obtained, by the number of all such possible choices.

The final result of this section, which we have to keep in mind, reads: *Our estimated number $r_k(m)$ does not exceed the sum of the numbers of solutions in integers z_i, $|z_i| \leq c(n)N^{(n-1)/n}$, with multiplicities $\lambda_i \leq c(n)N^{(2^s-n+1)/n}$, of equations of the form*

$$\sum_{i=1}^{k_0} h_{wk_0+i} z_{wk_0+i} = 0, \tag{35}$$

where w runs through the values $0, 1, \ldots, 2^s-1$, and the numbers h_r $(1 \leq r \leq 2^s k_0)$ assume, independently of one another, all integers in the interval $\langle -2N^{1/n}, +2N^{1/n}\rangle$.

And so we see that we have now obtained an estimate for $r_k(m)$, in whose formulation the given polynomial $f(x)$ does not appear, which lends this estimate a very general character.

§7

CONCLUSION

Now that we have reduced the problem to an estimation of the number of solutions of linear equations which are independent of the special form of the polynomial $f(x)$, we quickly reach our goal with the aid of Lemma 3.

Denote by K any particular combination of the numbers h_i,

62

$|h_i| \leqq 2N^{1/n}$ $(1 \leqq i \leqq k/2)$, and by $U_w(\mathrm{K})$ the number of solutions of equation (35) for this fixed combination K and for a certain prescribed w, where we are concerned with those solutions z_i which satisfy the inequalities $|z_i| \leqq c(n)N^{(n-1)/n}$, with multiplicities $\lambda_i \leqq c(n)N^{(2^s-n+1)/n}$. Then, according to the final result in the preceding section,

$$r_k(m) \leqq \sum_{\mathrm{K}} \{ \sum_{w=0}^{2^s-1} U_w(\mathrm{K}) \},$$

where the summation over K extends over all admissible combinations K of the numbers h_i. This can be written

$$r_k(m) \leqq \sum_{w=0}^{2^s-1} \{ \sum_{\mathrm{K}} U_w(\mathrm{K}) \}.$$

It is immediately evident, however, that for different w the sums $\sum_{\mathrm{K}} U_w(\mathrm{K})$ do not differ from one another at all (because for different w the equations (35) do not differ from one another in any respect). We can therefore write

$$r_k(m) \leqq 2^s \sum_{\mathrm{K}} U_0(\mathrm{K}) = c(n) \sum_{\mathrm{K}} U_0(\mathrm{K}).$$

Here $U_0(\mathrm{K})$ is the number of solutions of the equation

$$h_1 z_1 + h_2 z_2 + \ldots + h_{k_0} z_{k_0} = 0 \tag{36}$$

for the given combination K of the numbers h_i, $|h_i| \leqq 2N^{1/n}$ $(1 \leqq i \leqq k/2)$, where $|z_i| \leqq c(n)N^{(n-1)/n}$ and the z_i have multiplicities $\lambda_i \leqq c(n) \cdot N^{(2^s-n+1)/n}$. Let us denote by $U_0^*(\mathrm{K})$ the number of solutions of the same equation under the assumption that all z_i are simple. Then clearly

$$U_0(\mathrm{K}) \leqq \{ c(n) N^{\frac{2^s-n+1}{n}} \}^{k_0} U_0^*(\mathrm{K}),$$

or, recalling that $k_0 = 2n$,

$$U_0(\mathrm{K}) \leqq c(n) N^{2(2^s-n+1)} U_0^*(\mathrm{K}),$$

and hence

$$r_k(m) \leqq c(n) N^{2(2^s-n+1)} \sum_{\mathrm{K}} U_0^*(\mathrm{K}). \tag{37}$$

Now let us note the following. Every K represents a certain admissible combination of the values of all h_i $(1 \leq i \leq k/2)$; the number $U_0^*(K)$, however, is completely determined by the values of the first $k_0 = 2n$ of these values $(1 \leq i \leq 2n)$, because they alone appear in equation (36). Of course when we choose a certain fixed combination K, we thereby also uniquely define a certain combination K′ of the values $h_1, h_2, \ldots, h_{2n}$. But if, conversely, a certain combination K′ of the numbers $h_1, h_2, \ldots, h_{2n}$ is selected, there corresponds to it not the single combination K, but rather as many as there are ways of choosing the remaining "supplements" h_i $(2n < i \leq k/2)$. Since every h_i must belong to the interval $\langle -2N^{1/n}, +2N^{1/n} \rangle$, it is evident that to a combination K′ there correspond at most

$$c(n)\,(N^{1/n})^{(k/2)-2n} = c(n)\,N^{(k/2n)-2}$$

combinations K. Hence

$$\sum_{K} U_0^*(K) \leq c(n)\,N^{(k/2n)-2} \sum_{K'} U_0^*(K'),$$

where $U_0^*(K')$ is the number of solutions in integers z_i, $|z_i| \leq c(n) \cdot N^{(n-1)/n}$ $(1 \leq i \leq 2n)$ of equation (36) for the given combination K′ of the numbers h_i, $|h_i| \leq 2N^{1/n}$ $(1 \leq i \leq 2n)$, and the summation is to be extended over all such combinations. From (37) we therefore obtain*

$$(38)\quad r_k(m) \leq c(n) N^{2(2^s-n+1)} N^{(k/2n)-2} \sum_{K'} U_0^*(K') = c(n) N^{2(2^{s+1}-n)} \sum_{K'} U_0^*(K').$$

Finally, $\sum_{K'} U^*(K')$ is immediately estimated with the help of Lemma 3, where we have to put $l = 2n$, $A = 2N^{1/n}$, $B = c(n)N^{(n-1)/n}$; you can easily verify that all the hypotheses of Lemma 3 are satisfied. On applying this lemma we find that

$$\sum_{K'} U_0^*(K') \leq c(n)\,(AB)^{2n-1} = c(n) N^{2n-1}.$$

At last, inequality (38) yields

$$r_k(m) \leq c(n) N^{2(2^{s+1}-n)} \cdot N^{2n-1} = c(n) N^{2 \cdot 2^{s+1}-1} = c(n) N^{\frac{k}{n}-1},$$

which completes the proof of the fundamental lemma and thereby also of Hilbert's theorem.

This proof, so exquisitely elementary, will undoubtedly seem very complicated to you. But it will take you only two to three

*Recall that $k = 2n \cdot 2^{s+1}$.

64

weeks' work with pencil and paper to understand and digest it completely. It is by conquering difficulties of just this sort, that the mathematician grows and develops.

⁂

Andrey Nikolaevich Kolmogorov

Born – 25 April 1903 in Tambov, Tambov Province, Russia
Died – 20 Oct 1987 in Moscow, Russia

Andrey Kolmogorov was best known for his development of the probability theory that played an important role in making the subject an integral part of analysis. His work influenced several branches of modern mathematics.

Kolmogorov enrolled in the Moscow State University at age 17 in the year 1920. At that time, ideas in the theory of functions of real variables were in line with the basic directions of the Moscow mathematical school. In the autumn of 1921, he worked intensively on complex problems — the theory of trigonometrical series and operations on sets. He completed a study in the theory of operations on sets in the spring of 1922, which was later published in 1928.

In 1925, Kolmogorov graduated from the faculty of physics and mathematics of Moscow State University. He was appointed the post of research associate in the faculty, and began to gain keen interest in the probability theory during this period of time. His important work "*General Theory of Measure and Probability Theory*" (published in 1929), gave the first description of an axiomatic construction of probability theory based on measure theory. In 1933, Kolmogorov expanded the paper into the monograph "*Grundbegriffe der Wahrscheinlichkeitsrechnung*", which was translated into English under the title "*Foundations of the Theory of Probability*" in the year 1950.

Kolmogorov became a professor of Moscow State University in 1931. Two years later, he took on the role of director of the Institute of Mathematics. During his appointment, he elaborated on the principles of stochastic processes, which included the Markov processes, in the monograph "*Analytical Methods of Probability Theory*". He also made contributions to areas concerning aspects of functional analysis — a comparatively recent branch of mathematics in which the traditional techniques of algebra and calculus are applied to entire collections of functions. Also not to be disregarded was the work he did on the studies of topology and turbulent flow of fluids.

Kolmogorov's specialized field was in the study of probability theory. He formulated two systems of partial differential equations bearing his name, which describe transition probabilities controlling a Markov process. The Kolmogorov equations provide a mean of dealing with problems in the theories of Brownian motion and diffusion. His work opened a brand new era in the development of probability theory and its application in physics, chemistry, civil engineering, and biology.

In succession to the fundamental study of probability, Kolmogorov also did a fundamental study of Markov chains that embodied many new ideas in the theory of dynamic systems and served as a starting point for numerous works of other mathematicians.

Kolmogorov's investigations in problems of random stationary processes, which he related to the analysis of locally isotropic turbulent flow, were also of big significance. The founder of cybernetics (the theory of control and communication as applied to animals and machines) Norbert Wiener, who had independently investigated many aspects of stationary processes, pointed out that the investigations by Kolmogorov were relevant to the statistical theory of information upon which the science of cybernetics in part depended.

In topology, Kolmogorov put forward certain considerations relating to an operator called the nabla operator and applied these to complexes, and later to any topological space. Related algebraic structures called nabla groups turned out to be very effective and provided suitable instrumentation in the study of numerous problems of topology, including continuous maps. Based on these studies, Kolmogorov presented a concept of a homological ring, an important idea in topology. He formulated the duality law (on equivalent representations of certain mathematical properties) which proved to be very important for the period from 1935 to 36. In this context, the duality law concerned closed sets situated in a topological space that is locally bicompact and fully regular.

During the 1930s, Kolmogorov published articles entitled "*On Topological Group Formulation of Geometry*" and "*On Formulation of Projective Geometry*". There were also articles on areas of functional analysis and the optimum approximation of functions.

Later, Kolmogorov's interests broadened into problems of mathematical logic and the foundations of mathematics. In 1938, he published a large article "*Mathematics*" in the first edition of the *Bolshaya Sovyetskaya Entsiklopediya* ("Great Soviet Encyclopaedia"), in which he introduced the development of mathematics from ancient to modern times and interpreted it in terms of dialectical materialism.

In 1939, Kolmogorov was elected an Academician of the Academy of Sciences of the USSR shortly after which he was appointed an academician-secretary of the department of physical and mathematical sciences of the academy. This department had been established to unite and direct scientific work of the most prominent researchers in this field.

In the mid-1950s Kolmogorov turned his attention to information theory, dynamic-system theory, interconnections of information theory with function theory, classical mechanics, the 13th Hilbert Problem (representation of functions of a large number of variables by functions of a lesser number of arguments), and complex theory.

Throughout the latter part of his life, Kolmogorov devoted tremendous amount of energy into the mathematical education of schoolchildren. He became chairman of the Commission for Mathematical Education under the Presidium of the Academy of Sciences of the USSR. Under his leadership, a new state program of training in mathematics in the Soviet general school was developed and introduced.

45. LOCAL STRUCTURE OF TURBULENCE IN AN INCOMPRESSIBLE VISCOUS FLUID AT VERY LARGE REYNOLDS NUMBERS *

§1. We denote by

$$u_\alpha(P) = u_\alpha(x_1, x_2, x_3, t), \quad \alpha = 1, 2, 3,$$

the velocity components at time t at a point with rectangular Cartesian coordinates x_1, x_2, x_3. When studying turbulence it is natural to regard the velocity components $u_\alpha(P)$ at each point $P = (x_1, x_2, x_3, t)$ of the region G under consideration belonging to the four-dimensional space (x_1, x_2, x_3, t) as *random variables* in the sense of probability theory (for this approach, see the paper by Millionshchikov [1]).

Denoting the mathematical expectation of a random variable A by $\overline{A}$, we assume that

$$\overline{u_\alpha^2} \text{ and } \overline{(\partial u_\alpha / \partial x_\beta)^2}$$

are finite and bounded in each bounded subregion of the region G.

We introduce new coordinates in the four-dimensional space (x_1, x_2, x_3, t):

$$y_\alpha = x_\alpha - x_\alpha^{(0)} - u_\alpha(P^{(0)})(t - t^{(0)}), \quad s = t - t^{(0)}, \tag{1}$$

where

$$P^{(0)} = (x_1^{(0)}, x_2^{(0)}, x_3^{(0)}, t^{(0)})$$

is a fixed point belonging to G. Note that the coordinates y_α of a point P depend on the random variables $u_\alpha(P^{(0)})$ and therefore they are themselves random variables. In the new coordinates the velocity components are

$$w_\alpha(P) = u_\alpha(P) - u_\alpha(P^{(0)}). \tag{2}$$

Suppose that for fixed values of the components $u_\alpha(P^{(0)})$ the points $P^{(k)}$ $(k = 1, 2, \ldots, n)$ with coordinates $y_\alpha^{(k)}$ and $s^{(k)}$ in the coordinate system (1) lie in the region G. Then we can consider the $3n$-dimensional conditional probability distribution F_n of the variables

$$w_\alpha^{(k)} = w_\alpha(P^{(k)}), \quad \alpha = 1, 2, 3; \ k = 1, 2, \ldots, n,$$

* *Dokl. Akad. Nauk SSSR* **30**:4 (1941), 299–301.

for given

$$u_\alpha^{(0)} = u_\alpha(P^{(0)}).$$

In general, the distribution F_n depends on the parameters $x_\alpha^{(0)}, t^{(0)}, u_\alpha^{(0)}, y_\alpha^{(k)}$ and $s^{(k)}$.

Definition 1. A turbulence is said to be *locally homogeneous* in a region G if for any fixed n, $y_\alpha^{(k)}$ and $s^{(k)}$ the distribution F_n does not depend on $x_\alpha^{(0)}, t^{(0)}$, or $u_\alpha^{(0)}$ as long as all the points $P^{(k)}$ lie in G.

Definition 2. A turbulence is said to be *locally isotropic* in a region G if it is locally homogeneous and the distributions indicated in Definition 1 are invariant with respect to rotations and reflections of the original coordinate axes x_1, x_2, x_3.

As compared to the notion of *isotropic turbulence* introduced by Taylor, the above definition of locally isotropic turbulence is *narrower* in the sense that it requires that the distributions F_n be independent of $t^{(0)}$, that is, stationary in time, and is *broader* in the sense that the constraints are imposed only on the distributions of the velocity differences and not on the distribution of the velocities themselves.

§2. The isotropy hypothesis in Taylor's sense is in good agreement with experiment in the case of turbulence generated by a flow passing through a grid (see [3]). In the majority of other situations of practical significance, this hypothesis can only be regarded as a rather crude approximation to reality, even for small regions G and at very large Reynolds numbers.

By contrast, in our opinion, it seems very likely that in the case of an arbitrary turbulent flow with sufficiently large Reynolds number [1]

$$\mathrm{Re} = LU/\nu$$

in sufficiently small regions G of the four-dimensional space (x_1, x_2, x_3, t) not lying close to the boundaries of the flow or other singularities of it, the hypothesis of local isotropy holds with high accuracy. Here, by "small regions" are meant those whose linear dimensions and time scales are small relative to L and

$$T = U/L,$$

[1] Here L and U are the typical length and velocity scales for the flow as a whole.

respectively.

Naturally, this hypothesis, which we have put forward in such a general and somewhat ambiguous form, cannot be proved rigorously.[2]

To make its experimental verification possible for individual special cases, we present below a number of consequences of the hypothesis of local isotropy.

§3. We denote by y a vector with components y_1, y_2, y_3 and consider the random variables

$$w_\alpha(y) = w_\alpha(y_1, y_2, y_3) = u_\alpha(x_1 + y_1, x_2 + y_2, x_3 + y_3, t) - u_\alpha(x_1, x_2, x_3, t). \quad (3)$$

By virtue of the hypothesis of local isotropy, the distributions of (3) do not depend on x_1, x_2, x_3 or t. For the first moments of the variables $w_\alpha(y)$,

---

[2] Here we present only some general consideration in favour of the suggested hypothesis. For very large Re a turbulent flow can be represented in the following way. The mean flow (characterized by the mathematical expectations $\bar{u}_\alpha$) is accompanied by "first-order fluctuations" imposed on it, consisting of chaotic motions of individual fluid volumes with diameters of the order of $l^{(1)} = l$ (where l is the Prandtl mixing length) relative to one another; we denote by $v^{(1)}$ the order of the velocities of these relative motions. In turn, the first-order fluctuations become unstable at very large Re and there appear "second-order fluctuations" imposed on the former, the latter having mixing length $l^{(2)} < l^{(1)}$ and relative velocities $v^{(2)} < v^{(1)}$. This process of successive refinement of turbulent fluctuations goes on until for fluctuations of sufficiently high order n the Reynolds number $\mathrm{Re}^{(n)} = l^{(n)} v^{(n)} / \nu$ turns out to be sufficiently small, so that the effect of viscosity on the nth-order fluctuations is appreciable and prevents the formation of $(N+1)$th-order fluctuations imposed on the former.

From the energetic standpoint, the process of turbulent mixing can be thought of in the following manner: the first-order fluctuations absorb the energy of the mean motion and transfer it subsequently to higher-order fluctuations; as to the energy of the smallest fluctuations, it is dissipated and transformed into thermal energy due to viscosity.

By virtue of the chaotic mechanism of transfer of motion from lower-order fluctuations to higher-order ones, it is natural to assume that, within the limits of space regions which are small as compared to $l^{(1)}$, small-scale higher-order fluctuations are subject to an approximately spatially isotropic statistical regime. Within short time intervals this regime can naturally be regarded as being stationary, even when the flow as a whole is non-stationary.

Since at very large Re the velocity component differences

$$w_\alpha(P) = u_\alpha(P) - u_\alpha(P^{(0)})$$

at nearby points P and $P^{(0)}$ of the four-dimensional space (x_1, x_2, x_3, t) are determined almost exclusively by higher-order fluctuations, the presented scheme leads to the hypothesis of local isotropy in small regions in the sense of the definitions in §§1, 2.

local isotropy implies that

$$\overline{w_\alpha(y)} = 0. \tag{4}$$

Therefore we proceed to the study of the second moments[3]

$$B_{\alpha\beta}(y^{(1)}, y^{(2)}) = \overline{w_\alpha(y^{(1)}) w_\beta(y^{(2)})}. \tag{5}$$

Local isotropy implies that

$$B_{\alpha\beta}(y^{(1)}, y^{(2)}) = \tfrac{1}{2}[B_{\alpha\beta}(y^{(1)}, y^{(1)}) + B_{\alpha\beta}(y^{(2)}, y^{(2)}) - \\ - B_{\alpha\beta}(y^{(2)} - y^{(1)}, y^{(2)} - y^{(1)})] \tag{6}$$

This formula allows us to confine ourselves to second moments of the form $B_\alpha(y, y)$. For these we have

$$B_{\alpha\beta}(y, y) = \overline{B}(r) \cos\theta_\alpha \cos\theta_\beta + \delta_{\alpha\beta} B_{nn}(r), \tag{7}$$

where $r^2 = y_1^2 + y_2^2 + y_3^2$, $y_\alpha = r\cos\theta_\alpha$, $\delta_{\alpha\beta} = 0$ for $\alpha \neq \beta$, and $\delta_{\alpha\beta} = 1$ for $\alpha = \beta$;

$$\overline{B}(r) = B_{dd}(r) - B_{nn}(r), \tag{8}$$

$$B_{dd}(r) = \overline{[w_1(r, 0, 0)]^2}, \quad B_{nn}(r) = \overline{[w_2(r, 0, 0)]^2}. \tag{9}$$

For $r = 0$ we have

$$B_{dd}(0) = B_{nn}(0) = \frac{\partial}{\partial r} B_{dd}(0) = \frac{\partial}{\partial r} B_{nn}(0) = 0, \tag{10}$$

$$\frac{\partial^2}{\partial r^2} B_{dd}(0) = 2\overline{\left(\frac{\partial w_1}{\partial y_1}\right)^2} = 2a^2, \quad \frac{\partial^2}{\partial r^2} B_{nn}(0) = 2\overline{\left(\frac{\partial w_2}{\partial y_1}\right)^2} = 2a_n^2. \tag{11}$$

Formulas (6)–(11) have been derived without using the assumption that the fluid is incompressible. This assumption implies the equation

$$r \partial B_{dd}/\partial r = -2\overline{B}, \tag{12}$$

making it possible to express B_{nn} in terms of B_{dd}. Formulas (12) and (11) imply that

$$a_n^2 = 2a^2. \tag{13}$$

[3] All results in §3 are completely analogous to those obtained in [1, 2, 4] for the case of isotropic turbulence in Taylor's sense.

Further, it is easy to show that (under the assumption of incompressibility) the average energy dissipation rate per unit mass is given by

$$\bar{\epsilon} = \nu\left\{2\overline{\left(\frac{\partial w_1}{\partial y_1}\right)^2} + 2\overline{\left(\frac{\partial w_2}{\partial y_2}\right)^2} + 2\overline{\left(\frac{\partial w_3}{\partial y_3}\right)^2} + \overline{\left(\frac{\partial w_2}{\partial y_1} + \frac{\partial w_1}{\partial y_2}\right)^2} + \right.$$
$$\left. + \overline{\left(\frac{\partial w_3}{\partial y_2} + \frac{\partial w_2}{\partial y_3}\right)^2} + \overline{\left(\frac{\partial w_1}{\partial y_3} + \frac{\partial w_3}{\partial y_1}\right)^2}\right\} = 15\nu a^2. \tag{14}$$

§**4.** Consider the transformation of coordinates

$$y'_\alpha = y_\alpha/\eta, \quad s' = s/\sigma. \tag{15}$$

In the new coordinates the velocities, the kinetic viscosity, and the average energy dissipation rate per unit mass are expressed by the following formulas:

$$w'_\alpha = w_\alpha\sigma/\eta, \quad \nu' = \nu\sigma/\eta^2, \quad \bar{\epsilon}' = \bar{\epsilon}\sigma^3/\eta^2. \tag{16}$$

We now state the following hypothesis.

The first similarity hypothesis. *The distributions F_n for locally isotropic turbulence are uniquely determined by the parameters ν and $\bar{\epsilon}$.*

For

$$\eta = \lambda = \sqrt{\nu/a} = \nu^{3/4}/\bar{\epsilon}^{1/4}, \tag{17}$$
$$\sigma = 1/a = \sqrt{\nu/\bar{\epsilon}} \tag{18}$$

the transformation of coordinates (15) results in the parameters $\nu' = 1$, $\bar{\epsilon}' = 1$.

Therefore by virtue of the above similarity hypothesis, the corresponding function

$$B'_{dd}(r') = \beta_{dd}(r') \tag{19}$$

must be the same in all cases of locally isotropic turbulence. The formula

$$B_{dd}(r) = \sqrt{\nu\bar{\epsilon}}\beta_{dd}(r/\lambda) \tag{20}$$

in combination with the earlier derived expressions shows that in the case of locally isotropic turbulence the second moments $B_{\alpha\beta}(y^{(1)}, y^{(2)})$ can be uniquely expressed in terms of ν, $\bar{\epsilon}$ and the universal function β_{dd}.

§5. To determine the behaviour of the function $\beta_{dd}(r')$ for large r' we introduce one more hypothesis.

The second similarity hypothesis.[4] *If the absolute values of the vectors $y^{(k)}$ and their differences $y^{(k)} - y^{(k')}$ (where $k' \neq k$) are large relative to λ, the distributions F_n are uniquely determined by the parameter $\bar{\epsilon}$ and do not depend on ν.*

We put

$$y''_\alpha = y'_\alpha / k^3, \quad s'' = s'/k^2, \tag{21}$$

where y'_α and s' are determined according to formulas (15), (17) and (18). Since

$$\bar{\epsilon}' = \bar{\epsilon}'' = 1,$$

for any k, by virtue of the above hypothesis, for r' large relative to $\lambda' = 1$ we have

$$B''_{dd}(r'') \approx B'_{dd}(r'') = \beta_{dd}(r'/k^3).$$

On the other hand, formula (20) implies that

$$B''_{dd}(r'') = (1/k^2) B'_{dd}(r') = (1/k^2)\beta_{dd}(r').$$

Hence, for large r' we have

$$\beta_{dd}(r'/k^3) \approx (1/k^2)\beta_{dd}(r),$$

whence

$$\beta_{dd}(r') \approx C(r')^{2/3}, \tag{22}$$

where C is an absolute constant. By virtue of (17), (20) and (22) for r large relative to λ we have

$$B_{dd}(r) \approx C\bar{\epsilon}^{2/3} r^{2/3}. \tag{23}$$

From (23) and (12) it can easily be derived that for r large relative to λ,

$$B_{nn}(r) \approx 4/3 B_{dd}(r). \tag{24}$$

---

[4] In terms of the schematic description of turbulence presented in Footnote 2, λ is the scale of the smallest fluctuations, whose energy is dissipated and transformed directly into thermal energy due to viscosity. The meaning of the second similarity hypothesis is that for fluctuations of intermediate orders for which $l^{(k)}$ is greater than λ the mechanism of energy transfer from larger fluctuations to smaller ones does not depend on the viscosity.

We note that instead of formula (24), by virtue of (13) for r small as compared to λ the relation

$$B_{nn}(r) \approx 2B_{dd}(r) \tag{25}$$

holds.

28 December 1940

References

1. M.D. Millionshchikov, 'Degeneration of homogeneous isotropic turbulence in viscous incompressible fluids', *Dokl. Akad. Nauk SSSR* **22**:5 (1939) 236–240 (in Russian).
2. G.I. Taylor, 'Statistical theory of turbulence' I–IV, *Proc. Roy. Soc. London* **A 151** (1935), No. 874, 421–478.
3. S. Goldstein, ed., *Modern developments in fluid mechanics*, Vol. 1, §95, Oxford Univ. Press, 1938.
4. Th. von Karman, 'The fundamentals of the statistical theory of turbulence', *J. Aeronaut. Sci.* **4**:4 (1937), 131–138.

46. ON THE DEGENERATION OF ISOTROPIC TURBULENCE IN AN INCOMPRESSIBLE VISCOUS FLUID *

As in [1, 2], we assume that the velocity components

$$u_\alpha(P,t) = u_\alpha(x_1, x_2, x_3, t)$$

at the point $P = (x_1, x_2, x_3)$ at time t are random variables and denote by $\overline{A}$ the mathematical expectation of a random variable A.

In the case of isotropic turbulence in Taylor's sense (see [3, 4]) we have

$$\bar{u}_\alpha = 0, \tag{1}$$

and the second moments

$$b_{\alpha\beta} = \overline{u_\alpha(x_1^{(1)}, x_2^{(1)}, x_3^{(1)}, t) u_\beta(x_1^{(2)}, x_2^{(2)}, x_3^{(2)}, t)} \tag{2}$$

are given by the formulas (see [1, 5])

$$b_{\alpha\beta} = \bar{b}(r,t) \cos\theta_\alpha \cos\theta_\beta + \delta_{\alpha\beta} b_{nn}(r,t), \tag{3}$$

where

$$r^2 = (x_1^{(2)} - x_1^{(1)})^2 + (x_2^{(2)} - x_2^{(1)})^2 + (x_3^{(2)} - x_3^{(1)})^2,$$

$$x_\alpha^{(2)} - x_\alpha^{(1)} = r\cos\theta_\alpha, \quad \delta_{\alpha\beta} = 0 \text{ for } \alpha \neq \beta, \quad \delta_{\alpha\beta} = 1 \text{ for } \alpha = \beta;$$

$$\bar{b}(r,t) = b_{dd}(r,t) - b_{nn}(r,t); \tag{4}$$

$$b_{dd}(r,t) = \overline{u_1(x_1, x_2, x_3, t) u_1(x_1 + r, x_2, x_3, t)}; \tag{5}$$

$$b_{nn}(r,t) = \overline{u_2(x_1, x_2, x_3, t) u_2(x_1 + r, x_2, x_3, t)}; \tag{6}$$

$$\partial b_{dd}/\partial r = -(2/r)\bar{b}. \tag{7}$$

Formula (7), together with

$$b_{dd}(0,t) = b_{nn}(0,t) = \overline{[u_\alpha(x_1, x_2, x_3, t)]^2} = b(t), \tag{8}$$

* *Dokl. Akad. Nauk SSSR* **31**:6 (1941), 538–541.

makes it possible to express b_{nn} in terms of b_{dd}. Thus, all the second moments can be found from a single function $b_{dd}(r,t)$.

Von Karman and Howarth [5] derived the equation

$$\frac{\partial b_{dd}}{\partial t} + 2\Big(\frac{\partial d_{nnd}}{\partial r} + \frac{4}{r} b_{nnd}\Big) = 2\nu\Big(\frac{\partial^2 b_{dd}}{\partial r^2} + \frac{4}{r}\frac{\partial b_{dd}}{\partial r}\Big) \tag{9}$$

for $b_{dd}(r,t)$, where

$$b_{nnd}(r,t) = \overline{u_2^2(x_1,x_2,x_3,t)u_1(x_1+r,x_2,x_3,t)}. \tag{10}$$

Loitsyanskii [6] showed that (9) implies that

$$\Lambda = \int_0^\infty b_{dd}(r,t) r^4 dr \tag{11}$$

does not depend on time.

As the "turbulence scale" we take the length

$$L = (\Lambda/b)^{1/5}. \tag{12}$$

Using L and the average value of the velocity components

$$\mathbf{u} = \sqrt{b}, \tag{13}$$

we introduce the Reynolds number

$$\mathrm{Re} = L\mathbf{u}/\nu = (\Lambda^{1/5}/\nu)\mathbf{u}^{3/5}. \tag{14}$$

Since $\mathbf{u} \to 0$ as $t \to \infty$, (14) implies that the Reynolds number Re is small for large t. At this terminal stage of degeneration of isotropic turbulence it is legitimate to apply the theory neglecting the third moments. As was shown in [1], in this case

$$b_{dd}(r,t) = \frac{k}{(4\nu t)^{5/2}} \exp\Big(-\frac{r^2}{8\nu t}\Big), \tag{15}$$

$$b_{nn}(r,t) = \frac{k}{(4\nu t)^{5/2}} \Big(1 - \frac{r^2}{8\nu t}\Big) \exp\Big(-\frac{r^2}{8\nu t}\Big) \tag{16}$$

and consequently,

$$\mathbf{u} = At^{-5/4}, \tag{17}$$

$$L = B\sqrt{t}. \tag{18}$$

It is easy to express k, A and B in terms of ν and Λ.

For *large* Re, assuming that $b_{dd}(r,t)$ has the form

$$b_{dd}(r,t) = b\psi(r/L), \tag{19}$$

von Karman found (see [5]) that [1]

$$\mathbf{u} = \mathbf{u}_0(t-t_0)^{-p}, \quad L = L_0(t-t_0)^{1-p}, \tag{20}$$

but he did not determine the exponent p.

Below we present a new derivation of formula (20), which implies that

$$p = 5/7. \tag{21}$$

We first show that for sufficiently large Re the assumption (19) implies

$$db/dt = -Kb^{3/2}L^{-1} \tag{22}$$

where K is an absolute constant. Formula (22) can be justified in different ways. If the assumption made in my paper [2] is adopted, then the following argument can be used. In isotropic turbulence the average energy per unit mass is equal to $(3/2)b$. Therefore the average energy dissipation rate per unit mass is

$$\bar{\epsilon} = -\tfrac{3}{2}db/dt. \tag{23}$$

Since using the notation $B_{dd}(r)$ in [2] we can write

$$b_{dd}(r) = b - \tfrac{1}{2}B_{dd}(r), \tag{24}$$

formula (23) in [2] implies that for r small relative to L and large relative to λ we have

$$b_{dd}(r) \approx b\Big(1 - \frac{C}{2b}\bar{\epsilon}^{2/3}r^{2/3}\Big). \tag{25}$$

For small ρ we derive the expression [2]

$$\psi(\rho) \approx 1 - g\rho^{2/3} \tag{26}$$

[1] Equation (77) in [6] was integrated incorrectly. Actually, the exponent in formula (78) must be equal to $AB/[5+AB]$ and that in formula (81) must be equal to $-5/[5+AB]$. The value $p = 5/7$ corresponds to $AB = 2$.

[2] We assume that relation (26) remains valid for arbitrarily small ρ. However, it has a real meaning only for ρ large relative to λ/L, since the assumption (19) itself can naturally be adopted only for r large relative to λ.

for the function $\psi(\rho)$, where

$$g = (C/2)\bar{\epsilon}^{2/3}L^{2/3}b^{-1}.$$

Consequently,

$$\bar{\epsilon} = (2gb/C)^{3/2}L^{-1}. \tag{27}$$

Putting

$$K = \tfrac{2}{3}(2g/C)^{3/2} \tag{28}$$

we obtain formula (22) from (23) and (27). Formula (28) shows how the coefficient K is expressed in terms of the absolute constant C introduced in [2] and the parameter g determined from (26) taking into account the form of the function $\psi(\rho)$.

From (12) and (22) we find that

$$db/dt = -K\Lambda^{-1/5}b^{17/10}. \tag{29}$$

Integration of equation (29) results in

$$b = (10/7K)^{10/7}\Lambda^{2/7}(t-t_0)^{-10/7}. \tag{30}$$

Relations (30), (13) and (12) imply

$$u = (10/7K)^{5/7}\Lambda^{1/7}(t-t_0)^{-5/7}, \tag{31}$$

$$L = (7K/10)^{2/7}\Lambda^{1/7}(t-t_0)^{2/7}. \tag{32}$$

A comparison of (31) and (32) with (20) shows that we actually have $p = 5/7$.

4 March 1941

References

1. M.D. Millionshchikov, 'Degeneration of homogeneous isotropic turbulence in viscous incompressible fluids', *Dokl. Akad. Nauk SSSR* **22**:5 (1939), 236–240 (in Russian).
2. A.N. Kolmogorov, 'Local structure of turbulence in an incompressible viscous fluid at very large Reynolds numbers', *Dokl. Akad. Nauk SSSR* **30**:4 (1941), 299–303 (in Russian). (See No. 45.)

3. G.I. Taylor, 'Statistical theory of turbulence', I–IV, *Proc. Roy. Soc. London* **A151** (1935), No. 874, 421–478.
4. S. Goldstein, ed. *Modern developments in fluid mechanics*, Vol. 1, §91, Oxford Univ. Press, 1938.
5. Th. von Karman and L. Howarth, 'On the statistical theory of isotropic turbulence', *Proc. Roy. Soc. London* **A164** (1938), No. 917, 192–245.
6. L.G. Loitsyanskii, 'Some basic regularities of an isotropic turbulent flow', *Trudy TsAGI* **440** (1939), 3–23.

47. DISSIPATION OF ENERGY IN ISOTROPIC TURBULENCE *

In my paper [1] the definition of the notion of local isotropy was stated and the functions

$$B_{dd}(r) = \overline{[u_d(M') - u_d(M)]^2}, \quad B_{nn}(r) = \overline{[u_n(M') - u_n(M)]^2} \tag{1}$$

were introduced, where r denotes the distance between the points M and M', $u_d(M')$ and $u_d(M)$ are the velocity components at the points M and M' in the direction of $\overline{MM'}$, and $u_n(M)$ and $u_n(M')$ are the velocity components at M and M' respectively, in a direction perpendicular to $\overline{MM'}$.

In what follows we also need the third order moments

$$B_{ddd}(r) = \overline{[u_d(M') - u_d(M)]^3}. \tag{2}$$

Locally isotropic turbulence in an incompressible fluid is described by the equation

$$4\bar{\epsilon} + \left(\frac{dB_{ddd}}{dr} + \frac{4}{r}B_{ddd}\right) = 6\nu\left(\frac{d^2B_{dd}}{dr^2} + \frac{4}{r}\frac{dB_{dd}}{dr}\right), \tag{3}$$

which is analogous to the well-known von Karman equation for isotropic turbulence in Taylor's sense. Here $\bar{\epsilon}$ denotes the average energy dissipation rate per unit mass. Equation (3) can be written as

$$\left(\frac{d}{dr} + \frac{4}{r}\right)\left(6\nu\frac{dB_{dd}}{dr} - B_{ddd}\right) = 4\epsilon \tag{4}$$

and, by virtue of the condition $dB_{dd}(r)/dr\big|_{r=0} = B_{ddd}(0) = 0$, it yields

$$6\nu dB_{dd}/dr - B_{ddd} = (4/5)\bar{\epsilon}r. \tag{5}$$

As is well known, for small r we have

$$B_{dd} \approx (1/15\nu)\bar{\epsilon}r^2, \tag{6}$$

that is, $$6\nu dB_{dd}/dr \approx (4/5)\bar{\epsilon}r.$$

Thus, for small r the second term on the left-hand side of equation (5) is infinitesimally small relative to the first term. Conversely, for large r the first term is negligibly small as compared to the second term, that is, we can assume that

$$B_{ddd} \approx -(4/5)\bar{\epsilon}r. \tag{7}$$

It is natural to assume that for large r the ratio

$$S = B_{ddd} : B_{dd}^{3/2}, \tag{8}$$

* *Dokl. Akad. Nauk SSSR* **32**:1 (1941), 19–21.

that is, the *skewness* of the distribution of the difference

$$\Delta u_d = u_d(M') - u_d(M),$$

remains constant. Under this assumption, for large r we have

$$B_{dd} \approx C\bar{\epsilon}^{2/3} r^{2/3}, \tag{9}$$

where

$$C = (-4/5S)^{2/3}. \tag{10}$$

Relation (9) was derived in [1] on the basis of somewhat different considerations. [1]

In [1] we introduced the *local scale* of turbulence

$$\lambda = (\nu^3/\bar{\epsilon})^{1/4} \tag{11}$$

and justified the assumption that

$$B_{dd}(r) = \sqrt{\nu\bar{\epsilon}}\beta_{dd}(r/\lambda), \quad B_{nn}(r) = \sqrt{\nu\bar{\epsilon}}\beta_{nn}(r/\lambda), \tag{12}$$

where β_{dd} and β_{nn} are universal functions such that

$$\beta_{dd}(\rho) \approx (1/15)\rho^2, \quad \beta_{nn}(\rho) \approx (2/15)\rho^2 \tag{13}$$

for small ρ and

$$\beta_{dd}(\rho) \approx C\rho^{2/3}, \quad \beta_{nn}(\rho) \approx (4/3)C\rho^{2/3} \tag{14}$$

for large ρ. In isotropic turbulence in Taylor's sense the laws of locally isotropic turbulence must hold at distances much less than the *integral scale* of turbulence L (for a precise definition see my paper [3]). In this case the correlation coefficients

$$R_{dd}(r) = \overline{(u_d(M')u_d(M))} : b, \quad R_{nn}(r) = \overline{(u_n(M')u_n(M))} : b, \tag{15}$$

where b is the average value of the square of the velocity components, are related to $B_{dd}(r)$ and $B_{nn}(r)$ by

$$B_{dd} = 2b(1 - R_{dd}), \quad B_{nn} = 2b(1 - R_{nn}). \tag{16}$$

By (16) and (12), for r small relative to L we must have

$$1 - R_{dd} \approx \frac{\sqrt{\nu\bar{\epsilon}}}{2b}\beta_{dd}\left(\frac{r}{\lambda}\right), \quad 1 - R_{nn} \approx \frac{\sqrt{\nu\bar{\epsilon}}}{2b}\beta_{nn}\left(\frac{r}{\lambda}\right). \tag{17}$$

[1] A.M. Obukhov obtained relation (9) independently, by calculating the balance in the spectral distribution of the fluctuation energy (see [2]).

If r is small relative to L and large relative to λ, then by (14) and (11),

$$1 - R_{dd} \approx (1/2)C\bar{\epsilon}^{2/3}b^{-1}r^{2/3}, \tag{18^1}$$

$$1 - R_{nn} \approx (2/3)C\bar{\epsilon}^{2/3}b^{-1}r^{2/3}. \tag{18^2}$$

Formulas (18) make it possible to determine the constant C from experimental data The correlation coefficients R_{dd} and R_{nn} were measured most thoroughly by Dryden, Schubauer, Mock, and Scramstad [4]. Representing formula (18^2) in the form

$$1 - R_{nn} \approx 2C(kr)^{2/3}, \quad k = \bar{\epsilon} : (3b)^{3/2}, \tag{19}$$

we have calculated the values of the coefficient k corresponding to turbulence at a distance 40M from a grid with a mesh size M equal to 1″, 3.25″, and 5″ using the empirical formula (17) in [4] and putting $b = \sqrt{\bar{u}^2}$ and $\bar{\epsilon} = (3/2)Ud\sqrt{\bar{u}^2}/dx$ in the notation of [4]:

M (inch)	1	3.25	5
k (cm^{-1})	0.197	0.065	0.042

With these values of k, the k curves in Figure 5 in [4], taking into account the correction for the wire length, are in good agreement with formula (19) for values of r not very large relative to L and

$$C = 3/2. \tag{20}$$

The local scale λ is so small that under the conditions of the experiments described in [4] the deviations from relations (18) for small r cannot be detected.[2]

[2] In this connection we note that the attempt of Millionshchikov [5] to apply the theory of isotropic turbulence neglecting third order moments to experimental data in [4] is based on a misunderstanding. It can easily be seen that under the conditions of the experiments in [4] the terms involving second order moments in the equations relating second order moments to third order ones (for example, in equation (3)) are *much less* than those involving third order moments.

In Figure 3 in [5], the comparison with the theoretical curve of R_{nn} (obtained by neglecting third order moments) using the experimental data in [4] was performed incorrectly since: 1) the data under consideration relate to a definite spectral component of fluctuations and not to the total fluctuations; and 2) the experimental data do not correspond to the condition $8\nu t = 7.56$ which determines the theoretical curve.

In his later studies [6] (see also M.D. Millionshchikov, *Izv. Akad. Nauk SSSR Ser. Geogr. i Geofiz.* 5:4/5 (1941), 433–446 (in Russian)) Millionshchikov himself estimated the errors due to the neglect of third order moments, using subtle considerations involving third and fourth order moments.

The curves in Figure 28 in [4] cannot be used directly to determine C, since they do not include the correction for the wire length. However, for r small relative to L they confirm with sufficient accuracy the relation

$$(1 - R_{nn})/(1 - R_{dd}) = 4/3 \tag{21}$$

implied by (18). Including the correction for the wire length, the authors of [4] estimated the ratio on the left-hand side of (21) as 1.28 (see [4, p. 29]), which appears to be sufficiently close to the theoretical value 4/3 if we take into account the restricted accuracy of an experiment.

References

1. A.N. Kolmogorov, 'Local structure of turbulence in an incompressible viscous fluid at large Reynolds numbers', *Dokl. Akad. Nauk SSSR* **30**:4 (1941), 299–303 (in Russian). (See No. 45.)
2. A.M. Obukhov, 'On the energy distribution in the spectrum of a turbulent flow', *Dokl. Akad. Nauk SSSR* **32**:1 (1941), 22–24 (in Russian).
3. A.N. Kolmogorov, 'On the degeneration of isotropic turbulence in an incompressible viscous fluid', *Dokl. Akad. Nauk SSSR* **31**:6 (1941), 538–541 (in Russian). (See No. 46.)
4. H.L. Dryden, G.B. Schubauer, W.C. Mock and H.K. Scramstad, 'Measurements of intensity and scale of wind-tunnel turbulence and their relation to the initial Reynolds number of spheres', *Nat. Adv. Com. Aeronaut.* **581** (1937).
5. M.D. Millionshchikov, 'Damping of fluctuation velocity in aerodynamical tunnels', *Dokl. Akad. Nauk SSSR* **22**:5 (1939), 241–242 (in Russian).
6. M.D. Millionshchikov, 'On the influence of third order moments in isotropic turbulence', *Dokl. Akad. Nauk SSSR* **32**:9 (1941), 615–617 (in Russian).

48. EQUATIONS OF TURBULENT MOTION IN AN INCOMPRESSIBLE FLUID *

The general pattern of turbulent motion can be described (according to Taylor and Richardson) in the following way. The mean flow is accompanied by turbulent fluctuations imposed on it and having different scales, beginning with maximal scales of the order of the "external scale" of turbulence L (the "mixing length") to smallest scales of the order of the distances λ at which the effect of viscosity becomes appreciable (the "internal scale" of turbulence). In regions whose dimensions are small relative to L the velocity field is smooth. Most large-scale fluctuations receive energy from the mean flow and transfer it to fluctuations of smaller scales. Thus, there appears a flux of energy transferred continuously from fluctuations of larger scales to those of smaller scales. Dissipation of energy, that is, transformation of energy into heat, occurs mainly in fluctuations of scale λ. The amount of energy w dissipated in unit time per unit volume is the basic characteristic of turbulent motion for all scales.

The most characteristic properties of turbulence are those relating to motion in regions whose dimensions are small as compared to L. The local structure of turbulence (that is, its structure at scales $\ll L$) can be regarded as being spatially homogeneous and isotropic. Its study is carried out by analogy with Taylor's and von Karman's study of isotropic turbulence. However, in this case one should consider moments of various orders, defined as average values of products of components of vector velocity differences at two different points in space. By virtue of the isotropy, these moments are functions depending only on the distance r between the points (for $r \ll L$). For example, the second order moments B_{dd} and B_{nn} are the average values of the squares of differences of the projections of the velocities at two points on the line joining the points and the direction perpendicular to the line, respectively. For $r \ll L$ these moments turn out to be proportional to r^2, while for $r \gg \lambda$ (and $r \ll L$) B_{dd} and B_{nn} are proportional to $r^{2/3}$. [1] The internal scale of turbulence λ itself is proportional to $\nu^{3/4}(\rho/w)^{1/4}$, where ν is the viscosity of the fluid and

* *Izv. Akad. Nauk SSSR Ser. Fiz.* **6**:1–2 (1942), 56–58. Summary of the paper presented at the General Meeting of the Division of Physics and Mathematics of the USSR Academy of Sciences; 26–28 January 1942, Kazan'.

[1] *Dokl. Akad. Nauk SSSR* **30**:4 (1941), 299–303 (in Russian). (No. 45 in the present edition).

ρ is its density. These results are in good agreement with measurements of the velocity correlation performed by Dryden et al.

A somewhat less thorough mathematical investigation can be carried out for a homogeneous and isotropic (at all scales) turbulent motion in the case when there is no mean flow; such a motion must damp out with time. In addition to Taylor's and von Karman's results, the conservation law found by L.G. Loitsyanskii[2] can be applied to show[3] that in this case the fluctuation velocity of motion decays inversely proportionally to $t^{5/7}$. The scale of motion L increases proportionally as $t^{2/7}$.

Based on the above local properties of turbulence, we can construct (using a cruder approximate assumption) a complete system of equations of turbulent motion. These equations are written as

$$\frac{D\bar{v}_i}{Dt} = F_i - \frac{\partial}{\partial x_i}\left(\frac{\bar{p}}{\rho} + b\right) + A\sum_j \frac{\partial}{\partial x_j}\left[\frac{b}{\omega}\left(\frac{\partial \bar{v}_i}{\partial x_j} + \frac{\partial \bar{v}_j}{\partial x_i}\right)\right], \tag{1}$$

$$\frac{D\omega}{Dt} = -\frac{7}{11}\omega^2 + A'\sum_j \frac{\partial}{\partial x_j}\left[\frac{b}{\omega}\frac{\partial \omega}{\partial x_j}\right], \tag{2}$$

$$\frac{Db}{Dt} = -b\omega + \frac{A}{3}\frac{b}{\omega}\bar{\epsilon} + A''\sum_j \frac{\partial}{\partial x_j}\left[\frac{b}{\omega}\frac{\partial b}{\partial x_j}\right]. \tag{3}$$

Here D/Dt is the symbol of the total time derivative, F_i is an external force, $\bar{v}_i$ is the velocity of the mean motion, and $\bar{p}$ is the average value of the pressure; $b = (1/3)\sum_j \overline{v_j'^2}$ is equal to one third of the mean square of the fluctuation velocity; ω is a certain average "frequency", defined as $\omega = c\sqrt{b}/L$ where c is a constant, and $\epsilon = \sum_{j,i}\left(\frac{\partial \bar{v}_i}{\partial x_j} + \frac{\partial \bar{v}_j}{\partial x_i}\right)^2$. Finally, A, A' and A'' are numerical constants which must be determined once and for all by comparing an individual solution of the equations with experimental data. The ratio b/ω in equations (1), (2) and (3) plays the role of a "diffusion coefficient" for the average velocity, the "frequency" ω, and the fluctuation velocity b, respectively. The first term in (3) is equal to two thirds of the turbulent dissipation of energy related to a unit mass of the fluid $(2w/3\rho)$, and the second term is equal to two thirds of the energy transferred by turbulent fluctuations from the mean motion. The ratio of the coefficients in the first terms in (1) and (2) is determined by

---

[2] *Trudy TsAGI* **440** (1939), 3 (in Russian).

[3] *Dokl. Akad. Nauk SSSR* **31**:6 (1941), 538–541 (in Russian) (No. 46 in the present edition).

the above-mentioned Loitsyanskii relation. Equations (1)–(3) determine the average velocity vector $\bar{v}_i$, the fluctuation velocity b, and the scale of motion L.

The solution of these equations meets with severe difficulties. They can be integrated in finite form for fluid motion between movable parallel planes, the result being in agreement with well-known results of von Karman.

In conclusion, some preliminary results on numerical integration of equations (1)–(3) for a turbulent flow in a tube of a circular cross-section were presented.

Discussion. L.D. Landau noted that A.N. Kolmogorov had, for the first time, suggested a correct interpretation of the local structure of a turbulent flow. As to the equations of turbulent motion, L.D. Landau said that, in his opinion, they must necessarily take into account the fact that the presence of the curl of the velocity in a turbulent flow is confined to a finite region in space; qualitatively, correct equations must result in this very kind of vortex distribution.

P.L. Kapitsa also took part in the discussion.

26 January 1942

Lev Semenovich Pontryagin

Born – 3 Sept 1908 in Moscow, Russia
Died – 3 May 1988

Lev Pontryagin's father was a civil servant and their family was not well-off. Pontryagin's mother, Tat'yana Andreevna, who played a crucial role in Pontryagin's path to becoming a mathematician, used her sewing skills to help out the family's finance. Even then, there wasn't enough money in the family to provide a good education for young Pontryagin. He attended the town school where the standard of education was well below that of better schools that were financially out-of-reach for him.

At age 14, Pontryagin met with an accident. An explosion left him blinded for life. That would have meant an end for his education and career, had his mother not assumed complete responsibilities to all his needs and devoted herself to guide him onto the road of success. For many years, she worked as Pontryagin's personal secretary, reading scientific works to him, writing down the formulas in his manuscripts, correcting his work and so on. In order to do all this, she even had to learn to read foreign languages. Tat'yana Andreevna saw to her son's every need and took excellent care of him. She, with no mathematical training or knowledge, read mathematics papers to Pontryagin by describing the appearances of "mysterious math symbols". For example, an intersection point is a "tails down" while a union symbol became a "tails up". With her determination and extreme efforts, Tat'yana Andreevna allowed Pontryagin to become a mathematician against all odds.

It was quickly apparent to his lecturers that Pontryagin was an exceptional student when he entered the University of Moscow in 1925. He could not make notes, but was able to remember the most complicated manipulations with symbols. Even more remarkable, was the fact that this blind student could "see" the depth of meaning of the topics presented to him more clearly than his fellow students. Of all the advance courses he took, Pontryagin liked Aleksandrov's courses the most. He was strongly influenced by Aleksandrov and for many years, the area Pontryagin's work was determined by the direction of Aleksandrov's research.

In 1927, Pontryagin's father passed away. By then, 19 year-old Pontryagin had begun to produce important results on Alexander's duality theorem. His main tool was to use link numbers introduced by Brouwer. In 1932, he proved the duality between the homology groups of bounded closed sets in Euclidean space and the homology groups in the complement of the space, by producing the most significant duality results.

Pontryagin graduated from the University of Moscow in 1929, he was then appointed to the Mechanics and Mathematics Faculty. He became a member of Steklov Institute in 1934. And in 1935, he became head of the Department of Topology and Functional Analysis in the Institute.

Pontryagin worked on problems in Topology and Algebra. He described his work as "problems where these two domains of mathematics come together". The significance of his work lies in its effect on the development of topology, and the fact that Pontryagin's theorem enabled him to construct a general theory of characters for commutative topological groups.

In 1900, one of the 23 problems Hilbert posed was to prove his conjecture that any locally Euclidean topological group can be given the structure of an analytic manifold so as to become a Lie group. This became known as the Hilbert's Fifth Problem. In the year 1934, Pontryagin was able to prove Hilbert's Fifth Problem for abelian groups using the characters on locally compact abelian groups that he had introduced.

Pontryagin's most important books were on topological groups. These books were "classics" that exert a formative influence on the scientific outlook of whole generations of mathematics and remain significant even after decades.

Pontryagin attended Cartan's lecture when Cartan visited Moscow and lectured in the Mechanics and Mathematics Faculty in the year 1934. The lecture was in French, and Pontryagin had to listen to a whispered translation from Nina Bari who sat beside him. Cartan talked about the problem of calculating the homology groups of the classical compact Lie groups. He explained some ideas of how he thought this might be achieved. But by the following year, Pontryagin solved the problem completely using a different approach to the one Cartan suggested. He used ideas introduced by Morse on equipotential surfaces.

In 1952, the direction of Pontryagin's research changed completely. He turned to study applied mathematics problem, in particular, areas in differential equation and control theory. The shift in his focus was not totally sudden as it appeared. From the 1930s Pontryagin had been friendly with the physicist A.A. Andronov and had regularly discussed with him problems in the theory of oscillations and the theory of automatic control on which Andronov was working. He published a paper with Andronov on dynamical system in 1932. Around the time of Andronov's death in 1952, Pontryagin shifted focus in his work. A year later, there was an English translation. Then in 1962, Pontryagin received a Lenin Prize for his book. Subsequently, he produced a series of papers on differential games that extended his work on control theory.

Many mathematical concepts were associated with Pontryagin's name. For example, the essential tool of cobordism theory was the Pontryagin–Thom construction.

Pontryagin was well-honoured for his work. In 1939, he was elected to the Academy of Science and became its full member in 1959. He was the first recipient of the Stalin Prizes (later called the States Prizes) in 1941. In 1970, he was elected the Vice-President of the International Union.

Characteristic Cycles of Manifolds

Doklady Akademii Nauk SSSR (1942) 35(2): 35–39

Let R^{k+l} be $(k+l)$-dimensional Euclidean space, and $H(k,l)$ the manifold of all oriented k-dimensional planes in the space R^{k+l} that pass through a fixed point O. Let M be a k-dimensional oriented manifold with a continuous tangent in R^{k+l}. To each point x of the manifold M^k let us put into correspondence that plane $T(x)$ in $H(k,l)$ which is parallel to the tangent to the manifold M at the point x. Evidently, under a continuous differentiable deformation of M^k in R^{k+l}, the mapping T changes continuously, and consequently, the homotopy type of this mapping remains invariant.

Hence it follows that for $l \geqslant k+2$, the homotopy type of the mapping T is a topological invariant of the oriented manifold M^k (by a topological transformation of a manifold M we mean here a topological differentiable transformation). Thus, in particular, the homomorphisms of the Betti group of the manifold M^k into the Betti groups of the manifold $H(k,l)$ under the mapping T are certain topological invariants of the manifold M^k. These homomorphisms, by virtue of the Poincare–Veblen theorem, can be characterized as follows. Let z_j^{m-r} $(j=1,2,\ldots,p^r)$ be an $(m-r)$-dimensional basis of homology of the manifold $H(k,l)$, where $m=kl$ is the dimension of the manifold $H(k,l)$. The intersection of the manifold $T(M^k)$ with the cycle z_j^{m-r} is a cycle y_j^{k-r} of dimension $(k-r)$ in the manifold $T(M^k)$. Then $T^{-1}(y_j^{k-r}) = x_j^{k-r}$ is a cycle in the manifold M^k. It can be readily verified that for $l \geqslant k+2$, the homology class X_j^{k-r} containing the cycle x_j^{k-r} is uniquely defined by the homology class of the cycle z_j^{m-r} and does not depend on the location of M^k in R^{k+l}. Thus, each homology class X_j^{k-r} is a topological invariant of the oriented manifold M^k. Moreover, a knowledge of the homology classes X_j^{k-r} $(j = 1, 2, \ldots, p^r)$ is equivalent to knowledge of the homomorphism of the r-dimensional Betti group of the manifold M^k into the r-dimensional Betti group of the manifold $H(k, l)$ under the mapping T. The cycles x_j^{k-r} are called the *characteristic cycles* of the manifold M^k. Characteristic cycles are rather typical features of the topological structure of a manifold and are encountered, as we shall show below, in different problems. Study of characteristic cycles requires, first of all, a knowledge of the homology of $H(k, l)$.

Let $\omega(p)$ be an integral-valued non-decreasing function of p $(p = 1, 2, \ldots, k)$, satisfying the condition $0 \leqslant \omega(p) \leqslant l$. In the space R^{k+l}, take a sequence of planes $0 \subset R_1 \subset R_2 \subset \ldots \subset R_k$, the dimension of the plane R_p being $p + \omega(p)$. Let $Z(\omega(p))$ denote the set of all planes R_ξ^k in $H(k, l)$ satisfying the condition that the intersection of R_ξ^k with R_p is of dimension not less than p. The set $Z(\omega(p))$ is a pseudomanifold of dimension $s = \Sigma_{p=1}^k \omega(p)$. If it is orientable, it can be regarded as a cycle in $H(k, l)$, and if not, as a cycle modulo 2. The cycle $Z(\omega(p))$ depends on the function $\omega(p)$; its dependence on the planes R_p is unimportant from the viewpoint of homology because by rotating the space R^{k+l}, any system of planes R_p can be converted into any other system with the same function $\omega(p)$.

The number p is called *a jump* of the function ω, if $\omega(p+1) - \omega(p) > 0$. Let $p_1, p_2, \ldots, p_{n-1}$ be an increasing sequence of jump points. Put

$$p_0 = 0, \quad p_n = k, \quad \alpha_i = p_i - p_{i-1} \quad (i = 1, 2, \ldots, n);$$

$$\beta_1 = \omega(1), \quad \beta_{i+1} = \omega(p_i + 1) - \omega(p_i) \quad (i = 1, 2, \ldots, n-1).$$

The numbers α_i and β_i uniquely define the function $\omega(p)$. Furthermore, they are essential because the set $Z(\omega(p))$ is uniquely defined by the planes R_p for $p = p_i$ $(i = 1, 2, \ldots, n)$. Applying the aforementioned condition on the dimension of intersection to the intersection of R_ξ^k with the planes R_{p_i} only, we obtain the set $Z(\omega(p))$. The following two sequences

$$\alpha_1, \beta_2, \alpha_2, \beta_3, \ldots, \alpha_{n-1}, \beta_n; \quad \alpha_1, \beta_2, \alpha_2, \beta_3, \ldots, \beta_n, \alpha_n \qquad (1)$$

play an important part; if $\omega(k) = l$, the first sequence is to be taken, and if $\omega(k) < l$, the second sequence. If $\alpha_1 \geqslant 2$ and $\Sigma_{p=1}^k (p) \geqslant m - (l-2)$, denote the function $\omega(p)$ by $\omega^*(p)$. Thus, the dimension of $Z(\omega^*(p))$ is equal to $(m-r)$, where $r \leqslant l-2$. Let $\omega'(p)$ denote those functions $\omega^*(p)$ for which all the numbers in the sequence (1) are even, and let $\omega''(p)$ stand for those functions $\omega^*(p)$ for which the sequence (1) contains odd numbers, and let β be the first odd number in this sequence.

Theorem 1. All pseudomanifolds $Z(\omega'(p))$ of dimension $(m-r)$ are orientable and when each is taken with a certain orientation, they form an $(m-r)$-dimensional independent basis of the weak homology of the manifold $H(k,l)$. The pseudomanifold $Z(\omega''(p))$ is not orientable, and if its boundary is taken in a certain triangulation, we obtain a cycle $2Y(\omega''(p))$. The set of all cycles $Z(\omega'(p))$ and $Y(\omega''(p))$ of dimension $(m-r)$ form an $(m-r)$-dimensional canonical basis of homology in the manifold $H(k,l)$. Hence, all torsion coefficients of dimension $(m-r)$, where $r \leqslant l-1$, of the manifold $H(k,l)$ are equal to two, and the pseudomanifolds $Z(\omega''(p))$, which are cycles modulo 2, are independent in the sense of homology modulo 2.

Proof. From the set $Z(\omega(p))$ take a plane R_0^k whose intersection with R_{p_i} is of dimension p_i $(i=1,2,\ldots,n)$. Let $U(\omega(p))$ denote the closure of the set of those R_ξ^k in $Z(\omega(p))$ for which the orthogonal projection of R_ξ^k on R_0^k is positive, and let $\bar{U}(\omega(p))$ denote a similar set with a negative projection. Now let us impose an additional condition for the choice of the planes R_{p_i} and R_0^k, so that U and $\bar{U}$ should depend only on the function $\omega(p)$, but not on the random choice of these planes. For this purpose, in the space R^{k+l} fix an orthogonal system of coordinate axes taken in a definite order: (1), (2), $\ldots$, $(k+l)$. The plane defined by the first $p_i + \omega(p_i)$ coordinates axes shall always be taken as the plane R_{p_i}. And R_0^k is always taken to be the plane defined by the axes:

$$(\beta_1+1),\ldots,(\beta_1+\alpha_1),(\beta_1+\beta_2+\alpha_1+1),\ldots,(\beta_1+\beta_2+\alpha_1+\alpha_2),$$
$$(\beta_1+\beta_2+\beta_3+\alpha_1+\alpha_2+1),\ldots,(\beta_1+\ldots+\beta_n,\alpha_1+\ldots+\alpha_n),$$

with an orientation given by the order already specified. In the pseudomanifolds U and $\bar{U}$ with boundaries thus uniquely defined by the function $\omega(p)$, let us introduce a certain definite orientation. If $\omega(p)$ is a given function, then let $\omega_i(p)$ be a function that coincides everywhere with $\omega(p)$, except at the point $p_{i-1}+1$, $(i=1,2,\ldots,n)$ where it is equal to $\omega(p)-1$; if $\omega_1(p)<0$, the corresponding term in (2) is discarded. Thus, we have

$$\Delta U(\omega(p)) = \sum_{i=1}^{n} [(-1)^{\beta_1+\ldots+\beta_i+\alpha_i+\ldots+\alpha_n} U(\omega_i(p)) - \bar{U}(\omega_i(p))](-1)^{q_i(\omega)};$$

$$\Delta \bar{U}(\omega(p)) = \sum_{i=1}^{n} [(-1)^{\beta_1+\ldots+\beta_i+\alpha_i+\ldots+\alpha_n} \bar{U}(\omega_i(p)) - U(\omega_i(p))](-1)^{q_i(\omega)}.$$

Here $q_i(\omega)=\Sigma_{p=1}^{n}\,\omega(p)$. Each linear form in various U and $\bar{U}$ is called a *U-complex*. It turns out that every cycle in $H(k,l)$ is homologous to some

U-cycle, and each U-cycle which is homologous to zero in $H(k, l)$ bounds some U-complex. From this, by virtue of (2), the proof of Theorem 1 can be obtained.

Now put $Z(\omega(p)) = U(\omega(p)) + (-1)^{\beta_1 + k}\, \bar{U}(\omega(p))$. Thus, we assign a definite orientation to each pseudomanifold $Z(\omega(p))$, so that $Z(\omega'(p))$ turns out to be a cycle, while $Z(\omega''(p))$ acquires a definite boundary. In particular, the manifold $H(k, l) = Z(l)$ itself acquires an orientation. It can be readily found that the intersection of $H(k, l)$ with $Z(1 + \omega(p))$ in $H(k, l + 1)$ is $\pm Z(\omega(p))$. Thus we eliminate the apparent dependence of the characteristic cycle on the number l. In order to emphasize this independence, let us introduce a function $\varphi(p) = l - \omega(p)$ in place of the function $\omega(p)$, and let $X(\varphi(p))$ denote the intersection of the manifold $T(M^k)$ with $Z(l - \varphi(p))$ transferred into M^k with the help of the mapping T^{-1}. The dimension of $X(\varphi(p))$ is equal to $k - r(\varphi)$, where $r(\varphi) = \Sigma_{p=1}^{k} \varphi(p)$. Special mention should be made of the case where $r(\varphi) = k$. The cycle $X(\varphi(p))$ in this case is zero dimensional, and the number of its points is a numerical invariant of the manifold M^k. In particular, $X(1)$, i.e., $X(\varphi(p))$ when $\varphi(p) \equiv 1$, gives the Euler characteristic for even k. Other types of invariants also exist, for instance, when $k = 4$, the function defined by the conditions

$$\varphi(1) = \varphi(2) = 2, \quad \varphi(3) = \varphi(4) = 0$$

gives a zero-dimensional cycle $X(\varphi(p))$.

We now give examples where characteristic cycles are encountered in certain problems.

Let f be a differentiable mapping of the oriented manifold M^k into a q-dimensional Euclidean space R^q. Using the local coordinates $t_1, t_2, \ldots, t_k$ in M^k, we can analytically express it as follows:

$$y_i = f_i(t_1, \ldots, t_n) \quad (i = 1, \ldots, q).$$

The functional matrix $\|\partial y_i / \partial t_j\|$ has, in general, a rank equal to the lesser of the two numbers k and q. Let $C(k, q, \rho)$ denote the set of those points in M^k for which this rank is less than the number just indicated or less than or equal to ρ.

Theorem 2. The critical set $C(k, q, \rho)$ is a characteristic cycle $X(\varphi(p))$, where $\varphi(p)$ is given by the conditions: $\varphi(1) = \ldots = \varphi(k - \rho) = q - \rho$, $\varphi(k - \rho + 1) = \ldots = \varphi(k) = 0; r(\varphi) = (k - \rho)(q - \rho)$.

Proof. Let R^{k+l} be a Euclidean space with sufficiently large l and a fixed orthogonal coordinate system. Take the subspace of R^{k+l} given by the last q axes as R^q, and the normal to R^q as R^{k+l-q}. Now imbed M^k into R^{k+l} in such a manner that its orthogonal projection in R^q coincides with the mapping f. It is easily seen that a point x in M belongs to $C(k, q, \rho)$, if and only if the plane $T(x)$ parallel to the tangent intersects R^{k+l-q} along a plane of dimension $\leqslant (k-\rho)$. Theorem 2 follows directly from this assertion as a corollary.

Let M^k be a differentiable oriented manifold. In each of its tangential planes T_x, take $s < k$ vectors $v_1(x), \ldots, v_s(x)$ continuously dependent on x. And let $V(k, s)$ denote the set of all those points for which these vectors are linearly dependent. Stiefel investigated this problem.

Theorem 3. The set $V(k, s)$ is a characteristic cycle $X(\varphi(p))$, where $\varphi(p)$ is given by the conditions:

$$\varphi(1) = \ldots = \varphi(k-s+1) = 1; \quad \varphi(k-s+2) = \ldots = \varphi(k) = 0;$$
$$k - r(\varphi) = (s-1).$$

Proof. Assume that M^k sits in R^{k+l}. Let R^s denote a plane in R^{k+l} spanned by the last s axes, and R^{k+l-s} the normal to this plane. In the plane R^s take mutually orthogonal vectors $v_1, \ldots, v_s$. If the tangent T_x is in general position with R^{k+l-s}, the intersection of $T(x)$ with R^{k+l-s} is of dimension equal to $(k-s)$. Let $R^s(x)$ denote the normal to this intersection in $T(x)$. Projection of the plane $R^s(x)$ on the plane R^s is a one-to-one correspondence. Consequently, the vectors $v_1, \ldots, v_s$ uniquely define the vectors $v_1(x), \ldots, v_s(x)$ in $R^s(x)$. Thus, x in M^k will belong to $V(k, s)$, if and only if the intersection of $T(x)$ with R^{k+l-s} is of dimension greater than or equal to $(k - s + 1)$. Theorem 3 follows directly from this assertion as a corollary.

Received 16 March 1942

Characteristic Cycles

Doklady Akademii Nauk SSSR (1945) 47(4): 246–49

Whitney, in his papers [1, 2], introduced the concept of a fibre-bundle and studied mainly the case where fibres are spheres. My paper [3] also deals with a fibre-bundle for a case where a bundle consists of tangents to an oriented manifold. My technique is stronger than the Whitney method in that I succeeded in constructing more characteristic cycles than Whitney did. In the present note I demonstrate that my method is applicable to the general case as well.

Definitions 1 and 2, due to Whitney, will be supplemented by several remarks.

Definition 1. Let A be a topological space with a given topological group Γ of homeomorphisms, B a complex of simplices $T_1, T_2, \ldots$, and P a topological space. The space P is called a *fibre-bundle* or *skew product*, $P = P(A, B)$ of A and B, if:

(1) to every point $y \in B$, there corresponds some subset $A_y \subset P$; to distinct points y and z in B there correspond disjoint subsets A_y and A_z;

(2) to each simplex T_α, there corresponds some homeomorphic mapping ξ_α of the topological product $A \cdot T_\alpha$, i.e., a function $\xi_\alpha(x, y) = \xi_{\alpha y}(x)$ $(x \in A, y \in T)$ is defined so that $\xi_{\alpha y}$ is the homeomorphic mapping of the space A onto A_y;

(3) if $y \in T_\alpha \cap T_\beta$, then $\xi_{\alpha y}^{-1}\xi_{\beta y} = \eta_{\alpha\beta y} \in \Gamma$ and $\eta_{\alpha\beta y}$ is a continuous mapping of $T_\alpha \cap T_\beta$ into the group Γ.

Evidently, if $y \in T_\alpha \cap T_\beta \cap T_\gamma$, then

$$\eta_{\alpha\beta y}\eta_{\beta\gamma y} = \eta_{\alpha\gamma y}. \tag{1}$$

Assume that all the simplices $T_1, T_2, \ldots$ are oriented and let $\epsilon_{\alpha\beta}$ denote the coefficient with which the simplex T_β enters into the boundary of the simplex T_α. If $\epsilon_{\alpha\beta} \neq 0$, take $\eta_{\alpha\beta y} = \zeta_{\alpha\beta y}$. Let $\epsilon_{\alpha\beta} \neq 0$, $\epsilon_{\alpha\gamma} \neq 0$, $\epsilon_{\beta\delta} \neq 0$, $\epsilon_{\gamma\delta} \neq 0$; then for $y \in T_\delta$, from (1) it follows that

$$\zeta_{\alpha\beta y}\zeta_{\beta\delta y} = \zeta_{\alpha y\gamma}\zeta_{\gamma\delta y}. \tag{2}$$

If we put $\Delta_\Gamma T_\alpha = \Sigma_\beta\, \epsilon_{\alpha\beta}\zeta_{\alpha\beta y} T_\beta$, then (2) can be rewritten as

$$\Delta_\Gamma \Delta_\Gamma T_\alpha = 0.$$

It is a simple matter to verify that $\eta_{\alpha\beta y}$ satisfying (1) is uniquely determined by $\zeta_{\alpha\beta y}$ satisfying (2). The following existence theorem is easily proved.

(A) Let A be a topological space with a given topological group Γ of homeomorphisms, B a complex consisting of the simplices $T_1, T_2, \ldots,$ where $\Delta T_\alpha = \Sigma_\beta\, \epsilon_{\alpha\beta} T_\beta$ and $\zeta_{\alpha\beta y} \in \Gamma$ are continuous functions of $y \in T_\beta$ satisfying (2), defined for α, β for which $\epsilon_{\alpha\beta} \neq 0$. Then there exists a skew product $P(A, B)$ for which $\xi_{\alpha y}^{-1}\xi_{\beta y} = \zeta_{\alpha\beta y}$.

Definition 2. Let $P(A, B)$ and $P'(A, B)$ be two skew products with the same group Γ. A homeomorphic mapping f of the space P' onto the space P is called an *isomorphic mapping* of $P'(A, B)$ onto $P(A, B)$, if $f(A'_y) = A_y$ and for $y \in T_\alpha$, $\xi_{\alpha y}^{-1} f \xi'_{\alpha y} = \tau_{\alpha y} \in \Gamma$.

From the last assertion, it follows that for $y \in T_\beta$,

$$\zeta_{\alpha\beta y} - \tau_{\alpha y}^{-1}\zeta'_{\alpha\beta y}\tau_{\beta y}. \tag{3}$$

The following theorem is easily proved.

(B) Two skew products $P(A, B)$ and $P'(A, B)$ with defining functions $\zeta_{\alpha\beta y}$ and $\zeta'_{\alpha\beta y}$ are isomorphic, if and only if there exist continuous functions $\tau_{\alpha y} \in \Gamma$ of $y \in T_\alpha$ satisfying the conditions (3).

Whitney noted that if $P(A, B)$ is a skew product and f is a continuous mapping of some complex C into B, there exists a skew product $P(A, C, f)$ of A and C with the same group Γ as the initial product $P(A, B)$, where $P(A, C, f)$ is defined uniquely up to an isomorphism. It is a simple matter to demonstrate that if two mappings f and g are equivalent, then $P(A, C, f)$

and $P(A, C, g)$ are isomorphic. But, often isomorphism of the products $P(A, C, f)$ and $P(A, C, g)$ implies equivalence of the mappings f and g. Thus we arrive at the following

Definition 3. A skew product $P(A, B)$ with the group Γ is called *n-dimensionally universal* for the space A with the group Γ, provided:

(1) the skew product $P(A, C)$ with the group Γ, where C is of a dimension not greater n, is always isomorphic to $P(A, C, f)$, where f is a continuous mapping of C into B;

(2) two skew products $P(A, C, f)$ and $P(A, C, g)$ are isomorphic, if and only if the mappings f and g of C into B are equivalent.

Definition 4. Using the notation of Definition 3, assume that B is a b-dimensional oriented manifold and let z denote some $(b-r)$-dimensional cycle in B. Let f be a continuous mapping of C into B, and T some r-dimensional oriented simplex in C. Let $u_z^r(T)$ denote the index of intersection of $f(T)$ with z. We can easily show that u_z^r is an r-dimensional ∇-cycle in C, whose homology class is defined by the homology class of the cycle z and the class of the mapping f. Thus, since $P(A, B)$ is universal, the ∇-cycle u_z^r is an invariant of the skew product $P(A, C, f)$. The cycle u_z^r is called a *characteristic cycle* of the skew product $P(A, C, f)$.

Let $H_l = H(k, l)$ be the manifold of all k-dimensional oriented planes of $(k+l)$-dimensional Euclidean space R^{k+l} that pass through the origin of coordinates in R^{k+l}. Let A be some k-dimensional oriented Euclidean space R^k, and let Γ be the group of all isometric mappings of R^k onto itself that preserve the origin and orientation in R^k. We now define the skew product $P(R^k, H_l)$: a point in the space P is a pair (R_y^k, x), where $R_y^k \in H_l$, and $x \in R_y^k$. Then the set of all pairs (R_y^k, x) with a fixed R_y^k is a fibre A_y. Whitney noted that $P(R^k, H_2)$ satisfies the condition 1 of Definition 3 when $l = n$. In Theorem 1 we give another proof of this assertion, demonstrating at the same time that $P(R^k, H_2)$ is universal.

Theorem 1. The skew product $P(R^k, H_2)$ is an $(l-1)$-dimensional universal product for the space R^k with the rotation group Γ.

Proof. We show that $P(R^k, H_l)$ is universal for sufficiently large l; the dimension can be lowered without any difficulty. Let C be a complex consisting of the simplices $T_1, T_2, \ldots$, and C^r a subcomplex of C that consists of all simplices of dimension not greater r. Let $P(R^k, C)$ be some skew product with defining functions $\zeta_{\alpha\beta y} \in \Gamma$, $y \in T_\beta$ (see (A)). We shall construct a continuous mapping f of the complex C into H_l, such that $P(R^k, C, f)$ is isomorphic to $P(R^k, C)$, $f(y) = \bar{R}^k_y \in H_l$. We shall also construct an isometric mapping $\bar{\xi}_{\alpha y}$, $y \in T_\alpha$ of the space R^k onto the space $\bar{R}^k_y$ so that $\bar{\xi}^{-1}_{\alpha y}\bar{\xi}_{\beta y} = \zeta_{\alpha\beta y}$. Suppose that we have already constructed f and $\bar{\xi}_{\alpha y}$ for C^r, $f(y) = \bar{R}^k_y \subset R^{k+l}$ for $y \in C^r$. We assume that $R^{k+l} \subset R^{2k+l}$ and that $\bar{R}^k$ is the orthogonal complement of R^{k+l} in R^{2k+l}. Let T_α be an arbitrary $(r+1)$-dimensional simplex in C with center p. Put $f(p) = \bar{R}^k$, and arbitrarily define an isometric mapping $\bar{\xi}_{\alpha p}$ of the space R^k onto $\bar{R}^k$. Let S_α be the boundary of the simplex T_α, then the point $z \in T_\alpha$ in the simplex T_α can be defined as the pair (y, t), where $y \in S_\alpha$, and t is a real parameter $0 \leqslant t \leqslant \pi/2$, where $(y, 0) = p$, $(y, \pi/2) = y$, $z = (y, t)$. Let T_β be an arbitrary r-dimensional face of the simplex T_α, then $\bar{\xi}_{\beta y}$ is well defined. Let us define the function $\bar{\xi}_{\alpha z}$, $z = (y, t)$, $y \in T_\beta$ as follows:

$$\bar{\xi}_{\alpha z}(x) = \bar{\xi}_{\alpha p}(x)\cos(t) + \bar{\xi}_{\beta y}(\zeta^{-1}_{\alpha\beta y}(x))\sin(t), \quad x \in R^k, \tag{4}$$

and the function f as follows

$$f(z) = \bar{\xi}_{\beta z}(R^k). \tag{5}$$

If y belongs simultaneously to two r-dimensional faces T_β and T_γ of the simplex T_α, putting $T_\delta = T_\beta \cap T_\gamma$, we find that, by virtue of (2), the relation (4) uniquely defines the mapping $\bar{\xi}_{\alpha z}$, irrespective of the choice of T_β or T_γ.

Now let f and g be two continuous mappings of C into H_l, such that the products $P(R^k, C, f)$ and $P(R^k, C, g)$ are isomorphic. This implies that there exists an isometric mapping h_y of the plane $f(y)$ onto the plane $g(y)$ which depends continuously on $y \in C$. Let $R^{k+l} \subset R^{2k+2l}$, $\bar{R}^{k+l}$ the orthogonal complement of R^{k+l} in R^{2k+2l}, and let φ be some isometric mapping of R^{k+l} onto $\bar{R}^{k+l}$. Then φg is a mapping of C into H_{k+2l} isotopic to g, because φ can be realized by continuously rotating R^{k+l} in R^{2k+2l}. If $x \in f(y)$, take

$$\psi_{yt}(x) = \varphi(h_y(x))\cos(t) + x\sin(t), \quad \psi_t(y) = \psi_{ty}(f(y)). \tag{6}$$

Clearly, $\psi_t(y)$ is a k-dimensional plane in R^{2k+2l}; therefore ψ_t is a mapping of C into H_{k+2l}. Thus, from (6), we have $\psi_0 = \varphi g$, $\psi_{\pi/2} = f$, and as g and φg are equivalent, the mappings f and g are equivalent in H_{k+2l}. This completes the proof of Theorem 1.

The homology of the manifold $H(k, l)$ is discussed in detail in [3]. In studying the homotopic properties of $H(k, l)$ the following theorem is helpful.

Theorem 2. For $n \leqslant l-1$, the n-dimensional homotopy group of the manifold $H(k,l)$ is isomorphic to the $(n-1)$-dimensional homotopy group of the manifold Γ_k, where Γ_k is the group of positive rotations of a k-dimensional Euclidean space.

Proof. Just as in Theorem 1, we give the proof for sufficiently large l. Let R_0^k and R_1^k be two orthogonal k-dimensional subspaces of R^{k+l}. Assume that the group Γ_k acts in R_1^k and let φ be some isometric mapping of R_1^k onto R_0^k. Let S^{n-1} be an oriented $(n-1)$-dimensional sphere and ψ_y be an element of the group Γ_k which depends continuously on $y \in S^{n-1}$. For $x \in R_1^k$, suppose that

$$\theta(x,y,t) = \varphi(x)\cos(t) + \psi_y(x)\sin(t); \quad \theta(y,t) = \theta(R_1^k,y,t). \quad (7)$$

Obviously, $\theta(y,t)$ is a k-dimensional plane in R^{k+l}, and $\theta(y,0) = R_0^k$, $\theta(y,\pi/2) = R_1^k$. Thus, $\theta(y,t)$ is a continuous mapping of the direct product $S^{n-1}\cdot T$ into H_l, where T is the interval $0 \leqslant t \leqslant \pi/2$. Under this mapping both the bases $S^{n-1}\cdot 0$ and $S^{n-1}\cdot \pi/2$ of the cylinder $S^{n-1}\cdot T$ are mapped into points, and, consequently, we have a continuous mapping of the sphere S^n into H_l. The correspondence thus established between the mapping of S^{n-1} into Γ_k and the mapping of S^n into H_l gives an isomorphism of the corresponding homotopy groups. The proof proceeds in a manner similar to the proof of Theorem 1.

It should be noted that universal skew products can be easily constructed for any compact Lie group Γ, so $P(R^k, H_2)$ is no exception in this respect.

References

1. H. Whitney, *Bull. Am. Math. Soc.* **43**, 785 (1937).
2. H. Whitney, *Proc. Nation. Acad.* **26**, 148 (1940).
3. L. Pontryagin, *Doklady AN USSR*, **XXXV**, N 2 (1942).

Received 9 February 1945

Rough Systems*

Doklady Akademii Nauk SSSR (1937) 14(5): 247–350

Consider a dynamical system defined by two first order equations:

$$\frac{dx}{dt} = P(x,y); \quad \frac{dy}{dt} = Q(x,y), \tag{A}$$

where x and y are Cartesian coordinates; $P(x,y)$ and $Q(x,y)$ are analytic functions of x,y.

We shall confine ourselves to systems of the type (A) for which there exists a so-called "contactless cycle", i.e., a simple closed curve g with a continuous tangent such that all trajectories passing through the points on this curve intersect it and none of the trajectories touch it. The plane domain inside the curve g is called *domain G*. Without loss of generality, we may assume that the phase trajectories intersecting the curve g enter the domain G with increasing time t.

Along with the system (A), consider a modified system:

$$\frac{dx}{dt} = P(x,y) + p(x,y); \quad \frac{dy}{dt} = Q(x,y) + q(x,y) \tag{B}$$

where $p(x,y)$ and $q(x,y)$ are also analytic functions of x and y.

Definition I. The system (A) is said to be "rough" (in contrast to systems which are "non-rough") in the domain G, if for any $\eta > 0$, we can find

* In collaboration with A. Andronov. (Communicated by Academician L. I. Mandel'shtam on 2 January 1936).

an $\epsilon > 0$, such that for arbitrary analytic functions $p(x,y)$ and $q(x,y)$, satisfying the conditions

$$|p(x,y)| < \epsilon, \quad |q(x,y)| < \epsilon;$$

$$|p'_x(x,y)| < \epsilon, \quad |p'_y(x,y)| < \epsilon, \quad |q'_x(x,y)| < \epsilon, \quad |q'_y(x,y)| < \epsilon,$$

in the domain G, there exists such an one-to-one and bi-continuous transformation T of G into itself, under which

(1) corresponding points are at a distance less than η;

(2) points on the same trajectory of the system (B) correspond to points on the same trajectory of the system (A), and vice versa.*

Necessary conditions for the system (A) to be rough in the domain G may be formulated in the following three theorems.

Theorem 1. If the system (A) is rough in the domain G, then in the domain G the system (A) can only have equilibrium states for which the real parts of the roots of the characteristic equation are nonzero.

In other words, in the domain G the system (A) cannot have an equilibrium state $x = x_0, y = y_0$, for which

(a) $\Delta = \begin{vmatrix} P'_x(x_0,y_0) P'_y(x_0,y_0) \\ Q'_x(x_0,y_0) Q'_y(x_0,y_0) \end{vmatrix} = 0,$ or

(b) for $\Delta > 0, \sigma = -[P'_x(x_0,y_0) + Q'_y(x_0,y_0)] = 0.$

---

* This definition of roughness of a system may be regarded as the definition of stability of a set of trajectories of a dynamical system with respect to sufficiently small changes in the right-hand sides of Eq. (A). This type of stability is of great interest in physics. In particular, the condition of stability for periodic motion with respect to sufficiently small changes in the right-hand sides of the system (A) was postulated and used by Andronov [1] to prove that only limit cycles can describe real auto-oscillating processes in systems of the type (A).

In this connection it should be mentioned that for the motion of a point on trajectories, Bohl [2] gave a different definition of stability of a dynamical system with respect to small changes in the right-hand sides of the equations. Such stability, in the sense of Bohl, imposes rather stringent conditions on the initial system. For instance, an initial system of the type (A) may be stable in the sense of Bohl, only if the domain G contains only one equilibrium state and only if all other motions tend to this equilibrium state when $t \to +\infty$.

Theorem 2. If the system (A) is rough in the domain G, then in the domain G the system (A) can only have periodic motions for which the characteristic exponent is not zero.

In other words, in the domain G the system (A) cannot have a periodic motion $x = \varphi(t), y = \psi(t)[\varphi(t+\tau) = \varphi(t); \psi(t+\tau) = \psi(t)]$ for which

$$h = \frac{1}{\tau}\int_0^\tau [P'_x(\varphi, \psi) + Q'_y(\varphi, \psi)]\, dt = 0.$$

Theorem 3. If the system (A) is rough in the domain G, then in the domain G the system (A) can only have separatrices which do not join one saddle point to another.

In other words, in the domain G the system (A) cannot have separatrices which join one saddle point to another.

A system (A) is said to satisfy the conditions Γ in the domain G, if inside a contactless cycle g the system (A) has

(1) only equilibrium states for which $\Delta \neq 0$, and if $\Delta > 0$, then $\sigma \neq 0$;

(2) only periodic trajectories for which $h \neq 0$, and

(3) only separatrices which do not join one saddle point to another.

By virtue of the results obtained by Bendixson [3], a system (A) satisfying the conditions Γ in the domain G can have in the domain G only trajectories of the following eleven types:

I.	Equilibrium states:	Nodes (foci) $[\Delta > 0, \sigma \neq 0]$.	1
		Saddle points $[\Delta < 0]$.	2
II.	Limit cycles	$[h \neq 0]$	3
III.	Separatrices:	Leaving a node (focus) or entering a node (focus).	4
		Unwinding from a limit cycle or tending to a limit cycle.	5
		Entering into the domain G.	6
IV.	Trajectories whose limiting cycles are either nodes (foci) or limiting cycles inside the domain G:	Leaving a node (focus) and entering a node (focus).	7
		Unwinding from a limit cycle and tending to a limit cycle.	8
		Leaving a node (focus) and tending to a limit cycle (or vice versa).	9

V.	Trajectories entering the domain G that are not separatrices:	Tending to a node (focus).	10
		Tending to a limit cycle.	11

Definition II. A trajectory is said to be positively (negatively) stable in the sense of Lyapunov in the domain G [4], if, for any point M_0 ($M_0 \subset G$), the corresponding motion $M = \varphi(t - t_0, M_0)$ satisfies the following two conditions:

(1) for all $t > t_0$ $[t < t_0]$, $M \subset G$;

(2) for any $\epsilon > 0$, a $\delta(\epsilon, M_0)$ can be found such that for all $t > t_0$ $[t < t_0]$, $\rho[\varphi(t - t_0, M_0), \varphi(t - t_0, M_0')] < \epsilon$ as soon as $\rho[M_0, M_0'] < \delta$.*

A trajectory is called positively (negatively) unstable in the sense of Lyapunov† in the domain G, if it satisfies the condition 1 but does not satisfy the condition 2.

Definition III. A trajectory is said to be singular, if it is positively or negatively unstable in the sense of Lyapunov.

A trajectory is said to be ordinary, if it is

(1) either positively or negatively stable in the sense of Lyapunov,

(2) or positively stable in the sense of Lyapunov and leaves the domain G when $t \to -\infty$.

Theorem IV. For a system (A) satisfying the conditions Γ in the domain G, trajectories of the type I, II, III are singular and all other trajectories are ordinary.

Theorem V. In a dynamical system (A) for which the conditions Γ are satisfied in the domain G, the totality of singular trajectories subdivides the domain G into a finite number of connected components ("cells") covered by ordinary trajectories.‡ These components fall into two classes: a class of components adjacent to a contactless cycle g and a class of internal components. In each internal component every trajectory is positively or negatively stable in the sense of Lyapunov; the boundary of each such component contains a singular trajectory positively stable in the sense of Lyapunov which is an "attraction element" or a "sink", and a

---

* Here ρ denotes Euclidean distance (Ed.).

† For the sake of brevity, we shall henceforth omit the phrase "in the domain G".

‡ Note that there are only a finite number of possible topological types of cells taken together with their boundaries.

singular trajectory negatively stable in the sense of Lyapunov which is a "repulsion element" or a "source".

In each component adjacent to a contactless cycle, every trajectory is positively stable in the sense of Lyapunov; the boundary of each such component contains a singular trajectory positively stable in the sense of Lyapunov which is a "sink".

Let there be given two dynamical systems (A_1) and (A_2) of the type (A), and accordingly two partitions of the phase plane into the trajectories S_1 and S_2. Assume that the conditions Γ are satisfied in the domains G_1 and G_2.

We say the partitions of the phase plane into trajectories S_1 and S_2 are of identical qualitative structure in the domains G_1 and G_2, if there exists a one-to-one and bicontinuous transformation T of the domain G_1 into the domain G_2, for which points of the domain G_2 lying on the same trajectory in the partition S_2 correspond to points of the domain G_1 on the same trajectory in the partition S_1, and vice versa.

By virtue of the theorem stated below, we can assert that in a system of the type (A), for which the conditions Γ are satisfied in the domain G, the qualitative structure of the partitions of these domains into trajectories is determined by the singular trajectories, the direction of motion on closed singular trajectories and by the type of stability of the limiting singular trajectories.

Theorem VI. The qualitative structures of the partitions S_1 and S_2, for which the conditions Γ are satisfied inside the corresponding contactless cycles, are identical in the domains G_1 and G_2, if and only if there exists a one-to-one and bicontinuous transformation T_1 of the domain G_1 into the domain G_2, under which points of G_2 lying on a singular trajectory of partition S_2 correspond to points of G_2 lying on a singular trajectory of partition S_1, and vice versa, and when

(1) to all sources and sinks correspond sources and sinks (or vice versa sinks and sources);

(2) the direction of rotation (in time) over all closed singular trajectories is conserved (or reversed).

Under the conditions of this theorem, points (1) and (2) can be reformulated as follows:

In each component of the partition S_1 there exists an ordinary trajectory whose points under the transformation T_1 are mapped into the points of an ordinary trajectory in the corresponding component of the partition S_2, and vice versa.

Sufficient conditions for the system (A) to be rough in the domain G are formulated in the following:

Theorem VII. If, in the domain G, the system (A) satisfies the conditions Γ, the system is then rough in the domain G.

Evidently, rough systems do exist.

Theorems IV, V, and VI relating to a system of the type (A) satisfying the conditions Γ in the domain G may be formulated for systems of the type (A) rough in the corresponding domains G.

References

1. A. Andronow, *C. R.* **189**, 559 (1929).
2. P. Bohl, *J. f. reine u. angew. Mathem.* **144**, 284 (1914).
3. I. Bendixson, *Acta Mathem.* **24**, 1 (1901).
4. A. Markoff, *Math. ZS.* **36**, 708 (1933).

Institute of Physics and
Institute of Mathematics,
Moscow State University.

Received 2 January 1937

Homotopic Classification of the Mappings of an $(n+2)$-Dimensional Sphere into an n-Dimensional Sphere

Doklady Akademii Nauk SSSR (1950) 70(6): 957–59

I have already published a note [1] on this subject. However a mistake was made, which led to an erroneous result. Using the same method as in [1], but correcting the mistake, I arrive at the following result.

Theorem. For $n \geqslant 2$, the number of classes of mappings of an $(n+2)$-dimensional sphere S^{n+2} into an n-dimensional sphere S^n is two.

Proof. Since there are no more than two classes of mappings of S^{n+2} into S^n, I only prove that the number of these classes is not less than two.

(a) We assume that the spheres S^n and S^{n+k} (starting from c) we take $k=2$), are ordinary metric spheres in a Euclidean space. Continuous mappings and deformations can be approximated by analytic mappings; in the sequel we shall only deal with such mappings. Under an analytic mapping of some manifold into S^n, a point of the manifold is called *proper*, if the rank of the functional matrix of this mapping at this point is equal to n. We interpret the deformation f_t, $0 \leqslant t \leqslant 1$ as a mapping F of the topological product $T = S^{n+2} \times I$ into S^n, where I is the segment $0 \leqslant t \leqslant 1$. Let $p \in S^n$, and take $f_t^{-1}(p) = P_t$ and $F^{-1}(p) = Q$. The analyticity of the mappings f_0, f_1, and F imply the existence of a point p for which the sets P_0, P_1, and Q consist of only proper points, and are, in particular, analytic manifolds (Q has the boundary $P_0 \times O \cup P_1 \times I$). Let $y = (x, t) \in Q$ $(x \in S^{n+2}, t \in I)$ and $\theta(y) = t$, where $\theta(y)$ is an analytic function on Q. We can modify the deformation f_t, without changing this deformation at $t = 0$ and 1, so that all the critical points of the function $\theta(y)$ are non-degenerate.

(b) Let R^n denote the tangent to S^n at the point p, and $d_1, \ldots, d_n$ some linearly independent sequence of vectors in R^n. Let t be a fixed *noncritical* value of the function $\theta(y)$, R^{n+k} the tangent to the sphere S^{n+k} at the point $x \in P_t$, and N_x^n the normal in R^{n+k} to P_t at the point x. Since (x, t) is a proper point of the mapping F, and t is a non-critical value of the function $\theta(y)$, then x is a proper point of the mapping f_t and the corresponding linear mapping g of the space N_x^n into R^n is non-degenerate. Put $g^{-1}(d_i) = e_i(x)$ $(i = 1, \ldots, n)$.

(c) Let M denote the manifold of all sequences $x; e_0, e_1, \ldots, e_n$, where $x \in S^{n+2}$ and $e_0, e_1, \ldots, e_n$ are linearly independent vectors tangent to S^{n+2} at the point x. The one-dimensional Betti group of the manifold M contains only two elements: 0 and 1.

(d) Let $k = 2$ and t be a fixed non-critical value of the function $\theta(y)$ (see (a)); then P_t is a closed orientable (possibly, disconnected) surface. A finite set of pairwise disjoint simple closed oriented smooth curves is called a *curve on P_t*. The number of components of a curve C is denoted by $\nu(C)$. Let $x \in C$ and $e_0(x)$ be the unit vector tangent to the curve C at the point x. Put $\omega(x) = \{x; e_0(x), e_1(x), \ldots, e_n(x)\}$ (see (b)). This defines a mapping ω of the curve C into the manifold M (see (c)) and with it the homology class $\varphi(C)$ of the curve $\omega(C)$ in M is defined, i.e., 0 or 1 (see (c)). Put

$$\gamma(C) = \nu(C) + \varphi(C) \pmod 2. \tag{1}$$

Clearly, when the curve C is deformed, the functions $\nu(C)$ and $\varphi(C)$, consequently $\gamma(C)$, do not change. It is even true that $\gamma(C)$ is determined by the homology class c modulo 2 of the curve C on P_t: $\gamma(C) = \gamma(c)$. Furthermore, if c_1 and c_2 are two homology classes, then

$$\gamma(c_1 + c_2) = \gamma(c_1) + \gamma(c_2) + I(c_1, c_2), \tag{2}$$

where $I(c_1, c_2)$ is the intersection index (mod 2) of the classes c_1 and c_2.*

To prove this, let us define the operation of addition of curves, the result of which is again a curve. Let C_1 and C_2 be two curves on P_t in general position, i.e., they only have a finite number of intersection points and are not tangent at any of these points. The sum of the curves $C_3 = C_1 + C_2$ differs from the set-theoretic sum $C_1 \cup C_2$ of two oriented curves C_1 and C_2 only in the neighbourhood of their intersection. Figure 1 shows how we can pass from the union $C_1 \cup C_2$ to the sum $C_3 = C_1 + C_2$ in the neighbourhood of an intersection point.

* The mistake committed in [1] is that the term $I(c_1, c_2)$ was omitted in (2).

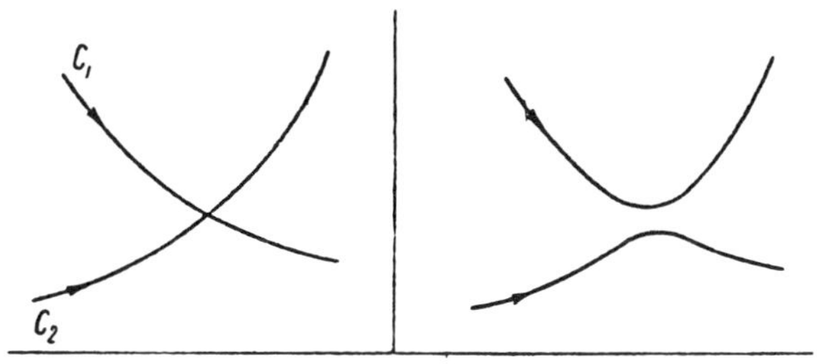

Fig. 1.

Evidently, the homology classes (mod 2) c_1, c_2, c_3 of the curves C_1, C_2, C_3 are connected by the relation

$$c_3 = c_1 + c_2. \tag{3}$$

Furthermore, it is readily seen that

$$\varphi(C_1) + \varphi(C_2) = \varphi(C_1 + C_2),$$

$$\nu(C_1) + \nu(C_2) = \nu(C_1 + C_2) + I(C_1 + C_2) \quad (\text{mod } 2).$$

Hence, from (1), we get

$$\gamma(C_1 + C_2) = \gamma(C_1) + \gamma(C_2) + I(C_1, C_2). \tag{4}$$

It now remains to demonstrate the homological invariance of the function $\gamma(C)$. Suppose that $C_1 \sim C_2$ (mod 2), then the last term in (4) vanishes and so it suffices to prove that $\gamma(C_3) = 0$ if $C_3 \sim 0$ (mod 2) (see (3)). From $C_3 \sim 0$ (mod 2) it readily follows that C_3 is a level curve of some smooth real-valued function $\zeta(x)$ defined on P_t, all of whose critical points are non-degenerate and whose values at different critical points are distinct. Let us denote by $C(s)$ the level curve of this function that corresponds to a non-critical value s. It is a simple matter to prove that $\gamma(C(s))$ does not depend on s. For this, it suffices to examine the passage of the parameter s through a critical value. When the parameter s passes through a critical value we find that the functions $\nu(C(s))$ and $\varphi(C(s))$ both change by one, so that $\gamma(C(s))$ does not at all change. Since, for sufficiently large s, the set $C(s)$ is empty, we have $\gamma(C_3) = \gamma(C(s)) = 0$.

(e) Let $2q$ denote the rank of the one-dimensional Betti group B (mod 2) of the surface P_t. Let us now examine the question: Does there exist in B a linearly independent system of elements $c_1, \ldots, c_q$ satisfying the conditions:

$$I(c_i, c_j) = 0, \quad \gamma(c_i) = 0 \quad (i, j = 1, \ldots, q).$$

Let us define a function $\Gamma(f_t)$ as follows: $\Gamma(f_t) = 0$ if such a system exists and $\Gamma(f_t) = 1$ if such a system does not exist. This function is defined only for non-critical values of t; it is readily verified that this function does not change when passing through critical values. And it is obvious that it remains constant on other intervals of variation of t. Further, $\Gamma(f_t) = 0$, if the surface P_t is homeomorphic to a sphere; therefore for the mapping f_0, which is homotopic to zero, we have $\Gamma(f_0) = 0$. Thus, Γ is an invariant of the homotopy class and it is zero for the zero class.

(f) We now construct a mapping $f = f_0$ of the sphere S^{n+2} into S^n for which $\Gamma(f) = 1$. For this purpose, it is sufficient to give a surface $P = P_0$ in S^{n+2} and to define at each point $x \in P$ a sequence of linearly independent vectors $e_1(x), \ldots, e_n(x)$ orthogonal to P such that the above construction gives $\Gamma = 1$. We shall take the sphere S^{n+2} to be a Euclidean space R^{n+2} with a point at infinity. In R^{n+2} choose a subspace R^3 and assume that P is an ordinary metric torus in R^3 with the usual cyclic coordinates α and β. Let $u_1(x)$ be the unit normal to P at $x = (\alpha, \beta) \in P$ in R^3 and $u_2, \ldots, u_n$ an orthonormal system of vectors in R^{n+2} orthogonal to R^3. Put

$$e_1(x) = u_1(x) \cos(\alpha - \beta) - u_2 \sin(\alpha - \beta),$$

$$e_2(x) = u_1(x) \sin(\alpha - \beta) + u_2 \cos(\alpha - \beta),$$

$$e_i(x) = u_i \quad (i = 3, \ldots, 4).$$

It is readily verified that $\Gamma = 1$.

References

1. L. S. Pontryagin, *Doklady AN USSR* **19**, N 5, 957 (1938).
2. H. Freudenthal, *Comp. Math.* **5**, 299 (1937).

Received 4 January 1950

Israil Moiseevich Gelfand

Born – 2 Sept 1913 in Krasnye Okny, Odessa, Ukraine

Israil Gelfand went to Moscow at the age of 16, in 1930, before completing his secondary education. In 1932 Gelfand was admitted as a research student under Kolmogorov's supervision. His work was in functional analysis and he was fortunate to be in a strong school of functional analysis so he received much support from other mathematicians such as A. E. Plessner and I. A. Liusternik. Gelfand presented his thesis *Abstract functions and linear operators* in 1935 which contains important results, but is perhaps even more important for the methods that he used, studying functions on normed spaces by applying linear functional to them and using classical analysis to study the resulting functions.

Gelfand's next major achievement was the theory of commutative normed rings that he created and studied in his D.Sc. thesis submitted in 1938. During the time that he was carrying out this work for his D.Sc., Gelfand taught at the Academy of Science of the USSR. He held a post at the Academy from 1935 until 1941 when he was appointed as professor at Moscow State University. In joint work with Naimark in the early 1940s, Gelfand worked on non-commutative normed rings with an involution. They showed that these rings could always be represented as ring of linear operators on a Hilbert space.

One important area which he started work on in the early 1940s was the theory of representations of non-compact groups. This work followed on from the representation theory of finite groups by Frobienius and Schur and the representation theory of compact groups by Weyl. Another important area of his work is that on differential equations where he worked on the inverse Sturm–Liouville problem. He saw the importance of the work of Sobolev and Schwartz on the theory of generalised functions and distributions, and he developed this theory in a series of monographs. He worked on computational mathematics, developing general methods for solving the equations of mathematical physics by numerical means. In this area he also worked on difference operators.

Gelfand's work on group representations led him to study integral geometry (a term due to Blaschke) which in turn he was led to by a study of the Radon transform. Between 1968 and 1972 Gelfand produced a series of important papers on the cohomology of infinite dimensional Lie algebras. From 1958 onwards Gelfand became interested in problems in biology and medicine. In 1960, together with Fomin and other scientists, he set up the Institute of Biological Physics of the Academy of Sciences of the USSR. In particular he became interested in cell biology and also became interested in experimental work as well as theoretical work, which was his first interest.

In 1973 Gelfand was awarded the Order of Lenin for the third time. This is only one of a very large number of honours that have been given to Gelfand over many years of outstanding contributions. He was president of the Moscow Mathematical Society during 1968–1970. He was elected an honorary member of the American National Academy of Science, the American Academy of Science and Arts, the Royal Irish Academy, the American Mathematical Society, and the London Mathematical Society. He has been awarded many honorary doctorates including one from the University of Oxford.

In 1989–1990 Gelfand taught at Harvard University and in 1990 he also taught at Massachusetts Institute of Technology. In that same year, 1990, he emigrated to the United States where he became Distinguished Visiting Professor at Rutgers. He also holds a chair in the departments of mathematics and biology at the Center for Mathematics, Science and Computer Education in the Institute for Discrete Mathematics and Computer Science at Rutgers University.

4.

On normed rings

Dokl. Akad. SSSR **23** (1939) 430–432. Zbl. **21**:294

(Communicated by A. N. Kolmogoroff, Member of the Academy, 27. III. 1939)

Definition 1. The set R of elements $x, y, \ldots$ is called a normed ring, if: a) it is a complete normed space with multiplication by complex numbers; b) together with the elements x and y exists their product xy possessing the usual algebraical properties and continuous at least with respect to one of its factors; c) it contains a unit element e.

It may be shown that in R exists a norm equivalent to a given one and possessing the following properties:

$$\|xy\| \leqslant \|x\| \cdot \|y\|, \quad \|e\| = 1.$$

In what follows we consider only commutative rings.

Definition 2. A set of elements $x, y, \ldots \subset R$ is called an ideal I, if it possesses the following properties: a) if $x \in I$, $y \in I$, then $\alpha x + \beta y \in I$, where α and β are arbitrary complex numbers; b) if $x \in I$ and z is an arbitrary element of the ring R, then $xz \in I$.

The whole ring and the nul ideal, i. e. the ideal consisting of only one element 0, are trivial ideals. Every ideal different from these two we shall call non-trivial.

Definition 3. A non-trivial ideal not contained in any other non-trivial ideal we call a maximal ideal.

Theorem 1. *Every maximal ideal is closed (i. e. the set of elements contained in this ideal is closed in R).*

Theorem 2. *Every non-trivial ideal is contained in a maximal ideal.*

The ring of residues R/I, where I is a closed ideal, may be again made a normed ring. As is known, the elements of R/I are classes X, Y, $\ldots$, consisting of such elements x', x'', $\ldots$ that $x' - x'' \in I$. The norm of the class X is defined in the following way: $\|X\| = \inf |x|$, where $x \in X$. It may be shown that in the ring of residues all axioms of a normed ring are satisfied.

Theorem 3. *The ring of residues to a maximal ideal is the corpus of complex numbers.*

Thus, to every element $x \in R$ and every maximal ideal M there corresponds a complex number $x(M)$, namely the number, which is correlated to x by the homomorphism $R \sim R/M$. Hence follows

Theorem 4. *To every element $x \in R$ there corresponds a function*

$x(M)$ defined on the set $\mathfrak{M}$ of all maximal ideals. To the sum of elements corresponds the sum of functions, to the product of elements — the product of functions.

The function $x(M)$ is bounded. In fact, $|x(M)| \leqslant \|x\|$. It may be shown that $\sup|x(M)| = \lim_{n\to\infty} \sqrt[n]{\|x^n\|}$. Hence we obtain

Theorem 5. *If in the ring R there are no elements different from zero, for which $\sqrt[n]{\|x^n\|} \to 0$, then our ring is isomorphic to the ring of functions defined on the set of maximal ideals.*

We shall show that these functions $x(M)$ are continuous. To this end we must topologize the set $\mathfrak{M}$ of maximal ideals. We shall define the topology by a system of neighbourhoods.

Definition 4. A set K of elements of the ring R is called a set of generators of the ring R, if the smallest closed ring containing K is the whole ring R.

Definition 5. By the neighbourhood of the maximal ideal M_0 we shall mean the set of maximal ideals satisfying the inequalities

$$|x_i(M) - x_i(M_0)| < \varepsilon, \qquad (i = 1, 2, \ldots, n),$$

where ε and n are arbitrary and $x, \ldots, x_n$ are arbitrary elements of the set K of generators of the ring R.

The so obtained topology in $\mathfrak{M}$ does not depend on the choice of K. The set $\mathfrak{M}$ of maximal ideals in this topology proves to be a bicompact Hausdorff space and the functions $x(M)$ are continuous on it. As the result we obtain the following fundamental

Theorem 6. *Every abstract normed ring R may be homomorphically mapped in the ring of continuous functions defined on the Hausdorff bicompact space of maximal ideals of the ring R. For the isomorphism it is necessary and sufficient that the conditions of Theorem 5 should be satisfied.*

In some cases the topological properties of the set $\mathfrak{M}$ may be stated more precisely, namely:

1. If the ring R is separable, then the set $\mathfrak{M}$ is metrisable.
2. If the ring R is generated by a system of n generators, then the set $\mathfrak{M}$ is homeomorphic to a compact subspace of the n-dimensional complex space.

Theorem 7. *If in the ring R for each element x there is a complex-conjugated element, i. e. such an element y that $y(M) = \overline{x(M)}$, whatever be M from $\mathfrak{M}$, then each function continuous on the set $\mathfrak{M}$ is the limit of a uniformly convergent sequence of functions corresponding to the elements of the ring R*.*

Hence it follows that if in the ring R a norm is introduced in such a way that from the uniform convergence of the functions $x_n(M)$ follows the convergence of the norms of the elements x_n (for which it is, for instance, sufficient to demand that $\|x^2\| = \|x\|^2$), then the ring R consists of all functions continuous on $\mathfrak{M}$ (we assume that the conditions of theorem 7 are satisfied in the ring R).

Thus, in many cases the aggregate of functions $x(M)$ uniquely determines the ring. This follows from

* This theorem was proved by the author together with G. Šilov. The proof will be given in their common paper.

T h e o r e m 8. *Let there be given two normed rings R_1 and R_2. Let the intersection of all maximal ideals in each of them be the nul ideal. Then from the algebraic isomorphism of the rings R_1 and R_2 follows their continuous isomorphism.*

The applications of the above results will be given elsewhere. A detailed exposition will be published in the «Recueil Mathématique de Moscou».

Institute of Mathematics.
Moscow State University.

Received
27. III. 1939.

7.

On identities for eigenvalues of a second order differential operator

Usp. Mat. Nauk **11** (1) (1956) 191–198. Zbl. **70**: 83

The paper [1] contains a formula that enables one to compute the sum $\sum_{n=1}^{\infty} (\lambda_n - n^2 - a_0)$, where λ_0 are eigenvalues of the equation

$$-y'' + p(x)y = \lambda y, \qquad y(0) - y(\pi) = 0$$

and $a_0 = \int_0^{\pi} p(x)dx$. For a simple proof of this formula see [2].

This formula is an analogue of the trace formula for differential operators*: it generalizes to differential operators an assertion that the sum of eigenvalues of a matrix is equal to its trace, i.e. to the sum of its diagonal elements. However for a matrix we can compute not only the sum of eigenvalues, but the sum of powers of eigenvalues as well. In this paper we get the similar result for a second order differential operator. Corresponding results were obtained by L.A. Dikij and by myself approximately at the same time; however the proofs are different. We think that both methods are of some interest, so we have decided to publish both proofs. In essence each proof depends on some summation method, or as one usually says in physics, on some method of regularization of the sum

$$\sum_{n=1}^{\infty} \lambda_n^k. \tag{1}$$

The regularization method accepted in this paper may be called "the Hadamard regularization". This method is reduced to the subtraction of some terms from λ_n^k so that the series (1) turns into a convergent one. (Hadamard used similar regularization for the computation of divergent integrals, see [4]). The regularization method used by Dikij, is similar to M. Riesz's method [5] for the computation of divergent integrals; it is reduced to the analytic continuation of $\sum \lambda_n^{-s}$ as a function on s.

1. Let us consider a differential equation

$$-y'' + p(x)y = \lambda y \tag{2}$$

---

* One should mention here an interesting paper by M.G. Krein [3], where he obtained some relations from traces in a class of operators in the Hilbert space. It would be interesting to study the relation of our results with those of M.G. Krein.

with boundary conditions $y(0)=y(\pi)=0$. The function $p(x)$ is supposed to be smooth. To make the formulae simpler we assume that in some neighbourhoods of the points $x=0$ and $x=\pi$ the function $p(x)$ is identically zero (one could assume instead that $p(x)$ and the boundary conditions are periodic). In the general case the final formulae are somewhat more complicated. The symbolic form of the equation (2) is

$$(-D^2+p)y=\lambda y$$

where D is the differentiation operator and p is the operator of multiplication by $p(x)$. Let us consider now the operator $(-D^2+p+\zeta E)^{-1}$. This is an integral operator whose kernel is the Green's function of the operator $-y''+p(x)y+\zeta y$, $y(0)=y(\pi)=0$ and eigenvalues are $(\lambda_n+\zeta)^{-1}$, λ_n being eigenvalues of the equation (2). The trace of any integral operator equals the sum of its eigenvalues, so

$$\operatorname{Sp}(-D^2+p+\zeta E)^{-1}=\sum_{n=1}^{\infty}\frac{1}{\lambda_n+\zeta}. \tag{3}$$

We will find asymptotic expansion of the right-hand side and left-hand side of this equality in powers of ζ^{-1}. These series contain integral and half-integral powers of ζ^{-1}. Half-integral powers are related to asymptotics of eigenvalues, and integral powers give the desired identities.

2. Let us consider first the right hand side of (3). It is known that eigenvalues λ_n of (2) have the following asymptotics:

$$\lambda_n=n^2+c_0+\frac{c_2}{n^2}+\dots. \tag{4}$$

Let us write down $(\lambda_n+\zeta)^{-1}$ as an asymptotic series in ζ

$$(\lambda_n+\zeta)^{-1}=\sum_{k=0}^{\infty}(-1)^k\frac{\lambda_n^k}{\zeta^{k+1}}. \tag{5}$$

With this formula we cannot yet write down asymptotic expansion of $\sum_{n=1}^{\infty}(\lambda_n+\zeta)^{-1}$ in powers of ζ^{-1}, because the series $\sum_{n=1}^{\infty}\lambda_n^k$ is divergent. It appears however that the final expression for the asymptotics of $\sum_{n=1}^{\infty}(\lambda_n+\zeta)^{-1}$ is similar to the one obtained by formal summation of equalities (5) for all n. Namely, integral powers of ζ^{-1} in the asymptotic series for $\sum_{n=1}^{\infty}(\lambda_n+\zeta)^{-1}$ are multiplied by coefficients that are equal to regularized sums $\sum_{n=1}^{\infty}\lambda_n^k$ (see below). The final expression contains also half-integral powers $\zeta^{-n+1/2}$; one can see that these powers are closely related to the regularization. Let us give precise statements.

The regularized sum $\sum_{n=1}^{\infty}\lambda_n^k$ (k is fixed) with λ_n having asymptotics (4), is the sum of terms that are obtained from λ_n^k by the subtraction of the divergent terms of asymptotics. For simplicity we will substract also half of the constant term of the asymptotic expansion. The regularized sum will be denoted by $\sum_{n=1}^{\infty}{}' \lambda_n^k$. So, for example,

$$\sum_{n=1}^{\infty}{}' \lambda_n = \sum_{n=1}^{\infty} (\lambda_n - n^2 - c_0) - \tfrac{1}{2} c_0,$$

$$\sum_{n=1}^{\infty}{}' \lambda_n^2 = \sum_{n=1}^{\infty} (\lambda_n^2 - n^4 - 2n^2 c_0 - 2c_1 - c_0^2) - \frac{2c_2 + c_0^2}{2}.$$

To describe half-integral coefficients of the expansion we will consider the inversion of the asymptotic series (4). Namely, if $\lambda_n = n^2 + c_0 + \frac{c_2}{n^2} + \ldots \equiv \varphi(n^2)$, where $\varphi(n^2)$ denotes this asymptotic series, then its inversion gives

$$n^2 = \psi(\lambda) \equiv \lambda + b_0 + \frac{b_1}{\lambda} + \ldots. \tag{6}$$

Theorem 1. *Let λ_n have asymptotic expansion (4). Then the sum $\sum_{n=1}^{\infty} (\lambda_n + \zeta)^{-1}$ has the following asymptotic expansion in powers of ζ^{-1}:*

$$\sum_{n=1}^{\infty} (\lambda_n + \zeta)^{-1} = \sum_{k=0}^{\infty} (-1)^k \frac{A_k}{\zeta^{k+1}} + \pi \frac{d}{d\zeta} \sqrt{-\psi(-\zeta)} \tag{7}$$

where the asymptotic power $\psi(\zeta)$ is defined by (6), and coefficients A_k are given by

$$A_k = \sum_{n=1}^{\infty}{}' \lambda_n^k.$$

Proof. Let us first substitute λ_n into its approximate asymptotic expression:

$$\tilde{\lambda}_n = \tilde{\varphi}(n^2) = n^2 + c_0 + \frac{c_2}{n^2} + \ldots + \frac{c_{2m}}{n^{2m}}.$$

Later m will tend to infinity. It is clear that when m increases, the inverse asymptotic series $\tilde{\psi}(\lambda)$ coincides with the asymptotic series $\psi(\lambda)$ up to terms of higher and higher order. Let us decompose the rational function in n^2

$$\frac{1}{n^2 + c_0 + \frac{c_2}{n^2} + \ldots + \frac{c_{2m}}{n^{2m}} + \zeta}$$

into elementary fractions:

$$\frac{1}{\tilde{\lambda}_n + \zeta} = \sum_{i=0}^{m} \frac{a_i(\zeta)}{n^2 - \alpha_i(\zeta)} \tag{8}$$

where $\alpha_i(\zeta)$ are roots of the equation

$$x^{m+1} + (c_0 + \zeta) x^m + c_2 x^{m-1} + \ldots + c_{2m} = 0. \tag{9}$$

One can easily see that the roots of the equation (9) have the following behavior for $\zeta\to\infty$: one of the roots, say $\alpha_0(\zeta)$ tends to infinity and has the asymptotic behavior

$$\alpha_0(\zeta)=-\zeta+d_0+\frac{d_1}{\zeta}+\dots$$

and all other roots tend to zero, and their asymptotic expansions contain only negative (not necessary integral) powers of ζ. It is clear that $\alpha_0(\zeta)=\tilde{\psi}(-\zeta)$. Let us note also that $a_0(\zeta)$ is the residue of the function

$$\frac{1}{x+c_0+\frac{c_2}{x}+\dots+\frac{c_{2m}}{x^m}+\zeta}$$

in the point $x=\tilde{\psi}(-\zeta)$, i.e.

$$a_0(\zeta)=\frac{1}{\tilde{\varphi}'[\tilde{\psi}(-\zeta)]}=\tilde{\psi}'(-\zeta).$$

Therefore

$$\frac{1}{\tilde{\lambda}_n+\zeta}-\frac{\tilde{\psi}'(-\zeta)}{n^2-\tilde{\psi}(-\zeta)}=\sum_{i=1}^{m}\frac{a_i(\zeta)}{n^2-\alpha_i(\zeta)}. \tag{10}$$

As $\alpha_i(\zeta)\to 0$ for $\zeta\to\infty$ $(i=1,2,\dots,m)$, to obtain the asymptotic of the sum in the right hand side of (10), one has to write

$$\frac{1}{n^2-\alpha_i(\zeta)}=\frac{1}{n^2}+\frac{\alpha_i(\zeta)}{n^4}+\frac{\alpha_i^2(\zeta)}{n^6}+\dots$$

so

$$\sum_{n=1}^{\infty}\frac{a_i(\zeta)}{n^2-\alpha_i(\zeta)}=a_i(\zeta)\sum_{n=1}^{\infty}\frac{1}{n^2}+\alpha_i(\zeta)a_i(\zeta)\sum_{n=1}^{\infty}\frac{1}{n^4}+\dots. \tag{11}$$

To get an asymptotic expansion of the expression

$$\sum_{n=1}^{\infty}\sum_{i=1}^{\infty}\frac{a_i(\zeta)}{n^2-\alpha_i(\zeta)}$$

in powers of ζ, one has to sum up m series of the form (11) and collect terms with given powers of ζ. As $\alpha_i(\zeta)$ begins from a negative power of ζ, any given power of ζ appears in only a finite number of terms in the series (11). However, it would be difficult to get asymptotics from this formula. The only statement we need is this:

If we perform the asymptotic expansion in powers of ζ (for $\zeta\to\infty$) of the right hand side (so also of the left hand side) of (10), and sum up over n all coefficients at a fixed power ζ^{-k}, then the corresponding series would be convergent.

Let us obtain now the asymptotic expansion of the left hand side of (10). Let us expand each term in series on ζ^{-1} (bearing in mind that $\psi(-\zeta)=\zeta+\dots$ for large ζ):

$$\frac{1}{\lambda_n+\zeta}=\frac{1}{\zeta}-\frac{\lambda_n}{\zeta^2}+\frac{\lambda_n^2}{\zeta^3}-\dots$$

$$\frac{\psi'(-\zeta)}{n^2-\psi(-\zeta)}=-\frac{\psi'(-\zeta)}{\psi(-\zeta)}+n^2\frac{\psi(-\zeta)}{\psi^2(-\zeta)}+\dots$$

Combining coefficients in ζ^{-n}, we get

$$\frac{1}{\lambda_n+\zeta}-\frac{\psi'(-\zeta)}{n^2-\psi(-\zeta)}=\sum_{k=0}^{\infty}\frac{(-1)^k}{\zeta^{k+1}}(\lambda_n^k-\dots),$$

where points denote combinations of nonnegative powers of n^2. We have proved already that the sum over n of the coefficients at $\frac{1}{\zeta^{k+1}}$ converges. So the series

$$\sum_{n=1}^{\infty}(\lambda_n^k-\dots) \tag{12}$$

converges. Therefore the combinations of nonnegative powers of n^2, denoted earlier by points, are initial terms in the asymptotic expansion of λ_n^k; in fact only after their substraction does the series (12) become convergent. So

$$\sum_{n=1}^{\infty}\left(\frac{1}{\lambda_n+\zeta}-\frac{\psi'(-\zeta)}{n^2-\psi(-\zeta)}\right)\sim\sum_{k=0}^{\infty}\frac{(-1)^k}{\zeta^{k+1}}\left[\sum_{n=1}^{\infty}(\lambda_n^k-\dots)\right],$$

where in $\sum_{n=1}^{\infty}(\lambda_n^k-\dots)$ each summand is λ_n^k minus the divergent asymptotic terms. To find asymptotic expansion of

$$\sum_{n=1}^{\infty}\frac{\psi'(-\zeta)}{n^2-\psi(-\zeta)}$$

we will use the formula

$$\sum_{n=1}^{\infty}\frac{1}{z^2+4\pi^2n^2}=-\frac{1}{2z^2}+\frac{1}{4z}\frac{e^z+1}{e^z-1}.$$

Then

$$\sum_{n=1}^{\infty}\frac{\tilde{\psi}'(-\zeta)}{n^2+[-\tilde{\psi}(-\zeta)]}=\frac{\pi\tilde{\psi}'(-\zeta)}{2\sqrt{-\tilde{\psi}(-\zeta)}}\cdot\frac{e^{2\pi\sqrt{-\tilde{\psi}(-3)}}+1}{e^{2\pi\sqrt{-\tilde{\psi}(-\zeta)}}-1}$$

$$+\frac{\tilde{\psi}'(-\zeta)}{2\psi(-\zeta)}\approx\frac{\pi\tilde{\psi}'(-\zeta)}{2\sqrt{-\tilde{\psi}(-\zeta)}}+\frac{\tilde{\psi}'(-\zeta)}{2\tilde{\psi}(-\zeta)}.$$

The last equality is an asymptotic one up to exponentially decreasing terms of the form $e^{-\zeta}$. Then

$$\sum_{n=1}^{\infty}\frac{1}{\tilde{\lambda}_n+\zeta}\approx\sum_{k}(-1)^k\frac{(\lambda_n^k-\dots)}{\zeta^{k+1}}+\frac{\tilde{\psi}'(-\zeta)}{2\tilde{\psi}(-\zeta)}+\frac{\pi\tilde{\psi}'(-\zeta)}{2\sqrt{-\tilde{\psi}(-\zeta)}}.$$

Expanding $\dfrac{\tilde{\psi}'(-\zeta)}{2\tilde{\psi}\cdot(-\zeta)}$ in powers of ζ we get as coefficients free terms of asymptotics of $\tilde{\lambda}_n^k$ divided by two; these are terms that have to be substracted from the sum as was mentioned before. So

$$\sum_{n=1}^{\infty}\frac{1}{\tilde{\lambda}_n+\zeta}=\sum_{k=0}^{\infty}(-1)^k\frac{\sum_{n=1}^{\infty}{}'\,\lambda_n^k}{\zeta^{k+1}}+\pi\frac{d}{d\zeta}\sqrt{-\tilde{\psi}(-\zeta)}$$

To replace $\tilde{\lambda}_n$ to λ_n we use the fact that the series $\sum\limits_n(\lambda_n^k-\tilde{\lambda}_n^k)$ converges for $k<m$ and $\sum\limits_n(\lambda_n^k-\tilde{\lambda}_n^k)=\sum\limits_n{}'\,\lambda_n^k-\sum\limits_n{}'\,\tilde{\lambda}_n^k$

$$\begin{aligned}\sum_{n=1}^{\infty}\frac{1}{\lambda_n+\zeta}&=\sum_{n=1}^{\infty}\left(\frac{1}{\lambda_n+\zeta}-\frac{1}{\tilde{\lambda}+\zeta}\right)+\sum_{n=1}^{\infty}\frac{1}{\tilde{\lambda}_n+\zeta}\\&=\sum_{k=0}^{m-1}(-1)^k\frac{\sum_n(\lambda_n^k-\tilde{\lambda}_n^k)}{\zeta^{k+1}}+O\left(\frac{1}{\zeta^{m+1}}\right)+\sum_{k=0}^{m-1}(-1)^k\frac{\sum_n{}'\,\tilde{\lambda}_n^k}{\zeta^{k+1}}\\&\quad+\pi\frac{d}{d\zeta}\sqrt{-\tilde{\psi}(-\lambda)}=\sum_{k=0}^{m-1}(-1)^k\frac{\sum_n{}'\,\lambda_n^k}{\zeta^{k+1}}+\pi\frac{d}{d\zeta}\sqrt{-\tilde{\psi}(-\zeta)}+O\left(\frac{1}{\zeta^{m+1}}\right).\end{aligned}$$

As m is arbitrary, we can take $m\to\infty$ thus proving our assertion.

3. Let us express now the left-hand side of the formula (3) in terms of $p(x)$ and its derivatives. Let us assume for the simplicity that $p(x)=0$ in some neighborhoods of the end points of the interval $[0,\pi]$. We have

$$\begin{aligned}(-D^2+p+\zeta E)^{-1}&=(-D^2+\zeta E)^{-1}[E+p(-D^2+\zeta E)^{-1}]^{-1}\\&=(-D^2+\zeta E)^{-1}-(-D^2+\zeta E)^{-1}p(-D^2+\zeta E)^{-1}\\&\quad+(-D^2+\zeta E)^{-1}p(-D^2+\zeta E)^{-1}p(-D^2+\zeta E)^{-1}+\dots.\end{aligned}$$

This is an asymptotic series in ζ (and even convergent in norm topology). We have to find out the asymptotic expansion of each summand. The operator $(-D^2+\zeta E)$ is an integral operator in the interval $(0,\pi)$. Its kernel is the Green's function of the differential operator $-y''+\zeta y$, $y(0)=y(\pi)=0$. It has the form

$$G(x_1,x_2)=\begin{cases}\dfrac{1}{\sqrt{\zeta}\,\operatorname{sh}\pi\sqrt{\zeta}}\operatorname{ch}\sqrt{\zeta}x_1\cdot\operatorname{ch}\sqrt{\zeta}(\pi-x_2), & (x_1<x_2)\\[2ex] \dfrac{1}{\sqrt{\zeta}\,\operatorname{ch}\pi\sqrt{\zeta}}\operatorname{ch}\sqrt{\zeta}(\pi-x_1)\cdot\operatorname{ch}\sqrt{\zeta}x_2, & (x_1>x_2).\end{cases}$$

Therefore this kernel is the sum of $\dfrac{1}{2\sqrt{\zeta}}e^{-\sqrt{\xi}|x_1-x_2|}$ and of some function that decreases exponentially as $\zeta\to\infty$ when x_1 and x_2 lie inside the interval $(0,\pi)$. As we have to integrate the product of the kernel with the function $p(x)$, equal to 0 near endpoints of the interval, the second summand is not essential for the asymptotic behavior. So

$$\mathrm{Sp}(-D^2+p+\zeta E)^{-1}=\mathrm{Sp}(-D^2+\zeta E)^{-1} \\ -\mathrm{Sp}\{(-D^2+\zeta E)^{-1}p(-D^2+\zeta E)^{-1}\}+\dots,$$

$$\mathrm{Sp}(-D^2+\zeta E)^{-4}=\sum_{n=1}^{\infty}\frac{1}{n^2+\zeta}\approx\frac{\pi}{2\sqrt{\zeta}}+\frac{1}{2\zeta},$$

$$\begin{aligned}\mathrm{Sp}\{(-D^2+\zeta E)^{-1}p(-D^2+\zeta E)^{-1}\}&=\frac{1}{4\zeta}\int_0^\pi\int_0^\pi e^{-\sqrt{\zeta}|x_2-x_1|}e^{-\sqrt{\zeta}|x_1-x_2|}p(x_2)dx_1dx_2\\&=\frac{1}{4\zeta}\int_0^\pi p(x_2)\int_0^\pi e^{-2\sqrt{\zeta}|x_2-x_1|}dx_1dx_2\\&=\frac{1}{4\zeta\sqrt{\zeta}}\int_0^\pi p(x)dx.\end{aligned}$$

All other traces can be expanded in the same way

$$\mathrm{Sp}\{(-D^2+\zeta E)^{-1}p(-D^2+\zeta E)^{-1}p\dots p(-D^2+\zeta E)^{-1}\}$$
$$=\frac{1}{2^n(\sqrt{\zeta})^n}\int_0^\pi\dots\int_0^\pi e^{-\sqrt{\zeta}(|x_1-x_2|+|x_2-x_3|+\dots+|x_n-x_1|)}p(x_2)p(x_3)\dots p(x_n)dx_1\dots dx_n.$$

For the asymptotic behavior only the neighbourhood of the point $x_1=\dots=x_n$ is crucial. So we expand $p(x_{i+1})$ in powers of $x_{i+1}-x_1=\xi_i$:

$$p(x_{i+1})=\sum\frac{D^k}{k!}p(x_1)\xi_i^k.$$

Let us change the variables $\sqrt{\zeta}\,\xi_i=\eta_i$. Then the coefficients are

$$\frac{1}{(\sqrt{\zeta})^{k_1+k_2+\dots+k_n+2n-1}}\int\dots\int e^{-(|\eta_1|+|\eta_1-\eta_2|+\dots+|\eta_{n-1}|)}d\eta_1\dots d\eta_{n-1}. \tag{13}$$

This implies that non-zero coefficients may appear only for odd powers of $\sqrt{\zeta}$, because an integral changing sign for $\eta_i\to-\eta_i$ should be equal to zero.

Theorem 2. *Let us consider a differential operator*

$$(-D^2+p(x)+\zeta E)y \qquad \left(D=\frac{d}{dx}\right)$$

with boundary conditions $y(0)=y(\pi)=0$. Let us assume that $p(x)$ is smooth and equal to zero near points $x=0$ and $x=\pi$. Then we have the following asymptotic equality

$$\mathrm{Sp}(-D^2+p+\zeta E)^{-1}=\frac{\pi}{2\sqrt{\zeta}}+\frac{1}{2\zeta}+\sum_{k=1}^{\infty}\frac{c_k}{\zeta^{k+1/2}}. \tag{14}$$

The coefficients c_k are linear combinations of expressions of the form

$$\int_0^x p^{(k_1)}(x)\,p^{(k_2)}(x)\dots p^{(k_n)}(x)\,dx,$$

and

$$k_1+k_2+\dots+k_n+n=2k.$$

Theorem 1 gives the asymptotic expansion of the left hand side of (3), and Theorem 2 gives that of the right hand side of (3). Comparing coefficients for integral powers of ζ, we get the following theorem.

Theorem 3. *Let the function $p(x)$ in the equation $-y''+p(x)y=\lambda y$, $y(0)=y(\pi)=0$, be smooth and equal to zero near $x=0$ and $x=\pi$; then*

$$\sum_{n=1}^{\infty}{}' \lambda_n^k=0 \tag{15}$$

here λ_n are eigenvalues and $\sum'$ is the regularized sum of eigenvalues defined on the page 512.

The above method can be easily generalized to the case when $p(x)$ has singularities in some points, or is not equal to zero near endpoints. The expression in the right hand side of (15) in this case will, in general be nonzero.

References

1. Gelfand, I.M. and Levitan, B.M.: On a simple identity for eigenvalues of second order differential operator. Dokl. Akad. Nauk SSSR **88** (4) (1953) 593–596.
2. Dikij, L.A.: On a formula of Gelfand and Levitan, Usp. Mat. Nauk **8** (2) (1953).
3. Krein, M.G.: On trace formulas in perturbation theory. Mat. Sb. **33** (3) (1953) 597–626.
4. Hadamard, J.: Le problème de Cauchy et les èquations aux dérivées partielles linéaires hyperboliques, Paris, 1932.
5. Riesz, M.: Intégrale de Riemann-Liouville et problèmes de Cauchy, Acta Math. **81** (1949) 1–223.

Received March 24, 1955

Sergei Lvovich Sobolev

Born – 6 Oct 1908 in St Petersburg, Russia
Died – 3 Jan 1989 in Leningrad (now St Petersburg), Russia

Sergei Sobolev's father, Lev Aleksandrovich Sobolev, was an important lawyer and barrister. His mother, Nataliya Georgievna, played an important role in Sobolev's upbringing, particularly after the death of Sobolev's father when Sobolev was 14 years old.

After graduating from high school in 1925, Sobolev entered the Physics and Mathematics Faculty of Leningrad State University where Smirnov who had returned to Leningrad three years earlier quickly spotted his talents. Sobolev became interested in differential equations, a topic that would dominate his research throughout his life, and even at this stage in his career he produced new results which he published.

In 1932 the Steklov Institute of Physics and Mathematics was divided into separate Departments of Mathematics and of Physics. Vinogradov headed the Mathematics Department and invited Sobolev to join the Department. By this time, however, Sobolev had already published a number of profound papers in which he put forward a new method for the solution of an important class of partial differential equations.

Working with Smirnov, Sobolev studied functionally invariant solutions of the wave equation. These methods allowed them to find closed form solutions to the wave equation describing the oscillations of an elastic medium. The methods also led them to a complete theory of Rayleigh surface waves and Sobolev went on to solve problems on diffraction. Sobolev was honoured for this outstanding work by election as Corresponding Member of the Academy of Sciences of the USSR in 1933.

In 1934, the Departments of the Steklov Institute of Physics and Mathematics were split into two independent Institutes, the Steklov Mathematical Institute and the Lebedev Physical Institute. In the same year the Steklov Mathematical Institute was moved from Leningrad to Moscow and Sobolev went with the new Institute to Moscow. By 1935 Sobolev was head of the Department of the Theory of Differential Equations at the Institute.

While working in Moscow, Sobolev built on the standard variational method for solving elliptic boundary value problems by introducing these Sobolev function spaces. He gave inequalities on the norms on these spaces that were important in the theory of embedding function spaces. He applied his methods to solve difficult problems in mathematical physics.

In 1939 Sobolev was elected a full member of the Academy of Sciences of the USSR. He was only 31 years of age at the time of his election and that was a remarkable achievement. It made him the youngest full member of the Academy of Sciences and in fact he remained the youngest member for quite a few years.

At the beginning of World War II, the Steklov Mathematical Institute was moved from Moscow to Kazan. In the October of that year Sobolev was appointed as Director of the Institute and in the spring of 1943 he supervised the move of the Institute back to Moscow. Sobolev became one of the first recipients of the Stalin prize (later called a State prize) in the first presentation of these prizes in 1941. His period as Director of the Institute ended in February 1944.

In the early 1950s Sobolev's work turned towards computational mathematics and in 1952 he became head of the first department of computational mathematics in the Soviet Union when he organized the first such department at Moscow State University. However in 1956 he joined with a number of colleagues in proposing ways in which the large areas of Russia in the east could be opened up with educational initiatives. The scheme was to set up a number of Institutes for Scientific research to balance the large number of high quality educational establishments in the east of the Soviet Union. After the plan was approved, Sobolev spent some time in Moscow recruiting staff and organizing the establishment of an Institute in Novosibirsk.

In 1958 Sobolev was part of the Soviet delegation to the International Mathematical Union, the delegation being led by Vinogradov, and Sobolev attended the International Congress at Edinburg that year and gave an invited address on partial differential equations. From 1960 until 1978 Sobolev, in addition to his work at the Institute of Mathematics of the Siberian Branch of the Academy of Sciences, was a professor at Novosibirsk University. During this period he published his famous text *Applications of functional analysis in mathematical physics which appeared* in 1962.

During the 1960s much of Sobolev's research was directed towards numerical methods, in particular to interpolation. Although interpolation for functions of a single variable was well worked out, the problem of interpolation in many dimensions was largely unsolved. Sobolev applied his theories of generalized functions and embeddings of function spaces to cubature formulas, the multi-dimensional analogues of quadrature formulas for functions of one variable. Again a major text by Sobolev *Introduction to the theory of cubature formulas* has been extremely influential in this area.

Sobolev received many honours for his fundamental contributions to mathematics. He was elected to many scientific societies, including the Academy of Sciences of the USSR, the Academie des Sciences de France, and the Academia Nazionale dei Lincei. He was awarded many prizes, including three State Prizes and the 1988 M. V. Lomonosov Gold Medal from the Academy of Sciences of the USSR.

ON A THEOREM OF FUNCTIONAL ANALYSIS

S. L. SOBOLEV

In this paper we give the proof of a theorem of functional analysis which is important in various questions of the theory of partial differential equations. In an earlier paper of the author it was established that the space $L_2^{(\nu)}$, i.e., the space of functions of n variables whose derivatives up to the order ν are square-integrable, is a subspace of the space $C^{(\nu-[\frac{n}{2}]-1)}$, i.e., the space of functions which have continuous derivatives up to the order $\nu-[\frac{n}{2}]-1$ inclusive [1, 2].

Convergence of a certain sequence in $L_2^{(\nu)}$ automatically induces convergence in $C^{(\nu-[\frac{n}{2}]-1)}$.

This theorem made it possible to prove certain new properties of semi-harmonic functions and also to introduce certain improvements into the theory of non-linear hyperbolic equations [3]. The results of the author were generalized by V. I. Kondrašov [4], who established that $L_p^{(\nu)}$ is a subspace of $C^{(\nu-[\frac{n}{p}]-1)}$. The present paper is a supplement and development of this result. We prove in it that the space $L_p^{(\nu)}$ lies in every $L^{(\nu-l)}_{\frac{1}{\frac{1}{p}-\frac{l}{n}}}$. In a subsequent paper we shall show how this result may be used in the theory of non-linear hyperbolic partial differential equations.

This first chapter contains a generalization of an important inequality due to F. Riesz. In the second chapter we apply this inequality to the proof of a theorem. A brief exposition of the results of the present paper is contained in a note by the author [5] in the Doklady Akademii Nauk.

CHAPTER I

Generalization of some inequalities of F. Riesz

§1. **Generalization of Riesz's lemma.** Let there be given in n-dimensional space three functions

$$\phi_1(x_1, \cdots, x_n),\ \phi_2(x_1, \cdots, x_n),\ \phi_3(x_1, \cdots, x_n), \tag{1.1}$$

taking only the two values 0 and 1, i.e., characteristic functions of three sets E_1, E_2, and E_3. Let

$$\text{mes } E_1 < \infty,\ \text{mes } E_2 < \infty,\ \text{and mes } E_3 < \infty. \tag{1.2}$$

Let us consider the integral

$$I = \overbrace{\int \cdots \int}^{2n} \phi_1(x_1, \cdots, x_n)\phi_2(y_1, \cdots, y_n)$$
$$\cdot \phi_3(x_1 - y_1, \cdots, x_n - y_n)\, dx_1 \cdots dx_n\, dy_1 \cdots dy_n. \tag{1.3}$$

Let E_1^* be the sphere in n-dimensional space with center at the origin whose volume equals mes E_1,

E_2^* such a sphere with volume mes E_2, and

E_3^* the sphere with volume mes E_3.

Let us denote by ϕ_1^*, ϕ_2^* and ϕ_3^*, respectively, the characteristic functions of E_1^*, E_2^*, and E_3^*. Let

$$I^* = \overbrace{\int \cdots \int}^{2n} \phi_1^*(x_1, \cdots, x_n)\phi_2^*(y_1, \cdots, y_n)$$
$$\cdot \phi_3^*(x_1 - y_1, \cdots, x_n - y_n)\, dx_1 \cdots dx_n\, dy_1 \cdots dy_n. \tag{1.4}$$

Then the inequality

$$I \leq I^* \tag{1.5}$$

holds.

Inequality (1.5) for the case $n = 1$ was proved by F. Riesz [6]. In the case when $n > 1$ we shall prove this inequality using complete induction and applying a procedure used by L. A. Ljusternik [7] for the proof of another theorem.

Suppose that our inequality has been established for any number of dimensions up to $n - 1$ inclusive. We shall show that it is true for n dimensions. We represent the integral I in the

$$I = \int_{-\infty}^{+\infty} \int_{-\infty}^{+\infty} dx_1\, dy_1 \overbrace{\int \cdots \int}^{2n-2} \phi_1(x_1, \cdots, x_n)\phi_2(y_1, \cdots, y_n)$$
$$\cdot \phi_3(x_1 - y_1, \cdots, x_n - y_n)\, dx_2 \cdots dx_n\, dy_2 \cdots dy_n. \tag{1.6}$$

By assumption this integral can only increase if we replace the functions $\phi_1(x_1, \cdots, x_n)$, $\phi_2(y_1, \cdots, y_n)$ and $\phi_3(x_1 - y_1, \cdots, x_n - y_n)$ for every fixed pair of values x_1, y_1 by new functions $\phi_1'(x_1, \cdots, x_n)$, $\phi_2'(y_1, \cdots, y_n)$, and $\phi_3'(x_1 - y_1, \cdots, x_n - y_n)$ which are characteristic functions of $(n-1)$-dimensional spheres E_1', E_2', and E_3' with centers at the origin and whose volumes coincide with the $(n-1)$-dimensional measure of the intersection of the sets E_1, E_2, and E_3 with the plane $x_1 = \text{const}$, $y_1 = \text{const}$, $x_1 - y_1 = \text{const}$. The description of the substitution bears the name of Schwartz's rounding operation [8]. Thus the integral does not decrease as a result of the application to E_1, E_2, and E_3 of Schwartz's rounding operation with respect to the axis x_1 (correspondingly, y_1 or $x_1 - y_1$).

It is easy to prove that an application to these sets of Steiner's symmetrization operation with respect to the plane $x_1 = 0$ ($y_1 = 0$ or $x_1 - y_1 = 0$) likewise does not decrease I. Indeed, let us represent the integral I in the form

$$I = \overbrace{\int \cdots \int}^{2n-2} dx_2 \cdots dx_n \, dy_2 \cdots dy_n \int_{-\infty}^{+\infty} \int_{-\infty}^{+\infty} \phi_1(x_1, \cdots, x_n) \phi_2(y_1, \cdots, y_n) \cdot \phi_3(x_1 - y_1, \cdots, x_n - y_n) \, dx_1 \, dy_1. \tag{1.7}$$

The inner integral, and with it I, is not decreased by this operation. A consecutive application of Schwartz's rounding operation and then Steiner's symmetrization operation we shall call, following Ljusternik, an L_0-operation with respect to the axis x_1 (y_1 or $x_1 - y_1$).

If in place of x_1 we choose any other line, then we may construct another operation L_0.

We shall now apply to each of the sets E_1, E_2, and E_3 in turn the operations L_0 and L_0'. As proved in the cited paper of L. A. Ljusternik, the sequences of sets

$$\left.\begin{array}{l} E_1, E_1', E_1'', \cdots, \\ E_2, E_2', E_2'', \cdots, \\ E_3, E_3', E_3'', \cdots \end{array}\right\} \tag{1.8}$$

thus obtained converge to the spheres E_1^*, E_2^*, and E_3^*.

At the same time the integral I transforms to I', I'', etc., such that

$$I \leq I' \leq I'' \leq \cdots, \tag{1.9}$$

$$\lim_{n \to \infty} I^{(n)} = I^*. \tag{1.10}$$

Hence inequality (1.5) follows immediately.

§2. **Generalization of Riesz's inequality.** Let the functions

$$\phi_1(x_1, \cdots, x_n), \phi_2(y_1, \cdots, y_n) \text{ and } \phi_3(x_1 - y_1, \cdots, x_n - y_n) \tag{2.1}$$

be any positive summable functions of their arguments. Let us form the integral

$$I = \overbrace{\int \cdots \int}^{2n} \phi_1(x_1, \cdots, x_n) \phi_2(y_1, \cdots, y_n) \cdot \phi_3(x_1 - y_1, \cdots, x_n - y_n) \, dx_1 \cdots dx_n \, dy_1 \cdots dy_n. \tag{2.2}$$

Now let $\phi_1^*(x_1, \cdots, x_n)$ be a function of $r = \sqrt{\sum_{i=1}^{n} x_i^2}$ alone which is non-increasing and also such that

$$\mathrm{mes}(l_1 \leq \phi_1^*) = \mathrm{mes}(l_1 \leq \phi) \tag{2.3}$$

for all values of l_1. Let us construct the functions ϕ_2^* and ϕ_3^* analogously. Let

$$I^* = \overbrace{\int \cdots \int}^{2n} \phi_1^*(x_1, \cdots, x_n)\phi_2^*(y_1, \cdots, y_n) \cdot\phi_3^*(x_1 - y_1, \cdots, x_n - y_n)\,dx_1 \cdots dx_n\,dy_1 \cdots dy_n. \tag{2.4}$$

Then

$$I \leq I^*. \tag{2.5}$$

The proof of this inequality is not difficult if we make use of a method due to F. Riesz. Let us replace each function ϕ_i by the function

$$\bar{\phi}_i = \sum_{j=1}^{n} a_j f_j^{(i)} \leq \phi_i,$$

where f assumes only the two values 0 and 1. Taking sufficiently many such terms, we can make the integral

$$\bar{I} = \overbrace{\int \cdots \int}^{2n} \bar{\phi}_1(x_1, \cdots, x_n)\bar{\phi}_2(y_1, \cdots, y_n) \cdot\bar{\phi}_3(x_1 - y_1, \cdots, x_n - y_n)\,dx_1 \cdots dx_n\,dy_1 \cdots dy_n \tag{2.6}$$

as close as desired to I:

$$I - \epsilon < \bar{I} \leq I. \tag{2.7}$$

By the lemma of Riesz which has been proved

$$\bar{I} \leq \bar{I}^*, \tag{2.8}$$

where $\bar{I}^*$ is obtained from $\bar{I}$ by replacing the functions $\bar{\phi}_i$ by ϕ_i^*. But $\bar{I}^* \leq I^*$, hence

$$I - \epsilon < \bar{I} \leq \bar{I}^* \leq I^*. \tag{2.9}$$

Because of the arbitrariness of ϵ we obtain (2.5) directly.

§3. An auxiliary lemma. Let $\phi_1(r_1)$ and $\phi_2(r_2)$ be arbitrary positive, non-increasing functions of the quantities

$$r_1 = \sqrt{(x_1 - a)^2 + \sum_{i=2}^{n} x_i^2}, \qquad r_2 = \sqrt{(x_1 + a)^2 + \sum_{i=2}^{n} x_i^2}. \tag{3.1}$$

Let us consider the integral

$$N = \overbrace{\int \cdots \int}^{n}_{\substack{r_1 \leq 2a \\ r_2 \leq 2a}} \phi_1(r_1)\phi_2(r_2)\,dx_1 \cdots dx_n \tag{3.2}$$

and try to estimate it.

To this end let us consider two domains

$$\left.\begin{array}{l} D_1(a \le r_1 \le 2a;\ 0 \le r_2 \le 2a), \\ D_2(a \le r_2 \le 2a;\ 0 \le r_1 \le 2a) \end{array}\right\} \tag{3.3}$$

and estimate separately the integrals over each of these domains. It is obvious that

$$N \le \overbrace{\int \cdots \int}^{n}_{D_1} \phi_1 \phi_2 \, dx_1 \cdots dx_n + \overbrace{\int \cdots \int}^{n}_{D_1} \phi_1 \phi_2 \, dx_1 \cdots dx_n = N_1 + N_2. \tag{3.4}$$

Since the estimation of the integrals N_1 and N_2 will be essentially the same, we shall consider either one of them, for instance N_1. For this purpose we change the variables of integration in it. First of all, we introduce in place of $x_1, \cdots, x_n$ cylindrical coordinates by the formula

$$x_i = \rho \lambda_i, \quad i = 2, 3, \cdots, n, \tag{3.5}$$

where

$$\rho = \sqrt{\sum_{i=1}^{n} x_i^2},$$

and $\lambda_1, \lambda_2, \cdots, \lambda_n$ is an arbitrary radius-vector of the unit sphere of $(n-1)$-dimensional space. We will have

$$N_1 = \chi_{n-1} \iint\limits_{\substack{0 \le r_1 \le 2a \\ 0 \le r_2 \le 2a}} \phi_1(r_1) \phi_2(r_2) \rho^{n-2} \, d\rho \, dx_1. \tag{3.6}$$

The domain of integration in the plane of ρ, x_1 is represented in Figure 1. The surface of the sphere in $(n-1)$-dimensional space is indicated by χ_{n-1}.

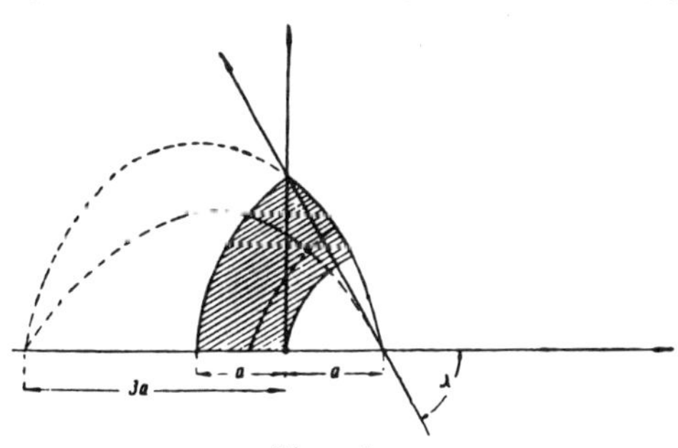

Figure 1

Now let us introduce in place of ρ, x_1 the new variables

$$r_1 \text{ and } \lambda = \operatorname{arc\,tg}\frac{\rho}{a-x_1} + \operatorname{arc\,tg}\frac{\rho}{3a+x_1}. \tag{3.7}$$

The geometrical meaning of the variable λ is easily clarified. Let us describe a circle through the given point x_1, ρ and the points $(a, 0)$ and $(-3a, 0)$. Draw the tangent to this circle at the point $(a, 0)$. The acute angle between the direction of the axis x_1 and the direction of this tangent will be equal to the value of λ at the given point.

The coordinates we have introduced have the property that the variable r_2 inside the domain D_1 is a decreasing function of r_1 for constant λ. Indeed, any circle with center at $(-a, 0)$ intersects the line $\lambda = \text{const}$ in two points which are symmetric with respect to the line $x_1 = -a$. Consequently, this line $\lambda = \text{const}$ enters every such circle to the right of the line $x_1 = -a$ and exits from it to the left of this line. Since all of the domain D_1 lies to the right of the line $x_1 = -a$, our assertion is obvious.

Let us note further two simple estimates in the domain D_1:

$$C_1 a \le \frac{\rho}{\sin\lambda} \le C_2 a, \quad C_3 \le a\left|\frac{D(r_1, \lambda)}{D(x_1, \rho)}\right| \le C_4, \tag{3.8}$$

where C_1, C_2, C_3, and C_4 are certain constants which do not need to be evaluated. Indeed,

$$\sin\lambda = \frac{4a\rho\sqrt{[(a-x_1)^2+\rho^2][(3a+x_1)^2+\rho^2]}}{(a-x_1)^2(3a+x_1)^2}, \tag{3.9}$$

from which the first of the inequalities (3.8) follows readily. The second inequality (3.8) follows from the formula

$$\frac{D(r_1, \lambda)}{D(x_1, \rho)} = \frac{6a(3a+2x_1)}{r_1[(3a+x_1)^2+\rho^2]}. \tag{3.10}$$

Using inequality (3.8) we give the integral N_1 the estimate

$$N_1 \leq C_5 a^{n-1} \int_0^{2a} \left\{ \int_0^{\frac{\pi}{2}} \phi_2(r_2) \sin^{n-2}\lambda \, d\lambda \right\} \phi_1(r_1) dr_1. \tag{3.11}$$

As was already noted above, r_2 is a decreasing function of r_1 for a given λ. Therefore, the integral

$$\int_0^{\frac{\pi}{2}} \phi_2(r_2) \sin^{n-2}\lambda \, d\lambda \tag{3.12}$$

will be an increasing function of r_1. Now let us make use of the well-known inequality of Čebyšev

$$\int_a^b \phi(x)\psi(x)\,dx \leq \frac{1}{b-a} \int_a^b \phi(x)\,dx \int_a^b \psi(x)\,dx, \tag{3.13}$$

which is valid if $\phi(x)$ is a non-decreasing and $\psi(x)$ a non-increasing function;

$$\phi(x) > 0, \quad \phi(x) > 0. \tag{3.14}$$

Applying (3.13) to the estimate of N_1, we obtain

$$N_1 \leq C_6 a^{n-2} \int_a^{2a} \phi_1(r_1)\,dr_1 \int_a^{2a} \left\{ \int_0^{\frac{\pi}{2}} \phi_2(r_2) \sin^{n-2}\lambda \, d\lambda \right\} dr_1. \tag{3.15}$$

By again making use of the inequalities (3.8), but in the other side, we will obtain moreover

$$\int_0^{2a} \left\{ \int_0^{\frac{\pi}{2}} \phi_2(r_2) \sin^{n-2}\lambda \, d\lambda \right\} dr_1 \leq \frac{C_7}{a^{n-1}} \iint_{D_1} \phi_2(r_2)\,\rho^{n-2}\,d\rho\,dx_1. \tag{3.16}$$

Introducing the coordinates r_2 and $\theta_2 = \text{arc tg}\,\frac{\rho}{x+a}$ in place of x_1 and ρ, we will thus obtain

$$\int_a^{2a} \left\{ \int_0^{\frac{\pi}{2}} \phi_2(r_2) \sin^{n-2}\lambda d\lambda \right\} dr_1 \leq \frac{C_8}{a^{n-1}} \int_0^{2a} \phi_2(r_2)\,r_2^{\,n-1}\,dr_2. \tag{3.17}$$

Note further that

$$\int_a^{2a} \phi_1(r_1)\,dr_1 \leq \frac{C_9}{a^{n-1}} \int_a^{2a} \phi_1(r_1)\,r_1^{\,n-1}\,dr_1 \leq \frac{C_9}{a^{n-1}} \int_0^{2a} \phi_1(r_1)\,r_1^{\,n-1}\,dr_1. \tag{3.18}$$

Substituting these estimates into (3.15), we will have

$$N_1 \leq \frac{C_{10}}{a^n} \int_0^a \phi_1(r_1) r_1^{n-1} dr_1 \int_0^a \phi_2(r_2) r_2^{n-1} dr_2. \tag{3.19}$$

Clearly, the integral N_2 admits the same estimate. Thus, finally,

$$\underbrace{\int \cdots \int}_{\substack{r_1 \leq 2a \\ r_1 \leq 2a}}^{n} \phi_1(r_1)\phi_2(r_2)\, dx_1 \cdots dx_n \leq \frac{L}{a^n} \int_0^{2a} \phi_1(r_1) r_1^{n-1} dr_1 \int_0^{2a} \phi_2(r_2) r_2^{n-1} dr_2. \tag{3.20}$$

This estimate is the aim of the preceding considerations.

§4. Generalization of the second integral inequality of Riesz. Let the functions $\phi(x_1, \cdots, x_n)$ and $\psi(y_1, \cdots, y_n)$ be pth power and qth power summable, respectively, over the entire space, where $p > 1$ and $q > 1$, $\frac{1}{p} + \frac{1}{q} > 1$. Then the integral

$$J = \overbrace{\int \cdots \int}^{2n} \frac{\phi(x_1, \cdots, x_n), \psi(y_1, \cdots, y_n)}{r^\lambda} dx_1 \cdots dx_n dy_1 \cdots dy_n, \tag{4.1}$$

where

$$r = \sqrt{\sum_{i=1}^n (x_i - y_i)^2} \text{ and } \lambda = n\left(2 - \frac{1}{p} - \frac{1}{q}\right) = n\left(\frac{1}{p'} + \frac{1}{q'}\right), \tag{4.2}$$

$$\frac{1}{p} + \frac{1}{p'} = 1, \quad \frac{1}{q} + \frac{1}{q'} = 1,$$

has meaning and satisfies the inequality

$$\overbrace{\int \cdots \int}^{2n} \frac{\phi(x_1, \cdots, x_n)\psi(y_1, \cdots, y_n)}{r^\lambda} dx_1 \cdots dx_n dy_1 \cdots dy_n$$

$$\leq K\left(\overbrace{\int \cdots \int}^{n} |\phi|^p dx_1 \cdots dx_n\right)^{\frac{1}{p}} \left(\overbrace{\int \cdots \int}^{n} |\psi|^q dy_1 \cdots dy_n\right)^{\frac{1}{q}}, \tag{4.3}$$

where the constant K depends on p and q, but does not depend on the functions ϕ and ψ.

For the proof we may assume from the beginning that $\phi > 0$ and $\psi > 0$. Moreover, by the previous lemma, we may suppose from the beginning that ϕ depends only on $r_1 = \sqrt{\sum_{i=1}^n x_i^2}$ and that ψ depends only on $r_2 = \sqrt{\sum_{i=1}^n y_1^2}$.

We break up the integral J into three terms. Let

$$J_1 = \overbrace{\int\cdots\int}^{2n}_{\substack{r>r_1\\ r>r_2}} \frac{\phi\psi}{r^\lambda}\, dx_1 \cdots dx_n dy_1 \cdots dy_n,$$

$$J_2 = \overbrace{\int\cdots\int}^{2n}_{\substack{r_1>r\\ r_1>r_2}} \frac{\phi\psi}{r^\lambda}\, dx_1 \cdots dx_n dy_1 \cdots dy_n, \tag{4.4}$$

$$J_3 = \overbrace{\int\cdots\int}^{2n}_{\substack{r_2>r\\ r_2>r_1}} \frac{\phi\psi}{r^\lambda}\, dx_1 \cdots dx_n dy_1 \cdots dy_n.$$

Then

$$J = J_1 + J_2 + J_3.$$

We shall estimate each of these integrals separately. It is obvious that the estimation of J_2 and J_3 will be analogous, and, therefore, it is sufficient to estimate one of them. Let us begin the investigation with the estimation of the integral J_2. Carrying out the outer integration with respect to $x_1, \cdots, x_n$ and the inner with respect to $y_1, \cdots, y_n$, we will give this integral the form

$$J_2 = \chi_{n-1} \int_0^\infty r_1^{n-1} \phi(r_1) \Big[\overbrace{\int\cdots\int}^{n}_{\substack{r<r_1\\ r_2<r_1}} \frac{\psi(r_2)}{r^\lambda}\, dy_1 \cdots dy_n \Big]\, dr_1. \tag{4.5}$$

The inner integral is easily transformed to that which was investigated by us in the previous paragraph. Estimating it and using the fact that $\lambda < n$, we will have

$$\overbrace{\int\cdots\int}^{n}_{\substack{r<r_1\\ r_2<r_1}} \frac{\psi(r_2)}{r^\lambda}\, dy_1 \cdots dy_n \leq \frac{L}{r_1^n} \int_0^{r_1} \psi(r_2)\, r_2^{n-1}\, dr_2 \int_0^{r_1} r^{n-1-\lambda}\, dr$$

$$\leq L_2 r_1^{-\lambda} \int_0^{r_1} \psi(r_2)\, r_2^{n-1}\, dr_2, \tag{4.6}$$

or

$$J_2 \leq L_3 \int_0^\infty \phi(r_1)\, r_1^{n-\lambda-1}\, dr_1 \int_0^{r_1} \psi(r_2)\, r^{n-1}\, dr_2. \tag{4.7}$$

Now let us denote

$$r_1^{-\frac{n}{q'}} \cdot \frac{1}{q} \int_0^r \psi(r_2)\, r_2^{n-1}\, dr_2 = \Psi(r_1). \tag{4.8}$$

We shall show that $\int_0^\infty \Psi(r_1)^q\, dr_1$ exists and give an estimate for this integral.

Since $\psi(r_2)$ is qth power summable, the integral

$$\int_0^\infty [\psi(r_2)]^q r_2^{n-1}\, dr_2 \leq C \overbrace{\int \cdots \int}^{n} [\psi(r_2)]^q\, dy_1 \cdots dy_n = CB^q \tag{4.9}$$

exists. Let us rewrite this inequality in the form

$$\int_0^\infty [\psi(r_2) r_2^{\frac{n-1}{q}}]\, dr_2 \leq CB^q. \tag{4.10}$$

Let

$$\psi(r_2) r_2^{\frac{n-1}{q}} = \chi(r_2). \tag{4.11}$$

Then

$$\Psi(r_1) = r_1^{-\frac{n}{q'} - \frac{1}{q}} \int_0^{r_1} \chi(r_2) r_2^{\frac{n-1}{q'}}\, dr_2. \tag{4.12}$$

Now let us make use of an inequality due to Hardy ([9], p. 245):

$$\int_0^\infty x^{-\alpha q} \left[\int_0^x \chi(y) y^{\alpha-1}\, dy\right]^q dx \leq \left(\frac{q}{\alpha q - 1}\right)^q \int_0^\infty [\chi(y)]^q dy, \tag{4.13}$$

which is valid for $\alpha q > 1$, $q > 1$.

If we set

$$\alpha = \frac{n-1}{q} + 1 = \frac{n}{q'} + \frac{1}{q}$$

in inequality (4.13), then we will obtain directly

$$\int_0^\infty [\psi(r_2)]^q\, dr_2 \leq C_{12} B^q. \tag{4.14}$$

Returning to the estimate of J_2 and denoting

$$\phi(r_1) r_1^{\frac{n-1}{p'}} = \Phi(r_1), \tag{4.15}$$

we will have

$$\int_0^\infty \Phi(r_1)^p\, dr_1 \leq C_{13} A^p, \text{ where } A^p = \overbrace{\int \cdots \int}^{n} |\phi|^p\, dx_1 \cdots dx_n, \tag{4.16}$$

and, moreover,

$$J_2 \leq C_{14} \int_0^\infty \Phi(r_1)\Psi(r_1) r_1^{n-1-\lambda-\frac{n-1}{p}+\frac{n}{q'}+\frac{1}{q}} dr_1$$

$$= C_{14} \int_0^\infty \Phi(r_1)\Psi(r_1) r_1^{\frac{1}{q}-\frac{1}{p'}} dr_1. \quad (4.17)$$

This inequality may be rewritten in the form

$$J_2 \leq C_{14} \int_0^\infty [\Psi(r_1) r_1^{\frac{1}{q}}]^{\frac{p'-q}{p'}} \Phi(r_1)[\Psi(r_1)]^{\frac{q}{p'}} dr_1. \quad (4.18)$$

Note further that

$$r_1^{\frac{1}{q}} \Psi(r_1) = r_1^{-\frac{n}{q'}} \int_0^{r_1} r_2^{\frac{n-1}{q'}} \chi(r_2)\, dr_2$$

$$\leq r_1^{-\frac{n}{q'}} \left[\int_0^{r_1} r_2^{n-1} dr_2\right]^{\frac{1}{q'}} \left[\int_0^{r_1} (\chi(r_2))^q dr_2\right]^{\frac{1}{q}} \leq C_{15} B, \quad (4.19)$$

from which, using Hölder's inequality,

$$J_2 \leq C_{16} B^{\frac{p'-q}{p'}} \int_0^\infty \Phi(r_1)[\Psi(r_1)]^{\frac{q}{p'}} dr_1 \leq C_{16} A B^{\frac{p'-q}{p'}} B^{\frac{q}{p'}} = C_{16} AB. \quad (4.20)$$

An analogous estimation gives

$$J_3 \leq C_{16} AB. \quad (4.21)$$

Now let us estimate the integral J_1. Introducing cylindrical coordinates and transforming this integral similarly to the previous one, we will obtain

$$J_1 \leq C_{17} \int_0^\infty r^{n-1-\lambda} \left\{ \underbrace{\int \cdots \int}_{\substack{n \\ r_1 < r \\ r_2 < r}} \phi(r_1)\psi(r_2)\, dx_1 \cdots dx_n \right\} dr. \quad (4.22)$$

The inner integral is estimated analogously to the previous one:

$$\underbrace{\int \cdots \int}_{\substack{n \\ r_1 < r \\ r_2 < r}} \phi(r_1)\psi(r_2)\, dx_1 \cdots dx_n \leq \frac{L}{r^n} \int_0^r \phi(r_1) r_1^{n-1} dr_1 \int_0^r \psi(r_2) r_2^{n-1} dr_2. \quad (4.23)$$

Setting

$$r^{-\frac{n}{p'}-\frac{1}{p}}\int_0^r \phi(r_1)r_1^{n-1}dr_1 = \Phi(r),$$
$$r^{-\frac{n}{q'}-\frac{1}{q}}\int_0^r \psi(r_2)r_2^{n-1}dr_2 = \Psi(r), \tag{4.24}$$

we will have

$$\int_0^r [\Phi(r)]^p dr \leq C_{18}A, \tag{4.25}$$

$$\int_0^r [\Psi(r)]^q\, dr \leq C_{19}B, \tag{4.26}$$

similar to what we had before. The estimate (4.23) gives

$$J_1 \leq C_{20}\int_0^\infty r^{\frac{1}{p}+\frac{1}{q}-1}\Phi(r)\Psi(r)\,dr = C_{20}\int_0^\infty r^{\frac{1}{q}-\frac{1}{p'}}\Phi(r)\Psi(r)\,dr. \tag{4.27}$$

Moreover, on the basis of (4.19),

$$r^{\frac{1}{q}}\Psi(r) \leq C_{15}B,$$

from which it is easy to obtain, as in the former case,

$$J_1 \leq C_{21}AB. \tag{4.28}$$

Combining formulas (4.20), (4.21), and (4.28), we obtain

$$J \leq KAB,$$

i.e., the desired formula (4.3).

§5. Fundamental theorem about integrals of potential type. Now let there be given some function $f(x_1, \cdots, x_n)$ which is pth power summable, $p > 1$. Let us agree for brevity to write it in the form $f(P)$. In addition, let $dx_1 \cdots dx_n = dP$. Now let

$$\overbrace{\int\cdots\int}^{n} |f(P)|^p\, dP = B^p. \tag{5.1}$$

We shall call the function

$$U(Q) = \overbrace{\int\cdots\int}^{n} r^{-\lambda} f(P)\, dP, \quad \lambda < \frac{n}{p'}, \tag{5.2}$$

the λ-potential of f, where by Q is denoted some point with coordinates $y_1, \cdots, y_n$ and

$$r = \sqrt{\sum_{i=1}^{n} (x_i - y_i)^2}. \tag{5.3}$$

Then the following theorem is valid.

Theorem. *The function $U(Q)$ is qth power summable on the entire space $y_1, \cdots, y_n$, where*

$$\frac{1}{q} = \frac{\lambda}{n} - \frac{1}{p'} \tag{5.4}$$

(it is obvious that q is greater than unity). Moreover, the inequality

$$\overbrace{\int \cdots \int}^{n} [U(Q)]^q \, dQ \leq KB^q \tag{5.5}$$

holds. *

---

* A weaker theorem which, otherwise, is sufficient for many applications may be proved much more easily:

Theorem. *Let*

$$U_1(Q) = \underbrace{\overbrace{\int \cdots \int}^{n}}_{r \leq R} r^{-\lambda} f(P) \, dP. \tag{5.6}$$

The function $U_1(Q)$ is q^*th power summable on the entire space $y_1, \cdots, y_n$ for any q^* where $p < q^* < q$. The inequality

$$\overbrace{\int \cdots \int}^{n} [U_1(Q)]^{q^*} \, dQ \leq KB^{q^*} \tag{5.7}$$

holds, where the constant K depends on R, p, q, q^*, and n.

In addition $p = 1$, as we shall see below, is not excluded.

This theorem is proved with the aid of the repetition of the proof of an integral inequality of Young ([9], p. 201). Let $\epsilon < n\left(\frac{1}{q^*} - \frac{1}{q}\right)$. Then by Hölder's inequality,

$$\begin{aligned} U_1(Q) &= \underbrace{\overbrace{\int \cdots \int}^{n}}_{r \leq R} r^{-\frac{n}{p'} + \epsilon} [f(P)]^{1 - \frac{p}{q^*}} \left(r^{-\frac{n}{q} + \epsilon} f(P)^{\frac{p}{q^*}}\right) dP \\ &< \left(\underbrace{\overbrace{\int \cdots \int}^{n}}_{r \leq R} r^{-n + \epsilon p'} \, dP\right)^{\frac{1}{p'}} \left(\underbrace{\int \cdots \int}_{r \leq R} [f(P)]^p \, dP\right)^{\frac{1}{p} - \frac{1}{q^*}} \\ &\quad \cdot \left(\underbrace{\overbrace{\int \cdots \int}^{n}}_{r \leq R} r^{-\frac{nq^*}{q} - q^*} [f(P)]^p \, dP\right)^{\frac{1}{q^*}}, \end{aligned} \tag{5.8}$$

from which

On the basis of a well-known theorem of functional analysis, for the proof of inequality (5.5) it is sufficient to establish that whatever be the given function $\omega(Q)$, q'th power summable, where $\frac{1}{q}+\frac{1}{q'}=1$,

$$\overbrace{\int\cdots\int}^{n}|\omega|^{q'}\,dQ = C^{q'}, \tag{5.12}$$

the inequality

$$\overbrace{\int\cdots\int}^{n} U(Q)\,\omega(Q)\,dQ \leq CKB \tag{5.13}$$

holds. Substituting for $U(Q)$ its representation, we will have

$$\overbrace{\int\cdots\int}^{2n}\frac{\omega(Q)\,f(P)}{r^{\lambda}}\,dP\,dQ \leq CKB. \tag{5.14}$$

On the basis of the generalized Riesz's theorem (see §4) this inequality is immediately apparent. Our theorem is proved.

CHAPTER II

Relation between the spaces $L_p^{(\nu)}$

§6. A new definition of derivative. Let $\phi(x_1, \cdots, x_n)$ be some function of the variables $x_1, \cdots, x_n$ which is summable in the domain D. Let D' be any other domain lying along with its boundary inside D.

Let us consider any function $\psi(x_1, \cdots, x_n)$ having continuous derivatives

---

$$[U_1(Q)]^{q^*} \leq MB^{\left(1-\frac{p}{q^*}\right)q^*}\overbrace{\int\cdots\int}^{n}_{r\leq R} r^{-\frac{nq^*}{q}-\epsilon q^*}[f(P)]^p\,dP, \tag{5.9}$$

and, moreover,

$$\overbrace{\int\cdots\int}^{n}_{r\leq R}[U_1(Q)]^{q^*}\,dQ \leq MB^{q^*-q}\overbrace{\int\cdots\int}^{n}[f(P)]^p\left[\overbrace{\int\cdots\int}^{n}_{r\leq R} r^{-\frac{nq^*}{q}-\epsilon q^*}\,dQ\right]dP, \tag{5.10}$$

from which

$$\overbrace{\int\cdots\int}^{n}[U_1(Q)]^{q^*}\,dQ \leq M_1B^{q^*}, \tag{5.11}$$

as was to be proved.

up to the order α inclusive and vanishing outside of D'. If the function ϕ itself has continuous derivatives up to the order α, then the equation

$$\overbrace{\int \cdots \int}^{n}_{D} \left[\phi \frac{\partial^{\alpha} \psi}{\partial x_1^{\alpha_1} \cdots \partial x_n^{\alpha_n}} + (-1)^{\alpha-1} \psi \frac{\partial^{\alpha} \phi}{\partial x_1^{\alpha_1} \cdots \partial x_n^{\alpha_n}} \right] dx_1 \cdots dx_n = 0 \qquad (6.1)$$

is valid.

We shall consider this equation as the definition of the derivative and call derivative of ϕ,

$$\frac{\partial^{\alpha} \phi}{\partial x_1^{\alpha_1} \cdots \partial x_n^{\alpha_n}},$$

such a function of the variables $x_1, \cdots, x_n$ which is summable over any bounded subset of the domain D and satisfies condition (6.1) for all ψ.

It is obvious that the derivative thus defined may only be unique, for supposing the existence of two such derivatives

$$\left[\frac{\partial^{\alpha} \phi}{\partial x_1^{\alpha_1} \cdots \partial x_n^{\alpha_n}} \right]_1 \text{ and } \left[\frac{\partial^{\alpha} \phi}{\partial x_1^{\alpha_1} \cdots \partial x_n^{\alpha_n}} \right]_2,$$

we will have for any ψ

$$\overbrace{\int \cdots \int}^{n}_{D} \psi \left\{ \left[\frac{\partial^{\alpha} \phi}{\partial x_1^{\alpha_1} \cdots \partial x_n^{\alpha_n}} \right]_1 - \left[\frac{\partial^{\alpha} \phi}{\partial x_1^{\alpha_1} \cdots \partial x_n^{\alpha_n}} \right]_2 \right\} dx_1 \cdots dx_n = 0, \qquad (6.2)$$

from which it follows that these two derivatives are equal to each other everywhere except, perhaps, a set of measure zero.

Comparatively simple examples show that the derivative in our sense may exist when, for instance, the function does not have a derivative almost everywhere, and conversely the existence of the derivative almost everywhere does not necessitate the existence of the derivative in our sense.

In fact, let

$$\phi = f_1(x_1) + f_2(x_2), \qquad (6.3)$$

where neither $f_1(x_1)$ nor $f_2(x_2)$ is differentiable. Then the derivative $\frac{\partial^2 \phi}{\partial x_1 \partial x_2}$ does not exist from the classical point of view. Nevertheless, by our definition

$$\frac{\partial^2 \phi}{\partial x_1 \partial x_2} = 0 \qquad (6.4)$$

at all points. Indeed,

$$\overbrace{\int\cdots\int}^{n}_{D} \phi \frac{\partial^2 \psi}{\partial x_1 \partial x_2}\, dx_1 \cdots dx_n = \overbrace{\int\cdots\int}^{n}_{D} f_1(x_1) \frac{\partial}{\partial x_2} \left[\frac{\partial \psi}{\partial x_1}\right] dx_1 \cdots dx_n$$

$$+ \overbrace{\int\cdots\int}^{n}_{D} f_2(x_2) \frac{\partial}{\partial x_1} \left[\frac{\partial \psi}{\partial x_2}\right] dx_1 \cdots dx_n = 0, \tag{6.5}$$

since both terms may be integrated by parts. On the other hand, let

$$\phi = F(x_1), \tag{6.6}$$

where F has a summable derivative almost everywhere, but is not absolutely continuous. Then, according to our definition, the derivative does not exist, of which the reader may easily convince himself.

For discovering the existence of the derivative it is useful to point out one indication based on the theory of approximations.

Theorem. *If one can approximate a given summable function ϕ by a sequence of continuously differentiable functions*

$$\phi_k(x_1, \cdots, x_n) \quad (k = 0, 1, \cdots) \tag{6.7}$$

such that for any function ψ, continuous and vanishing outside the domain D' which is interior to the domain D, the equation

$$\lim_{k\to\infty} \overbrace{\int\cdots\int}^{n}_{D} \phi_k \psi \, dx_1 \cdots dx_n = \overbrace{\int\cdots\int}^{n} \phi\psi \, dx_1 \cdots dx_n \tag{6.8}$$

holds, and if, in addition, for any bounded subset D' of the domain D

$$\overbrace{\int\cdots\int}^{n}_{D'} \left| \frac{\partial^{\alpha} \phi_k}{\partial x_1^{\alpha_1} \cdots \partial x_n^{\alpha_n}} \right|^p dx_1 \cdots dx_n \leq A, \tag{6.9}$$

where $p > 1$, then the derivative $\dfrac{\partial^{\alpha}\phi}{\partial x_1^{\alpha_1}\cdots\partial x_n^{\alpha_n}}$ exists and satisfies the condition

$$\overbrace{\int\cdots\int}^{n}_{D'} \left| \frac{\partial^{\alpha} \phi}{\partial x_1^{\alpha_1} \cdots \partial x_n^{\alpha_n}} \right|^p dx_1 \cdots dx_n \leq A. \tag{6.10}$$

For the proof we shall use the theorem on weak compactness of the unit sphere of the space L_p of functions which are pth power summable (see [10], p. 126).

By this weak compactness there exists a function $\omega_{a_1,\cdots,a_n}$ and a subsequence

$$\frac{\partial^a \phi_{k_i}}{\partial x_1^{a_1}\cdots\partial x_n^{a_n}}, \quad i = 1, 2, \cdots, \tag{6.11}$$

such that for any function ψ which is p'th power summable on D', $p' = \frac{p}{p-1}$, the equation

$$\lim_{i\to\infty} \overbrace{\int\cdots\int_{D'}}^{n} \psi \frac{\partial^a \phi_{k_i}}{\partial x_1^{a_1}\cdots\partial x_n^{a_n}} dx_1 \cdots dx_n = \overbrace{\int\cdots\int_{D'}}^{n} \psi\,\omega_{a_1,\cdots,a_n} dx_1\cdots dx_n \tag{6.12}$$

is valid. Moreover,

$$\overbrace{\int\cdots\int_{D'}}^{n} |\omega_{a_1,\cdots,a_n}|^p dx_1\cdots dx_n \le A. \tag{6.13}$$

Let us write the identity

$$\overbrace{\int\cdots\int_{D'}}^{n} \psi \frac{\partial^a \phi_{k_i}}{\partial x_1^{a_1}\cdots\partial x_n^{a_n}} dx_1\cdots dx_n$$

$$= (-1)^a \overbrace{\int\cdots\int_{D'}}^{n} \phi_{k_i} \frac{\partial^a \psi}{\partial x_1^{a_1}\cdots\partial x_n^{a_n}} dx_1\cdots dx_n, \tag{6.14}$$

where ψ is a continuously differentiable function equal to zero outside D'. Passing to the limit in equation (6.14) we obtain

$$\overbrace{\int\cdots\int_{D'}}^{n} \psi\,\omega_{a_1,\cdots,a_n} dx_1\cdots dx_n = (-1)^a \overbrace{\int\cdots\int_{D'}}^{n} \phi \frac{\partial^a \psi}{\partial x_1^{a_1}\cdots\partial x_n^{a_n}} dx_1\cdots dx_n. \tag{6.15}$$

Hence it follows that the function $\omega_{a_1,\cdots,a_n}$ is the derivative of ϕ, as was to be proved.

§7. Fundamental theorem for the case of continuous functions. Let us consider some domain D in the space $x_1, \cdots, x_n$. Let this domain be such that any of its interior or boundary points can be reached by the point of a moving spherical sector of constant magnitude and shape. (D may be unbounded.)

Let us consider in this domain some function $\phi(x_1, \cdots, x_n)$ with respect to which we shall make the following assumptions:

(1) *the function* $|\phi(x_1, \cdots, x_n)|^{q^*}$ *is summable on* D, *where* $q^* \ge 1$,

(2) *the function* $\phi(x_1, \cdots, x_n)$ *admits all derivatives of order* l *in the domain* D,

(3) *all derivatives of order l are pth power summable,* $p \geq 1$.

$$\overbrace{\int \cdots \int_D}^{n} \left| \frac{\partial^l \phi}{\partial x_1^{\alpha_1} \cdots \partial x_n^{\alpha_n}} \right|^p dx_1 \cdots dx_n \leq A^p, \tag{7.1}$$

where $l \leq n$.

In addition, several possibilities are meaningful:

$$1)\ 1 < p < \frac{n}{l}, \qquad 3)\ 1 < p = \frac{n}{l},$$
$$2)\ 1 = p < \frac{n}{l}, \qquad 4)\ 1 = p = \frac{n}{l}, \tag{7.2}$$
$$5)\ p > \frac{n}{l}.$$

The last case has been the object of special investigation. For $p = 2$ it was analyzed by the author [1, 2], and for the case of arbitrary p by V. I. Kondrašov, who used the same method. There it is proved that the function $\phi(x_1, \cdots, x_n)$ will be continuous and will satisfy Hölder's condition with certain positive exponents. Interesting results concerning this question were obtained by I. G. Petrovskiĭ and K. N. Smirnov [11].

Our task will be the consideration of the remaining cases. First of all, we shall study the first of these. Let us restrict ourselves only to those cases when $\frac{1}{q^*} > \frac{1}{p} - \frac{1}{n}$. Then we have the

Theorem. *If* $1 < p < \frac{n}{l}$, *then the function* ϕ *is qth power summable on the domain D, where*

$$\frac{1}{q} = \frac{1}{p} - \frac{l}{n}, \tag{7.3}$$

and, moreover,

$$\left[\overbrace{\int \cdots \int_D}^{n} |\phi|^q dx_1 \cdots dx_n\right]^{\frac{1}{q}} \leq NA^p + L\left[\overbrace{\int \cdots \int_D}^{n} |\phi|^{q^*} dx_1 \cdots dx_n\right]^{\frac{1}{q^*}}, \tag{7.4}$$

where the constants N and L depend on the shape of the domain D and on the exponents p, q^*, k, n and 1, but do not depend on the form of the function ϕ.*

---

*If instead of the fundamental theorem of the previous chapter we used its weakened form (see previous footnote), then we would obtain a weaker result in this case. We would prove the summability of the function ϕ only for the power $q^{**} < q$. This weak result is sometimes sufficient for the applications.

We shall carry out the proof of the theorem first of all for the case when ϕ is continuous with its derivatives up to the order l, and then pass over to the general case.

Let ϕ have these continuous derivatives. At each point $x_1^0, \cdots, x_n^0$ of the domain D let us consider the system of polar coordinates defined by the formulas

$$x_i - x_i^0 = r \sin\vartheta_1 \sin\vartheta_2 \cdots \sin\vartheta_{i-1} \cos\vartheta_i, \; i = 1, 2, \cdots, n-2,$$

$$x_{n-1} - x_{n-1}^0 = r \sin\vartheta_1 \sin\vartheta_2 \cdots \sin\vartheta_{n-2} \cos\phi,$$

$$x_n - x_n^0 = r \sin\vartheta_1 \sin\vartheta_2 \cdots \sin\vartheta_{n-2} \sin\phi.$$

Let the direction of the coordinate axes be chosen such that the spherical sector which reaches the point $x_1^0, \cdots, x_n^0$, has the equation

$$r = h, \quad \vartheta_1 = \alpha. \tag{7.5}$$

Consider the function

$$\psi(\xi) = \begin{cases} 1, & \xi \le \frac{1}{3}, \\ -\frac{1}{\pi} \operatorname{arc\,tg\,sh} \dfrac{\xi - \frac{1}{2}}{\left(\frac{2}{3} - \xi\right)\left(\xi - \frac{1}{3}\right)} + \frac{1}{2}, & \frac{1}{3} \le \xi \le \frac{2}{3}, \\ 0, & \xi \ge \frac{2}{3}. \end{cases} \tag{7.6}$$

Obviously, this function has continuous derivatives of any order everywhere.

Now let us construct the function Ω by the formula

$$\Omega = \phi \frac{\partial^{l-1}}{\partial r^{l-1}} \left[\frac{r^{l-1}}{(l-1)!} \psi\left(\frac{r}{h}\right)\right] - \frac{\partial \phi}{\partial r} \frac{\partial^{l-2}}{\partial r^{l-2}} \left[\frac{r^{l-1}}{(l-1)!} \psi\left(\frac{r}{h}\right)\right] + \cdots$$

$$\cdots + \frac{\partial^{l-1}\phi}{\partial r^{l-1}} (-1)^{l-1} \left[\frac{r^{l-1}}{(l-1)!} \psi\left(\frac{r}{h}\right)\right]. \tag{7.7}$$

It is easy to verify the formulas:

$$\Omega(0, \vartheta_1, \cdots, \vartheta_{n-2}, \phi) = \phi(x_1^0, \cdots, x_n^0), \quad \Omega(h, \vartheta_1, \cdots, \vartheta_{n-2}, \phi) = 0, \tag{7.8}$$

$$\frac{\partial \Omega}{\partial r} = \phi \frac{\partial^l}{\partial r^l} \left[\frac{r^{l-1}}{(l-1)!} \psi\left(\frac{r}{h}\right)\right] + (-1)^{l-1} \frac{\partial^l \phi}{\partial r^l} \left[\frac{r^{l-1}}{(l-1)!} \psi\left(\frac{r}{h}\right)\right]. \tag{7.9}$$

Thus we have the identity

$$\phi(x_1^0, \cdots, x_n^0) = \int_h^0 \frac{\partial \Omega}{\partial r}\, dr. \tag{7.10}$$

Let us multiply both sides of formula (7.10) by the function

$$\psi\left(\frac{\vartheta_1}{\alpha}\right) \sin^{n-2}\vartheta_1 \sin^{n-3}\vartheta_2 \cdots \sin\vartheta_{n-2}\, d\vartheta_1 \cdots d\vartheta_{n-2}\, d\phi$$

$$= \frac{1}{r^{n-1}} \psi\left(\frac{\vartheta_1}{\alpha}\right) \frac{D(x_1, \cdots, x_n)}{D(r, \vartheta_1, \cdots, \vartheta_{n-2}, \phi)} d\vartheta_1 \cdots d\vartheta_{n-2} d\phi \quad (7.11)$$

and integrate between the limits

$$0 \le \vartheta_1 \le \alpha,\ 0 \le \vartheta_i \le \pi\,(i = 2, 3, \cdots, n-2),\ -\pi \le \phi \le \pi. \quad (7.12)$$

The left-hand side is thus multiplied by a constant since it does not depend on $\vartheta_1, \cdots, \vartheta_{n-2}, \phi$. Dividing by this constant we will have

$$\phi(x_1^0, \cdots, x_n^0) = \overbrace{\int \cdots \int}^{n}_{V_0} \psi\left(\frac{\vartheta_1}{\alpha}\right) \frac{\partial \Omega}{\partial r} r^{-n+1} dx_1 \cdots dx_n \quad (7.13)$$

where V_0 is the sector reaching the point $x_1^0, \cdots, x_n^0$.

The function under the integral sign in formula (7.13) may be, on the basis of formula (7.9), somewhat simplified. If we bear in mind that the derivative

$$\frac{\partial^l \phi}{\partial r^l} \quad (7.14)$$

is a linear combination of various derivatives of order l of ϕ with respect to the variables $x_1, \cdots, x_n$ with bounded coefficients, then we obtain

$$\psi\left(\frac{\vartheta_1}{\alpha}\right) r^{-n+1} \frac{\partial \Omega}{\partial r} = \mu\phi + \sum \mu_{\alpha_1, \cdots, \alpha_n} \frac{\partial^l \phi}{\partial x_1^{\alpha_1} \cdots \partial x_n^{\alpha_n}}, \quad (7.15)$$

where the function μ is a bounded continuous function and the functions $\mu_{\alpha_1, \cdots, \alpha_n}$ satisfy the inequality

$$|\mu_{\alpha_1, \cdots, \alpha_n}| \le \frac{A}{r^{n-l}} \psi\left(\frac{r}{h}\right). \quad (7.16)$$

Inequality (7.16) is obvious; the continuity of the function μ follows from the fact that

$$\frac{\partial^l}{\partial r^l} \left[\frac{r^{l-1}}{(l-1)!} \phi\left(\frac{r}{h}\right) \right] = 0, \text{ if } r \le \frac{h}{3}. \quad (7.17)$$

The classical Minkowski inequality yields moreover

$$\left[\overbrace{\int\cdots\int}^{n}_{D} |\phi(Q)|^q dQ\right]^{\frac{1}{q}} \le \left[\overbrace{\int\cdots\int}^{n}_{D} \left(\overbrace{\int\cdots\int}^{n}_{\substack{r\le h\\ D}} |\phi||\mu| dP\right)^q dQ\right]^{\frac{1}{q}}$$

$$+ \Sigma\left[\overbrace{\int\cdots\int}^{n}_{D} \left(\overbrace{\int\cdots\int}^{n}_{\substack{r\le h\\ D}} \left|\frac{\partial^l\phi}{\partial x_1^{\alpha_1}\cdots\partial x_n^{\alpha_n}}\right| |\mu_{\alpha_1,\ldots,\alpha_n}| dP\right)^q dQ\right]^{\frac{1}{q}}. \qquad (7.18)$$

According to the fundamental theorem of §5 of Chapter I, the terms under the summation sign are estimated by means of the inequality

$$\left[\overbrace{\int\cdots\int}^{n}_{D} \left(\overbrace{\int\cdots\int}^{n}_{\substack{r\le h\\ D}} \left|\frac{\partial^l\phi}{\partial x_1^{\alpha_1}\cdots\partial x_n^{\alpha_n}}\right| |\mu_{\alpha_1,\ldots,\alpha_n}| dP\right)^q dQ\right]^{\frac{1}{q}} \le MA. \qquad (7.19)$$

In order to estimate the first term, note that

$$\left(\overbrace{\int\cdots\int}^{n}_{\substack{r\le h\\ D}} |\phi||\mu| dP\right)^q = \left(\overbrace{\int\cdots\int}^{n}_{\substack{r\le h\\ D}} |\phi||\mu| dP\right)^{q^*} \left(\overbrace{\int\cdots\int}^{n}_{\substack{r\le h\\ D}} |\phi||\mu| dP\right)^{q-q^*}. \qquad (7.20)$$

Moreover,

$$\overbrace{\int\cdots\int}^{n}_{\substack{r\le h\\ D}} |\phi||\mu| dP \le \left[\overbrace{\int\cdots\int}^{n}_{r\le h} |\mu|^{q^{*\prime}} dP\right]^{\frac{1}{q^{*\prime}}} \left[\overbrace{\int\cdots\int}^{n}_{\substack{r\le h\\ D}} |\phi|^{q^*} dP\right]^{\frac{1}{q^*}}$$

$$= M_2\left(\overbrace{\int\cdots\int}^{n}_{\substack{r\le h\\ D}} |\phi|^{q^*} dP\right)^{\frac{1}{q^*}}, \qquad (7.21)$$

from which

$$\left(\overbrace{\int\cdots\int}^{n}_{\substack{r\le h\\ D}} |\phi||\mu| dP\right)^q \le M_3 \overbrace{\int\cdots\int}^{n}_{\substack{r\le h\\ D}} |\phi|^{q^*} dP \left(\overbrace{\int\cdots\int}^{n}_{\substack{r\le h\\ D}} |\phi|^{q^*} dP\right)^{\frac{q-q^*}{q^*}}$$

$$\le M_3\left[\overbrace{\int\cdots\int}^{n}_{D} |\phi|^{q^*} dP\right]^{\frac{q-q^*}{q^*}} \overbrace{\int\cdots\int}^{n}_{\substack{r\le h\\ D}} |\phi|^{q^*} dP, \qquad (7.22)$$

$$\left[\overbrace{\int\cdots\int}^{n}_{D}\left(\overbrace{\int\cdots\int}^{n}_{\substack{r\leq h\\ D}}|\phi||\mu|\,dP\right)^{q} dQ\right]^{\frac{1}{q}}$$

$$\leq M_4\left[\overbrace{\int\cdots\int}^{n}_{D}|\phi|^{q^*}\,dP\right]^{\frac{q-q^*}{qq^*}}\left[\overbrace{\int\cdots\int}^{n}_{D}\left(\overbrace{\int\cdots\int}^{n}_{\substack{r\leq h\\ D}}|\phi|^{q^*}\,dP\right)dQ\right]^{\frac{1}{q}}$$

$$= M_4\left[\overbrace{\int\cdots\int}^{n}_{D}|\phi|^{q^*}\,dP\right]^{\frac{1}{q^*}-\frac{1}{q}}\left[\overbrace{\int\cdots\int}^{n}_{D}|\phi|^{q^*}\left(\overbrace{\int\cdots\int}^{n}_{\substack{r\leq h\\ D}}dQ\right)dP\right]^{\frac{1}{q}}$$

$$\leq M_5\left[\overbrace{\int\cdots\int}^{n}_{D}|\phi|^{q^*}\,dP\right]^{\frac{1}{q^*}}. \tag{7.23}$$

Combining (7.19), (7.23), and (7.18), we arrive at the desired inequality (7.4).

Thus, this inequality is established for continuous functions. Before going over to the general case, we shall make a certain digression, recalling several theorems relating to approximation of a given function by means of continuously differentiable ones.

§8. **Properties of averaged functions.** Let ϕ be some function summable on any finite subset of the space. We construct a new function

$$\phi_k(x_1, \cdots, x_n) = \chi k^n \overbrace{\int\cdots\int}^{n}\phi(y_1, \cdots, y_n)\psi(kr)\,dy_1\cdots dy_n, \tag{8.1}$$

where

$$r = \sqrt{\sum_{i=1}^{n}(x_i - y_i)^2}.$$

The function ψ was defined in the previous paragraph and the constant χ is defined by the formula

$$\chi = \frac{1}{\int\cdots\int\phi(r)dx_1\cdots dx_n}. \tag{8.2}$$

We shall call ϕ_k a sequence of averaged functions for ϕ. The properties of sequences of averaged functions were considered by us earlier [12]. However, as we plan to develop this question somewhat, we shall recall these properties again:

(1) *Almost everywhere*

$$\lim \phi_k(x_1, \cdots, x_n) = \phi(x_1, \cdots, x_n); \tag{8.3}$$

we shall prove that (8.3) holds at all points where

$$\lim \frac{\int\cdots\int\limits_{r\leq r} \phi(y_1, \cdots, y_n) dy_1 \cdots dy_n}{\int\cdots\int\limits_{r\leq r} dy_1 \cdots dy_n} = \phi(x_1, \cdots, x_n). \tag{8.4}$$

By a well-known theorem of Lebesgue (8.4) holds almost everywhere. Let us first prove the formula

$$\phi_k = \chi k^n \int_0^1 \left\{ \overbrace{\int\cdots\int}^{n}_{C(\psi>\lambda)} \phi(y_1, \cdots, y_n) \, dy_1 \cdots dy_n \right\} d\lambda, \tag{8.5}$$

where $C(\psi > \lambda)$ denotes the sphere in which $\psi(kr) > \lambda$. In order to prove this formula, note that the integral on the right-hand side of (8.1) may be represented with as much accuracy as desired in the form of a sum:

$$\overbrace{\int\cdots\int}^{n} \phi(y_1, \cdots, y_n) \, \psi(kr) \, dy_1 \cdots dy_n = \sum_{i=0}^{N} \lambda_i \left[\overbrace{\int\cdots\int}^{n}_{\psi(kr)\geq\lambda_i} \phi \, dy_1 \cdots dy_n \right.$$

$$\left. - \overbrace{\int\cdots\int}^{n}_{\psi(kr)\geq\lambda_{i+1}} \phi \, dy_1 \cdots dy_n \right] + \overbrace{\int\cdots\int}^{n}_{\psi(kr)=1} \phi \, dy_1 \cdots dy_n + \epsilon, \tag{8.6}$$

where

$$0 = \lambda_0 < \lambda_1 < \lambda_2 < \cdots < \lambda_{N-1} < \lambda_N = 1 \tag{8.7}$$

and the differences between the numbers λ_{i+1} and λ_i are uniformly as small as desired for sufficiently large N. Grouping the terms in (8.6), we obtain

$$\overbrace{\int\cdots\int}^{n} \phi(y_1, \cdots, y_n) \, \psi(kr) \, dy_1 \cdots dy_n = \epsilon(N) + \lambda_0 \overbrace{\int\cdots\int}^{n}_{r\leq k} \phi \, dy_1 \cdots dy_n$$

$$+ \Sigma(\lambda_i - \lambda_{i-1}) \overbrace{\int\cdots\int}^{n}_{\psi(kr)\geq\lambda_i} \phi \, dy_1 \cdots dy_n, \tag{8.8}$$

from which formula (8.5) is obtained by a simple passage to the limit. Analogously,

$$\frac{1}{k^n \chi} = \int_0^1 \left[\overbrace{\int\cdots\int}^{n}_{C(\psi(kr)\geq\lambda)} dy_1 \cdots dy_n \right]. \tag{8.9}$$

From formula (8.4) we will have

$$\overbrace{\int\cdots\int}^{n}_{r\leq r} \phi(y_1, \cdots, y_n) \, dy_1 \cdots dy_n = [\phi(x_1, \cdots, x_n) + \epsilon(r)] \overbrace{\int\cdots\int}^{n}_{r\leq r} dy_1 \cdots dy_n, \tag{8.10}$$

where

$$\epsilon(r) \underset{r \to 0}{\longrightarrow} 0.$$

Substituting (8.10) and (8.9) into (8.5), we obtain

$$\frac{\int_0^1 \left\{ [\phi(x_1, \cdots, x_n) + \epsilon_k(\lambda)] \overbrace{\int \cdots \int}^{n}_{\psi(kr) \geq \lambda} dy_1 \cdots dy_n \right\} d\lambda}{\int_0^1 \left[\overbrace{\int \cdots \int}_{\psi(kr) \geq \lambda} dy_1 \cdots dy_n \right] d\lambda}, \tag{8.11}$$

$\epsilon_k(\lambda) = \epsilon(r)$, where $\psi(kr) \geq \lambda$, in other words,

$$|\phi_k - \phi| \leq \frac{\int_0^1 \left[|\epsilon_k(\lambda)| \overbrace{\int \cdots \int}^{n}_{\psi(kr) \geq \lambda} dy_1 \cdots dy_n \right] d\lambda}{\int_0^1 \left[\overbrace{\int \cdots \int}^{n}_{\psi(kr) \geq \lambda} dy_1 \cdots dy_n \right] d\lambda}, \tag{8.12}$$

i. e., $\lim \phi_k = \phi$, as was to be proved.

(2) *If*

$$\overbrace{\int \cdots \int}^{n}_{\infty} |\phi|^p \, dx_1 \cdots dx_n < \infty, \tag{8.13}$$

$$p \geq 1,$$

then

$$\overbrace{\int \cdots \int}^{n}_{\infty} |\phi_k|^p \, dx_1 \cdots dx_n < \infty \tag{8.14}$$

and, more than that

$$\lim_{k \to \infty} \overbrace{\int \cdots \int}^{n}_{\infty} |\phi_k - \phi|^p \, dx_1 \cdots dx_n = 0. \tag{8.15}$$

It is obvious that property (2) holds if the function ϕ is bounded and differs from zero only on a finite domain; moreover, if

$$|\phi| \leq M,$$

then

$$|\phi_k| \leq \chi k^n \overbrace{\int \cdots \int}^{n} M\psi(kr) \, dy_1 \cdots dy_n = M. \tag{8.16}$$

(8.14) is thus obvious and (8.15) follows from Lebesgue's theorem on passing to the limit under the integral sign for bounded functions.

In the general case the function may always be represented in the form

$$\phi = \bar{\phi}^{(N)} + \bar{\bar{\phi}}^{(N)}, \tag{8.17}$$

where

$$\bar{\phi}^{(N)} = \phi \text{ for } |\phi| \le N, \ \bar{\phi}^{(N)} = N \text{ for } |\phi| > N; \tag{8.18}$$

obviously,

$$\overbrace{\int\cdots\int}^{n} |\bar{\bar{\phi}}^{(N)}|^p \, dx_1 \cdots dx_n < \epsilon. \tag{8.19}$$

Let us consider separately the functions $\bar{\phi}_k^{(N)}$ and $\bar{\bar{\phi}}_k^{(N)}$, the respective averages for $\bar{\phi}^{(N)}$ and $\bar{\bar{\phi}}^{(N)}$. Clearly,

$$\phi_k = \bar{\phi}_k^{(N)} + \bar{\bar{\phi}}_k^{(N)}. \tag{8.20}$$

Consider the integral

$$\left[\overbrace{\int\cdots\int_\infty}^{n} |\phi - \phi_k|^p \, dx_1 \cdots dx_n\right]^{\frac{1}{p}} \le \left[\overbrace{\int\cdots\int_\infty}^{n} |\bar{\phi}^{(N)} - \bar{\phi}_k^{(N)}|^p \, dx_1 \cdots dx_n\right]^{\frac{1}{p}}$$

$$+ \left[\overbrace{\int\cdots\int_\infty}^{n} |\bar{\bar{\phi}}^{(N)}|^p \, dx_1 \cdots dx_n\right]^{\frac{1}{p}} + \left[\overbrace{\int\cdots\int_D}^{n} |\bar{\bar{\phi}}_k^{(N)}|^p \, dx_1 \cdots dx_n\right]^{\frac{1}{p}}. \tag{8.21}$$

Choose the number N such that

$$\left[\overbrace{\int\cdots\int_\infty}^{n} \bar{\bar{\phi}}^{(N)p} \, dx_1 \cdots dx_n\right]^{\frac{1}{p}} < \frac{\epsilon}{3K}, \tag{8.22}$$

where

$$K = \chi \overbrace{\int\cdots\int_{r\le 1}}^{n} dx_1 \cdots dx_n > 1. \tag{8.23}$$

Let us estimate the integral

$$\overbrace{\int\cdots\int_D}^{n} |\bar{\bar{\phi}}_k^{(N)}|^p \, dx_1 \cdots dx_n. \tag{8.24}$$

We have

$$|\bar{\bar{\phi}}_k^{(N)}| = \chi k^n |\overbrace{\int \cdots \int}^{n}_{r \leq \frac{1}{k}} \bar{\bar{\phi}}^{(N)}(y_1, \cdots, y_n)\psi(kr)\,dy_1 \cdots dy_n|$$

$$\leq \chi k^n \left[\overbrace{\int \cdots \int}^{n}_{r \leq \frac{1}{k}} |\bar{\bar{\phi}}^{(N)}(y_1, \cdots, y_n)|^p\,dy_1 \cdots dy_n\right]^{\frac{1}{p}} \left[\overbrace{\int \cdots \int}^{n} |\psi(kr)|^{p'}\,dy_1 \cdots dy_n\right]^{\frac{1}{p'}}$$

$$\leq (\chi k^n)^{1 - \frac{1}{p'}} \left[\overbrace{\int \cdots \int}^{n}_{r \leq \frac{1}{k}} |\bar{\bar{\phi}}^{(N)}(y_1, \cdots, y_n)|^p\,dy_1 \cdots dy_n\right]^{\frac{1}{p}}, \tag{8.25}$$

where $\frac{1}{p'} + \frac{1}{p} = 1$, or

$$|\phi_k^{(N)}|^p \leq \chi k^n \overbrace{\int \cdots \int}^{n}_{r \leq \frac{1}{k}} |\bar{\bar{\phi}}^{(N)}(y_1, \cdots, y_n)|^p\,dy_1 \cdots dy_n. \tag{8.26}$$

Now let us estimate the integral

$$\overbrace{\int \cdots \int}^{n}_{\infty} |\bar{\bar{\phi}}_k^{(N)}|^p\,dx_1 \cdots dx_n$$

$$\leq \overbrace{\int \cdots \int}^{n}_{\infty} \chi k^n \left[\overbrace{\int \cdots \int}^{n}_{r \leq \frac{1}{k}} |\bar{\bar{\phi}}^{(N)}(y_1, \cdots, y_n)^p\,dy_1 \cdots dy_n\right] dx_1 \cdots dx_n$$

$$= \overbrace{\int \cdots \int}^{n}_{\infty} |\bar{\bar{\phi}}^{(N)}(y_1, \cdots, y_n)|^p \Big(\chi k^n \overbrace{\int \cdots \int}^{n}_{r \leq \frac{1}{k}} dx_1 \cdots dx_n\Big)\,dy_1 \cdots dy_n$$

$$\leq K \overbrace{\int \cdots \int}^{n}_{\infty} |\bar{\bar{\phi}}^{(N)}|^p\,dy_1 \cdots dy_n,$$

or, finally, by (8.22),

$$\overbrace{\int \cdots \int}^{n}_{\infty} |\bar{\bar{\phi}}_k^{(N)}|^p\,dx_1 \cdots dx_n \leq \frac{\epsilon}{3}. \tag{8.27}$$

Now taking k sufficiently large to have

$$\left[\overbrace{\int \cdots \int}^{n} |\bar{\phi}^{(N)} - \bar{\phi}_k^{(N)}|^p\,dx_1 \cdots dx_n\right]^{\frac{1}{p}} \leq \frac{\epsilon}{3}, \tag{8.28}$$

we obtain

$$\left[\overbrace{\int\cdots\int}^{n}_{\infty} |\phi - \phi_k|^p \, dx_1 \cdots dx_n\right]^{\frac{1}{p}} < \epsilon. \tag{8.29}$$

From formula (8.29) follows (8.14) and (8.15) in the general case.

Corollary. If for the function ϕ the integral

$$\overbrace{\int\cdots\int}^{n}_{D} |\phi|^p \, dx_1 \cdots dx_n$$

taken over any bounded domain D is bounded, then we may substitute for the infinite domain everywhere in the formulation of the theorem any finite domain. The proof is obvious.

(3) *The derivative of the averaged function is equal to the averaged function of the derivative if we understand the latter in the sense that it was indicated above.*

Indeed, using the fact that $\psi(kr)$ depends only on the differences of the coordinates $x_i - y_i$, we will obtain

$$\frac{\partial^\alpha \phi_k}{\partial x_1^{\alpha_1} \cdots \partial x_n^{\alpha_n}} = \chi k^n \overbrace{\int\cdots\int}^{n} \phi(y_1, \cdots, y_n) \frac{\partial^\alpha \psi(kr)}{\partial x_1^{\alpha_1} \cdots \partial x_n^{\alpha_n}} dy_1 \cdots dy_n$$

$$= \chi k^n \overbrace{\int\cdots\int}^{n} \phi(y_1, \cdots, y_n)(-1)^\alpha \frac{\partial^\alpha \psi(kr)}{\partial y_1^{\alpha_1} \cdots \partial y_n^{\alpha_n}} dy_1 \cdots dy_n,$$

or, accordi ng to the definition of the derivative,

$$\frac{\partial^\alpha \phi_k}{\partial x_1^{\alpha_1} \cdots \partial x_n^{\alpha_n}} = \chi k^n \overbrace{\int\cdots\int}^{n} \frac{\partial^\alpha \phi}{\partial y_1^{\alpha_1} \cdots \partial y_n^{\alpha_n}} \psi(kr) \, dy_1 \cdots dy_n, \tag{8.30}$$

as was to be proved.

(4) *If any derivative of the averaged functions*

$$\frac{\partial^\alpha \phi_k}{\partial x_1^{\alpha_1} \cdots \partial x_n^{\alpha_n}}$$

has over some domain D an integral uniformly bounded with respect to k:

$$\overbrace{\int\cdots\int}^{n} \left|\frac{\partial^\alpha \phi_k}{\partial x_1^{\alpha_1} \cdots \partial x_n^{\alpha_n}}\right|^p dx_1 \cdots dx_n \leq A, \tag{8.31}$$

where $p > 1$, then the derivative

$$\frac{\partial^\alpha \phi}{\partial x_1^{\alpha_1} \cdots \partial x_n^{\alpha_n}}$$

of the function ϕ exists and is pth power summable on D, satisfying in turn the inequality

$$\overbrace{\int \cdots \int}^{n}_{D} \left| \frac{\partial^\alpha \phi}{\partial x_1^{\alpha_1} \cdots \partial x_n^{\alpha_n}} \right|^p dx_1 \cdots dx_n \le A. \tag{8.32}$$

This property is a consequence of the general indicator of the existence of the derivative (see §6).

From (2) and (3) follows, moreover, the corollary:

$$\lim_{k \to \infty} \overbrace{\int \cdots \int}^{n}_{D} \left| \frac{\partial^\alpha \phi_k}{\partial x_1^{\alpha_1} \cdots \partial x_n^{\alpha_n}} - \frac{\partial^\alpha \phi}{\partial x_1^{\alpha_1} \cdots \partial x_n^{\alpha_n}} \right|^p dx_1 \cdots dx_n = 0. \tag{8.33}$$

It is obvious that the converse assertion is also valid. From (8.32) follows (8.31) and (8.33).

§9. **Proof of the fundamental theorem in the general case.** Now let ϕ be any function q^*th power summable on the domain D, and let the derivatives of order l of ϕ have bounded pth power integrals.

Let us construct a sequence of averaged functions ϕ_k. By what has been proved, for this sequence all of our hypotheses remain in force, since the averaged function will be q^*th power summable and its derivatives of order l pth power summable on the entire interior of the domain. We can establish inequality (7.4) for the averaged functions. After that a passage to the limit yields this inequality for the general case.

§10. **The remaining subcases of the fundamental theorem.** Let us return now to the fundamental theorem of §7 and consider the remaining possibilities. The second case

$$p = 1 < \frac{n}{l} \tag{10.1}$$

is easily considered if we make use of the weaker theorem indicated in the footnote to §5 in place of the generalized Riesz's inequality. In this case, repeating our considerations carried out in the first case, we can establish only the summability of the terms

$$\overbrace{\int \cdots \int}^{n}_{r \le h} \mu_{\alpha_1, \cdots, \alpha_n} \frac{\partial^l \phi}{\partial x_1^{\alpha_1} \cdots \partial x_n^{\alpha_n}} dy_1 \cdots dy_n \tag{10.2}$$

to any power q^{**} satisfying the condition

$$q^* \leq q^{**} < q. \tag{10.3}$$

Therefore inequality (7.4) is replaced by another one:

$$\left[\overbrace{\int \cdots \int_D}^{n} |\phi|^{q^{**}} dx_1 \cdots dx_n\right]^{\frac{1}{q^{**}}}$$

$$\leq LA + N\left[\overbrace{\int \cdots \int_D}^{n} |\phi|^{q^*} dx_1 \cdots dx_n\right]^{\frac{1}{q^*}}. \tag{10.4}$$

The next case presents an analogous picture. If

$$p = \frac{n}{l} > 1, \tag{10.5}$$

then $q = \infty$ and we may prove (10.4) for any positive q^{**}.

Finally, if

$$p = \frac{n}{l} = 1, \tag{10.6}$$

then all of the $\mu_{\alpha_1, \cdots, \alpha_n}$ will be bounded and, consequently, the function ϕ will be a bounded continuous function of all of its arguments if we impose certain new restrictions on the domain, which are similar to those that were considered by the author, [1] and [2].

It is not difficult to show by examples that in none of these cases can the exponent q be increased.

BIBLIOGRAPHY

[1] S. L. Sobolev, *On some estimates of families of functions having square-integrable derivatives*, Dokl. Akad. Nauk SSSR 1 (1936), 267–270. (Russian)

[2] ———, *Sur un problème limite pour les équations polyharmoniques*, Mat. Sb. (N. S.) 2 (44) (1937), 467–500. (Russian. French summary)

[3] F. Frankl, *Über das Anfangswertproblem für lineare und nichtlineare hyperbolische Differentialgleichungen zweiter Ordnung*, Mat. Sb. (N. S.) 2 (44) (1937), 793–814. (Russian. German summary)

[4] V. I. Kondrašov, *On some estimates of families of functions subject to an integral inequality*, Dokl. Akad. Nauk SSSR 18 (1938), 235–240. (Russian)

[5] S. L. Sobolev, *On a theorem of functional analysis*, Dokl. Akad. Nauk SSSR 20 (1938), 5–10. (Russian)

[6] F. Riesz, *Sur une inégalité intégrale (Extrait d'une lettre adressée à Hardy)*,

J. London Math. Soc. 5 (1930), no. 19, 162–168.

[7] L. A. Ljusternik, *The Brunn-Minkowski inequality for arbitrary measurable sets*, Dokl. Akad. Nauk SSSR 3 (1935), 55–58. (Russian)

[8] W. Blaschke, *Kreis und Kugel*, Viet, Leipzig, 1916; Reprint, Chelsea, New York, 1949.

[9] G. Hardy, J. Littlewood and G. Pólya, *Inequalities*, Cambridge Univ. Press, London, 1934.

[10] L. A. Ljusternik, *Fundamental concepts of functional analysis*, Uspehi Mat. Nauk 1 (1936), 77–140. (Russian)

[11] I. G. Petrovskiĭ and K. N. Smirnov, *On conditions of equicontinuity of families of functions*, Bjull. Moskov. Univ. Ser. A 1 (1938), no. 8. (Russian)

[12] *General theory of diffraction of waves on Riemann surfaces*, Trudy Fiz.-Mat. Inst. Steklov. 9 (1935), 39–108. (Russian)

Translated by:
J. R. Brown

Ivan Georgievich Petrovsky

Born – 18 Jan 1901 in Sevsk, Orlov guberniya, Russia
Died – 15 Jan 1973 in Moscow, USSR

Ivan Petrovsky was a student of Egorov at the University of Moscow. From this time he began to build up his collection of books. His personal library contained over 30 thousand books. It is remarkable that Petrovsky did not have a catalogue but knew his way round the library very well and was able to find his favourite books quickly.

Petrovsky taught at Moscow University and, from 1943, at the Steklov Institute. He became Rector of Moscow University and Vice-Director of the Steklov Institute.

Petrovsky's main mathematical work was on the theory of partial differential equations, the topology of algebraic curves and surfaces, and probability. Petrovsky also worked on the boundary value problem for the heat equation and this was applied to both probability theory and work of Kolmogorov.

P. S. Aleksandrov and O. A. Oleinik described his career as follows:- Petrovsky's knowledge was encyclopaedic. He had a thorough understanding of modern science and all its interconnections. He was perspicacious and far-sighted, was able to discern long-term trends, and always emphasized them. He also showed a wide general culture, knowledge of diverse branches of science, a deep understanding of state problems in many outstanding public and state activities.

I. G. PETROVSKIĬ

On Some Problems of the Theory of Partial Differential Equations

[Uspehi Matematicheskih Nauk (N.S.) 1, no. 3-4 (13-14), 44-70 (1946)]

I. Cauchy's Problem. Hyperbolic Equations

We consider an equation of the form

$$\frac{\partial^n u}{\partial t^n} = F\left(t, x_1, \ldots, x_p, u, \ldots, \frac{\partial^n u}{\partial t^{k_0}\,\partial x_1^{k_1} \ldots \partial x_p^{k_p}}\right),$$

whose right member depends on the independent variables $t, x_1, \ldots, x_p$, the unknown function u and its partial derivatives up to order n, $(k_0 < n)$, with respect to these independent variables. For the sake of simplicity we shall consider mostly one equation with one unknown function, although almost all results extend naturally to systems of equations.

Cauchy's problem consists in finding that solution of this equation which for $t = 0$ satisfies the conditions

$$\frac{\partial^k u}{\partial t^k} = \varphi_k(x_1, \ldots, x_p) \qquad k = 0, 1, \ldots, n-1.$$

The following theorem of Mme. Kovalevski is fundamental in this connection:

Let the function F be holomorphic with respect to all its arguments in a neighborhood of some fixed values (which for the sake of simplicity we shall assume to be zero) and let the functions ϕ_k be holomorphic in all their arguments. If then the functions ϕ_k as well as all their derivatives up to the nth order vanish when all x_k are equal to zero, then Cauchy's problem has one and only one holomorphic solution in a certain neighborhood of the origin.

N. A. Lednev proved recently that the requirement of analyticity of F with respect to t in Mme. Kovalevski's theorem can be omitted and replaced by the condition of differentiability with respect to t. Then the solution of Cauchy's problem will, in general, not be analytic in t.

In connection with Mme. Kovalevski's theorem the following question presents itself: Are there solutions of Cauchy's problem other than the holomorphic one given by her theorem, in a region G of the $(t, x_1, \ldots, x_p)$-space bounded by a domain G_0 in the

4

hypersurface $t = 0$, on which the initial conditions are prescribed? A first answer to this question is the theorem proved by Holmgren [1]* in 1901, guaranteeing the uniqueness of the solution of Cauchy's problem in the class of n times differentiable continuous functions when the differential equation is linear and has coefficients that are holomorphic in a certain region G (whose boundaries are given below).

On the other hand Hadamard showed, [2], that the uniqueness proof for the non-linear case can be reduced to the proof of this theorem for linear equations with non-analytic coefficients of sufficient smoothness, i.e., with a sufficient number of derivatives. Since then, all efforts to prove the uniqueness of the solution of Cauchy's problem within the class of non-analytic functions have been concentrated on the proof of this theorem for linear equations with non-analytic but smooth coefficients.

For 40 years no success was scored in this direction, except for the uniqueness proof for hyperbolic equations which will be mentioned below. It was not until 1940 that Carleman [3] proved that for linear partial differential equations of the form

$$\frac{\partial^n u}{\partial t^n} = \sum_{\substack{k=k_0+k_1\leq n \\ k_0<n}} a^{(k_0,k_1)} \frac{\partial^k u}{\partial t^{k_0}\, \partial x_1^{k_1}},$$

in two independent variables Cauchy's problem has a unique solution in the class of n times differentiable functions, provided that the equation

$$\lambda^n - \sum_{k_0+k_1=n} a^{(k_0,k_1)} \lambda^{k_0} = 0$$

has no multiple roots. Nothing more has been done in this direction. Thus the complete clarification of the uniqueness question for Cauchy's problem in the domain of non-analytic functions remains one of the fundamental and apparently most difficult problems in the theory of partial differential equations. In this question we presumably have to do with a situation which

* Numbers in brackets refer to the bibliography at the end of this article.

can often be found in the theory of partial differential equations: The theorem is easy to prove for one independent variable, i.e., for ordinary differential equations. It is more complicated when the number of independent variables is two (Carleman's case); but this latter problem is still much simpler than the case of more than two independent variables. The difficulty for an arbitrary number of independent variables is the same as for three. So far we have not said anything about the boundaries of the region in which Holmgren's and Carleman's theorems guarantee the uniqueness of the solution of Cauchy's problem. In order to describe these boundaries, we shall have to introduce the concept of "characteristic", which plays an important role in the whole theory of partial differential equations. For simplicity we limit ourselves to the case of a linear equation, i.e.,

$$\sum_{k=k_0+k_1+\cdots+k_p\leq n} a^{(k_0,k_1,\ldots,k_p)}(t,x_1,\ldots,x_p)\frac{\partial^k u}{\partial t^{k_0}\partial x_1^{k_1}\cdots\partial x_p^{k_p}} = f(t,x,\ldots,x_p), \quad (1)$$

The position of a plane

$$\alpha_0(t-t^0)+\alpha_1(x_1-x_1^0)+\ldots+\alpha_p(x_p-x_p^0)=0$$

through the point $P(t^0,x_1^0,\ldots,x_p^0)$ is called characteristic for the given equation at the point P, if

$$\Phi\equiv\sum_{k_0+\cdots+k_p=n} a^{(k_0,k_1,\ldots,k_p)}(P)\,\alpha_0^{k_0}\ldots\alpha_p^{k_p}=0.$$

A surface in the $(t,x_1,\ldots,x_p)$-space such that all its tangent (hyper)-planes have characteristic position, is called a "*characteristic surface*" or "*characteristic*". When we talk of characteristics in the following, we shall always mean *real* characteristics. It is easy to show that through any portion of a $(p-1)$-dimensional surface in the hyperplane $t=0$ there pass at most n real characteristics.

6

It may happen that several characteristics coincide; then they are called multiple. If the form Φ is definite, then the equation (1) has no real characteristics at all near the point P. Such an equation is called *elliptic*.

Let us now return to the question of the uniqueness of the solution of Cauchy's problem in the class of non-analytic functions. We assume that in the (hyper)-plane $t = 0$ initial values are prescribed on some region G_0 bounded by a sufficiently smooth $(p - 1)$-dimensional surface S. Through S we pass all possible real characteristics. By Holmgren's theorem a uniqueness domain of the solution of Cauchy's problem for linear equations with analytic coefficients is given by the domain which is bounded by G_0 on one side and whose other boundaries are defined by the condition that none of those characteristics passes through the domain.

For linear elliptic equations with analytic coefficients -- which, as we know, have no real characteristic -- Holmgren's theorem implies that two solutions of the equation which, together with their partial derivatives up to order $n - 1$, coincide on some $(p - 1)$-dimensional portion of a real surface, will coincide everywhere. Analogous consequences can be drawn from Carleman's theorem concerning non-analytic partial differential equations in two independent variables.

The question whether the unique regular solution can be continued beyond the characteristics is one of the fundamental problems of the theory of partial differential equations, and has not yet been solved even in the analytic case. All that exist are certain partial results.

Long ago, for example, Goursat [4] answered in the affirmative the question whether a non-unique continuation may exist in the analytic case, if the characteristics are not multiple. He obtained a similar answer for certain very special cases of multiple characteristics.

The uniqueness problem for Cauchy's problem has up to now been considered "in the small" only. A series of very interesting results has recently been obtained by A. D. Myshkis [4a] concerning the uniqueness of the solutions of systems of partial differential equations in the whole domain of existence of these solutions. He constructed examples which show that there exist real characteristics through which in a large domain the solution of the given system can be continued in one way only. It may be useful to remark here that it is unknown, even for most of the very simplest non-analytical equations, whether even one solution exists! An investigation of this question would be of the greatest importance.

The following remark by Hadamard [5] contains an important hint for the further study of Cauchy's problem. In the case of Laplace's equation

$$\frac{\partial^2 u}{\partial t^2}+\frac{\partial^2 u}{\partial x^2}=0,$$

to which the uniqueness theorem may be applied, it is possible to modify the solution of Cauchy's problem arbitrarily much and arbitrarily near the line $t = 0$ by changing the initial values of $u(t,x)$ and $u'_t(t,x)$ at $t = 0$ as little as we please. To do this, one adds to the previous solution

$$v(t, x)=\frac{1}{n^k}\sin nx\cdot\frac{e^{nt}-e^{-nt}}{2},$$

where n and k are positive constants. For n sufficiently large, the function v and all its partial derivatives with respect to t and x up to order $k - 1$ are arbitrarily small at $t = 0$, but for $t > \epsilon > 0$ the function $v(t,x)$ assumes arbitrarily large values. The solution of Cauchy's problem for such equations as that of Laplace is therefore without physical interest, even if the solution exists. For the initial conditions are usually obtained empirically and are, therefore, not known with absolute precision. But arbitrarily small errors in the determination of these initial conditions can distort the solution of Cauchy's problem to an arbitrarily large extent, even arbitrarily near the line on which the initial values are prescribed. For the same reason it is hopeless to attempt to find an approximate solution of Cauchy's problem for Laplace's equation with non-analytic initial conditions by means of approximation with analytic functions. On the

8

other hand, it is impossible to limit oneself to analytic initial conditions in physical investigations. Following Hadamard's ideas we shall say that Cauchy's problem is correctly posed in a certain domain G of the $(t,x_1,\dots,x_p)$-space, bounded by the region G_0 of the hyperplane $t = 0$, if the following two conditions are satisfied:

1. For every system of initial functions $\phi_k(x_1,\dots,x_p)$, $k = 0, 1,\dots, n-1$, there exists in G one and only one solution of Cauchy's problem provided the ϕ_k are differentiable up to a certain finite order L.

2. To every positive number ϵ there exists a positive η, such that if all ϕ_k and their derivatives up to order L are changed by amounts less than η, then the solution of Cauchy's problem in G will be changed by less than ϵ.*)

A class of equations for which it is generally known that Cauchy's problem is correctly posed are the so-called *hyperbolic equations*, [6]. For reasons of simplicity we limit the discussion to the linear equation

$$\frac{\partial^n u}{\partial t^n} = \sum_{\substack{k_0+\dots+k_p\le n\\ k_0<n}} a^{(k_0,k_1,\dots,k_p)}(t,x_1,\dots,x_p)\frac{\partial^k u}{\partial t^{k_0}\partial x_1^{k_1}\dots\partial x_p^{k_p}} + f(t,x_1,\dots,x_p) \quad (2)$$

with coefficients which are sufficiently often differentiable.

Consider in the $(\alpha_1,\alpha_2,\dots,\alpha_p)$-space the real algebraic surface

$$1 - \sum_{k_0+k_1+\dots+k_p=n} a^{(k_0,k_1,\dots,k_p)}(t,x_1,\dots,x_p)\,\alpha_1^{k_1}\dots\alpha_p^{k_p} = 0. \quad (3)$$

An equation (2) is called *hyperbolic* in a certain region of the $(t,x_1,\dots,x_p)$-space, if the points of the surface (3) form $[n/2]$ nested ovals containing the origin of the coordinate system in their interior. To this, in the case of odd n, has to be added a so-called "odd piece" consisting of a homeomorphic image of the projective hyperplane. It is assumed that these ovals and the

*)It would be of interest to clarify the connection between these two conditions. For ordinary differential equations it is known that the second condition is satisfied whenever the first is.

"odd piece" have no points of intersection. Then every real straight line through the origin of the coordinate system meets the surface (3) in n real distinct points. The ovals may extend to infinity. In order to explain what is meant by nested ovals in this case, we have to represent the projective space as a hypersphere in which opposite endpoints of the diameters are identified. To every oval corresponds a double cone which cuts out two ovals from the sphere. We say that one oval is contained in a second one, if the cone that belongs to the first oval lies inside the cone that belongs to the second oval. To the "odd piece" there corresponds on the hypersphere a surface which is homeomorphic to the intersection of the hypersphere with a hyperplane through the center of the sphere.

If two ovals of the surface (3) meet in even one point only, then -- as proved by the examples I constructed in [7] -- the coefficients of the lowest derivatives may have an influence on whether Cauchy's problem is posed correctly or not. If the coefficients of the equation (2) depend on t only, I have shown in [8] that a necessary condition for Cauchy's problem to be posed correctly is that the real surface (3) consist of $[n/2]$ possibly intersecting ovals, which contain the origin of the coordinate system, and -- if n is odd -- also of the "odd piece". John showed in [9], that the following is true for linear equations with analytic coefficients: If the surface (3) has any topological structure other than the one just described, then the initial values of Cauchy's problem for the equation (2) cannot be prescribed arbitrarily, not even on an arbitrarily small surface. This is true, even if the initial values are continuous together with their derivatives up to an arbitrarily high order L.

In 1938, S. A. Halpern [10] examined the question as to when Cauchy's problem is posed correctly for systems in which the number of equations exceeds the number of unknown functions.

We have not said anything yet concerning the size the number L, giving the order of differentiability the initial data, must have, if the existence of the solution of Cauchy's problem for hyperbolic equations is to be assured, As shown by Poisson's and Cauchy's formulas, the simplest equations of the form

$$\frac{\partial^2 u}{\partial t^2} = \frac{\partial^2 u}{\partial x_1^2} + \cdots + \frac{\partial^2 u}{\partial x_p^2} \tag{4}$$

require $L = [p/2] + 3$, if the solution is to be twice continuously differentiable. Schauder [11], who obtained for the first time solutions for quasi-linear equations, was not interested in this number. In F. J. Frankl's work [12], where an idea analogous to that of Schauder is used, the values $L = p + 3$ or $p + 4$ are obtained, according as p is even or odd. The same estimate is found in S. A. Christianovich [13]. S. L. Sobolev [14] found $L = [p/2] + 4$ for non-linear hyperbolic equations of the second order. His methods are interesting and will have many applications in the future. In my paper [6] on general hyperbolic equations of order n, L is $4p + 4 + n$. This is probably 8 times too large. It would be very interesting if L could be reduced to a minimal, actually necessary, size.

For the case $p = 1$, Cauchy's problem for hyperbolic equations was solved for the first time independently by Holmgren and N. M. Günther. V. I. Smirnov and S. L. Sobolev discovered and investigated in detail the class of functionally invariant solutions of hyperbolic equations of the second order with constant coefficients. These are solutions such that any function of them is again a solution of the same equation. These functionally invariant solutions can be applied in analogy to the method of D'Alembert for the equation of the vibrating string,

$$\frac{\partial^2 u}{\partial t^2} = \frac{\partial^2 u}{\partial x^2}.$$

This well known method is based on the two solutions $x + t$ and $x - t$ and the fact that all arbitrary functions of these are again solutions of the equation of the string.

I now pass on to the question of the character of the dependence of the solution of Cauchy's problem for hyperbolic equations on the initial data. The simplest examples show that the base of the characteristic cone with vertex at $P(t^0, x^0, \ldots x_p^0)$ may contain domains such that a change of the initial data there does not change the value of the solution of Cauchy's problem at the vertex of the cone. We shall call such domains *lacunae*. It is, for example, known that the value at P of the solution of Cauchy's problem for the wave equation (4) depends for odd $p > 1$ only on the initial values on the boundary of the base of the characteristic cone with vertex at P. But for even p, and also for $p = 1$ the quantity $u(P)$ depends on the initial data on the whole base of that cone. It follows that, if the values of u and u_t' for $t = 0$ are changed in some small neighborhood of the point $(x_1^0, \ldots, x_p^0)$, the value of the function u will for odd p and $t > 0$, be influenced only in those points of the $(x_1, \ldots, x_p)$ space which lie near the sphere of radius t and center at $(x_1^0, \ldots, x_p^0)$. In this way, a disturbance produced initially at the point $(x_1^0, \ldots, x_p^0)$ generates a spherical wave with center at that point. This wave has a well defined front and rear end. For even p and for $p = 1$, the initial disturbance at the point $(x_1^0, \ldots, x_p^0)$ creates an effect, generally speaking, at all points within the sphere of radius t and center at $(x_1^0, \ldots, x_p^0)$. Thus there is generated a wave with a precise front but a diffused rear. This is referred to as *diffusion* of the wave rear. There is no diffusion for odd $p > 1$..

Already at the beginning of this century Hadamard showed [15] that for linear hyperbolic equations of the second order with variable coefficients there is always diffusion of the waves for even p, provided the vertex of the cone is sufficiently close to the initial plane. For general linear hyperbolic equations with odd p the question of wave diffusion remained unsolved for a long time. Only in 1939 there appeared an article by Mathisson [16], where the following is proved: For $p = 3$ the only class of linear hyperbolic equations of second order without wave diffusion are those whichresult from (4) by the following operations:

1) Multiplication of both members of the equation by a function of $t, x_1, \ldots, x_p$,

2) Linear transformations of the unknown function,

3) Transition from one set of independent variables $t, x_1, \ldots, x_p$ to another.

Mathisson's methods are also applicable, in principle, to hyperbolic equations of the second order with any odd $p > 1$. But for $p > 3$ the calculations are more extensive and have not been carried out completely.

I have investigated analogous questions for general hyperbolic equations [17]. Consider a linear hyperbolic equation (3) with sufficiently smooth coefficients. The lateral surface of the characteristic cone with vertex at $P^*(t^*, x_1^*, \ldots, x_p^*)$ divides its base in the plane $t = t^0$, in general, into several domains. One such domain G_t^0 will be called a *lacuna*, if arbitrary but sufficiently smooth changes of the initial data, inside G_t^0 only, do not change the solution of Cauchy's problem for equation (2) at the point P^*. To fix the ideas we assume $t^0 < t^*$. Let t^0 be sufficiently close to t^*, and let $G^*_{t^0}$ be one of the domains into which the surface of the characteristic cone with vertex at P^* constructed for the linear equation with constant coefficients

$$\frac{\partial^n u}{\partial t^n} = \sum_{k_s} a^{(k_0, k_1, \ldots, k_p)}(t^*, x_1, \ldots, x_p) \frac{\partial^n u}{\partial t^{k_0} \partial x_1^{k_1} \ldots, \partial x_p^{k_p}} \qquad (5)$$

$$k_0 + k_1 + \ldots + k_p = n, \quad k_0 < n$$

divides the base of the cone in the plane $t = t^0$. Then there corresponds to $G^*_{t^0}$ a uniquely determined domain G_{t^0} constructed for the equation (2), such that $G^*_{t^0}$ is close to G_{t^0} (as $t^0 \to t^*$).

If the domain G_{t^0} corresponding to the domain $G^*_{t^0}$ is a lacuna for equation (2) for all t^0 sufficiently close to t^*, then the domain $G^*_{t^0}$ is a lacuna for equation (5). An analogous theorem can be obtained for non-linear hyperbolic equations. Hadamard's theorem asserting the occurrence of wave diffusion for all linear hyperbolic equations of second order with an even number of space dimensions is an obvious consequence of the preceding theorem. For wave diffusion is always present in second order linear hyperbolic equations with constant coefficients, which can be reduced to an equation of the type (4) by means of a linear transformation in the $(t, x_1, \ldots, x_p)$-space. It would be very important to find conditions necessary for the existence of lacunae not only for linear but also for non-linear hyperbolic equations, in the manner of Mathisson's work for linear second order equations.

All remarks below will be concerned with the investigation of lacunae in linear hyperbolic systems with constant coefficients of the form (5). We shall be interested only in *stable* lacunae, i.e., those which are not destroyed by any sufficiently small change in the coefficients of the equation (5). For that equation Herglotz [18] found in 1926 formulas giving a solution of Cauchy's problem.

According to him the characteristic cone K with vertex at the point $P(t^0, x_1^0, \ldots, x_p^0)$ can be obtained as envelope of that family of planes $(t - t^0) + \alpha_1(x_1 - x_1^0) + \ldots + \alpha_p(x_p - x_p^0) = 0,$ (6)

which passes through P and whose coefficients satisfy the equation

$$1 - \sum_{k_0 + \ldots + k_p = n} a^{(k_0, k_1, \ldots, k_p)} \alpha_1^{k_1} \ldots \alpha_p^{k_p} = 0. \quad (7)$$

For simplicity we write here $a^{(k_0, \ldots, k_p)}$ instead of $a^{(k_0, \ldots, k_p)}(t^*, x_1^*, \ldots, x_p^*)$.

Let us suppose that $x_k^0 = 0$, $k = 1, \ldots, p$, at $t = 0$; then (6) reduces to

$$-t^0 + \alpha_1 x_1 + \ldots + \alpha_p x_p = 0. \quad (8)$$

Equation (7) can be regarded as a tangential equation of the intersection between the lateral surface of the characteristic cone with vertex at P and the hyperplane $t = 0$. By virtue of the principle of duality every point $(x_1, \ldots, x_p)$ of that hypersurface corresponds to a plane of the form (8) in the $(\alpha_1, \ldots, \alpha_p)$-space.

In order to answer the question whether a given point $(x_1, \ldots, x_p)$ belongs to a stable lacuna in the base of the cone K we choose an arbitrary point A in the plane (8) of the $(\alpha_1, \ldots, \alpha_p)$-space. Through it we draw in that plane all possible straight lines with real coefficients, and consider their intersections with the surface (7). The real intersections (as well as the imaginary ones and their limits) form certain $(p - 2)$-dimensional cycles which we shall denote by C_{real} and C_{imag}, respectively. The cycle C_{real} is obviously independent of the choice pf the point A. An imaginary cycle C_{imag} is transformed into one that is homologous to it in the complex intersection of (7) and (8), if A' is changed.

Assuming that the surface (7) has no singular points, the following theorem can be formulated: *Let $n \leq p$; then a point belongs to a stable lacuna, if and only if for even p the corresponding cycles C_{real} (and for odd p the corresponding cycles C_{imag}) are homologous to zero in the complex intersection of (7) and (8).*

There are examples where such lacunae lie in the interior of the base of the characteristic cone.

For $n > p$ this condition is only necessary. If it is satisfied, one can construct in the complex surface (7) itself a $(p-1)$-dimensional cycle Σ such that the point $(x_1, x_2, \ldots, x_p)$ belongs to a lacuna, if and only if Σ is homologous to zero in the complex surface (7). We shall not describe here in more detail the construction of the cycle Σ. The existing examples for lacunae when $n > p$ are trivial in the sense that these lacunae form the exterior of the characteristic cone. It would be interesting to know if there exist other lacunae for $n > p$.

It would be important to be able to tell from the real intersection of the surface (7) and the plane (8) whether the corresponding cycles $C_{\mathrm{real}}\ (t,x)$ or $C_{\mathrm{imag}}\ (t,x)$ are homologous to zero in the complex intersection of (7) and (8). We know, e.g., that the cycles $C_{\mathrm{imag}}\ (t,x)$ are homologous to zero whenever the real intersection of (7) and (8) contains the greatest possible number -- i.e., $[n/2]$ -- of nested ovals. In other cases the solution of this problem is, as a rule, very difficult. A. M. Davydova has formulated an easily verifiable condition which is sufficient, but by no means necessary, for the absence of lacunae. The study of the topological properties of real algebraic curves and surfaces will probably play an important role in the investigation of the questions concerning the condition of homology to zero of the cycles $C_{\mathrm{real}}\ (t,x)$ or $C_{\mathrm{imag}}\ (t,x)$. How little progress has been made in this question can for example be seen from the fact that we do not yet know what shape the real algebraic surfaces of order 4 or the real algebraic curves of order 6 have. It should be added that my investigations on the conditions for the existence of lacunae refer only to the case that the surface (7) has no singular points, although such singular points exist in equations with important physical interpretations such as those of elasticity, crystal optics, and others. It is true, however, that it is possible, for all these equations, to investigate the question of the lacunae by means of a passage to the limit from equations without singularities. But it would be very important to have general conditions for the existence of lacunae in such equations.

I wish to discuss two more problems which are related to

Cauchy's problem. The first is Goursat's problem. In the simplest case it can be formulated as follows: Given a system of s sets of unknown functions containing, respectively, $n_1,\dots,n_k,\dots,n_s$ functions. The number of equations equals the number of unknown functions. Let us assume that the system -- which contains only derivatives of the first order -- is solved for the derivatives of the functions of the kth group with respect to x_k, $(k = 1,2,\dots,s)$. Find that solution of the system for which the functions of the kth group assume prescribed values on the coordinate plane $x_k = 0$, near the origin. E. Goursat [4] and S. L. Sobolev [19] have investigated this problem for $s = 2$ in the analytic case. By means of a new method Riquier succeeded in carrying through the discussion for $s > 2$ in the analytic case. L. A. Meltzer [20] has explored for $s = 2$ the question as to whether Goursat's problem is correctly posed. It would be of interest to extend these investigations to the case $s > 2$.

A further problem is the so-called "mixed problem", in which Cauchy's data are prescribed for the initial moment and boundary conditions at the boundaries of a certain domain in the $(x_1,x_2,\dots,x_p)$-space. For example, the study of the vibration of a membrane or an elastic plate of finite dimensions leads to this problem. The mixed boundary problem has been solved only for the simplest partial differential equations of the second order. In recent years there is an increasing concern with the very important question of the stability in the sense of Liapounoff of problems such as the "mixed problem". One is to investigate under what conditions a small change of the initial conditions will not produce arbitrarily large changes at some later moment. So far, only isolated problems, related to hydrodynamics, have been solved.

Taylor [21] has, for example, investigated the question of the motion of viscous fluids between rotating cylinders. S. L. Sobolev has studied the motion of a fluid inside a rotating top.

In all that has preceded we discussed the solutions of hyperbolic equations in the classical sense. By solution of a partial differential equation we meant a function with continuous derivatives of all those orders which occur in the equation. These derivatives must satisfy the given equation. Just as it is convenient in many questions of analysis to generalize the concept of number and to investigate irrational numbers together with the rational ones, so it is often useful to consider "generalized solutions" of partial differential equations. Such generalizations can be made in more than one way. It is, for example, often convenient to define "generalized solutions" as limits of uniformly convergent sequences of solutions in the usual sense. Then any function of the form $\phi(t+x) + \Phi(t-x)$ can be obviously called a generalized solution of the equation

$$\frac{\partial^2 u}{\partial t^2} = \frac{\partial^2 u}{\partial x^2}$$

even if that function does not possess continuous derivatives up to the second order. S. L. Sobolev has carried out systematic investigations of generalized solutions of hyperbolic equations.

II. Elliptic Equations

In order to simplify the presentation we limit ourselves to a consideration of the linear case. The system

$$\sum_{j,k_1,\ldots,k_p} a_{ij}^{(k_1,\ldots,k_p)}(x_1,\ldots,x_p)\frac{\partial^k u_j}{\partial x_1^{k_1}\ldots\partial x_p^{k_p}} = f_i(x_1,\ldots,x_p),$$
$$i,\ j=1,\ldots,N;\quad k=k_1+\ldots+k_p\leq n_j\ )$$

[n_j is the highest order of differentiation of the function u_j in the system] is called elliptic if the determinant

$$\left|\sum_{k_1+\ldots+k_p=n_j} a_{ij}^{(k_1,\ldots,k_p)}(x)\alpha_1^{k_1}\ldots\alpha_p^{k_p}\right|$$

is, for all values $x_1,\ldots,x_p$ under consideration, a definite form in $\alpha_1,\ldots,\alpha_p$, i.e., a form which for real $\alpha_1,\ldots,\alpha_p$ vanishes only if all α_j are zero. A single equation

$$\sum_{k=k_1+\ldots+k_p\leq n} a^{(k_1,\ldots,k_p)}(x_1,\ldots,x_p)\frac{\partial^k u}{\partial x_1^{k_1}\ldots\partial x_p^{k_p}} = f(x_1,\ldots,x_p)$$

is called elliptic if the form

$$\sum_{k_1+\ldots+k_p=n} a^{(k_1,\ldots,k_p)}(x)\,\alpha_1^{k_1}\ldots\alpha_p^{k_p}$$

is always definite.

The simplest elliptic equation is Laplace's equation

$$\frac{\partial^2 u}{\partial x_1^2}+\frac{\partial^2 u}{\partial x_2^2}=0. \tag{9}$$

The simplest elliptic system of the first order is the Cauchy-Riemann system

$$\left.\begin{aligned}\frac{\partial u}{\partial x}&=\frac{\partial v}{\partial y}\\ \frac{\partial v}{\partial x}&=-\frac{\partial u}{\partial y}\end{aligned}\right\}. \tag{10}$$

Every pair of sufficiently smooth functions u and v which satisfy the equations (10) also satisfy the equation (9).

The theory of the solutions u and v of the Cauchy-Riemann system is the theory of analytic functions. The properties of the solutions of all elliptic systems are to some extent similar to the properties of the functions u and v that satisfy the simplest elliptic system (10), i.e., the properties of analytic functions.

In this sense it can be said that the theory of analytic functions is a model for the theory of elliptic differential equations. To every branch of the theory of analytic functions corresponds -- or may be expected to correspond -- a branch of the theory of elliptic equations. We shall try to bring out this analogy in some of these branches.

Dirichlet's Problem or the First Boundary Problem.

Dirichlet's problem for elliptic equations of the second order is to find in the interior of a given domain a solution of this equation which assumes on the boundary the values of a given

18

bounded continuous function. The conditions for existence of a solution of Dirichlet's problem for Laplace's equation

$$\frac{\partial^2 u}{\partial x_1^2} + \ldots + \frac{\partial^2 u}{\partial x_p^2} = 0$$

have been thoroughly explored by many mathematicians, in particular, Perron and Wiener.

Let there be given an arbitrary domain (which we shall assume to be finite, in order to simplify matters) and on its boundary an arbitrarily prescribed continuous function f; then there exists a definite method for constructing in G a sequence of functions which converges in the interior of G to a certain harmonic function assuming the prescribed values at all "regular" boundary points. We shall not go into the necessary and sufficient conditions for regularity of a boundary point. We shall only remark that this property is "local", i.e., it is determined by the nature of the boundary near the point under consideration, and we shall mention some of the most easily understandable sufficient conditions for regularity.

For $p = 2$ one such condition is that the given boundary point P can be reached from the exterior by a Jordan arc none of whose points belongs to G. Lebesgue [22] and P. S. Urysohn [23] have proved that this condition is not sufficient for regularity of the boundary point if $p = 3$. When $p > 2$, it is sufficient for regularity of the boundary point P that the small cone with vertex at P and generated by revolving the curve

$$x_2 = \frac{x_1}{|\ln x_1|^{\frac{1}{p-3}}}, \quad x_3 = \ldots = x_p = 0, \text{ when } p > 3,$$

about its axis lies entirely outside the domains considered; when $p = 3$ we may rotate instead the curve $x_2 = x_1^k, x_3 \ldots = x_p = 0$, where k is an arbitrary positive number. Here we have taken the axis of the cone as OX-axis, and as origin of the coordinate system the point P. But if in the neighborhood of the point P the boundary of the domain lies inside a more pointed cone generated by the rotation about OX of the curve

$$x_2 = \frac{x_1}{|\ln x_1|^{\frac{1}{p-3}+\epsilon}}, \quad x_3 = \ldots = x_p = 0, \text{ when } p > 3,$$

or by the curve

$$x_2 = x_1^{|\ln x_1|^{\epsilon}}, \qquad x_3 = \ldots = x_p = 0, \text{ when } p = 3,$$

where ϵ is an arbitrarily small positive number, then the point P is irregular. Here we assume again that the origin of the coordinate system lies at P. For the doubly connected domain formed by the interior of a circle with the center removed the center is an irregular point. To M. A. Lavrentiev and M. V. Keldysh [24] we owe very interesting papers on the stability of the solutions of Dirichlet's problem for Laplace's equation. Thanks to the absence of a maximum in the interior of the domain in the case of harmonic functions it is easy to show that two bounded harmonic functions which differ on the boundary by less than ϵ cannot differ by more than ϵ in the interior of the domain. Thus there is complete stability in the dependence of the solution of Dirichlet's problem on the boundary values. It is quite different if the boundary curve is varied. Let us assume that the boundary of the domain G is everywhere regular and that on it a continuous function is prescribed. We extend this function into a neighborhood of the boundary. Consider two infinite sequences of domains, G_i^* and G_i^{**} $i = 1, 2, \ldots,$ such that for any i and j

$$G_i^* < G < G_j^{**}.$$

Let us assume that, as $i \to \infty$, the domains G_i^* and G_i^{**} tend to G in such a way that to every closed set A in the exterior of G a number i_A can be found that A is outside G_i^{**} whenever $i > i_A$. Analogously we assume that to every closed set a in the interior of G a number i_a can be found such that a is inside G_i^*, whenever $i > i_a$. We suppose that the boundaries of the domains G_i^* and G_i^{**} are of such a nature that Dirichlet's problem has always a solution u_i^* or u_i^{**}, respectively, no matter what continuous functions are prescribed at those boundaries. This will then in particular be true for those boundary values which coincide with the previously constructed continuation of the originally given function f. If, as $i \to \infty$, we have always $u_i^* \to u$, $(u_i^{**} \to u)$, where u is the solution of Dirichlet's problem for the originally given domain G and boundary function f, then we shall say that the solution of Dirichlet's problem is stable with respect to interior (exterior) domains. Surprisingly, M. A. Lavrentiev and M. V. Keldysh were able to construct an example which shows that Dirichlet's problem may be unstable with respect to exterior domains.

We may add one more remark concerning the solution of Dirichlet's problem for Laplace's equation: In recent years some progress has been achieved in solving it by means of the method of finite differences, which is very convenient for practical computations. Lusternik [25] was the first to prove rigorously for two independent variables the applicability of this method. A little later, Courant, Friedrichs and Lewy [25a] solved Dirichlet's problem for Laplace's equation by the method of finite differences for any number of independent variables; but they only proved that the harmonic function constructed by them satisfies the boundary conditions in the mean.

I have shown [25b] with the help of a procedure of S. N. Bernstein, that the method of finite differences yields a harmonic function which satisfies the boundary conditions in all regular points.

Neumann's problem, i.e., the problem of finding a harmonic function in the given domain which has a prescribed normal derivative on the boundary, has been explored for the simplest domains only. Its study is far from having reached the completeness of the investigations of Dirichlet's problem. The situation is even worse with respect to the mixed boundary problem in which the values of a linear combination of u and du/dn are prescribed on the boundary. Here d/dn designates differentiation in the normal direction.

The investigations of the solution of Dirichlet's problem for Laplace's equation can be extended without difficulty to elliptic equations of the type

$$\sum_{i,j=1}^{p} a_{ij} \frac{\partial^2 u}{\partial x_i \partial x_j} + \sum_{i=1}^{p} a_i \frac{\partial u}{\partial x_i} + au = f,$$

in which the coefficients and the right member depend on $x_1, \ldots, x_p$. The sign of the coefficient a plays a decisive role. If this sign is the opposite of the sign of the form

$$\sum a_{ij} \alpha_i \alpha_j,$$

then Dirichlet's problem has a solution under the same assumptions on the boundary of the domains as in the case of Laplace's equation. But if the sign of a coincides with the sign of that form, then it is not sufficient for the solvability of Dirichlet's problem to impose the same condition as for Laplace's equation concerning regularity of the boundary points, which, as mentioned above, is determined by local properties of the boundary. Sufficient conditions are obtained if, in addition, a certain smallness of the whole domain is required [26]. V. V. Nemytski [27] has proved that only the area of the domain matters in this context. For some domains with large area but very smooth boundary it can actually happen that for certain special boundary functions Dirichlet's problem has no solution. As far as I know, the statement that the boundary curve must "in the small" consist of regular points in the same sense as for Laplace's equation, if Dirichlet's problem is to be solvable for every continuous boundary function has been rigorously proved [47] only for self-adjoint equations, i.e., equations of the form

$$\sum_{i,j} \frac{\partial}{\partial x_i}\left(a_{ij} \frac{\partial u}{\partial x_j}\right) + au = f.$$

But there is every reason to assume that the same is true also for arbitrary linear elliptic equation of second order (Courant, Friedrichs, Lewy, A. I. Barabanoff.)

Assuming that the solution of Dirichlet's problem is unique, Adelson-Velski and Kronrod have shown recently that the solution of this problem for the linear elliptic equation of second order depends continuously on the coefficients of this equation. By making use of the metric determined by the coefficients of a linear second order differential equation with variable coefficients, Feller [25] has extended many theorems on harmonic functions to functions that solve these elliptic equations.

To S. N. Bernstein are due fundamental results concerning the existence of solutions of Dirichlet's problem for non-linear elliptic equations. Schauder and Leray have extended their important contributions to the case of more independent variables [28].

The following is a generalization of Dirichlet's problem to the polyharmonic equation, i.e., to the equation

$$\left(\frac{\partial^2}{\partial x_1^2}+\dots+\frac{\partial^2}{\partial x_p^2}\right)^m u=0.$$

To find a function u which satisfies this equation in the interior of the given domain and assumes on the boundary prescribed values with all its derivatives up to order $m-1$. For domains with sufficiently smooth $(p-1)$-dimensional boundaries and sufficiently smooth functions prescribed on them the existence of a solution of this problem has been known for a long time. S. L. Sobolev [29] has investigated the case when the boundary of the domain considered consists not only of $(p-1)$-dimensional surfaces but also of sets of lower dimension. The result was that on boundary sets of dimension $p-1-s$ the values of the required function and of its normal derivatives may be prescribed up to order $m-1-[(s+1)/2]$ only. The existence of a solution of such a problem has so far been established only for the case that the boundary conditions are to be satisfied "in the mean", in a certain sense. When $m-1-[(s+1)/2]<0$, then no arbitrary conditions may be prescribed on a piece of dimension $p-1-s$.

Z. Ya. Shapiro [30] and N. I. Simonov [31] have concerned themselves with a generalization of Dirichlet's problem to elliptic systems:

$$\sum_{j=1}^{2n}\sum_{k=1}^{3} a_{ij}^{k}\frac{\partial u_j}{\partial x_k}=0 \qquad i=1,\dots,2n.$$

Let there be given the elliptic system

$$\sum_{j=1}^{2n} l_\nu^j u_j = f_\nu(x_1, x_2) \qquad \nu = 1, \ldots, n,$$

with constant a_{ij}^k. There exists a unique solution of this system in the half-space $x_3 > 0$ satisfying at $x_3 = 0$ the conditions

$$\sum_{j=1}^{2n} l_\nu^j u_j = f_\nu(x_1, x_2), \qquad \nu = 1, \ldots, n.$$

Here the left members are n linearly independent linear forms in the u_j with constant coefficients. The functions $f(x_1, x_2)$ are assumed to be bounded and continuous on the whole (x_1, x_2)-coordinate plane. The solution $(u_1, \ldots, u_{2n})$ is to be such that the forms

$$\sum_{j=1}^{2n} l_\nu^j u_j$$

remain bounded in the whole half-space considered.

For a bounded domain in (x_1, x_2, x_3)-space all that has been done for this problem so far is to reduce it to a system of integral equations of Fredholm's type. In order to prove for arbitrary continuous boundary functions the existence of a solution of the boundary problem as formulated originally, there remains only to be proved the uniqueness of the solution of that problem; but this has not yet been possible.

The methods applied by Z. Ya. Shapiro can, in principle, be applied to any number of independent variables. They are based on the construction of so-called fundamental solutions (i.e. solutions with certain prescribed singularities) with the help of algebraic functions.

By using Fourier's method N. I. Simonov has solved analogous problems for half-spaces and for a strip between two parallel planes.

Still another approach to the solution of similar problems for elliptic equations will be mentioned below when we discuss parabolic equations.

Quasiconformal Mappings

M. A. Lavrentiev calls a mapping of the (x,y)-plane onto the (u,v)-plane *quasiconformal*, if u and v satisfy the linear system

$$\left.\begin{aligned}\frac{\partial u}{\partial x}&=a_1\frac{\partial u}{\partial y}+b_1\frac{\partial v}{\partial y},\\ \frac{\partial v}{\partial x}&=a_2\frac{\partial u}{\partial y}+b_2\frac{\partial v}{\partial y}.\end{aligned}\right\} \qquad (11)$$

If the coefficients of the equation (11) are assumed to be sufficiently smooth, a theorem analogous to Riemann's theorem on the possibility of mapping an arbitrary simply connected domain onto the circle is valid. As in Riemann's theorem the mapping is uniquely determined if the image of one point and the image of a direction through it are prescribed. M. A. Lavrentiev and his pupils (B.W. Shabat, Z. Ya. Shapiro) have investigated [32] the properties of such a mapping and have extended the mapping theorem, under certain conditions, to the case of simultaneous linear and non-linear elliptic equations.

As far as I know, little work has been done on mapping problems for a greater number of coordinates, even for linear elliptic systems with constant coefficients [33].

The Study of the Solutions in the Whole Space and near a Singular Point

To the first of these questions belongs Liouville's theorem on functions holomorphic on the whole plane, which states that such a function is a constant if it is bounded. For second order elliptic equations of the type

$$Ar+2Bs+Ct=0,$$

where A, B, C are bounded functions of x, y, z, p, q, r, s, t satisfying the conditions $AC < B^2$, S. N. Bernstein has shown that every solution which is bounded on the whole plane must be a constant. Closely connected with this result is the following geometric theorem: Let $z = f(x,y)$ be the equation of a surface such that $f(x,y)$ possesses first and second derivatives in the whole (x,y)-plane, and that its curvature is non-positive and not identically zero; then this surface cannot be contained between two planes $z = \pm h$. Recently, Adelson-Velski has somewhat generalized this result.

Picard's theorem stating that an analytic function assumes all but at most two values in the neighborhood of an essential singularity has been extended by H. Grötzsch [34] and M. A. Lavrentiev [32] to the solutions of certain elliptic systems with two independent variables. I do not know of any generalizations of Picard's theorem to more than two independent variables.

Analyticity, Smoothness and Uniqueness of the Solutions

It can be proved that if a solution of an analytic elliptic system is L times continuously differentiable, then that solution is analytic, provided L is sufficiently large [35]. There is every reason to assume that the value of L required in my paper, i.e., $L = 3p + 2 \max \{n_1, \ldots, n_N\} + 6$, can be considerably reduced. In some papers of S. N. Bernstein and in articles by other authors dealing with second order equations, L is equal to 3, in some cases even to 2. But if no smoothness of the solution is required beyond the existence at every point, of those derivatives which occur in the system under consideration, then the analyticity of such a solution cannot be guaranteed any longer. For example, the real part of the function

$$e^{-\frac{1}{(x+iy)^4}}$$

satisfies Laplace's equation everywhere, but it has a discontinuity at the point $x = y = 0$.

On the other hand, I have proved in [35] the following: Every system of linear equations of the type

$$\sum_{j,\, k_1+\cdots+k_p \leq n_j} a_{ij}^{(k_1, \ldots, k_p)} \frac{\partial^k u}{\partial x_1^{k_1} \ldots \partial x_p^{k_p}} = f_i(x_1, \ldots, x_p)$$
$$i, j = 1, \ldots, N$$

26

with constant coefficients has non-analytic solutions, if the determinant

$$\left| \sum_{k_1 + \ldots + k_p = n_j} a_{ij}^{(k_1, \ldots, k_p)} \alpha_1^{k_1} \ldots \alpha_\nu^{k_p} \right| \tag{12}$$

vanishes for even one set of real values of the α_k for which

$$\sum \alpha_k^2 > 0,$$

provided the determinant does not vanish identically.

It is easy to construct examples for which the determinant (12) vanishes identically for all α_k, and such that all sufficiently smooth solutions are analytic [36]. However, it is just as easy to construct examples of systems of the same kind which have arbitrarily smooth but non-analytic solutions. Thus the question of the analyticity of the solutions must remain open in the case that the determinant (12) vanishes identically for all α_k.

There is every reason to expect that all regular solutions of elliptic equations with sufficiently smooth coefficients are also smooth with their right members. As far as I know, this has up till now been proved only for the simplest elliptic equations. It is, for example, known that all regular solutions of linear elliptic equations of the second order have bounded derivatives up to order k, provided all coefficients and the right hand member have bounded derivatives up to order k. Adelson-Velski and Kronrod have shown that if the functions $u(x,y)$ possess total differentials and satisfy the elliptic system

$$\frac{\partial u}{\partial x} = A(x, y)\frac{\partial v}{\partial x} + B(x, y)\frac{\partial v}{\partial y},$$
$$\frac{\partial u}{\partial y} = C(x, y)\frac{\partial v}{\partial x} + D(x, y)\frac{\partial v}{\partial y},$$

then they assume their maximum and minimum on the boundary of the domain. Hence they conclude that these functions u and v must have bounded derivatives up to the kth order, whenever A, B, C, and D have bounded derivatives up to the kth order.

It is possible to link the question of the analyticity of the solution of elliptic equations with the question as to how much must be prescribed to determine a solution of elliptic equations. It follows from the above mentioned theorem on the analytic character of all solutions of an analytic elliptic equation that are sufficiently smooth, that such solutions are fully determined by the value of the function u and of all its partial derivatives at one point. Hence two such solutions coincide everywhere if they coincide on an arbitrarily small p-dimensional domain, or even if they coincide on an arbitrarily small piece of a real analytic $(p-1)$-dimensional surface. For linear elliptic systems with analytic coefficients the last result can also be obtained from Holmgren's theorem on uniqueness in Cauchy's problem, which I mentioned in the first part of this review. From Carleman's theorem the same result can be obtained for linear elliptic systems in two independent variables with sufficiently smooth coefficients.

In order to solve the system

$$\frac{\partial u}{\partial x}-\frac{\partial v}{\partial y}=\alpha u+\beta v,$$
$$\frac{\partial v}{\partial x}+\frac{\partial u}{\partial y}=\gamma u+\delta v,$$

where α, β, δ are continuous functions of x and y, Carleman [37] showed the following: If the regular function $u + iv$ vanishes on an infinite point set that converges to a regular limit point, then $u + iv$ is everywhere equal to zero.

The Work of L. Bers and A. Gelbart

In a very interesting article L. Bers and A. Gelbart studied the system

$$\left.\begin{aligned} G_1(x)\,u_x &= T_1(y)\,v_y, \\ G_2(x)\,u_y &= -T_2(y)\,v_x, \end{aligned}\right\} \qquad (\Sigma)$$

in which

1. G_i and T_i are real analytic functions which are defined for all

28

real x and y,

2. $G_1G_2T_1T_2 > 0$ (condition of ellipticity),

3. $G_i(+\infty)$, $G_i(-\infty)$, $T_i(+\infty)$, $T_i(-\infty)$ exist and are different from 0 and ∞.

The system

$$\left.\begin{aligned} \frac{u_x}{G_1(x)} &= T_1(y)\, u_{vk} \\ \frac{u_y}{G_1(x)} &= -\, T_2(y)\, v_y \end{aligned}\right\} \qquad (\Sigma')$$

is called conjugate to the system Σ.

Suppose

$$\frac{\Delta_\Sigma f(z)}{\Delta_\Sigma z} = R\,\{\Delta z^{-1}[G_1(x)\cdot\Delta_2 u + iT_1(y)\,\Delta_2 v]\}$$

$$+ I\,\{\Delta z^{-1}[\Delta_2 u/T_2(y) + i\Delta_1 v/G_2(x)]\},$$

where

$$\Delta_1 u = 2^{-1}[u(z+\Delta z) + u(z+\overline{\Delta z}) - 2u(z)],$$
$$\Delta_2 u = 2^{-1}[u(z+\Delta z) + u(z-\overline{\Delta z}) - 2u(z)].$$

(The expression $I(z)$ means iy, if $z = x + iy$.) The quantities $\Delta_1 v$ and $\Delta_2 v$ are defined analogously. The function $f(z)$ is called "Σ-monogenic", if the limit of

$$\frac{\Delta_\Sigma f(z)}{\Delta_\Sigma z},$$

as $\Delta z \to 0$ exists independently of the way in which Δz tends to zero. In that case we will call the limit the "Σ-derivative" and denote it by

$$\frac{d_\Sigma f(z)}{d_\Sigma z} = f^{[1]}(z).$$

Then the following theorem holds:

I. If $f = u + iv$ has a Σ-derivative, then u and v have first partial derivatives and

$$f^{[1]}(z) = G_1 u_x + \frac{iv_x}{G_1} = T_1 v_y - \frac{iu_y}{T_2}.$$

Hence, u and v satisfy the system Σ. Conversely, if u and v have continuous derivatives which satisfy the system Σ, then f has a Σ-derivative.

II. A function which is identically constant is Σ-monogenic.

III. If the Σ-derivative of a function $f = u + iv$ vanishes identically then f is a constant.

IV. If f and g are any Σ-monogenic functions, then $\alpha f + \beta g$ is a Σ-monogenic function for all real numbers α and β. But this is no longer true for complex α and β.

V. The derivative of a Σ-monogenic function is a Σ'-monogenic function, where Σ'-monogeneity is defined exactly like Σ-monogeneity, with the system Σ replaced by the system Σ'. It follows that the derivative of a Σ'-monogenic function is a Σ-monogenic function.

Let

$$f^{[2]}(z) = \frac{d_{\Sigma'} f^{[1]}(z)}{d_{\Sigma'} z}$$

and, more generally,

$$f^{[2k+1]}(z) = \frac{d_\Sigma f^{[2k]}(z)}{d_\Sigma z}, \qquad f^{[2k]}(z) = \frac{d_{\Sigma'} f^{[2k-1]}(z)}{d_{\Sigma'} z}.$$

Then it can be proved that, if for all $n \geq 0$

$$f^{(n)}(z_0) = 0,$$

then

$$f(z) \equiv 0.$$

The concept of a Σ-integral is defined analogously:

30

$$\int_L f(z)\,d_\Sigma z = \int_L (G_2 u dx - T_2 v dy) + i \int_L \left(\frac{v}{G_1} dx + \frac{u}{T_1} dy\right),$$

$$\int_L (\alpha f + \beta g)\,d_\Sigma z = \alpha \int_L f d_\Sigma z + \beta \int_L g d_\Sigma z,$$

where α and β are real.

If $f(z)$ is a Σ-monogenic function in a closed simply connected domain bounded by a rectifiable curve C, then

$$\int_C f(z)\,d_\Sigma z = 0.$$

It follows that the Σ-integral of a Σ-monogenic function is independent of the path of integration.

An indefinite Σ-integral of a Σ-monogenic function, i.e., a Σ-integral with a variable upper limit, is a Σ'-monogenic function. Σ-differentiation and Σ'-integration are mutually inverse.

With the help of Σ-and Σ'-integrals it is possible to define "formal powers", which are analogous to the powers of an ordinary complex number. We define

$$aZ^{(0)}(z_0, z) = a\tilde{Z}^{(0)}(z_0, z) = a \qquad (aZ \text{ and } a\tilde{Z} \text{ are inseparable symbols}),$$

$$aZ^{(n)}(z_0, z) = n\int_{z_0}^{z} (a\tilde{Z}^{(n-1)}(z_0, z))\,d_{\Sigma'} z \text{ — } \Sigma\text{-monogenic function},$$

$$a\tilde{Z}^{(n)}(z_0, z) = n\int_{z_0}^{z} (aZ^{(n-1)}(z_0, z))\,d_\Sigma z \text{ — } \Sigma'\text{-monogenic function},$$

$$aZ^{(n)}(z_0, z_0) = a\tilde{Z}^{(n)}(z_0, z_0) = 0.$$

Formal Power Series

The series

$$\sum_{n=0}^{\infty} a_n Z^{(n)}(a, z) \tag{13}$$

represent a Σ-monogenic function in every region where the series converges uniformly. If the ordinary power series

$$\sum_{n=0}^{\infty} a_n z^n \tag{14}$$

has a positive radius of convergence, then there exists a neighborhood of the point a at which the series (13) and all series resulting from it by termwise differentiation converge absolutely and uniformly in G. The authors did not investigate the size and shape of the region G.

Every Σ-mongenic function can be developed in a formal power series. It can be proved that this expansion is unique for a given a. The zeros of a Σ-monogenic function are isolated. A Σ-monogenic function is uniquely determined by its values on an infinite sequence of points with a regular finite limit point.

Formal Polynomials

A function of the form

$$M(z) = a_0 + a_1 Z^{(1)}(a, z) + \ldots + a_n Z^{(n)}(a, z)$$

with $a_n \neq 0$ is called a *formal polynomial of degree n*.

It can be proved that a formal polynomial of degree n has n roots.

The following interpolation theorem holds: Let there be given a number $b \neq 0$, a number a, n distinct constants $z_1, \ldots, z_n$ and n other constants $c_1, \ldots, c_n$. Then there exists a unique polynomial of the nth order,

$$M(z) = \sum_{k=0}^{n-1} a_k Z^{(k)}(a, z) + b Z^{(n)}(a, z),$$

such that

$$M(z_i) = c_i.$$

There is no need to emphasize how interesting it would be to have an extension of this theorem to more general elliptic systems.

32

L. Bers and A. Gelbart conclude their work with the following questions.

1. What is the relation between the domain of existence of the monogenic function determined by the series (13) and that of the corresponding holomorphic function represented by the series (14)?

2. Is it possible to approximate a Σ-monogenic function which is defined in a simply connected region by means of formal polynomials?

3. Under what conditions can a Σ-monogenic function be found which maps one given region onto another?

In an unpublished paper A. I. Markushevich proved the following theorem: If the coefficients of the system

$$a_1u_x + a_2v_x = b_1u_y + b_2v_y,$$
$$c_1u_x + c_2v_x = d_1u_y + d_2v_y$$

are sufficiently smooth, and if a solution exists, then the system can be transformed into the equivalent one

$$\left.\begin{aligned} \lambda_1u_x + \lambda_2v_x &= \mu_1u_y + \mu_2v_y, \\ \pi_1u_x + \pi_2v_x &= \rho_1u_y + \rho_2v_y, \end{aligned}\right\} \qquad (T)$$

where $\lambda_{ix} = \mu_{iy}$, $\pi_{ix} = \rho_{iy}$, $(i = 1,2)$. The system (T) expresses the independence of the path of integration of the integrals

$$U = \int_{(x_0, y_0)}^{(x,y)} (\rho_1u + \rho_2v)\,dx + (\pi_1u + \pi_2v)\,dy \quad \text{and}$$

$$V = \int_{(x_0, y_0)}^{(x,y)} (\mu_1u + \mu_2v)\,dx + (\lambda_1u + \lambda_2v)\,dy\,.$$

On the other hand, U and V are solutions of a certain linear system (T_1) of the type (T). Let us call (U,V) the "integral" belonging to the solution of the system (T). We shall assume that there exist a finite number of systems (T_j), $j = 0,1,\ldots,n-1$ such that the integral belonging to a solution of (T_j) satisfies the system (T_{j-1}), and that $T = T_0 = T_n$. Then we shall say that the system *(T)* is included in a cycle of order n. No example is yet known of a system which is included in a cycle of order higher than 2, but not in a cycle of order 2 or *1*. Similarly, no examples are known of systems which are not included in any cycle.

Starting with the system *(T)* formal powers and power series can be defined in analogy with the procedure of L. Bers and A. Gelbart. If the coefficients are analytic, it can be proved that all analytic solutions of the system (T_j) can be developed in series of formal powers.

Additional Problems

It would be interesting to study systems in which the number of equations exceeds the number of unknown functions, similar to the situation present in the theory of functions of two complex variables, where every variable satisfies two systems of Cauchy-Riemann equations; or similar to the way in which S. A. Halpern examined the generalization of hyperbolic systems to cases in which the number of equations is a multiple of the number of unknown functions. Just as we investigated concepts analogous to ordinary polynomials it would be interesting to study analogues of complex algebraic curves and surfaces.

III. Parabolic Equations

Let there be given the system

$$\frac{\partial^{n_l} u_l}{\partial t^{n_l}} = \sum_{j=1}^{N} \sum_{k_0} A_{lj}^{(k_0, k_1, \dots, k_p)} \frac{\partial^{k_0+\dots+k_p} u_j}{\partial t^{k_0} \partial x_1^{k_1} \dots \partial x_p^{k_p}} + f_l(t, x_1, \dots, x_p), \qquad (15)$$
$$l = 1, \dots, N.$$

So far, when we said that an equation under consideration contained derivatives from order n_j to order n_i it was implied that differentiation with respect to every variable was counted the same way. Now we shall modify this assumption. We shall now attach to differentiation with respect to t a certain weight b, so that one differentiation with respect to t will in a certain sense correspond to b-fold differentiation with respect to $x_1, x_2, \dots, x_p$. Accordingly, we shall suppose that the right member of the equation (15) contains only such derivatives that

34

$$k_0 b + k_1 + \ldots + k_p \leq n_j b \text{ and } k_0 b < n_j.$$

We shall call the system (15) *b-parabolic*, or simply *parabolic* [38], if for every real $a_1, \ldots, a_p$ for which $\Sigma a_k^2 = 1$ the real parts of the roots in λ of the determinant of the matrix

$$\left\| \begin{matrix} \lambda_1^{n_1} & & \\ & \ddots & \\ & & \lambda_N^{n_N} \end{matrix} \right\| - \left\| \sum_{bk_0 + k_1 + \ldots + k_p = bn_j} A_{ij}^{(k_0, k_1, \ldots, k_p)} \lambda^{k_0} (ia_1)^{k_1} \ldots (ia_p)^{k_p} \right\|$$

remain smaller than some constant negative number $-\delta$.

The following equations will, for example, be parabolic:

$$\frac{\partial u}{\partial t} = \sum_{k_1 + \ldots + k_p \leq b} A^{(k_1, \ldots, k_p)} \frac{\partial^{k_1 + \ldots + k_p} u}{\partial x_1^{k_1} \ldots \partial x_p^{k_p}} + f(t, x_1, \ldots, x_p),$$

provided b is even and the form

$$i^b \sum_{k_1 + \ldots + k_p = b} A^{(k_1, \ldots, k_p)} a_1^{k_1} \ldots a_p^{k_p}$$

is negative definite, i.e., it always remains negative, if

$$\sum a_k^2 > 0.$$

The simplest parabolic equation is the equation of heat conduction in a rod,

$$\frac{\partial u}{\partial t} = \frac{\partial^2 u}{\partial x^2}. \tag{16}$$

The simplest boundary problem for such an equation is the following Cauchy problem: Given the values of u for $t = 0$; to find for all $t > 0$ a solution of the equation which satisfies the given initial conditions. If the initial function $u(0,x) = f(x)$ is continuous and bounded, then it is known that Cauchy's problem has one and only one solution. It is given by Poisson's formula. It is evident from this formula that the value of $u(t,x)$ depends for arbitrarily small positive t on the values of $f(x)$ on the whole x-axis, contrary to what we saw in the case of hyperbolic equations. Heat spreads with infinite speed. From Poisson's formula it also follows that all bounded, continuous solutions of the equation of heat conduction which are defined on a whole strip $t_0 > t > 0$ are analytic in x for all positive t.

The uniqueness problem for the solution of Cauchy's problem in the case of the equation of heat conduction has been studied by A. N. Tikhonov [39]. He obtained the following result. Let us assume that the function $u(t,x)$ solves Cauchy's problem for the equation of heat conduction in an infinite strip $0 \leq t \leq t_0$. Let

$$\varphi(x) = \max_{0 \leq t \leq t_0} u(t, x).$$

If there exists a c such that

$$\varphi(x)\, e^{-cx^2} \longrightarrow 0,$$

as $x \to -\infty$, as well as as $x \to -\infty$, then $u(t,x)$ is uniquely defined by the initial function $f(x)$ and given by Poisson's integral. On the other hand, it is possible to find, for arbitrary $\epsilon > 0$ solutions of the equation of heat conduction which vanish for $t = 0$, for which

$$\varphi(x) \leqslant e^{x^{2+\epsilon}}.$$

I proved in [38] that Cauchy's problem is correctly posed for general parabolic systems whose coefficients depend only on t. Here again the initial values of the unknown function and its derivatives with respect to t up to order $n_i - 1$ are prescribed on the whole plane $t = 0$. Integrals similar to that of Poisson have been constructed for the solution of Cauchy's problem. It can be shown that all sufficiently smooth solutions of (15) are analytic in $x_1, \ldots x_p$, in every region of the $(t, x_1, \ldots, x_p)$-space in which all the functions $f_i(t, x_1, \ldots, x_p)$ are analytic in these variables.

S. Z. Bruck has generalized my results about the correct formulation of Cauchy's problem for linear parabolic systems to the case that only the coefficients of the highest derivatives depend on t alone while all the other coefficients may also depend on the other variables.

It would be interesting to find, for general parabolic systems, theorems on the uniqueness of the solution within the class of unbounded functions, in analogy to the theorem by A. N. Tikhonov.

The function $u(t,x)$ which solves the equation (16) on the whole strip $0 \leq t \leq t_0$ is not determined by its values on $t = t_0$, i.e. by the condition $u(t_0,x) = f_1(x)$. But A. N. Tikhonov showed that if we require, in addition, that $u(t,x)$ is to be bounded on this strip, then there cannot exist more than one solution of the equation of heat conduction satisfying this condition. From what was said above it follows that the function $f_1(x)$ must be analytic, if such a solution is to exist. But this condition is not sufficient to assure the existence of such solutions.

The first boundary problem for equation (16) is as follows: Let G be a region of the (t,x)-plane bounded by the curves

$$x = \varphi_1(t), \quad x = \varphi_2(t), \quad t = 0, \quad t = t_1 > 0, \quad \varphi_1(t) < \varphi_2(t);$$

to find a function $u(t,x)$ which satisfies the equation (16) in the region G and assumes given continuous boundary values on the curves

$$t = 0, \quad x = \varphi_1(t) \text{ and } x = \varphi_2(t).$$

If we assume that the functions $\phi_i(t)$ satisfy a Lipschitz condition, then the first boundary problem has a unique solution for any continuous prescribed values on the indicated parts of the boundary of G. This solution depends continuously on the prescribed boundary functions. In [40] I investigated a case where, at certain points, the functions $\phi_i(t)$ did not satisfy the Lipschitz condition. The first boundary problem for the equation of heat conduction can be solved very simply by the method of finite differences.

The general linear as well as non-linear parabolic equations with partial derivatives of second order and with an arbitrary number of independent variables have been very thoroughly investigated. The fundamental papers by Holmgren [41], E. E. Levi [42] and Gevray [43] are devoted to this problem.

I wish to state the following result of A. N. Tikhonov [44] concerning the differential equation

$$\frac{\partial u}{\partial t} = \frac{\partial^2 u}{\partial x_1^2} + \dots + \frac{\partial^2 u}{\partial x_p^2}. \tag{17}$$

Let $G_0^{t_1}$ be a domain of the $(t, x_1, \dots, x_p)$-space in which t ranges over all values between 0 and t_1, and whose projection on the $(x_1, \dots, x_p)$-plane is a certain region G. To find in G a solution of the equation (17) which assumes the values of a given continuous function f on the lateral surface and on the lower base of the cylinder described. It turns out that such a problem has a solution for every continuous function f, if and only if all the boundary points of the region G in the $(x_1, \dots, x_p)$-plane are regular points with respect to Dirichlet's problem for the Laplace equation.

If the values of the function f on the lateral surface of $G_0^{t_1}$ do not depend on t, then the solution $u(t, x_1, \dots, x_p)$ of the first boundary problem tends, as $t \to +\infty$, to a function $u(x_1, \dots, x_p)$, which satisfies Laplace's equation inside the region G and assumes on the boundary the values f, independent of the initial values of $u(t,x)$ at $t = 0$.

S. Z. Bruck solved the first boundary problem for linear parabolic systems of second order with constant coefficients and containing only the highest derivatives, provided the lateral surface of the region considered is formed by straight lines parallel to the Ot-axis. With the aid of the passage to the limit as $t \to +\infty$

described above he obtains from this a solution of the first boundary problem for those linear elliptic systems with partial derivatives of order 2 which result if we cancel terms containing the partial derivatives with respect to t in the system of S. Z. Bruck. But not all possible elliptic systems of second order can be obtained in this way. The question arises if there exist solutions of the first boundary problem for those linear elliptic systems of second order with constant coefficients and containing only the highest derivatives, which cannot be obtained from parabolic systems in this way. S. Z. Bruck solved the first boundary problem for parabolic systems also in the case that the coefficients depend on t.

If a and b are constant, the equation

$$\frac{\partial^2 u}{\partial z^2} - a\frac{\partial u}{\partial x} - b\frac{\partial u}{\partial y} + c\frac{\partial u}{\partial z} + du = f \qquad (18)$$

can be reduced to the equation

$$\frac{\partial^2 u}{\partial z^2} - \frac{\partial u}{\partial t} + c\frac{\partial u}{\partial z} + du = f$$

by the transformation $t = ax + by$. But if a and b depend on x, y, z, such a reduction is, in general, impossible. N. S. Piskunov [45] developed for equation (18) solutions of problems similar to the first boundary problem and to Cauchy's problem.

In conclusion we make the following remarks: Poincare said that boundary problems for partial differential equations involving non-analytic functions are posed correctly in a certain sense only if they are arrived at by physical considerations, and the methods used to solve these problems are also dictated by physics. This statement of Poincaré on the formulation of boundary value problems is still valid. If boundary problems that are not always posed correctly are formulated in physics, then they appear as immediate generalizations of such problems. In this way the hyperbolic, elliptic and parabolic equations have been explored up to a certain point. But there is, in addition, a sizable class of equations for which we do not know any correctly posed boundary problems. The so-called ultra-hyperbolic equation

$$\frac{\partial^2 u}{\partial x_1^2}+\ldots+\frac{\partial^2 u}{\partial x_p^2}=\frac{\partial^2 u}{\partial y_1^2}+\ldots+\frac{\partial^2 u}{\partial y_p^2}, \tag{19}$$

with $p \geq 2$ appears, for example, to be one of these.

Asgeirsson [46] has proved the following remarkable theorem for the equation (19): Let $u(x_1,\ldots,x_p,y_1,\ldots,y_p)$ be any solution of the equation (19) which is continuous with all its second order partial derivatives in the domain

$$\sum_i (x_i-a_i)^2+\sum_i (y_i-b_i)^2 \leqslant A.$$

Let ρ be an arbitrary positive number not exceeding A. We now consider two $(p-1)$-dimensional spheres with radius ρ and common center at the point $(a_1,\ldots,a_p,b_1,\ldots,b_p)$. Suppose the first of these, $S(x,\rho)$, lies in the plane

$$y_i=b_i, \quad i=1,\ldots,p,$$

and the second, $S(y,\rho)$, in the plane

$$x_i=a_i, \quad i=1,\ldots,p.$$

Then the arithmetic mean of u on $S(x,\rho)$ is, for every ρ, equal to the arithmetic mean of u on $S(y,\rho)$.

This theorem can easily be applied to the equation

$$\frac{\partial^2 u}{\partial x_1^2}+\ldots+\frac{\partial^2 u}{\partial x_p^2}=\frac{\partial^2 u}{\partial y_1^2}+\ldots+\frac{\partial^2 u}{\partial y_q^2}, \text{ where } q<p,$$

for to every solution $u(x_1,\ldots,x_p,y_1,\ldots y_q)$ of this equation corresponds that solution u of the equation (19) which is defined by the equation

$$u(x_1,\ldots,x_p,y_1,\ldots,y_q,y_{q+1},\ldots,y_p)=u(x_1,\ldots,x_p,y_1,\ldots,y_q).$$

40

In this way one obtains the theorems of the arithmetic mean for Laplace's equation (for $q = 0$) and for a hyperbolic equation (for $q = 1$). By using these latter theorems one can obtain formulas which are a solution of Dirichlet's problem for Laplace's equation and of Cauchy's problem for a hyperbolic equation.

In his recent papers M. G. Krein [48] exhibits methods by means of which it can be decided which problems are posed correctly, in some sense, for the different types of partial differential equations.

LITERATURE

1. Holmgren, Ofversigt af Kongl. Svenska Vetenskaps-Akad. Förhandlingar 58, 91-103 (1901). The contents of this paper are also included in Hadamard's book [2].

2. Hadamard, Lecons sur la propagation des ondes et les équations de l'hydrodynamique, Paris, 1903, note 1.

3. Carleman, Ark. Mat. Astr. Fys. 26B, no. 17 (1939).

4. Goursat, Leçons sur l'intégration des équations aux dérivées partielles du second ordre, v. 2, Hermann, Paris, 1898, note II.

4a. Myshkis, Rec. Math. [Mat. Sbornik] N.S. 19(61), 489-522 (1946).

5. Hadamard, Le problème de Cauchy et les équations aux dérivées partielles linéaires hyperboliques, Hermann, Paris, 1932, pp. 40-41.

6. Petrovsky, Rec. Math. [Mat. Sbornik] N.S. 2(44), 815-870 (1937).

7. Petrovsky, Bull. Univ. Moscou. Sér. Internat. Sect. A. 1, no. 7 (1938), pp. 65-67.

8. ibid.

9. John, Proc. Nat. Acad. Sci. U.S.A. 29, 98-104 (1943).

10. Halpern, Rec. Math. [Mat. Sbornik] N.S. 7(49), 111-141 (1940).

11. Schauder, Fund. Math. 24, 213-246 (1935).

12. Frankel, Rec. Math. [Mat. Sbornik] N.S. 2(44), 793-814 (1937).

13. Christianovich, ibid., 871-899 (1937).

14. Sobolev, C. R. (Doklady) Acad. Sci. URSS (N.S.) 10 (1936 I), 279-282.

15. Hadamard, Le problème de Cauchy et les équations aux dérivées partielles linéaires hyperboliques, Hermann, Paris, 1932, pp. 239-241.

16. Mathisson, Acta Math. 71, 249-282 (1939).

17. Petrovsky, C. R. (Doklady) Acad. Sci. URSS (N.S.) 38, 151-153 (1943).

18. Herglotz, Ber. Verh. Sächs. Akad. Wiss. Leipzig. Math.-Phys. Kl. 78, 93-126, 287-318 (1926).

19. Sobolev, Trav. Inst. Phys.-Math. Stekloff 5, 265-282 (1934).

20. Meltzer, C. R. (Doklady) Acad. Sci. URSS (N.S.) 30, 693-696 (1941); a detailed account has been printed in Rec. Math. [Mat. Sbornik] N.S. 18(60), 59-104 (1946).

21. Taylor, Philos. Trans. Roy. Soc. London. Ser. A. 223, 289-343 (1923);
Synge, Proc. Roy. Soc. London, Ser. A. 167, 250-256 (1938).

22. Lebesgue, C. R. Soc. Math. France 1913, 17.

23. Urysohn, Math. Z. 23, 155-158 (1925).

24. See the article by M. V. Keldysh, Uspehi Matem. Nauk no. 8, 171-231 (1941).

25. Lusternik, Rec. Math. [Mat. Sbornik] (1) 33, 173-201 (1926).

25a. Courant, Friedrichs and Lewy, Math. Ann. 100, 32-74 (1928).

25b. See the Bernstein reference in 35.

26. Bernstein and Petrovskiĭ, Uspehi Matem. Nauk no. 8, 12-13 (1941).

27. Niemytski, Rec. Math. [Mat. Sbornik] (1) 41, 438-452 (1934).

28. Leray and Schauder, Ann. Sci. Ecole. Norm. Sup. (3) 51, 45-78 (1934).

29. Sobolev, Rec. Math. [Mat. Sbornik] N.S. 2(44), 465-499 (1937).

30. Shapiro, C. R. (Doklady) Acad. Sci. URSS (N.S.) 46, 133-135 (1945).

31. Simonov, ibid. 44, 259-261 (1944).

32. Lavrentiev , Rec. Math. [Mat. Sbornik] (1) 42, 407-424 (1935); see also Z. Ya. Shapiro, C. R. (Doklady) Acad. Sci. URSS (N.S.) 30, 690-692 (1941).

33. Cf. Kreines, Rec. Math. [Math. Sbornik] N.S. 9(51), 713-720 (1941).

34. Grötzsch, Ber. Verh. Sächs. Akad. Wiss. Leipzig. Math.-Phys. Kl. 80, 503-507 (1928).

35. For a proof in the case $p = 2$ see Bernstein, Uspehi Matem. Nauk no. 8, 82-99 (1941); H. Lewy, Nachr. Ges. Wiss. Göttingen, Math. Phys. Kl. 1927, 178-186; for larger values of p a proof is given in my article in Rec. Math. [Mat. Sbornik] N.S. 5(47), 3-70 (1939).

36. An example is given on p. 67 of my paper mentioned in 35.

37. Carleman, C. R. Acad. Sci. Paris 197, 471-474 (1933).

38. Petrovsky, Bull. Univ. Moscou. Sér. Internat. Sect. A. 1, no. 7 (1938), chapter II.

39. Tikhonov, Rec. Math. [Mat. Sbornik] (1) 42, 199-216 (1935).

40. Petrovsky, Compositio Math. 1, 383-419 (1935).

41. Holmgren, Ark. Mat. Astr. Fys. 3, no. 1 (1906); no. 12 (1907); 4, nos. 14 and 18 (1908).

42. E. E. Levi, Ann. Mat. Pura Appl. (3) 14, 187-264 (1908).

43. Gevray, J. Math. Pures Appl. (6) 9, 305-471 (1913); 10, 105-148 (1914); Ann. Sci. École Norm. Sup. (3) 35, 129-190 (1918).

44. Tikhonov, Bull. Univ. Moscou. Sér. Internat. Sect. A. 1, no. 9 (1938), §1.

45. Piskunov, Rec. Math. [Mat. Sbornik] N.S. 7(49), 385-424 (1940).

46. Asgeirsson, Math. Ann. 113, 321-346 (1936); see also Courant and Hilbert, Methoden der mathematischen Physik, v. 2, Springer, Berlin, 1937, pp. 417-430.

47. Püschel, Math. Z. 34, 535-553 (1932).

48. M. Krein, C. R. (Doklady) Acad. Sci. URSS (N.S.) 48, 303-306 (1945).

49. Bers and Gelbart, Trans. Amer. Math. Soc. 56, 67-93 (1944).

Mark Grigorievich Krein

Born – 3 April 1907 in Kiev, Ukraine
Died – 17 Oct 1989 in Odessa, USSR

Mark G. Krein grew up in an atmosphere of persecution as the family were Jewish. This was highly significant for Krein's subsequent career, for discrimination against Jews in the Ukraine was bad and, by misfortune for Krein, was particularly bad in Odessa where he lived from the age of 17.

Krein showed a remarkable talent for mathematics at a young age. By the time he reached 14 years of age he was already attending research seminars in mathematics in Kiev. However, Krein never completed his undergraduate degree for he left his home in Kiev when he was 17 years old and ran away to Odessa. Despite the lack of an undergraduate degree, his talents were clearly visible to the mathematicians at Odessa University and, in 1926, Krein was accepted for doctoral studies under Chebotaryov.

In 1928 Chebotaryov left Odessa and became professor at Kazan University. Krein completed his doctorate at Odessa in the following year and remained on the staff at the university building up one of the most important centres for functional analysis research in the world. During this time he worked on topics such as Banach spaces, the moment problem, integral equations and matrices, and on spectral theory for linear operators.

Kolmogorov had laid the foundations for the study of extremal problems in 1935 and Krein began to work on extremal problems for the class of differentiable periodic functions. Together with N. I. Akhiezer, Krein made a major contribution to this field in 1937.

In 1941 Krein had to leave Odessa when the university was evacuated as the German armies advanced. He was appointed as professor of the theoretical mechanics at Kuibyshev Industrial Institute but he returned to Odessa in 1944. However, soon after taking up his post again he was dismissed. He was accused of Jewish nationalism, presumably for having too many Jewish students before the War. Not only was Krein dismissed but the whole of the functional analysis school at Odessa was closed down. Potapov, who had been one of Krein's non-Jewish students, tried hard to influence the university authorities to reverse their decisions. Krein was not reinstated, however, but held the chair of theoretical mechanics at Odessa Marine Engineering Institute from 1944. Also from 1944, he held a part-time post as head of the functional analysis and algebra department at the Mathematical Institute of the Ukrainian Academy of Science in Kiev. This latter position came to a sudden end in 1952 when he was dismissed for a second time. Officially the reason given was that he did not live in Kiev, but in reality it seems more likely that the accusation of Jewish nationalism in his classified file was again the reason.

Although the 1940s must have been difficult times for Krein, his mathematical research did not suffer. Among other important work, Krein wrote eight papers on harmonic analysis and representation theory in the 1940s. In these papers he studied the use of algebras of operators applied to obtain results on positive definite kernels and functions. He also studied analysis on homogeneous spaces and duality theorems.

From 1954 until his retirement Krein occupied the chair of theoretical mechanics at the Odessa Civil Engineering Institute. During the 1940s and 1950s there were many unsuccessful attempts to have Krein and his students reinstated at Odessa University but all attempts failed.

Despite the support of those around him, Krein did have to work at a mathematical disadvantage in one sense. However, for he was not allowed to travel abroad and so was unable to attend international conferences that many mathematicians feel to be almost essential for someone so preeminent at an international level. International recognition came to Krein despite his inability to visit centres of mathematical research around the world. In 1968 the American Academy of Arts and Sciences elected him an honorary member, and he was elected a Foreign Member of the National Academy of Sciences in 1979. Krein received many other honours, but the most prestigious award made to him was the Wolf Prize in Mathematics in 1982. After the years of discrimination on account of being Jewish it must have been particularly pleasing to receive this Prize from Israel.

Despite a life of persecution, in which he often feared arrest, Krein remained enthusiastic, friendly and kind, showing great mathematical generosity towards his students and colleagues. Towards the end of his life, however, the struggle which he continually had to make took its toll and he suffered from depression which became worst after the death of his wife.

On extreme points of regular convex sets

by

M. KREIN and D. MILMAN (Odessa).

Let E be a Banach space (a linear normed complete space) and let $\bar{E}$ be the space of linear functionals adjoint to it.

A set $K \subset \bar{E}$ is called *regularly convex* [1]) if for every $f_0 \in \bar{E}$ not belonging to K such an element $x_0 \in E$ can be found that

$$\sup_{f \in K} f(x_0) < f_0(x_0).$$

It is obvious that every regularly convex set is convex.

Let $f_0 \in \bar{E}$, $x_i \in E$ ($|x_i| \leqslant 1$, $i = 1, 2, \ldots, n$) and $\varepsilon > 0$; then by the neighbourhood $U(f_0; x_1, \ldots x_n, \varepsilon)$ we shall mean the set of all $f \in \bar{E}$ such that

$$|f_0(x_i) - f(x_i)| < \varepsilon \quad (i = 1, 2, \ldots, n).$$

All possible neighbourhoods $U(f_0; x_1, \ldots, x_n, \varepsilon)$, where $f_0 \in E$, $x_i \in \bar{E}$, $|x_i| \leqslant 1$ $(i = 1, 2, \ldots, n;\ n = 1, 2, \ldots,)$ and $\varepsilon > 0$, define in $\bar{E}$ a certain topology, which is called *weak topology* (*Tychonoff's topology*) [2]).

From Tychonoff's theorem on bicompactness of the topological product of segments, as it has been pointed out by Vera GANTMACHER and V. ŠMULYAN [3]), results the following proposition:

A. *A bounded convex set $K \subset \bar{E}$ is regularly convex if and only if it is bicompact in the weak topology.*

---

[1]) This definition has been lorrowed by us from the work of M. G. Krein and V. J. Šmulyan, On regularly closed sets etc. Annals of Mathematics 41 (1940).

[2]) A. Tychonoff, Über topologische Erweiterung von Raumen, Mathem. Annalen 102 (1929) 548.

[3]) V. Šmulyan, Sur les topologies différentes dans l'espace de Banach, Comptes Rendus de l'Acad. des Sc. de l'URSS, 23, 4 (1939).

A point of a convex closed set is called an *extreme point* if it is not an inner point of any segment belonging to the given set.

We now prove the theorem.

Theorem. *Let $K \subset \bar{E}$ be a bounded regularly convex set. Then the set S of extreme points of K is not empty and its regularly convex envelope* [4]) *coincides with K.*

Proof. According to proposition A, K is a bicompact set in the weak topology. To every element $x \in E$ corresponds a function $\varphi_x(f) = f(x)$ continuous on the bicompact set K.

Let $\{x_\alpha\}$ $(\alpha < \vartheta)$ be the set of all the elements of E with $|x| \leqslant 1$ well ordered in any way.

Correspondingly to the sequence $\{x_\alpha\}$ $(\alpha < \vartheta)$ we form a sequence of bicompact sets $\{K_\alpha\}$ $(\alpha < \vartheta)$, each one containing the following ($K_\alpha \supset K_\beta$ for $\alpha < \beta < \vartheta$), by induction.

Define K_1 as a set of those elements $f \in K$ on which the function $\varphi_{x_1}(f)$ reaches its maximum. The set K_1 is closed in the weak topology, and consequently is bicompact. Now let all K_α be defined for $\alpha < \xi$ $(\xi < \vartheta)$. If ξ is not a limiting number, then we denote by K_ξ the set of those $f \in K_{\xi-1}$ on which the function $\varphi_{x_\xi}(f)$ considered on $K_{\xi-1}$ reaches its maximum.

If ξ is a limiting number, then we denote by K'_ξ the intersection of all K_α, with $\alpha < \xi$; since $K_1 \supset K_2 \supset \ldots K_\alpha \supset K_{\alpha+1} \supset \ldots$ are bicompact, so K'_ξ is non-empty. K_ξ will now denote the set of all the points of K'_ξ, on which the function $\varphi_{x_\xi}(f)$ $(f \in K'_\xi)$ reaches its maximum.

Denote by P the non-empty intersection of all K_α $(1 \leqslant \alpha < \vartheta)$. If $g, f \in P$ then $g, f \in K_\alpha$ and consequently

$$f(x_\alpha) = g(x_\alpha) \quad (1 \leqslant \alpha < \vartheta),$$

whence $g = f$. Thus P consists of one point g. We shall prove that this point g is an extreme point of the set K. We assume the contrary, i. e. that with some $f_1, f_2, \in K$ $(f_1 \neq f_2)$ and some t $(0 < t < 1)$

$$g = t f_1 + (1 - t) f_2. \tag{1}$$

[4]) That is the smallest regularly convex set containing S.

Take the first K_ξ to which neither of the elements f_1, f_2 belong. Consider two cases. Let ξ not be a limiting number. Then $f_1, f_2 \in K_{\xi-1}$, and consequently

$$g(x_\xi) = \sup_{f \in K_{\xi-1}} f(x_\xi) \geqslant f_i(x_\xi) \qquad (i=1, 2),$$

the equality sign being excluded at least for one of f_1, f_2 (namely for that f_i which is not included in K_ξ). Whence

(2) $$g(x_\xi) = tg(x_\xi) + (1-t)g(x_\xi) > tf_1(x_\xi) + (1-t)f_2(x_\xi),$$

which contradicts to (1).

Let now ξ be a limiting number. Then $f_1, f_2 \in K_\alpha$ with $\alpha < \xi$, and consequently $f_1, f_2 \in K'_\xi$. Whence

$$g(x_\xi) = \sup_{f \in K'_\xi} f(x_\xi) \geqslant f_i(x_\xi) \qquad (i=1, 2),$$

the equality sign being, as before, excluded, and consequently (2) holds, which contradicts to (1).

Thus we have proved that the point g is an extreme point of the set K, and consequently the set S is not empty.

We now prove that the regularly convex envelope K' of the set S coincides with K. It is evident that $K' \subset K$. Assuming that K' does not coincide with K, we take an element $f_0 \in K - K'$. Since K' is regularly convex, there exists an element $x_0 \in E\,(|x_0| = 1)$ such that

(3) $$\sup_{f \in K'} f(x_0) < f_0(x_0).$$

Consider then the set K_0 of those $f \in K$ on which the function $\varphi_{x_0}(f) = f(x_0)$ $(f \in K)$ reaches its maximum. Evidently the set $K_0 \subset K$ is in the weak topology a certain convex bicompact set, and consequently is a regularly convex set. Whence, in virtue of the facts already proved, K_0 has an extreme point g_0, which is an extreme point of K (for it is easily seen that every extreme point of the set K_0 is also an extreme point of the set K); on the other hand, in virtue of (3) and the definition of the set K_0, the intersection of K_0 with K', and consequently, with S is empty. We have come to a contradiction, which completes our proof.

Corollary. *If a space E is regular (reflective), then any bounded convex closed set is the convex closed envelope of the set of its extreme points.*

M. Krein and V. Šmulyan [5]) have proved that if $S < \bar{E}$ is a bounded set, then its regularly convex envelope consists of those and only those $g \in \bar{E}$ that admit the representation

$$g(x) = M\{\varphi_x(f)\} \quad (x \in E,\ \varphi_x(f) = f(x)),$$

where $M\{\varphi\}$ is a certain mean value defined on the space of all bounded and continuous in the weak topology functions $\varphi(f)$ $(f \in S)$.

As it has been shown by A. Markoff [6]), to every mean value $M\{\varphi\}$ corresponds in a unique way an additive non-negative function $\mu(e)$ of sets $e \subset S$ $(\mu(S) = 1)$ possessing a number of properties and such that

$$M\{\varphi\} = \int \varphi(f)\, d\mu(e),$$

where the integral is understood in the sense of Fréchet-Stieltjes [7]).

Owing to all this, our theorem permits us to say that every point of a regularly convex space is, in a certain sense, the centre of gravity of masses, distributed on the extreme points of this set.

Notice that the unit sphere $|f| \leqslant 1$ of the adjoint space is regularly convex and therefore if $\bar{E}$ is infinite-dimensional, then the sphere has an infinite set of extreme points. Hence:

If the unit sphere of an infinite-dimensional space E has a finite number of extreme points, then E is not adjoint to any Banach space.

We shall now give two examples to which this remark is applicable.

1. Let Q be a topological space and let C_Q be a linear set of all the bounded continuous functions $\varphi(q)$ $(q \in Q)$, with the definition of the norm:

$$\|\varphi\| = \sup_{q \in Q} |\varphi(q)|.$$

---

[5]) See their work quoted in footnote [1]).

[6]) A. Markoff, On mean values and exterior densities, Recueil Mathématique 4 (46), 1 (1938).

[7]) For more details see A. Markoff [6]).

It is easily seen that in this case the point $\varphi(q)$ of the unit sphere $K(\|\varphi\| \leqslant 1)$ of the space C_Q is an extreme point for K if and only if $|\varphi(q)| \equiv 1$ $(q \in Q)$. Therefore, if the space Q is decomposed on α components [8]) then K has exactly 2^α extreme points.

In virtue of this, if α is a finite number and C_Q is infinite-dimensional (the latter, for instance, is carried out if Q contains an infinite number of points and is completely regular) [9]), then E is not adjoint to any Banach space.

2. Let Q be an arbitrary abstract set, and let $\mu(e)$ be an additive function of the subsets $e \in Q$, forming a certain Borel-corpus B. Let the corpus B besides that possess in respect to $\mu(e)$ the following property: if $\mu(e) > 0$ for a set $e \in B$, then there exists a sub-division of e: $e = e_1 + e_2$ $(e_1, e_2 \in B)$ such that $\mu(e_1) > 0$ and $\mu(e_2) > 0$.

Denote by L_Q^μ a linear set of all the functions $\varphi(q)$ $(q \in Q)$ measurable and absolutely integrable in respect to the function $\mu(e)$, the norm φ being defined by the equality:

$$\|\varphi\| = \int\limits_Q |\varphi(q)| \, d\mu(e).$$

It is easily seen that the unit sphere $\|\varphi\| \leqslant 1$ in the space L_Q^μ does not have extreme points and therefore L_Q^μ is not adjoint to any Banach space.

For the space (L) this result (in a more considerable general form) has been obtained by I. M. Gelfand [10]).

[8]) i. e. on disjointed closed connected parts.

[9]) Selim Krein has called our attention to the fact that in order that the space C_Q (with a finite α) should be infinite-dimensional, it is necessary and sufficient that Q should contain an infinite number of points and that the number of dimensions should be greater than α. The necessity of the conditions is obvious. To prove their sufficiency we show that if C_Q is finitely-dimensional, then the number of dimensions of C_Q exactly equals α. In fact, in this case the unit sphere K in C_Q contains exactly m (m is the number of dimensions of C_Q) linearly independent extreme points φ. But as we know, for every such point $\varphi(q) = \pm 1$ and consequently there is exactly α of such linearly independent points, and accordingly $m = \alpha$.

[10]) I. Gelfand, Abstrakte Funktionen und lineare Operatoren, Recueil Mathématique 4 (46), 2 (1938) p. 265.

(Received 10-th 4. 1940).

138

Про екстремальні точки регулярно конвексних множин

М. Крейн і Д. Мільман (Одеса).

(Резюме)

Нехай E є простір Банаха (тобто лінійний, нормований та повний простір) і $\bar{E}$ спряжений до нього простір лінійних функціоналів.

Множина $K \subset \bar{E}$ зветься регулярно конвексною [1]), якщо для кожного $f_0 \in \bar{E}$, не належного до множини K $(f \in K)$, знайдеться такий елемент $x \in E$, що

$$\sup_{f \in K} f(x_0) < f_0(x_0).$$

В цій статті встановлюється така

Теорема. *Якщо $K \subset \bar{E}$ є обмежена регулярно конвексна множина, то множина S екстремальних точок K не є пуста і, більш того, найменша регулярно конвексна множина, що містить S, співпадає з K.*

При цьому точка x даної конвексної множини C зветься екстремальною точкою C, якщо вона не є внутрішня точка жодного сегмента, що увіходить до C.

З теореми безпосередньо випливає

Висновок. *Якщо простір E є регулярний, то кожна обмежена, конвексна, замкнена множина є конвексна замкнена оболонка множини своїх екстремальних точок.*

Доведена теорема дозволяє вказати одну достатню ознаку того, щоб даний простір Банаха не був спряженим до жодного іншого простору Банаха.

ON THE THEORY OF WAVE OPERATORS AND SCATTERING OPERATORS

M. Š. BIRMAN AND M. G. KREĬN

In this paper wave operators are defined for pairs of unitary operators in Hilbert space. Existence of these wave operators is shown when the unitary operators differ by an operator of trace class. Existence of the wave operators for pairs of self-adjoint operators follows by the Cayley transform when the difference of their resolvents is of trace class.*

Scattering operators are constructed in a known way from wave operators and give rise in turn to S-matrices or scattering matrices. Some spectral properties of scattering matrices are given in the second part of the paper.

1. Let U_1 and U_2 be unitary operators in a Hilbert space $\mathfrak{H}$ and let P_k be the projection of $\mathfrak{H}$ into the absolutely continuous subspace $\mathfrak{H}_k$ for the operator U_k, $k = 1, 2$.

Theorem 1. *If the operator $V = U_2 - U_1$ is of trace class, then*

1) *the wave operators*

$$W_\pm(U_2, U_1) = \lim_{n\to\pm\infty} U_2^n U_1^{-n} P_1; \tag{1}$$

$$W_\pm(U_1, U_2) = \lim_{n\to\pm\infty} U_1^n U_2^{-n} P_2 \tag{2}$$

converge in the strong operator topology;

2) *the operator $W_\pm(U_2, U_1)$ maps $\mathfrak{H}_1$ isometrically onto $\mathfrak{H}_2$; the operator $W_\pm(U_1, U_2) = W_\pm^*(U_2, U_1)$ maps $\mathfrak{H}_2$ isometrically onto $\mathfrak{H}_1$;*

---

*Wave operators corresponding to pairs of abstract self-adjoint transformations have been studied in [1–6]. A brief summary of results appears in [6]. See this paper for the definition of a wave operator, absolutely continuous parts, and so on. We depart from [6] in denoting wave operators by $W_\pm$ rather than $U_\pm$. Trace class operators are defined, for instance, in [7]. This class was denoted $\mathfrak{S}$ in [8] by one of the authors. The trace class condition on the resolvents $R_z(H_2)$ and $R_z(H_1)$ means that for some, and hence every, regular point z the difference $R_z(H_2) - R_z(H_1)$ belongs to $\mathfrak{S}$.

3) *the absolutely continuous parts of the operators* U_1 *and* U_2 *are unitarily equivalent and in fact*

$$W_{\pm}(U_2, U_1)\, U_1 f = U_2 W_{\pm}(U_2, U_1)\, f$$

holds for every f *in* $\mathfrak{H}_1$.

The proof of the theorem follows the method of Kato [2, 3]. For one-dimensional perturbations the proof depends on a study of resolvents of U_1 and U_2. Finite dimensional V are then obtained by the "multiplication theorem" of wave operators [2, 5]. Approximation by finite dimensional perturbations is applied in the general case. This approximation by simple operators V_n must be made without destroying the unitary property of the operator $U_1 + V_n$. The possibility of such approximation was previously shown by one of the authors [8] in extending the "trace formula" to unitary operators.

2. Let H_1 and H_2 be self-adjoint operators in $\mathfrak{H}$. We say that they satisfy condition (A) if the difference of their resolvents is of trace class. They satisfy condition (B_n), n a positive integer, if they are strictly positive, the operator $H_2^{-1} - H_1^{-1}$ is completely continuous, and the operator $H_2^{-n} - H_1^{-n}$ is of trace class. The projection into the absolutely continuous subspace $\mathfrak{H}_k$ for H_k is denoted P_k, $k = 1, 2$.

Theorem 2. *Let* H_1 *and* H_2 *be self-adjoint operators in* $\mathfrak{H}$ *which satisfy condition* (A). *Then*

$$W_{\pm}(H_2, H_1) = \lim_{t \to \pm\infty} \exp(iH_2 t) \exp(-iH_1 t) P_1 \tag{3}$$

exists as a strong limit. If $U_k = (H_k - iI)/(H_k + iI)$ *is the Cayley transform of the operator* H_k, $k = 1, 2$, *then the difference* $U_2 - U_1$ *is of trace class and*

$$W_{\pm}(H_2, H_1) = W_{\pm}(U_1, U_2). \tag{4}$$

Additions to this theorem can clearly be made in analogy with 2) and 3) of Theorem 1. Kato's theorem [3] on trace class perturbations of self-adjoint operators is obtained from Theorem 1 applied to the operator $\exp(iH_k)$. Both theorems of Kuroda [4] on the existence of wave operators are contained in Theorem 2. To prove Theorem 2 it is sufficient to establish the existence of the weak limits in (3), together with the relations (4). Then because of Theorem 1, the operators $W_{\pm}(H_2, H_1)$ must be isometric and convergence must take place in the strong sense. This method of showing existence of strong limits (3) was previously developed by one of the authors [6] under additional invertibility hypotheses on the operators H_1 and H_2. In this case the relationship $W_{\pm}(H_2, H_1) = W_{\mp}(H_2^{-1}, H_1^{-1})$ is obtained along with formula (4). In the same paper the existence of strong limits of the form (3) was shown for operators H_1 and H_2, and for operators H_1^{-1} and H_2^{-1}, under the condition (B_n). In doing this it was shown that

$$W_{\pm}(H_2, H_1) = W_{\mp}(H_2^{-1}, H_1^{-1}) = W_{\mp}(H_2^{-n}, H_1^{-n}). \tag{5}$$

3. From the wave operators we construct the scattering operators

$$S = S_{21} = S(H_2, H_1) = W_+^*(H_2, H_1)\, W_-(H_2, H_1),$$

which are widely used in physics. The operator S is unitary in $\mathfrak{H}_1$ and commutes there with H_1. (With no loss of generality $\mathfrak{H}_1$ may be supposed separable since for perturbations of the types under consideration the orthogonal complement of the set of fixed elements of the operator S is separable.) From the multiplication theorem for wave operators one readily obtains the multiplication rule for scattering operators, which in the case of three operators is

$$S(H_4, H_1) = S_{41} = S'_{43} S'_{32} S_{21}, \tag{6}$$

where

$$S'_{k+1,k} = W^*_+(H_k, H_1)\, S(H_{k+1}, H_k)\, W_+(H_k, H_1) \quad (k = 2, 3)$$

are unitary operators in $\mathfrak{H}_1$ which commute with H_1. As in [9] write $\mathfrak{H}_1$ as a continuous direct sum of Hilbert spaces $\mathfrak{H}_\lambda$, taken over the ring of bounded functions in the absolutely continuous part of the operator H_1. Since the operator H_1 has an absolutely continuous spectral measure in $\mathfrak{H}_1$, the measure occurring in the direct sum representation of $\mathfrak{H}_1$ may be chosen as Lebesgue measure on the spectrum Λ of the operator H_1 in $\mathfrak{H}_1$. Since the operator S commutes with H_1, it gives rise in the family $\mathfrak{H}_\lambda$ to a family $S(\lambda)$ of unitary operators, defined for almost all λ in Λ. We call the family $S(\lambda)$ the scattering matrices corresponding to the operator S. If H_1 has finite multiplicity in $\mathfrak{H}_1$, every $\mathfrak{H}_\lambda$ is finite dimensional and $S(\lambda)$ may be represented by finite unitary matrix functions. Otherwise infinite matrix functions or kernels must be used to represent $S(\lambda)$ with respect to any chosen realization of the spaces $\mathfrak{H}_\lambda$. The multiplication rule (6) clearly carries over to scattering matrices:

$$S_{41}(\lambda) = S'_{43}(\lambda)\, S'_{32}(\lambda)\, S_{21}(\lambda). \tag{7}$$

4. In the proof of results given below, an essential role is played by the apparatus used in establishing the trace formula [10, 8]. Let us introduce as much as is necessary for later arguments. If condition (A) is satisfied, the representation

$$\operatorname{Sp}[R_z(H_2) - R_z(H_1)] = -\int_{-\infty}^{\infty} \frac{\xi(\lambda)\, d\lambda}{(\lambda - z)^2} \tag{8}$$

is applicable to the trace of the difference of the resolvents of the operators H_1 and H_2, for all nonreal values of z, where $(1+\lambda^2)^{-1}\xi(\lambda)$ belongs to $L_1(-\infty, \infty)$. The spectral displacement function $\xi(\lambda)$ is specified by (8) only within an added constant. In what follows $\xi(\lambda)$ is to be chosen, within an added integer, so that

$$\int_{-\infty}^{\infty} \frac{\xi(\lambda)\, d\lambda}{1+\lambda^2} = \frac{1}{2i} \operatorname{Sp} \ln (U_1^{-1} U_2), \tag{9}$$

where U_k is the Cayley transform of the operator H_k, $k = 1, 2$. If $H_2 = H_1 + V$ where V is of trace class, $\xi(\lambda)$ belongs to $L_1(-\infty, \infty)$ and condition (9) may be replaced by

$$\int_{-\infty}^{\infty} \xi(\lambda)\, d\lambda = \operatorname{Sp} V.$$

If V is one-dimensional, V and $\xi(\lambda)$ always have the same sign, and $\xi(\lambda)$ is bounded by 1. A spectral displacement function may also be defined as in [8] under the hypothesis (B_n), by a change of variable into the spectral displacement function for the operators H_1^{-n} and H_2^{-n}. In this case $\xi(\lambda)$ vanishes on the left of the spectrum of H_1 and H_2 and $\lambda^{-(n+1)}\xi(\lambda)$ belongs to $L_1(-\infty, \infty)$.

5. **Theorem 3.** *If the operators H_1 and H_2 satisfy either condition (A) or some condition (B_n), $n = 2, 3, \cdots$, then the scattering matrices are of the form*

$$S(\lambda) = I_\lambda + T(\lambda), \tag{10}$$

for almost all λ in Λ, where I_λ is the identity operator and $T(\lambda)$ is an operator of trace class in $\mathfrak{H}_\lambda$.

Furthermore,

$$\det S(\lambda) = e^{-2\pi i \xi(\lambda)}, \tag{11}$$

for almost all λ *in* Λ, *where* $\xi(\lambda)$ *is the spectral displacement function for* H_1 *and* H_2.

Let us give some explanations. If $H_2 = H_1 + V$ where V is of trace class, the transition from H_1 to H_2 can be made in steps, each being a one-dimensional disturbance. Such a transition corresponds to the nature of the scattering matrix, whose spectrum consists of simple eigenvalues of the form $\exp[-2\pi i \xi_k(\lambda)]$ whenever this number differs from unity. Here $\xi_k(\lambda)$ is the spectral displacement function for a one-dimensional disturbance at the kth stage of the process. As in [10] one has

$$\sum_{k=1}^{\infty} \int_{-\infty}^{\infty} |\xi_k(\lambda)| \, d\lambda \leqslant \mathrm{Sp}\, |V|. \tag{12}$$

Because of estimate (12) one can show that the above described infinite process of multiplying scattering matrices, constructed according to (7), converges in the trace norm for almost all λ in Λ, after deducting an identity matrix I_λ, to some operator $T(\lambda)$. It is then easily verified that $I_\lambda + T(\lambda)$ is the scattering matrix for the operator $S(H_2, H_1)$. Because of condition (A) (or alternatively condition (B_n)), formula (4) (or formula (5)) allows one to convert the problem into one for the Cayley transforms U_1 and U_2 (or for the operators H_1^{-n} and H_2^{-n}), after which the previous reasoning applies.

From (10) we obtain $S(\lambda)$ in the form

$$S(\lambda) = e^{-2\pi i K(\lambda)} \tag{13}$$

for almost all λ in Λ, where $K(\lambda)$ is an operator of trace class in $\mathfrak{H}_\lambda$. One may choose $K(\lambda)$ so that

$$\int_{\Lambda} \mathrm{Sp}\, |K(\lambda)| \, p(\lambda) \, d\lambda \leqslant C \; (< +\infty)$$

where 1) $p(\lambda) = 1$ and $C = \mathrm{Sp}|V|$ if $H_2 = H_1 + V$ where V is of trace class, or 2) $p(\lambda) = (1 + \lambda^2)^{-1}$ and $C = \mathrm{Sp}\,|U_2 - U_1|$ if condition (A) is satisfied, or 3) $p(\lambda) = n\lambda^{-(n+1)}$ and $C = \mathrm{Sp}\,|H_2^{-n} - H_1^{-n}|$ if condition (B_n) is satisfied.

The operator $K(\lambda)$ in the representation (13) may be chosen non-negative, in general, under the same conditions under which we were able in [10] to show that the function $\xi(\lambda)$ is bounded below. Namely this is true:

Theorem 4. *The trace class operator* $K(\lambda)$ *in the representation* (13) *may be chosen non-negative if condition* (A) *is satisfied and the following conditions:* 1) *the domain of* H_1 *coincides with the domain of* H_2; 2) *the quadratic form* $((H_2 - H_1) f, f)$, *which is defined for* f *in the common domain of* H_1 *and* H_2, *has only a finite number of negative terms; there exist constants* α *and* β, *where* $0 \leq \alpha < 1$, *such that* $\|(H_2 - H_1) f\| \leq \alpha \|H_1 f\| + \beta \|f\|$ *for all* f *in the domain of* H_1 *or* H_2.

These are the conditions used in [11, 4].

It is easily seen that $K(\lambda)$ may be chosen non-negative also if (B_n) is satisfied for some $n \geq 1$ and if the quadratic form $((H_1^{-n} - H_2^{-n}) f, f)$ has only a finite number of negative terms.

6. In conclusion we remark that for the radial Schroedinger equation a rigorous development of the incidence function $\pi\xi(\lambda)$ with bounded phase is found in [12]. In the three dimensional quantum scattering problem, the expression $\xi(\lambda)$ was previously used by B. S. Buslaev [13] as a scattering amplitude. The formula of B. S. Buslaev is interesting in that it is more easily made meaningful than the expression obtained from the above formula (11).

Leningrad State University
Odessa Institute of Construction Engineering

Received 27/DEC/61

BIBLIOGRAPHY

[1] M. Rosenblum, *Perturbation of the continuous spectrum and unitary equivalence*, Pacific J. Math. **7** (1957), 997–1010.

[2] T. Kato, *On finite-dimensional perturbations of self-adjoint operators*, J. Math. Soc. Japan 9 (1957), 239–249.

[3] ———, *Perturbation of continuous spectra by trace class operators*, Proc. Japan Acad. **33** (1957), 260–264.

[4] S. T. Kuroda, *Perturbations of continuous spectra by unbounded operators*. I, J. Math. Soc. Japan **11** (1959), 246–262.

[5] ———, *Perturbations of continuous spectra by unbounded operators*. II, ibid. **12** (1960), 243–257.

[6] M. Š. Birman, *Conditions for the existence of wave operators*, Dokl. Akad. Nauk SSSR **143** (1962), 506 = Soviet Math. Dokl. **3** (1962), 508.

[7] I. M. Gel'fand and N. Ja. Vilenkin, *Generalized functions*, No. 4, Fizmatgiz, Moscow, 1961. (Russian)

[8] M. G. Kreĭn, *Perturbation determinants and a formula for the traces of unitary and self-adjoint operators*, Dokl. Akad. Nauk SSSR **144** (1962), 268 = Soviet Math. Dokl. **3** (1962), 707.

[9] M. A. Naĭmark and S. V. Fomin, *Continuous direct sums of Hilbert spaces and some of their applications*, Uspehi Mat. Nauk **10** (1955), no. 2(64), 111–142. (Russian)

[10] M. G. Kreĭn, *On the trace formula in perturbation theory*, Mat. Sb. (N.S.) **33** (75) (1953), 597–626. (Russian)

[11] T. Kato, *Fundamental properties of Hamiltonian operators of Schroedinger type*, Trans. Amer. Math. Soc. **70** (1951), 195–211.

[12] V. S. Buslaev and F. D. Faddeev, *Formulas for traces for a singular Sturm-Liouville operator*, Dokl. Akad. Nauk SSSR **132** (1960), 13–16 = Soviet Math. Dokl. **1** (1960), 451.

[13] V. S. Buslaev, *Trace formulas for Schroedinger's operator in three-dimensional space*, ibid. **143** (1962), 1067.

Translated by:
L. de Branges

Lazar Aronovich Liusternik

Born – 31 Dec 1899
Died – 22 July 1981

Lazar Liusternik was a Corresponding Member of the Division of Physical-Mathematical Sciences since 4 Dec 1946.

ОБЗОРНЫЕ СТАТЬИ.

ТОПОЛОГИЯ И ВАРИАЦИОННОЕ ИСЧИСЛЕНИЕ.

Л. А. Люстерник.

1. Введение.

Лишь в исключительных случаях удаётся найти решение диференциального уравнения в виде окончательной формулы, с помощью которой можно и вычислять и исследовать характер решения. Это относится, в частности, и к уравнениям вариационного исчисления. Отсюда возникает задача качественного исследования соответствующего уравнения — исследования числа решений, возможности появления континуальных систем решений, исследования специальных видов решений (например, периодических), исследования устойчивости решений, экстремальных их свойств и т. д. При таких исследованиях весьма полезным орудием является топология. Настоящая статья посвящена вопросам качественного анализа уравнений вариационного исчисления с применением топологического аппарата. Она не претендует, конечно, на полноту освещения излагаемого вопроса.

Задача вариационного исчисления — нахождение экстремалей, т. е. кривых, для которых вариация некоторого функционала обращается в нуль, — является усложнением задачи нахождения критических точек функции, т. е. точек, в которых диференциал функции равен нулю. Поэтому мы остановимся прежде всего на связи топологии с этой более элементарной задачей.

Пусть в евклидовом n-мерном пространстве $(x_1, x_2, \ldots, x_n)$ задана функция точки $f(x_1, x_2, \ldots, x_n)$ и пусть A есть критическая точка для f, т. е. $df(A) = 0$ или, что то же самое, $\frac{\partial f(A)}{\partial x_i} = 0$, $i = 1, 2, \ldots, n$.

Свойство точки A быть критической для функции f не зависит от выбора системы координат: при переходе к новой системе координат $y_i (i = 1, 2, \ldots, n)$, $x_i = x_i(y_1, y_2, \ldots, y_n)$, в предположении, что якобиан преобразования в точке A отличен от нуля, частные производные $\frac{\partial f(A)}{\partial y_i}$, выражаясь линейно через $\frac{\partial f(A)}{\partial x_i}$, тоже равны нулю.

Пусть теперь функция f задана на римановом многообразии R_n. Предполагается, что R_n есть топологическое многообразие и что в некоторой окрестности U_A каждой точки A задана местная система координат $(x_1^A, x_2^A, \ldots, x_n^A)$, с началом координат в точке A.

Мы скажем, что многообразие R_n есть k раз диференцируемое (соответственно аналитическое) многообразие, если в пересечении окрестностей U_A и U_B, любых двух достаточно близких точек A и B из R_n координаты x_i^A являются k раз диференцируемыми (соответственно аналитическими) функциями координат x_i^B. Функция f точки на k раз диференцируемом многообразии R_n называется l раз диференцируемой ($l \leqslant k$), если в местных координатах она в окрестности каждой точки l раз диференцируема. Это свойство не зависит от выбора координат. Функция точки на аналитическом многообразии R_n называется аналитической, если она в окрестности каждой точки есть аналитическая функция местных координат. Точка A называется критической точкой диференцируемой функции f, если в точке A все частные производные $\frac{\partial f(A)}{\partial x_i}$ по местным координатам равны нулю. Свойство точки A быть критической не зависит от выбора местной системы координат.

Как известно, точки максимума и минимума функции являются критическими точками, но критические точки не обязательно являются точками максимума или минимума.

Введём следующие термины. Пусть $f(A)=c$, тогда совокупность точек R_n, в которых $f=c$, будем называть поверхностью уровня, содержащей точку A, и обозначать ($f=c$), а совокупность точек, в которых $f<c$, — областью меньших значений и обозначать ($f<c$). Аналогично определяется область бо́льших значений ($f>c$).

Если A критическая точка и $f(A)=c$, то c будем называть критическим значением. Совокупность критических точек, лежащих на поверхности уровня ($f=c$), будем называть критическим множеством. В дальнейшем без оговорок будем предполагать, что число критических значений функции конечно.

Пусть точка B есть некритическая точка функции f. С точностью до малых второго порядка, f можно представить в окрестности точки B в виде линейной функции местных координат

$$f(x_1, \ldots, x_n) \approx f(0, \ldots, 0) + \sum_i \frac{\partial f(B)}{\partial x_i} x_i.$$

Поверхность уровня ($f=f(B)$) в окрестности точки B «почти совпадает» с ($n-1$)-мерной гиперплоскостью $\sum_i \frac{\partial f(B)}{\partial x_i} x_i = 0$. В достаточно малой окрестности U_B точки B пересечение U_B с поверхностью уровня ($f=f(B)$) гомеоморфно окрестности точки в ($n-1$)-мерной гиперплоскости.

Примечание. Иногда вводят понятие топологической критической точки, т. е. точки, в которой окрестность поверхности уровня не гомеоморфна окрестности в ($n-1$)-мерной гиперплоскости, в отличие от аналитической критической точки — точки, в которой $df=0$. Всякая топологическая критическая точка будет и аналитической. Но аналитическая критическая точка может не быть топологической. Например, для функции, заданной на поверхности, роль поверхностей уровня играют линии уровня; точка возврата линии уровня будет аналитической, но не топологической критической точкой.

Пусть A есть критическая точка. С точностью до величин третьего порядка малости, функция f в окрестности точки A имеет, в местных координатах, вид:

$$f(x_1, \ldots, x_n) \approx c + {}^1/_2 \sum_{i,j} \frac{\partial^2 f(A)}{\partial x_i \partial y_j} x_i x_j,$$

где $c=f(A)$.

Поведение функции f в этой окрестности определяется в основном квадратичной формой $K = {}^1/_2 \sum_{i,j} \frac{\partial^2 f(A)}{\partial x_i \partial x_j} x_i x_j$. Если гессиан $\left|\frac{\partial^2 f(A)}{\partial x_i \partial x_j}\right|$ этой формы отличен от нуля, критическая точка A называется невырожденной. Преобразованием координат можно эту форму привести к сумме положительных и отрицательных квадратов:

$$-y_1^2-y_2^2-\dots-y_i^2+y_{i+1}^2+y_{i+2}^2+\dots+y_n^2.$$

Число i отрицательных квадратов в каноническом представлении формы K называется порядком критической точки. Точки минимума и максимума имеют порядки, равные соответственно 0 и n.

С точностью до величин высшего порядка малости поверхность уровня, проходящая через критическую точку A в окрестности этой точки, является конической гиперповерхностью

$$-\sum_{j=1}^{i} y_j^2+\sum_{k=i+1}^{n} y_k^2=0.$$

В случае $n=1$ невырожденная критическая точка может быть либо точкой минимума (при $i=0$, либо точкой максимума) при $i=1$.

В случае $n=2$ невырожденные критические точки A могут быть трёх типов.

1) $i=0$ (минимум). В некоторой окрестности точки A нет области меньших значений;

2) $i=1$ (седлообразная точка). A есть двойная точка линии уровня $(f=f(A))$. Линии уровня в области меньших и, соответственно, больших значений, расположены как системы сопряжённых гипербол. Область меньших значений в окрестности точки A не связна (части I, II этой области разделены линией уровня).

3) $i=2$ (максимум). Точка A — изолированная точка линии уровня. Область меньших значений в окрестности точки A есть вся окрестность без самой точки A. Линии уровня в этой области расположены как концентрические окружности с центром в A. Так как сама точка A не принадлежит области меньших значений, то в этой области в окрестности точки A линии уровня суть одномерные циклы, негомологичные нулю.

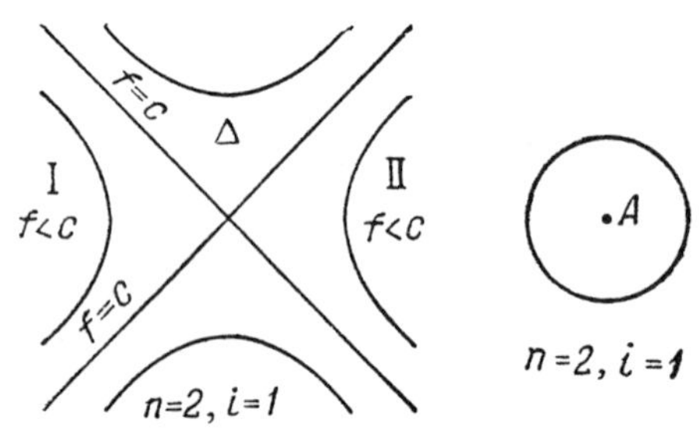

Черт. 1.

При $n=3$ могут иметь место четыре случая.

1) $i=0$ (минимум). Область меньших значений пуста в окрестности точки A.

2) $i=1$. Точка A есть обычная коническая точка поверхности уровня $(f=f(A))$. В окрестности точки A поверхности уровня в области меньших (соответственно больших) значений расположены как двуполостные (соответственно однополостные) гиперболоиды вокруг асимптотического конуса $(f=f(A))$. В окрестности точки A область меньших значений не связна (её части в полостях I и II конуса разделены конической поверхностью уровня).

3) $i=2$. Поверхности уровня в окрестности точки A в области меньших значений расположены как однополостные гиперболоиды вокруг асимптотического конуса, соответствующего поверхности уровня $(f=f(A))$. В окрестности точки A область меньших значений имеет одномерный цикл, негомологичный нулю.

Область бо́льших значений состоит из частей I и II в полостях конуса на черт. 2.

4) $i=3$ (максимум). Поверхности уровня в области меньших значений расположены как концентрические сферы с центром A. Каждая такая поверхность уровня в окрестности точки A в области меньших значений (с выключенной точкой A) есть двумерный цикл, негомологичный нулю.

Эти примеры показывают связь между порядком критической точки и топологической структурой области меньших значений в окрестности этой точки.

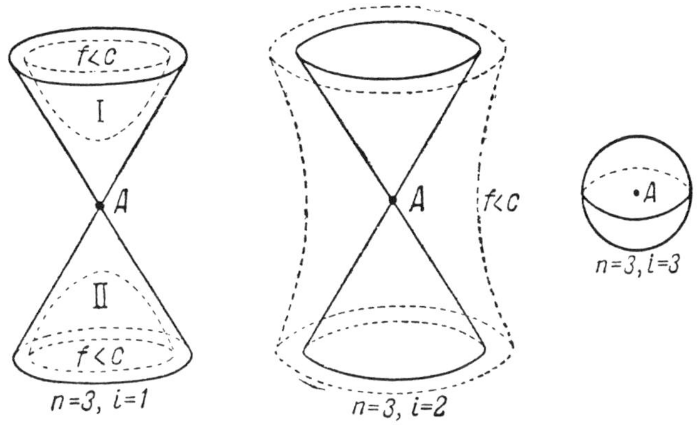

Черт. 2.

Пусть функция f в окрестности критической точки A в местных координатах задана в виде

$$f(y_1, y_2, \ldots, y_n) = c - \sum_{j=1}^{i} y_j^2 + \sum_{k=i+1}^{n} y_k^2.$$

Область меньших значений определится неравенством $f < c$.

Рассмотрим окрестность U_A точки A, заданную неравенством $\sum\limits_{i=1}^{n} y_i^2 \leqslant h^2$. Обозначим через D пересечение U_A с областью меньших значений $(f < c)$. Каждую точку $a(y_1, y_2, \ldots, y_n)$ области D спроектируем на точку $(y_1, y_2, \ldots, y_i, 0, 0, \ldots, 0)$. Последняя принадлежит той же области D, так как $-\sum\limits_{j=1}^{i} y_j^2 < 0$ и из $\sum\limits_{i=1}^{n} y_i^2 \leqslant h^2$ следует $\sum\limits_{j=1}^{i} y_j^2 \leqslant h^2$. Таким образом D спроектировано на часть D_1 линейного многообразия $y_{i+1} = y_{i+2} = \ldots = y_n = 0$. D_1 — i-мерный шар радиуса h с выключенным центром — точкой A. Области D и D_1 имеют одинаковые числа Бетти.

Именно, при $k \neq i-1$, $p_k = 0$ и $p_{i-1} = 1$ (это можно легко проверить на рассмотренных примерах для $n=2$, 3).

Таким образом порядок невырожденной критической точки можно определить топологически.

Будем в дальнейшем считать R_n замкнутым многообразием, функцию f трижды диференцируемой и обладающей лишь конечным числом критических значений.

Поскольку поведение функции в окрестности критической точки можно характеризовать топологически, естественно ожидать связи между топологической структурой многообразия и числом и характером критических точек функции, заданной на этом многообразии. Поясним сначала эту зависимость на простом примере. Пусть в трёхмерном пространстве (x, y, z) заданы сфера S_2 и тор T_2, ось которого параллельна плоскости $z=0$, и на этих поверхностях — функция точки $f=z$ (равная расстоянию точки от плоскости $z=0$). Линии уровня $(f=c)$ получаются пересечением плоскости $z=c$ соответственно с S_2 и T_2. Области меньших значений представляют собой части поверхностей S_2 и T_2, лежащие под этой плоскостью. Будем менять c и посмотрим, как при этом меняются линии уровня $(f=c)$ и области меньших значений $(f<c)$.

У сферы S_2 есть две точки a_1 и a_2 с горизонтальной касательной плоскостью; у тора T_2 четыре такие точки b_1, b_2, b_3, b_4. Эти точки и будут критическими точками функции $f=z$ соответственно на S_2 и T_2. Обозначим значения $f=z$ в этих точках на S_2 и T_2 через c_1 и c_2, где $c_1<c_2$, соответственно d_1, d_2, d_3 и d_4, где $d_1<d_2<d_3<d_4$. Точки a_1 и a_2, b_1 и b_4 суть изолированные точки соответственных линий уровня, b_2 и b_3 двойные точки. Точки a_1 и b_1 суть точки минимума (нулевого порядка), a_2 и b_4 точки максимума (второго порядка), b_2 и b_3 критические точки первого порядка.

На сфере S_2 при $c<c_1$ поверхности уровня $(z=c)$ и области меньших значений $(z<c)$ суть пустые множества; при $c_1<c<c_2$ линия уровня $(z=c)$ есть окружность, а область меньших значений $(z<c)$ гомеоморфна внутренности круга; при $c>c_2$, $(z=c)$ есть пустое множество, а $(z<c)$ — вся сфера. Итак, на S_2 линии уровня $z=c$ и области меньших значений $z<c$ два раза меняют свои топологические свойства при переходе через критические значения c_1 и c_2.

Для тора T_2, при $c<d_1$ линии уровня $(z=c)$ и области меньших значений $z<c)$ суть пустые множества; при $d_1<c<d_2$, линия уровня $(z=c)$ гомеоморфна окружности, а область меньших значений $z<c$ гомеоморфна внутренности круга; при $d_2<c<d_3$ линия уровня $(z=c)$ гомеоморфна паре окружностей, а область меньших значений $(z<c)$ цилиндру (содержит один негомологичный нулю одномерный цикл — меридиан); при $d_3<c<d_4$ линия уровня $(z=c)$ гомеоморфна окружности, а область меньших значений $(z<c)$ гомеоморфна тору с вырезанным кругом (содержит два независимых негомологичных нулю одномерных цикла — меридиан и экватор); при $c>d_4$ линия уровня $(z=c)$ есть пустое множество, область $(z<c)$ — весь тор. Топологическая структура $(z=c)$ и $(z<c)$ меняется на T_2 четыре раза, когда c проходит через критические значения d_1, d_2, d_3, d_4.

Этот пример нам иллюстрирует элементарную, но важную теорему: топологическая структура поверхности уровня $(f=c)$ и области меньших значений $(f<c)$ при изменении c не меняется, если c не проходит через критическое значение функции f. Точнее, если сегмент $[c', c'']$ не содержит критических значений функции f и $c'<\alpha<\beta<c''$, то $(f=\alpha)$ гомеоморфно $(f=\beta)$, а $(f<\alpha)$ гомеоморфно $(f<\beta)$.

В самом деле, введём в окрестности некоторой точки B местную евклидову систему координат $(x_1, x_2, \ldots, x_n)$. Если B есть некритическая точка, то не все частные производные $\frac{\partial f(B)}{\partial x_i}$ равны нулю. Следовательно, можно определить нормальное направление к поверхности уровня $(f=f(B))$ в точке B, именно, направление с угловыми коэфициентами, пропорциональными $\frac{\partial f(B)}{\partial x_i}$. Если сегмент $[\alpha, \beta]$ не содержит критических значений функции f, то в каждой точке B слоя

$\alpha-\varepsilon\leqslant f\leqslant\beta$, где число $\varepsilon>0$ достаточно мало, можно провести нормальное направление к поверхности уровня $(f=f(B))$. Этим определяется семейство нормальных траекторий к поверхностям уровня $(f=c)$, заполняющее слой. Эти нормальные траектории устанавливают гомеоморфное соответствие между точками поверхностей уровня $(f=\alpha)$ и $(f=\beta)$: именно, друг другу отвечают точки этих поверхностей уровня, лежащие на одной и той же нормальной траектории.

Далее, вследствие компактности пространства R_n и замкнутости множества критических точек, найдётся такое положительное число ε, что слой $\alpha-2\varepsilon\leqslant f<\beta$ не содержит критических точек функции f, иначе такие точки нашлись бы на $(f=\alpha)$. Поэтому весь слой $(\alpha-2\varepsilon<f<\beta)$ можно заполнить нормальными траекториями. Установим теперь гомеоморфное отображение области $(f<\beta)$ на область $(f<\alpha)$, оставив неизменной область $(f<\alpha-\varepsilon)$ и отобразив каждую дугу нормальной траектории, лежащую в слое $(\alpha-\varepsilon\leqslant f<\beta)$, на часть этой дуги, лежащую в меньшем слое $(\alpha-\varepsilon\leqslant f<\alpha)$. При этом мы имеем не только гомеоморфизм, но и изотопию областей $(f<\beta)$ и $(f<\alpha)$.

Из этого следует, что если область $(f<c')$ неизотопна $(f<c'')$, то в полуинтервале $c'\leqslant c<c''$ расположено критическое значение f; если $f<c+\varepsilon$ неизотопно $(f<c-\varepsilon)$ при сколь угодно малом $\varepsilon>0$, то c есть критическое значение f.

Пусть, например, дан k-мерный класс гомологий Z_k (т. е. класс всех гомологичных между собой k-мерных циклов в R_n). На каждом цикле z_k из Z_k функция f достигает максимума $\max\limits_{a\in z_k} f(a)$. Обозначим через x нижнюю границу этих максимумов по всем циклам z_k из Z_k: $c=\inf\limits_{z_k\in Z_k}\max\limits_{a\in z_k} f(a)$; c есть критическое значение функции f. В самом деле, при любом $\varepsilon>0$ множество $(f<c+\varepsilon)$ содержит цикл из класса Z_k, тогда как $(f<c-\varepsilon)$ не содержит ни одного цикла этого класса. Таким образом, эти области меньших значений не изотопны, ибо при изотопии циклы класса Z_k переходят в циклы того же класса. В силу предыдущего c есть критическое значение, — мы будем его называть критическим значением, порождаемым классом Z_k. В разобранном выше примере функции $f=z$ на торе T_2 критические значения d_2 и d_3 порождаются соответственно классами меридианов и экваторов.

Таким образом, гомологическая структура многообразия R_n существенно влияет на число и характер критических точек любой функции f на R_n. Однако критические точки определяются не только гомологическими свойствами многообразия, но и гомотопическими. Пусть $\mathfrak{M}$ есть некоторый класс всех континуумов в R_n, гомотопных между собой, т. е. преобразуемых один в другой непрерывной деформацией. Определим $C_{\mathfrak{M}}$ как нижнюю границу максимумов функции на всех континуумах m класса $\mathfrak{M}$. При любом ε множество $(f<C_{\mathfrak{M}}+\varepsilon)$ содержит континуум класса $\mathfrak{M}$, а множество $(f<C_{\mathfrak{M}}-\varepsilon)$ таких континуумов не содержит. Поэтому первое из этих множеств не изотопно второму, и $C_{\mathfrak{M}}$ есть критическое значение функции f [1]).

Как известно, Пуанкаре построил пример трёхмерного замкнутого многообразия F_3, гомологическая структура которого не отличается от гомологической

[1]) L. Lusternik et L. Schnirelmann. Sur un principe topologique d'analyse, C. R., 1929.

структуры трёхмерной сферы S_3 (все его одно- и двумерные циклы гомологичны нулю). Но, в отличие от S_3, фундаментальная группа многообразия F_3 нетривиальна, т. е. F_3 содержит класс L замкнутых кривых, не сводимых непрерывной деформацией к точке. На S_3 легко построить функцию, имеющую только две критические точки — максимума и минимума; на F_3 такой функции построить нельзя, — кроме значений максимума и минимума у функции f на F_3 появляется критическое значение c_L. Ниже мы вернёмся к этому примеру.

Определим теперь следующие топологические инварианты многообразия R_n:

1) $m(R_n)$ — наименьшее число геометрически различных критических точек, которые может иметь функция f на R_n;

2) $m^*(R_n)$ — наименьшее число различных невырожденных критических точек, которые может иметь функция f на R_n, если все её критические точки невырождены;

2') $m_i^*(R_n)$ — наименьшее число критических точек i-го порядка, которые может иметь функция f на R_n, если все критические точки f невырождены.

Если критические точки вырождены, то, вводя понятия кратности, их рассматривают как кратные критические точки. Инварианты $m^*(R_n)$, $m_i^*(R_n)$ оценивают тогда суммы кратностей критических точек или число аналитически различных критических точек функций f на R_n.

Исследования числа аналитически различных критических точек даны в работах Морса (M. Morse), числа геометрически различных критических точек — в работах московских математиков.

Допустим для простоты, что поверхность уровня $(f=c)$ содержит только одну критическую точку A. Вне некоторой окрестности T критической точки A через каждую точку поверхности уровня $(f=c)$ проходит ортогональная траектория. При достаточно малом ε слой $T_1=(c-\varepsilon\leqslant f\leqslant c+\varepsilon)$ можно разбить на сколь угодно малую окрестность T точки A и на часть $T_2=T_1-T$, наполненную ортогональными траекториями. Часть T_2 этого слоя можно «спроектировать» на поверхность уровня $(f=c-\varepsilon)$, заставляя каждую точку слоя двигаться по ортогональной траектории до этой поверхности уровня. Таким образом $(f<c+\varepsilon)-T$ изотопно множеству $(f\leqslant c-\varepsilon)$, тем более изотопно области $(f<c)$. От добавления части T_2 слоя T_1 все изотопные инварианты (например, числа Бетти) области $(f<c)$ не меняются. Лишь добавление окрестности T критической точки A может изменить эти инварианты. Область же $(f<c+\varepsilon)$ изотопна $(f<c)+T$.

Аналогичные рассуждения имеют место, если на поверхности уровня $(f=c)$ расположено критическое множество K, содержащее более чем одну точку. При достаточно малом ε область $(f<c+\varepsilon)$ изотопна $(f<c)+K_1$, где K_1 — окрестность критического множества K, сколь угодно малая при достаточно малом ε.

Поэтому основным орудием в исследовании поставленных вопросов являются аддиционные теоремы, изучающие изменение топологических инвариантов комплекса от добавления к нему другого комплекса (в зависимости от взаимного расположения этих комплексов).

В настоящее время исследованы аддиционные теоремы для чисел и групп Бетти. Важной очередной задачей является изучение аддиционных свойств других инвариантов, влияющих на критические точки (например, кольца пересечений).

2. Число аналитически различных критических точек.

Для того чтобы составить себе представление о том, как влияет прибавление окрестности критической точки на область меньших значений, ограничимся случаем точки первого порядка (разобранным ещё Биркгофом в 1917 г. [1])).

Пусть A критическая точка первого порядка функции f на R_n. Пусть далее K_1 есть некоторая окрестность точки A. Пересечение $K_2 = K_1 \cap (f < c)$ не связно. Пара точек A_1 и A_2, принадлежащих двум различным компонентам — множества K_2, образует негомологичный нулю нульмерный цикл (A_1, A_2) в K_2, их нельзя соединить дугой, целиком лежащей в K_2.

Рассмотрим отдельно два возможных случая:

1) Точки A_1 и A_2 образуют негомологичный нулю цикл во всей области меньших значений $(f < c)$, т. е. принадлежат разным компонентам области $(f < c)$. После прибавления окрестности K_1 точки A эти две компоненты сливаются в одну; число компонент области меньших значений $(f < c)$ — её нульмерное число Бетти — от добавления K_1 уменьшилось на 1 (черт. 3).

2) Цикл (A_1, A_2) гомологичен нулю во всей области меньших значений $(f < c)$, т. е. точки A_1, A_2 можно соединить путём l, целиком лежащим в $(f < c)$ (см. черт. 4). В окрестности K_1 их можно соединить новым путём l_2. Дуги l_1 и l_2 с общими концами порождают одномерный цикл $l = l_1 - l_2$ в $(f < c) + K$. Этот цикл не зависит от циклов, лежащих в $(f < c)$. Таким образом, мы имеем в этом случае появление «нового» одномерного цикла, т. е. увеличение одномерного числа Бетти на 1. При переходе через критическое значение c изменились числа Бетти области меньших значений: одномерное число увеличилось на 1 или нульмерное число уменьшилось на 1.

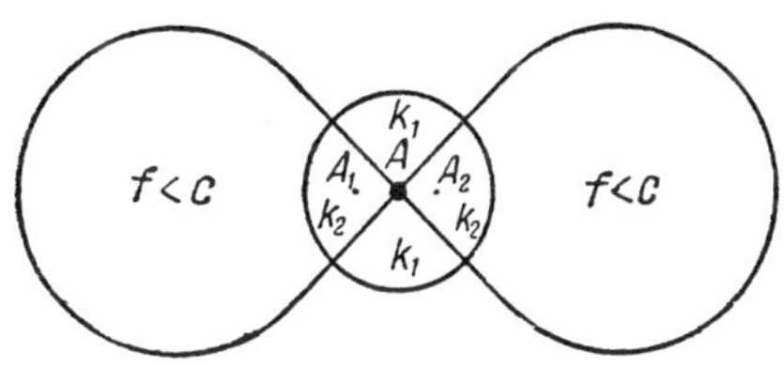

Черт. 3.

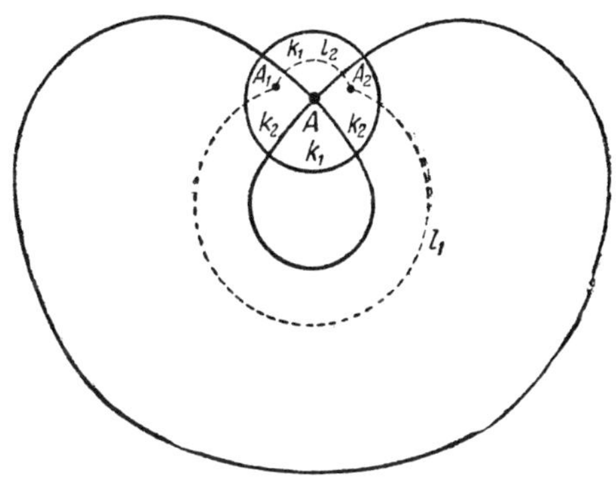

Черт. 4.

Этот результат был обобщён Морсом: при добавлении окрестности критической точки i-го порядка i-мерное число Бетти области меньших значений может возрасти на 1 (т. е. может появиться «новый» i-мерный цикл) или $(i-1)$-мерное число Бетти может уменьшиться на 1[2]).

Отсюда получаем оценку инварианта $m_i^*(R_n)$ наименьшего числа критических точек i-го порядка функции f на R_n (если все критические точки невырожденные):

$$m_i^*(R_n) \geq p_i(R_n),$$

[1]) Dynamical Systems, Transactions of the Amer. Math. Soc., 1917.

[2]) Изменение гомологической структуры поверхностей уровня при прохождении через критическую точку дано в работах Г. Чогошвили (ДАН, 1938) и Л. Эльсгольца (Матем. сборник, 1939).

где p_i есть i-мерное число Бетти R_n. В самом деле, область меньших значений ($f < c$) при возрастании c может приобрести новый i-мерный цикл лишь при прохождении через критическую точку i-го порядка функции f. При изменении c от минимума до максимума функции f, ($f < c$) должно по крайней мере $p_i(R_n)$ раз приобретать новые i-мерные циклы, т. е. функция f должна иметь не менее $p_i(R_n)$ критических точек i-го порядка.

Вместе с тем можно оценить и $m^*(R_m)$ — наименьшее число критических точек для функции f на R_n, у которой все критические точки невырожденные. Очевидно,

$$m^*(R_n) = \sum_i m_i^*(R_n) \geqslant \sum_i p_i(R_n),$$

т. е. число критических точек функции, у которой нет вырожденных критических точек, не меньше суммы чисел Бетти всех размерностей. Для тора T_2, например, $p_0(T_2) = 1$, $p_1(T_2) = 2$, $p_2(T_2) = 1$. Следовательно, если у функции f на T_2 все критические точки невырожденные, то число их не меньше $1 + 2 + 1 = 4$. Мы построили выше пример функции на T_2 с четырьмя критическими невырожденными точками, $m^*(T_2) = 4$.

В случае наличия вырожденных критических точек можно ввести понятие кратности критической точки. Ограничимся аналитическими функциями на аналитических многообразиях. Пусть A критическая точка функции f и $f(A) = c$. Пусть далее пересечение области меньших значений ($f < c$) с окрестностью точки A имеет локальные l-мерные числа Бетти, равные l_i. Мы скажем, что точка A есть критическая точка кратности $\sum_i l_i$, причём в этой точке совмещено l_i критических точек порядка i ($i = 0, 1, 2, \ldots, n$). Доказывается, что функции f можно дать сколь угодно малую вариацию δf так, чтобы функция $f + \delta f$ совпадала с f всюду, кроме некоторой окрестности A, причём в сколь угодно малой окрестности этой точки она имела бы ровно l_i невырожденных критических точек порядка i ($i = 0, 1, 2, \ldots, n$). Таким образом, топологическое понятие кратности критической точки имеет и аналитический смысл. Если не требовать, чтобы функция f имела только невырожденные критические точки, то под числом аналитически различных критических точек i-го порядка функции l_i следует понимать сумму кратностей критических точек i-го порядка l_i по всем критическим точкам функции f.

Аналогично, если мы имеем дело не с изолированной критической точкой, а с целым континуумом таких точек, то можно и для него ввести понятие кратности, позволяющее данный континуум считать эквивалентным некоторому конечному числу критических невырожденных точек разных порядков. Поэтому инвариант $m^*(R_n)$ можно определить как минимум сумм кратностей критических множеств по всем функциям, определённым на R_n.

Примечание 1. Оценка инвариантов $m_i^*(R_n)$, $m^*(R_n)$ через числа Бетти не является точной. Числа $m_i^*(R_n)$, $m^*(R_n)$ зависят не только от гомологических, но и от гомотопических свойств R_n, например, $m_i^*(R_n)$ зависит от фундаментальной группы. Обозначая через $f_1(R_n)$ наименьшее число образующих фундаментальной группы, можно доказать, как это сделал Л. Эльсгольц, что

$$m_1^*(R_n) \geqslant f_1(R_n).$$

Сфера S_3 и многообразие Пуанкаре имеют одинаковые, числа Бетти, в частности $p_1(F_3) = 0$. Между тем $m_1^*(F_3) > 0$.

Примечание 2. Числа $m_i^*(R_n)$ обладают свойством двойственности: $m_i^*(R_n) = m_{n-i}^*(R_n)$. В самом деле, число критических точек порядка i функции f на R_n равно числу критических точек порядка $n-i$ функции $-f$. Поэтому нижние границы числа критических точек порядков i и $n-i$ для всех функций на R_n совпадают.

3. Оценка числа геометрически различных критических точек.

Сумма чисел Бетти оценивает снизу сумму кратностей критических точек любой функции. Однако этим инвариантом нельзя оценить число $m(R_n)$ геометрически различных критических точек функции f на R_n.

Например, для тора T_2 сумма чисел Бетти равна 4. Однако легко построить функцию f на T_2 с числом критических точек, равным 3. Представим T_2 в виде прямоугольника с иденцифицированными противоположными сторонами (черт. 5). На этом чертеже изображены линии уровня функции f на T_2, имеющей три критические точки: точку минимума в A, точку максимума в B и кратную в C (двойную критическую точку первого порядка); итак, $m(T_2) \leqslant 3$.

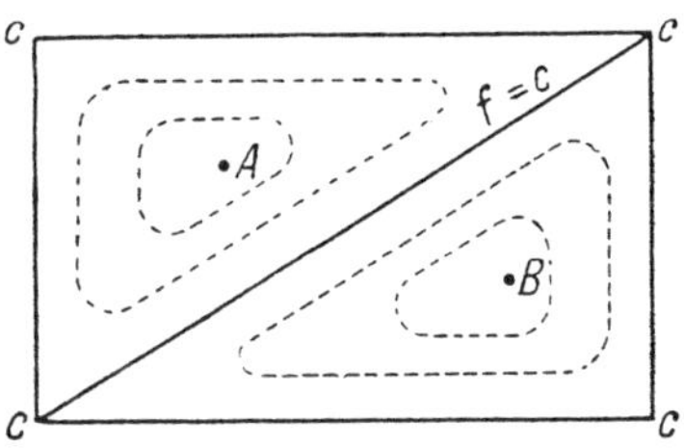

Черт. 5.

Докажем теперь, что $m(T_2) \geqslant 3$. Обозначим соответственно через c_1 и c_4 минимум и максимум функции f на T_2. Через c_2, c_3 обозначим нижние границы максимумов функции f, соответственно, на меридианах и экваторах, т. е. на негомологичных одномерных циклах. Имеем $c_1 \leqslant c_2 \leqslant c_4$ и $c_1 \leqslant c_3 \leqslant c_4$. Если $c_1 = c_4$, то функция постоянна, и все точки T_2 критические. Если $c_1 = c_2 = c$ (или $c_1 = c_3 = c$), то на T_2 имеется континуум критических точек функции f. В самом деле, из определения c_2 следует, что множество $f \leqslant c$ содержит цикл, гомологичный меридиану; так как c совпадает со значением минимума c_1, то $(f < c)$ пусто, и линия уровня $(f = c)$ содержит такой цикл, т. е. содержит целый меридиан. Но так как c есть значение минимума f, то всякая точка линии $(f = c)$ есть критическая точка. Аналогично, если $c_4 = c_2 = c$, а также если $c_4 = c_3 = c$ или $c_1 = c_3 = c$, то $(f = c)$ содержит континуальное критическое множество. Итак, $m(T_2) \geqslant 3$. Так как, с другой стороны, $m(T_2) \leqslant 3$, то $m(T_2) = 3$. Всякая функция на торе имеет не менее трёх критических точек.

Для оценки $m(R_n)$ введён ряд гомологических и гомотопических инвариантов. Вначале использовался гомотопический инвариант, названный «категория» [1]).

Замкнутое множество M называется множеством категории 1 на R_n, если M может быть стянуто в точку непрерывной деформацией в R_n. Замкнутое множество M называется множеством категории k на R_n, если оно может быть представлено как объединение k множеств категории 1 на R_n и не может быть составлено из меньшего числа таких частей. Будем это кратко записывать:

$$\operatorname{cat} M = k.$$

---

[1]) L. Lusternik, Ueber topologische Grundlagen der allgemeinen Eigenwerttheorie Monatsh. f. Math. u Physik., 1930.

Пример. Любая простая дуга на торе T_2 имеет категорию 1; меридиан или экватор на T_2 имеют категорию 2, так как они сами не могут быть стянуты к точке, но их можно разбить на две простые дуги. Сама поверхность T_2 не может быть разбита на две части, стягиваемые к точке, но может быть разбита на три такие части; категория T_2 на самом себе равна 3.

Категория обладает следующими свойствами:

1) $\operatorname{cat}(M+M_2) \leqslant \operatorname{cat} M_1 + \operatorname{cat} M_2$.

Это свойство непосредственно вытекает из определения категории.

2) Если $M_1 \subset M_2$, то $\operatorname{cat} M_1 \leqslant \operatorname{cat} M_2$.

3) При непрерывной деформации категория не понижается.

В самом деле, пусть множество A путём непрерывной деформации $D(A, B)$, переведено в множество B и $\operatorname{cat} B = k$. Это означает, что B может быть разбито на k частей $B_1, B_2, \ldots, B_k$, сводимых каждая непрерывной деформацией $D(B_i, b_i)$ к точке b_i, $i=1, 2, \ldots, k$. Обозначим через A_i, $i=1, 2, \ldots, k$, совокупность точек из A, которые при деформации $D(A, B)$ переходят в точки B_i, A_i сводимо деформацией $D(A, B) \times D(B_i, b_i)$ в точку b_i; следовательно, категория $A_i (i=1, 2, \ldots, k)$ есть 1. Так как $A = A_1 + A_2 + \ldots + A_k$, то $\operatorname{cat} A \leqslant k$.

4) Категория достаточно малой окрестности $S(A, \varepsilon)$ множества A совпадает с категорией A.

В случае $\operatorname{cat} A = 1$ это свойство вытекает из приводимой нами без доказательства теоремы о том, что деформацию множества A в точку можно распространить на некоторую замкнутую окрестность этого множества так, чтобы последняя сводилась к той же точке. Если $\operatorname{cat} A = k$, то $A = A_1 + A_2 + \ldots + A_k$, где $\operatorname{cat} A_i = 1$; каждое A_i можно окружить замкнутой окрестностью $\overline{S(A_i, \varepsilon)}$, где ε достаточно мало, так, что $\operatorname{cat}[\overline{S(A_i, \varepsilon)}] = 1$, $i=1, 2, \ldots, k$; отсюда $\operatorname{cat}[\overline{S(A, \varepsilon)}] \leqslant k$. Но так как $A \subset S(A, \varepsilon)$ и $\overline{S(A, \varepsilon)} = \sum_{i=1}^{k} \overline{S(A_i, \varepsilon)}$, то $\operatorname{cat}[\overline{S(A, \varepsilon)}] = k$.

Из этих свойств вытекает следующее важное предложение:

Если $\operatorname{cat} R_n = k$, то всякая функция f на R_n имеет не менее k критических точек. Другими словами, $m(R_n) \geqslant \operatorname{cat} R_n$[1]).

Рассмотрим множества $(f \leqslant c)$ при различных c. С изменением c категории этих множеств меняются лишь при прохождении c через критические значения функции f. С другой стороны, если $(f = c)$ содержит лишь изолированные критические точки $a_1, a_2, \ldots, a_l$, то при достаточно малом ε категория множества $(f \leqslant c + \varepsilon)$ совпадает, в силу предыдущего, с категорией множества $(f \leqslant c - \varepsilon) + T_1 + T_2, + \ldots + T_l$, где T_i есть некоторая сколь угодно малая окрестность критической точки a_i, $i=1, 2, \ldots, l$. Категория T_i равна 1, следовательно,

$$\operatorname{cat}(f \leqslant c+\varepsilon) \leqslant \operatorname{cat}(f \leqslant c-\varepsilon) + \operatorname{cat} T_1 + \operatorname{cat} T_2 + \ldots + \operatorname{cat} T_l, \quad \operatorname{cat} T_i = 1,$$

$$i = 1, 2, \ldots, l,$$

т. е. от включения окрестности T_i каждой изолированной критической точки

[1]) Другие гомотопические инварианты, оценивающие $m(R_n)$, даны в работах Л. Эльсгольца.

категория повышается не более чем на 1. Если критических точек лишь конечное число, то число этих критических точек не меньше категории всего R_n, что и требовалось доказать.

Уточним доказанную теорему.

Пусть категория R_n равна k. Обозначим через c_i нижнюю границу максимумов функции f на множествах категорий $\geqslant i$. Очевидно, $c_1 \leqslant c_2 \leqslant c_3 \leqslant \ldots \leqslant c_k$.

Возможны два случая:

1) все числа c_i различны: $c_1 < c_2 < \ldots < c_k$. Мы имеем k различных критических значений, а значит не менее k различных критических точек.

2) $c_i = c_{i+p} = c$,

т. е. слились $p+1$ критических поверхностей уровня $(f = c_i)$.

Теорема. *При $c_i = c_{i+p} = c$ поверхность уровня содержит множество A критических точек, категория которого не меньше $p+1$.*

Как мы уже выше указывали, множество $(f \leqslant c + \varepsilon)$ при достаточно малом ε изотопно $(f \leqslant c - \varepsilon) + T$, где T — сколь угодно малая замкнутая окрестность критического множества A. Отсюда

$$\operatorname{cat}(f \leqslant c + \varepsilon) \leqslant \operatorname{cat}(f \leqslant c - \varepsilon) + \operatorname{cat} T.$$

Но при достаточно малом ε окрестность T сколь угодно близка к A, и мы имеем

$$\operatorname{cat} T = \operatorname{cat} A.$$

Поэтому

$$\operatorname{cat}(f \leqslant c + \varepsilon) \leqslant \operatorname{cat}(f \leqslant c - \varepsilon) + \operatorname{cat} A. \tag{1}$$

Множество $(f \leqslant c - \varepsilon)$ имеет категорию не выше $i - 1$ (так как $c = c_i$ есть нижняя граница максимумов f на множествах категории $\geqslant i$, между тем максимум f на $(f \leqslant c - \varepsilon)$ равен $c - \varepsilon < c$; следовательно, это множество имеет категорию, меньшую i). Множество же $(f \leqslant c + \varepsilon)$ уже заключает множества категории $\geqslant (i + p)$, и его категория поэтому не меньше $i + p$. Сопоставляя это с (1), получим

$$\operatorname{cat} A \geqslant p + 1,$$

что и требовалось доказать.

Пример. Ниже мы покажем, что категория проективного n-мерного пространства равна $n+1$. Отсюда всякая функция f на n-мерном проективном пространстве имеет не менее $n+1$ критических точек (напомним, что n-мерное проективное пространство P_n есть пространство всех прямых в $(n+1)$-мерном эвклидовом пространстве E_{n+1}, проходящих через данную точку).

Пусть в E_n дано произвольное выпуклое тело S. Каждой прямой p, проходящей через данную точку (т. е. каждому элементу P_{n-1}), отвечает пара опорных плоскостей, перпендикулярных к этой прямой. Расстояние $\rho(p)$ (ширина S в направлении p) между этими плоскостями есть функция элемента p $(n-1)$-мерного проективного пространства P_{n-1}. В силу предыдущего на P_{n-1} существуют по крайней мере n элементов $p_1, p_2, \ldots, p_n$ таких, что в них $d\rho(p_i) = 0$.

Будем рассматривать хорды тела S, в концах которых тело S касается двух параллельных между собою опорных плоскостей. Если $d\rho(p) = 0$, то соответствующая хорда перпендикулярна к обеим параллельным между собою опорным плоскостям (т. е. хорда параллельна p). Итак, в теле существует не менее n различных хорд таких, что они перпендикулярны к опорным плоскостям, проведённым в их концах. Для n-мерного эллипсоида такими являются его главные оси. Если длины l таких хорд совпадают, то имеется уже $(l-1)$-мерный континуум таких хорд (как на эллипсоидах вращения).

Заметим, что категория не определяется полностью гомологическими свойствами многообразия. Так, Борсук доказал, что если фундаментальная группа нетривиальна, то категория многообразия не меньше 3. Поэтому $\operatorname{cat} F_3 \geqslant 3$ (где F_3 — сфера Пуанкаре), в то время как $\operatorname{cat} S_3 = 2$, хотя F_3 и сфера S_3 имеют одинаковую гомологическую структуру.

Для оценки инварианта $m(R_n)$ Л. Эльсгольц употребил другие инварианты, связанные с разбиением пространства на те или иные элементарные части[1]).

Перейдём теперь к гомологическим инвариантам, оценивающим число критических точек. Л. Г. Шнирельман, вычисливший сначала непосредственно категорию проективного пространства, нужную для ряда приложений, определил её также гомологическим образом, введя впервые[2]) такие гомологические инварианты, которые оказались тесно связанными с кольцом пересечений многообразия.

Исследования Л. Г. Шнирельмана положили начало ряду работ, посвящённых гомологическим оценкам числа критических точек. Так, С. Фролов и Л. Эльсгольц, упрощая конструкцию Л. Шнирельмана, ввели удобный для вычисления инвариант — «длину многообразия», (Матем. сборник, 1935 г.). Длиной многообразия R_n называется наибольшее число циклов $z_1, z_2, \ldots, z_i$ из R_n, отличных от самого R_n, пересечение которых негомологично нулю:

$$z_1 \times z_2 \times \ldots \times z_i \not\sim 0.$$

Длину R_n будем обозначать $\operatorname{long} R_n$.

П р и м е р ы. Длина тора T_2 равна 2. На T_2 можно найти два цикла — меридиан и экватор, пересечение которых (точка) негомологично нулю.

Длина n-мерного проективного пространства P_n равна n. В P_n можно найти n циклов — $(n-1)$-мерных проективных гиперплоскостей, — пересечение которых (точка) негомологично нулю.

Длина X_3 — топологического произведения окружности S_1 и двумерной сферы S_2 есть 2, ибо в X_3 имеются два цикла (множители S_1 и S_2), пересечение которых (точка) негомологично нулю.

Связь между категорией и длиной выражается неравенством:

$$\operatorname{cat} R_n \geqslant \operatorname{long} R_n + 1.$$

Это предложение основано на теореме Понтрягина. Пусть заданы в R_n k-мерный цикл z_k и замкнутое множество A такое, что для всякого $(n-k)$-мерного цикла z_{n-k} из A его индекс пересечения $H(z_k, z_{n-k})$ равен нулю. Тогда цикл z_k можно снять с A, т. е. существует цикл $\overline{z}_k$, гомологичный z_k и лежащий вне A. Если множество A имеет категорию 1, то все лежащие в нём циклы гомологичны нулю, и для любого цикла z_k удовлетворяются условия теоремы Понтрягина.

Таким образом, из теоремы Понтрягина сразу вытекает

С л е д с т в и е. Любой цикл можно снять с A.

Допустим теперь, что $\operatorname{cat} R_n = k$, $\operatorname{long} R_n \geqslant k$. Это значит:

$$R_n = M_1 + M_2 + \ldots + M_k, \quad \operatorname{cat} M_i = 1, \quad i = 1, 2, \ldots, k,$$

и на R_n существует k циклов $z_1, z_2, \ldots, z_k$ таких, что $z_1 \times z_2 \times \ldots \times z_k \not\sim 0$.

---

[1]) Л. Э л ь с г о л ь ц, Матем. сборник, 1939.

[2]) L. S c h n i r e l m a n n, Uber eine neue kombinatorische Invariante, Monatsh. f. Math. u. Phys., 1930.

В силу следствия из теоремы Понтрягина для любого M_i каждый цикл z_i ($i = 1, 2, \ldots, k$) можно снять с соответствующего M_i, т. е. можно найти цикл $\bar{z}_i \sim z_i$, где $\bar{z}_i$ не имеет общих точек с M_i. Имеем $\bar{z}_1 \times \bar{z}_2 \times \ldots \times \bar{z}_k \sim z_1 \times z_2 \times \ldots \times z_k \not\sim 0$. Следовательно, пересечение (теоретико-множественное) всех $\bar{z}_i$ не пусто. С другой стороны, каждое $\bar{z}_i$ лежит вне соответствующего множества M_i и их пересечение должно лежать вне $\sum_{i=1}^{k} M_i = R_n$, т. е. должно быть пустым. Полученное противоречие доказывает неравенство $\operatorname{long} R_n \leqslant \operatorname{cat} R_n - 1$, откуда

$$m(R_n) \geqslant \operatorname{cat} R_n \geqslant \operatorname{long} R_n + 1.$$

Что же касается оценок $m(R_n)$ сверху, то Гордон доказал, что на n-мерном многообразии R_n всегда можно построить функцию с $n+1$ критическими точками, т. е. $m(R_n) \leqslant n+1$.

Для проективного пространства P_n полученная нижняя оценка $m(P_n)$ совпадает с верхней. С одной стороны, $m(P_n) \leqslant n+1$. С другой стороны, $m(P_n) \geqslant \operatorname{cat} P_n \geqslant \operatorname{long} P_n + 1 = n+1$. Итак, $m(P_n) = \operatorname{cat}(P_n) = n+1$. Задача точного вычисления $m(R_n)$ в общем случае ещё не решена.

Уточним несколько этих результат. Будем рассматривать такие классы циклов X, в которые вместе с циклом x входят все циклы, гомологичные qx, где q — любое целое положительное число, если гомологии берутся по модулю 0 и $0 < q < p$, если гомологии берутся по простому модулю p.

Мы скажем, что класс k-мерных циклов Y подчинён классу X $(k+l)$-мерных циклов в R_n, если существует $(n-l)$-мерный цикл z такой, что для $x \in X z \times x \in Y$.

Пример. В трёхмерном проективном пространстве класс Y проективных прямых подчинён классу X проективных плоскостей. Здесь z — также проективная плоскость и пересечение проективных плоскостей $z \times x\, (x \in X)$ есть проективная прямая $y \in Y$.

Пусть теперь k-мерный класс Y подчинён $(k+l)$-мерному классу X посредством цикла z, т. е. если $x \in X$, то $z \times x \in Y$. Обозначим через c_1 и c_2 нижние границы максимумов функции f соответственно на циклах классов Y и X. Очевидно, $c_1 \leqslant c_2$.

Теорема. *Если $c_1 = c_2 = c$, то критическое множество на поверхности уровня $(f = c)$ содержит негомологичный нулю l-мерный цикл*[1]).

Таким образом, слияние поверхностей уровня $(f = c_1)$ и $(f = c_2)$ влечёт за собой существование на слившихся поверхностях уровня критического множества размерности не меньше l, т. е. не меньше, чем разность размерностей подчинённого и подчиняющего классов. Можно уточнить эту теорему: критическое множество A на $(f = c)$ содержит l-мерный цикл u_l, двойственный к z $(H(u_l, z) \neq 0)$.

В самом деле, допустим, что $H(u_l, z) = 0$ для любого l-мерного цикла u_l из A. Тогда, в силу теоремы Понтрягина, z можно снять с A, т. е. существует цикл $\bar{z}$, гомологичный z и лежащий вне A; $\bar{z}$ продолжает высекать из циклов x класса X циклы y класса Y, $(\bar{z} \times x) \in Y$, если $x \in X$.

При любом $\varepsilon > 0$ область $(f < c + \varepsilon)$ содержит цикл x класса X; область $(f < c + \varepsilon)$ изотопна $(f < c - \varepsilon) + A_1$, где A_1 — окрестность A, сколь угодно малая при достаточно малом ε. Мы выберем столь малое ε, чтобы соответствую-

[1]) Люстерник, Размерности критических множеств. ДАН, XXXIX, № 9, 1943.

щая окрестность A_1 множества A не имела общих точек с $\bar{z}$. Поскольку $(f < c - \varepsilon)$ изотопно $(f < c - \varepsilon) + A_1$, то в этой сумме заключён цикл x_2 класса X. Так как $\bar{z}$ лежит вне A_1, то и пересечение $y_1 = \bar{z} \times x_2$, являющееся циклом класса Y, лежит вне A_1. С другой стороны, $y_2 \subset x_2 \subset (f < c - \varepsilon) + A_1$; значит, y_2 лежит в $(f < c - \varepsilon)$. Максимум f на $y_2 \in Y$ не превосходит $c - \varepsilon$ и оказывается таким образом меньше c_1, что противоречит определению c_1 как нижней границы максимумов f на циклах класса Y.

Таким образом, устанавливается порядок следования критических значений c для разных классов гомологии. Если число критических точек конечно, то критическое значение, порождаемое циклами подчиняющих классов, строго больше критического значения, порождаемого циклами подчинённых классов. Пусть $\operatorname{long} R_n = k$, т. е. существуют k циклов $z_1, z_2, \ldots, z_k$, отличных от R_n, таких, что $z_1 \times z_2 \times \ldots \times z_k \nsim 0$. Мы имеем цепь из $k+1$ циклов R_n, z_1, $z_1 \times z_2$, $z_1 \times z_2 \times z_3$, $\ldots, z_1 \times z_2 \times z_3 \ldots \times z_k$, в которой каждый подчинён предыдущему. Таким образом, мы вновь доказали, что число геометрически различных критических точек $m(R_n)$ больше или равно $\operatorname{long} R_n + 1$.

Итак, исследование чисел Бетти определяет наличие критических значений, исследование кольца многообразия устанавливает порядок следования этих значений.

4. Вариационные задачи.

Перейдём теперь к собственно вариационному исчислению. Вместо n-мерных многообразий у нас будут функциональные пространства. Роль критических точек функций играют экстремали — элементы функционального пространства, для которых диференциал функционала (вариация) исчезает. Число экстремалей — критических элементов любого функционала — связано с топологической структурой этого пространства. При переходе от n-мерных пространств к функциональным необходимо обобщить для них те инварианты, которые служили для оценки числа критических точек в n-мерном случае.

Мы отвлекаемся от теоретико-функциональных трудностей, с которыми связан этот переход. Они аналогичны тем трудностям, с которыми связано доказательство существования простейшего критического элемента, дающего, например, минимум функционала. Мы опустим все эти рассмотрения, давая только намётки доказательств. Доказательства существования критических элементов (экстремалей) функционалов будут верными лишь при некоторых специальных ограничениях на функционалы, которые мы сейчас формулировать не будем.

Понятие категории логически непосредственно переносится на случай функциональных пространств. Множество M в пространстве L имеет категорию k, если оно может быть представлено в виде суммы k (но не меньшего числа) множеств, сводимых к точке путём деформации в L. Пусть, например, пространство L состоит из всех замкнутых кривых с непрерывной касательной на сфере S_2. Окрестность кривой q составляют близкие к q кривые в смысле близости первого порядка (т. е. с близкими направлениями касательных в соответственных точках). Все нулевые кривые, т. е. выродившиеся в точки, мы идентифицируем. Рассмотрим в L замкнутую лнию z (однопараметрическое семейство кривых — элементов L), именно, совокупность z_1 всех больших кругов с общим диаметром.

z_1 покрывает сферу S_2, и это свойство z_1 сохраняется при всех деформациях семейства (деформация множества M в L заключается в том, что каждая кривая q из M непрерывно деформируется, и достаточно близкие кривые из M остаются близкими во всё время деформации). При этом z_1 не может быть сведено к одному элементу из L, поэтому его категория больше 1 (нетрудно видеть, что она равна 2). Двумерное множество z_2 в L, состоящее из всех больших кругов на S_2 (представляющее собой проективную плоскость), имеет, как можно доказать, на L категорию 3. Наконец, трёхмерное множество z_3, образующее трёхмерное проективное пространство всех окружностей на S_2, имеет на L категорию 4.

Введём в S_2 произвольную риманову метрику и рассмотрим функционал $J(q)$ — длину кривой q в данной метрике на S_2. Критическими элементами этого функционала являются замкнутые геодезические (в данной метрике), критическими значениями — длины этих геодезических. Из того, что L содержит множество категории 4, следует, что существуют четыре критических элемента — замкнутых геодезических. Один из них — тривиальный — замкнутая геодезическая нулевой длины (выродившаяся в точку), отвечающая критическому значению 0. Три из них — настоящие замкнутые геодезические с длинами c_1, c_2, c_3, критическими значениями $J(q)$, где c_i нижняя граница максимумов длин на семействе кривых категории $\geqslant i$ $(i=2,\ 3,\ 4)$; при этом $c_2 \leqslant c_3 \leqslant c_4$. Если два из этих чисел совпадают, например, $c_2 = c_3 = c$ или $c_3 = c_4 = c$, то существует семейство геодезических категории 2 одинаковой длины c (пример: поверхность вращения). Если $c_2 = c_3 = c_4 = c$, то существует семейство замкнутых геодезических длины c категории 3 (как на сфере). Все замкнутые геодезические имеют в этом случае одинаковую длину c. Таким образом, на поверхности рода 0 имеются три замкнутых геодезических.

Дополнительные аналитические исследования позволяют доказать, что на поверхности рода 0 имеется три замкнутых не пересекающих самих себя геодезических. Для этого приходится рассматривать семейства линий, состоящие из несамопересекающихся кривых, и вводить только такие деформации этих семейств, которые переводят такие семейства опять в семейства несамопересекающихся кривых. Это связано с преодолением некоторых геометрических трудностей. Зато можно гарантировать, что каждая из полученных замкнутых геодезических не сводится к несколько раз повторённой одной из них. Окончательно можно сформулировать следующий результат:

На всякой трижды диференцируемой поверхности рода 0 существуют три замкнутых не пересекающих самих себя геодезических разной длины, или однопараметрическое семейство не пересекающих самих себя геодезических равной длины и одна такая геодезическая, или все геодезические суть замкнутые, не пересекающие самих себя кривые равной длины.

Подробное доказательство этой теоремы дано в мемуаре Л. Люстерника и Л. Шнирельмана — «Топологические методы в вариационных задачах», Труды Института математики и механики при I МГУ, 1930 (отдельное издание).

В данной задаче удалось определить категории семейств кривых z_1, z_2, z_3. Вообще определение категории множеств в функциональных пространствах весьма трудно. Неизвестно, например, существуют ли на сфере компактные семейства замкнутых кривых категории больше 4 (если такое семейство состоит из не пересекающих самих себя кривых, то его категория не больше 4, и, более того, любое такое семейство может быть деформировано в семейство окружностей, т. е. в часть z_3 [1]). Оценить снизу категории семейств дуг на сфере с общими концами удалось, только исследовав кольцо пересечений в функциональном пространстве таких дуг. (Оказалось, что существуют семейства таких дуг любой категории.)

Теория Морса была перенесена им на случаи вариационных задач. Гомологии (по модулю 2), которыми оценивалось число критических точек, непосредственно переносятся на случай функциональных пространств. Так же, как и в n-мерном случае, критические элементы (экстремали) классифицируются на вырожденные и невырожденные; невырожденные классифицируются по порядкам, вырожденные имеют кратность (эквивалентны некоторому числу невырожденных экстремалей разных порядков). Классификация критических элементов по порядкам производится на основе топологической структуры области меньших значений в окрестности критического элемента. Как и в n-мерном случае, порядок экстремали можно определять и аналитически, исследуя вторую вариацию окрестности экстремали: порядок есть число отрицательных собственных значений второй вариации [2]).

Как и в n-мерном случае, число экстремалей k-го порядка функционала J на функциональном пространстве R не меньше суммы чисел Бетти R всех измерений (при этом число экстремалей определяется как сумма кратностей всех экстремалей, геометрически они a priori могут все слиться в одну).

Морс определил числа Бетти по модулю 2 следующих функциональных пространств:

1) пространства R спрямляемых дуг на сфере с общими концами;

2) пространства L замкнутых спрямляемых кривых на сфере, с обобщением этих результатов на n-мерную сферу.

В применении к теории замкнутых геодезических это даёт оценку суммы кратностей числа таких геодезических — на двумерной поверхности рода 0 она равна трём (a priori эти геодезические могут сливаться). Для задачи о геодезических дугах, соединяющих данные точки a и b на поверхности S_2 рода 0, из результатов Морса следует существование бесконечной последовательности таких дуг, причём длины этих дуг неограниченно возрастают. Отсюда вытекает существование бесконечной последовательности геометрически различных геодезических дуг, соединяющих любые две точки на поверхности рода 0. Основные результаты Морса изложены в его книге «Calculus of variations in the large», 1934.

В «Трудах II Математического съезда» (Ленинград, 1934) напечатана обзорная статья Л. Люстерника и Л. Шнирельмана «Топологические методы в вариационных задачах», в которой приводятся основные теоремы как Морса, так и авторов обзора.

5. Верхние циклы и их произведения в функциональных пространствах.

Перейдём теперь к некоторым работам последнего времени. Как мы видели, число критических точек функций на многообразии самым тесным образом связано со свойствами кольца пересечений этого многообразия. Аналогично, число

---

[1]) L. Lusternik. Ueber topologische Eigenschaften der Kurvenfamilien auf Flächen. Математический сборник, 1931 г.

[2]) Порядок экстремали связан с числом сопряжённых точек на экстремали, как это доказал Морс. См., например, Лаврентьев и Люстерник, Курс вариационного исчисления, гл. XI.

экстремалей для некоторого функционала связано с кольцом пересечений («кольцом произведений») для функционального пространства, образованного допустимыми линиями соответственной вариационной задачи, определение же этого кольца связано с введением «верхних гомологий» данного функционального пространства.

Теория верхних гомологий, или ∇-гомологий (в отличие от обычных Δ-гомологий или нижних гомологий) разработана впервые А. Н. Колмогоровым и Александером и далее развита П. С. Александровым и Чехом. Мы отсылаем читателя к мемуару П. С. Александрова «О гомологических свойствах расположения комплексов и замкнутых множеств» [1]), где исследуются ∇-гомологии и соответственно ∇-произведения в локально бикомпактных пространствах. Как известно, классы допустимых линий для важнейших вариационных задач образуют компакты, а поэтому при исследовании их топологической структуры можно пользоваться некоторыми результатами уже развитой теории.

В n-мерном многообразии R_n верхние k-мерные образы суть по существу то же, что $(n-k)$-мерные обычные — теория верхних гомологий изучает по существу те же объекты, что и обычная теория нижних гомологий. В случае же бесконечномерных пространств k-мерный верхний цикл есть то, что можно назвать $(\infty-k)$-мерным подпространством. Мы приведём примеры верхних циклов в некоторых важных для приложений бесконечномерных пространствах.

1. *Гильбертово проективное пространство.* Если в гильбертовом пространстве идентифицировать диаметрально противоположные точки единичной сферы, то получим пространство, которое естественно назвать гильбертовым проективным пространством. Это пространство некомпактно, но из него можно выделить компактное бесконечномерное подпространство, если ограничиться, например, теми точками, у которых n-ые координаты x_n удовлетворяют неравенствам $x_n < \frac{1}{n}$ $n=1, 2, \ldots$). Обозначим этот компакт через P. Будем рассматривать в P гомологии по модулю 2.

Одномерный верхний цикл определяется функцией f, заданной для пар достаточно близких точек. Пусть A, B — две близкие точки в P. Каждая точка P определяется парой точек гильбертовой сферы: точка A парой точек $A_1(x_1, x_2, \ldots, x_n, \ldots)$, $A_2(-x_1, -x_2, \ldots, -x_n, \ldots)$, точка B парой точек $B_1(y_1, y_2, \ldots, y_n, \ldots)$ и $B_2(-y_1, -y_2, \ldots, -y_n, \ldots)$. Расстояние $\rho_P(A, B)$ между точками A и B в P определим как меньшее из двух расстояний

$$\rho(A_1, B_1) = \rho(A_2, B_2) \quad \text{и} \quad \rho(A_1, B_2) = \rho(A_2, B_1)$$

в гильбертовом пространстве. Пусть $\rho(A, B) \leqslant \varepsilon < {}^1/_2$, т. е. одно из расстоянии $\rho(A_1, B_1)$ или $\rho(A_1, B_2)$ не превосходит ε, например $\rho(A_1, B_2) \leqslant \varepsilon$: Тогда, при любом i, $|x_i - y_i| < \varepsilon$.

Определим теперь функцию $f_i^\varepsilon(A, B)$ от пар точек A, B из P, для которых $\rho(A, B) < \varepsilon$: именно $f_i^\varepsilon(A, B) = 0$, если i-ые координаты x_i и y_i (для которых $|x_i - y_i| \leqslant \varepsilon$) имеют разные знаки и $f_i = \rho(A, B)$, если при том же условии x_i, y_i имеют одинаковые знаки. Если $x_i = 0$ или $y_i = 0$, то $f_i^\varepsilon(A, B)$ не определено.

Эта функция определяет одномерный ∇-цикл, который мы обозначим через P_i. Для уяснения геометрической природы этого цикла введём понятие «носителя цикла»:

[1]) Известия Академии Наук СССР (серия математическая), 1942.

Пусть k-мерный ∇-цикл z определяется функциями f^{ε_n}, где f^{ε_n} заданы на системах из $k+1$ точек нашего пространства, попарные расстояния между которыми не превосходят $\varepsilon_n > 0$.

Обозначим через $\overline{Z}_n$ теоретико-множественную сумму тех систем точек, для которых $f^{\varepsilon_n} \neq 0$. Топологический предел Z множеств $\overline{Z}_n$ при $n \to \infty$ будем называть носителем цикла z.

В рассмотренном случае носителем P_i является «$(\infty - 1)$-мерное подпространство» $\overline{P}_i$, определённое уравнением $x_i = 0$. Носителями одномерных циклов в P являются $(\infty - 1)$-мерные проективные подпространства $\overline{P}_i$, определяемые одним линейным уравнением. Произведение ∇-циклов $P_{i_1} \times P_{i_2} \times \ldots \times P_{i_k}$ является k-мерным ∇-циклом $P_{i_1 i_2 \ldots i_k}$; его носителем является «$(\infty - k)$-мерное проективное подпространство», определённое k линейными условиями $x_{i_1} = 0$, $x_{i_2} = 0$, ..., $x_{i_k} = 0$. Двойственным в P к этому k-мерному ∇-циклу является k-мерный Δ-цикл — обычное k-мерное проективное пространство, заключённое в P (k-мерная единичная сфера с идентифицированными диаметрально противоположными точками). Пересечение $(k+l)$-мерного Δ-цикла — именно, $(k+l)$-мерного проективного подпространства — с описанным k-мерным ∇-циклом есть l-мерный Δ-цикл — обычное l-мерное проективное пространство.

2. *Пространство спрямляемых замкнутых кривых на двумерной сфере.* Рассмотрим на двумерной сфере S_2 совокупность L_N всех спрямляемых замкнутых кривых, длина которых не превосходит N. Число N будем полагать большим, чем длина большого круга на сфере. Расстояние $\rho(q_1, q_2)$ между кривыми q_1 и q_2 из l_N будем понимать в смысле Фреше, выбрав на q_1 и q_2 циклические параметры с общим периодом и определив для каждого такого выбора параметров максимум расстояний между точками обеих кривых с общими значениями параметров, мы обозначим через $\rho(q_1, q_2)$ нижнюю границу таких максимумов при всевозможных выборах параметров. Принимая $\rho(q_1, q_2)$ за расстояние между q_1 и q_2, мы превращаем L_N в компактное метрическое пространство. Рассмотрим в L_N гомологии по модулю 2.

Обозначим через z_3 совокупность всех окружностей на сфере; через z_2^a — совокупность таких окружностей, проходящих через некоторую точку a, через z_1^{ab} — совокупность таких окружностей, проходящих через точки a и b; z_3, z_2^a, z_1^{ab} суть Δ-циклы в L_N размерностей соответственно 3, 2 и 1.

Определим теперь в L_N одномерный ∇-цикл y_1^a следующим образом: пусть a произвольная точка S_2; выбрав достаточно малое число $\varepsilon > 0$, на каждую пару кривых q_1 и q_2, для которых $\rho(q_1, q_2) < \varepsilon$ и которые не проходят через a, натянем узкую плёнку[1]). Обозначим через $f_a^{\varepsilon}(q_1, q_2)$ степень покрытия такой плёнкой точки a по модулю 2 (она не зависит от выбора параметров на кривых q_1 и q_2). Функции $f_a^{\varepsilon}(q_1, q_2)$ определяют одномерный ∇-цикл на L_N, носителем которого является совокупность всех кривых из L_N, проходящих через точку a.

---

[1]) Такую плёнку можно получить, например, таким образом: выберем на q_1 и q_2 циклические параметры таким образом, чтобы точки с одинаковыми значениями параметров были удалены друг от друга на расстояние, меньшее ε, и соединим все возможные такие пары малыми дугами больших кругов; совокупность этих дуг и даст нужную нам плёнку.

Этот ∇-цикл двойственен Δ-циклу z_1^{bc}, индекс их пересечения равен 1. Геометрически это видно из того, что если a, b, c — различные точки, то пересечение носителя y_1^a с z_1^{bc} есть окружность q_{abc}, проходящая через точки a, b, c (т. е. один элемент L_N).

Произведение двух ∇-циклов y_1^a, y_1^b есть двумерный ∇-цикл y_2^{ab}, носителем которого является совокупность кривых из L_N, проходящих через две точки a и b. Этот цикл двойственен Δ-циклу z_2^c (пересечение носителя y_2^{ab} с z_2^c есть окружность q_{abc}). Наконец, произведение $y_1^a \times y_1^b \times y_1^c$ есть трёхмерный ∇-цикл y_3^{abc}, носитель которого есть совокупность кривых из L_N, проходящих через точки a, b, c; цикл y_3^{abc} двойственен z_3 (пересечение носителя y_3^{abc} и z_3 есть окружность q_{abc}).

В некоторых задачах приходится рассматривать вместо L_N его замкнутое подпространство L_N^*, элементами которого являются те из кривых L_N, которые не имеют самопересечений (но могут касаться самих себя), L_N^* есть замыкание в L_N множества всех кривых без двойных точек. Из двух кривых на черт. 6, в которых направление обхода указано стрелками, левая может быть элементом L_N^*, правая не может. Циклы z_3, z_2^a, $z_1{}^{ab}$ являются Δ-циклами также в L_N^*. Двойственные им ∇-циклы $\overline{y}_3^{abc}$ $\overline{y}_2^{ab}$, $\overline{y}_1^a$ в L_N^* определяются так, как y_3^{abc}, y_2^{ab}, y_1^a (носитель цикла $\overline{y}_1^a$ является пересечением носителя y_1^a с L_N^*, аналогично $\overline{y}_2^{ab}$, $\overline{y}_3^{abc}$).

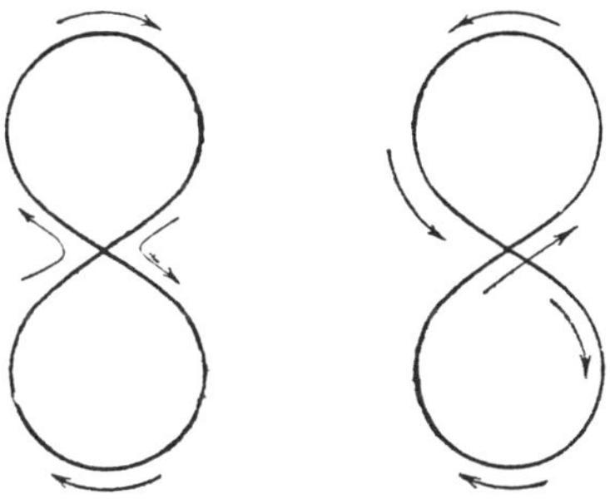

Черт. 6.

3. *Пространство допустимых линий вариационной задачи с фиксированными концами.* Пусть a и b произвольные точки сферы S_2. Обозначим через R совокупность всех спрямляемых кривых, соединяющих точки a и b сферы S_2. Расстояние $\rho(q_1, q_2)$ между дугами q_1 и q_2 — элементами R — определим по Фреше (аналогично расстоянию в L_N). Если обозначим через R_N совокупность всех дуг из R, длины которых не больше N, то R_N есть компакт. Само R некомпактно, но оно есть сумма компактов, $R = \sum R_N$ (т. е. есть множество типа F_σ). В дальнейшем гомологии в R будут обозначать гомологии в R_N при достаточно больших N. Гомологии будем понимать по произвольно большому простому модулю.

Определим в R одномерный ∇-цикл z_1^c. На достаточно близкие дуги q_1, q_2, для которых $\rho(q_1, q_2) < \varepsilon$, натягивается, как выше, узкая плёнка, границей которой является замкнутая кривая $q_1 - q_2$. Степень покрытия этой плёнкой точки c обозначим через $f_\varepsilon^c(q_1, q_2)$. Функция $f_\varepsilon^c(q_1, q_2)$ определяет одномерный ∇ цикл z_1^c. Носителем этого цикла является совокупность кривых из R, проходящих через точку c. Несколько более сложно определяется k-мерный ∇-цикл $z_k^{c_1, c_2, \ldots, c_k}$. Носитель его есть множество спрямляемых дуг, которые при движении по S_2 от точки a к b обходят в заданном порядке точки c_1, c_2, ... c_k[1]). Можно доказать, что всякий k-мерный ∇-цикл в R гомологичен $qz_k^{c_1, c_2, \ldots, c_k}$, где q — целое число.

[1]) Точное определение этих циклов см. в нашей заметке «Кольцо произведений в одном функциональном пространстве», ДАН, XXXVIII, № 2—3, 1943 г.

Кольцо произведений в R определяется следующей формулой:

$$z_k^{c_1, c_2, \ldots, c_k} \times z_l^{d_1, d_2, \ldots, d_l} = \sum \pm z_{k+l}^{e_1, e_2, \ldots, e_{k+l}}, \tag{2}$$

где $e_1, e_2, \ldots, e_{k+l}$ — перестановка элементов $c_1, c_2, \ldots, c_k, d_1, d_2, \ldots, d_l$ такая, что k элементов $c_1, c_2, \ldots, c_k$ сохраняют свой относительный порядок, равно как и l элементов $d_1, d_2, \ldots, d_l$. Такие перестановки назовём специальными. Например, из четырёх элементов c_1, c_2, d_1, d_2 можно образовать шесть специальных перестановок: $c_1c_2d_1d_2$, $c_1d_1c_2d_2$, $c_1d_1d_2c_2$, $d_1c_1c_2d_2$, $d_1c_1d_2c_2$, $d_1d_2c_1c_2$.

Сумма в правой части (2) берётся по всем специальным перестановкам $e_1, e_2, \ldots$ $\ldots, e_{k+l}$, причём знак перед её слагаемым — плюс или минус — зависит от чётности или нечётности соответствующей специальной перестановки. Например,

$$z_2^{c_1c_2} \times z_2^{d_1d_2} = z_4^{c_1c_2d_2d_2} - z_4^{c_1d_1c_2d_2} + z_4^{c_1d_1d_2c_2} + \\ + z_4^{d_1c_1c_2d_2} - z_4^{d_1c_1d_2c_2} + z_4^{d_1d_2c_1c_2}. \tag{3}$$

Обозначим через Z_k класс гомологии для $z_k^{e_1, e_2, \ldots, e_k}$ (все $z_k^{e_1, e_2, \ldots, e_k}$ при любых точках $e_1, e_2, \ldots, e_k$ гомологичны между собою). Тогда (3) запишется:

$$Z_k \times Z_l = C_{kl} Z_{k+l}, \tag{4}$$

где C_{kl} обозначает избыток чётных специальных перестановок из k элементов c_i и l элементов d_i над нечётными. Из (3), например, следует: $Z_2 \times Z_2 = 2Z_4$. Заметим, что $C_{2,n} \neq 0$ при любом n:

$$C_{1,\,2k} = 1; \quad C_{1,\,2k+1} = 0.$$

Формула (4) определяет кольцо произведений в R.

Построим теперь Δ-циклы в R, двойственные к Z_k[1]). Полагая a и b полюсами на сфере S_2, будем обозначать через (φ) меридиан долготы φ (полукруг, соединяющий a и b, плоскость которого образует угол φ с заданной начальной плоскостью). Обозначим через $(\varphi_1, \varphi_2, \ldots, \varphi_{2k+1})$ «полимеридиан» — кривую, которую получим, последовательно проходя от a к b и обратно, по меридианам (φ_1) (φ_2) $\ldots$ (φ_{2k+1}). Если a начало такой кривой, то b — её конец. Обозначим через T_{2k+1} совокупность всех полимеридианов $(\varphi_1, \varphi_2, \ldots, \varphi_{2k+1})$ при произвольных значениях цилиндрических координат $\varphi_1, \varphi_2, \ldots, \varphi_{2k+1}$; через T_{2k} — совокупность всех полимеридианов $(\varphi_1, \varphi_2, \ldots, \varphi_{2k}, \varphi^0_{2k+1})$ при произвольных значениях цилиндрических координат $\varphi_1, \varphi_2, \ldots, \varphi_{2k}$ и фиксированном значении $\varphi_{2k+1} = \varphi^0_{2k+1}$. Все T_n (n-мерные торы) суть n-мерные Δ-циклы в R. Индекс пересечения T_n и Z_n равен 1. Отсюда следует, что T_n, так же как и Z_n, негомологичны нулю (можно доказать, что всякий n-мерный Δ-цикл гомологичен qT_n).

Пересечения $Z_k \times T_{k+l}$ определяются по формуле

$$Z_k \times T_{k+l} = C_{k,\,l} T_l.$$

Из определения $C_{k,\,l}$ нетрудно заметить, что $C_{2,\,n} \neq 0$ при любом n, отсюда

$$Z_2 \times T_{k+2} = C_{2,\,k} T_k \not\sim 0.$$

Перенесём на случай функциональных пространств понятие подчинения циклов друг другу. Будем рассматривать классы Δ-циклов такие, что вместе с циклом x они содержат любой цикл, гомологичный qx, где $q > 0$, если гомологии берутся

[1]) См. нашу заметку «О семействах дуг с общими концами на сфере», ДАН, XXXIX, № 3, 1943.

по модулю 0 и $0<q<p$, если гомологии берутся по простому модулю p. Класс Δ-циклов Y назовём подчинённым классу Δ-циклов x, если существует такой ∇-цикл z, что если $x \in X$, то $(z \times x) \in Y$.

Пример. В гильбертовом проективном пространстве класс k-мерных проективных подпространств подчинён классу $(k+l)$-мерных подпространств.

В пространствах спрямляемых линий L циклы z_1^{ab}, z_2^a, z_3 последовательно подчинены друг другу.

В пространстве $R\ T_n$ всегда подчинён T_{n-2}, T_n подчинён T_{n+1}, если n чётное в силу формул (4).

6. Критические множества.

Пусть в компактном пространстве K задан функционал J. Будем называть «поверхностью уровня» $(J=c)$ функционала J множество точек, в которых $J=c$. Если любое замкнутое множество из $(J<c+\varepsilon)$ при достаточно малых ε может быть деформировано в замкнутое множество в $(J<c-\varepsilon)$, то c называется обыкновенным значением J, в противном случае — критическим. Пусть c — критическое значение. Будем называть критическим множеством на поверхности уровня $(J=c)$ непустое замкнутое множество M, обладающее следующими свойствами:

1) каково бы ни было число $\alpha>0$, можно найти такое число $\varepsilon>0$, что всякое замкнутое множество, лежащее в $(J<c+\varepsilon)$ и удалённое от M больше, чем на α, может быть деформировано в область $(J<c-\varepsilon)$.

2) M является неприводимым по отношению к этому свойству, т. е. никакая правильная часть множества M уже не обладает этим свойством.

Примечание. Во всём дальнейшем будем ограничиваться тем важнейшим для приложений случаем, когда для любого c существует замкнутое множество M', удовлетворяющее условию 1). Тогда всегда существует неприводимое множество M, пустое или непустое: при c — критическом значении существует непустое M — критическое множество.

Если X есть некоторый класс Δ-гомологий и c есть нижняя граница максимумов J на циклах класса X, то c есть критическое значение. В самом деле, каково бы ни было ε, множество $(J<c+\varepsilon)$ содержит цикл x_ε класса X. Если c не есть критическое значение, то x_ε при достаточно малом ε может быть деформировано в цикл $\bar{x} \subset (J<c-\varepsilon)$. Цикл $\bar{x}$ входит в класс X и потому максимум J на цикле $\bar{x}$ не меньше c, что противоречит соотношению $\bar{x} \subset (J<c-\varepsilon)$. Полученное противоречие доказывает предложение.

Примеры. 1. Если пространство K есть n-мерное многообразие, а J дважды диференцируемая функция на нём, то критическое множество есть множество точек, в которых $dy=0$ (см. ниже).

2. Пусть в n-мерном пространстве $(x_1, x_2, \ldots, x_n)$ задана $(n-1)$-мерная сфера $\sum x_i^2=1$. Квадратичная форма $F=\sum_{i,j} a_{ij}x_ix_j$, заданная на этой сфере, принимает равные значения в диаметрально противоположных точках сферы; можно идентифицировать противоположные точки и рассматривать F на $(n-1)$-мерном проективном пространстве P^{n-1}. Критические точки для F на P^{n-1} суть те, в которых $d(F-\lambda\sum x_i^2)=0$, где λ множитель Лагранжа. Каждой критической точке F на P^{n-1} отвечает пара точек сферы — концов единичных собственных векторов (отличающихся знаками). Собственные значения формы суть её критические значения.

Перейдём к гильбертовому проективному пространству. Рассмотрим, например, форму $F(\varphi)=\int_0^1\int_0^1 K(x, s)\,\varphi(x)\,\varphi(s)\,dx\,ds$, где $K(x, s)$ симметрическое ядро Фредгольма. Зададим F на единичной сфере $\int_0^1 \varphi^2\,dx=1$ гильбертова пространства. Так как $F(\varphi)=F(-\varphi)$, то F можно считать заданным на гильбертовом проективном пространстве. Критические значения формы F совпадут с собственными значениями соответствующего интегрального уравнения, а критические множества суть множества пар нормированных собственных функций, отвечающих данному собственному значению (и различающихся знаками). Отметим, что гильбертово проективное пространство P само не компактно. Но мы можем исследовать форму F на компактных частях P.

3. Пусть в R (см. выше) функционал J есть длина дуги. Критическое множество $(J=c)$ есть множество геодезических дуг длины c, соединяющих точки a и b на поверхности. В самом деле, множество дуг, в котором самые длинные кривые не являются геодезическими, может быть преобразовано деформацией в множество кривых, у которых максимум длин уменьшится.

4. Пусть в L (см. выше) функционал J означает длину замкнутой кривой. Критическое множество на $(J=c)$ есть множество замкнутых геодезических длины c. В пространстве L критическое множество на $(J=c)$ есть множество не пересекающих самих себя замкнутых геодезических длины c.

Перенесём теперь на случай компактных функциональных пространств [1]) теорему стр. 43.

Теорема. *Пусть X класс $(k+l)$-мерных Δ-циклов в R, Y — класс k-мерных Δ-циклов, подчинённый X, т. е. существует такой l-мерный ∇-цикл z^l, что если $x \in X$, то $x \times z^l \in Y$. Пусть, далее, c_1, c_2 — соответственно нижние границы максимумов функционала J на циклах классов Y, X. Тогда, если $c_1=c_2=c$, то критическое множество на $(J=c)$ содержит негомологичный нулю l-мерный Δ-цикл.*

Заметим, прежде всего, что теорема Понтрягина о снятии цикла распространена П. С. Александровым на случай локально бикомпактных пространств и, в частности, компактов (см. цитированную работу). Пусть M — замкнутое подмножество компакта K, z^k — верхний k-мерный цикл. Если для любого k-мерного Δ-цикла u_k из M индекс его пересечения с z^k есть нуль, $H(u_k, z^k)=0$, то z^k гомологичен $\bar{z}^{\kappa}$, носитель которого находится вне M. (В частности, если все k-мерные циклы из M гомологичны нулю, то z^k можно снять с M.)

В области $(J<c+\varepsilon)$ заключён цикл x_1 класса X. Класс Y подчинён X, т. е. существует l-мерный ∇-цикл z^l такой, что для $x \in X$, $z^l \times x \in Y$. Пусть M — критическое множество на $(J=c)$. Если в M все l-мерные Δ-циклы гомологичны нулю то ∇-цикл z^l можно снять с M, т. е. существует ∇-цикл z^l, гомологичный z^l, носитель которого расположен на расстоянии $\alpha>0$ от M. Существует, далее, число $\varepsilon>0$ такое, что всякое замкнутое множество из $(J<c+\varepsilon)$, лежащее на расстоянии α от M, может быть деформировано в $(J<c-\varepsilon)$. Пусть x есть цикл класса X, лежащий в $(J<c+\varepsilon)$.

[1]) Л. Люстерник. О числе решений вариационной задачи, ДАН, XL, № 6, 1943.

Пересечение $y = \bar{z}^l \times x$ есть k-мерный цикл класса Y. Он расположен в $(J < c + \varepsilon)$ на расстоянии $\geqslant \alpha$ от M. Поэтому y может быть деформирован в цикл y^*, расположенный в $(J < c - \varepsilon)$; y^* также принадлежит классу Y. Мы пришли к противоречию: максимум J на цикле y^* не больше $c - \varepsilon$, в то время как нижняя граница максимумов на циклах класса Y равна c. Тем самым теорема доказана.

Применения. 1. *Собственные значения интегральных уравнений*. Пусть

$$F(\varphi) = \int_0^1 \int_0^1 K(x, s)\, \varphi(x)\, \varphi(s)\, dx\, ds$$

на сфере $\int_0^1 \varphi^2 dx = 1$, т. е. в гильбертовом проективном пространстве P. Обозначим через $\{P^n\}$ класс всех n-мерных проективных пространств в P, через c_n верхнюю границу минимумов F на циклах класса $\{P^n\}$, c_n — критическое значение формы F — является её собственным значением. Пользуясь слабой непрерывностью формы $F(\varphi)$, можно все рассмотрения производить в компактных частях P. Так как класс $\{P^n\}$ подчинён $\{P^{n+k}\}$, то в случае совпадения $c_n = c_{n+k} = c$ на $(F = c)$ существует k-мерный цикл z_k критических элементов (собственных функций), причём z_k есть k-мерное проективное пространство. Мы используем лишь следующие свойства формы F: 1) F — чётная функция, т. е. $F(-\varphi) = F(\varphi)$ (что позволяет рассматривать её как функцию на проективном гильбертовом пространстве), 2) $F > 0$ на гильбертовой сфере, 3) F слабо непрерывна. Все рассмотрения остаются справедливыми для произвольных функционалов $G(\varphi)$, обладающих этими свойствами. Критические элементы суть те, на которых вариация $\delta \{G(\varphi) - \lambda \int_0^1 \varphi^2 dx\}$ обращается в нуль. $\lambda = c$ — критическое значение — собственное значение. Получаем класс нелинейных интегральных и, вообще, операторных уравнений, изученный автором [1]) и В. И. Соболевым [2]). Эти уравнения имеют бесконечное множество положительных собственных значений c_n, слияние которых $(c_n = c_{n+p} = c)$ приводит к бесконечному p-мерному множеству нормированных собственных функций, отвечающих кратному собственному значению c.

2. *Геодезические дуги с общими концами на поверхности S_2 рода* 0. Пусть a и b две точки поверхности S_2. В соответствующем пространстве R функционал J означает длину дуги. Обозначим через $[T_n]$ класс всех n-мерных циклов в R, гомологичных qT_n, $q \neq 0$, где T_n — определённый выше тор. Класс $[T_n]$ подчинён T_{n+2} при любом n и $[T_n]$ подчинён $[T_{n+1}]$ при чётном n. Обозначим через c_n нижнюю границу максимумов J на циклах класса (T_n). Так как критическое множество есть множество геодезических дуг данной длины, то существует хотя бы одна геодезическая дуга длины c, соединяющая точки a и b.

Пусть $c_{2k} = c_{2k+1} = c$. Так как класс (T_{2k}) подчинён классу T_{2k+1}, то должно существовать семейство геодезических дуг длины c, соединяющих точки a и b, содержащее негомологичный нулю одномерный цикл в R. Совокупность таких геодезических дуг покрывает всю поверхность S_2. Это значит, что при $c_{2k} = c_{2k+1} = c$ все геодезические, выходящие из точки a, проходят через точку b, причём точки

---

[1]) Л. А. Люстерник, Известия АН, серия матем., 1939.

[2]) В. И. Соболев, ДАН, 1940.

a и b ограничивают на этих геодезических дуги равной длины c (пример: точки a и b — диаметрально противоположные точки сферы).

Далее, числа c_k строго возрастают «через одно»: $c_k < c_{k+2}$. В самом деле, при $c_k = c_{k+2} = c$ существовало бы двумерное семейство геодезических длины c с концами в a и b, что невозможно.

3. *Замкнутые геодезические.* Рассмотрим в пространстве L^* замкнутых не пересекающих самих себя кривых на поверхности S_2 рода 0 функционал $J(q)$ — длину кривой. Критическое множество для J есть множество замкнутых геодезических, без самопересечений.

Обозначим через $[z_1]$, $[z_2]$, $[z_3]$ классы одно-, дву- и трёхмерных циклов, гомологичных соответственно z_1^{ab}, z_2^a, z_k. Эти классы последовательно подчинены друг другу. Обозначим через c_1, c_2, c_3 нижние границы максимумов на циклах классов $[z_1]$, $[z_2]$, $[z_3]$. Числа c_1, c_2, c_3 суть критические значения J, причём $c_1 \leqslant c_2 \leqslant c_3$. Так как классы $[z_1]$, $[z_2]$, $[z_3]$ последовательно подчинены друг другу, то равенство $c_1 = c_{i+p}$ влечёт за собою существование p-мерной системы замкнутых самонепересекающихся геодезических, содержащей негомологичный нулю p-мерный цикл в L. Приходим к сформулированной выше теореме о трёх геодезических[1]).

Аналогично, исследуя свойства произведения в пространстве замкнутых кривых на n-мерной сфере, можно доказать существование $(n+1)$-ой замкнутой геодезической на многообразии, гомеоморфном n-мерной сфере.

Подробное изложение результатов §§ 5 и 6 дано в мемуаре автора, выходящем в «Трудах Матем. ин-та им. Стеклова».

7. Метод Пуанкаре аналитического продолжения.

В заключение сделаем несколько замечаний о другом методе исследования диференциальных уравнений (не обязательно вариационного типа), восходящем также к Пуанкаре, — о методе аналитического продолжения. Этот метод был применён Пуанкаре для исследования замкнутых геодезических на выпуклых поверхностях. Грубо говоря, метод заключается в следующем. Исследуется система замкнутых выпуклых поверхностей, зависящая непрерывно от параметра t. Рассматривается, как меняются замкнутые геодезические с изменением t. При непрерывном изменении t, если не выполняются некоторые особые условия, замкнутые геодезические изменяются непрерывно, причём те из них, которые не пересекают самих себя, переходят опять в самонепересекающиеся. Особый случай наступает при слиянии двух таких геодезических: два вещественных периодических решения уравнения геодезической, сливаясь, переходят затем в пару таких же комплексных сопряжённых решений и обратно. Число вещественных решений не меняется по модулю 2. Так как на эллипсоиде существуют три замкнутые геодезические (три главных эллипса), то, соединяя эллипсоид с данной выпуклой поверхностью, непрерывно зависящей от параметра t системой поверхностей, получим, что на данной поверхности число замкнутых не пересекающих самих себя геодезических нечётное, т. е. отлично от нуля.

Это доказательство существования замкнутых геодезических на выпуклой поверхности не может считаться законченным: могут быть более сложные случаи слияния замкнутых геодезических, появление бесконечных систем их и т. д.

---

[1]) Л. Люстерник. Новое доказательство теоремы о трёх геодезических, ДАН, XLI, № 9, 1943.

Однако подобное доказательство можно сделать вполне строгим, применив тот метод «обхода особенностей», который применяется в алгебраическом случае. Пусть имеем систему уравнений

$$u_i(x_1, x_2, \ldots, x_n) = 0, \quad i = 1, 2, \ldots, n. \tag{5}$$

Мы будем искать вещественные решения системы (5). Если якобиан $J_n = \frac{D(u_1, u_2, \ldots, u_n)}{D(x_1, x_2, \ldots, x_n)} \neq 0$, то решения системы (5) непрерывно меняются с u_i. Если $J = 0$, то, вообще говоря, мы имеем слияние двух систем решений, и при непрерывном изменении u_i здесь может иметь место превращение пары вещественных систем решений в комплексные или обратно (т. е. изменение числа вещественных решений на 2). Более сложные особенности могут наступить, если, наряду с условием $J = 0$, удовлетворяются дополнительные, независимые от него, равенства. Системы, для которых это имеет место, назовём особыми. Если правые части уравнений (5) зависят от k параметров, то равенства, при которых для особых систем имеют место эти более сложные условия, определяют в k-мерном пространстве параметров не более чем $(k-2)$-мерное многообразие. Но так как такое многообразие не может разбивать пространства параметров (если последнее связно), то можно от любой системы (5) перейти непрерывно к любой другой, обходя многообразие особых систем (если обе системы не особые), причём число решений систем меняется на чётное число. Если одна неособая система имеет нечётное число вещественных решений, то то же имеет место и для любой неособой системы, т. е. все неособые системы имеют решения. Предельным переходом можно устанавливать наличие решения и на особых системах. Такие рассуждения применялись в различных алгебраических и геометрических исследованиях.

Но подобные рассуждения можно перенести и для случая функциональных пространств. В линейном бесконечномерном пространстве совокупность точек, определяемых уравнениями $l_i = 0$, где l_i — независимые линейные функционалы, $i = 1, 2, \ldots, n$, образует так называемое $(\infty - n)$-мерное линейное подпространство. Такое подпространство при $n \geqslant 2$ не разбивает всего пространства. Аналогично, если мы имеем пространство линейное в малом (аналог гладких k-мерных многообразий) и в нём подпространство, локально определяемое n независимыми линейными (точнее, диференцируемыми) условиями, то это линейное в малом подпространстве будем называть $(\infty - n)$-мерным. При $n \geqslant 2$ $(\infty - n)$-мерное линейное в малом подпространство не разбивает исходного пространства.

Пусть мы имеем некоторую систему уравнений $L = 0$ и пусть левые части этих уравнений образуют некоторое связное линейное в малом пространство $\overline{M}$. Рассмотрим в этом пространстве кривую, т. е. однопараметрическое семейство элементов L, где параметр t изменяется в сегменте $[0,1]$. С непрерывным изменением параметра решения уравнений $L_t = 0$ в общем случае изменяются непрерывно с сохранением числа решений. Случаю обращения в нуль якобиана $J_n = 0$ в конечномерной задаче в настоящей задаче отвечает некоторое условие $J = 0$, при котором эта непрерывность изменения решений может нарушаться (это будет, например, при наличии нулевого собственного значения в уравнении в вариациях для уравнения $L = 0$.) Если же имеет место условие $J = 0$, то здесь в общем случае имеем слияние двух решений соответствующего уравнения $L_t = 0$, и если оно наступает для $t = t_0$, то при перемене знака у $t - t_0$

уравнение $L_t = 0$, вообще говоря, теряет или приобретает два решения (пара вещественных решений становится комплексными сопряжёнными или обратно). Более сложные особенности наступают, когда кроме условия $J = 0$ удовлетворяются некоторые дополнительные условия.

Совокупность тех L, для которых удовлетворяются условие $J = 0$ и упомянутые дополнительные условия, образуют в пространстве уравнений некоторое подпространство $\overline{M}_1$ «особых уравнений». $\overline{M}_1$ не более чем $(\infty - 2)$-мерно, поэтому оно не разбивает пространства $\overline{M}$. Если L_0 и L_1 входят в $\overline{M} - \overline{M}_1$, то их можно соединить в $\overline{M} - \overline{M}_1$ кривой (т. е. однопараметрическим семейством L_t), где L и L_1 отвечают значениям параметра $t = 0$ и $t = 1$, лежащей в $\overline{M} - \overline{M}_1$. Вдоль этой кривой число решений уравнения $L_t = 0$ меняется на чётное число. Следовательно, числа решений $L_0 = 0$, $L_1 = 0$ совпадают по модулю 2. Число решений любых неособых уравнений $L = 0$, если для одного такого уравнения оно нечётно, будет нечётно. Любое особое уравнение может быть аппроксимировано неособым. Если при предельном переходе для уравнений $L_n = 0$, имеющих решения, предельное уравнение $L_\omega = 0$ тоже имеет решение, то тем самым можно доказать разрешимость уравнений $L = 0$ для всех L (неособых и особых).

Таковы в грубых чертах те рассуждения, которые приходится проводить для обоснования метода непрерывного продолжения.

Эти рассуждения применимы при непрерывном продолжении уравнений замкнутых геодезических от эллипсоида к любой поверхности рода 0. На трёхосном эллипсоиде существуют три замкнутые геодезические без двойных точек — три главных эллипса. Из них, как показал Пуанкаре, два устойчивых (наименьший и наибольший) и один неустойчивый (замкнутая геодезическая будет называться устойчивой, если достаточно близкие на некотором интервале геодезические будут на всём своём протяжении близки к ней). Пуанкаре показал в цитированной статье, что при слиянии двух замкнутых геодезических без двойных точек сливается одна устойчивая и одна неустойчивая и разность между числом тех и других не меняется.

Отсюда был сделан вывод (непрерывным продолжением от эллипсоида), что на всякой выпуклой поверхности существует нечётное число замкнутых самонепересекающихся геодезических, причём среди них число устойчивых на одно больше числа неустойчивых. С помощью рассуждений, аналогичных приведённым выше, доказательство Пуанкаре можно сделать совершенно точным. Сопоставляя полученные таким образом результаты с полученными топологическим методом, можно уточнить теорему о трёх замкнутых геодезических на выпуклых поверхностях: среди них имеется по крайней мере две устойчивые и одна неустойчивая. Поэтому приведённую выше теорему можно формулировать для выпуклых поверхностей следующим образом.

На всякой трижды диференцируемой выпуклой поверхности существуют или по крайней мере три замкнутые не пересекающие самих себя геодезических, из которых две устойчивые и одна неустойчивая, или континуальное семейство таких геодезических.

Vladimir Abramovich Rokhlin

Born – 23 Aug. 1919 in Baku (Azerbaijan)
Died – 1984 in Leningrad

In 1935 Vladimir Rokhlin entered the Moscow State University. His teachers were A. I. Plesner, L. S. Pontryagin, P. S. Alexandrov, A. N. Kolmogorov and I. M. Gelfand.

Rokhlin was a volunteer in the beginning of the war in 1941. He was a war prisoner in a German concentration camp and after the war he spent two years in a Soviet filtration camp.

During 1947–1949 he made fundamental contributions to measure theory and ergodic theory. He returned to these subjects after 1958 and made important works in the theory of entropy.

In the 50s he obtained his famous results in topology (homotopy groups of spheres, cobordisms of low dimensional manifolds, signature of 4-manifolds, characteristic classes).

For the last 25 years of his life Rokhlin was a professor of the Leningrad State University, where he created well-known mathematical schools. In this period his main interest was in characteristic classes, low dimensional topology and real algebraic geometry.

PROOF OF GUDKOV'S HYPOTHESIS

V. A. Rokhlin

1. Introduction. We know that a real algebraic curve of degree m can possess not more than 1/2 (m − 1) (m − 2) + 1 components in RP^2. Curves with this maximum number of components exist for any m, do not possess singularities (neither real nor imaginary), and are termed M curves. In the case of an even m, all the components of a nonsingular real algebraic curve of degree m are distributed bilaterally in RP^2 and are divisible by even (contained within an even number of the other components) and by odd (other components); we denote by p the number of even components, and by n the number of the odd components.

Recently, Arnol'd [1] proved that $p - n \equiv k^2 \bmod 4$ for any M curve of even degree 2k. In his paper, Arnol'd communicates that the stronger congruence $p - n \equiv k^2 \bmod 8$ has been formulated by Gudkov as a hypothesis. It follows from Gudkov's [2] results that this congruence holds for k = 3 (for k = 1, 2, the congruence is obvious).

In the present note, we show that $p - n \equiv k^2 \bmod 8$ for any M curve of any even degree 2k.

The proof rests upon the basic ideas set forth in Arnol'd's paper; however, use is made of more specialized facts in the topology of four-dimensional manifolds. Since all of these facts have not been published, they will be outlined in a separate section.

I would like to use this opportunity to thank V. I. Arnol'd for informing me about his work and for sharing his enthusiasm with me.

2. Brief Outline of Arnol'd's Paper. Let A be an arbitrary M curve of even degree 2k, and CA a complex curve defined by the same equation in CP^2 (so that $A = CA \cap RP^2$). We know that CA is an oriented closed surface of kind (k − 1) (2k − 1), that A divides CA in two, and that the two halves go into each other under complex conjugation. The projective plane RP^2 is also halved by curve A, A serving as the interface for the oriented portion RP^2_+ and the nonoriented portion RP^2_-. We will denote union of one of the halves of surface CA with RP^2_+ by $\mathfrak{A}$. The principal technical observation made by Arnol'd, which in the following will be referred to as Arnol'd's Lemma, is that this closed piecewise smooth surface realizes a zero of the group $H_2(CP^2; Z_2) = Z_2$ for an even k, and a value other than zero for an odd k.

Let Y be the branched double covering of manifold CP^2, branching along CA and possessing a natural complex structure, let τ: Y → Y be a nonidentical automorphism of the covering, and σ: Y → Y be a complex conjugation. The manifolds Fix τ and Fix σ of the fixed points of involutions τ and σ coincide with the inverse images of the manifolds CA and RP^2_+ for the covering Y → CP^2. From here, with the aid of Arnol'd's Lemma, it is not difficult to deduce that they realize the same element of group $H_2(Y; Z_2)$. Both of them are oriented, while the intersection indices of the elements t and s of the integral group $H_2(Y)$, which are realized by them, are defined by the formulas $t^2 = 2k^2$, $ts = 0$, $s^2 = 2(n - p)$. The class $w_2(y)$ is also easy to evaluate: it is zero for an odd k and is dual, in Poincaré's sense, to the class realized by the manifold Fix τ for an even k. Consequently, $(s + t)^2 \equiv 0 \bmod 8$ for an odd k, and $s^2 \equiv t^2 \bmod 8$ for an even k, which means that $p - n \equiv k^2 \bmod 4$.

3. Topological Regression. Let X be an oriented closed smooth four-dimensional manifold with $H_1(X; Z_2) = 0$ and F its orientable (closed, two-dimensional) submanifold that realizes an element of group $H_2(X; Z_2)$, which is dual to $w_2(X)$. The compact two-dimensional submanifold P of manifold X is termed the membrane on F, if the intersection $P \cap F$ consists of the edge ∂P along which P does not touch on F,

Moscow State University. Translated from Funktsional'nyi Analiz i Ego Prilozheniya, Vol. 6, No. 2, pp. 62-64, April-June, 1971. Original article submitted December 29, 1971.

and of a finite number of points at which the intersection is transverse. If we construct a vector field on ∂P, which touches on F and does not touch on P, the attempt at continuing the vector field without touching on P leads to a two-dimensional obstruction with an intergral index that is independent of the field selected on ∂P, and which is termed the index of membrane P. It appears that the modulo-2 reduced sum of this index and of the number of points of which the intersection (Int P) $\cap$ F is composed does not change when any other membrane with the same edge ∂P is substituted for membrane P and, moreover, is defined by the element of group $H_1(F; Z_2)$, which is realized by the edge ∂P. The function ψ: $H_1(F; Z_2) \to Z_2$ thus generated satisfies the relation $\psi(\alpha+\beta)=\psi(\alpha)+\psi(\beta)+\alpha\beta$, and defines in the conventional manner the invariant Arf, which in the following will be denoted by Arf (F). (If $\alpha_1, \ldots, \alpha_q;\ \beta_1, \ldots, \beta_q$ is the canonical basis in $H_1(F; \mathbf{Z}_2)$, then $\mathrm{Arf}(F)=\varphi(\alpha_1)\varphi(\beta_1)+\ldots+\varphi(\alpha_q)\varphi(\beta_q)$; the canonicality of the basis $\alpha_1, \ldots, \alpha_q;$ $\beta_1, \ldots, \beta_q$ means that $\alpha_1\beta_1=\ldots=\alpha_q\beta_q=1$, and that all the remaining intersection indices are zero.)

A fundamental property of the invariant Arf (F) is that it is related by the congruence $x(F)-\sigma(X) \equiv 8\,\mathrm{Arf}(F) \bmod 16$ to the normal Euler number x(F) of surface F and to the signature $\sigma(X)$ of manifold X. In particular, Arf (F) is an invariant of an integral class of intrinsic homologies realized by surface F, and is an invariant of a class of intrinsic homologies of the pair X, F. It can be defined even without the assumption that $H_1(X; Z_2) = 0$, however, this general definition is less effective and is not required in the following. In the simplest case where F is a sphere, the preceding congruence reduces to the known theorem of Kervaire and Milnor [3].

4. Proof of The Congruence $p-n \equiv k^2 \bmod 8$. It is essential for the following that manifold Y (see Sec. 2) be simply connected. We assume first that k is odd. Then the manifold $F = \mathrm{Fix}(\tau\circ\sigma)$ of the fixed points of involution $\tau\circ\sigma$: $Y \to Y$ is orientable, and from what was asserted in Sec. 2, it is not difficult to deduce that it realizes a zero of group $H_2(Y; Z_2)$. Since, $w_2(Y) = 0$ at the same time, the invariant Arf (F) is defined and satisfies the congruence $x(F)-\sigma(Y)\equiv 8\,\mathrm{Arf}(F) \bmod 16$, in which the Euler number x(F) (as can be easily computed) is $2(p-n-1)$ and $\sigma(Y)$, as follows from the formula $\sigma(Y) = 2-2k^2$ (see, for example, [4]), or from equality $w_2(Y) = 0$, is divisible by 16. Thus, the congruence being proved is equivalent to the equality Arf (F) = 0. This equality is a corollary of the almost obvious fact that group $H_1(F; Z_2)$ possesses a canonical basis $\alpha_1, \ldots, \alpha_p;\ \beta_1, \ldots, \beta_p$ with $\psi(\beta_1)=\ldots=\psi(\beta_p)=0$; the classes $\alpha_1, \ldots, \alpha_p$ are realized by the inverse images of the even components of curve A in the case of the covering $Y \to CP^2$, while the classes $\beta_1, \ldots, \beta_p$ are realized by the inverse images of the p sections of manifold RP^2_-, which transform its components into topological circles and which join the even and odd components of curve A (that $\psi(\beta_i) = 0$ follows from symmetry considerations).

We now assume that k is even. We realize class $t_1 = t/k \in H_2(Y)$ (image of the generating group $H_2(CP^2)$ for the inverse Hopf homomorphism that corresponds to the covering $Y \to CP^2$) of submanifold F_1 that does not intersect Fix σ, and set $F = F_1 \cup \mathrm{Fix}\,\sigma$. Submanifold F realizes an element of group $H_2(Y; Z_2)$, that is dual to $w_2(Y)$, so that invariant Arf (F) is again defined and satisfies the congruence $x(F)-\sigma(Y) = 8\,\mathrm{Arf}(F) \bmod 16$. Simple calculations show that $x(F) = 2(n-p+1)$, and that Arf (F) = 0. By substituting these values together with the value $\sigma(Y) = 2-2k^2$ into the preceding congruence, one can see that $p-n \equiv k^2 \bmod 8$. It is noteworthy that this reasoning holds also for an odd k; the proof of the equality Arf (F) = 0, however, is more complicated.

5. Concluding Remark: Arnol'd's Theorem and the Generalized Whitney Theorem. It is striking that the passage from manifold CP^2 to its branched double covering, proposed by Arnol'd (and used in Sec. 4), is quite unnecessary for proving Arnol'd's congruence $p-n \equiv k^2 \bmod 4$, and that this congruence may be derived directly from Arnol'd's lemma by making use of an elementary theorem in four-dimensional topology. This theorem generalizes the well-known Whitney theorem about the normal Euler numbers of closed surfaces in R^4 (see [5]), and states that if F is a (closed two-dimensional) submanifold of an oriented smooth connected four-dimensional manifold X, which realizes an element of group $H_2(X; Z_2)$ that is dual to $w_2(X)$, then the Euler characteristic $\chi(F)$, the normal Euler number $x(F)$, and the signature $\sigma(X)$ are related by the congruence $2\chi(F)+x(F)\equiv\sigma(X) \bmod 4$ (the submanifold F is not assumed to be orientable, and the Whitney theorem results for $X = S^4$). In order to prove Arnol'd's congruence for an odd k, it is sufficient to apply this theorem to a surface $\mathfrak{A}$ smoothed by natural means (CA and RP^2 do not touch on each other!), taking into consideration that for this surface $\chi=[2-(p+n)]+(p-n)=2-2n$, and $x = 2k^2 + (n-p)$, and that $\sigma(CP^2 = 1$. For an even k, the proof is carried out in the same fashion, except that $\mathfrak{A}$ is replaced by the union of the half-surface CA with RP^2_-.

LITERATURE CITED

1. V. I. Arnol'd, "Concerning the arrangement of the ovals of real plane algebraic curves, the involutions of four-dimensional smooth manifolds, and the arithmetic of integral quadratic forms," Funkts. Analiz, 5, No. 3, 1-9 (1971).
2. D. A. Gudkov, "Topology of sixth-order curves and fourth-order surfaces," Uchenye Zapiski Gor'k. Un-ta, No. 87, 3-153 (1969).
3. M. A. Kervaire and J. W. Milnor, "On 2-spheres and 4-manifolds," Proc. Nat. Acad. Sci. USA, 47, No. 10, 1651-1657 (1961).
4. V. A. Rokhlin, "Two-dimensional submanifolds of four-dimensional manifolds," Funkts. Analiz, 5, No. 1, 48-60 (1971).
5. H. Whitney, "On the topology of differentiable manifolds," Univ. of Michigan Lectures (1941), pp. 101-141.

CONGRUENCES MODULO 16 IN HILBERT'S SIXTEENTH PROBLEM

V. A. Rokhlin

In recent papers Arnol'd [1] and the author [2] have proved the Gudkov congruence, which determines modulo 8 the difference between the number of even and the number of odd ovals of a plane M-curve of even degree. In the present article the author extends this congruence to more complex algebraic varieties, both one-dimensional and multidimensional.

§1. The Fundamental Varieties

1.1. Varieties A, B, B_+, and B_-. In what follows we assume as given homogeneous real polynomials $P_1, \ldots, P_s$ of degrees $m_1, \ldots, m_s$ in q + 1 variables. We denote by A the subset defined in the real projective space RP^q by the set of equations

$$P_1(x_0, \ldots, x_q) = 0, \ldots, P_s(x_0, \ldots, x_q) = 0, \tag{1}$$

and by B the subset of that space defined by the truncated set of equations

$$P_1(x_0, \ldots, x_q) = 0, \ldots, P_{s-1}(x_0, \ldots, x_q) = 0. \tag{2}$$

The set A constitutes our fundamental object for even n = q − s, and the pair B, A is the fundamental object for odd n. It is assumed that A is nonempty and that the system (1) does not have any singularities on A, and it is assumed for odd n, in addition, that the system (2) does not have any singularities on B. Consequently, A is a smooth closed variety of dimension n, and if n is odd, then B is a smooth closed variety of dimension n + 1 containing A as a subvariety. If n is odd and m_s is even, then A partitions B into compact subvarieties defined by the inequalities $P_s(x_0, \ldots, x_q) \geq 0$, $P_s(x_0, \ldots, x_q) \leq 0$, with common edge A. We denote these compact subvarieties by B_+ and B_-.

The varieties A, B, B_+, and B_- can be either orientable or nonorientable. In the case when n is odd and m_s is even we need to deal with the endomorphism $\frown \omega$: $H_*(B_-, A; Z_2) \to H_*(B_-, A; Z_2)$, which is defined by the formula $\xi \mapsto \xi \frown \omega$, where

$$\omega = \begin{cases} 0, & \text{if } m_s \equiv 0 \mod 4, \\ w_1(B_-), & \text{if } m_s \equiv 2 \mod 4. \end{cases}$$

We denote by a the dimension (over Z_2) of the intersection of the kernel of this endomorphism with the kernel of the boundary homomorphism ∂: $H_*(B_-, A; Z_2) \to H_*(A; Z_2)$. Clearly, $\omega = 0$ if n = 1. It is also obvious that if $\omega = 0$, then

$$a = \dim(\operatorname{Ker}\partial) = \dim H_*(B_-; Z_2) - \frac{1}{2}\dim H_*(A; Z_2).$$

1.2. Varieties CA and CB. We denote the sets defined by the systems (1) and (2) in the complex projective space CP^q by CA and CB. The system (1) can have singularities on CA, and the system (2) on CB, but any singularities can be removed by a small variation of the coefficients of the polynomials $P_1, \ldots, P_s$. It is assumed in this and the following section that there are no singularities, so that CA and CB are smooth closed varieties of dimension 2n and 2n + 2 with natural orientations and CA is contained in CB as a subvariety. The complex conjugate $CP^q \to CP^q$ [defined by the formula $(z_0, \ldots, z_q) \to (\bar{z}_0, \ldots, \bar{z}_q)$] induces in CA and CB smooth involutions, for which A and B serve as fixed-point varieties. We denote these involutions by the symbol conj.

Leningrad State University. Translated from Funktsional'nyi Analiz i Ego Prilozheniya, Vol. 6, No. 4, pp. 58-64, October-December, 1972. Original article submitted June 21, 1972.

Unlike the variety A and the pair B, A, the variety CA and the pair CA/CB, are determined by the numbers q; $m_1, \ldots, m_s$ up to diffeomorphism. In particular, the numbers q; $m_1, \ldots, m_s$ determine homologies of the variety CA. These homologies are known [10] and are computed without major difficulty. We agree to denote by $\chi(X)$ the Euler characteristic of the variety X and we define the polynomial $\chi_s^q(m_1, \ldots, m_s)$ with $0 \leq s \leq q$ by the formulas

$$\chi_s^q(m_1, \ldots, m_s) = \begin{cases} q+1, & \text{if } s=0, \\ m_1 m_2 \ldots m_q, & \text{if } 0 < s = q, \end{cases}$$

$$\chi_s^q(m_1, \ldots, m_s) = m_s \chi_{s-1}^{q-1}(m_1, \ldots, m_{s-1}) - (m_s - 1)\chi_s^{q-1}(m_1, \ldots, m_s), \quad \text{if } 0 < s < q.$$

Regarding CA as an m_s-sheeted cyclic ramified covering of a hyperplane cross section of CB (with ramification along a hyperplane cross section of CA itself), we can easily show by induction on q that $\chi(CA) = \chi_s^q(m_1, \ldots, m_s)$. Also, from Lefschetz's theorem on hyperplane cross sections we infer by induction on q that if $0 \leq r \leq 2n$ and $r \neq n$, then the integer-valued homological groups $H_r(CA)$ are trivial for odd r and isomorphic to Z for even r. All that remains is to note that the foregoing then determines the group $H_n(CA)$; it is free, and its rank is equal to $\chi_s^q(m_1, \ldots, m_s)$ for even n and to $(n + 1) - \chi_s^q(m_1, \ldots, m_s)$ for odd n.

In the case of even n we also need the signature $\sigma(CA)$ of CA. It can be computed by means of the well-known "functional equation" of Hirzebruch ([3], Theorem 9.3.1). We define the polynomial $\sigma_s^q(m_1, \ldots, m_s)$ with $0 \leq s \leq q$ and $q - s \equiv 0 \bmod 2$ by the formulas

$$\sigma_s^q(m_1, \ldots, m_s) = \begin{cases} 1, & \text{if } s=0, \\ m_1 m_2 \ldots m_q, & \text{if } 0 < s = q, \end{cases}$$

$$\sigma_s^q(m_1, \ldots, m_s) = m_s \sigma_{s-1}^{q-1}(m_1, \ldots, m_{s-1}) - \sum_{\mu=1}^{m_s} \sigma_{s+1}^{q-1}(m_1, \ldots, m_{s-1}, \mu - 1, \mu), \quad \text{if } 0 < s < q.$$

Applying the Hirzebruch theorem in succession, we can show that $\sigma(CA) = \sigma_s^q(m_1, \ldots, m_s)$.

1.3. Variety Y. It is assumed in this section that the number m_s is even. We denote by Y a two-sheeted ramified covering of the variety CB with ramification along CA. This covering is a smooth closed variety of dimension 2n + 2 and is also determined by the numbers q; $m_1, \ldots, m_s$ up to diffeomorphism. It has natural orientation and natural complex structure. The involution conj: CB → CB is covered by two smooth involutions of Y, which are mapped one into the other by the nonidentical covering automorphism of the Y → CB. For one of these involutions the fixed-point variety lies over B_+, and for the other it lies over B_-. We denote by conj the involution for which it lies over B_-.

We shall be concerned with the variety Y only for odd n, and we shall have need of the signature $\sigma(Y)$. It can be computed as follows. We denote by $c_k(m)$ the coefficient of t^{2k} in the power decomposition of the function $\frac{(1+t)^m + (1-t)^m}{(1+t)^m - (1-t)^m} t$ [in particular, $c_0(m) = 1/m$ and $c_1(m) = (m^2 - 1)/3m$] and we put

$$\tau_s^q(m_1, \ldots, m_s) = \sum_{k=0}^{(n+1)/2} c_k(m_s/2)\, \sigma_s^{q-2k+1}(m_1, \ldots, m_s).$$

We also denote by Z the subvariety of CP^{q+1} defined by the set of equations

$$P_1(x_0, \ldots, x_q) + x_{q+1}^{m_1} = 0, \ldots, P_s(x_0, \ldots, x_q) + x_{q+1}^{m_s} = 0. \tag{3}$$

We can treat Z as an m_s-sheeted ramified covering of CB with ramification along CA and, hence, as an $(m_s/2)$-sheeted cyclic ramified covering of Y. Applying to the covering Z → Y the signature formula deduced by Hirzebruch [5] from the Atiyah-Singer formula ([4], §6), we see that $\sigma(Y) = \tau_s^q(m_1, \ldots, m_s)$.

§2. Application of the Smith Sequence

2.1. Preliminaries. The Smith sequence we need has the form

$$\ldots \xrightarrow{\gamma_{r+1}} H_r(X', \Phi; \mathbf{Z}_2) \oplus H_r(\Phi; \mathbf{Z}_2) \to H_r(X; \mathbf{Z}_2) \xrightarrow{\beta_r} H_r(X', \Phi; \mathbf{Z}_2) \xrightarrow{\gamma_r} H_{r-1}(X', \Phi; \mathbf{Z}_2) \oplus H_{r-1}(\Phi; \mathbf{Z}_2) \ldots,$$

in which X is a space with involution, X' is the corresponding factor space, and Φ is a fixed-point involution set regarded as a subset of space X'. We are interested only in the case in which X is a smooth variety and the involution is smooth. In this case homologies may be regarded as singular, the homologies of the pair X', Φ can be identified with the homologies of the pair U, ∂U, where U is the complement of a regular neighborhood of the set Φ in X', and the component $H_r(X',\Phi; \mathbf{Z}_2) \to H_{r-1}(X',\Phi;\mathbf{Z}_2)$ of the homomorphism γ_r can be defined by the formula $\xi \mapsto \xi \frown x$, where x is the characteristic class of the covering $X\setminus\Phi \to X'\setminus\Phi$. The component of γ_r is a boundary homomorphism, whereas the definition of the homomorphisms α_r and β_r is irrelevant for our purpose, and all we need to know is that the automorphism $H_r(X;\mathbf{Z}_2) \to H_r(X;\mathbf{Z}_2)$, induced by involution is identical on Im α_r and that the sequence is exact. Its exactness implies that

$$\dim H_*(X;\ \mathbf{Z}_2) = \dim H_*(\Phi;\ \mathbf{Z}_2) + 2\sum \dim(\operatorname{Ker}\gamma_r), \tag{4}$$

and it follows from this equality, in turn, that

$$\dim H_*(\Phi;\ \mathbf{Z}_2) \leqslant \dim H_*(X;\ \mathbf{Z}_2). \tag{5}$$

It is also clear that if the equality holds in (5), then Ker $\gamma_r = 0$, $\beta_r = 0$, and Im $\alpha_r = H_r(X;\ Z_2)$ for any r, so that the action of the involution in $H_*(X;\ Z_2)$ is identical. See [6, 7] for the details of this matter.

2.2. Application to the Involution conj: CA $\to$ CA. As noted by Thom [8], the application of inequality (5) to the involution conj: CA $\to$ CA yields the inequalities

$$\dim H_*(A;\mathbf{Z}_2) \leqslant \begin{cases} \chi_s^q(m_1,\dots,m_s), & \text{if n is even,} \\ 2(n+1) - \chi_s^q(m_1,\dots,m_s), & \text{if n is odd} \end{cases} \tag{6}$$

(see 1.2). For n = 1 this inequality reduces to the classical Harnack inequality; the number of components of the curve A does not exceed the genus of the surface CA plus one. As we are aware, for plane curves, i.e., for q = 2, s = 1 the Harnack inequality is sharp, in other words, for any m_1 there exist in RP^2 curves of degree m_1, i.e., so-called M-curves, for which the number of components is equal to the indicated upper bound. In more complex situations the sharpness problem for inequality (6) poses considerable difficulties.

2.3. Application to the Involution conj: Y $\to$ Y. The total Betti number dim $H_*(Y;\ Z_2)$ is determined by the numbers q; $m_1, \dots, m_s$ (see 1.3) and is denoted by $b_s^q(m_1, \dots, m_s)$. It can be computed according to formula (4) by putting Y in the role of X and putting the nonidentical covering automorphism Y $\to$ CB in the role of involution; here X' = CB and Φ = CA. The computation relies on the fact that the inclusion homomorphism $H_r(CA;\ Z_2) \to H_r(CB;\ Z_2)$ is an isomorphism for r < n and is trivial for $r \geqq n$ (see [10]). It yields

$$\dim(\operatorname{Ker}\gamma_r) = \begin{cases} 0 & \text{for } r \neq n+1, \\ \dim H_{n+1}(CB;\mathbf{Z}_2) = \chi_{s-1}^q(m_1,\dots,m_{s-1}) - (n+1) & \text{for } r = n+1, \end{cases}$$

so that $b_s^q(m_1,\dots,m_s) = 2\chi_{s-1}^q(m_1,\dots,m_{s-1}) - \chi_s^q(m_1,\dots,m_s)$. More complete information (irrelevant for our subsequent purpose) on the homologies of the variety Y is contained in the fact that $H_r(Y) \cong H_r(CB)$, if $r \neq n+1$.

We denote the fixed-point variety of the involution conj: Y $\to$ Y by F. The total Betti number dim $H_*(F;\ Z_2)$ can be computed according to (4) by putting F in the role of X and putting the transformation F $\to$ F induced by the nonidentical covering automorphism Y $\to$ CB in the role of involution. Here X' = B_-, Φ = A, and the characteristic class x, as shown by a straightforward computation, is determined by the formula

$$x = \begin{cases} 0, & \text{if } m_s \equiv 0 \mod 4, \\ w_1(B_-\setminus A), & \text{if } m_s \equiv 2 \mod 4. \end{cases}$$

Consequently, dim $H_*(F;\ Z_2) = 2a + \dim H_*(A;\ Z_2)$, and if n = 1 or $m_s \equiv 0$ mod 4, then dim $H_*(F;\ Z_2)$ = 2dim $H_*(B_-;\ Z_2)$.

The foregoing results enable us to represent inequality (5) as applied to the involution conj: Y $\to$ Y, i.e., the inequality dim $H_*(F;\ Z_2) \leq \dim H_*(Y;\ Z_2)$, in the form

$$2a + \dim H_*(A;\ \mathbf{Z}_2) \leqslant b_s^q(m_1,\ \dots,\ m_s), \tag{7}$$

and if n = 1 or $m_s \equiv 0$ mod 4, in the form

$$\dim H_*(B_-;\ \mathbf{Z}_2) \leqslant \frac{1}{2}\, b_s^q(m_1,\dots,m_s). \tag{8}$$

§ 3. Principal Results

3.1. General Formulations. The principal results of the article are comprised in two theorems.

A. If n is even and

$$\dim H_*(A;\, \mathbf{Z}_2) = \chi_s^q(m_1, \ldots, m_s), \tag{9}$$

then

$$\chi(A) \equiv \sigma_s^q(m_1, \ldots, m_s) \bmod 16. \tag{10}$$

B. If n is odd, m_s is even, and

$$2a + \dim H_*(A;\, \mathbf{Z}_2) = b_s^q(m_1, \ldots, m_s), \tag{11}$$

then

$$2\chi(B_-) \equiv \tau_s^q(m_1, \ldots, m_s) \bmod 16. \tag{12}$$

3.2. Case q = 2, s = 1 (Plane Curves). In this case one of the varieties B_+ or B_- is orientable, while the other is nonorientable; suppose that B_- is nonorientable. Then

$$\dim H_*(B_+;\, \mathbf{Z}_2) = \frac{1}{2}\dim H_*(A;\, \mathbf{Z}_2),$$

$$\dim H_*(B_-;\, \mathbf{Z}_2) = \frac{1}{2}\dim H_*(A;\, \mathbf{Z}_2) + 1,\ a = 1$$

($1/2 \dim H_*(A;\, Z_2)$ is the number of components of curve A), and since $b_1^2(m_1) = m_1^2 - 3m_1 + 6$, we find that inequalities (7) and (8) are equivalent to the Harnack inequality and equality (11) connotes that A is an M-curve.

The components of curve A separate into even (those situated within an even number of other components) and odd (all others); the number of even components is denoted by P, and the number of odd components by N. Clearly, $\chi(B_-) = N - P + 1$, and since $\tau_1^2(m_1) = 2 - m_1^2/2$, the congruence (12) can be represented in the form

$$P - N \equiv (m_1/2)^2 \bmod 8.$$

In the case q = 2, s = 1, therefore, Theorem B reduces to the Gudkov congruence.

3.3. Case q = 3, s = 1 (Surfaces in $\mathbf{RP}^3$). In this case Theorem A states that

$$\chi(A) \equiv (4m_1 - m_1^3)/3 \bmod 16 \tag{13}$$

if

$$\dim H_*(A;\, \mathbf{Z}_2) = m_1^3 - 4m_1^2 + 6m_1.$$

For even m_1 the congruence (13) connotes that $\chi(A) \equiv 0 \bmod 16$. For example, $\chi(A) \equiv 0 \bmod 16$ for every surface A of degree 4 with $\dim H_*(A;\, Z_2) = 24$, and this is the theorem stated (in equivalent form) as the hypothesis of D. A. Gudkov.

We note that the weaker congruence $\chi(A) \equiv 0 \bmod 8$ was recently verified for even m_1 and under the same condition $\dim H_*(A;\, Z_2) = m_1^3 - 4m_1^2 + 6m_1$ by Kharlamov [9]. This congruence [which is analogous to the congruence $P - N \equiv (m_1/2) \bmod 4$ established by Arnol'd [1] for M-curves] was in fact what motivated the author to generalize the Gudkov congruence in Theorems A and B.

3.4. Case q = 3, s = 2 (Curves in $\mathbf{RP}^3$). In this case Theorem B states that if m_2 is even and

$$\dim H_*(B_-;\, \mathbf{Z}_2) = \frac{m_1}{2}(2m_1^2 + m_1m_2 + m_2^2 - 8m_1 - 4m_2 + 12),$$

then

$$\chi(B_-) \equiv \frac{m_1}{12}(16 - 4m_1^2 - 3m_2^2) \bmod 8.$$

3.5. Two Lemmas. We denote by u(X), where X is a smooth closed variety of even dimension 2k, the k-dimensional homological Wu class of this variety, i.e., an element of the group $H_k(X;\, Z_2)$ for which the index of intersection with any class $\xi \in H_k(X;\, Z_2)$ is equal to the self-intersection index of class ξ. Those elements of the integer-valued group $H_k(X)$ which map into u(X) under the coefficient homomorphism $H_k(X) \to H_k(X;\, Z_2)$ are called integer-valued homological Wu classes.

In the ensuing Lemmas CA and Y, as in §§1 and 2, it is assumed that the system (1) does not have singularities on CA and that the system (2) does not have singularities on CB.

LEMMA CA. For even n the variety CA has an integer-valued homological Wu class v with $conj_*(v) = (-1)^{n/2}v$.

For the proof it suffices to note that the integer-valued homological Wu class of the variety CA can be found among the classes delineated in CA by elements of the group $H_{q+s}(CP^q)$. The latter statement is readily deduced from the well-known Wu formulas, which permit the class u(CA) to be expressed in terms of the Stiefel-Whitney classes $w_r(CA)$, and the fact that the classes $w_r(CA)$ (as implied by the multiplicative theorem for Stiefel-Whitney classes) belong to the images of the inclusion homomorphisms $H^r(CP^q; Z_2) \to H^r(CA; Z_2)$.

LEMMA Y. For odd n the variety Y has an integer-valued homological Wu class v with $conj_*(v) = (-1)^{(n+1)/2}v$.

Proof. We put $u_1 = D_Y \circ p^* \circ D_{CB}^{-1}(u(CB))$, where D_{CB} and D_Y are Poincaré isomorphisms and p is the projection of the variety Y onto CB. The classes $D_Y^{-1}(u(Y))$ and $D_Y^{-1}(u_1)$ assume identical values on the image of any element of the group $H_{n+1}(Y\setminus CA; Z_2)$ under the inclusion homomorphism $H_{n+1}(Y\setminus CA; Z_2) \to H_{n+1}(Y; Z_2)$, whence it follows that the class $u(Y) - u_1$ is represented as the image of a certain class $u_0 \in H_{n+1}(CA; Z_2)$ under the inclusion homomorphism $i_*: H_{n+1}(CA; Z_2) \to H_{n+1}(Y; Z_2)$. Consequently, $u(Y) = i_*(u_0) + D_Y \circ p^* \circ D_{CB}^{-1}(u(CB))$, and all that remains is to note that the classes u_0 and u(CB) are represented as the image (under coefficient homomorphisms) of certain classes $v_0 \in H_{n+1}(CA)$, $v_1 \in H_{n+1}(CB)$ with $conj_*(v_0) = (-1)^{(n+1)/2}v_0$, $conj_*(v_1) = (-1)^{(n+1)/2}v_1$: this fact is obvious for u_0, and for u(CB) it follows from Lemma CA as applied to CB.

3.6. Proof of Theorem A. It suffices to analyze the case in which the system (1) does not have singularities on CA. We denote by G_+ the subgroup of $H_n(CA)$ composed of homological classes invariant under the automorphism $conj_*: H_n(CA) \to H_n(CA)$, and by G_- the subgroup composed of classes $\xi \in H_n(CA)$ with $conj_*(\xi) = -\xi$. According to Sec. 2.1, it follows from condition (9) that the automorphism $conj_*: H_n(CA; Z_2) \to H_n(CA; Z_2)$ is identical, and it follows from the latter, in turn, that the classes $\xi - conj_*(\xi)$ and $\xi + conj_*(\xi)$ are divided by 2 in $H_n(CA)$ for every $\xi \in H_n(CA)$. As a result, the formula

$$\xi = \frac{1}{2}[\xi + \mathrm{conj}_*(\xi)] + \frac{1}{2}[\xi - \mathrm{conj}_*(\xi)]$$

shows that $H_n(CA) = G_+ \oplus G_-$, and it is clear that this direct decomposition is orthogonal on the intersection index. Consequently, the quadratic form of the variety CA (self-intersection index) decomposes into an orthogonal sum of quadratic forms induced by it on G_+ and G_-, whereupon it is obvious that the discriminants of these forms are equal to ±1 and the sum of their signatures is equal to $\sigma(CA)$.

Let σ_+ and σ_- be those signatures. By virtue of the At'ya-Singer-Hirzebruch formula (see [4], §6) the normal Euler number of the variety A in CA is equal to $\sigma_+ - \sigma_-$, and since multiplication by i takes a tangential fibering of A into a normal fibering, this number is equal to $(-1)^{n/2}\chi(A)$. Therefore,

$$\chi(A) = \begin{cases} \sigma(CA) - 2\sigma_-, & \text{if} \quad n \equiv 0 \mod 4, \\ \sigma(CA) - 2\sigma_+, & \text{if} \quad n \equiv 2 \mod 4, \end{cases}$$

where in the first case, due to Lemma CA, the group G_+ contains the integer-valued Wu class of CA, while in the second case, due to Lemma CA, G_- contains the integer-valued Wu class of CA. Consequently, in the first case the quadratic form of CA is even on G_-, so that σ_- is divided by 8, while in the second case the quadratic form of CA is even on G_+, so that σ_+ is divided by 8, and in both cases $\chi(A) \equiv \sigma(CA) \mod 16$.

3.7. Proof of Theorem B. It suffices to consider the case in which the system (1) does not have singularities on CA and the system (2) does not have singularities on CB. We denote by G_+ the subgroup of $H_{n+1}(Y)$/Torsion composed of elements invariant under the automorphism $conj_*: H_{n+1}(Y)/\text{Torsion} \to H_{n+1}(Y)/\text{Torsion}$, and by G_- the subgroup composed of elements $\xi \in H_{n+1}(Y)$/Torsion with $conj_*(\xi) = -\xi$. [Actually, Tors $H_{n+1}(Y) = 0$.] It follows from condition (11) that the automorphism $conj_* = H_{n+1}(Y; Z_2) \to H_{n+1}(Y, Z_2)$ is identical, and it follows from this fact, in turn, that $H_{n+1}(Y)/\text{Torsion} = G_+ \oplus G_-$ (see 3.6). This direct decomposition is orthogonal, so that the discriminants of the forms induced by the quadratic form of Y on G_+ and G_- are equal to ±1, and the sum of the signatures of those forms is equal to $\sigma(Y)$.

Let σ_+ and σ_- be the indicated signatures. The normal Euler number of the variety F in Y is equal to $\sigma_+ - \sigma_-$ and it is also equal to $(-1)^{n+1}\chi(F)$ (see 3.6), while $\chi(F) = 2\chi(B_-) - \chi(A) = 2\chi(B_-)$. Therefore,

$$2\chi(B_-) = \begin{cases} \sigma(Y) - 2\sigma_-, & \text{if } n+1 \equiv 0 \mod 4, \\ \sigma(Y) - 2\sigma_+, & \text{if } n+1 \equiv 2 \mod 4, \end{cases}$$

and all that remains is to use Lemma Y exactly as Lemma CA was used in Sec. 3.6.

LITERATURE CITED

1. V. I. Arnol'd, "Decomposition of the ovals of real plane algebraic curves, involutions of four-dimensional smooth varieties, and the arithmetic of integer-valued quadratic forms," Funktsional. Analiz i Ego Prilozhen., 5, No. 3, 1-9 (1971).
2. V. A. Rokhlin, "Proof of the Gudkov hypothesis," Funktsional. Analiz i Ego Prilozhen., 6, No. 2, 62-64 (1972).
3. F. Hirzebruch, Neue Topologische Methoden in der Algebraischen Geometrie (2nd ed.), Springer (1962).
4. M. F. At'ya and I. M. Singer, "Index of elliptic operators, Part III," Usp. Matem. Nauk, 26, No. 1, 127-182 (1969).
5. F. Hirzebruch, "The signature of ramified coverings," Global Analysis, Papers in Honour of Kodaira, Tokyo (1969).
6. Seminar on Transformation Groups, Chap. III, Ann. Math. Studies, No. 46, Princeton (1960).
7. G. E. Bredon, "Cohomological aspects of transformation groups," Proc. Conf. Transformation Groups, New Orleans, 1967, Springer (1968).
8. R. Thom, "Sur l'homologie des variétés algebriques reeles," Differential and Combinatorial Topology, A Symposium in Honor of Marston Morse, Princeton Univ. Press (1965), pp. 255-265.
9. V. M. Kharlamov, "Maximum number of components of a fourth-degree surface in RP^3," Funktsional. Analiz i Ego Prilozhen., 6, No. 4, 101 (1972).
10. J. Fary, "Cohomologie des variétés algebriques," Ann. Math., 65, No. 1, 21-73 (1957).

Petr Sergeevich Novikov

Born – 15 Aug 1901 in Moscow, Russia
Died – 9 Jan 1975 in Moscow, Russia

Petr Novikov was the son of Sergi Novikov, a Moscow merchant. He attended school in Moscow then, in September 1919, he entered the Faculty of Physics and Mathematics at Moscow University. However, even before Novikov entered university, the Russian nation had been plunged into civil war. The Red Army had been formed in February 1918 with Trotsky as its leader. The Red Army opposed the White Army formed of anticommunists led by former imperial officers. In the spring of 1920, with the Civil war still raging, Novikov joined the Red Army. He served with this army until July 1922 when he returned to Moscow University to complete his studies.

He graduated in 1925 then, remaining at Moscow University, he undertook research under Luzin's supervision. Novikov graduated in 1929 and then taught at the Moscow Chemical Technology Institute until he joined the Department of Real Function Theory at the Steklov Institute in 1934. He was awarded his doctorate in 1935 and, in 1939, he was promoted to full professor. Novikov married Ludmila Vsvolodovna Keldysh in 1935. They had five children; one of their sons Sergi Novikov was awarded a Fields Medal in 1970.

Novikov headed the Department of Analysis at Moscow State Teachers Training Institute from 1944. In 1957 Novikov set up a new department at the Steklov Institute, namely the Department of Mathematical Logic, and he was appointed as the first head of that department. He held the two posts, one at the Moscow State Teachers Training Institute and the other at the Steklov Institute, until he retired in 1972 and 1973 respectively.

After early work on set theory, influenced by Luzin and his school, he began to publish results in mathematical physics from 1938. Perhaps his most fundamental result in this area was that "...*any two solids having the same constant density must coincide if they both are star-shaped relative to a common point and have the same external gravitational potential*".

He began to study mathematical logic and the theory of algorithms just before 1940. He studied consistency of arithmetic, proving that formal arithmetic with recursive definitions is consistent. He also examined the consistency of certain propositions in Gödel's system of axiomatic set theory.

Novikov showed, in 1952, that the word problem for groups is insoluble. The word problem asks the fundamental question of whether there is an algorithm to determine whether a word in a group given by a presentation consisting of a finite number of generators and relations is trivial. The problem was first posed by Dehn in 1912 and Novikov was able to show that no such algorithm exists in general. Research into questions of this type is still of major importance in combinatorial group theory. Novikov was awarded the Lenin Prize in 1957 for this outstanding piece of work. In fact Boone published another proof of this result in 1957, the same year that Novikov received his prize.

The word problem was not the only problem of major importance in combinatorial group theory that Novikov solved. Jointly with Adian he showed that the problem of the finiteness of periodic groups proposed by Burnside in 1902 had a negative solution. Although in 1959 Novikov announced that for every $n > 71$ there exists a finitely generated infinite group with every element of order dividing n, his proof was not quite correct.

To be more precise, the Burnside problem asks whether, for fixed d and n, the group $\mathrm{B}(d, n)$ having d generators and in which every element x satisfies $xn = 1$, is finite. Novikov's argument of 1959 was correct in general terms but the details were not, and in putting the arguments right it was found that one required larger values of n. In 1968 Novikov and Adian jointly published a proof $\mathrm{B}(d, n)$ is infinite for every $d > 1$ and every $n > 4380$. They continued to work on improving the result and, in 1979, published a book *The Burnside problem and identities in groups* which they improved the result to $n > 664$.

There is still a large gap, however, between those values of n for which $\mathrm{B}(d, n)$ is known to be finite and those for which it is known to be infinite. It is really easy to show the $\mathrm{B}(d, 2)$ is finite. Burnside himself showed that $\mathrm{B}(d, 3)$ is finite, Sanov showed $\mathrm{B}(d, 4)$ is finite and Marshall Hall showed $\mathrm{B}(d, 6)$ is finite. However, it is still an open question as to whether B(2, 5) is finite.

Appendix: Petr Sergeevich Novikov's son, Sergi Petrovich Novikov

Born on March 20, 1938 in Gorki (Nizni Novgorod), Novikov enrolled as a research student in 1960 at the Stelov Institute of Mathematics, where his supervisor was M. M. Postnikov; since 1963 he has been on the staff there. He was awarded the degree of Ph. D there in 1964 and that of Doctor of Science in 1965. In 1966, he was elected Corresponding Member of the Academy of Sciences of the USSR, and in 1981 a full member. Since 1984 he has been the head of the Department of Geometry and Topology of the Mathematical Institute of the Academy of Sciences and in charge of the problem committee of *Geometriya I topologiya* (Geometry and Topology) at the Mathematics Division of the Academy of Sciences of the USSR. He has been head of Mathematics Division at the L. D. Landau Institute of Theoretical Physics of the Academy of Sciences since 1971, where he works closely with the physicists. During the period 1985–1966 Novikov served as President of the Moscow Mathematical Society, and during 1986–1990 he was also a Vice-President of the International Association in Mathematical Physics.

In the Gorbachev–Elzyn era, Novikov started to visit western countries more actively. Before that it was very difficult for Soviet scientists to attend scientific conferences abroad. In 1970 Novikov was not permitted to attend a ceremony at the International Congress of Mathematicians in Nice where he was awarded the Fields Medal for his work in topology because he had signed some letters defending dissidents. In 1991 he was able to work for the first time in his life for a half year in Paris at the Laboratory of Theoretical Physics, Ec. Norm. Superior de Paris. Beginning in 1992 he regularly worked at the University of Maryland at College Park as a Visiting Professor. In September 1996 he became a full time professor at the University of Maryland at College Park in the Department of Mathematics and the Institute for Physical Science and Technology (IPST). He also continues to work in Moscow for a period during the winter and summer.

Доклады Академии Наук СССР
1952. Том LXXXV, № 4

МАТЕМАТИКА

П. С. НОВИКОВ

ОБ АЛГОРИТМИЧЕСКОЙ НЕРАЗРЕШИМОСТИ ПРОБЛЕМЫ ТОЖДЕСТВА

(Представлено академиком И. М. Виноградовым 31 V 1952)

1. Пусть G — произвольная группа, заданная конечным числом образующих, связанных конечным числом определяющих отношений. Проблема тождества ставится следующим образом: требуется указать алгоритм, позволяющий устанавливать для любых двух элементов группы, равны они между собой или нет.

В некоторых частных случаях эту проблему удается решить. Магнус (3) построил алгоритм для случая произвольной группы с одним определяющим соотношением. В ряде работ В. А. Тартаковского (5–9) построены алгоритмы, применимые к более широким классам групп. Однако в общем случае вопрос о существовании алгоритма для решения проблемы тождества оставался до настоящего времени открытым. Нами доказана следующая теорема.

Теорема. *Существуют группы, для которых невозможно построить алгоритм, решающий проблему тождества.*

Понятие алгоритма понимается в смысле определения, принятого в современной теории алгоритмов. Мы разрабатываем способ, позволяющий свести вопрос о тождестве элементов группы к ранее решенным задачам о несуществовании алгоритма. Наиболее удобно для этого воспользоваться системой продукций Поста. Мы сводим сначала вопрос об алгоритмической неразрешимости проблемы тождества для групп с конечным числом определяющих соотношений к тому же вопросу для некоторых систем специального вида — с бесконечным числом определяющих соотношений, которые мы называем круговыми системами. Затем вопрос об алгоритмической неразрешимости проблемы тождества для круговых систем мы сводим к тому же вопросу для системы продукций Поста.

2. Определение круговых систем. Мы рассматриваем конечный алфавит и слова в нем, к которым присоединено пустое слово, обозначаемое 1. Система задается конечным числом соотношений, которые мы назовем исходными равенствами:

$$A_i = B_i, \tag{1}$$

где A_i и B_i — слова. Для каждой буквы x существует обратная x^{-1} и в (1) существуют все равенства вида $xx^{-1} = 1$ и $x^{-1}x = 1$. Для обозначения равенства мы употребляем тот же знак $=$, что и в исходных равенствах.

Равенство между словами определяется следующими условиями: 1) равенство рефлективно, симметрично и транзитивно; 2) для любой буквы a и любого слова A имеет место $aA = Aa$; 3) каковы бы ни были слова A' и A'' (может быть, пустые), имеет место $A'A_iA'' = A'B_iA''$, где A_i и B_i — слова из соотношения (1).

3. Мы будем говорить, что круговая система $\mathfrak{B}$ представима посредством группы $\mathfrak{A}$, если в $\mathfrak{A}$ можно алгоритмически выделить подмножество $\mathfrak{A}'$, которое можно алгоритмически привести во взаимно-однозначное соответствие с системой $\mathfrak{B}$ (иными словами, равные слова системы $\mathfrak{B}$ переходят в равные элементы $\mathfrak{A}'$ и обратно).

Теорема 1. *Для каждой круговой системы $\mathfrak{B}$ с конечным числом образующих и конечным числом исходных равенств существует группа с конечным числом образующих и конечным числом определяющих соотношений, посредством которой можно представить круговую систему $\mathfrak{B}$.*

Построение группы $\mathfrak{A}$ для данной круговой системы $\mathfrak{B}$. Пусть $\mathfrak{B}$ определена равенствами (1) и $a, b, c, \ldots$ — ее образующие. Среди букв группы $\mathfrak{A}$ имеются буквы: опорные p_1, p_2, p_3; основные $a_j, b_j, c_j, \ldots$ с индексами; сигнальные μ_i, λ_i, ν_i с индексами. Основные и сигнальные буквы делятся на 4 рода, отмечаемые индексами 0, 1, 2, 3. Основные буквы каждого рода — все буквы из $\mathfrak{B}$, но снабженные индексами.

Слова группы $\mathfrak{A}$, которые представляют систему $\mathfrak{B}$, имеют вид

$$A_0p_1A_1p_2A_2p_3A_3. \tag{2}$$

При этом слово A_0 отличается от слова A системы $\mathfrak{B}$ только индексами при буквах. Слово A_j состоит только из основных букв j-го рода. Слова (2) являются с точностью до различающих типы индексов и индексов у p центрально-симметричными относительно p_2, а части слова (2) $A_0p_1A_1$ и $A_2p_3A_3$ — центрально-симметричными относительно букв p_1 и p_3. Слово (2) соответствует слову A системы $\mathfrak{B}$.

Определяющие соотношения группы $\mathfrak{A}$ обладают в известном смысле подобной же симметрией. Так, если это соотношение содержит только буквы нулевого и первого рода, а среди опорных букв только p_1, то существует соответствующее ему определяющее соотношение, получаемое из данного заменой p_1 на p_3, букв нулевого рода — соответствующими буквами третьего рода, букв первого — буквами второго, а также изменением порядка написания входящих в соотношение слов на обратные. Мы напишем поэтому все определяющие соотношения, содержащие центральную опорную букву p_2. Остальные определяющие соотношения напишем только для образующих первых двух родов, как основных, так и сигнальных, и буквы p_1. Основные буквы каждого рода j разделяются на два подрода, отмечаемые штрихами. Вместе с основной буквой x_j в группе существуют основные буквы x'_j. Для произвольной основной буквы x_0 нулевого рода имеются сигнальные буквы того же рода μ_0^{1,x_0}, μ_0^{2,x_0} и λ_0^{1,x_0,y_0}, $\lambda_0^{1,x_0,y_0,z_0}$, λ_0^{2,x_0,y_0} и $\lambda_0^{2,x_0,y_0,z_0}$. Имеем следующие соотношения:

1) $x_0 = \mu_0^{1,x_0}\mu_0^{2,x_0}$; 2) $\mu_0^{1,x_0}\mu_0^{2,x_0} = \mu_0^{2,x_0}\mu_0^{1,x_0}$;

3) $\mu_0^{2,x_0}y_0 = y'_0\mu_0^{2,x_0}\lambda_0^{2,x_0,y_0}$; 4) $\mu_0^{1,x_0}y'_0 = y_0\mu_0^{1,x_0}\lambda_0^{1,x_0,y_0}$,

где y_0 — произвольная буква нулевого рода, отличная от x_0. Пусть z_0 и u_0 — любые другие буквы того же рода (может быть, совпадающие с x_0 или y_0).

5) $\lambda_0^{2,x_0,y_0}z_0 = z_0\lambda_0^{2,x_0,y_0}\lambda^{2,x_0,y_0,z_0}$; 6) $\lambda_0^{2,x_0,y_0,z_0}u_0 = u_0(\lambda_0^{2,x_0,y_0,z_0})^2$;

7) $\lambda_0^{1,x_0,y_0}z'_0 = z'_0\lambda_0^{1,x_0,y_0}\lambda_0^{1,x_0,y_0,z_0}$; 8) $\lambda^{1,x_0,y_0,z_0}u'_0 = u'_0(\lambda_0^{1,x_0,y_0,z_0})^2$;

9) $\lambda_0^{2,x_0,y_0,z_0} p_1 \lambda_1^{2,x_1,y_1,z_1} = p_1$; 10) $\lambda_0^{2,x_0,y_0} p_1 \lambda_1^{2,x_1,y_1} = p_1$;

11) $\lambda^{1,x_0,y_0,z_0} \mu_0^{2,x_0} p_1 \mu_1^{2,x_1} \lambda_1^{1,x_1,y_1,z_1} = \mu_0^{2,x_0} p_1 \mu_1^{2,x_1}$;

12) $\lambda_0^{1,x_0,y_0} \mu_0^{2,x_0} p_1 \mu_1^{2,x_1} \lambda_1^{1,x_1,y_1} = \mu^{2,x_0} p_1 \mu_1^{2,x_1}$;

13) $x_1 = \mu_1^{2,x_1} \mu_1^{1,x_1}$, $\mu_1^{2,x_1} \mu_1^{1,x_1} = \mu_1^{1,x_1} \mu_1^{2,x_1}$,

где x_1, y_1 и z_1 — буквы первого рода, соответствующие буквам x, y, z системы $\mathfrak{B}$.

14) $y_1 \mu_1^{2,x_1} \nu_1^{2,x_1,y_1} = \lambda_1^{2,x_1,y_1} \mu_1^{2,x_1} y_1'$; 15) $u_1' \lambda_1^{1,x_1,y_1,z_1} = (\lambda_1^{1,x_1,y_1,z_1})^2 u_1'$;

16) $y_1' \mu_1^{1,x_1} \nu_1^{1,x_1,y_1} = \lambda_1^{1,x_1,y_1} \mu_1^{1,x_1} y_1$; 17) $\nu_1^{2,x_1,y_1} z_1' = z_1' \nu_1^{2,x_1,y_1} \nu_1^{2,x_1,y_1,z_1}$;

18) $z_1 \lambda_1^{2,x_1,y_1} = \lambda_1^{1,x_1,y_1,z_1} \lambda_1^{2,x_1,y_1} z_1$; 19) $\nu_1^{2,x_1,y_1,z_1} u_1' = u_1' (\nu_1^{2,x_1,y_1,z_1})^2$;

20) $z_1' \lambda_1^{1,x_1,y_1} = \lambda_1^{1,x_1,y_1,z_1} \lambda_1^{1,x_1,y_1} z_1'$; 21) $\nu_1^{1,x_1,y_1} z_1 = z_1 \nu_1^{1,x_1,y_1} \nu_1^{1,x_1,y_1,z_1}$;

22) $u_1 \lambda_1^{2,x_1,y_1,z_1} = (\lambda_1^{2,x_1,y_1,z_1})^2 u_1$; 23) $\nu_1^{1,x_1,y_1,z_1} u_1 = u_1 (\nu_1^{1,x_1,y_1,z_1})^2$.

Пусть $A^{(i)} = B^{(i)}$ — произвольное определяющее соотношение круговой системы $\mathfrak{B}$. Тогда слово $A_k^{(i)}$ состоит из тех же букв, что и $A^{(i)}$, только снабженных индексом k ($k = 0, 1, 2, 3$); при этом, если k четно, то буквы $A_k^{(i)}$ выписаны в том же порядке, как у $A^{(i)}$, а если k нечетно, то в обратном порядке. Такой же смысл имеет обозначение $B_k^{(i)}$. Тогда имеем следующее определяющее соотношение:

24) $A_0^{(i)} p_1 A_1^{(i)} = B_0^{(i)} p_1 B_1^{(i)} \rho_1^{(i)}$ и $\rho_1^{(i)} x_1 = x_1 (\rho_1^{(i)})^2$.

Кроме того, имеется столько же соотношений, написанных по симметрии для букв второго и третьего рода. Последние из них имеют вид

24') $A_2^{(i)} p_3 A_3^{(i)} = \rho_2^{(i)} B_2^{(i)} p_3 B_3^{(i)}$ и $x_2 \rho_2^{(i)} = (\rho_2^{(i)})^2 x_2$.

Наконец, имеются еще соотношения:

а) $\nu_1^{1,x_1,y_1,z_1} p_2 \nu_2^{1,x_2,y_2,z_2} = p_2$ и $\nu_1^{1,x_1,y_1} p_2 \nu_2^{1,x_2,y_2} = p_2$;

б) $\nu_1^{2,x_1,y_1,z_1} \mu_1^{1,x_1} p_2 \mu_2^{1,x_2} \nu_2^{2,x_2,y_2,z_2} = \mu_1^{1,x_1} p_2 \mu_2^{1,x_2}$ и $\nu_1^{2,x_1,y_1} \mu_1^{1,x_1} p_2 \mu_2^{1,x_2} \nu_2^{2,x_2,y_2} =$
$= \mu_1^{1,x_1} p_2 \mu_2^{1,x_2}$;

в) $\rho_1^{(i)} p_2 \rho_2^{(i)} = p_2$.

Здесь x_2, y_2 и z_2 — буквы второго рода, соответствующие буквам x, y, z круговой системы $\mathfrak{B}$, а $\mu_2^{1,x_2}, \lambda_2^{1,x_2,y_2}, \lambda_2^{1,x_2,y_2,z_2}$ и $\mu_2^{2,x_2}, \lambda_2^{2,x_2,y_2}, \lambda_2^{2,x_2,y_2,z_2}$ — буквы, соответствующие по симметрии буквам $\mu_1^{1,x_1}, \lambda_1^{1,x_1,y_1}, \lambda_1^{1,x_1,y_1,z_1}$ и $\mu_1^{2,x_1}, \lambda_2^{2,x_1,y_1}, \lambda_1^{2,x_1,y_1,z_1}$.

Пусть A — произвольное слово системы $\mathfrak{B}$, а $A_0 p_1 A_1 p_2 A_2 p_3 A_3$ — соответствующее слово в группе $\mathfrak{A}$.

Теорема 2. *Слово группы $\mathfrak{A}$ вида $A_0 p_1 A_1 p_2 A_2 p_3 A_3$ равно слову $B_0 p_1 B_1 p_2 B_2 p_3 B_3$ тогда и только тогда, когда $A = B$ в круговой системе $\mathfrak{B}$.*

После этого вопрос о разрешимости задачи тождества слов в группах сводится к той же задаче для круговых систем.

4. Равенство слов в круговой системе можно свести к задаче равенства слов в системе Поста ([10]). Эта система состоит из определенных правил переработки любого слова, или продукций. Правила преобразования исходного слова A следующие.

Дается конечное число пар слов (P_i, Q_i). Каждое слово вида $P_i X$ преобразуется в слово $X Q_i$ и обратно. Эта операция называется продукцией. Пусть A — некоторое слово. Путем продукций можно, исходя из A, получить совокупность слов. Все эти слова называются равными слову A. Известно ([10]), что можно построить такую конкретную систему пар (P_i, Q_i) и указать такое слово A, что не существует алгоритма, позволяющего решать вопросы равенства между словами в данной системе. Напишем теперь исходные равенства, определяющие круговую систему, для которой вопрос об отыскании алгоритма для решения задачи равенства слов прямо сводится к такому же вопросу для системы Поста. Пусть Ω — некоторая система Поста и A — исходное слово системы Поста, а (A_i, B_i) — пары слов, определяющих продукции. Мы строим круговую систему $\mathfrak{B}$, алфавит которой содержит каждую букву системы, но в двойном числе, так что имеются буквы двух родов, отмеченные индексами 1 и 2.

Эти буквы назовем основными буквами системы $\mathfrak{B}$. Две буквы p_1 и p_2 назовем опорными и, наконец, буквы $l_1^{(i)}, q_1^{(i)}, r_1^{(i)}$ и $l_2^{(i)}, q_2^{(i)}, r_2^{(i)}$ — сигнальными (здесь i пробегает столько же чисел, сколько есть пар (A_i, B_i)).

Пусть x_1 и x_2 — произвольные основные буквы соответственно первого и второго рода, а $(A_1^{(i)}, B_1^{(i)})$ и $(A_2^{(i)}, B_2^{(i)})$ — i-е пары слов, буквы которых соответственно отмечены индексами 1 и 2.

1) $A_1^{(i)}p_1A_2^{(i)} = q_1^{(i)}l_1^{(i)}p_1l_2^{(i)}q_2^{(i)}$; 2) $x_1q_1^{(i)} = (q_1^{(i)})^2x_1$;

3) $q_2^{(i)}x_2 = x_2(q_2^{(i)})^2$; 4) $x_1l_1^{(i)} = l_1^{(i)}x_1$;

5) $l_2^{(i)}x_2 = x_2l_2^{(i)}$; 6) $q_1^{(i)}p_2q_2^{(i)} = p_2$;

7) $r_1^{(i)}p_1r_2^{(i)} = p_1$; 8) $q_1^{(i)}l_1^{(i)}p_2l_2^{(i)}q_2^{(i)} = B_1^{(i)}p_2B_2^{(i)}$.

К этим восьми равенствам присоединяются все равенства вида $uu^{-1} = 1$ и $u^{-1}u = 1$, где u — произвольная буква системы $\mathfrak{B}$.

Рассмотрим слово

$$p_1A_1p_2A_2, \tag{3}$$

где A_1 и A_2 — слова, отвечающие слову A системы Ω, только записанные во взаимно-противоположном порядке, так что A_2 есть зеркальное отражение A_1 относительно центра p_2.

Пусть A имеет вид A_ix; тогда преобразование слова $p_1A_1^{(i)}x_1p_2x_2A_2^{(i)}$ после исчезновения сигнальных букв приводит к слову $p_1x_1B_1^{(i)}p_2B_2^{(i)}x_2$.

Это значит, что происшедшее при этом преобразование части данного слова, являющегося словом A_1, есть продукция, определяемая парой $(A_1^{(i)}, B_1^{(i)})$. То же справедливо и для симметрического слова A_2.

Теорема 3. *Всякое преобразование слова* (3) *путем равенств* (2), *после которого исчезают все сигнальные буквы и отсутствуют буквы в отрицательных степенях, приводит к слову* $p_1B_1p_2B_2$, *где* B_1 *преобразуется из* A_1 *в результате применения продукций системы* Ω.

Этим заканчивается доказательство сводимости задачи об алгоритмической разрешимости проблемы тождества в теории групп к такому же вопросу для систем продукций. Но так как можно построить конкретную систему продукций, для которой не может быть алгоритма, решающего задачу равенства слов, то можно построить и конкретную группу, для которой проблема тождества неразрешима.

Поступило
29 V 1952

ЦИТИРОВАННАЯ ЛИТЕРАТУРА

[1] А. И. Мальцев, Матем. сборн., **6** (48), 331 (1939). [2] А. И. Мальцев, там же, **8** (50), 251 (1940). [3] В. Магнус, Усп. матем. наук, **8**, 365 (1941). [4] В. А. Тартаковский, ДАН, **58**, № 8, 1605 (1947). [5] В. А. Тартаковский, ДАН, **58**, № 9, 1909 (1947). [6] В. А. Тартаковский, Матем. сборн., **25** (67), 3 (1949). [7] В. А. Тартаковский, там же, **25** (67), 251 (1949). [8] В. А. Тартаковский, Изв. АН СССР, сер. матем., **13**, № 6, 283 (1949). [9] В. А. Тартаковский, Матем. сборн., **30** (72), 39 (1952). [10] E. Post, Bull. Am. Math. Soc., **50**, No. 5 (1944). [11] Н. Г. Чеботарев, Усп. матем. наук, **8**, 336 (1941).

Доклады Академии наук СССР
1959. Том 127, № 4

МАТЕМАТИКА

Член-корреспондент АН СССР П. С. НОВИКОВ

О ПЕРИОДИЧЕСКИХ ГРУППАХ

Мы рассматриваем группы с конечным числом образующих, определенные тождеством $X^n = 1$, где X — произвольный элемент группы, n — натуральное число.

Эти группы обычно называются свободными периодическими. Число n назовем показателем периодичности.

Теорема. *Всякая свободная периодическая группа с показателем периодичности, не меньшим 72, бесконечна.*

Обозначим через $\mathfrak{A}_n$ периодическую группу с показателем периодичности n и с некоторым конечным числом образующих. Число n мы будем предполагать во всех дальнейших рассуждениях фиксированным и не меньшим чем 72. Число образующих группы $\mathfrak{A}_n$ больше или равно 2 и также фиксировано. Пусть $\mathfrak{A}$ — свободная группа с тем же числом образующих, что и группа $\mathfrak{A}_n$.

Все дальнейшие определения и построения относятся главным образом к группе $\mathfrak{A}$; в иных случаях мы будем делать соответствующие оговорки. Единичный элемент группы $\mathfrak{A}$ мы будем обозначать через 1.

В основе доказательства приведенной выше теоремы о бесконечности периодических групп с показателем $\geqslant 72$ лежит понятие Ω_q-преобразования.

Мы будем рассматривать Ω_q-преобразования двух типов: круговые и линейные.

Определение круговых Ω_q-преобразований. Пусть q—натуральное число, удовлетворяющее неравенству $6q \leqslant n$. Назовем круговым Ω_q-преобразованием, во-первых, переход от слова вида AX^tB к слову $AX^{t-n}B$, где $q \leqslant t \leqslant n-q$, и во-вторых, циклические перестановки слов.

Определение линейного Ω_q-преобразования. Назовем линейным Ω_q-преобразованием переход от слова AX^tB к слову $AB^{t-n}B$, где $q \leqslant t \leqslant n-q$. В случае, когда либо A либо B пустое, линейным Ω_q-преобразованием называется тот же переход при более слабых условиях: $q/2 \leqslant t \leqslant n-q$. В дальнейшем мы будем предполагать, что $q \geqslant 12$.

Определение периодического и условно периодического слова. Периодическим словом назовем слово вида X^t, где t — некоторое натуральное число; X будем называть периодом слова X^t, а t—его показателем. Назовем условно-периодическим словом произведение $X_1 X_2 \ldots X_t$, если каждое X_i линейными Ω_q-преобразованиями можно перевести в слово вида $u_i X v_i$ так, что $v_i u_{i+1} = 1$ для $i = 1, 2, \ldots, n-1$ и $v_t u_1 = 1$.

Назовем слова X_i условными периодами слова $X_1, \ldots, X_t$, а число t — показателем этого слова,

Если слово A имеет вид $A_1 A_2$, где A_1 и A_2 — непустые слова, то A_1 и A_2 будем называть отрезками слова A.

Определение приведенного слова. Слово A называется приведенным, если оно не сократимо и не содержит никакого отрезка, приводимого круговыми Ω_q-преобразованиями к слову, равному 1, и никакого отрезка, равного условно периодическому слову с показателем, большим чем $n-q$.

Теорема 1. *Каждое слово X группы $\mathfrak{A}$ можно круговыми Ω_q-преобразованиями перевести в приведенное слово.*

Это последнее мы будем называть приведенной формой слова X. Пусть R_x — класс приведенных форм слова X.

Теорема 2. *Если хотя бы одно слово, содержащееся в R_x, не равно* 1, *то и каждое не равно* 1.

Определение основного условно-периодического слова. Условно-периодическое слово $X_1 X_2 \ldots X_t$ назовем основным, если каждый условный период X_i является непустым приведенным словом и t — наибольший из возможных показателей этого слова.

Пусть $X_1 X_2 \ldots X_t$ и $Y_1 Y_2 \ldots Y_t$ — основные условно-периодические слова с одинаковым показателем. Назовем их эквивалентными, если существуют такие слова $u_1, v_1, u_2, v_2, \ldots, u_t, v_t$, что $u_i Y_i v_i \in R_{v_i}$ и $u_i v_{i+1} = 1$, $u_t v_t = 1$.

Введенное отношение эквивалентности рефлексивно, симметрично и транзитивно.

Обозначим через $L_{X_1 \ldots X_t}$, где $X_1 \ldots X_t$ — основное условно-периодическое слово, класс основных условно-периодических слов, эквивалентных слову $X_1 \ldots X_t$.

Теорема 3. *Если A — ссновное условно-периодическое слово, то A^{-1} также условно-периодическое слово с тем же показателем.*

Теорема 4. *Если AB — основное условно-периодическое слово, то BA также основное условно-периодическое слово.*

Теорема 5. *Каждое условно-периодическое слово $X_1 \ldots X_t$ содержит периодическое слово с показателем, не меньшим чем* $\min[t, q/2]$.

Это утверждение прямо вытекает из определения условно-периодического слова. Если в слове $X_1 \ldots X_t$ все слова X_i различны, то его можно перевести в периодическое слово X^t так, что каждое X_i переводится в $u_i X v_i$ линейными Ω_q-преобразованиями. Но в таком случае хотя бы одно слово X_i должно содержать условно-периодическое слово с показателем, не меньшим чем $q/2$.

Из последней теоремы вытекает, что основное условно-периодическое слово с показателем t также содержит хотя бы одно периодическое слово с показателем, не меньшим чем $\min[t, q\ 12]$.

Теорема 6. *Пусть A и B — основные условно-периодические слова с одинаковым показателем, не меньшим q, причем A имеет вид A_1C, а B имеет вид CB_1. Если C содержит не менее четырех условных периодов слова A и не менее четырех условных периодов слова B, то классы L_A и L_B совпадают.*

Теорема 7. *Пусть A и B — основные условно-периодические слова с показателями, не меньшими q, и классы L_A и L_B различны. Если слово A имеет вид A_1C, а слово B имеет вид $C^{-1}B_1$ и C^{-1} содержит не менее q условных периодов слова B, то приведенная форма слова A_1B_1 принадлежит классу L_A.*

Две последние теоремы свазаны с тем, что все основные условно-периодические слова с показателями, не меньшими чем q, могут быть частично упорядочены следующим образом: будем говорить, что основное условно-периодическое слово B предшествует основному условно-периодическому слову A, если классы L_A и L_B различны, слово A имеет вид $A_1 C A_2$, а слово B имеет вид $B_1 C B_2$, где C содержит не менее q условных периодов слова B.

Заметим, что условно-периодическое слово с показателем n равно 1 в группе $\mathfrak{A}_n$. Так как все периодические слова с показателем n также являются условно-периодическими, то совокупность всех равенств $X_1 \ldots X_n = 1$, где $X_1 \ldots X_n$ — произвольное условно-периодическое слово образует систему определяющих соотношений для группы $\mathfrak{A}_n$. Можно, однако, доказать более сильное утверждение.

Теорема 8. *Совокупность всех равенств вида $X_1 \ldots X_n = 1$, где $X_1 \ldots X_n$ — произвольное основное условно-периодическое слово, является системой определяюших соотношений группы $\mathfrak{A}_n$.*

Всякое равное единице слово какой-либо группы, заданной определяющими соотношениями вида $A_i = 1$, может быть получено, по терминологии Тартаковского, композицией слов A_i, обратных им слов и циклических перестановок этих слов. Это значит, что каждое слово Дика может быть получено из пустого слова, во-первых, преобразованиями по схеме $XY \to XuY$, где u есть либо A_i, либо обратное ему слово, либо циклическая перестановка слов A_i и A_i^{-1}, и, во-вторых, сокращениями.

Из работ Тартаковского ([1]) можно извлечь следующий результат:

Теорема 9. *Пусть группа определена конечным или бесконечным числом определяющих соотношений вида*

$$A_{1i}\, A_{2i} \ldots A_{ti} = 1, \quad t \geqslant 6.$$

Предположим, что среди слов $A_{1i} \ldots A_{ti}$ содержатся все их циклические перестановки и обратные слова и что ни одно из двух слов A_{ri} и A_{sj} не может полностью исчезнуть при сокращении слова $A_{ri}\, A_{sj}$. Тогда каждое непустое и несократимое слово Дика A может быть получено такой композицией слов $A_{1i} \ldots A_{ti}$, что после сокращений в слове A сохранится хотя бы один из отрезков A_{ki}.

Теорема 10. *Каждое непустое и несократимое слово Дика группы $\mathfrak{A}_n$ может быть получено путем такой композиции основных условно-периодических слов $A_{1i} \ldots A_{ni}$ с показателем n и периодами A_{ji}, что каждые два условно-периодических слова, входящих в композицию, сокрашаются взаимно не более чем на q условных периодов.*

Идея доказательства этой теоремы заключается в перестройке произвольной композиции непустого и несократимого слова в такую, которая удовлетворяет требованиям теоремы. Каждому несократимому слову приписывается некоторым образом индекс. Затем берется произвольная композиция непустого и несократимого слова Дика A группы $\mathfrak{A}_n$, составленная из основных условно-периодических слов $A_{1i} \ldots A_{ni}$.

Докажем, что два таких слова A' и A'' сократятся между собой более чем на q условных периодов.

Применяя лемму Тартаковского, мы можем перестроить композицию слова A так, что во вновь полученной композиции вставки слов A' и A'' следуют непосредственно друг за другом. Тогда возможно либо, используя теорему 7, заменить композицию слов A' и A'' одним условно-периодическим словом, либо элиминировать композицию слов A' и A'', заменяя эти условно-периодические слова другими условно-периодическими словами с меньшим индексом.

Из теорем 5, 9 и 10 следует:

Теорема 11. *Всякое непустое и несократимое слово Дика содержит отрезок, являющийся периодическим словом с показателем, не меньшим чем $q/2$.*

Так как $q \geqslant 12$, то этот показатель не меньше 72.

Для того чтобы придти к окончательному результату, достаточно использовать одну работу Аршона ([2]), в которой строится последовательность из трех символов такая, что в ней не содержится никакого периодического отрезка с периодом, не меньшим 2.

Из сказанного выше следует, что каждая группа $\mathfrak{A}_n$ для $n \geqslant 72$ с 3 образующими бесконечна.

Однако, следуя способу Аршона, можно построить последовательность из 2 символов, не содержащую периодического отрезка, с показателем, не меньшим 3. Отсюда можно заключить, что группа $\mathfrak{A}_n$ с 2 образующими для $n \geqslant 72$ также бесконечна.

Поступило
27 V 1959

ЦИТИРОВАННАЯ ЛИТЕРАТУРА

[1] В. А. Тартаковский, Изв. АН СССР, сер. матем., **13**, № 6, 483 (1949); Матем. сборн., **30** (72), 39 (1952). [2] С. Е. Аршон, Матем. просвещ., № 2, 24 (1934).

Nikolai Nikolaevich Bogoliubov

Born – 21 Aug 1909
Died – 13 Feb 1992

Nikolai Bogoliubov was a Corresponding Member of the Division of Physical–Mathematical Sciences since 1946. He was also an Academician of the Division of Physical–Mathematical Sciences since 1953.

Awards:
M.A. Lavrent'ev Gold Medal
– For the work *On stochastic processes in dynamic systems* (1983);

M.V. Lomonosov Gold Medal
– For outstanding achievements in mathematics and theoretical physics (1984);

A.M. Lyapunov Gold Medal
– For a series of works in the problems of stability, critical phenomena and phase transitions in the theory of systems of many interacting particles (1989).

MATHEMATICAL PROBLEMS OF QUANTUM FIELD THEORY

N.N. BOGOLYUBOV

Quantum field theory, which studies the interactions of elementary particles, was born comparatively long ago: at the end of the twenties and the beginning of the thirties. In accordance with the situation then prevailing, it was formulated as the quantum electrodynamics of the interactions of electrons, positrons, and photons.

Many elementary particles have now been discovered; it has become clear that their interactions fall into three classes: 1) strong, or nuclear, interactions, 2) electromagnetic, 3) weak interactions, differing sharply in their intensities.

The range of application of quantum field theory has thus been quite essentially extended. I should like to emphasize that in spite of the successes achieved in the consideration of various problems of this theory, the fundamental difficulties that we encountered right at the very start of work on quantum electrodynamics, those associated with the need to combine the principles of relativity theory with those of quantum mechanics, remain unresolved to this day. In the various proposed approaches the form in which they are expressed is sometimes changed, but in essence things remain just as they were. In general outline one may say that the nature of these fundamental difficulties is rooted in the basic ideas of quantum field theory themselves – the concepts of the point character of the particles and the localization of their interactions: the various fields interact at one and the same point. The experimentally observed structure of the elementary particles is regarded as secondary, a dynamical effect. Now one says that the "original" or "bare" particle is surrounded as it were by a cloud of virtual particles, responsible for the observed structure of the real particles.

It is not difficult to understand why, because of the point character of the "bare" particles, generalized functions now turn up in the theory. On the other hand, the equations of motion are non-linear, due to the presence of the local interaction, and we are faced with the necessity of applying non-linear operations to generalized functions. Herein lies the deep physical reason for those mathematical difficulties of principle upon which the development of quantum field theory hinges.

It may turn out that things can easily be put right, and that the notion of the point character of the "bare particles" is excessively idealized. It might seem that from the physical point of view it would be more realistic to introduce into the theory particles having some spatial extension l, to introduce, as one says, an elementary length. But since we need to take the requirements of relativity theory into account, we must now also assign to the particle a "temporal extension" $\Delta t \sim \frac{l}{e}$. But what does the presence of a temporal extension Δt signify? It signifies that in an interval of time of duration one cannot speak of the relation "earlier-later". This, in its turn, leads to a deviation, "microscopic" so to speak, it is true, from the law of causality. Such an infringement of causality "in the small" may also lead, because of the presence of the corresponding groups, to an infringement of causality "in the large", which is inadmissible.

Let us now comprehend these simple and intuitive ideas as applied to the formulation of the basic equations of quantum field theory. We shall discuss the very first idea, which however still retains its value up to the present day.

As we know, classical field theory can be based upon the variational principle

$$\delta \int L\left(\ldots, \varphi_\alpha, \ldots, \frac{\partial \varphi_\alpha}{\partial x_\beta}, \ldots\right) dx = 0$$

with a local Lagrangian L, depending on the field functions φ at points x of space-time and their partial derivatives, the Lagrangian having to possess the property of invariance relative to the Lorentz group and respective other groups.

This variational principle leads to equations of motion for the field functions. In quantum theory we keep, essentially, the same equations, but the meaning of the field functions is radically altered. Namely, in the classical theory the values of the components of the φ's are ordinary real or complex numbers. In quantum theory however, we have to do with fields of particles that may be created and annihilated; the field functions $\varphi(x)$ must now characterize the creation and annihilation of particles at points of space-time.

Thus, the $\varphi(x)$ are considered as operators, acting on the "state vectors" – elements of a certain linear space, the space of states of the dynamical system in question. Therefore in quantum theory the equations of motion are supplemented by the associated commutation relations. Typical equations of motion considered in quantum field theory have the form

$$D\varphi + k_0\varphi = \sum g_0 \prod \varphi.$$

Here for the Bose-field D is the D'Alembertian operator, $k_0 = m_0^2$ is the square of the mass of the "bare" particle; for the Fermi-field, D is the Dirac operator, $k_0 = m_0$; $\Pi\varphi$ is the product of two or three fields at the same point x of space-time; finally, the g_0 are the "seeding" coupling constants. The left-hand sides of the above equations depend on the Lagrangian L_0 for free, non-interacting particles (a quadratic form), and the right-hand sides on the Lagrangian L_{int} of the interaction, chosen in the form of a sum of products of field operators at one and the same point (localization of the interactions). The commutation relations express the commutator or anti-commutator

$$[\varphi_\alpha(x_1); \quad \varphi_\beta(x_2)]$$

for two arbitrary points x_1, x_2 lying in one space-like surface, in terms of three-dimensional δ-functions. They were first obtained by a passage to the Hamiltonian formalism and by employment of the commutation relations, generally accepted in ordinary non-relativistic quantum mechanics, between the "coordinates" and the "momenta". The presence of δ-functions in the commutation relations indicates that the $\varphi(x)$ can be regarded only as singular or generalized operators. On the other hand, their product enters into the equations of motion. In view of this, the problem of a mathematically correct approach to the accepted equations is exceptionally complicated. It is not clear whether in general they have any sense whatever, and whether one can attempt to consider them as a logically non-contradictory basis for field theory with the aid of the introduction of an appropriate definition of the concept of the multiplication of generalized operators. If one endeavours to regularize the commutation relations (renunciation of the point character of the particles) or to replace the local products

$$\prod \varphi(x)$$

by non-local expressions

$$\int K(x, x_1, x_2, \ldots)\, \varphi(x_1)\, \varphi(x_2) \ldots dx_1\, dx_2 \ldots$$

(passage to non-local interactions), then the equations of motion become inconsistent with the commutation relations. This situation simultaneously reflects the difficulties associated with the violation of the law of causality involved in a departure from point localization, of which we have already spoken above.

We now note that in quantum field theory a basic rôle is played by the scattering operator, usually called the "S-matrix". This characterizes the passage from the initial state of the dynamical system at the time $t = -\infty$, when only infinitely separated and therefore non-interacting particles are present, to the final state at the time $t = +\infty$, when the particles are again infinitely separated.

We cannot, in general, consider the equations of motion for the field operators, but must deal only with the operator S. Formally it may be expressed in terms of the interaction Lagrangian by means of the well-known formula

$$S = T\left(\exp i \int L_{\text{int}}(x)\, dx \right).$$

Here in L_{int} the dependence of the field operators $\varphi_x(x)$ on x is taken to be the same as for the free, non-interacting fields. It should be emphasized that this formula has only a purely symbolic, intuitive character, and cannot be considered as a mathematical definition. The T-exponent is simply a symbolic notation for the series

$$\sum_{n \geqslant 0} \frac{i^n}{n!} \int \ldots \int T\,\{L_{\text{int}}(x_1) \ldots L_{\text{int}}(x_n)\}\, dx_1 \ldots dx_n,$$

in which the T-product

$$T\{L_{\rm int}(x_1)\ \ldots\ L_{\rm int}(x_n)\}$$

is "defined" as the product of the operators $L(x_j)$, arranged from left to right in increasing order of the time coordinates t_j. In other words, we have here only the formal writing of a perturbation theory series "in powers of the interaction".

We mention that perturbation theory appeared right at the start of work on the creation of quantum electrodynamics. This was then completely natural, since in electrodynamics we indeed have a small parameter – a dimensionless coupling constant, known as the fine structure constant, which is equal to 1/137. It turned out however that only the first approximation in perturbation theory led to results that were in satisfactory agreement with experiment. The construction of the higher approximations not only did not give improvements on the first approximation, but led to mathematical absurdities, divergent expressions. Namely, in these higher approximations one always encountered integrals, over momentum space, of functions not decreasing sufficiently rapidly at infinity, and therefore one began to refer to the ultra-violet catastrophe in perturbation theory.

Analyzing the physical reasons for this situation, research workers came to understand that the basic difficulty is rooted in the very concepts of the point character of particles and localization of their interactions. In view of the fact that it is very easy to criticize such concepts as excessively idealized and non-physical, the view became popular that success in the resolution of these difficulties must come from new physical ideas, and a great number of exploratory papers with this orientation began to appear.

At the present time it is interesting to emphasize that the first success here came not from new physical ideas but from a purely formal approach, now known under the name of the renormalization, or subtraction, procedure. I shall now say a few words about this procedure. Denote by m_0 and α_0 the mass and dimensionless charge of the electron (the fine structure constant) that occur in the Lagrangian. Then the complete mass and charge of the electron, observed experimentally, are represented by expansions:

$$m = m_0 + m_1 + m_2 + \ldots,$$
$$\alpha = \alpha_0 + \alpha_1 + \alpha_2 + \ldots$$

Thus m_0, α_0 may be interpreted as the "seeding values", or values corresponding to the "bare particle", and the terms $m_1 + m_2 + \ldots$, $\alpha_1 + \alpha_2 + \ldots$ as the field additions to the mass and charge, caused by the interactions of the electron-positron and electromagnetic fields. We note, incidentally, that all the m_1, m_2, $\ldots$, α_1, α_2, $\ldots$ contain divergent integrals.

Now suppose that we wish to deal only with the real mass and charge, and correspondingly wish to reconstruct the perturbation theory expansion in such a way that in the first and any subsequent approximation the mass and charge are exactly equal to m, α. For this purpose we need only substitute the systems

$$m_0 = m - m_1 - m_2 - \ldots,$$
$$\alpha_0 = \alpha - \alpha_1 - \alpha_2 - \ldots$$

in the original Lagrangian and successively define $m_1, m_2, \ldots, \alpha_1, \alpha_2, \ldots$ in the appropriate way. The required quantities, for example the elements of the scattering matrix, are then represented in the form of series in powers of α.

A remarkable property of such expansions was observed: there were no ultra-violet divergences in their terms. They, as it were, cancel each other out, in fact the divergences due to the terms of the Lagrangian proportional to α are successively compensated for by divergences arising from the terms of the Lagrangian containing m_1, α_1 and so on. This unexpected fact was first noticed in the second approximation during the computation of a number of concrete effects, and was afterwards also established in the general case. We now know how to formulate a perturbation theory in which any fixed approximation is free from divergences. This certainly represents only a very partial success, but it is of great practical value in quantum electrodynamics, because the way was thereby opened to the computation of the radiative corrections for important physical quantities to any order in powers of α. The numerous computations that have been carried out (in particular of the Lamb shift, displacement of energy levels, anomalous magnetic moment of the electron and so on) are in striking agreement with experiment, and in a number of cases a higher degree of accuracy has been attained than that hitherto reached in experiments.

The renormalization procedure within perturbation theory was also formally extended to the case of the calculation of strong interactions. But here the situation is completely different than in electrodynamics. The generalization of the renormalization technique to the theory of strong interactions has more the character of a purely formal achievement, without *direct* practical results, since here we do not have a small parameter, and the very idea of an approach based on perturbation theory appears ineffective.

Nevertheless it must be acknowledged that the working out of a general scheme of perturbation theory made an important contribution so to speak to our "general education", and was of great heuristic value. Some very useful concepts and methods were developed in connection with this theory. Let us take, for example, the introduction of the Green's function and the diagram technique. The diagram technique is now widely used even also in qualitative investigations. Not only theorists but also experimentalists have recourse to Feynman diagrams in the discussion of experimental situations.

We also mention that the consideration of a general schema of perturbation theory made if possible to elucidate the mathematical nature of the success of the renormalization procedure in eliminating the divergences from the separate terms. In this connection we remark that the n-th term of perturbation theory contains the T-product

$$T\{L_{\text{int}}(x_1) \ldots L_{\text{int}}(x_n)\}.$$

Expanding it as a sum of a product of coefficient functions and simple operator functions, we observe that the coefficient functions are of the form

$$\Pi D(x_i - x_j)$$

where the D's are the Green's functions for the corresponding free fields. They are generalized functions of a very special type.

In addition, there are a number of algebraic relations between the coefficient functions, expressing the properties of unitarity and causality.

If the Fourier representations for the functions D are substituted into the above product and a formal integration over the momentum variables is carried out, we immediately obtain the usual ultra-violet divergences.

One can however make a perfectly correct definition of the product of generalized functions D of the type studied. For this purpose one can start from the fact that this product is completely determined as a functional on a subspace of the corresponding space C_∞ of functions $f(x_1, \ldots, x_n)$ in which $f(x_1, \ldots, x_n)$ and a sufficient number of its partial derivatives vanish when any two of the arguments $x_1, \ldots, x_n$ coincide. Then the technique of extending linear functionals, with the preservation of the above-mentioned algebraic relations, is employed.

Thus, the product of generalized functions in question is now rigorously defined as a functional over the whole of C_∞. It turns out that this procedure is completely equivalent to renormalization. One may say that in order to eliminate the ultra-violet divergences in the separate terms of the perturbation theory it is sufficient just to define correctly the T-products:

$$T\,\{L_{\text{int}}(x_1) \,\ldots\, L_{\text{int}}(x_n)\}.$$

We have already said that in the theory of strong interactions there is no small parameter, and therefore perturbation theory cannot be employed directly.

It should however be emphasized that also in theories in which there is a small parameter, say quantum electrodynamics, the situation becomes very complicated in the passage to high enough energies. For example, the n-th term of the expansion of the single-particle Green's function is proportional not simply to α^n, but to the product $\alpha^n \left| \ln \frac{k^2}{m^2} \right|$, where k^2 is the relativistic square of the 4-momentum, and therefore as $k \to \infty$ we shall no longer have a small parameter. Hence it is clear that the problem of convergence of the series of perturbation theory presents very exceptional difficulties. Having eliminated the divergences in the individual terms of these series, we do not have even the first intuitive idea of how to sum the whole series.

Thus, the question whether in general quantum electrodynamics even exists as a theory, free of internal contradictions, remains open.

In the consideration of the problem of approximate solutions the question may arise, why such great value is attributed to the small parameter. Indeed, in principle other methods of obtaining approximate solutions are possible, not depending on a small parameter. But the fact is that so far it has been possible to formulate the renormalization procedure only in application to perturbation theory, and we do not know how to remove the divergences in the other approaches.

In other words, we know how to define correctly the T-products only of a finite number of factors, and so far we cannot arrive at a definition of transcendental T-operations. In this situation various directions of research have been proposed, giving rise to a great many papers. Thus, a number of simplified models for quantum theory have been studied.

Various methods of summation of the "principal" diagrams have been developed. Unfortunately the concept of "principal" diagram still has to a marked degree a purely subjective character. Great attention has been paid to the application of continuous integration.

The fact is that it has proved possible to express the quantum-field Green's functions by means of functional quadratures of the Green's functions for free non-interacting particles in an external classical field. These functions are the ordinary Green's functions for linear partial differential equations of hyperbolic type. The external field is regarded as their "functional argument". The investigation of these ordinary Green's functions does not present any difficulties of principle. The deep difficulties of principle appear when we pass to the performance of the integration with respect to a functional argument – the external classical field.

This integration must be made with respect to a measure having a certain resemblance to the Wiener measure employed in the theory of stochastic processes. The essential difference consists in that under the sign of the exponent there stands not a real expression, as with Wiener, but a complex one. In spite of a great number of published papers, investigations in this direction have not yet left the very first stage of a more or less formal approach. So far it is not clear what will happen to the divergences. Can one in general assign a definite mathematical sense to this functional quadrature? Neither have qualitative properties been successfully studied on this basis. For example, the spectral representations, easily established by different arguments, have not been obtained starting from functional quadratures.

Great value, theoretical as well as practical, is attached to investigations connected with the dispersion relations.

In essence, a dispersion relation consists in the Cauchy formula for a function of a complex variable regular in the whole complex plane (without essential singularities at infinity) with the exception of cuts along the real axis. This Cauchy formula is called a dispersion relation when it is formulated for the function representing a matrix element of the scattering operator.

At present the main attention is focussed on processes in which there are two particles at the beginning and the end. For such processes the associated scattering amplitude is expressed in terms of functions depending on two variables; the energy relative to some frame of reference may be taken as one of them, and the momentum transfer as the other.

Consider the dispersion relation with respect to the energy variable for the process of scattering of two particles. With its aid, by passing to the real axis, we can at the same time express the real part of the scattering amplitude in terms of its imaginary part. It should be noted here that the effective total cross-section is given in terms of the imaginary part of this amplitude, and the effective differential cross-section for scattering through a given angle in terms of its modulus. Thus, the real and imaginary parts of the amplitude can be found experimentally, at any rate in the physically meaningful range of values for the energy.

Since in the theory of strong interactions the derivation even of approximate quantitative results is very non-trivial, the exact equations that follow from the dispersion relations have attracted great interest. In the cases when the cuts include only physically attainable energies, these relations can be verified experimentally. Such is the case, for example, for the forward scattering of pions by nucleons.

In order to make use of the dispersion relations it was first of all necessary to establish the corresponding properties of analyticity for the elements of the scattering matrix under consideration. In this connection the question arose of which formulation of quantum field theory could be employed for the purpose.

The fact is that although there exist many approaches in quantum field theory, yet non of them so far yields this mathematical formulation, although the question is one of establishing a completely definite mathematical fact - the analyticity property. On the other hand, it already became clear from the first preliminary ideas on the foundation of the dispersion relations that they depend only on very general principles, and that the specific peculiarities of the dynamical system under investigation are irrelevant. It was therefore natural that in this situation interest in the construction of an axiomatic approach should develop.

Interest was here directed towards the distillation of an extract from quantum field theory, an extract whose formulation would be given by a series of general axioms expressing the most basic notions. Several such axiomatic systems have now been constructed. The attempt has been made in them to formulate mathematically certain very deep and apparently essential requirements, namely:

1. Invariance with respect to some fundamental group, naturally including the Lorentz group.
2. The spectral condition, as it is called. It is assumed that there exists a complete system of states of the dynamical system with positive energies; the only state with zero energy is the vacuum state.
3. Unitarity of the scattering operator.
4. The condition of localization and micro-causality. This last condition requires the explicit introduction of space-time, which can be done in various ways. We consider one of them here.

Let us introduce the classical external field ("arbitrary" functions of space-time) together with the corresponding quantized fields, and regard the S-operator as a functional of these classical fields:

$$S=S(\varphi).$$

Then the dynamical quantities will become expressible in terms of the variational derivatives:

$$\frac{\delta S}{\delta\varphi_{x_1}\ldots\delta\varphi_{x_s}}S^{+}.$$

The condition is imposed that their averages are generalized functions of slow growth.

The condition of micro-causality is represented in the form

$$\frac{\delta}{\delta\varphi_y}\left(\frac{\delta S}{\delta\varphi_x}S^{+}\right)=0 \quad \text{for} \quad y\lesssim x.$$

On the basis of these conditions we can establish the appropriate properties of analyticity of the functions of two complex variables appearing in the scattering amplitude of the two particles.

The region of analyticity obtained here is rather restricted, but completely adequate for the establishment of the dispersion relations in a number of physically important situations. Dispersion relations have been much applied in very varied questions, in particular in the approximate calculation of the different processes of collisions of particles.

We wish, however, to emphasize that the greatest interest in the situation would arise if the dispersion relations accessible to experimental test were contradicted as the result of such an experiment. For this would signify that experiment had come into contradiction with the most fundamental notions of quantum field theory, since these relations are a direct consequence of them.

We also have such a situation with what are called the asymptotic relations between the effective cross-sections, which are established, starting from the properties of analyticity of the scattering amplitude, with the aid of the Phragmén-Lindelöf theorem.

In my lecture I have not been able to deal with a new and extremely rapidly developing and promising line of research, in which the groups SU_3 and SU_6 are applied to the study of the properties of elementary particles and resonances. Such a group approach has already given many important results. So far it is still in essence of a purely kinematical nature. It would now be very interesting to construct corresponding dynamical models and to elucidate in them the nature of the group laws and regularities that have been established.

Translated by R.O. Davies

Alcksandr Danilovich Aleksandrov

Born – 4 Aug 1912 in Volyn, Ryazan, Russia
Died – 27 July 1999

Aleksandr Aleksandrov's father was the headmaster of a secondary school in St Petersburg and his mother was a teacher at the same school. In fact, although he was born in the village of Volyn, he lived in St Petersburg from a very young age. St Petersburg changed its name to Petrograd when Aleksandrov was two years old. Aleksandrov attended school in Petrograd but while he was at school the name of the city in which he lived changed yet again and so by the time he left school in 1928 it was a Leningrad school from which he graduated.

When Aleksandrov left school he did not intend to study mathematics but rather his interests were in physics. Therefore when he entered Leningrad University in 1929 he set out on a theoretical physics course in the Faculty of Physics. In 1930, while still only 18 years of age, he began original work on optics in the Optics Institute. However Aleksandrov was taught mathematics in the Faculty of Physics by B. N. Delone. Delone's interests in the geometry of numbers and the structure of crystals soon began to attract Aleksandrov at least as much as his work in physics which was supervised by V. A. Fok.

In 1932 Aleksandrov moved from the Optics Institute to Physics Research Institute of Leningrad University where he worked on the theoretical side of the subject. He graduated with a degree in theoretical physics in 1933 and continued his research, working with two supervisors in Fok and Delone. The influence of these two are clearly seen in Aleksandrov's first few publications which appeared in 1933 and 1934 and represented research largely carried out while he was still an undergraduate.

While continuing to work at the Physics Research Institute, Aleksandrov began to teach in the Faculty of Mathematics and Mechanics from 1933. Two years later he presented a thesis on geometry for a Master's degree which was on the topic of mixed volumes of convex bodies and he published six papers over the next couple of years on the results of this thesis. This work generalizes classical problems in differential geometry. Aleksandrov's doctoral thesis (more of a habilitation thesis) was presented in 1937 and in it he studied the topics of additive set functions and the geometrical theory of weak convergence.

Aleksandrov was appointed as Professor of Geometry at Leningrad University in 1937. He was also appointed to the Steklov Mathematical Institute of the Russian Academy of Sciences which, as a part of the Steklov Institute of Physics and Mathematics had been set up in the early 1920s. In 1934 the Steklov Mathematical Institute had been set up but moved to Moscow and Delone, Aleksandrov's supervisor, had moved with it to head the Algebra Department. In 1940 the Leningrad Branch of the Mathematical Institute was founded and it had among its members Aleksandrov, Kantorovich, Linnik, and Faddeev. However the Moscow part of the Steklov Mathematical Institute was moved to Kazan at the beginning of World War II and, in 1942 Aleksandrov went to Kazan to continue his research within the Mathematical Institute.

In 1944 Aleksandrov returned to the University of Leningrad where he was Professor of Geometry. In 1952 he became Rector of the University of Leningrad. It was a period in which he worked hard to recreate the mathematical activity in Leningrad which had been associated with the St Petersburg Mathematical Society.

The St Petersburg Mathematical Society was founded in 1890 and was the third oldest mathematical society in Russia (Moscow founded 1867 and the Kharkov founded 1879 are older). It had ceased to exist in 1917 due to the Revolution, but was recreated after initiatives from Steklov as the Petrograd Physical and Mathematical Society in 1921. Both of Aleksandrov's supervisors, Fok and Delone, played major roles in the Physical and Mathematical Society. However the Society was again closed down due to political pressure. Then Smirnov organised the Leningrad Mathematical Seminar in 1953 which went some way to filling the gap left but both Aleksandrov and Smirnov worked hard to restart the Leningrad Mathematical Society. They succeeded in 1959 when the Leningrad Mathematical Society again began to hold meetings.

In 1964 Aleksandrov left Leningrad and moved to Novosibirsk where he was appointed as Head of the Department of Geometry of the University of Novosibirsk. He also became Head of the Department of Geometry of the Mathematical Institute of the Siberian Branch of the Academy of Sciences of the USSR.

Aleksandrov's work in physics did not stop in his student days. He published on optics, quantum mechanics, and relativity. He often lectured on the history of mathematical ideas, a topic which greatly fascinated him. In addition he wrote encyclopaedia articles and wrote chapters on methodology.

Aleksandrov received many awards for his major contributions to geometry. In 1942 he received the State Prize for his work in geometry, then in 1946 he was elected a Corresponding Member of the Academy of Sciences. In 1951 he received the international Lobachevsky Prize.

Comptes Rendus (Doklady) de l'Académie des Sciences de l'URSS
1937. Volume XIV, № 4

MATHEMATIK

NEUE UNGLEICHUNGEN FÜR DIE MISCHVOLUMEN KONVEXER KÖRPER

Von A. ALEXANDROV

(Vorgelegt von I. M. Vinogradow, Mitglied der Akademie, d. 10. XI. 1936)

Es seien im n-dimensionalen Raume N Richtungen $\bar{n}_1, \ldots, \bar{n}_N$ derart gegeben, dass die zu ihnen senkrechte Ebenen $P_1, \ldots, P_N$ ein konvexes Vielflach Π begrenzen. Die Ebene P_i bezeichnen wir als wesentlich Begrenzende, wenn sie auf Π eine Fläche von nicht-nulltem $(n-1)$-dimensionalem Volumen gibt. Zwei Vielecke, die ein und dasselbe Normalensystem $\bar{n}_1, \ldots, \bar{n}_N$ haben, werden wir als analoge bezeichnen, wenn die Gerade P_i für das eine sowohl wie für das andere wesentlich Begrenzende ist. Zwei Vielflache heissen analog, wenn sie das gleiche Normalensystem besitzen und alle ihre Flächen von nicht-nulltem $(n-1)$-dimensionalem Volumen sind und deren entsprechende Flächen analog sind. Die Abstände H_i vom Ursprung O bis zu den begrenzenden Flächen P_i nennen wir Stützzahlen des Vielflaches. Wenn wir als Ursprung in P_i die Projektion des Ursprungs O auf dieselbe nehmen, dann ist der Abstand von demselben bis zum Schnitt von P_i mit P_k (Θ_{ik} bedeutet der Winkel zwischen $\bar{n}_i$ und $\bar{n}_k$):

$$H_{ik} = \frac{H_k - H_i \cos\Theta_{ik}}{\sin\Theta_{ik}}. \tag{1}$$

Das sind die Stützzahlen der i-ten Fläche.

Der Mischung der analogen Vielflache entspricht das Summieren ihrer Stützzahlen. Für die Vielecke ist diese Behauptung offenbar. Aus (1) ist klar, dass bei dem Summieren der Stützzahlen der Vielflache sich auch die Stützzahlen ihrer Flächen summieren. Ausserdem ist bekannt, dass die Fläche eines gemischten Vielflaches das Resultat der Mischung paralleler Flächen ist. Dabei ist unsere Behauptung, wenn sie für $(n-1)$-dimensionale Vielflache richtig ist, auch für n-dimensionale richtig.

Es sei V das Volumen eines Vielflachs, das durch Mischung analoger Vielflache $H^{(1)}, \ldots, H^{(n)}$ erhalten worden ist, und F_i — die Volumen seiner Flächen.

Dann erhalten wir, wenn wir mit $V(H^{(j_1)} \ldots H^{(j_n)})$ die Mischvolumen bezeichnen:

$$V = \sum_{j_1 \ldots j_n} \lambda_{j_1} \ldots \lambda_{j_n} V(H^{(j_1)} \ldots H^{(j_n)}) = \frac{1}{n} \sum_{i=1}^{N} \Big(\sum_{k=1}^{n} \lambda_k H_i^{(k)} \Big) F_i, \tag{2}$$

und da seinerseits

$$F_i = \sum_{j_1 \ldots j_{n-1}} \lambda_{j_1} \ldots \lambda_{j_{n-1}} F_i (H^{(j_1)} \ldots H^{(j_{n-1})}), \tag{3}$$

$$V = \sum_{j_1 \ldots j_n} \lambda_{j_1} \ldots \lambda_{j_n} \sum_{i=1}^{N} H_i^{(j_n)} F_i (H^{(j_1)} \ldots H^{(j_{n-1})}). \tag{4}$$

Wir zeigen, dass unabhängig von der Reihenfolge der Indices $j_1 \ldots j_n$:

$$V(H^{(j_1)} \ldots H^{(j_n)}) = \frac{1}{n} \sum_{i=1}^{N} H_i^{(j_n)} F_i (H^{(j_1)} \ldots H^{(j_{n-1})}). \tag{5}$$

Wir nehmen an, dass dies zutrifft für den $(n-1)$-dimensionalen Raum. Dann ist

$$F_i (H^{(j_1)} \ldots H^{(j_{n-1})}) = \frac{1}{n-1} \sum_{k=1}^{N} H_{ik}^{(j_{n-1})} F_{ik} (H^{(j_1)} \ldots H^{(j_{n-2})}), \tag{6}$$

$$F_i (H^{(j_1)} \ldots H^{(j_n)}) = \frac{1}{n-1} \sum_{k=1}^{N} H_{ik}^{(j_n)} F_{ik} (H^{(j_1)} \ldots H^{(j_{n-2})}). \tag{7}$$

Indem wir hier die Werte H_{ik} aus (1) einsetzen, (6) mit $H_i^{(j_n)}$ und (7) mit $H_i^{(j_{n-1})}$ multiplizieren, und über i summieren, erhalten wir:

$$\sum_{i=1}^{N} H_i^{(j_n)} F_i (H^{(j_1)} \ldots H^{(j_{n-1})}) = \sum_{i=1}^{N} H_i^{(j_{n-1})} F_i (H^{(j_1)} \ldots H^{(j_n)}). \tag{8}$$

Hieraus ergibt sich im Zusammenhang mit Formel (4) Formel (5). Man kann leicht zeigen, dass der Beweis für die Ungleichung:

$$V(H^{(1)} \ldots H^{(n-1)} H^{(n)})^2 \geqslant V(H^{(1)} \ldots H^{n-1} H^{n-1}) V(H^{(1)} \ldots H^{(n)} H^{(n)}), \tag{9}$$

gleichwertig (im Falle der analogen Vielflache) ist dem Beweis, dass unter der Bedingung:

$$V(H^{(1)} \ldots H^{(n-1)} Z) = 0, \tag{10}$$

die quadratische Form der Veränderlichen Z_i:

$$V(H^{(1)} \ldots H^{(n-2)} ZZ) \leqslant 0, \tag{11}$$

wo das Gleichheitszeichen nur bei einer linearen Stützfunktion Z steht.

Die Überführung der Form (10) in die kanonische Form unter der Bedingung:

$$\sum_{i=1}^{N} Z_i^2 \frac{F_i (H^{(1)} \ldots H^{(n-1)})}{H_i^{(n-1)}} = 1 \tag{12}$$

führt zur Lösung der Gleichungen:

$$F_i (H^{(1)} \ldots H^{(n-2)} Z) = \lambda \frac{F_i (H^{(1)} \ldots H^{(n-1)})}{H_i^{(n-1)}} Z_i. \tag{13}$$

Die Lösung für $\lambda = 1$, $Z = H^{(n-1)}$ spielt wegen der Bedingung (10) keine Rolle. Für $\lambda = 0$, wenn $Z = H - H^{(n-2)}$, wo H ein analoges Vielflach ist, erhalten wir:

$$F_i(H^{(1)} \dots H^{(n-2)} H) = F_i(H^{(1)} \dots H^{(n-2)} H^{(n-2)}). \tag{14}$$

Unser Theorem führt für den Fall von Vielecken zum Brunn-Minkowskischen Theorem. Ist das Theorem richtig für $(n-1)$-dimensionale Vielflache, so erhalten wir:

$$F_i H^{(1)} \dots H^{(n-2)} H) \geqslant F_i(H^{(1)} \dots HH), \tag{15}$$

oder

$$F_i(H^{(1)} \dots H^{(n-3)} ZZ) \leqslant 0. \tag{16}$$

Aus (8) erhalten wir, dass

$$\sum_{i=1}^{N} H_i^{(n-2)} F_i(H^{(1)} \dots ZZ) = \sum_{i=1}^{N} Z_i F_i(H^{(1)} \dots H^{(n-2)} Z) = 0, \tag{17}$$

wegen (13) für $\lambda = 0$, und da $H_i^{(n-1)} > 0$, so ist nach (16):

$$F_i(H^{(1)} \dots H^{(n-3)} ZZ) = 0. \tag{18}$$

Folglich ist Z eine lineare Stützfunktion, so dass das Gleichheitszeichen in (11) unter der Bedingung (10) nur für ein solches Z möglich ist.

Für $H^{(1)} = \dots = H^{(n-2)}$ stellt die Ungleichung (9) die bekannte Minkowskische Ungleichung dar. Daher ist in diesem Falle $\lambda = 1$ der einzige positive λ-Wert. Aber wir haben gezeigt, dass $\lambda = 0$ immer denselben Lösungen von (13) entspricht.

Daher sehen wir, wenn wir stetig von gleichen $H^{(1)}, \dots, H^{(n-2)}$ zu verschiedenen übergehen, dass dies auch für den allgemeinen Fall richtig ist, woraus auch unser Theorem folgt.

Die Übertragung der Ungleichung (9) auf beliebige konvexe Körper gewinnt man durch folgendes Theorem: Man kann beliebige n nichtentartete konwexe Körper mit erwünschter Genauigkeit durch analoge Vielflache approximieren. Die Erbringung des Beweises dieser einfachen Satzes unterlassen wir (¹).

Mathematisches Institut der Universität.
Leningrad.

Eingegangen
d. 10. XI. 1936.

ZITIERTE LITERATUR

¹ Bonnesen u. Fenchel, Theorie der konvexen Körper, Springer (1934).

ALEKSANDER DANILOVICH ALEKSANDROV
(On his fiftieth birthday)

August 4th, 1962 was the fiftieth birthday of Aleksander Danilovich Aleksandrov, the leading contemporary geometer. His ideas, results and methods have enriched geometry and increased its influence on other branches of mathematics.

Aleksandrov was born in the village of Volyn' in the Ryazan district, but from early childhood lived in Leningrad where his father was head-master and his mother a teacher at a secondary school. He left school in 1928, and in 1929 entered the Faculty of Physics at the University of Leningrad from which he graduated four years later, specializing in theoretical physics. In 1930 he began original work as a scientific and technical worker in the Optics Institute, and in 1932 moved to the theoretical division of the Physics Research Institute at the University of Leningrad. Here he worked until 1936 and, after 1933, also taught in the Faculty of Mathematics and Mechanics of the University. In physics Aleksandrov was a pupil of V.A. Fok. His work in physics deals with optics [9], [17], quantum mechanics [4], [5], [90], [91], relativity [79], [97], [98], [117], and also with problems of a methodological character [73], [74], [99], [112], [148], [149], [150].

In mathematics Aleksandrov was a pupil of B.N. Delone, who was then teaching mathematics in the Faculty of Physics. Under his influence Aleksandrov completed his first geometrical work while still a student. In 1935 he presented his thesis for the Candidate's degree in geometry, and from 1936 his work was entirely mathematical. In 1937 he presented his doctoral thesis. He became acting Professor of Geometry at the University of Leningrad in 1937, and from 1938 worked with his old colleagues in the Leningrad branch of the V.A. Steklov Mathematical Institute of the Academy of Sciences of the USSR. Between 1942 and 1944 he went to Kazan with the Institute where he continued intensive scientific research. He was awarded the State Prize for his work in goemetry in 1942. Since 1944 he has been Professor of Geometry at the University of Leningrad. In 1946 he was elected a Corresponding Member of the Academy of Sciences of the USSR. In 1951 he was awarded the international Lobachevskii Prize. Sine 1952 he has been Rector of the University of Leningrad.

Aleksandrov's early work in geometry was concerned with mathematical crystallography and regular decompositions of space [1], [2], [3], [6], [10]. (Related questions were touched on later in [14] and [101].) Then followed his first work on the theory of surfaces [7], [8], [11], substantial advances in the theory of polyhedra [13], [23], [31] and his first investigations on general convex surfaces [18], [21], [22], [24], [26]. He also perfected analytical methods of investigation. An important series of papers on mixed volumes [12], [15], [16], [19], [20], [25] formed the subject of his Candidate's dissertation, and his doctoral dissertation was concerned with additive set functions and the geometrical theory of weak convergence [27], [30], [37]; [32], [55], [66] are also concerned with this topic. In his work on mixed volumes Aleksandrov was already developing the first generalized solutions of classical problems of differential geometry in the large.

This was followed by an immense series of papers on the theory of convex surfaces [28], [29], [33], [34], [36], [38 - 51], [54], [60], [64], [66], [78], [105], [110], [118], the general theory of curves [52], [56], [177], and the foundations of the general theory of non-regular surfaces [60], [62], [63], [69], [70], [76], [77], [82], [96], [103], [120], [126], [128], [157], [176], [177]. From these subjects Aleksandrov turned to related questions of geometry and partial differential equations [102], [119], [121], [127], [135 - 139], [145 - 147], [153 - 156], [163 - 166], [178].

The results of these geometrical investigations form an extensive and profound new theory. The paramount significance of this theory became indisputable later when it was applied to many difficult concrete problems (both by Aleksandrov and other authors).

To see the importance of Aleksandrov's work in contemporary geometry, let us recall the position of the metric theory of surfaces at the beginning of the century. Methods for local problems concerned with regular surfaces were already fully developed; they could be reduced to problems of analysis in a sufficiently general form and solved by analytical methods. The position of global problems, or problems of geometry in the large, was quite different. These problems are concerned with the relation between properties of the surface as a whole and its differential properties at each of its points. Both geometry and analysis seemed unable to solve such problems. For example, the classical problem of the unique definition of the ovaloid and only partially solved after persistent attempts by first-class mathematicians (Liebmann, Minkowski, Hilbert, Weyl, Blaschke). At the beginning of the thirties Cohn-Vossen finally gave the first solution of the problem, proving that isometric ovaloids were equal, but this was not a definitive solution, because the ovaloids were required to be three times differentiable. Cauchy, among other, had already proved the equality of isometric closed convex polyhedra. It seemed obvious that these two results were special cases of a general theorem on the equality of arbitrary (in general, non-regular) ovaloids. No method of approaching such a theorem was known. On the other hand, there seemed to be no method of proving the possibility of an isometric mapping (deformation) of an ovaloid after cutting off some part of it. What is more, there was no way of estimating the degree of deformability of a partial ovaloid. If sufficient regularity is assumed, these problems can easily be expressed as differential equations, but some of these equations are non-linear, and the theory of such equations was in a rudimentary state. In this case analysis was unable to help geometry.

It should be noted that these geometrical problems were being studied by a wide circle of mathematicians who made great efforts to develop the necessary methods of analysis. For example we might mention the work on Weyl's problem on the existence of closed convex surfaces with a given metric (defined on a topological sphere). The solution of this problem was outlined in general terms by Weyl himself in 1916, and completed in the thirties by H. Lewy, using very delicate results of his own in the analytic theory of the Monge-Ampere equation. However in Lewy's work the analysis was developed independently of the geometry and applied as a ready-made tool from outside. In most other papers in this direction the matter was reduced to a device of solving individual problems. The search for such methods, usually highly ingenious, was rarely successful and few results were obtained with their help. In fact, no one at that time penetrated to the essentials of the problem, and many other problems of deformation of surfaces and geometry in the large remained inaccessible. It was quite clear that completely new methods of attack were needed before further development of this branch of geometry was possible. Aleksandrov approached the search for such methods in the general spirit of the development of contemporary mathematics.

It is well known that the advances in general analysis are essentially due to extending the notion of the "objects" studied. Aleksandrov's achievement is to carry out a similar extension in geometry and to leave analysis and the theory of differential equations behind. He approached the differential geometry of surfaces in this way, extending the class of regular convex surfaces to the class of all convex surfaces (such a surface being defined as the boundary of an arbitrary convex body). In order to solve concrete problems Aleksandrov had to replace the Gaussian geometry of regular surfaces by a much more general theory. In the first place the intrinsic properties (i.e., those properties that appear as a result of measurements carried out on the surface) of an arbitrary convex surface had to be studied, and methods found for the proof of theorems on the connection between intrinsic and exterior properties of convex surfaces. Aleksandrov constructed a theory of intrinsic geometry of convex surfaces on this basis. Because of the depth of this theory, the importance of its applications and the breadth of its generality, Aleksandrov comes second only to Gauss in the history of the development of the theory of surfaces. Aleksandrov constructed the intrinsic geometry of convex (and also arbitrary) surfaces from the general concept of a metric space. Let F be a connected metric surface so that any two points X and Y can be joined by a rectifiable arc in F. Then an intrinsic metric is defined on F by letting $\rho(X,Y)$ be the greatest lower bound of the lengths of the arcs joining X and Y. The surface F with the metric $\rho(X,Y)$ forms a two-dimensiona metric space, and it is this space that is studied in the theory of the intrinsic geometry of F. Now let R be a given abstract two-dimensional metric space. Among all such spaces Aleksandrov singled out the class of spaces with intrinsic metric, on which the distance between two arbitrary points is the greatest lower bound of the length of the arcs joining X

and Y. In order that R be isometric to an arbitrary surface, its metric must necessarily be intrinsic. He constructed the geometry of the space R with an intrinsic metric. In R it is natural to define the concepts of geodesic (between two points), triangle, polygon and polygonal arc (constructed from geodesics). The definition of the angle between two curves is non-trivial. The concept of angle is basic to the theory, and was made by Aleksandrov the object of a very deep study.

After the concept of angle has been introduced, the excess of a triangle, $\alpha + \beta + \gamma - \pi$, where α, β, γ are the angles between its sides, is defined. This makes it possible to generalize the definition of positive curvature for two-dimensional spaces: R is called a space with positive curvature if every point of R has a neighbourhood such that the excess of every triangle contained in it is non-negative. Every convex surface is a surface of positive curvature.

The construction of a theory of curvature is related to many important problems in the theory of surfaces. Aleksandrov regards curvature as an additive set function. It is natural to take the area of the spherical image of a set M on a convex surface as its external curvature. Curvature in the intrinsic geometry of convex surfaces was defined by Aleksandrov firstly for three "fundamental" sets: (i) the curvature of an open triangle was taken as its excess; (ii) the curvature of an open geodesic was zero; (iii) the curvature of a point was $2\pi - \theta$, where θ is the whole angle round the point. From this, curvature as an additive set function is uniquely defined on all Borel sets. Aleksandrov proved the theorem: the curvature of every Borel set on a convex surface is the area of its spherical image. In this way Gauss's *theorema egregium* was generalized to arbitrary convex surfaces. It also established a fundamental connection between the intrinsic geometry and the external properties of the surface. A number of important concrete conclusions followed from this. In particular, the specific curvature of a region on a convex surface (i.e., the ratio of the area of the spherical image to the area of the region itself) is invariant under isometric mappings of the convex surfaces. Thus, the class of convex surfaces with bounded specific curvature, introduced by Aleksandrov, could be defined in intrinsic terms. Aleksandrov proved that every complete convex surface with bounded specific curvature (whose metric is not everywhere euclidean) is smooth, and an incomplete surface can depart from smoothness only by having a rectilinear edge whose ends lie on the boundaries of the surface. This was the first result on the degree of regularity of a convex surface and its dependence on the regularity of its intrinsic metric.

In contrast to the analytic methods of the Gaussian theory of surfaces, Aleksandrov applied methods of a geometrical character. The basic method is that of approximation by the so-called manifolds with polyhedral metric. (In external geometry this corresponds to approximating to a general convex surface by convex polyhedra.) This simple and natural approximation enables the solution of a problem for polyhedra to be generalized by passage to the limit. Thus the unified theory of curvature for convex surfaces which we mentioned above was created.

Let P be a closed convex polyhedron. We imagine that it is divided by polygonal arcs into parts each of which may be developed and placed on a plane polygon; let these plane polygons be $G_1, \ldots, G_n$. The system of polygons $G_1, \ldots, G_n$ with a given identification of their vertices and sides is called a development of the polyhedron P. It is obvious that a development satisfies the following conditions: (i) the complex $G_1, \ldots, G_n$ is homeomorphic to a sphere; (ii) the identified sides are equal; (iii) at the identified vertices, the sum of the angles is less than or equal to 2π. Now suppose that we are given a complex of euclidean polygons $G_1, \ldots, G_n$, satisfying the above conditions; in other words, we are given the development of an abstract polyhedron with a metric of positive curvature. Aleksandrov proved the "sewing theorem" for a polyhedron which states: every development may be sewn together to form a closed convex polyhedron (the polygons $G_1, \ldots, G_n$ must in general be folded along a definite, but previously unknown, line). This theorem solved Weyl's problem on polyhedral metrics; it was proved by a completely general method developed by Aleksandrov, based on the topological theorem of the invariance of domain. The same method was used by Aleksandrov to solve Minkowski's problem on the existence of a convex polyhedron with given area and directions of its faces, to prove a theorem on the existence of a polyhedron with given curvatures and many other propositions. This method indicates a clear way to look for further theorems of this type.

On the basis of the sewing theorems, it was remarkably easy to solve Weyl's problem under very general assumptions about the given metric. In fact, Aleksandrov proved that if a two-dimensional space of positive curvature is homeomorphic to a sphere, then it is isometric with some closed convex surface. Aleksandrov's solution of Weyl's problem illustrated the power of purely geometric methods. Further, he used the proof of this theorem as a stepping-stone in a profound study of problems of deformation of convex surfaces. What we have in mind is his "sewing method" based on his theorem known as the "general sewing theorem", which greatly generalizes the sewing theorem for polyhedra. We shall give an outline of this in very general terms.

Let $G_1, \ldots, G_n$ be regions in a two-dimensional space of positive curvature. We imagine that these regions are cut out from the space and certain portions of their boundaries are identified ("sewn") in such a way that the result is a two dimensional manifold. If the identified portions are of equal length, then an intrinsic metric is defined in a natural way on this manifold. Aleksandrov formulated necessary and sufficient conditions for the manifold to be of positive curvature; these conditions are simple, natural and effective. In particular, if the sewn manifold is homeomorphic to a sphere, then it may be realized as a closed convex surface. He also proved an extraordinarily general theorem on the existence of convex surfaces sewn together from pieces of abstractly given manifolds or pieces of given convex surfaces. By using the sewing method Aleksandrov proved the local theorem that each point of a two-dimensional manifold R has a neighbourhood isometric with a convex surface if and only if R is a space of positive curvature. He solved the fundamental problem of the intrinsic geometry of convex surfaces by giving an exhaustive intrinsic characterization of convex surfaces. The sewing method can be used to prove that after an arbitrary piece is cut from an ovaloid, the remaining portion admits deformations (it is sufficient to sew the ovaloid together *in abstracto* and to change the form of the sewn-up piece.) For example, it can be shown that bending a half-ellipsoid may "close" it and turn it into a closed convex surface. In principle, the sewing method, together with theorems on realizability of a convex, abstractly given surface with abstractly given convex metric, enables all the basic problems on the bending of surfaces to be solved. The method has also successfully been applied to other problems, in particular the solution of various extremal problems. There are many effective examples of the variety of applications of Aleksandrov's sewing method.

Aleksandrov's ideas and methods were an essential preparation for the researches of A.V. Pogorelov who completely solved the problem of the unique definition of the ovaloid and then that of the connection between the regularity of convex surfaces and the regularity of their metrics. As a consequence of this, the sewing method became a method of classical differential goemetry. More precisely, if the sewing method is applied to the deformation of a regular surface of positive curvature, then the surface obtained after deformation will be regular: from an analytic surface an analytic surface is obtained. Now analysis was no longer being applied to geometry, but geometry was being used to obtain conclusions of an analytic character.

Aleksandrov's work was not confined to convex surfaces. Using methods of his own, he created a general theory of metric manifolds, and in particular of generalized Riemannian manifolds [176]. The two-dimensional manifolds of bounded curvature introduced by Aleksandrov exhaust the class of all metric two-dimensional manifolds that allow approximation by Riemannian manifolds without increase of general integral curvature. The generality of his methods is evident from the fact that they can be applied equally to regular surfaces, arbitrary polyhedral surfaces, to all convex surfaces and to a remarkably wide class of non-convex non-regular surfaces. Many problems in the classical theory of surfaces can be simply and elegantly solved by this theory. The close connection between the theory of surfaces and partial differential equations is well-known. The problems of geometry in the large for the classical, regular, surfaces may be formulated as problems on the solubility and uniqueness of solution of the corresponding differential equations, and conversely, the solution of geometrical problems gives results about the corresponding differential equations. In particular, Aleksandrov's results on the deformation of convex surfaces provide various theorems on the solubility of boundary-value problems for the so-called Darboux equations related to equations of Monge-Ampère type.

The new synthetic methods of research in geometry in the large throw light on new directions of purely geometrical research into general Monge-Ampère equations. Aleksandrov took the initiative in this in connection with his theorem on the existence of convex surfaces with given curvature. These researches were successfully carried out by Aleksandrov and his pupils for a number of years and led to the construction of an elegant and concise theory of Monge-Ampère equations and their higher-dimensional analogues.

Aleksandrov made notable investigations on a general method of attacking questions of uniqueness for convex bodies and surfaces and for the solutions of differential equations. This led to theorems about the equality of bodies or surfaces satisfying certain conditions of a local character. Aleksandrov both developed earlier methods and created new ones, in particular a method based on a very general form of the maximum principle. Aleksandrov's results in this field are notable for their great generality as well as for a striking simplicity and completeness.

Aleksandrov has a remarkable capacity for scientific work and great versatility. His work has not only been published in separate notes, but also in the form of detailed monographs [3], [60], [75], [176], [177], and unified series of articles ([15], [16], [19], [20], [25]), ([27], [30], [37]), ([119], [127], [136] - [138], [145], [154]), ([139], ·[146], [147], [155], [156], [163], [164]).

Aleksandrov is a very active and demanding supervisor and has built up a fine school of geometers. His pupils carry on teaching and research in many towns at home and abroad. He has given original courses of lectures not only on all types of geometry, both general and specialized, but also on the history of mathematics, or, more precisely, the history of mathematical ideas. He is continually active in spreading scientific ideas. His survey articles include [57] - [59], [61], [92], and the reports [103], [117], [128], [157], [178]. He is a member of the editorial board of several journals and is one of the editors of, and author of a number of articles [83], [84], [92] - [94], [104], [111], [112], [129] in, the Great Soviet Encyclopedia. He is also the chief editor of the Leningrad University Gazette to which he devotes special attention. He has taken part in many discussions on methodology, and is editor and author of several chapters [113] - [115] in a monograph of methodological character, and the author of a great many interesting methodological and controversial articles, many of which, like his scientific work, have been translated intc many languages. Aleksandrov invariably takes part in Youth meetings, Komsomol debates, is a warm supporter of his country and Soviet science and its honoured representative at scientific congresses in many countries.

Aleksandrov is also very active in public and adminstrative work. He is a member of the local and regional Leningrad district committee of the Soviet Communist Party, vice-president of the Italy-USSR Friendship Society, for a number of years was chairman of the Leningrad division of the Society for the Dissemination of Political and Scientific Knowledge, a Deputy of the City Soviet, Deputy of the Supreme Soviet of the Russian Socialist Federal Soviet Republic, President of the Fourth All-Union Mathematical Congress, and finally for the last ten years has been Rector of the University of Leningrad. In spite of all these duties, Aleksandrov also finds the energy to go mountaineering and is a Master of Sport and spent his fiftieth birthday as a member of a mountaineering expedition in Pamir.

N.V. Efimov, V.A. Zalgaller, A.V. Pogorelov.

List of Aleksandrov's Published Work

1933

1. A theorem on convex polyhedra. Trudy Mat. Inst. Steklov, 4, 87.
2. An elementary proof of the existence of a centre of symmetry in a three-dimensional convex parallelohedron. Trudy Mat. Inst. Steklov, 4, 89-99.

1934

3. *Matematicheskie osnovy strukturnogo analiza kristallov* (Mathematical foundations of the structural analysis of crystals). Moscow-Leningrad, Gostekhizdat, 328 pp. (with B.N. Delone and N.N. Padurov).
4. On the calculation of the energy of a bi-valent atom by Fok's method. Zhurn. eksperiment. i teoretich. fiz., 4, 326-341.
5. Remark on the commutation rule in Schrödinger's equation. Dokl. Akad. Nauk SSSR, 4, 198-200.
6. The derivation of a four-dimensional non-normal parallelohedron. Izv. Akad. Nauk SSSR, ser. fiz.-mat., No. 6, 803-817.

1935

7. A new proof of the non-deformability of the surface of a ball. Dokl. Akad. Nauk SSSR, 1, 353-356.

1936

8. On infinitesimal deformations of non-regular surfaces. Mat. Sb. 1(43), 307-322.
9. The dispersion of light in an infinite flat layer. Trudy Gos. Optich. Inst., 11., 56-71 (with N.G. Boldyrev).
10. On four-dimensional non-normal parallelohedra. In *Trudy II Vsesoyuznogo matematicheskogo s" ezda* (Proceedings of the Second All-Union Mathematical Congress), Leningrad, 24th-30th June 1934, Vol. 2, Sectional Reports. Moscow-Leningrad. Page 21.

1937

11. On the question of the existence of a convex body in which the sum of the principal radii of curvature is a given positive function satisfying closure conditions. Dokl. Akad. Nauk SSSR 14, 59-60.
12. New inequalities for mixed volumes of convex bodies. Dokl. Akad. Nauk SSSR, 14, 155-157.
13. An elementary proof of Minkowski's theorem and some other theorems on convex polyhedra. Izv. Akad. Nauk, Ot. Mat. Enzh. Nauk, Ser. Mat. No. 4, 597-608.
14. On the decomposition and covering of surfaces. Mat. Sb. 2(44), 307-18.
15. On the theory of mixed volumes of convex bodies, I. Mat. Sb. 2(44), 947-72.
16. On the theory of mixed volumes of convex bodies, II. New inequalities between mixed volumes and their applications. Mat. Sb. 2(44), 1205-38.
17. Errors in colorimetric measurements and the metrics of coloured space. zh. eksperiment. i. teoretich. fiz. 7, 785-791.

1938

18. A general uniqueness theorem for closed surfaces. Dokl. Akad. Nauk SSSR, 19, 233-236.
19. On the theory of mixed volumes of convex bodies, III. The extension of two theorems of Minkowski on convex polyhedra to arbitrary convex bodies. Mat. Sb. 3(45), 27-46.
20. On the theory of mixed volumes of convex bodies, IV. Mixed discriminants and mixed volumes. Mat. Sb. 3(45), 227-251.
21. On a class of closed surfaces. Mat. Sb. 4(46), 69-77.

1939
22. On uniqueness theorems for closed surfaces. Dokl. Akad. Nauk SSSR, 22, 99-102.
23. The application of the theorem on the invariance of domain to existence proofs. Izv. Akad. Nauk SSSR Ser. Mat, No. 3, 243-256.
24. On convex surfaces with flat-edged shadows. Mat. Sb. 5(47), 309-16.
25. On surface functions of a convex body (a note to the paper "On the theory of mixed volumes of convex bodies"). Mat. Sb. 6 (48), 167-174.
26. The existence almost everywhere of the second differential of a convex function and some connections between it and properties of convex surfaces. Leningrad Gos. Univ. Uch. Zap. Ser. Mat. Nauk, No. 37, pt. 6, 3-35.

1940
27. Additive set-functions in abstract spaces, I. (In English, with Russian summary). Mat. Sb. 8 (50), 307-48.

1941
28. The existence of convex polyhedra and convex surfaces with given metric. Dokl. Akad. Nauk SSSR, 30, 103-106.
29. The intrinsic geometry of an arbitrary convex surface. Dokl. Akad. Nauk SSSR, 32, 467-470.
30. Additive set-functions in abstract spaces, II-III. (In English, with Russian summary) Mat. Sb. 9 (51), 563-628.
31. The theory of polyhedra. Sovetskaya nauka. No. 4, 97-117.

1942
32. On groups with invariant measure. Dokl. Akad. Nauk SSSR, 34, 7-11.
33. The existence and uniqueness of a convex surface with given integral curvature. Dokl. Akad. Nauk SSSR, 35, 143-147.
34. The smoothness of a convex surface with bounded Gaussian curvature. Dokl. Akad. Nauk SSSR, 36, 211-216.
35. On the extension of a Hausdorff space to an H-closed space. Dokl. Akad. Nauk SSSR, 37, 138-141.
36. The existence of convex polyhedra and convex surfaces with given metric. Mat. Sb. 11 (53), 1-2, 15-65.

1943
37. Additive set-functions in abstract spaces, IV. (In English, with Russian summary.) Mat. Sb. 13 (55), 2-3, 169-238.

1944
38. The intrinsic metric of a convex surface in a space of constant curvature. Dokl. Akad. Nauk SSSR, 45, 3-6.
39. Synthetic methods in the theory of surfaces. In *Nauchnaya Sessiya, posvyashchennaya 125-letiyu Leningradskogo Universiteta* (Scientific Meeting in commemoration of the 125th anniversary of Leningrad University), Reports of the mathematical section. Leningrad. 9-10.

1945
40. Isoperimetric inequalities on curved surfaces. Dokl. Akad. Nauk. SSSR, 47, 239-242.
41. Curves on convex surfaces. Dokl. Akad. Nauk SSSR, 47, 319-322.
42. On triangles on curved surfaces. Dokl. Akad. Nauk SSSR, 50, 19-22.
43. The curvature of convex surfaces. Dokl. Akad. Nauk SSSR, 50, 23-26.
44. Convex surfaces as surfaces of positive curvature. Dokl. Akad. Nauk SSSR, 50, 27-30.
45. An isoperimetric problem. Dokl. Akad. Nauk SSSR, 50, 31-34.
46. Complete convex surfaces in Lobachevskian space. Izv. Akad. Nauk SSSR Ser. Mat.

9, 113-120.
47. The intrinsic geometry of convex surfaces. In *Nauchnaya Sessiya . . .* (see [39]), page 7.

1946
48. On the metric of convex surfaces in a space of constant curvature. Dokl. Akad. Nauk SSSR, 51, 407-410.
49. On the gluing of convex surfaces. Dokl. Akad. Nauk SSSR, 54, 99-102.
50. The intrinsic geometry of convex surfaces. Uspekhi Mat. Nauk, 1, 13-14.
51. The foundations of the intrinsic geometry of surfaces. Nauchnyi Byulleten' Leningradskogo Gos. Univers., No. 7, 3-4.
52. The theory of curves based on approximation by polygonal arcs. In *Nauchnaya Sessiya Leningradskogo Universiteta* (Scientific meeting at Leningrad University), Reports of the Mathemetical Section. P. 11-12.
53. What is topology? Mat. v shkole, No. 1, 7-19.

1947
54. The securing method in the theory of surfaces. Dokl. Akad. Nauk SSSR, 57, 863-865.
55. On the work of S.E. Cohn-Vossen. Uspekhi Mat. Nauk, 2, 107-141.
56. The theory of curves based on approximation by polygonal arcs. Uspekhi Mat. Nauk, 2, 182-184.
57. Geometry and topology in the Soviet Union, I. Uspekhi Mat. Nauk, 2, 3-58.
58. Geometry and topology in the Soviet Union, II. Uspekhi Mat. Nauk, 5, 9-92.
59. Geometry in Leningrad University. Vestnik Leningrad. Univ., No. 11, 124-148.

1948
60. *Vnutrennyaya geometriya vypuklykh poverkhnostei* (The intrinsic geometry of convex surfaces). Moscow-Leningrad, Gostekhizdat, 387 pp. (translated into German with additions.)
61. Geometry "in the large". In *Matematika v SSSR za tridtsat' let 1917-1947* (Mathematics in the USSR in the thirty years 1917-1947). Moscow-Leningrad, Gostekhizdat, 919-938.
62. The foundations of the intrinsic geometry of surfaces. Dokl. Akad. Nauk SSSR, 60, 1483-1486.
63. Curves in manifolds of bounded curvature. Dokl. Akad. Nauk SSSR, 63, 349-352.
64. A generalization of a theorem of Herglotz. (Introductory paragraph to N.V. Efimov's article "Qualitative questions in the theory of the deformation of surfaces"). Uspekhi Mat. Nauk 3, 89-98.
65. On formalism in the mathematical sciences. Vestnik Leningrad. Univ., No. 12, 137-144.
66. Additive domain functions in the theory of convex surfaces. Leningrad Gos. Univ. Uch. Zap. Ser. Mat. Nauk, pt. 15, 82-100.
67. The intrinsic geometry of surfaces. In *Nauchnaya Sessiya . . .* (see [52]), p. 6-7.
68. Review of V.F. Kagan's book "*Osnovy teorii poverkhnostei v tenzornom izlozhenii*" (Foundations of the theory of surfaces in terms of tensors). Part 2, Moscow-Leningrad, Gostekhizdat. Sovetskaya Kniga, No. 9, 31-34.

1949
69. Quasigeodesics. Dokl. Akad. Nauk SSSR, 69, 717-720.
70. On surfaces representable by the difference of convex functions. Izv. Akad. Nauk Kazakh. SSR Ser. Mat. Mekh., 3, 3-20.
71. On the foundations of differential geometry and their presentation. Uspekh. Mat. Nauk 4, 139-170. (Also translated into Rumanian.)
72. Review of V.F. Kagan's book *Osnovy teorii poverkhnostei v tenzornom izlozhenii* (Foundations of the theory of surfaces in terms of tensors),

Part 1, Moscow-Leningrad, Gostekhizdat, 1947, Part 2, 1948. Uspekhi Mat. Nauk 4, 213-217.

73. Against idealism and muddle in the concepts of quantum mechanics. Vestnik Leningrad. Univ. No. 4, 48-68.
74. The principle of indeterminacy and the Party spirit in science. (a short report of a philosophical seminar, the content of the seminar: "A discussion of the philosophical content of the principle of indeterminacy in quantum mechanics"). Gazeta Leningrad. Universiteta, January 12th, p. 2-3.

1950

75. *Vypuklye mnogogranniki* (Convex polyhedra). Moscow-Leningrad, Gostekhizdat, 428 pp. (Also translated into German.)
76. Quasigeodesics on manifolds homeomorphic to the sphere. Dokl. Akad. Nauk SSSR, 70, 557-560.
77. Surfaces representable as the differences of convex functions. Dokl. Akad. Nauk. SSSR, 72, 613–616.
78. The unique definition of convex surfaces of revolution. Mat. Sb. 26 (68), 183-204, (with A.V. Pogorelov).
79. On the Lorentz transformation. Uspekhi Mat. Nauk, 5, 187.
80. On some general questions of research and teaching of mathematics. Vestnik Leningrad. Univ. No. 1, 3-20.
81. Leninist dialectic and mathematics. Vestnik Leningrad. Univ., No. 4, 24-30.

1951

82. A theorem on triangles in metric spaces and some of its applications. Trudy Mat. Inst. Steklov, 38, 5-23.
83. Intrinsic geometry. In *Bol'shaya Sovetskaya Entsiklopediya*(Great Soviet Encyclopedia), 2nd edition, Vol. 8, 298.
84. Convex body (geometrical). Ibid., Vol. 9, 457-458.
85. Review of *Entsiklopediya elementarnoi matematiki* (Encyclopedia of elementary mathematics), edited by P.S. Aleksandrov, A.I. Markushevich, and A. Ya. Khinchin. Book 1, Arithmetic, 448 pp., Book 2, Algebra, 424 pp., Moscow-Leningrad, Gostekhizdat, 1951. Sovetskaya Kniga, No. 5, 19-25.
86. On logic. Voprosy Filosofii, No. 3, 152-163.
87. Leninist dialectic and mathematics. Priroda, No. 1, 5-15. (Also translated into Czech and Chinese.)
88. On idealism in mathematics. Priroda, No. 7, 3-11. (Also translated into Czech and Chinese.)
89. On idealism in mathematics (conclusion), Priroda, No. 8, 3-9. (Also translated into Czech and Chinese.)

1952

90. On Einstein's paradox in quantum mechanics. Doklady Akad. Nauk SSSR, 84, 253-256.
91. On the concept of wave function. Dokl. Akad. Nauk SSSR, 85, 291-294.
92. Geometry. In *Bol'shaya Sovetskaya Entsiklopediya* (Great Soviet Encyclopedia), 2nd edition, Vol. 10, 533-550. (Also in Chinese, Polish.)
93. Geometry of convex bodies. Ibid., 551-552.
94. Efimov, Nikolai Vladimirovich. Ibid., Vol. 15, 566.
95. To train fully-fledged scientists. Gazeta Leningrad. Univ., November 13th, 1.

1953

96. The estimation of the length of a curve on a surface. Dokl. Akad. Nauk SSSR, 93, 221-224, (with V.V. Strel'tsov).
97. On the essence of the theory of relativity. Vestnik Leningrad. Univ., No. 8, 103-128.
98. Remarks on the foundations of the theory of relativity. Vestnik Leningrad.

Univ., No. 11, 95-110, (with V.V. Ovchinnikova).
99. Concerning certain views of the theory of relativity. Voprosy filosofii, No. 5, 225-245.
100. Problems of the new academic year. Gazeta Leningrad. Univ., September 4th, p. 1.

1954
101. On the filling of space by polyhedra. Vestnik Leningrad. Univ., No. 2, 33-43.
102. Some theorems on second-order partial differential equations. Vestnik Leningrad. Univ., No. 8, 3-17.
103. Synthetic methods in the theory of surfaces. (In English). In *Convegno internazionale di geometria differenziale Italia, 20-26 Settembre 1953*, Rome, Cremonese, 162-175.
104. Lobachevskian geometry. In *Bol'shaya Sovetskaya Entsiklopediya* (Great Soviet Encyclopedia), 2nd edition, Vol. 25, 317-320.
105. On the conditions for non-deformability of convex surfaces with a boundary. In *Nauchnaya Sessiya Leningradskovo Universiteta* (Scientific Meeting at Leningrad University), Reports of the mathematical section. Leningrad. 45-46.
106. The ascent to the highest point of the earth. Priroda, No. 8, 62-72, (with V.P. Berkov).
107. On the condition and measures for improvement of ideological work in the university. Gazeta Leningrad. Univ., October 1st, 2-3.
108. Happy new academic year! Gazeta Leningrad. Univ., September 3rd, 1.
109. Happy new year, dear friends. Gazeta Leningrad. Univ., January 1st 1.

1955
110. On the non-deformability of convex surfaces. Vestnik Leningrad. Univ., No. 8, 3-13 (with E.P. Sen'kin).
111. Riemannian Geometry. In *Bol'shaya Sovetskaya Entsiklopediya* (Great Soviet Encyclopedia), 2nd edition, Vol. 36, 520-523 (with Yu. F. Borisov).
112. The theory of relativity (epistemological significance). Ibid., Vol. 31, 411-413.

1956
113. A general view of mathematics. In *Matematika, ee soderzhanie, metody i znachenie* (Mathematics, its content, methods and significance). Vol. 1, 5-78. Moscow, Fizmatgiz. (Also in Chinese and Rumanian; English edition in preparation).
114. Curves and surfaces. Ibid. Vol. 2, 97-152.
115. Abstract spaces. Ibid. Vol. 3, 93-180.
116. Theorems of Minkowski and A.D. Aleksandrov. In *Vypuklye figury i mnogogranniki* (Convex figures and polyhedra), by L.A. Lyusternik, Moscow, Gostekhizdat, 149-170.
117. The space-time of the theory of relativity. (In English.) In *Fünfzig Jahre Relativitätstheorie*, Basel (Birkhäuser), 44-45.
118. Supplement to "On the non-deformability of convex surfaces". Vestnik Leningrad. Univ., No. 1, 104-106 (with E.P. Sen'kin).
119. Uniqueness theorems for surfaces "in the large", I. Vestnik Leningrad. Univ., No. 19, 5-17.
120. On a generalization of Riemannian geometry. In *Trudy III Vsesoyuznogo matematicheskogo s"ezda* (Proceedings of the third All-Union mathematical congress), Moscow (Academy of Sciences Publishing House), Vol. 2, 138.
121. Uniqueness theorems for differential equations and surfaces. In *Nauchnaya Sessiya Leningradskogo Universiteta* (Scientific Meeting at Leningrad University), Reports of the mathematical section. Leningrad, 4-7.
122. The most important means of developing scientific creation. Vestnik vysshei shkoly, No. 7, 18-25.
123. On mathematical education in the USSR. (In English). The Math. Student,

24, 99-108.
124. In a country of great people (Indian impressions). Vechernaya Leningrad. August 4th, 4.
125. A school of creative thought. Literaturnaya Gazeta, September 4th, 2.

1957

126. Ruled surfaces in metric spaces. Vestnik Leningrad. Univer. No. 1, 5-26.
127. Uniqueness theorems for surfaces "in the large", II. Vestnik Leningrad. Univ., No. 7, 15-44.
128. Uber eine Verallgemeinerung der Riemannschen Geometrie. Deutsch. Akad. Wiss. Berlin Schr. Forschungsinst. Math., No. 1, 33-84.
129. Elementary geometry. In *Bol'shaya Sovetskaya Entsiklopediya*, (Great Soviet Encyclopedia), 2nd edition, Vol, 48, 645-648.
130. Dialectic and science. Vestnik Akad. Nauk SSSR, No. 6, 3-17.
131. The education of students, a very important political problem. Vestnik vysshei shkoly, No. 3, 12-19.
132. The abominable snowman - myth or reality? Literaturnaya gazeta, March 21st, 2 (with E. Simonov).
133. In the first ranks of native science. Pravda, June 22nd, 2.
134. New landmarks of history. Izvestiya, October 18th, 2.

1958

135. Dirichlet's problem for the equation Det $\|z_{ij}\| = \varphi(z_1, \ldots, z_n, z, x_1, \ldots, x_n)$, I. Vestnik Leningrad. Univ., No. 1, 5-24.
136. Uniqueness theorems for surfaces "in the large", III. Vestnik Leningrad. Univ., no. 7, 14-26.
137. Uniqueness theorems for surfaces "in the large", IV. Vestnik Leningrad. Univ., No. 13, 27-34 (with Yu. A. Volkov).
138. Uniqueness theorems for surfaces "in the large", V. Vestnik Leningrad Univ., No. 19, 5-8.
139. Investigations into the maximum principle, I. Izv. Vyssh. Uchebn. Zaved. Matematika, No. 5(6), 126-157.
140. Remember the demands of life, Izvestiya, August 9th, 2.
141. The way to higher education. Izvestiya, December 10th, 4.
142. Our seven-year plans. Gazeta Leningrad. Univ., March 31st, 3.
143. To educate skilfully, creatively. Gazeta Leningrad. Univ., March 10th, 1-2.
144. The main thing is creative activity. Sovetskaya Rossiya, December 12th, 2.

1959

145. Uniqueness theorems for surfaces "in the large", VI. Vestnik Leningrad. Univ., No. 1. 5-13.
146. Investigations into the maximum principle, II. Izv. Vyssh. Uchebn. Zaved. Matematika, No. 3(10), 3-12.
147. Investigations into the maximum principle, III. Izv. Vyssh. Uchebn. Zaved. Matematika, No. 5(12), 6-32.
148. The philosophical content and significance of the theory of relativity, Voprosy filosofii, No. 1, 67-84.
149. The philosophical content and significance of the principle of relativity, In *Filosofskie problemy sovremennogo estestvoznaniya* (Philosphical problems of contemporary science), Moscow, (Academy of Sciences Publishing House), 93-136.
150. The theory of relativity as a theory of absolute space-time. In *Filosofskie voprosy sovremennoi fiziki* (Philosphical questions of contemporary physics) Moscow (Academy of Sciences Publishing House), 269-323.
151. Education in the USSR. (In English). In *Proceedings of the Fourth Canadian Mathematical Congress*, Toronto (University of Toronto Press), 14-19.
152. First steps. Vechernaya Leningrad, March 10th, 2.

1960

153. Some estimates connected with Dirichlet's problem. Dokl. Akad. Nauk SSSR, 134, 1001-1004.
154. Uniqueness theorems for surfaces "in the large", VII. Vestnik Leningrad. Univ., No. 7, 5-13.
155. Investigations into the maximum principle, IV. Izv. Vyssh. Uchebn. Zaved. Matematika, No. 3(16), 3-15.
156. Investigations into the maximum principle, V. Izv Vyssh. Uchebn. Zaved. Matematika, No. 5(18), 16-26.
157. Modern development of surface theory. (In English.) In *Proceedings of the International Congress of Mathematicians, Edinburgh, 1958*, 1-18. (Also translated into Russian.)
158. Lenin's role in the development of science. Voprosy folosofii, No. 8, 35-45.
159. Nikolai Vladimirovich Efimov (on his fiftieth birthday). Uspekhi Mat. Nauk, 15, 175-177 (with A.V. Pogorelov).
160. The most important problem in the building of Communism. Gazeta Leningrad. Univ., October 11th, 2.
161. On the relation of biology to physics and chemistry. Gazeta Leningrad. Univ., May 9th, 2.
162. Your hand, colleague! Komsomol'skaya Pravda, December 22nd, 2.

1961

163. Investigations into the principle of the maximum, IV. Izv. Vyssh. Uchebn. Zaved. Matematika, No. 1(20), 3-20.
164. On the maximum principle. In *Nekotorye problemy matematiki i mekhaniki* (Some problems of mathematics and mechanics), Novosibirsk, (Siberian branch of the Academy of Sciences of USSR), 25-41.
165. A condition for the equality of closed convex surfaces. Vestnik Leningrad. Univ., No. 7, 5-7.
166. The theory of surfaces and partial differential equations. Fourth All-Union Mathematical Congress, annotations to the plenary reports. Leningrad. 3-4.
167. A discussion of the report of M.V. Keldysh on the First All-Union Conference of Scientists, 12-14th June, 1961. Vestnik Akad. Nauk SSSR, No. 7, 42-43.
168. A fundamental link, the high school. Pravda, February 8th, 3.
169. The principal wealth. Izvestiya, March 23rd, 3.
170. As a measure of Communism. Gazeta Leningrad. Univ., October 13th, 1-2.
171. A matter of the first importance. Gazeta Leningrad. Univ., June 23rd, 2.
172. Daring, enquiring, curious. Gazeta Leningrad. Univ., June 2nd, 1.
173. In defence of philosophy. Gazeta Leningrad. Univ., November 10th, 2-3.
174. Let more be obsessed. Komsomol'skaya Pravda. November 22nd, 2-3, Gazeta Leningrad. Univ., December 8th, 2-3.

1962

175. On the deformation of polyhedra with rigid boundaries. Vestnik Leningrad. Univ., No. 13, 138-141. (with S.M. Vladimirov).
176. Two-dimensional manifolds with bounded curvature (foundations of the intrinsic geometry of surfaces). Trudy Mat. Inst. Steklov, 63, 262 pp., (with V.A. Zalgaller).
177. *Obshchaya teoriya neregulyarnykh krivykh v trekhmernom prostranstve* (General theory of non-regular curves in three-dimensional space). Novosibirsk, Publication of the Siberian branch of the Academy of Scicnces of the USSR. 240 pp. (with Yu. G. Reshetnyak).
178. Theory of surfaces and partial differential equations. *Trudy IV Vsesoyuznogo matematicheskogo s"ezda* (Proceedings of the IVth All-Union Mathematical Congress) Vol. 1, (with A.V. Pogorelov).
179. Diamonds must be cut. Leningradskaya pravda, May 20th, 2.
180. Forward, with enthusiasm! Izvestiya, July 28th, 3.

Leonid Vitalyevich Kantorovich

Born – 19 Jan 1912 in St Petersburg, Russia
Died – 7 April 1986 in USSR

Leonid Kantorovich studied mathematics at Leningrad State University, receiving his doctorate in mathematics in 1930 at the age of eighteen. From 1934 to 1960 he was a professor of mathematics at Leningrad University. He held the chair of mathematics and economics in the Siberian branch of the USSR Academy of Sciences (1961–1971), then directed research at Moscow's Institute of National Economic Planning (1971–1976).

Kantorovich's background was entirely in mathematics but he showed a considerable feel for the underlying economics to which he applied the mathematical techniques. He was one of the first to use linear programming as a tool in economics and this appeared in a publication *Mathematical methods of organising and planning production* which he published in 1939.

Kantorovich introduced many new concepts into the study of mathematical programming such as giving necessary and sufficient optimality conditions on the base of supporting hyperplanes at the solution point in the production space, the concept of primal-dual methods, the interpretation in economics of multipliers, and the column-generation method used in linear programming.

One of his most fundamental works on economics was *The best use of economic resources* which he wrote in 1942 but was not published until 1959. In this work Kantorovich applies optimisation techniques to a wide range of problems in economics.

He also proposed a theory to handle the economics of technological innovations. This had three components namely the effect on the producer, the effect on the consumer and, the novel part of the theory, the effect derived from the increasing economic potential arising from the innovation.

Kantorovich was a joint winner of the 1975 Nobel Prize for economics for his work on the optimal allocation of scarce resources.

The work of Kantorovich that we have looked at up until now has been on the application of mathematical methods, particularly mathematical programming to economics. It is this work for which Kantorovich is most famous, but he also worked in many other areas of mathematics. These other areas include functional analysis and numerical analysis. Within these topics he published papers on the theory of functions, the theory of complex variables, approximation theory in which he was particularly interested in using Bernstein polynomials, the calculus of variations, methods of finding approximate solutions to partial differential equations, and descriptive set theory.

From 1929 he worked on the theory of analytic sets and the Baire classification of functions. This work continued through the early 1930s then in the late 1930s he studied ordered topological vector spaces. However later in his career he also became interested in computer architecture. His remarkable contributions to mathematics, economics and computers were published in over 300 papers and books.

Comptes Rendus (Doklady) de l'Académie des Sciences de l'URSS
1940. Volume XXVIII, № 3

MATHEMATICS

A NEW METHOD OF SOLVING OF SOME CLASSES OF EXTREMAL PROBLEMS

By L. KANTOROVITCH

(Communicated by S. N. Bernstein, Member of the Academy, 8. V. 1940)

In many extremal problems the extremum point lies always on the boundary of the fundamental region and never inside this region; it is then often difficult to effect all the necessary computations and find the extremum point. Thus, in the simplest problem of this kind, viz. that of finding the maximum (or the minimum) of a linear function f in the region bounded by a polygon, the extrema are always reached at the vertices of the polygon so that in order to find the required extremal point we have only to compare the values of f in all these vertices. If, however, the number of vertices is very great, this comparison can become practically impossible.

I shall propose here a new method of solving problems of this kind, which has certain likeness to Lagrange's method of finding relative extrema. This method, although it is based upon very simple considerations, is in many cases much more effective than usual methods. § 1 contains the exposition of our method; in § 2 we give some examples of problems which may be treated with success by this method.

§ 1. Theorem 1. Let E be a linear normed space, $A \subset E$—a convex weakly compact subset of E, and F—a finite functional defined in A and possessing the following properties: 1) it is weakly continuous from above, i. e. $x_n \dashrightarrow x$ implies $\overline{\lim} F(x_n) \leqslant F(x)$; 2) it does not reach its maximum in the interior of A, i. e. if x is an interior point of A, then in every neighbourhood of x we can find such x' that $F(x') > F(x)$.

Then there exist a point x_0 (lying on the boundary of A) at which F reaches its maximum, and a linear functional $f_0(x)$ which reaches its maximum in A at the same point x_0.

Proof. Let M be the exact upper bound of F on A. We can find then for every positive integer n such point $x_n \in A$ that $F(x_n) > M - \frac{1}{n}$. The sequence $\{x_n\}$ contains a weakly convergent subsequence $x_{n_k} \dashrightarrow x_0$. By condition 1) of the theorem, we have $F(x_0) = M$, and by condition 2) x_0 lies on the boundary of A. Further, according to a theorem of S. Mazur (1), there exists a linear functional $f_0(x)$ such that the set A lies wholly on one side of the plane $f_0(x) = f_0(x_0)$. Obviously we can make (by changing the sign of f_0, if needed) that $f_0(x) \leqslant f_0(x_0)$ for any $x \in A$.

Our theorem is thus proved in all its parts.

Fundamental Corollary. Let f be a linear functional and denote by H_f the set of all points of A where it reaches its maximum, and by $p(f)$ the maximum of F upon H_f. Then

$$M = \max_{x \in A} F(x) = \max_{f} \{p(f)\}. \tag{1}$$

In fact, on the one hand, we always have $p(f) \leqslant M$ and accordingly $\max p(f) \leqslant M$; on the other hand, according to theorem 1, $p(f_0) = F(x_0) = M$, and consequently $\max \{p(f)\} \geqslant M$.

The equality (1) will serve us as a starting point for the proposed method of finding a boundary extremum. This method may be described as follows. Chosing a linear functional f we determine for this functional first the set H_f and then the number $p(f)$. Then we vary f in such manner that $p(f)$ becomes greater and approaches its maximum. An essential advantage of this method in comparison with the direct variation of x is that we are certain to move along the boundary of A. Besides, with this method it is often easier to find the right direction of variation. The question arises how to verify that we have really reached the maximum. A criterion is in many cases furnished by the following theorem.

Theorem 2. Let besides all conditions of theorem 1, the following condition be satisfied: the set $F(x) \geqslant C$ is convex whatever be C. Suppose further that there exists a point x_0 and a linear functional f_0 satisfying the following condition: 1) $f_0(x) \leqslant f_0(x_0)$ for any $x \in A$, and either 2) on the plane $f_0(x) = f_0(x_0)$ the inequality $F(x) \leqslant F(x_0)$ subsists, or 2′) on the surface $F(x) = F(x_0)$ we have $f(x) \geqslant f(x_0)$; then the functional F reaches maximum in the point x_0. Furthermore, the point x_0 and the functional f_0 satisfying the above conditions (1 and 2 or 1 and 2′) do always exist.

Proof. It follows from the conditions of the theorem that the plane $f_0(x) = f_0(x_0)$ passes between the convex sets A and $\underset{x}{\varepsilon}(F(x) \geqslant F(x_0))$. Hence it is clear that in all points of A we have $F(x) \leqslant F(x_0)$, i. e. x_0 is a maximum point for $F(x)$. The existence of the point x_0 and the functional f_0 follows from a theorem of Eidelheit[2] stating that there always exists a plane passing between two disjoint convex sets.

Note that in the conditions of theorem 2 we can state somewhat more precisely what has been already said about the direction of variation of f. Viz., if the point x_0 and the functional f_0 are not the required maximal point and functional, then the plane $f_0(x) = f_0(x_0)$ is a tangent plane for A, but not for the set $\underset{x}{\varepsilon}(F(x) \geqslant F(x_0))$. Let $f_1(x) = f_1(x_0)$ be the tangent plane to set $\varepsilon(F(x) \geqslant F(x_0))$ at the point x_0. Then for the next approximation we must substitute $f_0 + \varepsilon f_1$ for f_0.

We shall make now some remarks showing that the suggested method remains applicable in some cases not covered by theorems 1 and 2.

Remark 1. If the existence of a finite extremum for F is already established, then the conditions that A must be weakly compact and F weakly semicontinuous are superfluous.

Remark 2. The condition of convexity of A can also be weakened. Viz., it is sufficient to require that for every boundary point x of A there should exist a functional f such that

$$\overline{\lim_{\substack{x' \to x \\ x' \in A}}} \frac{f(x' - x)}{\| x' - x \|} \leqslant 0. \tag{2}$$

In this case H_f is the set of points x satisfying (2).

§ 2. We shall give now some examples of problems to which our method can be applied.

I. Let $\{\alpha_i(t)\}$ (where $i=1, 2, \ldots, m$) be a set of functions, defined and integrable (in Lebesque's sense) in the interval $(a; b)$. The problem is to find measurable functions $h_i(t)$ $(i=1, 2, \ldots, m)$ satisfying the following conditions:

1) $h_i(t) \geqslant 0$ $(i=1, 2, \ldots m;\ a \leqslant t \leqslant b)$

2) $h_1(t)+h_2(t)+\ldots+h_m(t)=1 \quad (a \leqslant t \leqslant b)$

3) the number $\min\limits_{i=1}^{m} \int\limits_a^b \alpha_i(t)\, h_i(t)\, dt$

takes its maximum possible value [i. e. if $\{\bar{h}_i\}_{i=1,2,\ldots,m}$ is a set of functions satisfying (1) and (2), then

$$\min \int\limits_a^b \alpha_i(t)\, \bar{h}_i(t)\, dt \leqslant \min \int\limits_a^b \alpha_i(t)\, h_i(t)\, dt].$$

We can apply theorems 1 and 2 if we take for A the set of systems $(z_1, \ldots, z_m)$ of numbers, where $z_i = \int\limits_a^b \alpha_i(t)\, \bar{h}_i(t)\, dt$ and the functions $\bar{h}_i$ $(i=1, 2, \ldots, m)$ satisfy conditions (1) and (2); and for $F(z_1, z_2, \ldots, z_m)$, the number $\min(z_1, z_2, \ldots, z_m)$. From theorems 1 and 2 follows that there exist such numbers $\lambda_1, \ldots, \lambda_m$ that the linear function

$$\sum_{i=1}^{m} \lambda_i z_i = \int\limits_a^b \Sigma \lambda_i \alpha_i(t)\, \bar{h}_i(t)\, dt \tag{3}$$

reaches its maximum together with F. But it is clear that if the system $\{h_i\}$ gives the maximum value to this functional, then these functions h_i must almost everywhere satisfy the condition:

$$h_i(t)=0, \quad \text{if } \lambda_i \alpha_i(t) < \max \lambda_j \alpha_j(t). \tag{4}$$

If the numbers λ_i are known, then we can easily find the functions $h_i(t)$ such that condition (4) be satisfied and, besides, that these functions take only the values 0 and 1 and that

$$\int\limits_a^b h_1(t)\, \alpha_1(t)\, dt = \int\limits_a^b h_2(t)\, \alpha_2(t)\, dt = \ldots = \int\limits_a^b h_m(t)\, \alpha_m(t)\, dt. \tag{5}$$

Conversely, if the numbers λ_i are such that the functions $h_i(t)$ can be defined satisfying conditions (4) and (5), then these functions give an extremum to the functional F. Thus, the problem of finding the system of functions $\{h_i(t)\}$ is reduced to an easier problem of finding the system of numbers $\{\lambda_i\}$. As to this last problem, a good method of successive approximations can be given for its solution.

II. The preceding problem can also be solved if the functions h_i have to satisfy a supplementary condition of the form

$$R = \int\limits_a^b \Sigma \beta_i(t)\, h_i(t)\, dt \leqslant B.$$

In fact it suffices in this case to take for F the expression

$$F(z_1, z_2, \ldots, z_m, R) = \min z_i + \partial(R)$$

where $\partial(R)=0$ if $R \leqslant B$, and $\partial(R) = -\infty$ if $R > B$.

Applying theorems 1 and 2, we find that F reaches its maximum together with the functional $\Sigma\lambda_i z_i - \mu R$.

III. Let $\alpha(x, y)$ be a function positive and continuous in the rectangle $[a, b; c, e]$. The problem is to find a non-negative function Φ of elementary rectangle d such that

$$1. \quad \int_a^b \Phi(d_x) = 1 \qquad 2. \quad \int_c^e \alpha(x, y)\,\Phi(d_y) = C$$

where the constant C takes the greatest possible value compatible with condition 1 (condition 1 means that whatever be γ, δ such that $c \leqslant \gamma < \delta \leqslant e$, we should always have $\Phi[(a, b; \gamma, \delta)] = 1$).

If we consider the space of functions

$$z(x) = \int_c^e \alpha(x, y)\,\Phi(d_y)$$

then the problem is to find the maximum of the functional $F(z) = \min\limits_{a \leqslant x \leqslant b} z(x)$. To this problem we can again apply our method.

Besides a great many problems similar to those considered above, the method is also applicable to some problems of more complicated nature. We shall give two examples of such problems taking them from the theory of approximations.

IV. An approximate solution of a system of non-compatible equations. The problem [cf. e. g. (3), 4th paper] is to find such numbers $x_1, \ldots, x_n$ that setting

$$\begin{aligned} a_{11}x_1 + \ldots + a_{1n}x_n - c_1 &= z_1 \\ \cdots\cdots\cdots\cdots&\cdots\cdots \\ a_{m1}x_1 + \ldots + a_{mn}x_n - c_m &= z_m \end{aligned}$$

the expression $\Phi(z) = (|z_1|^p + \ldots + |z_m|^p)^{\frac{1}{p}}$ would take its minimum value.

If $p = \infty$, we have the best Tchebyshev's approximation. If $p = 2$, we have the best approximation in the sense of the method of minimum squares of Gauss, and the solution of the problem in this case is well known. Let us set $y_i = z_i^2$ and $F(y_1, \ldots, y_m) = \Phi(\sqrt{y_1}, \ldots, \sqrt{y_m})$. Then remark 2 above makes it possible to apply our method. Applying it we see that the required extremum of F is reached together with the extremum of the functional $\Sigma\lambda_i y_i = \Sigma\lambda_i z_i^2$. This last extremum can be easily found if we know the numbers λ_i. The numbers λ_i can be found by the method of successive approximations. In particular in Tchebyshev's case (i. e. when $p = \infty$) these numbers must be varied in such a way that if $|z_{i_0}|$ is greater than all other $|z_i|$, then the corresponding number λ_{i_0} must be magnified.

V. The same considerations may be applied to the problem of finding a polynomial of a given degree giving the best approximation of a given function either in Tchebyshev's or Jackson's sense, viz., we must find such weight $p(x)$ that the polynomial giving the best mean approximation with the weight $p(x)$ is the required one.

Leningrad State University.

Received
11. V. 1940.

REFERENCES

[1] S. Mazur, Studia Mathem., IV. [2] Eidelheit, ibidem, VI. [3] Н. Ахиезер и М. Крейн, О некоторых проблемах теории моментов, Харьков (1938).

Comptes Rendus (Doklady) de l'Académie des Sciences de l'URSS
1942. Volume XXXVII, № 7—8

MATHEMATICS

ON THE TRANSLOCATION OF MASSES

By L. KANTOROVITCH

(Communicated by S. L. Soboleff, Member of the Academy, 29.IX.1942)

R will denote a compact metric space, though some of the following definitions and results are valid in more general spaces.

Let $\Phi(e)$ be a mass distribution, i. e. a set function possessing the following properties: 1) $\Phi(e)$ is defined on Borel sets in R, 2) $\Phi(e)$ is non-negative, $\Phi(e) \geqslant 0$, 3) $\Phi(e)$ is absolutely additive, i. e. if $e = e_1 + e_2 + \ldots$, $e_i e_k = 0 \, (i \neq k)$, then $\Phi(e) = \Phi(e_1) + \Phi(e_2) + \ldots$ Let further $\Phi'(e')$ be another mass distribution, such that $\Phi(R) = \Phi'(R)$. Under the translocation of masses we shall understand the function $\Psi(e, e')$ defined on the pairs of B—sets $e, e' \in R$ and such that: 1) $\Psi(e,e')$ is non-negative and absolutely additive in each of its arguments, 2) $\Psi(e, R) \equiv \Phi(e)$; $\Psi(R, e') \equiv \Phi'(e')$.

Let a continuous non-negative function $r(x, y)$ be given that represents the work expended in transferring a unit mass from x to y.

By the work required for transferring the given mass distributions will be understood

$$W(\Psi, \Phi, \Phi') = \int_R \int_R r(x, x') \Psi(de, de') = \lim_{\lambda \to 0} \sum_{i,k} r(x_i, x'_k) \Psi(e_i, e'_k)$$

where e_i are mutually disjoint and $\sum_1^n e_i = R$; e'_k are also mutually disjoint and $\sum_1^m e'_k = R$, $x_i \in e_i$, $x'_k \in e'_k$, λ being the maximum of diameter e_i $(i = 1, 2, \ldots, n)$ and diameter e'_k $(k = 1, 2, \ldots, m)$. This integral does evidently exist.

$$W(\Phi, \Phi') = \inf_{\Psi} W(\Psi, \Phi, \Phi')$$

is said to be the minimal translocation work. Since the totality of functions $\{\Psi\}$ is compact, there exists a function Ψ_0 that realizes this minimum, i. e. such that

$$W(\Phi, \Phi') = W(\Psi_0, \Phi, \Phi').$$

This function, it is true, is not unique. Such a translocation is said to be minimal translocation.

We shall further say that the translocation Ψ from x to y is not zero, and write $x \to y$ if $\Psi(U_x, U_y) > 0$ for any neighbourhoods U_x and U_y of the points x and y. The translocation Ψ will be called potential if there exists a function $U(x)$ such that: 1) $|U(x) - U(y)| \leqslant r(x, y)$; 2) $U(y) - U(x) = r(x, y)$ when $x \to y$.

Theorem. *The translocation Ψ is minimal if and only if it is potential.*

Sufficiency of the Condition. Let Ψ_0 be a potential translocation and U the corresponding potential. According to property 2) of U

$$W(\Psi_0, \Phi, \Phi') = \int_R \int_R r(x, y)\, \Psi_0(de, de') = \int_R \int_R [U(y) - U(x)]\Psi_0(de, de') =$$
$$= \int_R \int_R U(y)\, \Psi_0(de, de') - \int_R \int_R U(x)\, \Psi_0(de, de') =$$
$$= \int_R U(y)\, \Phi'(de') - \int_R U(x)\, \Phi(de),$$

while for any other function Ψ

$$W(\Psi, \Phi, \Phi') = \int_R \int_R r(x, y)\, \Psi(de, de') \geqslant \int_R \int_R [U(y) - U(x)]\, \Psi(de, de') =$$
$$= \int_R U(y)\, \Phi'(de') - \int_R U(x)\, \Phi(de),$$

i. e. $W(\Psi, \Phi, \Phi') \geqslant W(\Psi_0, \Phi, \Phi')$ and Ψ_0 is thus minimal.

Necessity of the Condition. Let Ψ_0 be a minimal translocation. Consider a set of points $\xi_0, \xi_1, \ldots$ dense in R. Denote by D_n the minimal point set containing ξ_n and such that $x \in D_n$ and $x \to y$, or $y \to x$, imply $y \in D_n$. It is obvious that if $y \in D_n$, there exists then a system of points x_i, y_i such that $\xi_0 = x_0 \to y_1 \cdot x_1 \to y_1, x_2 \to y_2, \ldots, x_{n-1} \to y_n, x_n \to y_n\, (y_n = y)$ (or a similar chain with the inverse direction of arrows at the beginning or at the end). In our case we set

$$U(y) = \sum_1^n r(x_{i-1}, y_{i-1}) - \sum_1^n r(x_i, y_i).$$

It is readily verified that U does not depend on the choice of the joining chain and that properties 1) and 2) of the potential are valid for U, if $x, y \in D_0$. Indeed, it is easy to show that if one of these circumstances failed to take place, Ψ_0 might be replaced by another translocation involving less work, which contradicts the assumed minimality of Ψ_0.

Suppose the function U to be already defined on the domains D_0, $D_1, \ldots, D_{n-1}$.

If the point r_n belongs to $D_0 + D_1 + \ldots + D_{n-1}$, then U is defined in r_n and on the whole D_n as well. Otherwise we define $V(x)$ on the domain D_n in the same way as U on D_0, provided that r_0 is now replaced by r_n. Further we choose the number μ satisfying the inequalities

$$\inf_{\substack{x \in D_0 + \ldots + D_{n-1} \\ y \in D_n}} \{U(x) - V(y) - r(x, y)\} \leqslant \mu \leqslant$$
$$\leqslant \inf_{\substack{x \in D_0 + \ldots + D_{n-1} \\ y \in D_n}} \{U(x) - V(y) + r(x, y)\}.$$

The existence of such μ is again implied by the minimality of Ψ_0. Then we put for $x \in D_n$: $U(x) = V(x) + \mu$. The function U is thus defined on the set $D_0 + D_1 + \ldots$ As the latter is everywhere dense in R, U is defined on the whole space R according to the property 2), so that properties 1) and 2) are preserved. We see that Ψ_0 is a potential translocation.

The theorem just demonstrated makes it easy for one to prove that a given mass translocation is or is not minimal. He has only to try and construct the potential in the way outlined above. If this construction turns out to be impossible, i. e. the given translocation is not minimal, he at least will find himself in the possession of the method how to lower the translocation work and eventually come to the minimal translocation.

It is interesting to investigate the space of mass distributions with the metric defined as $W(\Phi, \Phi')$ [in the case, when $r(x, y)$ coincides with the distance $\rho(x, y)$]. This method of metrization of the space considered seems to be the most natural in some respects.

Finally, we give two practical problems, for the solution of which our theorem may be applied.

Problem 1. *Location of consumption stations with respect to production stations.* Stations $A_1, A_2, \ldots, A_m$, attached to a network of railways deliver goods to an extent of $a_1, a_2 \ldots, a_m$ carriages per day respectively. These goods are consumed at stations $B_1, B_2, \ldots, B_n$ of the same network at a rate of $b_1, b_2, \ldots, b_n$ carriages per day respectively ($\Sigma a_i = \Sigma b_k$). Given the costs $r_{i,k}$ involved in moving one carriage from station A_i to station B_k, assign the consumption stations such places with respect to the production stations as would reduce the total transport expenses to a minimum.

The solution of this and more complicated problems of the same type the reader will find exposed in detail in a paper by L. Kantorovitch and M. Govurin which is soon to be published.

Problem 2. *Levelling a land area.* Given the relief of the locality, namely, the equations $z = f(x, y)$ and $z = f_1(x, y)$ for the earth surface, before and after the levelling $\left[\text{under the condition } \int\int f(x, y)\,dx\,dy = \right.$ $\left.= \int\int f_1(x, y)\,dx\,dy\right]$ and the cost of transport of 1 m^3 of earth from point (x, y) to point (x_1, y_1); find a plan of transporting the earth masses with the least total expense.

Higher Engineering-Technical School.
Yaroslavl.

Received
29. IX. 1942.

Anatoly Ivanovich Malcev

Born – 27 Nov 1909 in Misheronsky (near Moscow), Russia
Died – 7 July 1967 in Novosibirsk, USSR

Anatoly Malcev was born on 27 Nov 1909 in Misheronsky, a place near Moscow. His family background was not well-to-do, his father being only a glass-blower. However, poverty did not hide the shine of his mathematical abilities. Even while at secondary school, his teachers were already convinced of his destiny to become an outstanding mathematician.

In 1927, Malcev graduated from school and entered the Moscow State University to study mathematics. He began a teaching career in a secondary school in Moscow a year before he graduated from the university in 1930. After graduating, he continued teaching in the secondary school for an entire year until 1932, when he was appointed as an assistant at the Ivanovo Paedagogical Institute in Ivanova, north-east of Moscow.

Malcev only had a first degree when he started teaching in Ivanovo, but thanks to the good rail link between Ivanovo and Moscow, he was able to make frequent trips to discuss his research findings with Kolmogorov. His first publications on logic and model theory were results of the work that he started on all by himself. These publications later influenced Robinson's study on non-standard analysis and ideas from these paper reappeared in Robinson's work.

Kolmogorov invited Malcev to join his graduate programme at Moscow University and he held a scholarship there for a year while continuing his teaching post at Ivanovo Paedagogical Institute. Malcev had always considered himself to be Kolmogorov's student, and during his scholarship year, Kolmogorov directed him to certain algebra problems.

In 1937, answering a question posed to him by Kolmogorov, Malcev published a paper on the embeddability of a ring in the field. The original problem was posed by Van Der Waerden whether there existed rings without zero-dividers which could not be embedded in the field. Malcev answered the question by constructing a ring whose multiplicative semigroup was not embedded in a group. Step-by-step, he was led into investigating the existence of rings whose multiplicative semigroup was embeddable in a group yet the ring still was not embeddable in a field. Two years later, Malcev gave sufficient and necessary conditions for a semigroup to be embeddable in a group.

Malcev also wrote a dissertation on *Torsion free abelian groups of finite rank* in 1937. Then between the years 1939 and 1941, he did his doctorate at the Steklov Institute of the Academy of Science, during which he continued as a lecturer at Ivanovo Paedagogical Institute. He received his degree of Doctor of Science for a dissertation *Structure of isomorphic representable infinite algebras and groups* in 1941.

Malcev became a professor at Ivanovo Paedagogical Institute in 1944. He continued his work in group theory and proved important results on linear groups, in particular, on linear soluble groups.

Malcev also studied Lie groups and topological algebras, producing a synthesis of algebra and mathematics logic. For example, in 1944 he wrote on semisimple subgroups of Lie groups, and in 1957 he wrote on free topological algebras. In 1946, he was awarded a State Prize for his work on Lie group. He also created a synthesis of the theory of algebras and algorithms called constructive algebras.

In 1960, Malcev was appointed to a chair in mathematics at the Mathematics Institute at Novosibirsk. He also became chairman of the Algebra and Logic department at Novosibirsk State University. Malcev founded the Siberian section of the Mathematics Institution of the Academy of Science, a logic-algebraic school with many members and directed the world famous seminar Algebra I Logika. He founded the specialized journal Algebra I Logika in 1962 and became its editor since then. He was also the founder of the Siberian Mathematical Society and its first president.

Malcev started working on problems of decidability of elementary theories of various algebraic structures in the early 1960s. He showed the undecidability of the elementary theory of finite groups, free nilpotent groups, free soluble groups and many others. He carried investigations on the undecidability of the elementary theory of classical linear group and proved that the class of locally free algebras had a decidable theory. An undergraduate text Malcev written in 1948 *Foundation of linear algebra*, giving a superb understanding on the subject, was translated to English in 1962.

Malcev died at a regrettable age of 57 in 7 July 1967, while taking part in the Novosibirsk Topology Conference that he helped to organize. He delivered his final lecture at this conference just before his death. The lecture was on algebra, now called Malcev algebra, and was a natural generalization of Lie algebra. Malcev first introduced these algebras in 1955, and he gave a survey of his work on this topic over the 12 preceding years.

Malcev received world-wide recognition and many honours for his innovative work. In 1958, he was elected as a member of the USSR Academy of Science. He also received a Lenin Prize for his series of papers on the applications of mathematical logic to algebra in 1964.

FREE TOPOLOGICAL ALGEBRAS

A. I. MAL'CEV

This paper deals with the fundamentals of the theory of free topological algebraic systems, generalizing the theory of the free topological groups of A. A. Markov [13] and related results of other authors.

Nearly ten years ago A. A. Markov [13] had developed the theory of the free topological groups over a completely regular space. In papers by M. I. Graev [5], [6], T. Nakayama [15], S. Kakutani [8] the proofs of the basic results of this theory have been substantially simplified; moreover, these authors obtained a number of new theorems on the free topological groups, throwing more light on their structure and going within a certain measure parallel to the results of the theory of abstract free groups. It is known, however, that the theory of the free abstract groups, rings, and other algebraic systems is more naturally placed within the framework of a general theory of algebraic systems. This fact, as well as the inner development of certain questions of the theory of the topological groups itself, seemed to encourage an attempt to create a general theory of free topologized algebraic systems. The fundamentals of such a theory will be given in the present paper.

Apart from the publications mentioned above, this paper is also a sequel to the paper [12]. For the convenience of the reader proofs will be carried out in detail also in such cases where they differ only in form from the proofs of the corresponding facts of the theory of the topological groups.

'*Algebra*' means here 'universal algebra' in the sense of G. Birkhoff [3], i.e., an arbitrary set of elements on which there is defined a finite number of one-valued operations, to be applied to finitely many elements. As distinguished from the general algebras, ordinary algebras over fields will be called linear algebras. An algebra is said to be topological if the set of its elements is a topological space and if the basic operations are continuous in this space.

In §1 we give the definition of a topological algebra with a given generating topological space and a prescribed system of defining relations; we prove its existence and uniqueness. It is also shown that a topological algebra over a topological space, given by the defining relations, is finitely generated by the elements of that space.

The main result of §2 is the proof of a generalized theorem of v. Dyck from which follows in particular that the classes of algebras with interchangeable congruences take a special position also in the theory of free topological algebras (cf., [12]). Hereafter free algebras and their free unions are defined. Analogous concepts have been studied by P. Sikorski [19].

§3 has an auxiliary character. Known facts concerning convergence in Hausdorff spaces are recalled and a method is considered to introduce a topology by means of limit operations which are essential for the subsequent discussion.

In the theory of free topological algebras it is important to find a free topology. A certain transfinite process for this purpose is studied in §4. There we also investigate more in detail the properties of a primitive topology obtained by this transfinite process.

In §5 we prove the theorems of M. I. Graev on the determination of a free topology of a group over a bicompact space for arbitrary algebras. Moreover, the free topology of an algebra with a locally compact generating space will be explained in §5. The last result is apparently new also for topological groups.

In §§6, 7, 8 we investigate the algebraic structure of special classes of free topological algebras. To begin, in §6 we study the classes with the simplest type of defining identities and here the topology and the algebraic structure of the free algebras are made evident.

In §7 it is shown how to derive known results of the theory of the free topological groups from the general results of the present work; it is also proved that the free nilpotent topological groups of different degrees are the factor groups of a free topological group with respect to the corresponding members of its lower central series.

In §8 we establish the algebraic structure of extended classes of free topological rings, including classes of non-associative rings, associative rings, Lie rings, etc. We also study the structure of free topological Boolean rings.

We conclude with the enumeration of some unsolved problems.

§1. Defining relations

If nothing is stated to the contrary all topological spaces occurring in the following are supposed to be Hausdorff spaces. In particular a Hausdorff space will be called a *topological algebra* if it is provided with a finite number of continuous operations $\phi_i(x_1, \cdots, x_{n_i})$ $(i = 1, \cdots, s)$, the fundamental operations of the algebra. A meaningful expression, consisting of the letters $x, \cdots z$, parentheses, and symbols of the fundamental operations of the algebra is called a *polynomial* in the letters $x, \cdots, z$. Algebras with one and the same set of fundamental operations, all elements of which satisfy a certain fixed system of identities of the form $f = g$ where f and g are polynomials, are said to belong to one and the same *primitive class* $\mathfrak{S}$ (cf., [12]).

Let be given a primitive class of algebras, characterized by a set of identities $\mathfrak{S}$ and a certain topological space X.

Definition. A topological algebra A of the class $\mathfrak{S}$ with a prescribed continuous mapping σ of the space X into A is said to be defined in $\mathfrak{S}$ by the space X and the relations

$$f_\lambda(x_{\lambda 1}, \cdots, x_{\lambda m_\lambda}) = g_\lambda(x_{\lambda 1}, \cdots, x_{\lambda m_\lambda}) \qquad (x_{\lambda i} \in X),$$

where f_λ and g_λ are certain polynomials if

F1) *in* A *one has the equations:* $f_\lambda(x^\sigma_{\lambda 1}, \cdots, x^\sigma_{\lambda m_\lambda}) = g_\lambda(x^\sigma_{\lambda 1}, \cdots, x^\sigma_{\lambda m_\lambda})$;

F2) A *is topologically generated by the images of the elements of* X, *i.e., there is in* A *no closed subalgebra different from* A *that contains* X^σ;

F3) *for each continuous mapping* γ *of the space* X *into an arbitrary topological algebra* C *of the class* $\mathfrak{S}$ *where the relations*

$$f_\lambda(x^\gamma_{\lambda 1}, \cdots, x^\gamma_{\lambda m_\lambda}) = g_\lambda(x^\gamma_{\lambda 1}, \cdots, x^\gamma_{\lambda m_\lambda}),$$

are satisfied, there exists a continuous homomorphism α *of the algebra* A *into the algebra* C *in correspondence with the mappings* σ, γ, *i.e., such that* $x^{\sigma\alpha} = x^\gamma$ *for all* $x \in X$.

In the usual way one can now prove

Theorem 1. *For each topological space* X *of any class* $\mathfrak{S}$ *and an arbitrary system of polynomials* f_λ, g_λ *there exists an algebra* A *with the properties* F1, F2, F3 *which is uniquely defined by these properties up to a topological isomorphism over the image of the space* X *in the algebra.*

Let us begin with the uniqueness. Suppose that the algebras A, B with the mappings σ, τ of the space X into them satisfy the conditions F1, F2, F3. According to F3 one can find a homomorphism α of the algebra A into B and a homomorphism β of the algebra B into A such that

$$x^{\sigma\alpha\beta} = x^{\tau\beta} = x^\sigma, \quad x^{\tau\beta\alpha} = x^{\sigma\alpha} = x^\tau.$$

Hence the homomorphisms α, β are mutually inverse and therefore onevalued on the sets X^σ, X^τ. Let us denote by A^*, B^* the subalgebras, generated finitely (algebraically) by the elements of the sets X^σ, X^τ in the algebras A, B respectively. Let

$$a^\alpha = b, \quad a = f(x^\sigma_1, \cdots, x^\sigma_n), \quad x_i \in X,$$

where f is a certain polynomial. Because of the homomorphy of the mapping α we have

$$b = f(x^\tau_1, \cdots, x^\tau_n),$$

whence

$$b^\beta = f(x^{\tau\beta}_1, \cdots, x^{\tau\beta}_n) = a,$$

i.e., the mappings α, β are mutually inverse on A^* and B^*. Since the closure of a subalgebra is a subalgebra, it follows that, by F2, the sets A^*, B^* are dense in A, B. Making use of the continuity of α and β we derive easily that α and β

are mutually inverse and therefore one-to-one correspondences on A and B. Thus α maps A isomorphically on B such that the image of any element of X into A is mapped into the image of the same element into B.

The existence of an algebra A follows immediately from the properties of direct products of topological algebras. We recall the corresponding definitions. Let A_α $(\alpha \in \Sigma)$ be a system of topological algebras of a fixed class $\mathfrak{S}$. The set of all the functions $h(\alpha)$, defined on Σ with values in $\bigcup A_\alpha$, and satisfying the conditions $h(\alpha) \in A_\alpha$, is defined as the cartesian product $H = \prod A_\alpha$. In H one introduces the Tichonov topology by distinguishing as open the cartesian products of open subsets chosen in a finite number of spaces, to be multiplied into the other spaces, and also any unions of such products.

Finally by definition

$$\phi_i(h_1, \cdots, h_{n_i}) = h \qquad (h_i, h \in H),$$

where

$$h(\alpha) = \phi_i(h_1(\alpha), \cdots, h_{n_i}(\alpha)), \qquad \alpha \in \Sigma.$$

Thus the set H is turned at the same time into a topological algebra of the same primitive class $\mathfrak{S}$ as the given algebras A_α. Subsequently the algebra H will be called the direct topological product of the given topological algebras.

Turning now to the existence proof we denote by $\mathfrak{m}$ the cardinal number of the given space X and out of the set of all topologically non-isomorphic algebras of the cardinal number $\leq 2^{2^{\mathfrak{m}+\aleph_0}}$ we choose all those algebras A_α that satisfy the conditions F1, F2. We denote by σ_α the continuous mapping of X into A_α which is mentioned in F1, F2. The mappings σ_α generate the natural mapping σ of the space X into the direct product $H = \prod A_\alpha$ which mapping is defined by

$$x^{\sigma}(\alpha) = x^{\sigma_\alpha} \qquad (\alpha \in \Sigma).$$

It is easily seen that this mapping is also continuous. The subalgebra A which is topologically generated by the elements of X^σ in the algebra H, has the properties F1, F2. But A also satisfies the condition F3. Indeed if B is any algebra having the properties F1, F2, then its cardinal number cannot be greater than $2^{2^{\mathfrak{m}+\aleph_0}}$,* because B contains a dense subalgebra B^*, algebraically generated by the elements of X^τ, and consequently of cardinal number $\leq \mathfrak{m} + \aleph_0$. Therefore among the factor algebras A_α there is an algebra A_λ, isomorphic with B over X. The projection of $a \to a(\lambda)$ of the algebra A onto the factor A_λ represents the required homomorphism of A onto $A_\lambda = B$, corresponding to the mappings $X \to A$

* A. A. Markov informed me that in a so far unpublished paper M. Ya. Perel'man has proved the theorem that is being used here, viz., that the cardinal number of a Hausdorff space H is never greater than 2^{2^a} if it contains a dense subset A with the cardinal number a. He also constructed examples which show that the number 2^{2^a} can really be reached.

and $X \to B$.

Theorem 2. *The algebra A, given in a class $\mathfrak{S}$ by the generating topological space X and some defining relations, is algebraically generated by the images of the elements of X, i.e., each element of A can be expressed in form of a polynomial of images of elements of X.*

For the proof we denote by A^* the set of those elements of A that can be written in the form of finite polynomials in the images of elements of X. Retaining the topology which A^* has as subset of the topological space A, we turn A^* into a *topological* algebra whereby the space X is continuously mapped into A^* using the same mapping σ as in the case of the given algebra A. With regard to F3 a continuous homomorphism τ of the algebra A into the algebra A^* can be found, corresponding to the mappings $X \to A$, $X \to A^*$. As A^* is generated algebraically by the images of the elements of X, so will τ be a mapping of A onto the whole algebra A^*, whereby for $a \in A^*$ simply $a^\tau = a$. Since A^* is dense in A, there is for every $a \in A$ a directed set $\{a_\lambda\}$ of elements of A^*, which tends to a (cf., §3). The continuity of τ entails that

$$a^\tau = \lim \{a_\lambda^\tau\} = a,$$

whence $A = A^*$.

The problem of a clear description of a topological algebra, given by its generating space X and the defining relations S in the class $\mathfrak{S}$, splits up naturally into the following parts:

A) Is A isomorphic to an abstract algebra $\tilde{A}$, given in the class $\mathfrak{S}$ by the same defining relations S and the same generating space X as the algebra A itself?

B) Does A contain the space X topologically, i.e., is the canonical mapping σ of the space X into A a homeomorphism?

C) Indicate a topology in A, in some sense naturally evident.

We point out that a positive answer to question B) has been included as an axiom in the definition of a free topological group in earlier papers on the theory of free topological groups [13], [5]. Therefore the existence proof has been given in these papers for completely regular spaces X. In our definition there is no such axiom. Hence the corresponding algebras do always exist, do not, however, always contain topologically the space X.

We shall formulate sufficient conditions for positive answers to the questions A), B) in the form of the following remarks which are immediate generalizations of known results of the theory of topological groups:

Remark 1. *The topological algebra A with the defining relations S and the generating space X is isomorphic to an abstract algebra $\tilde{A}$ with the same relations* *and generating set, if for every two polynomials f, g of elements of X with distinct images in A there is a continuous mapping of the space X into a suitable topological algebra B of the primitive class to be considered here, such that all defining relations for the images of the elements of X are satisfied in B and* $f \neq g$ *in B.*

Remark 2. *The algebra A contains the space X topologically if there is a system* $\{\alpha\}$ *of continuous mappings of the space X into suitable topological algebras* A_α*, leaving invariant the defining relations, such that for every neighborhood U of an arbitrary point* $x \in X$ *there is a finite system of neighborhoods* $U_1, \cdots, U_n$ *of the points* $x^{\alpha_1}, \cdots, x^{\alpha_n}$*, for which the intersection* $\bigcap U_i^{\alpha_i^{-1}}$ *of the originals is contained in U. If moreover* X^α *proves to be closed in every algebra* A_α*, then X will be contained in A as a closed subset.*

The statement of the first remark follows immediately from the definition of an algebra with given space X and defining relations S.

In order to prove the second remark we consider the construction of an algebra A by means of the direct product H as described above. All the algebras A_α enter the combination H as factors with mappings α. One has only to make sure that the canonical mapping of X into H will under the given conditions be a homeomorphism. In the case that all the A_α are homeomorphic a simple proof of this statement has been given in the paper [18]. In the general case the proof is the same.

So far we have considered arbitrary *Hausdorff* spaces. As, however, all discussions have been based only on properties of direct topological products, the preceding definitions and theorems remain correct if we replace the Hausdorff spaces by any other classes of spaces which satisfy the following two conditions (cf., [8]):

K_1) *The direct product of K-algebras is a K-algebra;*

K_2) *Every subalgebra of a K-algebra is a K-algebra.*

Such classes are, for instance, the regular spaces, the completely regular spaces (cf., [1]), with functionally separable pairs of points and others. Moreover, for the construction of a class of the above mentioned kind we may take any set of topological spaces and then consider all those spaces which can be obtained from the given spaces by means of the operations K_1 and K_2. In particular if we take as initial spaces those with a discrete topology, then the operations K_1, K_2 lead to a class of *algebraic* topologies as studied in the paper [7]. Topological spaces of this class are regular and completely separable in the sense that for an arbitrary system of points $x_1, \cdots, x_n$ of such a space X there is a decomposition of X in n mutually non-intersecting closed sets of which each contains one of the given points.

From the remarks 1, 2 it follows now immediately that *for algebras which are generated by a regular or a completely regular space X the problem* A) *is solved positively for arbitrary* $\mathfrak{S}$ *and* S, *and the problem* B) *is answered positively under the condition that the abstract algebra* $\tilde{A}$ *contains* X.

As in the case of the class K of all Hausdorff algebras this statement holds true also for the class of the regular and completely separable algebras. In the following, however, we shall discuss only algebras which are free in the class of the Hausdorff spaces.

§2. Free topological algebras. Free unions of topolical algebras

A topological algebra A is called a *free topological algebra* over the generating topological space X in the primitive class $\mathfrak{S}$, if A is defined in $\mathfrak{S}$ only by the space X, the set S of defining equalities being empty. In the abstract case the role of the free algebras is stressed by the fact that every abstract algebra of a class $\mathfrak{S}$ with generating set X is isomorphic to the factor algebra of a free algebra over X in $\mathfrak{S}$ with respect to a suitable congruence. In the topological case one can only state that every topological algebra of the class $\mathfrak{S}$ over X is the continuous homomorphic image of a free topological algebra over X (cf., Theorem 3, below). For the construction of a *topological* factor algebra the homomorphism has to be not only continuous, but also *open* (cf., [12]). As an example we consider the class $\mathfrak{S}$ of the semigroups, i.e., the algebras with one binary associative multiplication operation. Let X be a connected topological space having more than one point. We choose in X a certain point e and denote by B the semigroup with the relations $xy = e$ $(x, y \in X)$ consisting of the points of the space X, and by A a free topological semigroup over X. Obviously B is a topological semigroup. As elements of A appear the finite sequences $(a_1, \cdots, a_m)$ of points of X which multiply according to the law

$$(a_1, \cdots, a_m)(a_{m+1}, \cdots, a_n) = (a_1, \cdots, a_n),$$

and as neighborhood of the element $(a_1, \cdots, a_m) \in A$ (cf., §7) appears the system of all sequences of the form

$$(x_1, \cdots, x_m), \quad x_i \in U_i \quad (i = 1, \cdots, m),$$

where U_i are some fixed neighborhoods of the points a_i in X respectively. The canonical homomorphic mapping $(a_1, \cdots, a_n) \rightarrow a_1, \cdots, a_m$ of the algebra A onto the algebra B will be continuous, but not open, since the open set (X, X) will be mapped into the point $XX = e$ which does not represent an open set.

Nevertheless there exists a large system of classes $\mathfrak{S}$ for which the canonical homomorphism will be open.

Theorem 3. *Let A be a free topological algebra over the topological space X with the defining relations S in the class of identities $\mathfrak{S}$; let B be a topolo-*

gical algebra over a space Y with the defining relations T in the class of identities $\mathfrak{T}$. Further we suppose that a continuous mapping τ of the space X onto the space Y has the property that every identity of S, written in the corresponding elements of Y, is an identity in T, and that every identity of $\mathfrak{S}$ is contained in $\mathfrak{T}$. Then the mapping τ can be continued as a continuous homomorphism τ^* of the algebra A onto B. This homomorphism will be introduced as open if in the class $\mathfrak{S}$ all congruences are interchangeable, if B contains Y topologically, and if the mapping τ of the space X onto Y is a homeomorphism.*

The possibility to continue τ into a continuous homomorphism τ^* follows from the fundamental definition of §1, since it is possible to replace those identities of $\mathfrak{T}$ which are superfluous in comparison with $\mathfrak{S}$, by an additional system of individual relations to be included into T.

For the proof that τ^* is open let us denote by θ the congruence relation on A which corresponds to the homomorphic mapping τ^* of the algebra A onto the algebra B. Since the homomorphism τ^* is continuous and the congruences on $\mathfrak{S}$ are interchangeable, it follows from Theorems 10 and 11 of the paper [12] that the congruence θ is continuous and complete. Therefore we may consider the topological factor algebra A/θ with the open continuous homomorphism $A \to A/\theta$ and the continuous, both-ways one-valued homomorphism $A/\theta \to B$ (cf., [12]). By supposition B contains Y. Denoting by σ the canonical mapping of X into A we have the continuous mappings $X \to X^\sigma \to Y \to X$, whence it is evident that σ is a homeomorphism, i.e., that A contains X. Under the mapping $A \to A/\theta$ distinct elements of X have distinct images in A/θ. Moreover, this mapping is continuous and open. Thus the image X_θ of the set X in A/θ is contained topologically in A/θ. Under the mapping $Y \to X_\theta$ all defining relations of T will be satisfied for the corresponding elements of X_θ. Hence the mapping $Y \to X_\theta$ may be continued into a continuous homomorphism ϕ of the algebra B onto A/θ. This is evident for the inverse mapping $A/\theta \to B$ and therefore ϕ is a both-ways continuous isomorphism between the algebras A/θ and B, q.e.d.

Consequence. *Every topological algebra B, defined in the primitive class $\mathfrak{S}$ with interchangeable congruences contained in B, with generating space X and a system T of defining relations, is topologically isomorphic with the factor algebra of a free topological algebra A over X in the class $\mathfrak{S}$ with respect to a suitable continuous and complete congruence relation.*

For the proof one has to take in Theorem 3 as S the empty set and to put $X = Y$, so that τ is the identity mapping.

As a special case we point out that *every topological algebra B that belongs*

* Interchangeability of congruences in a class $\mathfrak{S}$ of algebras is to be understood in the sense of the paper [12].

to the primitive class $\mathfrak{S}$ with interchangeable congruence relations, appears as a continuous and open homomorphic image of a free topological algebra A in the class $\mathfrak{S}$ over the space B.

This statement is verified from the preceding consequence by identifying X and B and by taking in place of T all the relations between the elements of B.

Now we introduce some definitions. Suppose that the topological algebra A is finitely generated by the elements of one of its subsets, say X. We denote by S the system of all the relations between the elements of X in the algebra A and we consider the algebra B that has X as generating space and S as defining relations. According to remark 2 the algebra B contains X topologically and the continuous homomorphism of B onto A which appears as continuation of the identity mapping of X onto X, will be an algebraic isomorphism between B and A. Thus we may take the algebra B to be the same as the algebra A, only provided with another topology which we shall call *free with respect to* X. From the basic definition of §1 it follows that the identity mapping of the algebra A with the free topology with respect to X onto the algebra A with an arbitrary other topology, compatible with the given topology in the set X, will be continuous. This property is characteristic for the free topology.

Let X_α be topological spaces any two of which have no common points. Let $X = \bigcup X_\alpha$ be their union. A topology of X is called a *free topology* of X if unions of open subsets of the spaces X_α are open subsets of X in this topology.

We consider topological algebras A_α of one and the same class $\mathfrak{S}$ which have no common elements; we assume that A_α, A_β possess topologically isomorphic subalgebras $A_{\alpha\beta}$, $A_{\beta\alpha}$. The isomorphism of $A_{\alpha\beta}$ onto $A_{\beta\alpha}$ we denote by $\sigma_{\alpha\beta}$. The topological algebra, defined in the class $\mathfrak{S}$ by the space $\bigcup A_\alpha$ with its free topology and the defining relations $a^{\sigma_{\alpha\beta}} = a$ $(a \in A_{\alpha\beta})$, will be called the *free topological union of the algebras A with the identified subalgebras* $A_{\alpha\beta}$, $A_{\beta\alpha}$.

In the case that the subalgebras $A_{\alpha\beta}$ are empty, the free union with the identified subalgebras is called simply the *free union*. If the set of operations is empty, then we obtain the definition of the free union of topological spaces with identified subspaces.

The free topological union of algebras can also be characterized as follows: An algebra A is free topological union of the algebras A_α in the class $\mathfrak{S}$ if and only if there exist continuous homomorphisms σ_α of the algebras A_α into A with the following properties:

1) A is topologically generated by the elements of $\bigcup A_\alpha^{\sigma_\alpha}$;

2) for every selection of continuous homomorphisms τ_α of the algebras A_α into some algebra B of the given class $\mathfrak{S}$ there is a continuous homomorphism τ of the algebra A into B such that $a^{\sigma_\alpha\tau} = a^{\tau_\alpha}$ for $a \in A_\alpha$.

From Theorem 2 follows immediately that a free topological union is finitely generated by the images of the elements of the united algebras.

It is easy to give an example where the canonical mappings σ_α of the given algebras into their free union are no isomorphisms. In the general case we have

Theorem 4. *Suppose that each of the given topological algebras A_α of the class $\mathfrak{S}$ can be imbedded topologically into a topological algebra C_α of the same class, which contains a subalgebra consisting of a single element. Then the canonical mappings of the given algebras into their free union will be topological isomorphisms. The images of the algebras A_α will be closed subalgebras in their free union if the A_α are closed in C_α.*

Indeed, let each of the given algebras A_α be imbedded into a topological algebra C_α with a one-element subalgebra $\{e_\alpha\}$. We form the topological direct product $H = \Pi C_\alpha$ and consider the mapping ϕ_α of the algebra C_α into H, defined by

$$e^{\phi_\alpha}(\lambda) = e_\lambda, \qquad a^{\phi_\alpha}(\alpha) = a \qquad (\lambda \neq \alpha,\ a \in C_\alpha).$$

Obviously ϕ_α is a topological isomorphism of A_α onto the closed subalgebra $H_\alpha = C_\alpha^{\phi_\alpha}$ in H. In conformity with the characteristic property of the free union the homomorphisms ϕ_α can be continued into continuous homomorphisms of the whole free union into the algebra H. Since the C_α, H_α are topologically isomorphic and the A_α closed in C_α, the free union will contain A_α as a closed subalgebra.

§3. L-operations

The parallelism between the theory of abstract algebras and the theory of topological algebras becomes more perceptible if we look at the topological space as a discrete algebra with a system of partial operations, viz., limiting processes. The difficulty cousists in the fact that these operations depend on an infinity of variables and, in particular, that their properties are not described by means of identities. Nevertheless the juxtaposition of the operator point-of-view in the topological space with the ordinary point of view is of interest and will be briefly shown in the present section.

Let A be some set, Σ an auxiliary set of indices, which is partially ordered such that any two elements in Σ have a greater one in common. Any mapping of the set Σ into A: $\alpha \to a_\alpha$ is called a *direction* $[a_\alpha]$ in A with the carrier Σ. The direction $[a_\alpha]$ in the topological space A is said to be converging to the point a, symbolically $\lim [a_\alpha] = a$, if for every neighborhood U of the point a there is an index $\alpha \in \Sigma$ such that all a_λ are contained in U if $\lambda > \alpha$. In Hausdorff spaces a direction can converge to not more than one point. In T_0 or T_1-spaces (cf., [1]) directions may exist which converge to several points. The set F in a topological space A is closed if and only if all limit points of directions of elements of F belong to F (cf., [2], p. x).

The subset Σ' of the elements of a partially ordered set will be called a *cofinal subset* if for every $\alpha \in \Sigma$ there is an α' in Σ' which is greater than α (cf., [2], p. xii, translator's note).

In Hausdorff spaces the operation lim has the following obvious properties:

L1. *For every* Σ *the operation* lim *is a one-valued partial operation.*

L2. *If for all* $\alpha \in \Sigma$ *one has* $a_\alpha = a$, *then* $\lim [a_\alpha] = a$.

L3. *If* $\{\alpha_i\}$ *is a cofinal subset of* $\{\alpha\}$ *and* $\lim [a_\alpha]$ *exists, then also* $\lim [a_{\alpha_i}]$ *exists and coincides with* $\lim [a_\alpha]$.

It has been mentioned that, from the algebraic point of view, taking the limit of a direction can be considered as a partial operation with an infinity of arguments, or more accurately, as a system of partial operations which are distinguished by the carriers. The conditions L2, L3 have the same form as the identical relations joining the partial operations in the primitive classes. Thus we give the whole question another turn and assume that in a certain set A to certain directions $[a_\alpha]$ is associated one definite element $L[a_\alpha]$ each, from A, and that the operations defined this way satisfy the conditions L1 - L3. In this case we shall agree to call A an *L-algebra.* Subsets of A which are closed with respect to the L-operations, in particular subalgebras which are L-algebras in A, will be called *L-closed.* From L3 follows that the union of two L-closed sets is L-closed, and from L2 we conclude that every point for itself is an L-closed set. Since the intersection of a system of L-closed sets is L-closed, one can say that the L-closed sets form a system of all closed sets with respect to a definite topology in A which we shall call an *L-topology*. Making use of the L-topology one can define the notion of limit $\lim [a_\alpha]$ of a direction $[a_\alpha]$ in A.

If the operation L *has been defined for a direction* $[a_\alpha]$, *then* $\lim [a_\alpha] = L[a_\alpha]$ *where* lim *is to be taken in the sense of the* L*-topology.*

Indeed let U be a neighborhood of the point $a = L[a_\alpha]$ and F be the completion of U in A. We choose from the direction $[a_\alpha]$ those elements a_{α_i}, which are contained in F. If $\{\alpha_i\}$ is a cofinal subset of $\{\alpha\}$, then by axiom L3 the expression $L[a_{\alpha_i}]$ is defined and equal to a. Because F is closed we have $a \in F$ in contradiction to the condition $a \in U$. Hence $\{\alpha_i\}$ cannot be a cofinal subset of $\{\alpha\}$; an α can therefore be found such that for all $\lambda > \alpha$ we have $a_\lambda \in U$, q.e.d.

The theory of abstract algebras with generating set and defining relations, well known for algebras with partial operations with a finite number of arguments can without much difficulty be applied to algebras with partial operations on an infinity of arguments. Thus for the definition of a topological algebra with generating space X it might be possible to try to replace the topology on X by the system of L-operations and consider an algebra, part of whose operations would be finite and part of them infinite (new L-operations). However, in the development of this program there arise difficulties because the axioms L1-L3 do not guaranty a Hausdorff L-topology. This difficulty could be avoided if the properties of the limit processes in Hausdorff spaces could be completely characterized by axioms of the type L1-L3. But the complete axiomatics, as indicated for instance by Birkhoff (cf., [4]), contains axioms of an essentially different type. The axiomatics, given by Čogošvili [17] for other classes of topological spaces, also contains axioms of an undesirable form. Therefore in the following discussion we shall use L-operations only as a convenient method for the definition of a topology and as a suggestive design for definitions.

Further we need the following property of an L-topology which establishes an analogy between algebraic homomorphism and continuity.

Let X and Y be topological spaces in which the topology is defined by L-operations. A mapping σ of X into Y is continuous if from every equation $L[x_\alpha] = x$ in X follows the equation $L[x_\alpha^\sigma] = x^\sigma$ in the space Y.

Indeed let F_1 be some closed subset of Y and F its complete original in X. If $[x_\alpha]$ is a direction in F and $L[x_\alpha] = x$, then by supposition

$$L[x_\alpha^\sigma] = x^\sigma,$$

whence $x^\sigma \in F_1$, $x \in F$, F closed in X and consequently σ is continuous.

§4. The primitive topology of algebras

Let A be a topological algebra which is finitely generated by the elements of one of its subsets, say X. In §2 we have defined the free topology of the algebra A with respect to X. Independent of the prescribed topology in A one can obtain this topology from the topology of the space X by the following transfinite method.

The set U of elements of A is said to be X_0-open if for every polynomial f, and among these the polynomial $f(x) = x$, and for each system of elements $x_1, \cdots, x_n$ in X, for which $f(x_1, \cdots, x_n) \in U$ one can find neighborhoods $U_1, \cdots, U_n$ of those elements of X such that $f(U_1, \cdots, U_n) \subseteq U$. Evidently the intersection of a finite system, and the union of any system of X_0-open sets are X_0-open sets. Therefore the X_0-open sets define a special topology in A, which we shall call the *X_0-topology* or the *primitive* topology of A.

Further we define by induction the X_λ-topology of A for transfinite indices λ. Suppose that for some transfinite λ the X_λ-topology has been defined. The set U in A will be called $X_{\lambda+1}$-open, if for every fundamental operation $f(x_1, \cdots, x_n)$ and every $a_1, \cdots, a_n$ in A for which $f(a_1, \cdots, a_n) \in U$, X_λ-neighborhoods $U_1, \cdots, U_n$ of the points $a_1, \cdots, a_n$ of A respectively can be found such that $f(U_1, \cdots, U_n) \subseteq U$. Further, if the X_λ-topologies have been defined for all λ under a bound α, then the set U will be called X_α-open when it is X_λ-open for all $\lambda < \alpha$.

A topologization π_1 of a set A will be called 'greater', or *of higher degree free*, than the topologization π_2 of the same set if every π_2-closed subset of A is also π_1-closed (cf., [1]). We shall see that in this sense the topologizations X_λ are monotonically decreasing with increasing λ, and that all of them contain the given topology of A and, more generally, every admissible topologization of A; thus without change in the topology of X the fundamental operations are made continuous.

Let τ be the smallest transfinite number for which $X_\tau = X_{\tau+1}$. Then X_τ is the required topology of the algebra A that is free with respect to X.

In fact from the relation $X_\tau = X_{\tau+1}$ follows that the fundamental operations of the algebra A are continuous in the topology X_τ. The topology X_τ of X coincides with the first primitive topology of A. Hence X_τ is the greatest of the admissible topologizations of A, q.e.d.

The topologies X_λ can also be determined in another way. Let A be again a topological algebra which is finitely generated by the subspace X. In the *abstract* algebra A we introduce the L_0-operation by putting

$$L_0[a_\alpha] = a \quad (a_\alpha \in A),$$

if there exists a polynomial $f(x_1, \cdots, x_n)$ and directions $[x_{1\alpha}], \cdots, [x_{n\alpha}]$ in X with a common carrier $\{\alpha\}$, such that

$$\begin{aligned} &f(x_{1\alpha}, \cdots, x_{n\alpha}\} = a_\alpha, \quad f(x_1, \cdots, x_n) = a, \\ &\lim [x_{i\alpha}] = x_i, \quad x_i \in X \quad (i = 1, \cdots, n). \end{aligned} \tag{1}$$

The lim-operation is here to be understood in the sense of the given topology of X. It is evident that the L_0-operation defined in this way satisfies the axioms L1-L3 and thus defines in A a certain L_0-topology which we may call the topology L_0. For transfinite λ we define the topologies L_λ by induction. Using the known L_λ-topology we introduce the $L_{\lambda+1}$-operation in A by putting $L_{\lambda+1}[a_\alpha] = a$ if there is a polynomial $f(x_1, \cdots, x_n)$ and directions $[a_{1\alpha}], \cdots, [a_{n\alpha}]$ in A with common carrier such that

$$f(a_{1\alpha}, \cdots, a_{n\alpha}) = a_\alpha, \quad \lim [a_{i\alpha}] = b_i, \quad f(b_1, \cdots, b_n) = a \quad (i = 1, \cdots, n),$$

where lim is to be taken in the sense of the L_λ-topology. The $L_{\lambda+1}$-topology thus defined in A will be the required topology $L_{\lambda+1}$. If α is the bound and L_λ known for $\lambda < \alpha$, then closed sets in the L_α-topology are by definition those which are closed in all L_λ-topologies where $\lambda < \alpha$.

For every order number λ the topology L_λ coincides with the topology X_λ of the algebra A.

We prove first that the topologies L_0 and X_0 coincide. Let U be an X_0-open set. By definition L_0-non-closedness of $F = \text{compl } U$ would mean that for certain directions $[x_{1\alpha}], \cdots, [x_{n\alpha}]$ in X and for a suitable polynomial f the following relations are valid

$$\lim [x_{i\alpha}] = x_i \quad (i = 1, \cdots, n), \tag{2}$$

$$f(x_{1\alpha}, \cdots, x_{n\alpha}) \in F, \quad f(x_1, \cdots, x_n) \in U. \tag{3}$$

Because U is X_0-open one can find in X neighborhoods $U_1, \cdots, U_n$ of the points $x_1, \cdots, x_n$ for which $f(U_1, \cdots, U_n) \subseteq U$. But then by (2) for all α which are greater than a certain λ, we have $x_{i\alpha} \in U_i$ $(i = 1, \cdots, n)$ and therefore $f(x_{1\alpha}, \cdots, x_{n\alpha}) \in U$ in contradiction to (3).

Let now F be an L_0-closed set. We shall prove that $U = \text{compl } F$ is X_0-open. If this is not so then a polynomial f and points $x_1, \cdots, x_n$ in X can be found such that $f(x_1, \cdots, x_n) \in U$ and for every system of neighborhoods $U_1, \cdots, U_n$ of the given points in X the set $f(U_1, \cdots, U_n)$ will have a non-empty intersection with F, and therefore there will exist points $u_1 \in U_1, \cdots, u_n \in U_n$ such that $f(u_1, \cdots, u_n) \in F$. Let us denote by Σ the set of all systems $(U_1, \cdots, U_n)$ of neighborhoods of the points $x_1, \cdots, x_n$. We order these systems by putting

$$(U_1, \cdots, U_n) \leq (V_1, \cdots, V_n),$$

if $U_1 \subseteq V_1, \cdots, U_n \subseteq V_n$. The choice of the elements $u_1, \cdots, u_n$ depends on the system $\alpha = (U_1, \cdots, U_n)$ and thus we denote these elements by $u_{1\alpha}, \cdots, u_{n\alpha}$. It is readily seen that

$$\lim [u_{i\alpha}] = x_i \quad (i = 1, \cdots, n),$$

and hence the operation L will for the direction $[f(u_{i\alpha}, \cdots, u_{n\alpha})]$ have the value $f(x_1, \cdots, x_n)$. As all the elements of the direction are contained in the L_0-closed set F, we conclude that $f(x_1, \cdots, x_n) \in F$ which is in contradiction to the supposition.

By transfinite induction one proves that L_λ coincides with X_λ.

The X_0-topology of an algebra A can be characterized in a typically intuitive way as the free union of topologies of an essential sequence of closed subsets of A. For this purpose we shall define as the *length* of a polynomial f of the relevant variables $x_1, \cdots, x_n$ the number of these variables which are actually used in writing down the polynomial. Thus if f, g are binary fundamental operations, then the polynomial $f(x, g(x, y))$ has the length 3. Let X be a subset of elements of an algebra A. By X^s we shall denote the set of all those elements of A which can be written as values of polynomials of length $\leq s$ of elements of X.

Theorem 5. *Let A be an abstract algebra which is generated by one of its subsets, say X. In some way we topologize X and we call a direction $[a_\alpha]$ in A*

L_0-converging to $a \in A$ if for a suitable polynomial f and suitable elements $x_{i\alpha} \in X$ the relations (1) are satisfied. Suppose that the L-operation defined in this way is one-valued and that in the corresponding L_0-topology of the algebra A the sets X^s ($s = 1, 2, \cdots$) are closed. A set F in A is closed if and only if all intersections $F \cap X^s$ are closed.

The condition is obviously necessary. Let us prove that it is sufficient. Suppose that the intersections $F \cap X^s$ are closed, and that for a certain direction $[a_\alpha]$ in F we have $L_0[a_\alpha] = a$, so that for a certain polynomial f and for suitable $x_{i\alpha} \in X$ the equalities (1) are satisfied. If t denotes the length of f, then we have

$$a_\alpha \in X^t \cap F,$$

whence in view of the supposed closedness of the set $X^t \cap F$ we obtain

$$a \in X^t \cap F,$$

i.e., F is closed.

Theorem 6. *Let A be an abstract free algebra with the free generating set X in a certain primitive class $\mathfrak{S}$. By an arbitrary method we topologize X and we introduce in A an L_0-operation in the same way as in Theorem 5. Then this operation will be one-valued and in the L_0-topology of the algebra A all sets X will be closed.*

Let for instance

$$L_0[a_\alpha] = a, \quad L_0[a_\alpha] = b$$

for a certain direction $[a_\alpha]$ in A. This means that for suitable polynomials f, g and for suitable $x_{i\alpha} \in X$ we have

$$a_\alpha = f(x_{1\alpha}, \cdots, x_{m\alpha}), \quad \lim x_{i\alpha} = x_i \quad f(x_1, \cdots, x_m) = a, \tag{4}$$

$$a_\alpha = g(x_{m+1\alpha}, \cdots, x_{n\alpha}), \quad \lim x_{i\alpha} = x_i \quad g(x_{m+1}, \cdots, x_n) = b. \tag{5}$$

Consider a number α such that for all $i, j = 1, \cdots, n$ from $x_i \neq x_j$ follows $x_{i\alpha} \neq x_{j\alpha}$. Since the algebra A is free we conclude that all relations between its generating elements are identities in A. In particular the equality

$$f(x_{1\alpha}, \cdots, x_{m\alpha}) = g(x_{m+1\alpha}, \cdots, x_{n\alpha})$$

will be an identity. Therefore one can replace in it the elements $x_{1\alpha}, \cdots, x_{n\alpha}$ by $x_1, \cdots, x_n$, because then the identical elements will be replaced by identical elements. In the end we obtain the equality $a = b$ which proves the operation L_0 to be one-valued.

In order to prove the set X^s to be closed, let us suppose that for a certain direction $[a_\alpha]$ in X^s the equalities (4) are satisfied, whereby the length of the polynomial f may also be greater than s. The conditions $a_\alpha \in X^s$ indicate that for suitable polynomials g_α of length not exceeding s

$$a_\alpha = g_\alpha(x_{m+1\alpha}, \cdots, x_{n_\alpha \alpha}) \quad (x_{j\alpha} \in X, \ n \leq s). \tag{6}$$

Now there is only a finite number of fundamental operations; thus among the polynomials $g_a(y_1, \cdots, y_{n_a})$, where the $y_1, \cdots, y_{n_a}$ are auxiliary symbols, there is only a finite number of distinct ones and therefore there exists a cofinal subdirection $[a_{a_\beta}]$ for which all the polynomials $g_{a_\beta}(y_1, \cdots, y_n)$ are mutually identical. Let

$$g_{a_\beta}(y_1, \cdots, y_n) = g(y_1, \cdots, y_n) \quad (n = n_{a_\beta}).$$

Then the equalities (4), (6) yield the relations

$$f(x_{1a_\beta}, \cdots, x_{ma_\beta}) = g(x_{m+1a_\beta}, \cdots, x_{na_\beta}), \tag{7}$$

which are identities in A. Choosing again $\gamma = a_\beta$, so that from $x_i \neq x_j$ follows $x_{i\gamma} \neq x_{j\gamma}$ $(i, j = 1, \cdots, m)$ and replacing on both sides of the identity (7) the arguments $x_{1\gamma}, \cdots, x_{m\gamma}$ by $x_1, \cdots, x_m$ we obtain relations of the form

$$a = g(z_1, \cdots, z_n) \quad (z_j \in X),$$

which show that a is contained in X^s.

Theorem 7. *If the free topological algebra A with the generating space X has an X_0-topology, then the subalgebra B which is finitely generated by some closed subspace Y of the space X, will be closed in A and will have a Y_0-topology; in this way it will appear as free topological algebra over Y.*

Indeed every subset F of elements of B, closed in B in the sense of the X_0-topology, is evidently closed in the sense of the Y_0-topology. Conversely if F is closed in the Y_0-topology and for a certain direction $[b_a]$ in F we have $L[b_a] = c$ where the operation L is taken in the sense of the X_0-topology, then

$$b_a = f(x_{1a}, \cdots, x_{na}), \quad \lim x_{ia} = x_i, \quad f(x_1, \cdots, x_n) = c \quad (x_i, x_{ia} \in X)$$

and at the same time

$$b_a = \phi_a(y_{1a}, \cdots, y_{m_a a}) \quad (y_{ja} \in Y).$$

Now we choose a cofinal subdirection $[b_\beta]$ in the direction $[b_a]$ such that for each i either all the elements $x_{i\beta}$ are contained in Y or none of them is contained in Y. Because A is free, the equalities

$$f(x_{1\beta}, \cdots, x_{n\beta}) = \phi_\beta(y_{1\beta}, \cdots, y_{m_\beta \beta})$$

are identities in A. Therefore we can replace all the elements $x_{i\beta}$ on the left side which are not contained in Y, by an arbitrary fixed element $y \in Y$ without destroying these equalities. But then also the corresponding elements on the left side of the equality $f(x_1, \cdots, x_n) = c$ can be replaced by the same element y. After this change all the arguments on the left side belong to Y, since those which have not been changed will be equal to limits of directions of Y, and as such belong to Y because Y is closed in X. This shows that $c \in B$ and $c = L[b_\beta]$, where L is taken in the sense of the Y_0-topology. Since F is Y_0-closed it follows that $c \in F$, i.e., the set F is X_0-closed.

§5. Algebras with a bicompact generating space

We now consider important cases in which a free topology is known to appear as an L_0-topology.

Lemma. *Suppose that the topological algebra A is finitely generated by the elements of its subset X which appears as union of a chain of nested bicompact subsets $X_1 \subseteq X_2 \subseteq \cdots$. The subset F in A is said to be ω-closed if for an arbitrary polynomial $f(x_1, \cdots, x_n)$ all the intersections of the form $F \cap f(X_i, \cdots, X_i)$ $(i = 1, 2, \cdots)$ are closed. The corresponding topology, called ω-topology will be a Hausdorff topology and the fundamental operations of the algebra A will be continuous in the ω-topology.*

Indeed since X_i is bicompact, the same is true of the set $f(X_i, \cdots, X_i)$ which therefore is closed in A. Hence follows that all the closed subsets of A will be ω-closed and in particular the ω-topology will be a Hausdorff topology.

In order to prove the ω-continuity of the fundamental operations of A we enumerate all the polynomials $f_1, f_2, \cdots, f_m, \cdots$ and put

$$Y_m = f_1(X_m, \cdots, X_m) \cup \cdots \cup f_m(X_m, \cdots, X_m) \quad (m = 1, 2, \cdots).$$

Clearly all the Y_m are bicompact, $Y_1 \subseteq Y_2 \subseteq \cdots$, and the ω-closedness of a set F is equivalent with the closedness of all the intersections $F \cap Y_m$ $(m = 1, 2, \cdots)$.

Let W be an ω-open set, and for some fundamental operation f and certain $a_1, \cdots, a_n$ in A let $f(a_1, \cdots, a_n) \in W$. By p we denote the smallest number such that $a_1, \cdots, a_n$ are contained in Y_p; by induction we construct the sets

$$U_m(a_1), \cdots, U_m(a_n) \qquad (m = p, p + 1, \cdots)$$

with the following properties

1. $a_i \in U_m(a_i)$, $U_m(a_i) \subseteq Y_m$, $U_m(a_i)$ open in Y_m;
2. $\overline{U_m(a_i)} \subseteq U_{m+1}(a_i)$;
3. $f(\overline{U_m(a_1)}, \cdots, \overline{U_m(a_n)}) \subseteq W$,

where the upper bar indicates a closure operation in the given topology of A.

We introduce the notation $W_m = W \cap Y_m$; then for a suitable q we have

$$f(Y_p, \cdots, Y_p) \subseteq Y_q, \quad f(a_1, \cdots, a_n) \in W_q.$$

As W_q is open in Y_q, one can find a set V, open in A, such that

$$V \cap Y_q = W_q.$$

Now $f(a_1, \cdots, a_n) \in V$ and the operation f is continuous in A; hence there are neighborhoods $V_1, \cdots, V_n$ of the elements $a_1, \cdots, a_n$ in A for which

$$f(V_1, \cdots, V_n) \subseteq V.$$

Putting

$$V_i' = V_i \cap Y_p \qquad (i = 1, \cdots, n),$$

we have

$$f(V_1', \cdots, V_n') \subseteq V \cap Y_q \subseteq W.$$

Finally, denoting by $U_p(a_i)$ the neighborhood of a_i in Y_p whose closure is contained in V_i', we get the original sets $U_p(a_1), \cdots, U_p(a_n)$ thus having the properties 1, 3.

Suppose that we know already that $U_m(a_1), \cdots, U_m(a_n)$ have the three properties 1, 2, 3, for all $m \leq s$. We construct $U_{s+1}(a_i)$ $(i = 1, \cdots, n)$. Let q be an integer such that

$$f(Y_{s+1}, \cdots, Y_{s+1}) \subseteq Y_q,$$

and consider an arbitrary system of points

$$a_1' \in \overline{U_s(a_1)}, \cdots, a_n' \in \overline{U_s(a_n)}.$$

Then $f(a_1', \cdots, a_n') \in W_q$; hence there is an open set V such that $V \cap Y_q = W_q$, and neighborhoods $V_1, \cdots, V_n$ of the points $a_1', \cdots, a_n'$ in A such that

$$f(V_1, \cdots, V_n) \subseteq V.$$

Putting

$$V_i' = V_i \cap Y_{s+1}$$

we have

$$f(V_1', \cdots, V_n') \subseteq W$$

where $V_1', \cdots, V_n'$ are open in Y_{s+1}. The neighborhoods $V_1', \cdots, V_n'$ depend on the choice of $a_1', \cdots, a_n'$. Fixing $a_1', \cdots, a_{n-1}'$ and substituting an arbitrary a_n' from $\overline{U_s(a_n)}$ we see that $\overline{U_s(a_n)}$ will be covered by neighborhoods V_n'. Because of the bicompactness of $\overline{U_s(a_n)}$ a finite number of neighborhoods among the V_n' can be found which cover $\overline{U_s(a_n)}$. Let us denote these neighborhoods by $V_{n1}', \cdots, V_{nr}'$, and let $V_{1i}', \cdots, V_{n-1i}'$ be the corresponding neighborhoods of the points $a_1', \cdots, a_{n-1}'$ for V_{ni}' $(i = 1, \cdots, r)$. Putting

$$V_n'' = \bigcup V_{ni}', \qquad V_\alpha'' = \bigcap V_{\alpha i}' \qquad (i = 1, \cdots, r;\ \alpha = 1, \cdots, n-1),$$

we see that for each system of points

$$a_1' \in \overline{U_s(a_1)}, \cdots, a_{n-1}' \in \overline{U_s(a_{n-1})}$$

there exist sets $V_1'', \cdots, V_n''$ open in Y_{s+1}, for which

$$f(V_1'', \cdots, V_n'') \subseteq W, \qquad \overline{U_s(a_n)} \subseteq V_n'', \qquad a_\alpha' \in V_\alpha'' \qquad (\alpha = 1, \cdots, n-1).$$

Fixing $a_1', \cdots, a_{n-2}'$ and substituting an arbitrary a_{n-1}' from $\overline{U_s(a_{n-1})}$ we obtain a system of sets V_{n-1i}'' open in Y_{s+1} which cover the bicompact set $\overline{U_s(a_{n-1})}$. From these we choose a finite number $V_{n-11}'', \cdots, V_{n-1t}''$ that covers $\overline{U_s(a_{n-1})}$ and denoting by $V_{1\alpha}'', \cdots, V_{n-2\alpha}''$ the corresponding neighborhoods of the points $a_1', \cdots, a_{n-2}'$ in Y_{s+1} for $V_{n-1\alpha}''$ and putting

$$V_{n-1}'' = \bigcup V_{n-1\alpha}'', \qquad V_i''' = \bigcap V_{i\alpha}'' \qquad (i \neq n-1;\ \alpha = 1, \cdots, t),$$

we see that for every system of points

$$a_1' \in \overline{U_s(a_1)}, \cdots, a_{n-2}' \in \overline{U_s(a_{n-2})}$$

there exist sets $V_1''', \cdots, V_n'''$ open in $\underline{Y_{s+1}}$ and such that

$$f(V_1''', \cdots, V_n''') \subseteq W, \quad \overline{U_s(a_\lambda)} \subseteq V_\lambda''', \quad a_i' \in V_i'''$$

$$(i = 1, \cdots, n-2; \ \lambda = n-1, n).$$

By continuing this process we find eventually sets $V_1^{(n+1)}, \cdots, V_n^{(n+1)}$ open in Y_{s+1}, which contain $\overline{U_s(a_1)}, \cdots, \overline{U_s(a_n)}$ and for which

$$f(V_1^{(n+1)}, \cdots, V_n^{(n+1)}) \subseteq W. \tag{8}$$

As a bicompact space Y_{s+1} is normal and thus there are sets $U_{s+1}(a_1), \cdots, U_{s+1}(a_n)$ open in Y_{s+1} and containing the sets $\overline{U_s(a_i)}$ respectively; moreover

$$\overline{U_{s+1}(a_i)} \subseteq V_i^{(n+1)} \quad (i = 1, \cdots, n).$$

In view of (8) the sets $U_{s+1}(a_i)$ satisfy the conditions 1, 2, 3, and consequently we have proved the existence of a system of sets $U_m(a_i)$ $(m = p, p+1, \cdots)$ that have these three properties.

Put

$$U(a_i) = \bigcup_m U_m(a_i) \qquad (i = 1, \cdots, n).$$

From the conditions 2, 3 follows that

$$f(U(a_1), \cdots, U(a_n)) \subseteq W. \tag{9}$$

On the other hand

$$U(a_i) \cap Y_s = \bigcup_m (U_m(a_i) \cap Y_s) = U_s(a_i) \cup (U_{s+1}(a_i) \cap Y_s) \cup \cdots.$$

All the terms $U_{s+t}(a_i) \cap Y_s$ are intersections of Y_s with a set $U_{s+t}(a_i)$ open in Y_{s+t}; they are open in Y_s. Hence $U(a_i) \cap Y_s$ is open in Y_s and $U(a_i)$ is ω-open. Thus from (9) we conclude that the operation f is continuous in the ω-topology

Finally let the ω-topology in X coincide with the topology defined in X as the space of the given topological algebra A. We shall show that in this case the ω-topology coincides with the X_0-topology of A. As the X_0-topology contains each admissible topology of A with respect to X, we conclude that the X_0-topology contains the initial topology of A and thus is Hausdorffian. By supposition we can say that the sets X_i, and together with them the sets Y_i are bicompact in the X_0-topology and therefore also closed. Consequently every X_0-closed set will also be ω-closed, i.e., the X_0-topology is contained in the ω-topology. The converse has already been established; thus in the present case the X_0-topology coincides with the ω-topology.

Consequence 1. *If a topological algebra A is generated finitely by the elements of its bicompact subspace X, then the free topology of A with respect to X is its X_0-topology.*

For the proof it is sufficient to put $X = X_1 = X_2 = \cdots$ in Theorem 8. Then one can take $Y_i = X_i$ $(i = 1, 2, \cdots)$. In the end the ω-topology of A will coincide with the X_0-topology.

Consequence 2. *If the free topological algebra A of the primitive class $\mathfrak{S}$ over the locally compact space X satisfying the second axiom of countability contains X topologically and is algebraically free over X, then the topology of A is its X_0-topology.*

Indeed let $U_1, U_2, \cdots$ be an open basis of X such that the closures $\overline{U}_i$ are compact. Put

$$V_m = U_1 \cup U_2 \cup \cdots \cup U_m, \quad X_m = \overline{V}_m \quad (m = 1, 2, \cdots).$$

All the X_m are bicompact; $X = \bigcup X_m$ and the set $F \subseteq X$ is closed in X if and only if all the intersections $F \cap X_m$ $(m = 1, 2, \cdots)$ are closed. We show that

$$f(X_m, \cdots, X_m) \cap F \subseteq X_m \cap F. \tag{10}$$

If that were not so one could find an element $x \in X$ for which

$$x = f(x_1, \cdots, x_n), \quad x \overline{\in} X_m, \quad x_1, \cdots, x_n \in X_m. \tag{11}$$

Since A is algebraically free over X the relation (11) is an identity; further we can without changing the right side substitute instead of x an arbitrary element of X, which means that in this case X consists of one single element. This contradicts the condition $x \overline{\in} X_m$. The relation (10) now shows that F is ω-closed if and only if F is closed in X.

We indicate one more proposition which follows immediately from consequence 1.

Theorem 9. *If the topological algebra A is defined by a bicompact generating space X and an arbitrary system of defining relations, then its space is normal.*

By definition the topology of the algebra A is a free topology with respect to the canonical image X^σ in its space X. But the space X is bicompact and so is therefore also its image X^σ. By consequence 1 the topology of the algebra A is an X_0^σ-topology. This means that the set $F \subseteq A$ is closed if and only if the intersections $F \cap (X^\sigma)^m$ $(m = 1, 2, \cdots)$ are closed. With other words this means that the space of A is a free union of an increasing chain of topological spaces $X^{\sigma 1} \subseteq X^{\sigma 2} \subseteq \cdots$ each of which is closed in the following one, all the subspaces $X^{\sigma m}$ being bicompact, and therefore normal. However, it is well known (cf., [5], p. 302) that the free union of an increasing chain of normal spaces is normal. Thus the algebra A is normal.

§6. Simpler types of free topological algebras

An identity of the form $f(x_1, \cdots, x_n) = g(x_1, \cdots, x_n)$ will be said to be *homogeneous* if each of the symbols x_i occurs in each of the polynomials f, g once and only once. As examples of primitive classes of algebras with homogeneous defining identities we may take the following: the absolutely free class where the system of identities is empty; the class of the semigroups with the one identity $x(yz) = (xy)z$; the class of the commutative semigroups with the two identities $x(yz) = (xy)z$, $xy = yx$; etc. It is readily seen that in a free algebra with homogeneous identities each identity with repeated arguments can be obtained from a homogeneous identity by a suitable identification of arguments.

Theorem 10. *A free topological algebra A with homogeneous defining identities over the topological space X will be algebraically free over X and the topology of A is an X_0-topology. The subalgebra B of the algebra A, finitely generated by the elements of some subspace Y of X, is a free topological algebra over Y.*

Let A be an abstract free algebra over X. As in Theorem 6 we introduce an X_0-topology in A. One has to show that this is a Hausdorff topology and that the fundamental operations of the algebra A are continuous in this topology. We show first that the sets of the form $\phi(U_1, \cdots, U_m)$ are open in the X_0-topology if ϕ is an arbitrary polynomial and $U_1, \cdots, U_m$ are open subsets of the space X. For this purpose we apply the initial definition of an X_0-open set as indicated in §4, and we suppose that $\psi(x_1, \cdots, x_n) \in \phi(U_1, \cdots, U_m)$ whence

$$\psi(x_1, \cdots, x_n) = \phi(x_{n+1}, \cdots, x_{n+m}) \quad (x_{n+i} \in U_i). \tag{12}$$

In accordance with a remark made above one can find a homogeneous identity

$$f(z_1, \cdots, z_s) = g(z_1, \cdots, z_s),$$

from which the identity (12) and be obtained by means of substitution of suitable elements $x_1, \cdots, x_n$ instead of $z_1, \cdots, z_s$. Let for instance x_i be substituted in place of $z_{m_i}, z_{m_i+1}, \cdots, z_{m_{i+1}-1}$ and

$$x_{n+m_i+\lambda} - x_i \qquad (i - 1, \cdots, n; \quad m_1 = 1, \quad \lambda = 0, 1, \cdots, m_{i+1} - m_i - 1).$$

Putting

$$V_i = U_{m_i} \cap \cdots \cap U_{m_{i+1}-1},$$

we get: $x_i \in V_i$, V_i open in X and

$$\psi(V_1, \cdots, V_n) \subset f(V_1, \cdots, V_n) = g(V_1, \cdots, V_n) \subset \phi(U_1, \cdots, U_m),$$

which means that $\phi(U_1, \cdots, U_m)$ is X_0-open. Let us consider now any two different elements

$$a = \phi(x_1, \cdots, x_k, x_{k+1}, \cdots, x_m), \quad b = \psi(x_1, \cdots, x_k, x_{m+1}, \cdots, x_n)$$

in A where $x_1, \cdots, x_n$ are different, where, however, $x_1, \cdots, x_k$ may be absent. By $U_1, \cdots, U_n$ we denote a set of mutually non-intersecting neighborhoods of the elements $x_1, \cdots, x_n$ in the space X. Then $P = \phi(U_1, \cdots, U_m)$ and $Q = \psi(U_1, \cdots, U_n)$ will be non-intersecting neighborhoods of a and b. Indeed if P, Q have a common element c such that $c = \phi(u_1, \cdots, u_m)$ $(u_i \in U_i)$ and $c = \psi(u_1, \cdots, u_n)$, then

$$\phi(u_1, \cdots, u_m) = \psi(u_1, \cdots, u_n). \tag{13}$$

Because the algebra A is free, we know that (13) is an identity in A and instead of $u_1, \cdots, u_n$ we may write $x_1, \cdots, x_n$ respectively. Thus we obtain the equality

$$\phi(x_1, \cdots, x_m) = \psi(x_1, \cdots, x_n)$$

contradicting the condition $a \neq b$. Consequently $P \cap Q = 0$ and the space of A is Hausdorffian. Finally let $a_1, \cdots, a_m$ be elements of A and with the fundamental operation ϕ let $a = \phi(a_1, \cdots, a_m)$. Let U be an X_0-open subset which contains a. Putting $a_i = f_i(x_{i1}, \cdots, x_{in})$ we shall have

$$a = \phi(f_1, \cdots, f_m) = F(x_{11}, x_{12}, \cdots, x_{mn}) \quad (x_{ij} \in X).$$

Because U is X_0-open it follows that for suitable sets $U_{11}, \cdots, U_{mn}$ which are open in the space X and contain the points $x_{11}, \cdots, x_{mn}$ respectively, we have $F(U_{11}, \cdots, U_{mn}) \subset U$ or $\phi(U_1, \cdots, U_m) \subset U$, where the sets $U_i = f_i(U_{i1}, \cdots, U_{in})$ according to what has been proved, are neighborhoods of the elements $a_1, \cdots, a_m$ in the algebra A. Thus we have proved the continuity of the operations and at the same time the first two statements of the theorem. The third statement follows immediately from the relation

$$f(U_1, \cdots, U_n) \cap B = f(U_1 \cap Y, \cdots, U_n \cap Y),$$

where f is any polynomial and $U_1, \cdots, U_n$ are arbitrary subsets of the space X.

A more delicate example is furnished by the free topological *semilattices*, i.e., algebras with one binary operation of addition and three defining identities:

$$x + x = x, \qquad x + y = y + x, \qquad x + (y + z) = (x + y) + z.$$

An algebraically free semilattice A with the generating set X is realized by the system of all the non-empty finite subsets of X where the ordinary union of subsets is taken as operation of addition. Thus the problem becomes the topologization of the system of all non-empty finite subsets of a topological space X such that the union operation on subsets becomes continuous and the topology of A is in correspondence with the topology of the space X.

One of these topologizations is the "ordinary" topologization: As neighborhood of the set consisting of the elements $x_1, \cdots, x_m$ appears a system of unions of non-empty finite subsets of certain fixed neighborhoods $U_1, \cdots, U_m$ of the points $x_1, \cdots, x_m$ in the space X. In virtue of the remark 1 of §1 one can say that *a free topological semilattice, generated by some topological space X, contains X and is algebraically free over X.*

This "ordinary" topology of a semilattice will in general not be its free topology. In order to verify this it will be sufficient to compare the "ordinary" topology with the X_0-topology. In fact, in the present case every polynomial in $x_1, \cdots, x_m$ has the form

$$k_1 x_1 + \cdots + k_m x_m, \quad x_i \in X$$

where

$$kx = x + x + \cdots + x.^{*}$$

Thus, by §4, a system U in A is X_0-open if and only if for every set $u = \{x_1, \cdots, x_m\}$ which is contained in U, and for every natural number k there are neighborhoods $U_1, \cdots, U_m$ of the points $x_1, \cdots, x_m$ in X such that a finite set, containing not more than k points from each neighborhood U_i, belongs to U.

Now let X be the system of the numbers $1, 2, \cdots, n, \cdots, \omega$ with its natural topology. The space X is bicompact and therefore as free topology of the semilattice A we shall have its X_0-topology. Let us denote by U the system of all the finite subsets of A which contain ω, and in addition all those subsets $x_1, \cdots, x_m$ not containing ω, whose elements, however, satisfy the condition

$$x_1 \cdots x_m \leq 2^m.$$

It is readily seen that U is an X_0-open system containing ω. At the same time U contains no "ordinary" neighborhood of the point ω in A.

§7. β-classes. Free topological groups

A primitive class of algebras, characterized by a set of identities $\mathfrak{S}$, will be called a *β-class*, if every free topological algebra of this class over an arbitrary *bicompact* space X contains X topologically and is algebraically free over X.

A free topological algebra A of the β-class which is generated by a completely regular space X, contains X as a closed subset and is algebraically free over X. A subalgebra B which is finitely generated in A by the elements of a closed subspace $Y \subset X$, is closed in A.

For the proof we imbed X into a bicompact space Z and arrange it so that $Y = Z \cap F$ where F is a closed subset of X. By supposition the free topological algebra D over Z of the class under consideration contains Z and is algebraically free over Z.

Let us denote by A_1, B_1, C the subalgebras of the algebra D, which are finitely generated by the elements of the sets X, Y, F respectively. Since D is algebraically free over Z, it follows that

$$A_1 \cap Z = X, \qquad A_1 \cap C = B_1. \tag{14}$$

But Z is closed in D and, by Theorem 7 and Theorem 8, consequence 1, the subalgebra C is closed in D. Thus from the relations (14) follows that the set X is closed in the algebra A_1, and B_1 is a closed subalgebra of the algebra A_1. Moreover A_1 and B_1 are algebraically free over X and Y respectively as subalgebras of the algebraically free algebra D (cf., [11]). When we pass over from the constructed topology of the algebra A_1 to the free topology of A_1 with respect to X, then all the closed sets in A_1 remain closed in the free topology of A_1, q. e. d.

*Translator's note: By defining relations $kx = x$.

A homomorphic mapping of an algebra A into an algebra B is called a *representation* of A in B. A system of representations ψ_i of the algebra A in the algebras B_i is said to be *complete*, if for every two distinct elements of the algebra A one can find a representation ψ_i, in which these two elements have distinct images in B_i. Further we recall that a topological space A is said to be *linearly connected* if for every two of its points, say a, b, one can find a continuous mapping of the interval $[0, 1]$ into A whereby 0 is mapped into a and 1 into b.

Lemma. *For a primitive class to be a β-class it is sufficient that the abstract free algebras of this class with a finite number of generators admit a complete system of representations in topological linearly connected algebras.*

In fact, let X be a bicompact space, A a free topological algebra over X, of the class under consideration. We take the two polynomials f, g of the elements $x_1, \cdots, x_m$ in X, whose values are different in the abstract free algebra S of the class $\mathfrak{S}$ with the generators $x_1, \cdots, x_m$. Let ψ be a representation of the algebra S in a suitable linearly connected algebra B whereby $f^\psi \neq g^\psi$. We choose in B an auxiliary point b distinct from $x_1^\psi, \cdots, x_m^\psi$ and we denote by ψ_i the continuous mapping of the interval $[0, 1]$ into B, by which the numbers 0, 1 are mapped into the elements b and x_i^ψ $(i = 1, \cdots, m)$ respectively. Now we separate in the space X the points $x_1, \cdots, x_m$ by the non-intersecting neighborhoods $U_1, \cdots, U_m$ and construct the continuous functions $\phi_i(x)$ mapping X onto the interval $[0, 1]$ in such a way that the completion of U_i in X is mapped into 0 at the point x_i into 1. Further putting $\phi(x) = \psi_i[\phi_i(x)]$ for $x \in U_i$ and $\phi(x) = b$ for $x \,\overline{\in}\, \bigcup U_i$ we obtain a continuous mapping of X into the algebra B whereby the points $x_1, \cdots, x_m$ are mapped into the elements $x_1^\psi, \cdots, x_m^\psi$ respectively. As A is topologically free over X we conclude that the mapping ϕ is continued into a homomorphism of A into B. But

$$f^\phi = f(x_1^\phi, \cdots, x_m^\phi) = f^\psi, \quad g^\phi = g^\psi,$$

and therefore we have also in the algebra A: $f \neq g$, which means that A is algebraically free over X. The statement that A contains X topologically is then an immediate consequence of the remark 2 in §1.

Coming to the applications we point out first of all that the lemma, proved just now, implies the central result of Markov's theory: *the class of the groups is a β-class.*

Indeed it is well known that the group of all non-singular matrices of the second order over the field of the complex numbers is linearly connected and contains free subgroups with a countable number of generators. By the lemma this guarantees that the class of the groups is a β-class.

Still in the way of an example we consider the class $\mathfrak{N}_s$ of the nilpotent groups of a given degree of nilpotency. This means that we consider groups whose elements satisfy an identity of the form

$$x_1 \circ x_2 \circ \cdots \circ x_{s+1} = e, \qquad (15)$$

where $x \circ y = x^{-1} y^{-1} xy$ and in the "product" (15) parentheses are put in ad libitum. The class $\mathfrak{N}_1$ is evidently the class of the abelian groups.

Theorem 11. *The nilpotent groups of a given degree of nilpotency form a β-class. The members of the lower central series $G \supset G^{(2)} \supset G^{(3)} \supset \cdots$ of a free topological group G over a completely regular space X are closed in G and the factor group $G/G^{(s+1)}$ is isomorphic to a free nilpotent topological group of degree s over X.*

It is, indeed, well known that the free nilpotent groups with a finite number of generators are groups without torsion and therefore belong to the nilpotent Lie groups which admit isomorphic representations by matrices (cf., [9]). Thus the free nilpotent groups with a finite number of generators admit of isomorphic representation in a suitable group of matrices. According to the lemma we conclude that the nilpotent groups of a given degree form a β-class. The second statement of Theorem 11 follows immediately from the consequence of Theorem 3, because congruences on groups are interchangeable, and continuous complete congruences correspond to a decomposition into remainder classes with respect to a closed normal divisor.

§8. Free topological rings

We shall consider rings that are not necessarily associative. The ring R is said to be of zero characteristic if from $mx = my$ follows $x = y$ when m is a positive integer and $x, y \in R$. A class of rings whose elements satisfy – apart from the ring axioms – a certain system of identities $\mathfrak{S}$, are said to be 'of primitive class of characteristic zero' if a free ring in this class with a countable number of generating elements has characteristic zero, and consequently the same is true for every free ring in this class $\mathfrak{S}$.

Theorem 12. *Every primitive class of rings with zero characteristic is a β-class.*

Let R be a free ring of the given class with a finite number of generating elements $x_1, \cdots, x_m$. By supposition R has the characteristic zero and therefore, according to [10], all identities that characterize the given class, can be reduced to the multi-linear form. Moreover if the characteristic of R equals zero it follows (cf., [16]) that R can be imbedded as a subring into an algebra over the field of the rational numbers. This field can be extended into the field of the real numbers and in this way we obtain an imbedding of the ring R into an algebra A over the field of the real numbers. In view of the multi-linearity of the identities that define R, all these are also satisfied in the algebra A, which in this way will be a free linear algebra over the field of the real numbers in the class $\mathfrak{S}$. We denote by I_n

the system of all elements of A, of degree $> n$ with regard to the generators $x_1, \cdots, x_m$. According to [10], I_n is an ideal in A. The algebra of the remainders A/I_n is a finite-dimensional linear algebra over the field of the real numbers, and thus appears as an r-dimensional euclidean space with respect to its natural topology. So we have obtained a linear connected topological algebra A/I_n with the corresponding natural homomorphism ψ_n of the algebra A, including the ring R, into A/I_n. Arbitrary elements of the ring R of a degree not exceeding n with respect to the generators have distinct images in A/I_n. Therefore the representations ψ_n form a complete system. In virtue of the lemma we conclude that the class of rings under consideration is a β-class.

As a special case of Theorem 12 we obtain that *the classes of all (non-associative) rings, all associative rings, all Lie rings, all commutative rings, etc., appear as β-classes.*

Among the rings of characteristic different from zero of particular importance are the Boolean rings, i.e., associative rings with unit element in which the identity $x^2 = x$ is satisfied. This identity, as is well known, implies the identity $2x = 0$ (cf., [2], p. 154) which shows that a Boolean ring has the characteristic 2. By Stone's theorem the class of Boolean rings is rationally equivalent* to a class of Boolean algebras, i.e., algebras with three operations $+, \cdot, '$, semilattices with respect to addition and multiplication, and related by the identities

$$x(y + z) = xy + xz, \quad x + xy = x, \quad (xy)' = x' + y', \quad x + yy' = x, \quad x'' = x.$$

In view of this one can consider free topological Boolean algebras instead of free topological Boolean rings.

The class of the Boolean algebras, and consequently the class of the Boolean rings are β-classes.

Indeed from the representation theory of Boolean algebras (cf., [2], p. 141) it is known that each abstract Boolean algebra admits of a complete system of homomorphisms onto an algebra with two elements $0, 1$ which is contained in any Boolean algebra. Thus in order to apply the lemma it suffices to set in evidence only one topological, linearly connected Boolean algebra. Such an algebra $\mathfrak{B}$ is well known. As its elements appear the unions of a finite number of semi-segments of the form $(a, b]$ $(0 \leq a < b \leq 1)$, and also the empty set. The operations $+, \cdot, '$, in this algebra coincide with union, intersection, complementation (taking the complement) in the semi-segment $(0, 1]$. Let P, Q be elements of this algebra $\mathfrak{B}$; as distance $\rho(P, Q)$ we introduce the expression

$$\text{mes}\,(P \cup Q) - \text{mes}\,(P \cap Q).$$

*'Rationally equivalent' means that in the Boolean ring one can define derived operations with regard to which the Boolean ring becomes a Boolean algebra and through the original ring operations can again be expressed as derivatives with analogous properties.

The relations (cf., [2], p. 77)

$$\rho(P \cup Q, S \cup T) + \rho(P \cap Q, S \cap T) \leq \rho(P, S) + \rho(Q, T),$$
$$\rho(P', S') = \rho(P, S),$$

show that in the topology, defined by this metric, the algebra becomes a topological algebra and, as it is readily established, linearly connected. Thus the statement is proved.

With regard to the ring addition, defined by the formula

$$P \oplus Q = P'Q + PQ',$$

the algebra $\mathfrak{B}$ is an abelian group, all elements of which have the order 2. This group is linearly connected; it is the very first example of a connected topological periodic group. A general method for the construction of such groups has been considered by A. A. Markov [14].

Finally we point out a few questions whose solution would be of interest for the development of the theory of free topological algebras.

In §4 we have constructed a transfinite sequence of topologies X_λ $(\lambda \leq \tau)$. The question is whether an arbitrary transfinite τ can actually occur with various A and various X. In a series of cases it has been established that $\tau = 0$. In the case of the free topological groups where X is not a completely regular space, one has $\tau \neq 0$, because all the subspaces of a group space are completely regular. What value can τ have for free groups over a completely regular space?

Another series of questions is connected with the algebraic structure of the free topological algebras. A free topological group over a regular space, will it be algebraically free? Is it possible to indicate some necessary purely algebraic characteristics of the β-classes? We also point out the question as to a clear characterization of the free topology, which, of course, is closely connected with the question concerning the possible values of τ, but cannot be reduced to this question.

LITERATURE

[1] P. S. Alexandrov, *On the concept of space in topology*, Uspehi Mat. Nauk (N.S.) 2 (1948), no. 1(17), 5-57. (Russian)

[2] G. Birkhoff, *Lattice theory*, Amer. Math. Soc. Colloq. Publ. no. 25, revised ed., New York, 1948.

[3] G. Birkhoff, *On the structure of abstract algebras*, Proc. Cambridge Philos. Soc. **31** (1935), 433-454.

[4] G. Birkhoff, *Moore-Smith convergence in general topology*, Ann. of Math. (2) **38** (1937), 39-56.

[5] M. I. Graev, *Free topological groups*, Izv. Akad. Nauk SSSR. Ser. Mat. **12**

(1948), 279-324. (Russian) = Amer. Math. Soc. Transl. no. 35 (1951).

[6] M. I. Graev, *On free products of topological groups*, Izv. Akad. Nauk SSSR. Ser. Mat. **14** (1950), 343-354. (Russian)

[7] W. Hämisch, *Über die Topologie in der Algebra*, Math. Z. 60 (1954), 458-487.

[8] S. Kakutani, *Free topological groups and infinite direct products of topological groups*, Proc. Imp. Acad. Tokyo 20 (1944), 595-598.

[9] A. I. Mal'cev, *On a class of homogeneous spaces*, Izv. Akad. Nauk SSSR. Ser. Mat. **13** (1949), 9-32. (Russian) = Amer. Math. Soc. Transl. no. 39 (1951).

[10] A. I. Mal'cev, *On algebras defined by identities*, Mat. Sb. **26** (68) (1950), 19-33. (Russian)

[11] A. I. Mal'cev, *On a class of algebraic systems*, Uspehi Mat. Nauk (N.S.) **8** (1953), no. 1(53), 165-171. (Russian)

[12] A. I. Mal'cev, *On the general theory of algebraic systems*, Mat. Sb. (N.S.) **35** (77) (1954), 3-20. (Russian)

[13] A. A. Markov, *On free topological groups*, Izv. Akad. Nauk SSSR. Ser. Mat. **9** (1945), 3-64. (Russian. English summary) = Amer. Math. Soc. Transl. no. 30 (1950).

[14] A. A. Markov, *On the existence of periodic connected topological groups*, Izv. Akad. Nauk SSSR. Ser. Mat. **8** (1944), 225-232. (Russian. English summary) = Amer. Math. Soc. Transl. no. 30 (1950).

[15] T. Nakayama, *Note on free topological groups*, Proc. Imp. Acad. Tokyo **19** (1943), 471-475.

[16] L. Ya. Okunev, *The ring as an algebra over a field*, Mat. Sb. **40** (1933), 410-424. (Russian)

[17] G. S. Čogošvili, *On convergence spaces*, Mat. Sb. (N.S.) **9** (51) (1941), 377-381. (Russian. German summary)

[18] N. A. Shanin, *On imbedding in a power of topological space*, Izv. Akad. Nauk SSSR. Ser. Mat. **8** (1944), 233-242. (Russian. English summary)

[19] R. Sikorski, *Products of abstract algebras*, Fund. Math. **39** (1952), 211-228.

Translated by:

Hans W. E. Schwerdtfeger

Yuri Vladimirovich Linnik

Born – 21 Jan 1915 in Belaya Tserkov, Ukraine
Died – 30 June 1972 in Leningrad (now St Petersburg), Russia

Yuri Linnik's father was Vladimir Pavlovitch Linnik who was himself a famous scientist making major contributions to optics. Like his son, Vladimir was elected to the USSR Academy of Sciences.

After studying at secondary school Yuri Linnik worked as a laboratory assistant for a year in 1931. Then he entered Leningrad University (St Petersburg before 1914 and now St Petersburg again) to study mathematics and theoretical physics. He graduated in 1938, obtaining a doctorate there in 1940. In the same year he joined the Leningrad branch of the Steklov Institute for Mathematics.

From 1944 Linnik was professor of mathematics at Leningrad University in addition to his position in the Steklov Institute. He organised the chair of probability theory there and founded the Leningrad school of probability and mathematical statistics.

His main research topics were number theory, probability theory and mathematical statistics. He introduced ergodic methods into number theory in his first work. In a 1941 paper he introduced the large sieve method in number theory. In 1950 he introduced the ideas of probability into number theory and introduced the dispersion method in number theory.

Later Linnik made major contributions to probability with his work on limit theorems and was the first to use powerful techniques from analysis in mathematical statistics. He solved the Behren–Fisher problem and many other difficult problems of mathematical statistics.

Linnik wrote several important texts including *Characterisation Problems in Mathematical Statistics.*

Linnik was President of the Moscow Mathematical Society for many years. He was honoured by the national mathematical society of Sweden and was awarded an honorary degree from Paris. He was elected to the USSR Academy of Sciences in 1964.

AN APPLICATION OF THE THEORY OF MATRICES AND OF LOBATSCHEVSKIAN GEOMETRY TO THE THEORY OF DIRICHLET'S REAL CHARACTERS

By Y. V. LINNIK

[Received March 12, 1956]

1. Many difficult problems of the analytical theory of Dirichlet characters are not resolved up to the present. Several of these problems concern the summation of the character values and the distribution of these values on narrow intervals.

A survey of these problems and some theorems on the connections between the different problems may be found in the paper [5]. One of the most difficult of such problems is the problem of the least quadratic non-residue. (I. M. Vinogradov, 1918).

Let D be an odd number, and $\chi(n)$ a real non-principal character (mod D); we shall call a number m such that $\chi(m) = -1$ a *non-residue* (mod D). Let $N_{\min}(D)$ be the least non-residue amongst the numbers $1, 2, \ldots, D-1$; I. M. Vinogradov's hypothesis consists in the relation

$$\lim_{D\to\infty} \frac{\ln N_{\min}(D)}{\ln D} = 0. \tag{1.1}$$

This hypothesis is not completely proved so far* though there are many reasons for the probable truth of it (see, for instance, [3]). The hypothesis (1.1) is an easy consequence of the Riemann hypothesis for $L(s, \chi)$; one may prove even the following conditional theorem.

This paper was communicated by title to the International Colloquium on Zeta-functions held at the Tata Institute of Fundamental Research, Bombay, on February 14-21, 1956.

* In the *Duke Math. Journal*, 21 (1954) appeared N. C. Ankeney's paper "Quadratic residues", where the author claims to prove (1.1) for primes $D \equiv -1 \pmod 4$. Unluckily, his arguments are completely erroneous and give no information about least non-residues (see K. A. Rodosski's review [6]).

CONDITIONAL THEOREM 1. *Let $\psi(D) \to \infty$ be a monotonic (as slowly increasing as we please) positive function. Form the series* $L(s, \chi) = \sum_{n=1}^{\infty} \chi(n)\, n^{-s}$, *$\chi(n)$ being a real non-principal character* (mod D). *If the number of zeros of these L-series inside the corresponding semi-circles*

$$|s-1| \leqslant \frac{\psi(D)}{\ln D}, \quad \operatorname{Re} s < 1, \tag{1.2}$$

may be estimated as $o(\psi(D))$, then hypothesis (1.1) *is true.*

2. For the sums of the non-principal characters $\chi(n)$ (mod D) the following Vinogradov-Polya estimate(†) is known:

$$\sum_{n \leqslant x} \chi(n) = O(\sqrt{D}.\ln D) \tag{2.1}$$

for all values of x.

This estimate enables us to study the distribution of the character values on intervals of length exceeding $\sqrt{D}$. $\ln D$. The behaviour of the $\chi(n)$ values on more narrow segments escapes all efforts so far.

Very little was added to the result of I. M. Vinogradov [8]:

$$\limsup_{D\to\infty} \frac{\ln N_{\min}(D)}{\ln D} \leqslant \frac{1}{2\sqrt{e}}. \tag{2.2}$$

As is well known, Dirichlet's real characters $\chi(n)$ are connected with binary quadratic forms; therefore it is possible to study them by means of these forms.

Such an approach is considered in the present report; the theorems obtained are essentially theorems in quadratic form theory, but some of them are directly related to Dirichlet's series and real characters.

We shall consider classical integral properly primitive (*propriae primitivae*) binary quadratic forms $\phi(x, y) = ax^2 + 2bxy + cy^2$ with the determinant $-D = b^2 - ac < 0$; the class number of these

† An equivalent of this inequality as well as Polya's summation formula were found recently in G. F. Voronoi's scientific diaries for the year 1907 (see [9]).

forms will be denoted by $h(-D)$; the Lagrange reduced forms will be those under the conditions

$$c > a > 2\,|b|, \tag{2.3}$$

(the equality signs are allowed sometimes).

To each form $\phi(x, y) = ax^2 + 2bxy + cy^2$ we make correspond the integral matrix with trace zero:

$$L = \begin{bmatrix} b & -a \\ c & -b \end{bmatrix}, \tag{2.4}$$

so that

$$L^2 = -D, \tag{2.5}$$

where $D = ac - b^2 > 0$ is the determinant of the matrix.

The classical theorem of C. L. Siegel holds :

$$\ln h(-D) \sim \tfrac{1}{2} \ln D, \tag{2.6}$$

for $D \to \infty$.

By means of the machinery of the matrices (2.4) it is possible to prove a refinement of this theorem and to connect it with Lobatschevskian geometry.

3. To each form $\phi(x, y) = ax^2 + 2bxy + cy^2$, besides the matrix (2.4) we make correspond the point (a, b, c) on the half-hyperboloid H :

$$ac - b^2 = D\,;\; a > 0. \tag{3.1}$$

Introducing the normed coordinates

$$x_1 = \frac{a}{\sqrt{D}},\; x_2 = \frac{c}{\sqrt{D}},\; x_3 = \frac{b}{\sqrt{D}},$$

we obtain the normed hyperboloid H_0:

$$x_1 x_2 - x_3{}^2 = 1\,;\; x_1 > 0. \tag{3.2}$$

The hyperboloid H_0 may be considered as an interpretation of the Lobatschevsky plane, with the points (x_1, x_2, x_3), and the straight lines, the hyperbolic sections of H_0 by the planes $a_1x_1 + a_2x_2 + a_3x_3 = 0$. If we are given a finitely connected figure S_0 on H_0, with a piecewise smooth contour, then its Lobatschevsky area, which will be denoted by $\Lambda(S_0)$, is numerically equal to the Euclidean volume of the cone with the vertex (0, 0, 0) and base S_0.

The basic points (a, b, c) on H (with the reducibility conditions (2.3) and the condition: g.c.d. $(a, 2b, c) = 1$ (greatest common divisor $= 1$)), are projected from (0, 0, 0) on the triangle $\Delta_0 \subset H_0$. This triangle is bounded by the Lobatschevsky straight lines: $x_2 - x_1 = 0$; $x_1 - 2x_3 = 0$; $x_1 + 2x_3 = 0$, and has the angles

$$\frac{\pi}{3}, \frac{\pi}{3}, 0; \ \Lambda(\Delta_0) = \frac{2}{9}\pi.$$

The number of the projections of basic points (a, b, c) on Δ_0 equals $h(-D)$ and is governed by C. L. Siegel's theorem (2.6). One may prove a theorem on the asymptotic number of these projections inside rather arbitrary subsets of Δ_0.

Let a convex figure Σ be given on the initial hyperboloid H, with piecewise smooth contour and the projection Σ_0 on Δ_0 completely situated inside Δ_0.

Let $H(\Sigma)$ be the number of the basic integer points inside Σ. We are interested in the ratio

$$\frac{H(\Sigma)}{h(-D)}.$$

4. In what follows, $\alpha \leqslant 0.01$ is a small positive constant. D being given, we draw the Lobatschevsky straight line $x_2 - \lambda x_1 = 0$, where $\lambda = (\ln D)^2$.

It cuts off the basic triangle Δ_0, the quadrangle $G_0(\lambda)$; moreover

$$\Lambda(\Delta_0) \sim \Lambda(G_0(\lambda)), \ \text{for } D \to \infty.$$

THEOREM 2. *Let $D \geqslant 0$ be odd, and let there exist a prime $p \geqslant 3$ such that $(-D/p) = +1$. Let the figure Σ_0, the projection of $\Sigma \subset H$, be situated completely inside $G_0(\lambda)$. Let, moreover, the Lobatschevsky area of Σ_0 satisfy the condition*

$$\Lambda(\Sigma_0) > (\ln D)^{-\alpha}. \tag{4.1}$$

Then the following asymptotic relation holds :

$$H(\Sigma)/h(-D) = \Lambda(\Sigma_0)/\Lambda(\Delta_0)\,(1 + \eta\,(p, D)), \tag{4.2}$$

where $\eta(p, D) \to 0$ *for a fixed* p *and* $D \to \infty$.

As we have $\Lambda(\Delta_0) = 2\pi/9$, we obtain

$$H(\Sigma) = \frac{9}{2\pi}\,h(-D)\,\Lambda(\Sigma_0)\,(1 + \eta(p, D)). \tag{4.3}$$

Thus we obtain a complement to C. L. Siegel's theorem in terms of Lobatschevskian geometry.

It should be remarked that we use an auxiliary arbitrary prime p under the condition $(-D/p) = +1$. Apparently the use of this prime is not essential, it is a technical defect of the proof.

We must remark that Theorem 2 is not a direct consequence of the Riemann hypothesis for $L(s, \chi)$ with $\chi = (-D/n)$; $-D < 0$ being a fundamental discriminant. But if $L(s, \chi)$ is known not to have zeros for $\sigma > 9/10$, $|t| \leqslant \frac{1}{2}$ $(s = \sigma + it)$, Theorem 2 can be improved : instead of $\eta(p, D)$ in (4.2) and (4.3) one can take $\eta(D) \to 0$ for $D \to \infty$; no auxiliary prime p is needed then.

5. From formula (4.2), by a simple computation of area on the Lobatschevsky plane, we can deduce the theorems immediately connected with Dirichlet's L-series.

Let $h(-D, \gamma\sqrt{D})$ be the number of properly primitive forms (a, b, c) for which $a \leqslant \gamma\sqrt{D}$. Computing the area $\Lambda(\Sigma_0)$ in (4.3) we deduce

THEOREM 3. *For* $0 \leqslant \gamma \leqslant 1$, $\gamma \geqslant (\ln D)^{-\alpha}$ *and a fixed* p, *we have*

$$h(-D, \gamma\sqrt{D}) = \frac{3\gamma}{\pi}\,h(-D)\,(1 + \eta\,(p, D)), \tag{5.1}$$

where $\eta(p, D) \to 0$ *for* $D \to \infty$ *uniformly on* $\gamma \geqslant (\ln D)^{-\alpha}$.

For $1 \leqslant \gamma \leqslant \sqrt{(4/3)}$ *we have*

$$h(-D, \gamma\sqrt{D}) = f(\gamma)\, h(-D)\, (1 + \eta(p, D)), \tag{5.2}$$

where

$$f(\gamma) = \frac{6}{\pi} \text{ arc sin } \sqrt{(1-\gamma^{-2})} + \frac{3\gamma}{\pi} (1 - 2\sqrt{(1-\gamma^{-2})}).$$

For $\gamma > \sqrt{(4/3)}$ *we have (trivially)*

$$h(-D, \gamma\sqrt{D}) = h(-D).$$

Here $\eta(p, D) \to 0$, *for fixed* p *and* $D \to \infty$.

A part of Theorem 3 may be formulated in terms of L-series.

Let $D < 0$ be an odd fundamental discriminant and $\chi(n) = (-D/n)$ Kronecker's character.

Let

$$(-D/p) = +1;\ L(s, \chi) = \sum_{n=1}^{\infty} \chi(n)\, n^{-s};$$

$$\zeta(s)\, L(s, \chi) = \sum_{n=1}^{\infty} a_n\, n^{-s}, \quad (s > 1).$$

THEOREM 4. *For* $\gamma \geqslant (\ln D)^{-\alpha}$ *we have*

$$\sum_{n \leqslant \gamma\sqrt{D}} a_n = \gamma\sqrt{D}.\, L(1, \chi)\, (1 + \eta(p, D)), \tag{5.3}$$

where $\eta(p, D) \to 0$ *for fixed* p, $D \to \infty$ *uniformly on* $\gamma \geqslant (\ln D)^{-\alpha}$.

6. Relation (5.3) represents but a slight advance in the theory of the real characters, but the method of the analytical arithmetic of matrices (2.4) is also fit for more refined investigations. In particular, it gives some partial results on the hypothesis (1.1) for certain progressions modulo 16.

The study of the behaviour of characters on narrow segments of the type $(1, x)$ for $x = D^{\epsilon}$, $\epsilon > 0$ small and fixed, is reduced to the study of the distribution of the projections of integer points inside the zero angle of the Lobatschevskian triangle Δ_0.

We will not formulate the results thus obtained, which are cumbersome enough, except a theorem on the zeros of $L(s, \chi)$, though rather cumbersome also.

THEOREM 5. *Let* $-D < 0$ *be an odd fundamental discriminant*; $\chi(n) = (-D/n)$; K *a large positive constant. Consider all odd primes not exceeding* K:

$$3, 5, 7, \ldots, p_r \leqslant K. \tag{6.1}$$

Let $\rho > 0$ *be a small constant, and suppose for at least* 100ρ % *of primes* (6.1) *we have the condition*

$$(-D/p_i) = +1. \tag{6.2}$$

Then the series $L(s, \chi)$ *has no zeros in the domain*

$$\sigma > 1 - \frac{\psi_1(K)}{\ln D}; \quad \psi_2(K) \geqslant |t| \geqslant 1, \tag{6.3}$$

where $\psi_1(K)$ *and* $\psi_2(K) \to \infty$ *for* $K \to \infty$.

7. The methods of proof of theorems slightly feebler than those mentioned above (except Theorems 1 and 5) are exposed in detail in the paper [4].

We shall sketch here the principal ideas of these methods.

For a given odd $D > 0$, we consider the matrices (2.4) satisfying the conditions of reducibility (2.3) and primitivity: g.c.d. $(a, 2b, c) = 1$ (basic vector-matrices),

$$L_1, L_2, \ldots, L_h; \; h = h(-D); \quad L_i^2 = -D. \tag{7.1}$$

They form a part of the ring of the integral matrices $\begin{pmatrix} \alpha & \beta \\ \gamma & \delta \end{pmatrix}$.

If A is any non-singular matrix and $L' = ALA^{-1}$, then

$$\operatorname{tr}(L') = \operatorname{tr}(L) = 0; \, L'^2 = -D.$$

If the matrix L' is integral and basic, it must be one of the matrices (7.1).

If we have the equality $b + L_i = CA$ with an integer $b = b \begin{bmatrix} 1 & 0 \\ 0 & 1 \end{bmatrix}$ and integer matrices C and A, then $b + L'_i = b + ALA^{-1} = AC$, so that $L' = AL_i A^{-1}$ is an integer vector matrix. If ϵ is a unimodular matrix, then $\epsilon L'_i \epsilon^{-1}$ is also an integer vector matrix; it can be basic only for two particular values of ϵ, $\epsilon = \pm \epsilon_0$.

The connection between inequivalent matrices in this sense with the class number of the corresponding field (of any degree, not only quadratic ones) was established by A. Hurwitz some sixty years ago [1] and forms essentially the background of these investigations.

Let $(-D/p) = +1$; then it proves possible and expedient to consider the equalities

$$l + L_i = \Pi_{1i}\, \Pi_{2i} \dots \Pi_{si}\, V_i, \quad (i = 1, 2, \dots, h), \tag{7.2}$$

where Π_{ki} are integer matrices of determinant p, and l an integer under the condition

$$l^2 + D \equiv 0 \ (\text{mod } 0\, p^s).$$

The matrices Π_{ki} are so chosen in the " bundle " of matrices unimodularly associated from the right hand side, that if

$$Q_{\nu i} = \Pi_{1i}\, \Pi_{2i} \dots \Pi_{\nu i}, \tag{7.3}$$

then

$$Q_{\nu i}^{-1}\, L_i\, Q_{\nu i} = L'_i$$

is a basic vector matrix.

The matrices $Q_{\nu i}$ may be considered as operators generating the Lobatschevskian motions on the hyperboloid H. The application of these operators in a convenient way (see [4]) enables us to shift L_i into the given domain.

For the corresponding asymptotic computations some probability-theoretic limit theorems for Markov chains are applied (for analogous applications see the paper [2]).

REFERENCES

1. A. Hurwitz : Die unimodulären Substitutionen in einem algebraischen Zahlkörper, *Nachrichten von der k. Gesellschaft der Wiss. zu Göttingen*, Math.-Phys. Klasse (1895), 332-356.

2. Y. V. Linnik : An application of Markov chains in the quaternion arithmetic, *Uspehi Matem. Nauk*, 4 (62), 1954. (Russian).

3. Y. V. Linnik : A remark on the least quadratic non-residue, *Doklady Akademii Nauk SSSR*, 1942, *t.* 36, 4-5, 131-132 (Russian).

4. Y. V. Linnik : Asymptotic distribution of the reduced binary quadratic forms in connection with the Lobatschevskian geometry, I, II, III, *Viestnik Leningradskogo Universiteta*, N 2, 3-23; N 5, 3-32; N 8; 15-27, 1955. (Russian).

5. Y. V. Linnik and A. Renyi : On some hypotheses of the theory of Dirichlet's characters, *Izvestia Akademii Nauk SSSR, math. series*, 11, 1947, 539-546 (Russian).

6. K. A. Rodosski : Review of the paper: N. C. Ankeny, Quadratic residues, *Referativni Journal " Matematika "*, N 3, 1955, review N 1079 (Russian).

7. C. L. Siegel : Über die Klassenzahl quadratischer Zahlkörper, *Acta Arithm.* 1 (1935), 83-86.

8. I. M. Vinogradov : On the distribution of quadratic residues and non-residues, *Journal of the Physico-mathematical Society of Perm University*, 1919, 2, 1-16. (Russian).

9. G. F. Voronoi : *Collected Works*, Kiev, 1953, III, 44 and 210. (Russian).

University of Leningrad
U.S.S.R.

Andrei Andreyevich Markov

Born – 14 June 1856 in Ryazan, Russia
Died – 20 July 1922 in Petrograd (now St Petersburg), Russia

Andrei Markov was a graduate of Saint Petersburg University (1878), where he began to be a professor in 1886. Markov's early work was mainly in number theory and analysis, continued fractions, limits of integrals, approximation theory and the convergence of series.

After 1900 Markov applied the method of continued fractions, pioneered by his teacher Pafnuty Chebyshev, to probability theory. He also studied sequences of mutually dependent variables, hoping to establish the limiting laws of probability in their most general form. He proved the central limit theorem under fairly general assumptions.

Markov is particularly remembered for his study of Markov chains, sequences of random variables in which the future variable is determined by the present variable but is independent of the way in which the present state arose from its predecessors. This work launched the theory of stochastic processes.

In 1923 Norbert Wiener became the first to treat rigorously a continuous Markov process. The foundation of a general theory was provided during the 1930s by Andrei Kolmogorov.

Markov was also interested in poetry and he made studies of poetic style, interestingly Kolmogorov had similar interests.

Markov had a son (of the same name) who was born on September 9, 1903 and followed his father in also becoming a renowned mathematician.

THE THEORY OF ALGORITHMS

A. A. MARKOV

1. The term "algorithm" in mathematics is taken to mean a computational process, carried out according to a precise prescription and leading from given objects, which may be permitted to vary, to a sought-for result.

Euclid's algorithm for finding the greatest common divisor of two natural numbers is a typical example of an algorithm. The pair of natural numbers here plays the role of the original given objects. The prescription consists in the successive construction of a descending series of numbers, the first of which is the large of the two given numbers, the second is the smaller, the third is obtained as the remainder upon dividing the first by the second, the fourth as the remainder upon dividing the second by the third, and so on, until division can be carried out without remainder. Then the divisor in this division will be the sought-for result of the algorithm -- the greatest common divisor of the two given natural numbers.

We note here three distinguishing features, which we take as characteristic of algorithms:

a) The presence of a precise prescription, leaving no possibility of arbitrary choice, and in the known sense generally understood -- the algorithm is *determined*.

b) The possibility of starting from original given objects which can vary within known limits -- *applicability* of the algorithm.

c) The tendency of the algorithm to obtain a certain result, finally obtained for appropriate original given objects -- the *effectiveness* of the algorithm.

2. The description of algorithms just given does not pretend to mathematical exactitude. Up to recent times, however, mathematicians have contented themselves with the rather diffuse concept which corresponds to this description. This was permissible while the term "algorithm" was met with in mathematics only in statements of the following type: "for the solution of such-and-such problems there exists an algorithm, and here is what it consists of". No negative results, no theorems concerning the impossibility of algorithms could be obtained at this stage, precisely because of the vagueness of the concept of algorithm.

3. In recent times theories have been developed by a series of authors [7–11, 13] which, if not directly concerned with a sharpening of the concept of algorithm, in any case lead naturally to such a sharpening.

This gave the possibility of establishing a series of important negative results -- theorems of impossibility for algorithms. First of all the theorem of Church [12] on the unsolvability of the general "problem of solvability" of computing predicates relates to this program. As more concrete examples one can cite the proof of unsolvability of a series of problems of the general theory of associative systems [1, 2, 4, 5, 14] and the theory of integer matrices [3, 16].

4. In view of the fact that all of the theories referred to above lead to a sharpening of the concept of algorithm only in an indirect and rather complicated way, the author thought it appropriate to study a direct sharpening of this concept and to develop the general theory of algorithms on the basis of this sharpening. This article contains an outline of the theory constructed in this way.

5. *Elementary signs* are signs that we shall consider as not having parts. The content of this concept depends upon the conventions that are assumed. For example, the sign "a'" can be considered as an elementary sign and as consisting of the two parts "a" and "$'$", depending upon the conventions that are agreed upon.

6. In simultaneous consideration of any two elementary signs, we determine whether they are the *same* or *different*. These concepts are also conditional.

7. The possibility of determining when two elementary signs are the same permits us, applying an *abstraction of identification,* to speak of two identical elementary signs or of *one and the same* elementary sign. On this basis, we introduce the concept of an *abstract elementary sign,* that is, of an elementary sign, considered up to identity.

Concrete elementary signs will be considered as *representatives* of the corresponding abstract elementary signs. Two concrete elementary signs represent one and the same abstract elementary sign if and only if they are identical.

8. Lists of elementary signs are called *alphabets*. We shall call two alphabets *equal* if every elementary sign appearing in the first alphabet is identical with a certain elementary sign appearing in the second alphabet, and conversely. Alphabets considered up to equality will be called *abstract alphabets.*

An abstract alphabet is defined by those abstract elementary signs, representatives of which occur in a representative of the abstract alphabet. An abstract alphabet can hence be considered as a finite set of abstract elementary signs. Abstract elementary signs, *belonging* in this sense to a given abstract alphabet, will be called *letters of this abstract alphabet.*

In the sequel, when speaking of "alphabets", we shall usually mean "abstract alphabets".

9. Series of elementary signs will be called *words*. If the elementary signs constituting the work P are representatives of letters of an alphabet A, we shall say that P is a *word in this alphabet.*

For spelling words, we produce a series of requirements of intelligibility.

Since the order of the elementary signs that comprise a word plays an essential role, it must be clear; that is, for any two such elementary signs, it must be clear which of them stands to the left.

The beginning and the end of a word must be indicated clearly.

Finally, a very important requirement is the requirement that the representation of a word as a series of the elementary signs under consideration be unique -- that is, that there be no possibility of "variant readings". In order to guarantee the fulfillment of this requirement, the simplest method is to impose an appropriate limitation on the alphabets under consideration. It suffices, for example, to demand that the elementary signs representing letters of the alphabet should not be obtainable one from another by writing something on the right. We shall impose this limitation.

10. We say that two words P and Q are *equal* if they consist of identical elementary signs, arranged identically. To determine equality of words, there is a general effective method -- simultaneous comparison of the words from left to right sign by sign.

In considering words, we shall also apply abstract identification, which in the given case consists in speaking of two equal words as being one and the same word. This leads to the construction of the concept of an *abstract word,* words, considered up to equality. To every abstract word there correspond many concrete words, of which we say that they *represent* this abstract word. Two concrete words are equal if and only if they represent one and the same abstract word. An abstract word can be considered as constructed from abstract elementary signs -- from *letters of this word.*

11. Another abstraction, also continually applied in the sequel, is *abstraction of potential realizability.* This consists in departing from the real limits of our constructive possibilities and beginning to discuss arbitrarily long abstract words as if they were constructible. Their realizability is *potential*: their representatives could be practically realized if we had at our disposal sufficient time, space, and materials. Applying this abstraction, we shall in the sequel understand by "word" an abstract potentially realizable word.

12. Putting down an arbitrary representative of the word Q to the right of an arbitrary representative of the word P, we obtain a representative of a certain new word. It does not depend upon the choice of these representatives, which are defined completely by the words P and Q themselves. We call this word the *join* of the words P and Q and denote it by the symbol PQ. In like manner we define the join PQR of the three words P, Q, and R, and so on.

The abstraction of potential realizability gives the possibility of asserting the existence of the join of any two words.

13. It is convenient to introduce the "void word" in the alphabet under consideration, that is, the word having no letters at all. This can be done as follows. Suppose that the alphabet A is such that the elementary sign Λ does not represent a letter in it. We shall agree to call the corresponding abstract elementary sign the *void word* in the alphabet A.

Extending the concept of word in the alphabet A, we shall now call words the following: first, every word in A in the previous sense; second, the void word in A. We extend the definition of join to words in the new sense, setting

$$P\Lambda = P;\ \Lambda P = P$$

for every word P in A.[1)]

14. We shall say that a word P *begins* with the word Q if there exists a word S such that

$$P = QS;$$

we shall say that a word Q *occurs* in the word P if there exist words R and S such that

$$P = RQS. \tag{1}$$

If the word Q occurs in the word P, then it can happen that P admits several representations of the form (1). This is the state of affairs, for example, for the words "papagiglemma" and "pa", which the second occurs in the first.[2)] In such cases we speak of several "occurrences" of the word in the word, and we distinguish these occurrences from each other. It is expedient to give the following exact definition of occurrence.

We shall consider an alphabet A not containing the asterisk "$*$" among its letters. Let P and Q be letters in this alphabet. We shall call *occurrences* of the word Q in the word P words in the alphabet $A \cup \{*\}$[3)] of the form

$$R * Q * S,$$

where R and S are words in A satisfying condition (1). In other words, occurrences of Q in P are words in $A \cup \{*\}$, obtained from P by the insertion of two asterisks in such a way that Q stands between two of these asterisks.

For example, there are two occurrences of the word "pa" in the word "papagiglemma": *pa*pagiglemma and pa*pa*giglemma.

Q occurs in P if and only if there exists at least one occurrence of Q in P.

1) Here and in the sequel the sign of equality means identity of abstract words.

2) *Translator's note.* The expressions in quotation marks in the present sentence are transliterations of groups of letters in the Cyrillic alphabet. Thus we tamper with the original in selecting a different alphabet.

3) Here and in the sequel we denote by $\{a_1, \cdots, a_n\}$ the alphabet with the letters $a_1, \cdots, a_n$; by $A \cup B$ the result of combining the alphabets A and B.

Let $T = R * Q * S$ and $T_1 = R_1 * Q * S_1$ be two different occurrences of Q in P. Then, as is not hard to see, $R \neq R_1$ and either R_1 begins with R or R begins with R_1. In the first case, we say that T *precedes* T_1 and in the second case that T_1 precedes T. We order occurrences of Q in P in this way.

It is not hard to see that if there are occurrences of Q in P, there is among them one and only one that precedes all of the others. We call this the *first occurrence* of Q in P.

A general effective method can easily be given, which tells whether or not a given word Q occurs in an arbitrary other given word P, and giving the first occurrence of Q in P, in the case that Q occurs in P.

The void word occurs in every word and the first occurrence of the void word in the word P is the occurrence $**P$.

15. If $T = R * Q * S$ is an occurrence of Q in P, and if U is any word, then the construction of the word RUS will be called *substitution of the word U in place of the word Q* and this word itself will be called the *result* of this substitution.

The word $\mathfrak{A}P$ is the result of substitution of the word $\mathfrak{A}$ in place of the first occurrence of the void word in the word P.

For given $\mathfrak{A}$ and B, the substitution of the word B in place of an arbitrary occurrence of the word $\mathfrak{A}$ in an arbitrary word P is a "local" operation on P in the sense that it affects the given form of parts of the transformed word.

16. Every computational process applied in mathematics leads to a certain potentially realizable process of sequential transformation of words in an appropriate alphabet. Here by "words" one may understand just "abstract words", since all of these processes are determined up to equality of the words obtained. These processes are carried out in agreement with generally understood prescriptions, leaving no room for arbitrary choices. At the beginning of the process, we have certain "original given objects", expressed by the original word. In the course of the process we obtain in agreement with the given prescription a series of words in a sequence, which breaks off when the process terminates. We are interested in the word with which this series ends -- the result of the process.

Corresponding to this, we agree to understand by an *algorithm in the alphabet A* a generally understood prescription defining a potentially realizable process on abstract words in A, starting from an arbitrary word in A. We shall say that the algorithm is *applicable* to the word P, if, starting with this word and applying the algorithm, we finally obtain a word with which the process breaks off. Of this word, we shall then say that the algorithm *remakes* the word P into it. The word into which the algorithm $\mathfrak{A}$ remakes P will be denoted by the symbol $\mathfrak{A}(P)$.

We shall agree further to understand by an *algorithm over the alphabet A* an algorithm in an arbitrary alphabet containing A. If $\mathfrak{A}$ and $\mathfrak{B}$ are algorithms over A, then we shall say that they are *equivalent with respect to* A if whenever $\mathfrak{A}$ remakes an arbitrary word P in A into another word also in A, then $\mathfrak{B}$ remakes P into the same word, and conversely. We shall say that these algorithms are *completely equivalent with respect to* A if whenever $\mathfrak{A}$ remakes an arbitrary word P in A, then $\mathfrak{B}$ remakes P into the same word and conversely.

17. The concept of algorithm in a given alphabet just introduced is not sufficiently precise to define objects of a mathematical theory. It is subject to a sharpening, to which we shall immediately turn.

We demand first of all in guaranteeing general intelligibility of a prescription that it should stipulate the breaking up of the process of applying the algorithm into elementary steps of local character -- into substitution of given words in place of occurrences of other given words. We must give in this connection an enumeration of the types of such steps, where each type will be characterized by a pair of words in the alphabet under consideration: by the word the occurrence of which is replaced, and by the word which is substituted. Supposing that the arrow sign is not a letter of our alphabet, we agree to designate by a formula of the form

$$\mathcal{A} \rightarrow B \qquad (2)$$

a substitution of the word B in place of an occurrence of the word $\mathcal{A}$. The essential component part of the algorithm will then consist of a certain enumeration of such formulas.

It is now necessary to guarantee the determinacy of the process. Indeed, the formulas of substitution may be multitudinous. Thus it is needful to give instructions relating to the method of applying them in each case. Further, if we know that in a given case it is necessary to apply a substitution expressed by a given formula (2), then its left part $\mathcal{A}$ may occur several times in the word, and this means that we must state which of its occurrences is to be replaced by the right part B. Different variants of such instructions are possible. We content ourselves with the following.

We suppose first that the arrangement of the formulas of substitution in the list defines their order in the sense that every time we must take the first formula in the list that is applicable to the given word. Second, we will apply the corresponding substitution to the first occurrence of the left side of the formula in the word under consideration.

The process of successive substitutions, carried out on the basis of this prescription, has the following appearance. We start with an arbitrary word Q. In the given list of formulas (2), we seek the first of them, the left part of which occurs in Q. We seek the first occurrences of its left part in Q, and in place of this occurrence, we substitute the right side of the formula, which gives a new word Q_1.

We operate with Q_1 just as was done with Q, and so on.

This process may break off at a certain word Q_n such that none of the operations under consideration is applicable to it, that is, such that none of the left sides of the formulas of the list occurs in this word. This is the *natural termination* of the process. To guarantee the appropriate generality, it is further necessary to introduce a different type of termination of the process, in which we single out certain of the operations of substitution as *concluding* and break off the process whenever one of these operations is applied. In formulas expressing concluding operations, we shall add a period to the arrow. We shall place the period directly to the right of the arrow.

The list of formulas written in this way will be called the *scheme* of the given algorithm. The algorithms themselves of the kind described will be called *normal algorithms*. Every normal algorithm is completely defined by specifying the alphabet in which it acts and the scheme.

The definition of a normal algorithm is sufficiently precise.

18. We give two examples of normal algorithms.

Let $\mathfrak{A}_1$ be the normal algorithm in the alphabet $A_0 = \{a, b\}$ with the scheme

$$\{ \ \rightarrow \cdot\, a.$$ [1)]

It is not hard to see that $\mathfrak{A}(P) = aP$ for every word P in A_0. If we removed the period, we would obtain a normal algorithm applicable to no words at all.

Let $\mathfrak{A}_2$ be the normal algorithm in the alphabet $\{a, b, c, d, e\}$ with the scheme

$$\left\{\begin{array}{r}
\alpha\beta d \rightarrow \beta d\alpha \\
c\alpha \rightarrow \alpha d\alpha c \\
d \rightarrow e \\
e \rightarrow \\
c \rightarrow \cdot \\
\rightarrow c
\end{array}\right.$$

Here we apply the following abbreviated notation: α and β denote arbitrary letters of the alphabet A_0. To obtain a complete list, we must substitute in the first formula one of the letters a, b instead of α and one of these letters in place of β. This substitutions must be made in all possible ways, which gives four formulas. We write them in a definite order (the one selected is immaterial for the sequel). We then write two formulas, obtained from the second formula of the abbreviated scheme by substituting a and b in place of α. In all, the scheme contains 10 formulas of substitution and admits 48 variants. We single out one of them.

In application to the word $P = \alpha_1 \cdots \alpha_k$ in the alphabet A_0 ($\alpha_i = a$ or b), the

1) To the left of formulas of a scheme, written one below the other, we place a curly bracket combining them.

algorithm works in the following way:

$$
\begin{aligned}
P &\vdash cP \\
&= ca_1 \cdots a_k \\
&\vdash a_1 da_1 ca_2 \cdots a_k \\
&\vdash a_1 da_1 a_2 da_2 ca_3 \cdots a_k \\
&\vdash a_1 da_2 da_1 a_2 ca_3 \cdots a_k \\
&\cdots\cdots\cdots\cdots\cdots \\
&\vdash a_1 da_2 da_3 da_1 a_2 a_3 ca_4 \cdots a_k \\
&\cdots\cdots\cdots\cdots\cdots\cdots \\
&\vdash a_1 da_2 da_3 d \cdots a_k dPc \\
&\vdash a_1 ea_2 da_3 d \cdots a_k dPc \\
&\cdots\cdots\cdots\cdots\cdots \\
&\vdash a_1 ea_2 ea_3 e \cdots a_k ePc \\
&\vdash a_1 a_2 ea_3 e \cdots a_k ePc \\
&\cdots\cdots\cdots\cdots\cdots \\
&\vdash PPc \\
&\vdash .\, PP.
\end{aligned}
$$

Here the sign $\vdash$ is applied to emphasize the individual elementary steps. A period after this sign indicates the application of a concluding operation. We have in this way $A(P) = PP$ for word P in A_0, that is, A_2 doubles these words.

19. It is now natural to ask to what extent the exact concept of a normal algorithm corresponds to the general, not entirely precise, concept of an algorithm in a given alphabet, formulated above. As an answer to this question one can advance the following principle of normalization of algorithms: *every algorithm over the alphabet A is equivalent with respect to A to a certain normal algorithm over A.*

This principle cannot be proved, in view of the vagueness of the general concept of algorithm. It can, however, be supported by various arguments. The weightiest of these consists in the fact that all algorithms known in mathematics up to the present time are equivalent to normal algorithms. Taking into account that mathematics has existed for a fairly long time, we are compelled to admit that this argument is persuasive.

Another important argument in favor of the principle of normalization consists in the fact that different combinations of normal algorithms give algorithms equivalent to normal ones. We shall immediately formulate some theorems relating to this.

20. Theorem of composition. *If $\mathfrak{A}$ and $\mathfrak{B}$ are normal algorithms over an alphabet A, then one can construct a normal algorithm $\mathfrak{C}$ over A such that*

$$\mathfrak{C}(P) \approx \mathfrak{B}(\mathfrak{A}(P))$$

for words P in A.

The sign of *conditional equality* "$\approx$" applied here and in the sequel means that the expressions connected by it simultaneously (in the present case, for one and the same P) have a meaning and then express equal words.

Theorem of combination. *If* $\mathfrak{A}$ *and* $\mathfrak{B}$ *are normal algorithms in an alphabet* A, *and if* γ *is an elementary sign not belonging to* A, *then one can construct a normal algorithm* $\mathfrak{C}$ *over* $A \cup \{\gamma\}$ *such that*

$$\mathfrak{C}(P\gamma Q) \approx \mathfrak{A}(P)\,\gamma\mathfrak{B}(Q)$$

for words P *and* Q *in* A.

Theorem of ramification. *If* $\mathfrak{A}$, $\mathfrak{B}$, *and* $\mathfrak{C}$ *are normal algorithms over an alphabet* A *and* $\mathcal{A}$ *is a word in this alphabet, then one can construct a normal algorithm* $\mathfrak{D}$ *over* A *such that*

$$\mathfrak{D}(P) \approx \mathfrak{A}(P)$$

for words P *in* A *such that* $\mathfrak{C}(P) = \mathcal{A}$, *and*

$$\mathfrak{D}(P) \approx \mathfrak{B}(P)$$

for words P *in* A *such that* $\mathfrak{C}(P)$ *exists and* $\mathfrak{C}(P) \neq \mathcal{A}$.

Theorem of reiteration. *If* $\mathfrak{A}$ *is a normal algorithm in an alphabet* A *and* α *is a letter of this alphabet, then one can construct a normal algorithm* $\mathfrak{B}$ *over* A *with the following property: the equality*

$$\mathfrak{B}(P) = Q$$

obtains for a word P *in* A *and an arbitrary word* Q, *if and only if there exists a series of words* $P_0, \cdots, P_k$ *such that* $P_0 = P$, $P_k = Q$, $P_i = \mathfrak{A}(P_{i-1})$ $(i = 1, \cdots, k)$, *and such that* P_k *is the only one of the words* $P_1, \cdots, P_k$ *not containing an occurrence of* α.

21. In considering normal algorithms over a given alphabet A, we are often interested in them only so far as they remake words in this alphabet into words in the same alphabet, that is, up to equivalence with respect to A. In such considerations it is natural to try to make do with the smallest collection of "auxiliary" letters, that is, letters not belonging to A.

This is attained by means of the operation of *translation* from one alphabet into the other, defined in the following way. Suppose that the alphabet B is obtained from the alphabet A by adding letters $\gamma_1, \cdots, \gamma_n$; the alphabet C by adding the letters α and β to A. With every letter γ_i $(i = 1, \cdots, k)$, we associate the word

$$\Gamma_i = \alpha\overbrace{\beta \cdots \beta}^{i \text{ times}}\alpha.$$

The translation of a word P in B is defined as the word obtained from P as a result of replacing every letter γ_i appearing in P by the corresponding word Γ_i. The translation of every word in B is thus a word in C. In the sequel we shall denote the translation of the word P by the symbol $[P^\tau$. It is clear that

$$[PQ^\tau = [P^\tau[Q^\tau$$

for arbitrary words P and Q in B. It is also evident that a word in the alphabet B can be uniquely reconstructed from its translation.

The following theorem on "translation" of an algorithm is easily proved.

For every normal algorithm $\mathfrak{A}$ *in* B, *there exists a normal algorithm* $\mathfrak{B}$ *in* C *such that*

$$\mathfrak{B}([P^\tau) \approx [\mathfrak{A}(P)^\tau \ (P \text{ in } B).$$

Taking into account that $[P^\tau = P$ for every word P in A, we thus obtain the following result.

Every normal algorithm over an alphabet A *is equivalent with respect to* A *to a certain normal algorithm in the alphabet* $A \cup \{\alpha, \beta\}$, *where* α *and* β *are elementary signs not belonging to* A.

22. We shall consider normal algorithms in the alphabet A. Up to this time we have applied "schemes" for them -- columns of formulas of substitutions. It is now expedient to turn to a different method of giving normal algorithms. For this purpose we introduce new elementary signs α, β, and γ, different from each other, from the letters of the alphabet A and from the signs "$\rightarrow$" and "$\cdot$". We agree to describe every normal algorithm $\mathfrak{A}$ in the alphabet A by a word, obtained in the following fashion. The formulas of substitution of the algorithm $\mathfrak{A}$ are written one after the other in the order of sequence specified by the algorithm, with the arrow replaced by α, the period replaced by β, and addition on the right of the sign γ. The word obtained in this way will be called the *image* of the algorithm and will be denoted by the symbol $\overline{\mathfrak{A}}$. The image of a normal algorithm A in the alphabet A is a word in the alphabet $B = A \cup \{\alpha, \beta, \gamma\}$. As is not hard to see, it defines this algorithm. The following theorem can serve as the basis of many proofs of impossibility of algorithms.

Theorem on a universal algorithm. *Let* δ *be a sign that is not a letter of the alphabet* B (see above). *One can construct a normal algorithm* $\mathfrak{U}$ *over* $B \cup \{\delta\}$ *such that*

$$\mathfrak{U}(\overline{\mathfrak{A}}\delta P) \approx \mathfrak{A}(P)$$

for words P *in* A *and normal algorithms* $\mathfrak{A}$ *in* A.

For application of this theorem to proofs of impossibility it is expedient to go over to the specification of normal algorithms in A by words in the alphabet $A_0 = \{a, b\}$.

For this, we apply the operation of translation, analogous to the operation of translation defined above. With letters of the alphabet B we associate in a one-to-one way the words

$$aba,\ abba, \cdots, \overbrace{ab\cdots ba}^{n \text{ times}}, \tag{3}$$

where n is the number of letters of the alphabet B. We choose one such association. This gives the possibility of "translating" words in the alphabet B into the alphabet A_0. The translation of a word P in B is obtained simply by the process of replacing every letter of this word by the corresponding word from the series (3). Applying this operation to the image $\overline{\mathfrak{A}}$ of the normal algorithm $\mathfrak{A}$ in A, we obtain a word in the alphabet A_0, which we shall call the *transcription* of this algorithm. We will denote the transcription of the algorithm $\mathfrak{A}$ by the symbol $\mathfrak{A}^*$.

From the theorem on a universal algorithm, one easily infers the existence of a normal algorithm $\mathfrak{F}$ over the alphabet $A \cup \{a, b, \delta\}$ such that

$$\mathfrak{F}(\mathfrak{A}^* \delta P) \approx \mathfrak{A}(P)$$

for words P in A and normal algorithms $\mathfrak{A}$ in A. [1)]

23. We shall now consider normal algorithms over the alphabet A_0. In agreement with the theorem stated above, every such algorithm is equivalent to a certain normal algorithm in the alphabet $A_1 = \{a, b, c, d\}$. Every normal algorithm $\mathfrak{A}$ in A_1 has a transcription $\mathfrak{A}^*$ in the alphabet A_0, and there exists a normal algorithm $\mathfrak{F}$ over $A_2 = \{a, b, c, d, e\}$ such that $\mathfrak{F}(\mathfrak{A}^* eP) \approx \mathfrak{A}(P)$ for words P in A_0 and normal algorithms $\mathfrak{A}$ in A_1.

We shall say that a normal algorithm in A_1 is *self-applicable* if it is applicable to its own transcription.

Self-applicable algorithms exist. For example, the normal algorithm in A_1 with the scheme

$$\{ \;\rightarrow\cdot$$

is applicable to every word in A_1, in particular to its own transcription. There also exist non self-applicable normal algorithms in A_1. Such, for example, is the normal algorithm in A_1 with the scheme

$$\{ \;\rightarrow,$$

which is applicable to no word in A_1.

We shall prove first of all that *it is impossible to have a normal algorithm over A_0 that is applicable just to those transcriptions of normal algorithms in A that are transcriptions of non self-applicable algorithms.*

In fact, assume that $\mathfrak{H}$ is such an algorithm. In agreement with the theorem on translation of an algorithm, one can construct a normal algorithm $\mathfrak{K}$ in the alphabet A_1 which, just like $\mathfrak{H}$, is applicable just to those transcriptions of normal algorithms in A_1 that are transcriptions of non self-applicable algorithms. $\mathfrak{K}$ has the transcription $\mathfrak{K}^*$, which is a word in A_0. We shall see if $\mathfrak{K}$ can be self-

1) Here we suppose that the sign δ does not belong to the alphabet $A \cup \{a, b\}$.

applicable, that is, applicable to $\mathfrak{K}^*$.

If $\mathfrak{K}$ is applicable to $\mathfrak{K}^*$, then $\mathfrak{K}^*$ is the transcription of a non self-applicable algorithm, and hence $\mathfrak{K}$ in non self-applicable, that is, $\mathfrak{K}$ is not applicable to $\mathfrak{K}^*$. Thus $\mathfrak{K}$ is non self-applicable. It follows, however, from this that $\mathfrak{K}^*$ is the transcription of a non self-applicable algorithm, in view of which $\mathfrak{K}$ is applicable to $\mathfrak{K}^*$, that is, $\mathfrak{K}$ is self-applicable. Thus the assumption of the existence of an algorithm $\mathfrak{H}$ with the properties described above leads tc a contradiction. Such an algorithm is consequently impossible, which we wished to prove.

We shall now say of a normal algorithm over A_0 that it *identifies non self-applicability (self-applicability)* if it is applicable to the transcription of every normal algorithm in A_1 and remakes A^* into the void word if and only if the algorithm $\mathfrak{A}$ is non self-applicable (self-applicable).

The construction of a normal algorithm over A_0 that identifies non self-applicability would evidently give us a general effective method by means of which, for a given transcription, that is, for a given scheme of a normal algorithm in A_1, it would be possible in a finite number of steps to determine whether or not this algorithm is non self-applicable.

However, *a normal algorithm over A_0 that identifies non self-applicability is impossible.*

In fact, let us assume that $\mathfrak{H}$ is such an algorithm in the alphabet A containing A_0. We construct a normal algorithm $\mathfrak{L}$ in A with the abbreviated scheme

$$\{ \xi \rightarrow \xi,$$

where ξ denotes every letter of the alphabet A. This algorithm evidently is applicable to the void word and to no other word.

Applying the theorem of composition to the algorithms $\mathfrak{H}$ and $\mathfrak{L}$, we construct a normal algorithm $\mathfrak{K}$ over A_0 such that

$$\mathfrak{K}(P) \approx \mathfrak{L}(\mathfrak{H}(P))$$

for words P in A_0. This algorithm is applicable to the transcription of the normal algorithm $\mathfrak{A}$ in A_1 if and only if $\mathfrak{H}(\mathfrak{A}^*) = \Lambda$, that is, if and only if $\mathfrak{A}$ is non self-applicable. Such an algorithm $\mathfrak{K}$, however, is impossible in view of the preceding theorem. Consequently, the algorithm $\mathfrak{H}$ is also impossible, which we wished to prove.

It is easily inferred further from this, that *there is no normal algorithm over A_0 identifying self-applicability.*

We now remark that, in agreement with the conditional equality $\mathfrak{F}(\mathfrak{A}^* eP) \approx \approx \mathfrak{A}(P)$ established above, a normal algorithm $\mathfrak{A}$ in A_1 is self-applicable if and only if the algorithm $\mathfrak{F}$ is applicable to the word $\mathfrak{A}^* e \mathfrak{A}^*$. Recalling that a normal algorithm over A_2 remaking $\mathfrak{A}^*$ into $\mathfrak{A}^* e \mathfrak{A}^*$ can be constructed easily, we infer on the basis of the preceding that *there is no normal algorithm identifying non applicability (applicability) of the algorithm* F, *that is, a normal algorithm over* A_2 *applicable to every word in* A_2 *and remaking into the void word just those words in* A_2 *to which the algorithm is non applicable (applicable).*

Using the normal algorithm $\mathfrak{Z}$ in A_0 with the scheme

$$\begin{cases} a \rightarrow \\ b \rightarrow . \end{cases}$$

remaking every word in A_0 into the void word, and applying the theorem of composition to $\mathfrak{F}_0$ and $\mathfrak{Z}$, and then again making a "translation" in A_0, we obtain finally the following result.

There exists a normal algorithm $\mathfrak{F}_1$ *in* A_0 *such that there is no normal algorithm over* A_0 *applicable to every word in* A_0 *and remaking into the void word exactly those words in* A_0 *which* $\mathfrak{F}_1$ *does not remake into the void word.*

This result can serve as the starting point for proving the unsolvability of many algorithmic problems, in particular the problems referred to above of the theory of associative systems and the theory of integer matrices.

LITERATURE

[1] A. A. Markov, *On the impossibility of certain algorithms in the theory of associative systems*, Dokl. Akad. Nauk SSSR (N.S.) 55 (1947), 583–586.

[2] A. A. Markov, *The impossibility of certain algorithms in the theory of associative systems* II, Dokl. Akad. Nauk SSSR (N.S.) 58 (1947), 353–356.

[3] A. A. Markov, *On certain unsolvable problems concerning matrices*, Dokl. Akad. Nauk SSSR (N.S.) 57 (1947), 539–542.

[4] A. A. Markov, *The impossibility of certain algorithms in the theory of associative systems*, Dokl. Akad. Nauk SSSR (N.S.) 77 (1951), 19–20.

[5] A. A. Markov, *The impossibility of algorithms identifying certain properties of associative systems*, Dokl. Akad. Nauk SSSR (N.S.) 77 (1951), 953–956.

[6] A. A. Markov, *On an unsolvable problem concerning matrices*, Dokl. Akad. Nauk SSSR. (N.S.) 78 (1951), 1089–1092.

[7] S. C. Kleene, *General recursive functions of natural numbers*, Math. Ann. 112 (1936), 727–742.

[8] S. C. Kleene, *Recursive predicates and quantifiers*, Trans. Amer. Math. Soc., 53, 1943, 41–73.

[9] A. Church, *The calculi of* λ*–conversion*, Princeton University Press, Prince-

ton, N. J., 1941.

[10] A. Church, *An unsolvable problem of elementary number theory*, Amer. J. Math. 58 (1936), 345–363.

[11] A. Church, *A set of postulates for the foundation of logic*, Ann. of Math., (2) 33 (1932), 346–366.

[12] A. Church, *A note on the Entscheidungsproblem*, J. Symb. Logic 1 (1936), 40–41, 101–102.

[13] A. M. Turing, *On computable numbers with an application to the Entscheidungs problem*, Proc. London Math. Soc., (2) 42 (1936), 230–265; 43 (1937), 544.

[14] E. L. Post, *Recursive unsolvability of a problem of Thue*, J. Symb. Logic 12 (1947), 1–11.

Translated by:
Edwin Hewitt

A. MARKOFF

Foundations of the Algebraic Theory of Tresses

SUMMARY

1. We define the group of tresses of order n as the abstract group determined by the following relations between the generators σ_i $(i=1, \ldots, n-1)$:

$$\sigma_i \sigma_{i+1} \sigma_i = \sigma_{i+1} \sigma_i \sigma_{i+1} \qquad (i=1, \ldots, n-2),$$
$$\sigma_i \sigma_j = \sigma_j \sigma_i \qquad (i, j=1, \ldots, n-1;\ |i-j|>1).$$

This group is denoted by $\mathfrak{Z}_n$. Here n is a positive integer. The unit element of $\mathfrak{Z}_n$ is denoted by 1_n. The group $\mathfrak{Z}_1$ does not possess any generators, i. e. it consists of the unit element only.

2. We define the elements $s_{i,j}$ $(1 \leqslant i < j \leqslant n)$ of $\mathfrak{Z}_n$ inductively by the equalities:

$$s_{i,i+1} = \sigma_i^2 \qquad (1 \leqslant i \leqslant n-1),$$
$$s_{i,j+1} = [s_{i,j}, \sigma_j^{-1}] \qquad (1 \leqslant i < j \leqslant n-1).$$

Here and in the following the square brackets have the meaning

$$[x,y] = y^{-1} xy.$$

The following relations hold in $\mathfrak{Z}_n$

$$[s_{i,j}, \sigma_k^{\pm 1}] = s_{i,j} \qquad (k \neq i-1,\ i,\ j-1,\ j),$$
$$[s_{i,i+1}, \sigma_i^{\pm 1}] = s_{i,i+1},$$
$$[s_{i,j}, \sigma_{i-1}] = s_{i-1,j},$$
$$[s_{i,j}, \sigma_{i-1}^{-1}] = [s_{i-1,j}, s_{i-1,i}^{-1}],$$
$$[s_{i,j}, \sigma_i] = [s_{i+1,j}, s_{i,i+1}],$$
$$[s_{i,j}, \sigma_i^{-1}] = s_{i+1,j},$$
$$[s_{i,j}, \sigma_{j-1}] = s_{i,j-1},$$

$$[s_{i,j}, \sigma_{j-1}^{-1}] = [s_{i,j-1}, s_{j-1,j}^{-1}],$$
$$[s_{i,j}, \sigma_j] = [s_{i,j+1}, s_{j,j+1}].$$

The first of these relations is valid, whenever both its sides exist and the right — hand condition is satisfied. The other relations are valid, whenever they have sense.

These relations yield an algorithm for the solution of every problem of the following type: given two functions* f and g of $\frac{1}{2}n(n-1)$ and $n-1$ indeterminates respectively, to find a function h of $\frac{1}{2}n(n-1)$ indeterminates such that the relation

$$[f(s_{1,2}, \ldots, s_{n-1,n}), g(\sigma_1, \ldots, \sigma_{n-1})] = h(s_{1,2}, \ldots, s_{n-1,n})$$

be fulfilled in $\mathfrak{Z}_n$.

Corollary 1. The elements $s_{i,j}$ of the group $\mathfrak{Z}_n$ generate an invarint subgroup of $\mathfrak{Z}_n$.

Let $\mathfrak{N}_n$ denote the least invarint subgroup of $\mathfrak{Z}_u$ containing the elements

$$\sigma_1^2, \ldots, \sigma_{n-1}^2.$$

Theorem 1. The elements $s_{i,j}$ of $\mathfrak{Z}_n$ generate the group $\mathfrak{N}_n$.

3. If $z \in \mathfrak{Z}_n$, then $\breve{z}$ denotes the coset of $\mathfrak{N}_n$ in $\mathfrak{Z}_n$ which contains z.

Theorem 2. The group $\mathfrak{Z}_n/\mathfrak{N}_n$ is generated by the elements

$$\breve{\sigma}_i \, (i = 1, \ldots, n-1)$$

and can be determined by the relations:

$$\breve{\sigma}_i^2 = 1 \qquad (1 \leqslant i < n),$$
$$(\breve{\sigma}_i \breve{\sigma}_{i+1})^3 = 1 \qquad (1 \leqslant i < n-1),$$
$$\breve{\sigma}_i \breve{\sigma}_j = \breve{\sigma}_j \breve{\sigma}_i \qquad (1 \leqslant i < n, 1 \leqslant j < n, |i-j| > 1).$$

We define the elements $\sigma_{i,j} \, (1 \leqslant i \leqslant j \leqslant n)$ of the group $\mathfrak{Z}_n$ inductively by the equalities:

$$\sigma_{i,i} = 1_n \qquad (1 \leqslant i \leqslant n),$$
$$\sigma_{i,j+1} = \sigma_{i,j} \sigma_j^{-1} \qquad (1 \leqslant i \leqslant j < n).$$

* By a function of the indeterminates $x_1, \ldots, x_n$ we mean a formal product of powers of $x_1, \ldots, x_n$.

Theorem 3 (Ivanovsky). Every element of the group $\mathfrak{Z}_n$ given as a function of $\sigma_1, \ldots, \sigma_{n-1}$ is representable in the form

$$I_0 \qquad f(s_{1,2}, \ldots, s_{n-1,n}) \prod_{j=1}^{n} \sigma_{i_j, j},$$

where f is a function of $\frac{1}{2} n(n-1)$ arguments and i_j are positive integers such that $i_j \leqslant j$ $(j=1, \ldots, n)$.

The symbol $\prod_{j=1}^{m}$ used here is determined inductively by the equalities:

$$\prod_{j=1}^{0} x_j = 1, \quad \prod_{j=1}^{m} x_j = x_m \prod_{j=1}^{m-1} x_j \quad (m > 0).$$

Corollary 2. Every element of the group $\mathfrak{Z}_n/\mathfrak{N}_n$ given as a function of $\breve{\sigma}_1, \ldots, \breve{\sigma}_{n-1}$ is representable in the form

$$\breve{I}_0 \qquad \prod_{j=1}^{n} \breve{\sigma}_{i_j, j},$$

where i_j are positive integers such that $i_j \leqslant j$ $(j=1, \ldots, n)$.

Corollary 3. The index of $\mathfrak{N}_n$ in $\mathfrak{Z}_n$ does not exceed $n!$.

Let $\mathfrak{S}_n$ denote the symmetric group of degree n, i. e. the group of all permutations of the integers $1, \ldots, n$. Let further $\hat{\sigma}_i$ be the transposition $(i, i+1)$ considered as an element of the group $\mathfrak{S}_n$.

Theorem 4. The group $\mathfrak{Z}_n/\mathfrak{N}_n$ is isomorphic to $\mathfrak{S}_n$. An isomorphism arises if the element $\hat{\sigma}_i$ of $\mathfrak{S}_n$ is put into correspondence to the element $\breve{\sigma}_i$ of $\mathfrak{Z}_n/\mathfrak{N}_n$ $(i=1, \ldots, n-1)$.

Corollary 4. The group $\mathfrak{S}_n$ can be determined by the relations:

$$\begin{aligned} \hat{\sigma}_i^2 &= 1 && (1 \leqslant i < n), \\ (\hat{\sigma}_i \hat{\sigma}_{i+1})^3 &= 1 && (1 \leqslant i < n-1), \\ \hat{\sigma}_i \hat{\sigma}_j &= \hat{\sigma}_j \hat{\sigma}_i && (1 \leqslant i < n,\ 1 \leqslant j < n,\ |i-j| > 1). \end{aligned}$$

Corollary 5. The index of $\mathfrak{N}_n$ in $\mathfrak{Z}_n$ is $n!$.

Corollary 6. Every element of the group $\mathfrak{Z}_n/\mathfrak{N}_n$ is uniquely representable in the form $\breve{I}_0$.

Corollary 7. The elements

$$\prod_{j=1}^{n} \sigma_{i_j, j} \qquad (i_j \leqslant j \text{ for } j=1, \ldots, n)$$

form a system of representatives of the cosets of $\mathfrak{N}_n$ in $\mathfrak{Z}_n$.

Corollary 8. If

$$f'(s_{1,2}, \ldots, s_{n-1,n}) \prod_{j=1}^{n} \sigma_{i_j',j}$$

and

$$f''(s_{1,2}, \ldots, s_{n-1,n}) \prod_{j=1}^{n} \sigma_{i_j'',j}$$

are any two representations of the same element of $\mathfrak{Z}_n$ in the form I_0, then $i_j'=i_j''(j=1, \ldots, n)$.

4. The following relations of Burau are valid in $\mathfrak{N}_n$:

(48) $$s_{i,j}\, s_{k,l} = s_{k,l}\, s_{i,j} \qquad (j<k \text{ or } (i<k,\ l<j))$$

(49) $$s_{i,j}\, s_{i,k}\, s_{j,k} = s_{i,k}\, s_{j,k}\, s_{i,j},$$

(50) $$s_{i,k}\, s_{j,k}\, s_{i,j} = s_{j,k}\, s_{i,j}\, s_{i,k},$$

(51) $$s_{i,k}\, s_{j,k}\, s_{j,l}\, s_{j,k}^{-1} = s_{j,k}\, s_{j,l}\, s_{j,k}^{-1}\, s_{i,k} \qquad (i<j,\ k<l).$$

The abstract group with the generators $s_{i,j}\,(1 \leqslant i<j \leqslant n)$ determined by the relations (48)—(51) will be denoted by $\mathfrak{N}_n^*$.

The above relations imply a system of equalities of the form:

(59) $$[s_{j,l},\, s_{i,k}^{\varepsilon}] = f_{j,l,i,k,\varepsilon}(s_{1,l}, \ldots, s_{l-1,l})$$

where $\varepsilon = \pm 1$ and $\{j, l, i, k\}$ is an arbitrary system of indices, such that the left-hand side of (59) exists and $k<l$. This system of relations yields an algorithm for the solution of every problem of the following type: given a positive integer n and two functions f and g of $l-1$ and $\frac{1}{2}(l-1)(l-2)$ arguments respectively $(2<l \leqslant n)$, to find a function h of $l-1$ arguments such that the relation

$$[f(s_{1,l}, \ldots, s_{l-1,l}),\, g(s_{1,2}, \ldots, s_{l-2,l-1})] = h(s_{1,l}, \ldots, s_{l-1,l})$$

holds in $\mathfrak{N}_n^*$.

This algorithm enables us to represent every element of the group $\mathfrak{N}_n^*$ given as a function of $s_{i,j}$ with $j>k\,(1 \leqslant k \leqslant n)$ in the form

I_k $$\prod_{j=k+1}^{n} f_j(s_{1,j}, \ldots, s_{j-1,j}),$$

where f_j is a function of $j-1$ variables $(j=k+1, \ldots, n)$.

As every relation between the elements $s_{i,j}$ in $\mathfrak{N}_n^*$ is also valid in $\mathfrak{N}_n$, we obtain the following result:

Theorem 5 (Ivanovsky). Every element of the group $\mathfrak{N}_n$ given as a function of the elements $s_{i,j}$ with $j > k$ can be represented in the form I_k.

Corollary 9. Every element of the group $\mathfrak{N}_n$ given as a function of the elements $s_{i,j}$ can be represented in the form I_1.

Corollary 10. Every element of the group $\mathfrak{Z}_n$ given as a function of the elements σ_i can be represented in the form

$$\prod_{j=2}^{n} f_j(s_{1,j}, \ldots, s_{j-1,j}) \prod_{j=1}^{n} \sigma_{i_j,j}.$$

5. Let $\mathfrak{F}_n$ denote the free group with the free basis $\{t_1, \ldots, t_n\}$. Define the endomorphisms $\bar{\sigma}_i$, σ_i^* $(i=1, \ldots, n-1)$ of this group by the equalities:

$$\begin{aligned} \bar{\sigma}_i t_j &= t_j && (j \neq i, i+1), \\ \bar{\sigma}_i t_i &= t_{i+1}, \\ \bar{\sigma}_i t_{i+1} &= [t_i, t_{i+1}], \\ \sigma_i^* t_j &= t_j && (j \neq i, i+1), \\ \sigma_i^* t_i &= [t_{i+1}, t_i^{-1}], \\ \sigma_i^* t_{i+1} &= t_i. \end{aligned}$$

One can immediately verify that $\bar{\sigma}_i \sigma_i^* = \sigma_i^* \bar{\sigma}_i = 1$, where 1 denotes the identical automorphism of the group $\mathfrak{F}_n$.

Corollary 11. $\bar{\sigma}_i$ are automorphisms of the group $\mathfrak{F}_n$ and

$$\sigma_i^* = \bar{\sigma}_i^{-1} \qquad (i=1, \ldots, n-1).$$

The automorphisms $\bar{\sigma}_i$ generate a group of automorphisms which will be denoted by $\overline{\mathfrak{Z}}_n$.

It is easy to verify that

$$\begin{aligned} \bar{\sigma}_i \bar{\sigma}_{i+1} \bar{\sigma}_i &= \bar{\sigma}_{i+1} \bar{\sigma}_i \bar{\sigma}_{i+1} && (i=1, \ldots, n-2), \\ \bar{\sigma}_i \bar{\sigma}_j &= \bar{\sigma}_j \bar{\sigma}_i && (i, j=1, \ldots, n-1; |i-j| > 1 \end{aligned}$$

Corollary 12 (Artin). A homomorphism of the group $\mathfrak{Z}_n$ onto the group $\overline{\mathfrak{Z}}_n$ arises if the element $\bar{\sigma}_i$ of $\overline{\mathfrak{Z}}_n$ is put into correspondence to the element σ_i of $\mathfrak{Z}_n$ $(i=1, \ldots, n-1)$.

In what follows $\bar{z}$ will denote the image of the element z of $\mathfrak{Z}_n$ under the homomorphism just defined.

The following relations are valid:

$$\begin{aligned} \bar{s}_{i,j} t_k &= t_k && (k < i \text{ or } j < k), \\ \bar{s}_{i,j} t_i &= [t_i, t_j], \end{aligned}$$

$$\bar{s}_{i,j}\,t_j=[t_j,\ t_i\,t_j],$$
$$s_{i,j}^{-1}\,t_i=[t_i,\ t_j^{-1}\,t_i^{-1}],$$
$$\bar{s}_{i,j}^{-1}\,t_j=[t_j,\ t_i^{-1}].$$

6. We determine the endomorphisms α_j $(j=1, \ldots, n)$ of the group $\mathfrak{F}_n$ by the equalities

$$\alpha_j\,t_i=t_i \qquad (i\neq j),$$
$$\alpha_j\,t_j=1.$$

The following equalities take place:

$$\alpha_j\,\bar{\sigma}_i^{\varepsilon}=\bar{\sigma}_i^{\varepsilon}\,\alpha_j \qquad (i<j-1,\ \varepsilon=\pm 1),$$
$$\alpha_j\,\bar{s}_{i,j}^{\varepsilon}=\alpha_j \qquad (\varepsilon=\pm 1).$$

The following lemma will be very important:

Lemma 1 (Lemma 15 in the Russian text). Whatever be an integer $j\,(2\leqslant j\leqslant n)$ and a function f of $j-1$ arguments, we can choose an element w of the group $\mathfrak{F}_n$ such that

$$f(\bar{s}_{1,j}, \ldots, \bar{s}_{j-1,j})\,t_j=[t_j,\ w],$$
$$\alpha_j\,w=f(t_1, \ldots, t_{j-1}).$$

7. Let $\mathfrak{T}_i$ denote the totality of elements of the group $\mathfrak{F}_n$ conjugate to t_i.

It is easy to prove that the equality

$$f(\bar{\sigma}_1, \ldots, \sigma_{n-1})\,\mathfrak{T}_i=\mathfrak{T}_{f(\hat{\sigma}_1, \ldots, \hat{\sigma}_{n-1})\,i}$$

holds for every function f of $n-1$ variables. Using this equality and Theorem 4, one can easily prove.

Lemma 2 (Lemma 18 in the Russian text). If $z'\in\mathfrak{Z}_n$, $z''\in\mathfrak{Z}_n$ and $\overline{z'}=\overline{z''}$ then $\overleftarrow{z'}=\overleftarrow{z''}$.

8. The following lemma can be established by means of Lemmas 1 and 2:

Lemma 3 (Lemma 20 in the Russian text). If

$$z'=\prod_{j=2}^{n} f_j'(s_{1,j}, \ldots, s_{j-1,j})\prod_{j=1}^{n}\sigma_{i_j',j},$$

$$z''=\prod_{j=2}^{n} f_j''(s_{1,j}, \ldots, s_{j-1,j})\prod_{j=1}^{n}\sigma_{i_j'',j},$$

in $\mathfrak{Z}_n$ and $\bar{z}'=\bar{z}''$, then $f_j'=f_j''\ (j=2,\ldots,n)$ and $i_j'=i_j''\ (j=1,\ldots,n)$.

Hence follow the results:

Theorem 6 (Ivanovsky). Every element of the group $\mathfrak{Z}_n$ given as a function of σ_i can be uniquely represented in the form I.

Theorem 7 (Artin). The homomorphism of the group $\mathfrak{Z}_n$ onto the group $\bar{\mathfrak{Z}}_n$ carrying every σ_i into $\bar{\sigma}_i$ is an isomorphism.

Theorem 8 (Burau). The system of relations (48)—(51) determines the group $\mathfrak{N}_n$.

Theorems 6 and 7 yield two solutions of the problem of identity in the group of tresses of order n.

9. The results established throw light on the structure of the groups $\mathfrak{Z}_n$ and $\mathfrak{N}_n$. In fact, denote by $\mathfrak{N}_{n,l}$ the subgroup of the group $\mathfrak{N}_n$ generated by the elements $s_{i,j}$ with $j>l$. The above results imply

Theorem 9 (Ivanovsky). 1° The groups $\mathfrak{N}_{n,l}\ (1\leqslant l\leqslant n)$ are invariant subgroups of the group $\mathfrak{N}_n$;

2°. $(1_n)=\mathfrak{N}_{n,n}\subset\mathfrak{N}_{n,n-1}\subset\cdots\subset\mathfrak{N}_{n,1}=\mathfrak{N}_n$;

3°. The factor-group $\mathfrak{N}_{n,l-1}/\mathfrak{N}_{n,l}\ (2\leqq l\leqq n)$ is a free group with $l-1$ free generators.

Corollary 13. The group $\mathfrak{Z}_n$ possesses the normal series

$$(1_n)=\mathfrak{N}_{n,n}\subset\mathfrak{N}_{n,n-1}\subset\cdots\subset\mathfrak{N}_{n,1}=\mathfrak{N}_n\subset\mathfrak{Z}_n;$$

the factors $\mathfrak{N}_{n,l-1}/\mathfrak{N}_{n,l}$ of this series are free groups; the factor $\mathfrak{Z}_n/\mathfrak{N}_n$ is isomorphic to the symmetric group of degree n.

10. The following theorem reveals the connexion between the groups $\mathfrak{N}_n$ and $\mathfrak{N}_l$ for $l<n$.

Theorem 10. Let $n>l\geqslant 1$. There exists a unique homomorphism φ, of the group $\mathfrak{N}_n$ onto the group $\mathfrak{N}_l$ such that

$$\varphi_{n,l}\,s_{i,j}=\begin{cases}1_l & (1\leqslant i<j,\ l<j\leqslant n)\\ s_{i,j} & (1\leqslant i<j\leqslant l).\end{cases}$$

The kernel of this homomorphism is the group $\mathfrak{N}_{n,l}$.

Denote by $\mathfrak{N}_l^n$ the subgroup of the group $\mathfrak{N}_n$ generated by the elements $s_{i,j}$ with $1\leqslant i<j\leqslant l\ (1\leqslant l\leqslant n)$.

Corollary 14. The groups $\mathfrak{N}_l^n$ and $\mathfrak{N}_l$ are isomorphic. The transformation $\varphi_{n,l}$ maps $\mathfrak{N}_l^n$ isomorphically onto $\mathfrak{N}_l$.

Corollary 15. Every element of the group $\mathfrak{N}_n$ can be uniquely represented in the form xy, where $x\in\mathfrak{N}_{n,l}$ and $y\in\mathfrak{N}_l^n\ (1\leqslant l\leqslant n)$.

11. We define the elements $\rho_{i,j}\ (1\leqslant i\leqslant j\leqslant n)$ of $\mathfrak{Z}_n$ by induction as follows:

$$\rho_{i,i}=1 \qquad (1\leqslant i\leqslant n),$$

$$\rho_{i,j+1}=\sigma_j\,\rho_{i,j}\,\sigma_j \qquad (1\leqslant i\leqslant j<n).$$

These elements are connected with the elements $s_{i,j}$ by the equalities

$$s_{i,j} = \rho_{i,j}\, \rho_{i+1,j}^{-1} \qquad (1 \leqslant i < j \leqslant n).$$

Corollary 16. The elements $\rho_{1,j}, \ldots, \rho_{j-1,j}$ generate the same subgroup of the group $\mathfrak{N}_n$ as the elements $s_{1,j}, \ldots, s_{j-1,j}$ $(2 \leqslant j \leqslant n)$.

Corollary 17. The elements $\rho_{i,j}$ $(1 \leqslant i < j \leqslant n)$ generate $\mathfrak{N}_n$.

Theorem 11. Every element of the group $\mathfrak{Z}_n$ can be uniquely represented in the form

$$I' \qquad \prod_{j=2}^{n} f_j(\rho_{1,j}, \ldots, \rho_{j-1,j}) \prod_{j=1}^{n} \sigma_{i_j,j}.$$

Theorem 12. The group $\mathfrak{N}_n$ can be determined by the following relations between the generators $\rho_{i,j}$:

$$[\rho_{k,l}, \rho_{i,j}^{\varepsilon}] = \rho_{k,l} \qquad (j < k \text{ or } (k \leqslant i,\ j < l),\ \varepsilon = \pm 1),$$

$$[\rho_{k,l}, \rho_{i,j}] = \rho_{i,l}\, \rho_{j+1,l}^{-1}\, \rho_{j,l}\, \rho_{i,l}^{-1}\, \rho_{k,l}\, \rho_{j,l}^{-1}\, \rho_{j+1,l} \qquad (i \leqslant k \leqslant j < l),$$

$$[\rho_{k,l}, \rho_{i,j}^{-1}] = \rho_{j+1,l}\, \rho_{j,l}^{-1}\, \rho_{k,l}\, \rho_{i,l}^{-1}\, \rho_{j,l}\, \rho_{j+1,l}^{-1}\, \rho_{i,l} \qquad (i \leqslant k \leqslant j < l).$$

Mikhail Alekseevich Lavrentev

Born – 19 Nov 1900 in Kazan, Russia
Died – 15 Oct 1980 in Moscow, Russia

Mikhail Lavrentev studied at the University of Moscow, graduating in 1922. He was awarded doctorate by Moscow University in 1933. From 1933 he held the chair of Analysis and Theory of Functions at Moscow State University.

In 1934 the Steklov Mathematical Institute, together with some other institutions of the Academy of Sciences of the USSR, was moved from Leningrad to Moscow. Lavrentev headed the Department of the Theory of Functions of a Complex Variable at the Steklov Mathematical Institute.

From 1939 to 1941 and then again from 1944 to 1949, Lavrentev was the Director of the Institute of Mathematics of the Academy of Sciences of the Ukraine in Kiev. In 1945 he was also appointed Vice-President of the Academy of Sciences of the Ukraine. Then, in 1950, he became Director of the Institute of Mechanics and Computational Technology of the Ukraine.

Lavrentev moved to Novosibirsk when he was Vice-President of the Academy of Sciences of the USSR between 1957 snd 1975. During this time he was Head of the Siberian Division of the Academy of Sciences of the USSR. Then, in 1975, Lavrentev returned to Moscow where he worked at the Institute of Physics and Technology until his death.

Lavrentev is remembered for an outstanding book on conformal mappings and he made many important contributions to that topic. In the 1940s he developed the theory of quasiconformal mappings which gave a new geometrical approach to partial differential equations. One of the major areas to which he applied this work was to hydrodynamics.

The 1940s was a period of industrialization and construction and, after 1945, Lavrentev founded new areas of research in mechanics and applied physics which were aimed at laying the theoretical foundation necessary for the large construction projects of building dams, canals and bridges on the Volga, Dnieper and Don rivers.

He also applied the theory of complex variables to other topics, in particular to nonlinear waves. Other topics where he made substantial contributions where the theory of sets, the general theory of functions, and the theory of differential equations.

ON THE THEORY OF QUASI-CONFORMAL MAPPINGS OF THREE-DIMENSIONAL DOMAINS

By

M. A. LAVRENTIEFF

in Novosibirsk, U.S.S.R.

The theory of quasi-conformal mapping was established as an independent discipline about 30 to 35 years ago. Since that time the theory has undergone a broad development with a special flavor of its own, together with applications to closely related mathematical areas. In its applications to the theory of analytic functions, in particular to the classical type problem of Riemann surfaces, it owes much to the basic results of L. Ahlfors, who also contributed many important results concerning extremal problems in the theory. Other areas of fruitful applications concerned geometry and continuum mechanics. In the last decade have appeared many interesting papers on the theory of quasiconformal mappings in three-dimensional regions; the basic direction of these papers is the qualitative investigation of such mappings, and here it would be particularly interesting to connect this theory to the rapidly developing branch of classical topology—the theory of differentiable mappings. Also interesting and important is the construction of a general theory of quasi-conformal mappings in three-dimensional regions, mappings whose three components satisfy a certain system of differential equations of three unknown functions of three independent variables. Success in this direction would be of fundamental importance for several unsolved problems in continuum mechanics.

In this paper we shall mention three classes of mappings corresponding to certain systems of equations. The results mentioned are of a very special character, but they indicate certain directions in which the theory could develop.

We shall consider homeomorphic mappings of domains D in the space (x, y, z) onto domains Δ in the space (u, v, w):

$$u = u(x, y, z), \qquad v = v(x, y, z), \qquad w = w(x, y, z). \tag{1}$$

We shall assume that the functions u, v, w are continuously differentiable and that the functional determinant is strictly positive:

$$\begin{pmatrix} u & v & w \\ x & y & z \end{pmatrix} > 0.$$

Suppose that we are given a system of k equations, $k \geqq e$,

$$F_i\left(x, y, z, u, v, w, \frac{\partial u}{\partial x}, \frac{\partial u}{\partial y}, \ldots, \frac{\partial w}{\partial z}\right) = 0. \tag{2}$$

By the Riemann problem for (2) we shall mean the problem of constructing a mapping of the given domain D onto the given domain Δ satisfying the system (2). We shall also call such mappings the quasi-conformal mappings of D onto Δ corresponding to the system (2).

The general problem consists of finding a sufficiently wide class of equations (2) and a class of domains D and Δ to solve the Riemann problem. In the examples to be considered below we shall assume that the domain D is a layer

$$z_0(x, y) < z < z_1(x, y) \tag{3}$$

bounded by the surfaces $\Gamma_0 : z = z_0(x, y)$ and $\Gamma_1 : z = z_1(x, y)$, where the functions z_0 and z_1 have bounded derivatives up to and including the second order. Moreover, we assume that

$$e^{|x|+|y|}\left|z_0(x, y)\right| < K, \qquad e^{|x|+|y|}\left|z_1(x, y) - 1\right| < K,$$

where K is some constant. For the domain Δ we take the layer

$$0 < w < 1. \tag{4}$$

We shall restrict ourselves in the sequel to mappings (1) for which the point $[0,0,z_0(0,0)]$ goes into the point $[0,0,0]$, the point ∞ goes into the point ∞, and the x-axis goes into a curve which is asymptotic to the u-axis.

1. The theorem of Liouville. The first natural question concerns the number of equations of (2) for which the Riemann problem can be expected to be solvable for a sufficiently wide class of domains. Here we have the classical result of Liouville that if the mapping (that is, the functions u,v,w) is three times differentiable and conformal, then it reduces to the composition of a linear mapping and an inversion. This is not surprising since it follows from the conformality that the functions u,v,w must satisfy a system of five equations. We shall show, from the case of the mappings of layers, that the Liouville theorem is also valid, in a certain sense, if we discard one of the conditions of conformality. This will also give a proof of stability for Liouville's theorem.

Theorem *Let the system* (1) *effect a mapping of the domain* (3) *onto the layer* (4) *and satisfy the following conditions*:

(1) *All second partial derivations of u,v,w are uniformly continuous in D*;

(2) *An arbitrary infinitesimal cube in Δ with edges parallel to the axes corresponds to an infinitesimal parallelepiped in such a way that* (i) *the angles between the edges differ from $\pi/2$ by no more than ε*; (ii) *the face corresponding to the face parallel to the plane $w=0$ is a square to within ε* (i.e., *the ratio of the sides differs from unity by no more than ε*); (iii) *all surfaces $z=z_w(x,y)$ corresponding to the planes $w=$* const. *satisfy*

$$e^{|x|+|y|}\left|z_w(x,y)-w\right|<K.$$

Under these conditions there exists a continuous function $\phi(\varepsilon)$, with $\phi(0)=0$, depending only on the degree of continuity of $z_{xx},z_{xy},\cdots,w_{zz}$, and such that

$$\left|z_w(x,y)-z_w(0,0)\right|<\phi(\varepsilon),$$

where $z=z_w(x,y)$ is the surface corresponding to the plane $w=$ const. *If in addition, we assume that*

$$\left| z_w(0,0) - w \right| < \phi(\varepsilon),$$

then we have

$$\left| u(x,y,z) - x \right| < \psi(\varepsilon)$$
$$\left| v(x,y,z) - y \right| < \psi(\varepsilon)$$
$$\left| w(x,y,z) - z \right| < \psi(\varepsilon),$$

where $\psi(\varepsilon)$ is a continuous function depending on $\phi(\varepsilon)$ such that $\psi(0) = 0$.

Proof. Let us mention the basic idea of the proof. We shall show that the surface

$$z = z_w(x,y)$$

corresponding to $w = \text{const.}$ has at each point "almost" equal principal curvatures. In other words, it follows from the hypotheses of the theorem that there exists a function $\beta(\varepsilon)$, $\lim_{\varepsilon \to 0} \beta(\varepsilon) = 0$, such that for the principal curvatures K_1 and K_2 at each point of the surface, we have

$$K_2 - K_1 < \beta(\varepsilon).$$

Let us assume, to the contrary, that for fixed degree of continuity of the second derivatives and for arbitrarily small ε, there is a point A_0 at which

$$K_2 - K_1 \geqq a,$$

where a is some positive constant. But then, by the continuity of K_1 and K_2 in a sphere of radius $\kappa_0 = \kappa_0(a)$ with center A_0, we have

$$K_2 - K_1 \geqq a/2. \tag{5}$$

Let us consider now the surface $\Gamma: z = z_{w_0}(x,y)$ passing through the point. A_0 and the surface $\Gamma': z = z_{w_0+h}(x,y)$ passing through the point A_0' lying on the normal to Γ at a distance $r_0/2$ from A_0. By hypothesis, if we take a tube containing A_0 which makes an infinitesimal circular intersection with Γ

and whose generators are orthogonal to the surface $z = z_{w_0}(x, y)$, then the intersection of this tube with Γ should be almost circular; on the other hand, it follows from (5) that this intersection will be an ellipse such that the ratio of its semi-axes differs from unity by a finite quantity.

Thus, by the hypotheses of the theorem, every surface $\Gamma: z = z_w(x, y)$ must have almost equal principal curvatures at every point, so that, by a theorem in geometry [1], any such Γ is almost spherical. Since every such Γ lies in a layer of finite width, it follows that Γ is almost planar. The uniform closeness of Γ to some plane

$$z = \text{const}$$

is a consequence of condition (iii) of the theorem.

The following conjecture seems most plausible: Under the hypotheses of the theorem there exists an absolute constant K such that

$$|z_0(x, y)| < K\varepsilon, \qquad |z_1(x, y) - 1| < K\varepsilon.$$

This theorem has not been proved, nor is there a counter-example.

2. **Examples of systems of three equations.** The solvability of the Riemann problem, as well as the generalization of several properties of conformal mappings of planar regions to the corresponding quasiconformal mappings in space, can be achieved for mappings of the corresponding systems of three first-order equations related to potential theory.

Let Π be the parallelepiped in the domain D corresponding to an infinitesimal cube in Δ with edges parallel to the axes. Let a be the length of an edge of the cube, and let $\Delta\sigma_1$ and $\Delta\sigma_2$ be the areas of those faces of Π corresponding to the faces of the cube which are perpendicular, respectively, to the u- and w-axes. Let ΔV_1 and ΔV_2 be the lengths of the edges of Π corresponding to those edges of the cube which are parallel, respectively, to the u- and w-axes.

Let us set

$$\lim \frac{\Delta\sigma_1}{a} = \sigma_1, \qquad \lim \frac{\Delta\sigma_2}{a} = \sigma_2$$

$$\lim \frac{\Delta V_1}{a} = V_1, \qquad \lim \frac{\Delta V_2}{a} = V_2.$$

A large number of results can be obtained for mappings of layers which are similar to theorems in the theory of the conformal mapping of curvilinear strips; these results correspond to the following two classes of systems of three equations:

1) Two equations determine the orthogonality of four edges of Π by the face σ_1 and one equation corresponds to the condition

$$V_1 = f_1(\sigma_1),$$

where f_1 is a sufficiently smooth increasing function with $f_1(0) = 0$.

2) Two equations determine the orthogonality of four edges of Π by the face σ_2 and one equation corresponds to the condition

$$V_2 = f_2(\sigma_2),$$

where f_2 is a sufficiently smooth function with $f_2(0) = 0$.

The solution of the Riemann problem for each of the systems can be interpreted as the motion of an ideal fluid. Whenever f is of a specified type, the systems will define the motion of an ideal fluid in the layer (3). The Riemann problem is equivalent to determining the motion of a fluid or gas in the layers. The problem of finding one of the boundaries Γ of D for certain properties of the solutions on Γ corresponds to the classical problem, but the three-dimensional problems in hydrodynamics and aerodynamics with free boundaries are not yet sufficiently advanced, in particular, in the theory of jets and the theory of waves in heavy fluids.

Although not many concrete results have been obtained in the theory of quasiconformal mappings in space, it appears that this direction ought to lead to many significant results.

3. Harmonic mappings. In the course of more than half a century the attention of many leading mathematicians and schools of mathematics has been directed to various mixed boundary-value problems. Many important results could have been obtained if it had been possibly to carry over to higher dimensions the well-known results of Carleman for estimating the modulus of a harmonic function in terms of its modulus on a certain set of the boundary of the domain of the given function. It is possible to approach this problem from the point of view of quasi-conformal mappings of domains in space.

We now introduce a class of quasi-conformal mappings which is of interest both from the point of view of the Riemann problem and from the point of view of mixed boundary-value problems.

As before, we restrict ourselves to mappings of layers: mappings of the layer (3) onto the layer (4).

We shall call a mapping of D onto Δ harmonic if the functions u, v, w are the components of the gradient of a harmonic function which is regular in D. Harmonic mappings are quasi-conformal mappings corresponding to a system of four equations.

$$\frac{\partial u}{\partial y} = \frac{\partial v}{\partial x}, \qquad \frac{\partial v}{\partial z} = \frac{\partial w}{\partial x}, \qquad \frac{\partial w}{\partial y} = \frac{\partial v}{\partial z},$$
$$\frac{\partial u}{\partial x} + \frac{\partial v}{\partial y} + \frac{\partial w}{\partial z} = 0.$$

Despite the fact that a harmonic mapping is apparently overdetermined, there exists a sufficiently wide class of domains admitting harmonic mappings.

We remark first that it is possible to give a simple construction of the solution of the Riemann problem for a mapping of the layer (3) onto the layer (4).

Indeed, by the hypotheses of the problem, the function w must vanish on Γ_0 and must be 1 on Γ_1. Hence, if we denote by U the harmonic function such that

$$\frac{\partial U}{\partial z} = w, \tag{6}$$

we see that the harmonic function w has prescribed values on Γ_0 and Γ_1 and can be determined everywhere in D. If we integrate (6) we find U up to an arbitrary function $\phi(x, y)$:

$$U = \int_{z_0(x,y)}^{z} \frac{\partial U}{\partial z} dz + \phi(x, y) \equiv V(x, y, z) + \phi(x, y). \tag{7}$$

The function ϕ is determined by the condition that U be a harmonic function, that is, from the equation

$$\frac{\partial^2\phi}{\partial x^2} + \frac{\partial^2\phi}{\partial y^2} = -\frac{\partial^2 V}{\partial x^2} - \frac{\partial^2 V}{\partial y^2}. \tag{8}$$

If the harmonic function U defined by (7) exists, then the righthand side of (8) does not contain z, and we determine the unknown function ϕ from the solution of the Poisson equation (8).

If we consider the simplest case when Γ_0 and Γ_1 are the planes $z = 0$ and $z = 1$, then

$$\frac{\partial U}{\partial z} = z, \quad V = \frac{1}{2}z^2.$$

Equation (8) takes the form

$$\Delta\phi = -1, \tag{9}$$

and we obtain as the desired potential

$$V = -\tfrac{1}{2}x^2 + y^2 + \tfrac{1}{2}z^2.$$

The mapping will be realized by the system

$$u = -x, \quad v = 2y, \quad w = z.$$

This is the reason for assuming that, for harmonic mappings, the Riemann theorem should be valid for a very wide class of strip regions, but it yields an existence theorem only for strips which are sufficiently close (closeness of second order) to a unit strip $0 < z < 1$.

Theorem. *Let the boundaries of D, the surfaces Γ_0 and Γ_1, satisfy the following condition: the functions $z_0(x,y)$ and $z_1(x,y) - 1$, along with their first and second partial derivatives, do not exceed $a\, e^{-(x^2+y^2)}$ in modulus. Then for arbitrary Γ_0 and Γ satisfying this condition there exists, for sufficiently small a, a harmonic mapping of D onto the strip $0 < w < 1$.*

The mapping is unique if it is required that the point of Γ_0 *with coordinates* $x = y = 0$ *map into the point* $u = v_z - w = 0$, *and that the u-axis correspond to a curve which is asymptotic to the x-axis.*

Sketch of the basic steps of the proof. First, from the hypotheses, it is easy to show the existence of a harmonic function U for which $\frac{\partial U}{\partial z} = 0$ and $\frac{\partial U}{\partial z} = 1$ on Γ_0 and Γ_1, respectively, since at infinity U is asymptotically close to the function

$$U_0 = -\tfrac{1}{2}x^2 + y^2 + \tfrac{1}{2}z^2.$$

For sufficiently small a, equation (8) has the form

$$\Delta\phi = -1 + \alpha(x, y),$$

where $\alpha(x, y)$ is small along with a and tends to zero as $x^2 + y^2 \to \infty$ by the exponential law. Hence it is not difficult to show that grad U_0 is close to grad U, and that this gradient will effect the desired harmonic mapping of D onto the strip $0 < w < 1$.

We introduce a very interesting class of domains admitting a harmonic mapping of closed simply connected domains onto a ball. We are sure (not proved) that there does not exist a harmonic mapping of a sphere onto a sphere. But, for some prescribed correspondence of the axes, there exists a harmonic mapping of an ellipsoid onto the sphere. Hence a new formulation suggests itself: Can a given domain D (bounded by a smooth surface, simply connected and bounded) be subjected to a one-parameter deformation in such a way that, the resulting deformation of D can be mapped harmonically onto a sphere.

Reference

[1] A. V. Pogorelov, Nearly spherical surfaces, *Journal d'Analyse Mathématique*, Vol. 19. pp. 313–321.

The Academy of Sciences of the U.S.S.R.

(Received November 4, 1966)

Andrei Nikolaevich Tikhonov

Born – 30 Oct 1906 in Gzhatska, Smolensk, Russia
Died – 1993

Andrei Tikhonov attended secondary school as a day student and completed his school education in 1922. The very same year, he entered Moscow University and studied in the Mathematics Department of the Faculty of Mathematics and Physics. Tikhonov made remarkable progress and published his first papers in 1925 while still in the middle of his undergraduate course.

His first work was related to the results of Aleksandrov and Urysohn on the conditions for a topological space to be metrisable. He did not end there, but continued to do further investigations in topology. By 1926, he discovered the topological construction that was named after him today. The Tikhonov Topology defined the product of topological spaces. Tikhonov had given the right definition that was counterintuitive even to a topologist as great as Aleksandrov. His idea allowed him to push on and prove important topological results as the products of any set of compact topological space is compact.

There were not many mathematicians that gained worldwide recognition even before they started their research career, but this was essentially what happened to Tikhonov. He achieved results on the Tikhonov topology of products before he graduated in 1927. With this impressive record, he became a research student in Moscow University. Though Tikhonov was blessed with an intuitive grasp on topological ideas, he was equally bestowed with talents in other areas of Mathematics.

In fact in 1935, Tikhonov went from topology to functional analysis with his famous fixed-point theorem for continuous maps from convex compact subsets of locally convex topological spaces. The results he obtained are of importance in both topology and functional analysis. They were applied by Tikhonov to solve problems in mathematical physics.

In 1936, Tikhonov defended his habilitation thesis in the *Functional equations of Volterra type and their applications to mathematical physics.* The thesis extended Emile Picard's method of approximating the solution of a differential equation and gave applications to heat conduction, in particular, cooling, which obeys the law to Josef Stefan and Boltzmann. Tikhonov was appointed as a professor of Moscow University after successfully defending his thesis. Three years later, he was elected as a Corresponding Member of the Academy of Sciences of the USSR.

In mathematical physics, one characteristic of his research was, he would combine a concrete theme in natural science with investigations into a fundamental mathematical problem. He knew how to pick out a typical concrete physical problem in nature, and give it a clear mathematical formulation. Yet his mathematical investigations were never confined to the solution of a given concrete problem, but instead served as a starting point for stating a general mathematical problem that was a broad generalization of the first problem.

His work on mathematical physics continued throughout the 1940s and in 1953, he was awarded the State Prize for his contributions. In 1948 however, he began a study on a new type of problem when he considered the behavior of the solutions of the systems of equations with a small parameter in the term with the highest derivative. He wrote a series of fundamental papers introducing the topic. From then on, his students carried on with his work.

Tikhonov also made significant contributions on the area of computational mathematics. One of the most outstanding achievements in this area is the theory of homogeneous difference schemes, which Tikhonov developed in collaboration with Samarskii.

In 1960s, Tikhonov began to produce a series of important papers on ill-posed problems. He defined a class of regularizable ill-posed problems and further introduced the concept of a regularizing operator that was used as the solution to those problems. By combining his computing skills with solving problems of this type, Tikhonov gave implementations of algorithms to compute the operators. He was awarded the Lenin Prize for his work on ill-posed problems in 1966, during which he was also elected to full membership of the Academy of Sciences of the USSR.

Tikhonov's wide interest in Mathematics led him to hold a number of different chairs at Moscow University. In particular, he held a chair in the Mathematical Physics Faculty and also in the Computational Mathematics in the Engineering Mathematics Faculty. He was the Dean of the Faculty of Computing and Cybernetics at Moscow University and was also appointed as the Deputy Director of the Institute of Applied Mathematics of the Academy of Sciences, a position that he held for many years.

Tikhonov passed away in 1993 at a ripe old age of 87.

Über die topologische Erweiterung von Räumen.

Von

A. Tychonoff in Moskau.

Einleitung.

Die vorliegende Arbeit schließt an die abstrakt-topologischen Untersuchungen von Alexandroff und Urysohn an[1]) und behandelt das Problem der Einbettung von beliebigen topologischen Räumen in absolut abgeschlossene[2]) und insbesondere in bikompakte Räume. Ich beweise in erster Linie den folgenden

Satz I. *Zu jeder Kardinalzahl τ gibt es einen nur von dieser Kardinalzahl abhängenden bikompakten Raum R_τ von der Eigenschaft, daß jeder normale topologische Raum, der ein Umgebungssystem von einer Mächtigkeit $\leqq \tau$ besitzt, einer Teilmenge des Raumes R_τ homöomorph ist; dabei besitzt der Raum R_τ selbst ein Umgebungssystem von der Mächtigkeit τ. Wenn $\tau = \aleph_0$ ist, so ist R_τ dem Fundamentalquader des Hilbertschen Raumes homöomorph.*

(Man versteht bekanntlich unter dem Fundamentalquader des Hilbertschen Raumes die Gesamtheit aller Punkte $\xi = (x_1, x_2, \ldots, x_n, \ldots)$, deren Koordinaten x_n für jedes n den Ungleichungen $0 \leqq x_n \leqq \frac{1}{n}$ genügen.)

---

[1]) Siehe vor allem Alexandroff und Urysohn, Zur Theorie der topologischen Räume, Math. Annalen **92** (1924), S. 258, und Alexandroff, Über stetige Abbildungen kompakter Räume, Math. Annalen **96** (1926), S. 555; die Kenntnis dieser Arbeit (insbesondere auch die dort gebrauchte Terminologie) wird im folgenden vorausgesetzt; die erste dieser Arbeiten wird kurz durch „Alexandroff-Urysohn“, die zweite durch „Alexandroff“ zitiert. Wegen ausführlicher Darstellung der erwähnten Untersuchungen möge insbesondere auf „Mémoire sur les espaces compacts“ (Verh. K. Akademie Amsterdam, Deel XIV, No. 1 (1929)) derselben Verfasser hingewiesen sein.

[2]) Alexandroff-Urysohn, S. 261.

In dem letzten Teile des soeben ausgesprochenen Satzes sind insbesondere die beiden bekannten Urysohnschen Metrisationssätze[3]) (der Metrisationssatz für separable normale Räume, sowie der Metrisationssatz für kompakte topologische Räume) und auch der Urysohnsche Satz von der Einbettbarkeit jedes separablen metrischen Raumes in den Fundamentalquader des Hilbertschen Raumes[4]) enthalten.

Die Frage liegt nahe, ob die Normalität eines topologischen Raumes auch notwendig ist, um den Raum in einen bikompakten topologischen Raum einbetten zu können. In der vorliegenden Arbeit (§ 4) wird auf diese Frage eine negative Antwort gegeben, indem gezeigt wird, daß auch nichtnormale topologische Räume als Teilmengen von bikompakten Räumen auftreten können. Somit entsteht die Frage, notwendige und hinreichende Bedingungen aufzustellen, damit ein Raum in einen bikompakten topologischen eingebettet werden kann. Um diese Bedingung formulieren zu können, führen wir die folgende Definition[5]) ein.

Ein topologischer Raum heißt *vollständig regulär*, wenn zu jedem Punkte x_0 und zu jeder ihn nicht enthaltenden abgeschlossenen Menge A eine im ganzen Raume stetige Funktion $f(x)$ definiert werden kann, die im ganzen Raume der Bedingung

$$0 \leqq f(x) \leqq 1$$

genügt und die überdies in x_0 gleich 0 und in sämtlichen Punkten von A gleich 1 ist.

Jeder vollständig reguläre Raum ist, wie leicht ersichtlich, regulär, während jeder normale Raum vollständig regulär ist. Im § 4 der vorliegenden Arbeit wird gezeigt, daß es vollständig reguläre Räume gibt, die nicht normal, und daß es reguläre Räume gibt, die nicht vollständig regulär sind, womit die Klasse der vollständig regulären Räume als eine logisch berechtigte Klasse von Räumen erscheint. Ihre mathematische Berechtigung wird durch den folgenden Satz gegeben:

Satz II. *Ein Raum ist dann und nur dann einer Teilmenge eines bikompakten topologischen Raumes homöomorph, wenn er vollständig regulär ist.*

Was endlich den allgemeinen Fall topologischer Räume betrifft, so gilt der

Satz III. *Jeder topologische Raum R ist einer Teilmenge eines absolut abgeschlossenen Raumes homöomorph, der ein Umgebungssystem besitzt,*

---

[3]) Siehe Urysohn, Zum Metrisationsproblem, Math. Annalen **94** (1925), S. 309, wo auch andere Arbeiten über denselben Gegenstand angegeben sind.

[4]) Siehe Urysohn, Der Hilbertsche Raum ..., Math. Annalen **92** (1924), S. 302.

[5]) Diese Definition rührt von Urysohn her. Vgl. seine Arbeit: Über die Mächtigkeit der zusammenhängenden Mengen, Math. Annalen **94** (1925), S. 292.

welches dieselbe Mächtigkeit hat wie ein gegebenes Umgebungssystem des Raumes R.

In der Formulierung der obigen Sätze wird unter einem Umgebungssystem selbstverständlich ein volles (d. h. dem System aller im Raume vorhandener Gebiete gleichwertiges) System von Umgebungen verstanden. Ein solches Umgebungssystem soll im folgenden kurz eine *Basis* genannt werden.

Ich möchte zum Schluß noch erwähnen, daß ich bei der endgültigen Redaktion dieser Arbeit von Herrn Alexandroff unterstützt worden bin.

§ 1.

Konstruktion der Räume R_τ.

Es sei $\{J_\alpha\}$ eine Menge von der Mächtigkeit τ von abstrakt gegebenen zueinander fremden Einheitsstrecken[6]) $0 \leqq t \leqq 1$. Ein Punkt x des Raumes R_τ ist definitionsgemäß der Inbegriff $\{t_1, t_2, \ldots, t_\alpha, \ldots\}$ von „Koordinaten" t_α, wobei t_α ein Punkt von J_α, also eine reelle Zahl $0 \leqq t_\alpha \leqq 1$ ist. Die Umgebungsdefinition geschieht folgendermaßen: Es sei $x_0 = \{t_1^0, t_2^0, \ldots, t_\alpha^0, \ldots\}$ ein Punkt von R_τ. Wir wählen beliebig endlichviele $J_{\alpha_1}, J_{\alpha_2}, \ldots, J_{\alpha_k}$ und auf jedem dieser Intervalle J_{α_i} zwei rationale Zahlen $\tau'_{\alpha_i} < t^0_{\alpha_i} < \tau''_{\alpha_i}$; eine Umgebung von $x_0 = \{t_1^0, t_2^0, \ldots, t_\alpha^0, \ldots\}$ besteht dann definitionsgemäß aus allen Punkten $x = \{t_1, t_2, \ldots, t_\alpha, \ldots\}$, die den Bedingungen $\tau'_{\alpha_i} < t_{\alpha_i} < \tau''_{\alpha_i}$ genügen.

Dieses Umgebungssystem hat, wie leicht ersichtlich, die Mächtigkeit $\aleph_0 \cdot \tau = \tau$. Ein gleichwertiges Umgebungssystem erhält man, wenn man für τ'_{α_i} bzw. τ''_{α_i} die Punkte $t^0_{\alpha_i} - \varepsilon_i$ bzw. $t^0_{\alpha_i} + \varepsilon_i$ wählt (ε_i sind beliebige hinreichend kleine positive Zahlen). Wir werden kurz schreiben $t_{\alpha_i} \subset S(t^0_{\alpha_i}; \varepsilon_i)$. Wenn $\varepsilon_i = \frac{1}{n_i}$ ist, werden wir die soeben definierte Umgebung durch

$$U^{n_1 n_2 \ldots n_k}_{\alpha_1 \alpha_2 \ldots \alpha_k}(x_0)$$

bezeichnen.

Der mittels dieses Umgebungssystems definierte Raum genügt, wie man leicht zeigt, allen Hausdorffschen Axiomen; R_τ ist also ein topologischer Raum.

Es ist nun leicht zu zeigen, daß $R_{\aleph_0}$ dem kompakten, durch die Bedingungen $0 \leqq x_n \leqq \frac{1}{n}$ bestimmten Fundamentalquadern des Hilbertschen Raumes homöomorph ist. Um diese Homöomorphie festzustellen, lassen wir jedem Punkte $x = \{t_1, \ldots, t_n, \ldots\}$ des Raumes $R_{\aleph_0}$ den Punkt

[6]) α nimmt alle Werte von Null bis zur ersten Ordnungszahl $\omega^{(\tau)}$ von der Mächtigkeit τ an.

$x^* = \left\{t_1, \ldots, \frac{t_n}{n}, \ldots\right\}$ des Hilbertschen Raumes entsprechen. Die Abbildung ist eineindeutig. Um zu zeigen, daß sie beiderseits stetig ist, muß man folgende Behauptungen beweisen:

a) Zu jeder Umgebung U von x_0 im $R_{\aleph_0}$ gibt es eine Zahl ε von der Eigenschaft, daß aus

$$\varrho(x_0^*; x^*) < \varepsilon$$

im Hilbertschen Raume die Inklusion

$$x \subset U(x_0) \qquad (\text{in } R_{\aleph_0})$$

folgt.

b) Zu jeder Zahl $\varepsilon > 0$ gibt es eine Umgebung $U(x_0)$ von der Eigenschaft, daß aus

$$x \subset U(x_0) \qquad (\text{in } R_{\aleph_0})$$

die Ungleichung

$$\varrho(x_0^*; x^*) < \varepsilon$$

im Hilbertschen Raum folgt.

Beweis von a). Es sei eine Umgebung $U(x_0)$ durch die Bedingungen

$$t_{n_i} \subset S(t_{n_i}^{(0)}; \varepsilon_i) \qquad (i = 1, 2, \ldots, k)$$

gegeben. Man wähle N größer als alle n_i, und ε kleiner als alle $\frac{\varepsilon_i}{N}$.

Wenn nun

$$\varrho(x_0^*; x^*) = \sqrt{\sum_{n=1}^{\infty} \frac{(t_n^0 - t_n)^2}{n^2}} < \varepsilon$$

ist, so ist auch

$$\frac{|t_n^{(0)} - t_n|}{n} < \varepsilon.$$

Es ist also für jedes n_i

$$|t_{n_i}^0 - t_{n_i}| < \varepsilon n_i < \varepsilon N < \varepsilon_i$$

und somit $x \subset U(x_0)$.

Beweis von b). Es sei $\varepsilon > 0$ gegeben; man wähle N so groß, daß

$$\sum_{n=N+1}^{\infty} \frac{1}{n^2} < \frac{\varepsilon^2}{2}$$

ist. Dann ist für jeden Punkt x der durch die Bedingungen

$$t_i \subset S\left(t_i^0; \frac{\varepsilon\sqrt{3}}{\pi}\right) \qquad (i = 1, 2, \ldots, N)$$

bestimmten Umgebung U von x_0

$$\varrho(x_0^*; x^*) = \sqrt{\sum_{n=1}^{\infty} \frac{(t_n^0 - t_n)^2}{n^2}} < \sqrt{\sum_{n=1}^{N} \frac{(t_n^0 - t_n)^2}{n^2} + \frac{\varepsilon^2}{2}} < \sqrt{\frac{\varepsilon^2}{2} \cdot \frac{6}{\pi^2} \sum_{n=1}^{N} \frac{1}{n^2} + \frac{\varepsilon^2}{2}} < \varepsilon,$$

w. z. b. w.

§ 2.

Beweis der Bikompaktheit von R_τ.

Vorbemerkung 1. Es sei irgendein Raum R und eine (abstrakt gedachte) Menge M (beliebiger Mächtigkeit) gegeben. Es sei ferner $\mathfrak{a}$ eine eindeutige (im allgemeinen nicht eineindeutige) Abbildung von M auf eine echte oder unechte Teilmenge $\mathfrak{a}(M)$ von R. Dann ist für jeden Punkt x von $\mathfrak{a}(M)$ eine Kardinalzahl, nämlich die Mächtigkeit der Menge aller Originalpunkte von x bei der Abbildung $\mathfrak{a}$, bestimmt. Diese Kardinalzahl nenne ich das *Gewicht* des Punktes x. Die Summe von Gewichten aller Punkte einer Teilmenge von $\mathfrak{a}(M)$ soll das Gewicht dieser Teilmenge heißen.

Wenn ξ ein Punkt von R ist, so nennen wir ihn *Konzentrationspunkt von M (in bezug auf $\mathfrak{a}$)*, wenn, wie auch die Umgebung $U(\xi)$ von ξ gewählt sei, das Gewicht der Durchschnittsmenge $U(\xi)\cdot\mathfrak{a}(M)$ gleich der Mächtigkeit von M ist.

Die Menge aller Konzentrationspunkte von M in bezug auf $\mathfrak{a}$ ist eine in R abgeschlossene Menge, die wir durch $[M]_\mathfrak{a}$ bezeichnen werden.

Wenn R bikompakt ist, so besitzt jede unendliche Menge M in bezug auf jede Abbildung $\mathfrak{a}$ mindestens einen Konzentrationspunkt (Beweis mit Hilfe des Borel-Lebesgueschen Satzes ohne Schwierigkeit zu führen).

Im folgenden wird diese Bemerkung unter der Voraussetzung, daß R ein abgeschlossenes Intervall ist, benutzt.

Es sei E irgendeine unendliche Teilmenge von R_τ, deren Mächtigkeit $\mathfrak{m} \geqq \aleph_0$ sein möge. Indem man die Gesamtheit der ersten Koordinaten aller Punkte von E betrachtet, erhält man eine eindeutige Abbildung $\mathfrak{a}_1$ von E auf die Einheitsstrecke; wir bezeichnen durch t_1^0 einen (beliebig, aber eindeutig gewählten) Konzentrationspunkt von E in bezug auf $\mathfrak{a}_1$.

Es seien jetzt die reellen Zahlen

$$t_1^0,\ t_2^0,\ \ldots,\ t_\alpha^0,\ \ldots \qquad (0 \leqq t_\alpha \leqq 1)$$

für alle $\alpha < \alpha_0$ definiert (dabei ist α_0 eine beliebige Ordnungszahl unter ω^τ, und ω^τ die erste Ordnungszahl von der Mächtigkeit τ), und zwar so, daß folgende Bedingung erfüllt ist.

(A) Wenn E_n^α die Menge aller Punkte von E bedeutet, deren α-te Koordinate t_α der Ungleichung $|t_\alpha - t_\alpha^0| < \frac{1}{n}$ genügt, so hat der Durchschnitt eines beliebigen endlichen Systems

$$E_{n_1}^{\alpha_1}\cdot E_{n_2}^{\alpha_2}\cdot\ldots\cdot E_{n_k}^{\alpha_k} \qquad (\alpha_1 \leqq \alpha_2 \leqq \ldots \leqq \alpha_k < \alpha_0)$$

die Mächtigkeit $\mathfrak{m}$.

(Für $\alpha_0 = 2$ ist die Bedingung (A) erfüllt, weil dann

$$\alpha_1 = \alpha_2 = \ldots = \alpha_k = 1$$

ist und $E_{n_1}^{\alpha_1} \cdot E_{n_2}^{\alpha_2} \cdot \ldots \cdot E_{n_k}^{\alpha_k} = E_\nu^1$, wo ν die größte unter den Zahlen $n_1, n_2, \ldots, n_k$ ist; die Menge E_ν^1 hat offenbar die Mächtigkeit $\mathfrak{m}$, weil t_1^0 ein Konzentrationspunkt von E ist.)

Es sei jetzt $\mathfrak{a}$ die Abbildung von R_τ auf die Einheitsstrecke, die dadurch entsteht, daß man jedem Punkte von R_τ seine α_0-te Koordinate entsprechen läßt.

Man betrachte alle abgeschlossenen Mengen von der Gestalt

$$[E_{n_1}^{\alpha_1} \cdot E_{n_2}^{\alpha_2} \cdot \ldots \cdot E_{n_k}^{\alpha_k}]_\mathfrak{a} = F_{n_1 n_2 \ldots n_k}^{\alpha_1 \alpha_2 \ldots \alpha_k},$$

wobei wie immer $\alpha_1 \leqq \alpha_2 \leqq \ldots \leqq \alpha_k < \alpha_0$ ist.

Ich behaupte, daß je endlichviele unter den Mengen $F_{n_1 n_2 \ldots n_k}^{\alpha_1 \alpha_2 \ldots \alpha_k}$ einen nicht leeren Durchschnitt haben. Es seien in der Tat

$$F_{n_1^1 n_2^1 \ldots n_{k_1}^1}^{\alpha_1^1 \alpha_2^1 \ldots \alpha_{k_1}^1}, \quad F_{n_1^2 n_2^2 \ldots n_{k_2}^2}^{\alpha_1^2 \alpha_2^2 \ldots \alpha_{k_2}^2}, \quad \ldots, \quad F_{n_1^h n_2^h \ldots n_{k_h}^h}^{\alpha_1^h \alpha_2^h \ldots \alpha_{k_h}^h}$$

beliebig gegeben. Dann ist

$$(1) \qquad \prod_{i=1}^{h} F_{n_1^i n_2^i \ldots n_{k_i}^i}^{\alpha_1^i \alpha_2^i \ldots \alpha_{k_i}^i} = \prod_{i=1}^{h} \left[E_{n_1^i n_2^i \ldots n_{k_i}^i}^{\alpha_1^i \alpha_2^i \ldots \alpha_{k_i}^i} \right]_\mathfrak{a}.$$

Nun ist aber (wie die Mengen A und B gewählt sein mögen) stets

$$[A]_\mathfrak{a} \cdot [B]_\mathfrak{a} \supset [A \cdot B]_\mathfrak{a}.$$

Aus dieser Bemerkung und der Identität (1) folgt somit

$$(2) \qquad \prod_{i=1}^{h} F_{n_1^i n_2^i \ldots n_{h_i}^i}^{\alpha_1^i \alpha_2^i \ldots \alpha_{k_i}^i} \supset \left[\prod_{i=1}^{h} \prod_{j=1}^{k_i} E_{n_j^i}^{\alpha_j^i} \right]_\mathfrak{a}.$$

Es gilt ferner vermöge der Bedingung (A):

$$(3) \qquad \text{Mächt. von} \left[\prod_{i=1}^{h} \prod_{j=1}^{k_i} E_{n_j^i}^{\alpha_j^i} \right] = \mathfrak{m},$$

woraus folgt, daß

$$\left[\prod_{i=1}^{h} \prod_{j=1}^{k_i} E_{n_j^i}^{\alpha_j^i} \right]_\mathfrak{a}$$

und insbesondere

$$\prod_{i=1}^{h} F_{n_1^i n_2^i \ldots n_{k_i}^i}^{\alpha_1^i \alpha_2^i \ldots \alpha_{k_i}^i}$$

nicht leer ist.

Da je endlichviele unter den Mengen $F_{n_1 n_2 \ldots n_k}^{\alpha_1 \alpha_2 \ldots \alpha_k}$ $(\alpha_1 \leqq \alpha_2 \leqq \ldots \leqq \alpha_k < \alpha_0)$ einen nicht leeren Durchschnitt haben, so haben alle diese Mengen (die ja abgeschlossene Teilmengen der Strecke $0 \leqq t \leqq 1$ sind) mindestens einen Punkt gemein. Einen beliebig (aber eindeutig) gewählten unter den Punkten der Durchschnittsmenge aller $F_{n_1 n_2 \ldots n_k}^{\alpha_1 \alpha_2 \ldots \alpha_k}$ bezeichnen wir durch $t_{\alpha_i}^0$, und man zeigt leicht, daß die Bedingung (A) für alle $\alpha_1 \leqq \alpha_2 \leqq \ldots \leqq \alpha_k < \alpha_0 + 1$ erfüllt ist [7]).

Auf diese Weise bestimmt man für jede Ordnungszahl $\alpha < \omega^\tau$ einen Punkt t_α^0 der Strecke $0 \leqq t \leqq 1$. Man betrachte jetzt den Punkt x_0 von R_τ mit den Koordinaten

$$t_1^0,\ t_2^0,\ \ldots,\ t_\alpha^0,\ \ldots \qquad (\alpha < \omega^\tau):$$

$$x_0 = \{t_1^0,\ t_2^0,\ \ldots,\ t_\alpha^0,\ \ldots\}.$$

Ich behaupte, daß x_0 ein vollständiger Häufungspunkt der Menge E ist. In der Tat enthält eine beliebige Umgebung von x_0 $U_{\alpha_1 \alpha_2 \ldots \alpha_k}^{n_1 n_2 \ldots n_k}(x_0)$ die Menge $E_{n_1}^{\alpha_1} \cdot E_{n_2}^{\alpha_2} \cdot \ldots \cdot E_{n_k}^{\alpha_k}$, die (vermöge der Bedingung (A)) die Mächtigkeit $\mathfrak{m}$ hat, w. z. b. w.

§ 3.

Beweis der Sätze I und II.

Um die Sätze I und II zu beweisen, genügt es zu zeigen:

1. daß jede Teilmenge eines bikompakten Raumes vollständig regulär ist;

2. daß jeder vollständig reguläre Raum R mit einer Basis von der Mächtigkeit τ einer Teilmenge von R_τ homöomorph ist.

Die erste Behauptung folgt daraus, daß die vollständige Regularität eine im Urysohnschen Sinne transitive Eigenschaft ist, d. h. daß eine Teilmenge eines vollständig regulären Raumes, als Raum betrachtet, selbst vollständig regulär ist. Es sei in der Tat R eine Teilmenge eines vollständig regulären Raumes R^*. Ist nun x_0 ein Punkt und A eine zu x_0 fremde in R abgeschlossene Menge, so gibt es eine in R^* abgeschlossene Menge A^* derart, daß $A = R \cdot A^*$. Da R^* vollständig regulär ist, kann man dort eine stetige Funktion $f(x)$ erklären, die folgenden Bedingungen genügt:

$$\begin{cases} f(x_0) = 0, \\ f(x) = 1 & \text{für alle } x \subset A^*, \\ 0 \leqq f(x) \leqq 1 & \text{für alle übrigen Punkte.} \end{cases}$$

---

[7]) Es seien in der Tat

$$E_{n_1}^{\alpha_1},\ E_{n_2}^{\alpha_2},\ \ldots,\ E_{n_k}^{\alpha_k}$$

gegeben. $E^* = E_{n_1}^{\alpha_1} \cdot E_{n_2}^{\alpha_2} \cdot \ldots \cdot E_{n_k}^{\alpha_k}$ hat die Mächtigkeit $\mathfrak{m}$, und $t_{\alpha_0}^{(0)}$ ist ein Konzentrationspunkt dieser Menge in bezug auf $\mathfrak{a}$; daraus und aus der Definition von $E_{n_0}^{\alpha_0}$ folgt dann unmittelbar, daß bei jedem n_0 $E^* \cdot E_{n_0}^{\alpha_0}$ von derselben Mächtigkeit $\mathfrak{m}$ ist.

Die Menge A ist in der Menge A^* enthalten, also ist für alle ihre Punkte $f(x)=1$. Betrachtet man die Funktion $f(x)$ nur in Punkten von R, so genügt sie allen aufgestellten Bedingungen, womit die vollständige Regularität von R bewiesen ist.

Bemerkung 2. Ist R ein vollständig regulärer Raum, so gibt es für jeden Punkt x_0 und jede seine Umgebung $U(x_0)$ eine andere Umgebung desselben Punktes $V(x_0)$ derart, daß eine stetige Funktion $f(x)$ sich erklären läßt, so daß

$$\begin{cases} f(x)=0, & \text{wenn } x \subset \overline{V(x_0)}, \\ f(x)=1, & \text{wenn } x \subset R-U(x_0), \\ 0 \leqq f(x) \leqq 1 & \text{für alle übrigen Punkte von } R. \end{cases}$$

Um letztere Tatsache zu beweisen, erkläre man im Raume R eine stetige Funktion $f^*(x)$ dadurch, daß man

$$\begin{cases} f^*(x_0)=0, & \\ f^*(x)\ =1, & \text{wenn } x \subset R-U(x), \\ 0 \leqq f^*(x)\ \leqq 1 & \text{in allen übrigen Punkten von } R \end{cases}$$

setzt.

Die Funktion $f(x)$ ist stetig, es gibt also eine Umgebung von x_0, $V(x_0)$, für deren Punkte $f^*(x) < \frac{1}{2}$. Dann kann man eine Funktion $f(x)$ z. B. folgendermaßen definieren:

$$\begin{cases} f(x)=f^*(x) & , \text{ falls } f^*(x)=1, \\ f(x)=0 & , \text{ falls } f^*(x) \leqq \frac{1}{2}, \\ f(x)=2\left(f^*(x)-\frac{1}{2}\right), & \text{ falls } \frac{1}{2} < f^*(x) < 1. \end{cases}$$

Der Beweis der Behauptung 2 geschieht nunmehr mit Hilfe einer von Urysohn zum Beweis eines seiner Metrisationssätze angewandten Methode. Es sei R ein vollständig regulärer Raum mit einer Basis $\mathfrak{B}$ von der Mächtigkeit τ. Ein Paar zur Basis $\mathfrak{B}$ gehörender Umgebungen $(U, V)=\pi$ soll kanonisch heißen, falls $U \supset \overline{V}$ ist und eine stetige Funktion $f(x)$ im Raume R bestimmt werden kann, die in $\overline{V}$ verschwindet, in $R-U$ gleich 1 ist und überall in R der Ungleichung $0 \leqq f(x) \leqq 1$ genügt.

Es seien

$$\pi_1, \pi_2, \ldots, \pi_\alpha, \ldots \qquad (\alpha < \omega^\tau)$$

alle kanonischen Paare[8]) und

[8]) Die Mächtigkeit der Menge aller kanonischen Paare ist höchstens gleich der Mächtigkeit der Basis; sie kann aber auch nicht kleiner sein, weil (vermöge der Bemerkung 2) jedes U aus $\mathfrak{B}$ mindestens in einem π vorkommt; die Mächtigkeit aller kanonischen Paare ist also genau τ.

$$f_1, f_2, \ldots, f_\alpha, \ldots \qquad (\alpha < \omega^\tau)$$

die ihnen entsprechenden stetigen Funktionen.

Wir lassen nun einem Punkt $x \subset R$ den Punkt

$$f(x) = x^* = \{t_1, t_2, \ldots, t_\alpha, \ldots\}$$

entsprechen, wo $t_\alpha = f_\alpha(x)$. Die dadurch bestimmte Abbildung von R auf eine Teilmenge R^* von R_τ ist eineindeutig. Sind in der Tat x und y verschiedene Punkte von R, so gibt es zunächst eine Umgebung U, die x, aber nicht y enthält; vermöge der Bemerkung 2 gibt es sodann eine in U enthaltene Umgebung V von x, so daß U und V ein kanonisches Paar $\pi_\alpha = (U, V)$ bilden. Somit ist $f_\alpha(x) = 0$, während $f_\alpha(y) = 1$ ist; die Punkte x^* und y^* sind also verschieden.

Die Abbildung $x^* = f(x)$ ist *beiderseits* stetig.

1. Es sei x_0 ein Punkt,

$$f(x_0) = x_0^* = \{t_1^0, t_2^0, \ldots, t_\alpha^0, \ldots\}$$

sein Bildpunkt in R^*, $U(x_0^*) = U_{\alpha_1 \alpha_2 \ldots \alpha_k}^{n_1 n_2 \ldots n_k}$ eine beliebige Umgebung letzteren Punktes. Um die Stetigkeit der Funktion $x^* = f(x)$ nachzuweisen, genügt es eine den Punkt x_0 enthaltende offene Menge G zu finden, dessen Bild $f(G)$ in $U(x_0^*)$ enthalten ist. Man beachte zu diesem Zwecke die Menge $G_{\alpha_i}^{n_i}$ aller Punkte x von R, für die

$$(1_i) \qquad t_{\alpha_i}^0 - \frac{1}{n_i} < f_{\alpha_i}(x) < t_{\alpha_i}^0 + \frac{1}{n_i} \qquad (1 \leqq i \leqq k)$$

ist; da $f_\alpha(x)$ stetig ist, so ist G_α^n eine offene Menge; da außerdem $f_{\alpha_i}(x_0) = t_{\alpha_i}^0$ ist, so ist $x_0 \subset G_{\alpha_i}^{n_i}$. Es genügt jetzt

$$G = G_{\alpha_1}^{n_1} \cdot G_{\alpha_2}^{n_2} \cdot \ldots \cdot G_{\alpha_k}^{n_k}$$

zu setzen. G ist eine offene Menge und für $x \subset G$ sind alle Bedingungen (1_i) erfüllt, d. h. es ist $f(x) \subset U(x_0^*)$; da überdies $x_0 \subset G$ ist, ist unsere Behauptung bewiesen.

2. Um die Stetigkeit der Umkehrungsfunktion $x = f^{-1}(x^*)$ zu beweisen, zeigen wir, daß für jede $U(x_0)$ sich eine $U(x_0^*)$ finden läßt, für welche $f^{-1}(U(x_0^*)) \subset U(x_0)$. Letztere Bedingung heißt aber nichts anderes, als daß aus

$$y \subset R - U(x_0)$$
$$y^* \subset R^* - U(x_0^*)$$

folgen soll. Es ist nun $U(x_0)$ beliebig gegeben; man bestimme ein V so, daß $x_0 \subset V$ und (U, V) ein kanonisches Paar, und zwar $\pi_\alpha = (U, V)$ ist; dann ist aber $f_\alpha(x_0) = 0$; $f_\alpha(y) = 1$ für jedes $y \subset R - U$, d. h. die α-te Koordinate von x_0^* ist gleich Null, die α-te Koordinate eines jeden Punktes

y^* für $y \subset R - U$ ist dagegen gleich 1. Wenn jetzt $U(x_0^*) = U_\alpha^2$ gesetzt wird, so folgt aus $y \subset R - U(x_0)$, daß die α-te Koordinate von y^* gleich 1; also $> \frac{1}{2}$ ist; y^* liegt somit außerhalb von $U(x_0^*)$.

Die Homöomorphie der beiden Räume R und R^* ist hiermit nachgewiesen. Dadurch sind aber auch die beiden Sätze I und II bewiesen.

§ 4.

Konstruktion von Beispielen.

In diesem Paragraphen will ich zeigen:

1. daß es vollständig reguläre Räume gibt, die nicht normal sind, und
2. daß es reguläre Räume gibt, die nicht vollständig regulär sind.

Ich konstruiere zunächst einen Raum S folgendermaßen. Die Punkte x von S sind Paare (α, β), wobei α und β Ordnungszahlen sind, und zwar durchläuft α alle Werte $\leqq \omega_1$ und β alle Werte $\leqq \omega$. Wenn $x_0 = (\alpha_0, \beta_0)$ $(1 \leqq \alpha_0 \leqq \omega_1,\ 1 \leqq \beta_0 \leqq \omega)$ ein Punkt von S ist, so erhält man eine beliebige Umgebung $U_{\alpha'\beta'}(x_0)$ $0 \leqq \alpha' < \alpha_0$, $0 \leqq \beta < \beta_0$ dadurch, daß man alle $x = (\alpha; \beta)$ betrachtet, für die $\alpha' < \alpha \leqq \alpha_0$, $\beta' < \beta \leqq \beta_0$ ist.

Wie leicht ersichtlich, ist S ein bikompakter Raum.

Man bezeichne nun durch $\dot{S}$ den Raum, der aus S durch Tilgen des Punktes (ω_1, ω) entsteht. Als Teilmenge eines bikompakten Raumes S ist $\dot{S}$ vollständig regulär; andererseits ist aber $\dot{S}$ nicht normal, weil die beiden abgeschlossenen und zueinander fremden Teilmengen von $\dot{S}$

$$X = \left\{ \begin{matrix} (\alpha, \omega) \\ 1 \leqq \alpha < \omega_1 \end{matrix} \right\} \quad \text{und} \quad Y = \left\{ \begin{matrix} (\omega_1, \beta) \\ 1 \leqq \beta < \omega \end{matrix} \right\}$$

durch keine zwei zueinander fremde Gebiete getrennt werden können. Es gilt sogar die schärfere

Behauptung I. Wenn G ein die Menge $Y_n = \left\{ \begin{matrix} (\omega_1, \beta) \\ n \leqq \beta < \omega \end{matrix} \right\}$ enthaltendes Gebiet ist, so gibt es ein $\alpha_0 < \omega_1$ derart, daß $\bar{G}$ alle Punkte $X_{\alpha_0} = \{\alpha, \omega\}$, $\alpha_0 \leqq \alpha < \omega_1$, enthält.

In der Tat gibt es für jeden Punkt $x = (\omega_1, \beta)$ eine $U_{\alpha_\beta, \beta'}(x) \subset G$; also sind insbesondere alle $x = (\alpha, \beta)$ $\alpha > \alpha_\beta$ in G enthalten. Da die Menge aller α_β abzählbar ist, gibt es eine Zahl α_0, die kleiner als ω_1 und größer als alle α_β ist. Somit sind alle Punkte (α, β), $\alpha > \alpha_0$, $n \leqq \beta < \omega$ in G und alle Punkte (α, ω), $\alpha > \alpha_0$ in $\bar{G}$ enthalten, w. z. b. w.

Ein regulärer Raum ist (vermöge des Satzes II) dann und nur dann vollständig regulär, wenn er in einen bikompakten Raum eingebettet werden kann. Da jeder bikompakte Raum regulär ist, so folgt aus dem soeben

Gesagten, daß ein nicht bikompakter regulärer Raum, der in bezug auf reguläre Punkte abgeschlossen ist, sicher nicht vollständig regulär ist. Bevor wir einen solchen Raum konstruieren, machen wir die folgende

Behauptung II. Wenn $\dot{G}$ ein die Mengen

$$X_{\alpha_0} = \left\{ \begin{matrix} (\alpha, \omega) \\ \alpha_0 < \alpha < \omega_1 \end{matrix} \right\}, \quad Y_{n_0} = \left\{ \begin{matrix} (\omega_1, \beta) \\ n_0 < \beta < \omega \end{matrix} \right\}$$

enthaltendes Gebiet in $\dot{S}$ ist, so ist $\dot{S} - \dot{G}$ (als Relativraum in $\dot{S}$ betrachtet) bikompakt.

In der Tat ist

$$\dot{S} - \dot{G} = S - G$$

(wobei $G = \dot{G} + (\omega_1, \omega)$ ist), woraus unsere Behauptung folgt.

Es sei jetzt R^* ein topologischer Raum, der dadurch entsteht, daß man die Vereinigungsmenge von abzählbar vielen Exemplaren

$$\dot{S}^1, \dot{S}^2, \ldots, \dot{S}^k, \ldots$$

von $\dot{S}$ betrachtet. Wir bezeichnen die Punkte von $\dot{S}^k$ durch $(\alpha, \beta)^k$, die Mengen X_α und Y_n von $\dot{S}^k$ durch X_α^k bzw. Y_n^k usw. Man betrachte ferner die stetige Zerlegung[9]) von R^* in (höchstens aus zwei Punkten bestehende) Mengen X, wobei

X entweder ein Punkt $(\alpha, \beta)^k$, $\alpha \neq \omega_1$, $\beta \neq \omega$ von R

oder ein Punktepaar $[(\omega_1, \beta)^{2n-1}, (\omega_1, \beta)^{2n}]$, $n = 1, 2, \ldots$

oder ein Punktepaar $[(\alpha, \omega)^{2n}, (\alpha, \omega)^{2n+1}]$, $n = 1, 2, \ldots$

ist. Der durch diese Zerlegung induzierte Raum ist, wie leicht ersichtlich, ein regulärer Raum $\dot{R}^*$[10]). Jedem Punkte $(\alpha, \beta)^k$ von R^* entspricht ein Punkt $(\alpha, \beta)^{*k}$ von $\dot{R}^*$, ebenso entsprechen die Teilmengen $\dot{S}^k$ bzw. X_α^k bzw. Y_n^k von R^* Teilmengen $\dot{S}^{*k}$ bzw. X_α^{*k} bzw. Y_n^{*k} von $\dot{R}^*$.

Wir konstruieren endlich einen Raum

$$R = \dot{R}^* + \xi$$

dadurch, daß wir einen neuen Punkt ξ mittels der Umgebungserklärung

$$U_m(\xi) = \xi + \left(\dot{R}^* - \sum_{k=1}^{m} \dot{S}^{*k}\right) \qquad (m = 1, 2, \ldots)$$

einführen. Man zeigt ohne Schwierigkeit, daß R ein regulärer Raum ist.

Ich behaupte, daß R in bezug auf alle regulären Punkte abgeschlossen ist.

---

9) Alexandroff, Math. Annalen **96** (1926), S. 557 (§ 4).

10) Man könnte sagen, daß $\dot{R}^*$ aus R^* dadurch entsteht, daß man in R^* die Punkte $(\omega_1, \beta)^{2n-1}$ bzw. $(\alpha, \omega)^{2n}$ mit $(\omega_1, \beta)^{2n}$ bzw. $(\alpha, \omega)^{2n+1}$ identifiziert.

Um dies zu beweisen machen wir die folgende

Behauptung III. Es sei G ein Teilgebiet von $\dot{S}$, welches die Menge X_{α_0} enthält; es gibt dann eine natürliche Zahl n_0 von der Eigenschaft, daß $\bar{G}$ alle Punkte (ω_1, β), $n_0 \leqq \beta < \omega$ (d. h. die Menge Y_{n_0} enthält.

Es existiert in der Tat für jedes $x = (\alpha, \omega)$, $\alpha > \alpha_0$, eine $U_{\alpha', n_\alpha}(x) \subset G$; infolgedessen sind alle (α, β), $n_\alpha < \beta < \omega$, in G enthalten; da es nur abzählbar viele verschiedene n_α gibt, gibt es ein $n_0 < \omega$, so daß für unabzählbar viele α alle (α, β), $n_0 < \beta < \omega$, in G enthalten sind; dann sind aber alle (ω_1, β), $n_0 < \beta < \omega$, in $\bar{G}$ enthalten.

Es sei jetzt

$$R + \eta$$

ein regulärer Raum. Dann gibt es zwei Umgebungen $U(\xi) = U_N(\xi)$ und $V(\eta)$, so daß

$$\overline{U(\xi)} \cdot \overline{V(\eta)} = 0$$

ist; wir dürfen dabei stets voraussetzen, daß N gerade ist. Man konstruiere ferner

$$\overline{V_1}(\eta) \subset V(\eta), \ldots, \quad \overline{V_{m+1}(\eta)} \subset V_m(\eta), \ldots \quad (m = 1, 2, \ldots)$$

und setze

$$G_n = R - \overline{V_m(\eta)}.$$

Dann ist

$$(0) \qquad \bar{G}_m \subset R - V_m(\eta) \subset R - \overline{V_{m+1}(\eta)} \subset G_{m+1}.$$

Außerdem ist

$$(1) \qquad G_1 \supset U_N(\xi) \supset \sum_{k=N+1}^{\infty} \dot{S}^{*k},$$

so daß

$$V(\eta) \subset \sum_{k=1}^{N} \dot{S}^{*k} + \eta$$

ist.

Aus (1) folgt weiter, daß

$$G_1 \supset X^{*N+1},$$

also auf Grund der Behauptung III existiert ein n_1, so daß

$$\bar{G}_1 \supset Y_{n_1}^{*N+1} = Y_{n_1}^{*N}$$

und also (vermöge (0))

$$G_2 \supset Y_{n_1}^{*N}.$$

Aus der Behauptung I folgt sodann, daß es ein α_2 gibt, so daß

$$G_3 \supset \bar{G}_2 \supset X_{\alpha_2}^{*N} = X_{\alpha_2}^{*N-1}.$$

Daraus folgt weiter (nach Behauptung III), daß

$$G_4 \supset \bar{G}_3 \supset Y_{n_3}^{*N-1} = Y_{n_3}^{*N-2},$$

was (nach Behauptung I)

$$G_5 \supset \bar{G}_4 \supset X_{\alpha_4}^{*N-2}$$

ergibt.

Eine wiederholte Anwendung der beiden Behauptnngen I und III schließt mit

$$G_{N+2} \supset Y_{\alpha_N}^{*1} + X_{n_{N+1}}^{*1}.$$

Nun ist

$$V_{N+2} \subset R - G_{N+2}.$$

Da aber G_{N+2} sämtliche $X_{\alpha_{N-k}}^{*k}$ und $Y_{n_{N-k}}^{*k}$ $k \leqq N$ und sämtliche $\dot{S}^{*m}$ $m > N+1$ enthält, so ist $V_{N+2}(\eta)$ in einer nach Bemerkung II bikompakten Teilmenge von R^* enthalten, was aber nur dann möglich[11]) ist, wenn η in V_{N+2} und somit in $R+\eta$ isoliert ist; damit ist die Abgeschlossenheit von R^* in bezug auf reguläre Punkte bewiesen.

Es sei hierzu noch bemerkt, daß durch das soeben konstruierte Beispiel eine von Alexandroff und Urysohn (a. a. O. S. 266) gestellte Frage, ob jeder reguläre nicht absolut abgeschlossene Raum sich zu einem ebenfalls regulären Raum durch Hinzufügung eines nicht isolierten Punktes erweitern läßt, eine volle (und zwar negative) Antwort erhalten hat.

§ 5.

Beweis des Satzes III.

Der Beweis des Satzes III beruht auf der Betrachtung von Zerlegungen[12]) der Räume R_τ in zueinander fremde abgeschlossene Mengen: es wird sich nämlich zeigen, daß die absolut abgeschlossenen Räume, deren Existenz im Satze III behauptet wird, durch Zerlegungen der Räume R_τ bestimmt werden. Bei der Untersuchung dieser Zerlegungen werden wir folgende von Alexandroff a. a. O. bewiesene Tatsache benutzen: ein durch eine Zerlegung eines bikompakten Raumes R bestimmter Raum R^* ist ein topologischer Raum; wenn R eine Basis von einer Mächtigkeit $\leqq \mathfrak{m}$ besitzt, so gilt dasselbe auch für R^*. [13])

Es sei R ein topologischer Raum mit einer Basis $\mathfrak{B}$ von der Mächtigkeit $\leqq \tau$. Der Satz III wird bewiesen sein, wenn wir gezeigt haben werden, daß es eine Zerlegung von R_τ gibt, die einen absolut abgeschlossenen

---

[11]) Alexandroff und Urysohn, S. 263, Satz V.

[12]) Im Sinne von Alexandroff, Math. Annalen **96** (1926), § 3 (S. 556—557). Es sei hier ausdrücklich erwähnt, daß wir (der Alexandroffschen Definition entsprechend) beliebige allgemeine und nicht notwendig etwa stetige Zerlegungen betrachten.

[13]) Bei Alexandroff ist diese Behauptung explizite nur für stetige Zerlegungen und $\mathfrak{m} = \aleph_0$ (S. 262, Satz V und § 13) bewiesen. Der Beweis gilt aber wörtlich auch in unserem allgemeinen Falle.

Raum bestimmt, welcher eine dem Raume R homöomorphe Teilmenge enthält. Wir gehen jetzt zum Beweise letzterer Behauptung über.

Es sei X_α eine auf der Einheitsstrecke liegende abgeschlossene Menge. Im folgenden werden wir die Mengen $X = \{X_1, X_2, \ldots, X_\alpha, \ldots\}$ des Raumes R_τ betrachten, die aus allen Punkten $x = \{t_1, t_2, \ldots, t_\alpha, \ldots\}$ bestehen, für die $t_\alpha \subset X_\alpha$ ist. Die Mengen X sind abgeschlossen. In der Tat, wenn $x_0 = \{t_1^0, t_2^0, \ldots, t_\alpha^0, \ldots\}$ nicht zu X gehört, so ist wenigstens für ein α

$$\varrho(t_\alpha^0, X_\alpha) = \delta > 0$$

(die Entfernung auf J_α gemessen). Dann enthält aber seine, durch die Bedingung $t_\alpha \subset S\left(t_\alpha^0, \frac{\delta}{2}\right)$ bestimmte Umgebung keinen Punkt der Menge X; x_0 ist somit kein Häufungspunkt von X.

Jetzt wollen wir diejenigen Mengen X, die als Zerlegungseinheiten von R_τ auftreten sollen, definieren. Zu diesem Zwecke bestimmen wir für ein beliebiges Gebiet G des Raumes R zwei Funktionen $\varphi(x)$ und $\psi(x)$ durch die Vorschrift

$$\varphi_G(X) = \begin{cases} 0 & \text{falls} \quad x \subset \bar{G} \\ 1 & \text{falls} \quad x \subset G; \end{cases} \qquad \psi_G(X) = \begin{cases} 0 & \text{falls} \quad x \subset G \\ 1 & \text{falls} \quad x \subset R - G. \end{cases}$$

Für jeden Punkt des Raumcs ist $\varphi_G(x) \leqq \psi_G(x)$. Man sieht auch leicht, daß, wenn $\varphi_G(x_0) = \psi_G(x)$ ist, eine Umgebung $U(x_0)$ des Punktes x_0 existiert, für deren jeden Punkt $y \subset U(x_0)$

$$\varphi_G(y) = \psi_G(y) = \varphi_G(x_0) = \psi_G(x_0)$$

ist.

Es seien

$$U_1, U_2, \ldots, U_\alpha, \ldots \qquad (\alpha < \omega^{(\tau)})$$

die Elemente der Basis $\mathfrak{B}$ des Raumes R. Einfachheitshalber setzen wir

$$\varphi_\alpha(x) = \varphi_{u_\alpha}(x) \quad \text{und} \quad \psi_\alpha(x) = \psi_{u_\alpha}(x).$$

Es sei ferner $X_\alpha(x)$ die Gesamtheit aller der Bedingung $\varphi_\alpha(x) \leqq t \leqq \psi_\alpha(x)$ genügender reeller Zahlen und

$$X(x) = \{X_1(x), X_2(x), \ldots, X_\alpha(x), \ldots\}.$$

Wenn x und y verschiedene Punkte von R sind, so gibt es eine U_α so, daß $x \subset U_\alpha$, $y \subset R - \bar{U}_\alpha$, woraus folgt, daß $X(x)$ und $X(y)$ zueinander fremd sind.

Wenn die Vereinigungsmenge Ξ der (für alle Punkte x von R) bestimmten $X(x)$ nicht mit dem ganzen Raum R_τ identisch ist, betrachten wir alle Punkte von $R - \Xi$ als neue Zerlegungseinheiten, so daß schließlich der ganze Raum R_τ in zueinander fremde abgeschlossene Mengen zerlegt wird: $R_\tau = \sum X$. Diese Zerlegung ist durch den Raum R allein eindeutig bestimmt und soll deshalb als die Zerlegung $\zeta(R_\tau, R)$ bezeichnet werden. Der durch $\zeta(R_\tau, R)$ bestimmte Raum soll R_τ^R heißen. Die Zerlegung $\zeta(R_\tau, R)$ induziert aber auch eine, im allgemeinen unstetige Abbildung von R_τ auf R_τ^R, die wir durch φ bezeichnen werden: $R_\tau^R = \varphi(R_\tau)$. Die Punkte von R_τ^R werden wir durch x^* bezeichnen. Da jedem Punkte $x \subset R$ eine Menge X, d. h. ein Punkt $x^* = f(x)$ zugeordnet ist, sind wir im Besitze einer eineindeutigen Abbildung $x^* = f(x)$ des Raumes R auf eine (echte oder unechte) Teilmenge R^* von R_τ^R. Wir werden zeigen, daß diese Abbildung eine Homöomorphie ist. Dazu brauchen wir aber eine bequeme Basis des Raumes R_τ^R.

Es sei $G_{\alpha_1\alpha_2\ldots\alpha_k}^{n_1n_2\ldots n_k}(x_0)$ die Gesamtheit aller Punkte $x = \{t_1, t_2, \ldots, t_\alpha, \ldots\}$ von R_τ, für die $t_{\alpha_i} \subset S\left(X_{\alpha_i}(x_0), \frac{1}{n_i}\right)$ ist. $G_{\alpha_1\alpha_2\ldots\alpha_k}^{n_1n_2\ldots n_k}$ ist ein Gebiet. Es sei ferner $V_{\alpha_1\alpha_2\ldots\alpha_k}^{n_1n_2\ldots n_k}(x_0^*)$ die Gesamtheit aller x^*, die, als Mengen X betrachtet, in $G_{\alpha_1\alpha_2\ldots\alpha_k}^{n_1n_2\ldots n_k}$ liegen. Ich behaupte, daß die $V_{\alpha_1\alpha_2\ldots\alpha_k}^{n_1n_2\ldots n_k}$ eine Basis bilden. Um es einzusehen, genügt es zu zeigen, daß man für jedes Gebiet $G \supset X_0$ ein Gebiet $G_{\alpha_1\alpha_2\ldots\alpha_k}^{n_1n_2\ldots n_k}(x_0)$ finden kann von der Eigenschaft, daß

$$X_0 \subset G_{\alpha_1\alpha_2\ldots\alpha_k}^{n_1n_2\ldots n_k} \subset G \tag{1}$$

ist.

Um dieses $G_{\alpha_1\alpha_2\ldots\alpha_k}^{n_1n_2\ldots n_k}$ zu konstruieren, wähle man für jeden Punkt $x \subset X_0$ eine $U_{\alpha_1^x, \alpha_2^x, \ldots \alpha_k^x}^{2m_1^x, 2m_2^x, \ldots, 2m_k^x}(x)$, wobei die $\alpha_1^x, \alpha_2^x, \ldots, \alpha_k^x, m_1^x, m_2^x, \ldots, m_k^x$ durch die Bedingung

$$U_{\alpha_1^x, \alpha_2^x, \ldots, \alpha_k^x}^{m_1^x, m_2^x, \ldots, m_k^x}(x) \subset G$$

bestimmt sind. Auf diese Weise entsteht eine Überdeckung der abgeschlossenen Teilmenge X_0 von R_τ; nach dem Borel-Lebesgueschen Satze kann man diese Überdeckung durch eine endliche Teilüberdeckung ersetzen. Es seien $\alpha_1, \alpha_2, \ldots, \alpha_k$ alle bei den Elementen dieser Überdeckung auftretenden untere Indizes und $2m_i$ der größte unter den zu α_i gehörenden oberen Indizes. Dann ist

$$G_{\alpha_1, \alpha_2, \ldots, \alpha_k}^{2m_1, 2m_2, \ldots, 2m_k} \subset G.$$

Es sei in der Tat $\xi = \{t_1, t_2, \ldots, t_\alpha, \ldots\}$ irgendein Punkt von $G_{\alpha_1, \alpha_2, \ldots \alpha_k}^{2m_1, 2m_2, \ldots, 2m_k}$. Es ist $|t_{\alpha_i} - t_{\alpha_i}^0| < \frac{1}{2m_i}$, wobei

$$x_0 = \{t_1^0, t_2^0, \ldots, t_\alpha^0, \ldots\}$$

ein Punkt von X_0 ist; nun ist aber x_0 in einer bestimmten $U_{\alpha_1^x, \alpha_2^x, \ldots, \alpha_k^x}^{2m_1^x, 2m_2^x, \ldots, 2m_k^x}(x)$ mit $x = \{t_1^x, t_2^x, \ldots, t_\alpha^x, \ldots\}$ enthalten. Alle $\alpha_1^x, \alpha_2^x, \ldots, \alpha_k^x$ kommen unter den $\alpha_1, \alpha_2, \ldots, \alpha_k$ vor und es ist

$$|t_{\alpha_i^x}^x - t_{\alpha_i^x}| \leqq |t_{\alpha_i^x}^x - t_{\alpha_i^x}^0| + |t_{\alpha_i^x}^0 - t_{\alpha_i^x}| < 2 \cdot \frac{1}{2m_i^x} = \frac{1}{m_i^x}.$$

Daraus folgt, daß die Koordinaten von ξ allen die Umgebung $U_{\alpha_1^x, \alpha_2^x, \ldots, \alpha_k^x}^{m_1^x, m_2^x, \ldots, m_k^x}(x)$ bestimmenden Bedingungen genügen und also

$$\xi \subset U_{\alpha_1^x, \alpha_2^x, \ldots, \alpha_k^x}^{m_1^x, m_2^x, \ldots, m_k^x} \subset G$$

ist.

Wir wollen jetzt zeigen, daß die eineindeutige Abbildung

$$R^* = f(R)$$

in beiden Richtungen stetig ist.

Der Beweis vollzieht sich in zwei Schritten.

I. Es sei $U(x_0^*) = V_{\alpha_1 \alpha_2 \ldots \alpha_k}^{n_1 n_2 \ldots n_k}(x_0^*)$ beliebig gegeben. Man betrachte ein $X_{\alpha_i}(x_0)$. Wenn $\varphi_{\alpha_i}(x_0) = \psi_{\alpha_i}(x_0)$ ist, so gibt es (wie bei der Definition der Funktionen φ und ψ erwähnt wurde) eine $U(x_0)$ in R — nennen wir sie W_{α_i} — derart, daß für sämtliche $x \subset W_{\alpha_i}(x_0)$

$$\varphi_{\alpha_i}(x) = \psi_{\alpha_i}(x) = \varphi_{\alpha_i}(x_0) = \psi_{\alpha_i}(x_0)$$

ist, es ist m. a. W.

$$X_{\alpha_i}(x) = X_{\alpha_i}(x_0)$$

und also um so mehr

$$X_{\alpha_i}(x) \subset S\left(X_{\alpha_i}(x_0); \frac{1}{n_i}\right)$$

für alle $x \subset W_{\alpha_i}(x_0)$.

Wenn aber $X_{\alpha_i}(x_0)$ die ganze Einheitsstrecke ist, so ist für jeden Punkt $x \subset R$

$$X_{\alpha_i}(x) \subset S\left(X_{\alpha_i}(x_0); \frac{1}{n_i}\right)$$

und wir setzen dementsprechend $W_{\alpha_i}(x_0) = R$.

Es sei jetzt $U(x_0)$ eine im Gebiete

$$W_{\alpha_1}(x_0) \cdot W_{\alpha_2}(x_0) \cdot \ldots \cdot W_{\alpha_k}(x_0)$$

enthaltene Umgebung; offenbar ist dann für jeden Punkt $x \subset U(x_0)$ und jedes $i = 1, 2, \ldots, k$

$$X_{\alpha_i}(x) \subset S\left(X_{\alpha_i}(x_0); \frac{1}{n_i}\right);$$

diese Tatsache bedeutet aber gerade, daß

$$f(x) \subset V_{\alpha_1 \alpha_2 \ldots \alpha_k}^{n_1 n_2 \ldots n_k}(x_0^*),$$

sobald $x \subset U(x_0)$.

II. Es sei eine beliebige $U(x_0) = U_\alpha$ gegeben; um die Stetigkeit der Umkehrungsfunktion $f^{-1}(x^*)$ zu beweisen, genügt es zu zeigen, daß für jeden Punkt x^* von R^*, der in $V_\alpha^2(x_0^*)$ liegt, $f^{-1}(x^*) \subset U_\alpha$ ist. Nun ist aber $\varphi_\alpha(x_0) = \psi_\alpha(x_0) = 0$, also besteht $X_\alpha(x_0)$ aus dem einzigen Punkte 0. Wenn $f^{-1}(x^*) = x \subset R - U_\alpha$ wäre, wäre $\psi_\alpha(x) = 1$, also könnte auch x^* definitionsgemäß nicht zu $V_\alpha^2(x_0^*)$ gehören.

Die Homöomorphie zwischen R und R^* ist hiermit bewiesen. Da $R^* \subset R_\tau^R$ ist, haben wir noch zu zeigen, daß R_τ^R ein absolut abgeschlossener Raum ist. Dazu brauchen wir aber folgende zwei Bemerkungen zu machen:

1. Es gibt für jeden Punkt $x_0 = \{t_1^0, t_2^0, \ldots, t_\alpha^0, \ldots\}$ von R_τ abzählbare Punktmengen A, die zu diesem Punkte konvergieren, und zwar solche, daß unter den Koordinaten der Punkte von A weder 0 noch 1 vorkommt. Wir brauchen in der Tat z. B. die Punktmenge $x_n = \{t_1^{(n)}, t_2^{(n)}, \ldots, t_\alpha^{(n)}, \ldots\}$ $(n = 1, 2, \ldots)$ zu nehmen mit $t_\alpha^n = t_\alpha^0 - \frac{t_\alpha^0}{n+1}$, falls $t_\alpha^0 \neq 0$ und $t_\alpha^{(n)} = \frac{1}{n+1}$, falls $t_\alpha^{(0)} = 0$. Von einer solchen Folge sagen wir, sie konvergiere regelmäßig gegen x_0.

Sind alle Koordinaten eines Punktes x_0 von 0 und 1 verschieden, so ist $x_0 \subset R - \Xi$ und somit ist $\varphi(x_0) = x_0$; wenn x_n regelmäßig gegen x_0 konvergiert, so konvergiert auch $\varphi(x_n)$ gegen $\varphi(x_0)$ (weil ja auch $\varphi_n(x_n) = x_n$ ist.

2. Die den Raum R_τ^R bestimmende Zerlegung von R_τ ist im allgemeinen nicht stetig [14]). Es kann also vorkommen, daß für eine Umgebung $U(x_0^*)$ in R_τ^R die Menge $\varphi^{-1}(U)$ die Menge $X_0 = \varphi^{-1}(x_0^*)$ nicht im Innern enthält; dagegen existiert ein Teilgebiet G von R_τ von der Eigenschaft, daß

$$\varphi^{-1}(\overline{U(x_0^*)}) \supset G \supset X_0$$

ausfällt. Es genügt in der Tat, für G das der Definition von $U(x_0^*)$ zugrunde gelegtes Gebiet [15]) zu wählen. Es sei x ein beliebiger Punkt von G, $x_1, x_2, \ldots, x_n, \ldots$ eine gegen x regelmäßig konvergierende Folge. Da $\varphi(x_n) = x_n$ ist, so ist $\varphi(x_n) \subset U(x_0^*)$, also $\varphi(x) \subset \overline{U(x_0^*)}$.

Um jetzt die absolute Abgeschlossenheit des Raumes R_τ^R zu beweisen, genügt es nach einem Satze von Alexandroff und Urysohn [16]) zu zeigen, daß man aus jedem den Raum R_τ^R überdeckenden System von Umgebungen $\{U\}$ ein endliches Teilsystem $U_1, U_2, \ldots, U_N$ so wählen kann, daß

$$\overline{U}_1 + \overline{U}_2 + \ldots + \overline{U}_n = R_\tau^R$$

ist. Man betrachte zu diesem Zwecke die der Definition von $\{U\}$ zugrunde gelegten [17]) Gebiete G in R_τ. Da R_τ bikompakt ist und die $\{G\}$ eine

---

[14]) Aus dem Satze II und dem unter [13]) zitierten Satz von Alexandroff folgt vielmehr, daß unsere Zerlegung nur dann stetig sein kann, wenn der gegebene Raum regulär ist.

[15]) Alexandroff, a. a. O. § 3, S. 537.

[16]) Alexandroff und Urysohn, Zur Theorie der topologischen Räume, Math. Annalen **92** (1924), S. 262, Satz II.

[17]) Alexandroff a. a. O. § 3, S. 537.

Überdeckung von R_τ bilden, lassen sich endlichviele unter den $\{G\}$ — etwa $G_1, G_2, \ldots, G_N$ — so wählen, daß

$$G_1 + G_2 + \ldots + G_N = R_\tau$$

ist. Dann ist

$$\varphi(G_1) + \varphi(G_2) + \ldots + \varphi(G_N) = R_\tau^R;$$

nach dem soeben Bewiesenen ist aber $\varphi(G) \subset \overline{U}$, so daß

$$\overline{U}_1 + \overline{U}_2 + \ldots + \overline{U}_N$$

erst recht den ganzen Raum R_τ^R ausfüllt.

Dadurch ist die absolute Abgeschlossenheit von R_τ^R und mithin auch der Satz III vollständig bewiesen.

Ich möchte hervorheben, daß die von Alexandroff und Urysohn seit langem gestellte Frage, *ob jeder topologische Raum als überalldichte Teilmenge eines absolut abgeschlossenen Raumes betrachtet werden kann*, durch den soeben bewiesenen Satz keine Antwort erhalten hat[18]), da man nicht weiß, ob die Menge $\overline{R}^* \subset R_\tau^R$, als Raum betrachtet, absolut abgeschlossen ist. Auch die Frage, ob man alle topologischen Räume, die Basen von der Mächtigkeit τ besitzen, in *einen und denselben absolut abgeschlossenen Raum einbetten kann*, bleibt unentschieden.

[18]) Eine abgeschlossene Teilmenge eines absolut abgeschlossenen Raumes braucht nicht (wie elementare Beispiele zeigen) absolut abgeschlossen zu sein (siehe Alexandroff und Urysohn, Mémoire sur les espaces topologiques compacts, Verh. Kon. Ak. Amsterdam, Deel XIV, No. 1 (1929)).

(Eingegangen am 6. 1. 1929.)

Russian Math. Surveys 31:6 (1976), 13–15
From Uspekhi Mat. Nauk 31:6 (1976), 13–16

THE PRINCIPAL TOPOLOGICAL DISCOVERIES OF A. N. TIKHONOV

P. S. Aleksandrov

Tikhonov began his scientific career in a brilliant way at the age of 19, by obtaining several remarkable topological results, which soon made him world famous. The first of these was the discovery of the concept of the topological or the so-called Tikhonov or Cartesian product of an arbitrary family of topological spaces.

In spite of the flattery of associating the names of Descartes and Tikhonov in the term for the topological product, in this case it underrates rather than overrates Tikhonov's achievement. In fact, the great founder of the coordinate method and of analytic geometry, in his discussion of products of spaces, did not go beyond the representation of the plane and three-dimensional space as the product of two and three real lines. A natura generalization of this concept is the representation of the n-dimensional space as the product of n straight lines and of an n-dimensional cube contained in it as the product of n segments. This representation can be raised one step further by regarding a Hilbert parallelepiped of countable dimension as the product of segments $I_1, I_2, \ldots, I_n, \ldots$. But if we go further in this direction, proceeding from the natural to the transfinite numbers and attempting to define the product of $I_1, I_2, \ldots, I_n, \ldots, I_\alpha, \ldots$ where α ranges over all countable transfinite numbers, we obtain, as we now know, by no means what is needed. In fact, "what is needed" was discovered by Tikhonov, and this is what is now called the Tikhonov product topology. To find it, to *visualize* it, was a real discovery, the discovery a new continent in the ocean of set-theoretical mathematics, which is called a Tikhonov cube after the explorer.

Let us recall its precise definition. Let $\mathfrak{A} = \{\alpha\}$ be an infinite set whose elements α are called indices. With each index $\alpha \in \mathfrak{A}$ we associate a topological space $|X_\alpha|$, and the same space $|X_\alpha| = |X_{\alpha'}|$ can correspond to distinct indices α and α'. The pair consisting of an index α and its corresponding space $|X_\alpha|$ is denoted by X_α and is called the space labelled by the index α. For each α we choose a point $x_\alpha \in |X_\alpha|$ and we write X_α

instead of $|X_\alpha|$. So we obtain a function defined on $\mathfrak{A}$ that assigns to each $\alpha \in \mathfrak{A}$ the point $x_\alpha \in X_\alpha$. These functions are defined to be the points $x = \{x_\alpha\}$ of the product $\prod_{\alpha\in\mathfrak{A}} X_\alpha$.

We now define a topology on this set as follows. We choose a finite number of indices $\alpha_1 \in \mathfrak{A}, \ldots, \alpha_s \in \mathfrak{A}$ and select in each space $X_{\alpha_1}, \ldots, X_{\alpha_s}$ an open set $\Gamma_{\alpha_1}, \ldots, \Gamma_{\alpha_s}$. The set of all the points $x = \{x_\alpha\} \in \Pi X_\alpha$ for which $x_{\alpha_1} \in \Gamma_{\alpha_1}, \ldots, x_{\alpha_s} \in \Gamma_{\alpha_s}$ is called an elementary open set corresponding to the given choice of indices $\alpha_1, \ldots, \alpha_s$ and sets $\Gamma_{\alpha_1}, \ldots, \Gamma_{\alpha_s}$. These elementary sets and all their unions are *defined* to be open sets in the Tikhonov product $\prod_{\alpha\in\mathfrak{A}} X_\alpha$. Assuming that in each X_α a basis $\mathfrak{B}_\alpha$, is chosen, we can subject the sets $\Gamma_{\alpha_1}, \ldots, \Gamma_{\alpha_s}$ in the definition of the topology of ΠX_α to the condition that they belong to the corresponding bases $\mathfrak{B}_{\alpha_1}, \ldots, \mathfrak{B}_{\alpha_s}$. This implies at once the following result: if the cardinality of $\mathfrak{A}$ is τ and the weight[1] of each X_α does not exceed τ, then the weight of ΠX_α does not exceed τ.

It is also easy to see that if all the X_α are Hausdorff spaces, then so is their Tikhonov product.

If Tikhonov's first topological discovery is that of the concept of the topological product of an arbitrary family of topological spaces, his second discovery is the theorem asserting that the Tikhonov product of an arbitrary family of compact topological spaces is itself compact.

Suppose that the cardinality of the set $\mathfrak{A}$ above is an infinite cardinal number τ and that each space $X_\alpha = I_\alpha$ is a unit segment on the real line. Then the Tikhonov product $\prod_{\alpha\in\mathfrak{A}} X_\alpha$ is compact of weight τ; it is denoted by I^τ and is called the Tikhonov cube of weight τ.

A Hausdorff space is called normal if any two disjoint (that is, non-intersecting) closed sets A and B in this space can be *geometrically separated*, that is, they have disjoint neighbourhoods. A space is called regular if any closed set and a point outside it can be geometrically separated. Uryson proved the extremely important proposition (Uryson's famous lemma) that in a normal space X any two disjoint closed sets A and B can be *functionally separated,* in the sense that there one can construct a real function f, continuous on the whole of X, such that $0 \leqslant f(x) \leqslant 1$ for all $x \in X$ and $f(x) = 0$ for $x \in A$ and $f(x) = 1$ for $x \in B$. Since two closed sets can be functionally separated only if they can be geometrically separated, Uryson's lemma actually expresses the fact that in the definition of a normal space geometric separability can be replaced by functional separability. For regular spaces the corresponding assertion no longer holds; moreover, there even exists a regular space in which there are no continuous functions except constants. Therefore, we

[1] The weight of a space is the least cardinality of any basis of the space.

come to an essentially new class of topological spaces wider than normal but narrower than regular spaces, namely, the (Hausdorff) spaces considered by Tikhonov in which each point can be functionally separated from any closed set not containing it. These so-called completely regular or Tikhonov spaces turned out to be extremely important in topology as well as in all its various applications (in algebra, functional analysis, etc.). In particular, the space of any topological group is always completely regular (Pontryagin) but need not be normal (Markov).

Tikhonov's third topological discovery is the fact that all completely regular spaces and only they are subspaces of compact Hausdorff spaces. Furthermore, every Tikhonov space X of weight τ is homeomorphic to a set lying in the Tikhonov cube I^τ of the same weight τ, that is, from the topological point of view can be regarded as a subset of I^τ, and its closure in this cube is its so-called *compactification* (of weight τ), that is, a compact Hausdorff space containing X as a dense subset.

By establishing this remarkable result, Tikhonov founded the theory of compactifications, nowadays one of the most developed important and beautiful branches of general topology.

I would say that Tikhonov's theorem on the topological embedding of completely regular spaces in the cubes I^τ is also of great philosophical significance. The fact that these spaces and no others, can be regarded as subsets of the cubes I^τ, means no more and no less than that in the study of these and no other spaces the coordinate method in the widest sense of the word can be applied, a fact that is constantly used in the investigation of problems concerning Tikhonov spaces.

So we have come full circle back to Descartes.

Tikhonov's results are fundamental for the whole of topology. Moreover, statistics show that his theorem on the compactness of the product of compact spaces occupies among all topological theorems the first place as regards references in the general mathematical literature.

However, the significance of Tikhonov's topological results lies not only in their importance for topology and their wide applicability in other fields, but also in the fact that, even after half a century, they still inspire mathematicians to further major interesting research in mathematics. I hope that the subsequent articles in this issue of the Russian Mathematical Surveys will help to convince the reader about this fact.

I dedicate these lines to Andrei Nikolaevich Tikhonov on the occasion of his seventieth birthday, as an expression of my deep regard for this outstanding contemporary mathematician and of my affection for one of my first students.

Received by the Editors 2 July 1976

Translated by D. Mathon

Boris Nikolaevich Delone

Born – 15 March 1890 in St Petersburg, Russia
Died – 1980

Boris Delone graduated from Kiev University in 1913. At Kiev he was a student of Grave and he followed Grave's work in algebra and number theory. After graduating Delone taught in Kiev where he was a member of the Mathematical Society which had among its members Ch. T. Bialobzeski, P. V. Voronets, N. B. Delone, D. A. Grave, A. A. Friedmann, A. P. Kotelnikov, V. P. Linnik (I. V. Linnik's father) and O. Yu Schmidt.

Following the Revolution of 1917 there was a change in policy towards education which, certainly in the Ukraine, had to become more technology based and more practical. Algebra certainly did not fit into this new educational philosophy and Grave's algebra seminar was forced to close. Some mathematicians, such as Grave himself, changed to study applied mathematical topics. Delone, however, chose to continue to study algebra and so was forced, in the 1920s, to leave the Ukraine.

Delone moved to Petrograd in 1922. Petrograd was the name that St Petersburg had been given in 1914 and, two years after Delone began working there, in 1924, it was again renamed, this time to Leningrad. Delone worked at Leningrad University from 1922 until 1935.

The Institute of Physics and Mathematics had been established by Steklov in Petrograd (as it was called at the time) in 1921. In 1932 the Institute of Physics and Mathematics was divided into two independent Departments, the Mathematics Department headed by Vinogradov and the Physics Department headed by Vavilov. Vinogradov invited some outstanding mathematicians to join the new Mathematics Department including Delone.

In this new Mathematics Department, Delone became a colleague of Sergi Bernstein, Luzin, Smirnov, Kuzmin, N. S. Koshlyakov, Kochin, Sobolev and Faddeev.

The St Petersburg Mathematical Society was founded in 1890 but disbanded at the time of the 1917 Revolution. However the Society was reformed in 1921 as the Petrograd Physical and Mathematical Society and Delone joined in the following year. He played an active role in the Society along with other outstanding mathematicians such as Ya V.

Uspenskii, V. I. Smirnov, V. A. Steklov, A. A. Friedmann, V. A. Fok, A. S. Besicovitch, Sergi Bernstein, Ya D. Tamarkin, R. O. Kuzmin and B. G. Galerkin.

In 1934 the Division of Mathematical and Natural Sciences of the Academy of Sciences of the USSR decided to split the Departments of the Steklov Institute of Physics and Mathematics into independent Institutes, the Steklov Mathematical Institute and the Lebedev Physical Institute. Delone became the Head of the Algebra Department of the Steklov Mathematical Institute. However, the Steklov Mathematical Institute moved to Moscow and, in 1935 Delone moved to Moscow.

Delone was professor of mathematics at the University of Moscow from 1935 to 1942.

The mathematical topics that Delone studied include algebra, the geometry of numbers. He also did important work on the structural analysis of crystals.

THÉORIE DES NOMBRES. — *La solution générale de l'équation* $X^3\rho + Y^3 = 1$. Note de M. **Boris Delaunay**, présentée par M. Émile Picard.

C'est à Lagrange que nous devons la résolution générale des équations indéterminées du second degré à deux variables. Lagrange a fait la remarque qu'il paraît que la résolution des équations analogues, dont le degré dépasse le second, ne peut pas être trouvée, circonstance analogue à celle qui se présente dans la résolution des équations algébriques après le quatrième degré.

Jusqu'à nos jours nous savons encore bien peu sur ces équations. Il faut tout de même citer un théorème dû à Axel Thue, que l'équation $\varphi(x, y) = m$, où φ est une forme binaire, dont le degré est supérieur à 2, ne peut avoir un nombre infini de solutions en nombres entiers.

Après avoir étudié tout ce qu'on peut bien trouver sur ces matières et particulièrement un Mémoire du mathématicien russe Woronoï (*Sur une extension de l'algorithme des fractions continues*, 1896, en russe, Petrograd), qui contient une méthode sûre pour calculer les unités fondamentales de Dirichlet des corps algébriques du troisième degré, j'ai essayé de chercher la résolution générale des équations indéterminées binaires du troisième degré.

Voici un théorème aussi simple qu'élégant auquel je suis parvenu, au moins pour les équations de la forme $X^3\rho + Y^3 = 1$:

L'équation $X^3\rho + Y^3 = 1$, ρ *étant un nombre entier différent d'un cube, n'a pas de solution en nombres entiers* X *et* Y, *si l'unité fondamentale du corps* $\Omega(\sqrt[3]{\rho})$ *n'est pas de la forme* $B\sqrt[3]{\rho} + C$, *et a la seule solution* $X = B$, $Y = C$, *si elle a la forme* $B\sqrt[3]{\rho} + C$.

Je parle ici de l'unité fondamentale de Dirichlet $a(\sqrt[3]{\rho})^2 + b\sqrt[3]{\rho} + c$, dont les coefficients a, b et c sont des nombres entiers, n'ayant pas le même signe, et dont les puissances à exposants entiers positifs et négatifs représentent toutes les unités de la forme $A(\sqrt[3]{\rho})^2 + B\sqrt[3]{\rho} + C$, dont les coefficients A, B et C sont entiers.

On voit que tout revient à calculer l'unité fondamentale $a(\sqrt[3]{\rho})^2 + b\sqrt[3]{\rho} + c$, ce qu'on peut faire par la méthode de Woronoï.

THÉORIE DES NOMBRES. — *Sur les formes quadratiques binaires positives.* Note de M. **Gaston Julia**, présentée par M. G. Humbert.

Soit

$$ax^2 - 2bxy + cy^2$$

une forme positive *réduite*, on a

$$a, c > 0, \qquad ac - b^2 > 0, \qquad 2|b| \leqq a \leqq c.$$

On sait que les trois premiers minima propres de la forme sont

$$a, c \quad \text{et} \quad a - 2b + c$$

en supposant, ce qui ne diminue aucunement la généralité, $b \geqq 0$.

Dans sa Note du 17 mai 1915 [1], M. Humbert établit les propositions suivantes :

1° Le quatrième minimum est $a + 2b + c$, obtenu pour $x = 1$, $y = -1$;

2° Le cinquième minimum est $4a - 4b + c$, obtenu pour $x = 2$, $y = 1$;

3° Le sixième minimum est le plus petit des deux entiers $4a + 4b + c$ et $a - 4b + 4c$, obtenus respectivement pour $x = 2$, $y = -1$, et $x = 1$, $y = 2$;

4° Si le sixième minimum est le second, $a - 4b + 4c$, de ces entiers, le septième minimum sera le premier, $4a + 4b + c$.

Je vais donner de ces propositions une nouvelle démonstration.

---

[1] *Comptes rendus*, t. **160**, 1915, p. 647.

THÉORIE DES NOMBRES. — *Sur le nombre des représentations d'un nombre par une forme cubique binaire à discriminant négatif.* Note de M. **Boris Delaunay**, transmise par M. Hadamard.

1. Soit $Ax^3 + Bx^2y + Cxy^2 + Ey^3 = (A, B, C, E)$ une forme à discriminant D négatif, c'est-à-dire n'ayant qu'une racine réelle. Selon Lagrange (1768), la résolution de $(A, B, C, E) = \sigma$ se réduit au plus à la résolution de σ nouvelles équations $(A_i, B_i, C_i, E_i) = 1$. Une telle équation, ou bien n'a pas de solutions, ou bien, si l'on connaît une de ces solutions, peut se réduire à la forme $(1, n, -p, q) = 1$. Nous allons étudier cette dernière.

2. Soit $\alpha^3 = n\alpha^2 + p\alpha + q$, alors on a

$$(x + \alpha y)(x + \alpha' y)(x + \alpha'' y) = 1,$$

et tout revient à trouver toutes les unités de l'anneau $[1, \alpha, \alpha^2] = O(\alpha)$ qui sont *binomes*, c'est-à-dire n'ont pas de membre contenant α^2. Puisque $D < 0$ toutes les unités de $O(\alpha)$ sont des puissances à exposants positifs d'une unité fondamentale $0 < \varepsilon_0 < 1$ et de l'unité inverse ε_0^{-1}. En définissant dans l'espace les coordonnées $\xi = \frac{\omega' + \omega''}{2}$; $\eta = \frac{\omega' - \omega''}{2i}$; $\zeta = \omega$, où ω est un nombre de $O(\alpha)$, considérant la surface des unités $(\xi^2 + \eta^2)\zeta = 1$ et le *plan binome* $\xi + B\eta - \zeta = 0$, on reconnaît qu'il peut avoir seulement un nombre limité de solutions entre les unités inverses, qu'on peut toutes trouver. Soit $\varepsilon_0^{-m} = x + \alpha y$ la plus grande des solutions inverses, si nous transformons $(1, n, -p, q)$ par la substitution $\begin{pmatrix} x & \beta \\ y & \gamma \end{pmatrix}$ où $x\delta - y\beta = 1$, nous obtenons une nouvelle équation $(1, n', -p', q') = 1$ dont les solutions sont les solutions de $(1, n, -p, q) = 1$ divisées par ε_0^{-m} et qui, par contre, n'a pas de solutions inverses.

3. Considérons maintenant les puissances d'une unité binome

$$(b\alpha + c)^m = M\alpha^2 + P\alpha + Q;\ \frac{M}{b} = \frac{m(m-1)}{1.2} c^{m-2} + \frac{m(m-1)(m-2)}{1.2.3} bc^{m-3} n + \ldots;$$

soit b_1 un diviseur premier de b plus grand que 3, alors $|b_1^2| > 4$, $|b_1^3| > 5$, etc. et $M = 0$ est impossible. Si $|b_1| = 3$, alors $c^2 \equiv 1 \pmod 3$, $c \equiv 1$, $n \equiv 0 \pmod 3$ et $M = 0$ est impossible. Si $|b_1| = 2$, $M = 0$ est également impossible. Nous voyons que si $b \neq \pm 1$, aucune puissance de l'unité binome $b\alpha + c$ ne peut être binome.

4. Nous abordons maintenant la méthode qui nous conduira au but. Nous appelons cette méthode « algorithme de rehaussement ». Soit $(a\alpha^2 + b\alpha + c)^{m_i} = P_i\alpha + Q_i$, alors $\varepsilon_0'^{m_i} - \varepsilon_0''^{m_i} = P_i(\alpha' - \alpha'')$ et tous les P_i sont donc divisibles par $\frac{\varepsilon' - \varepsilon''}{\alpha' - \alpha''} = -a\alpha + b + an$; si $(a, b) = \delta$, alors les P_i sont divisibles par $\varkappa = \frac{1}{\delta^2} N(-a\alpha + b + an)$. Soit $\bar{\alpha} = \alpha\varkappa$, alors nous passons dans l'anneau $O(\bar{\alpha})$; toutes les unités de $O(\bar{\alpha})$ sont aussi puissances d'une unité fondamentale $\bar{\varepsilon}_0 = \varepsilon_0^{\nu}$, où ν est diviseur de $\varphi(\varkappa^2)$. Nous avons $(\bar{a}\bar{\alpha}^2 + \bar{b}\bar{\alpha} + \bar{c})^{\frac{m_i}{\nu}} = \bar{P}_i\bar{\alpha} + Q_i$, où $\bar{P}_i = \frac{P_i}{\varkappa}$. Nous répétons cette considé-

ration et ainsi de suite jusqu'à ce qu'un certain $\Delta^\star = -a^\star\alpha^\star + b^\star + a^\star n^\star$ ne soit une unité, alors $\varkappa^\star = 1$ et l'algorithme s'arrête.

Il y a à distinguer trois cas : I. Si $\Delta^\star$ ne devient nulle part unité, l'équation $(1, n, -p, q) = 1$ n'a pas de solutions (à part $x = 1, y = 0$); II. Si $\Delta^\star$ est ± 1, alors $a^\star = 0$, $b^\star = \pm 1$ et toutes les solutions doivent être puissances de $\pm\alpha^\star + c^\star$ de la forme $P_i^\star\alpha^\star + Q_i$, c'est-à-dire des puissances de $k\alpha + c^\star$ de la forme $P_i\alpha + Q_i$ (où $k = \varkappa\bar{\varkappa}\bar{\bar{\varkappa}}\ldots \neq 1$ autrement que quand le cas II se rencontre au début), ce qui est impossible d'après 3; il n'y a donc, dans ce cas, qu'une seule solution $x = c^\star$; $y = k$; III. $-a^\star\alpha^\star + b^\star + a^\star n^\star$ est une unité algébrique. Soit par exemple $-a\alpha + b + an$ une unité algébrique, c'est une unité binome, soit ε^μ; $\varepsilon^{\mu+1}$ est aussi binome. Si l'on transforme $(1, n, -p, q)$ par $\varepsilon^{\mu+1}$, on obtient une équation $(1, n', -p', q')$ qui a les solutions ε^{-1} et $\varepsilon^{-\mu-1}$ et dont la racine $\eta = \varepsilon^{-1}$, c'est-à-dire que $\eta^{\mu-1}$ est de la forme $P\eta + Q$. Nous voyons que si $-a^\star\alpha^\star + b^\star + a^\star n^\star$ était une unité algébrique, alors $\eta^{\star\mu+1} = P\eta^\star + Q$, ce qui serait encore impossible d'après 3, puisque $\alpha^\star = k\alpha$, où $k \neq \pm 1$ (autrement que le cas III arrive au début), et par suite $\eta^\star = k\beta + c'$.

C'est-à-dire que $-a\alpha + b + an$ dans notre algorithme ne peut être unité algébrique que dès le début. Dans ce dernier cas notre équation peut être ramenée à une équation de la forme $(1, n, -p, 1) = 1$ dont la racine est ε_0; nous nommons une telle équation *réversible*.

5. Soit τ le premier exposant pour lequel $\Delta_\tau = \dfrac{\varepsilon'^\tau - \varepsilon''^\tau}{\varepsilon' - \varepsilon''}$ n'est pas une unité, s'il y a alors une solution telle que $m \equiv r_i \pmod{\tau}$, nous transformons au moyen de cette solution et nous obtenons une nouvelle forme, laquelle, à la place des solutions $m \equiv r_i$, admet des solutions telles que $m \equiv 0 \pmod{\tau}$ et Δ_ε permet d'amorcer notre algorithme, et alors nous nous trouvons dans les cas I et II; pour cette raison il n'y a pas plus de $2\varepsilon - 1$ solutions. Une simple discussion montre qu'en général, pour les équations du type III, $\tau = 2$ et les seuls cas où $\tau > 2$ se réduisent aux formes $(1, 0, 1, 1)$ $\tau = 4$ et $(1, -1, 0, 1)$ $\tau = 6$ que nous avons examinées en détail. De tout cela nous déduisons le théorème : *L'équation* $(1, n, -p, q) = 1$ *n'a pas plus d'une ou deux solutions et seulement pour certaines équations il y en a* 3 *et* 4, *mais pas plus de* 4 *solutions* (la solution triviale $x = 1$, $y = 0$ non comptée). Ceci nous fait voir que pour le nombre des solutions de l'équation $(A, B, C, E) = \sigma$ $(D < 0)$ est ainsi assignée une limite qui dépend du nombre σ, mais non des coefficients de la forme; elle est en général $< 2\sigma_1$, où σ_1 est le nombre des racines de la congruence $Ax^3 + Bx^2 + Cx + E \equiv 0 \pmod{\sigma}$.

Mstislav Vsevolodovich Keldysh

Born – 10 Feb 1911
Died – 24 Jun 1978

Mstislav Keldysh's role in managing space researches in Russia was greatly important. In 1935 alone he published 4 papers on aerodynamics, 4 papers on hydrodynamics and 2 papers on mathematics. M.V. Keldysh carried out fundamental investigations in three directions — aerodynamics, hydrodynamics and aircraft structural vibrations.

For the first time M.V. Keldysh and F.I. Frank strictly established Zhukovsky's formulae for wing lift at subsonic speed with due account of air compressibility. In 1935 the work by M.V. Keldysh and F.I. Frank *Strict Ground for Zhukovsky's Propeller Theory* was published in mathematical journal. At present this theory is vital for heavy-loaded propellers.

The results of fundamental theoretical studies carried by M.V. Keldysh on hydrodynamics are the basis for a number of trends in this science. For example, some of these developments were used for investigation of tsunami.

For aircraft structural vibration M.V. Keldysh and group of scientists solved a difficult flutter problem and so doing helped aircraft designers to find flutter damping means. For his works on flutter M.V. Keldysh was awarded the State Prize.

The contribution of M.V. Keldysh to science was highly estimated by the State — Academician of USSR Academy of Sciences (1946), trice Hero of Socialist Labour (1956, 1961, 1971), State and Lenin Prize Winner, President of USSR Academy of Sciences (1961–1975), member of 16 foreign academies and scientific associations and honorary doctor of 6 world universities.

ON THE COMPLETENESS OF THE EIGENFUNCTIONS OF SOME CLASSES OF NON-SELFADJOINT LINEAR OPERATORS[1]

M. V. Keldysh

CHAPTER I

Some general properties of linear functional equations

In this chapter we state some general properties of linear functional equations which we shall need later. We do not claim complete novelty for all the propositions mentioned in this chapter, but some of them have apparently never been given in the form that we need.

All our discussions will take place in a Hilbert space $\mathfrak{H}$, although many of the results could be stated in more general form.

1. Following the usual definition, we shall call an element $x(\lambda)$ of a Hilbert space an *analytic function of the complex variable* λ *in some domain* D of the λ-plane if at each point of D the ratio

$$\frac{x(\lambda+h)-x(\lambda)}{h}$$

strongly converges to a limit $x'(\lambda)$.

A bounded linear operator $A(\lambda)$ is called an *analytic function of* λ *in the domain* D if at each point of D the ratio

$$\frac{A(\lambda+h)-A(\lambda)}{h}$$

converges in the norm of $\mathfrak{H}$ to a limit $A'(\lambda)$.

Later we shall use without proof a number of fairly obvious properties of analytic functions $x(\lambda)$ and $A(\lambda)$. In particular, for $x(\lambda)$ and $A(\lambda)$ Cauchy's theorem on the vanishing of an integral around a closed contour holds and they have a representation by a Cauchy integral. The maximum

---

[1] This article was written in 1950. It contains a detailed account of the first part of the article "On eigenvalues and eigenfunctions of some classes of non-selfadjoint equations", published by the author in Dokl. Akad. Nauk SSSR 77 (1951). The contents of the article were reported in 1951 at a meeting of the Moscow Mathematical Society. The author then put the manuscript at the disposal of a number of mathematicians. (*Ed.*)

of $\| x(\lambda) \|$ and $\|A(\lambda)\|$ is not attained inside the domain of analyticity. In a neighbourhood of an isolated singular point we have the expansions

$$x(\lambda) = \sum_{-\infty}^{+\infty} \frac{x_n}{(\lambda - c)^n}, \tag{1}$$

$$A(\lambda) = \sum_{-\infty}^{+\infty} \frac{A_n}{(\lambda - c)^n}, \tag{2}$$

that converge in norm, locally uniformly with respect to λ.

A singular point c is a pole if the expansions (1), (2) contain only finitely many terms with negative powers of $(\lambda - c)$. If in D the functions $x(\lambda)$ and $A(\lambda)$ have as singular points only poles, we call them *meromorphic* functions. Meromorphic functions $x(\lambda)$ and $A(\lambda)$ admit representations

$$x(\lambda) = \frac{x_1(\lambda)}{f(\lambda)}, \quad A(\lambda) = \frac{A_1(\lambda)}{F(\lambda)}, \tag{3}$$

where $f(\lambda)$ and $F(\lambda)$ are numerical functions, analytic in D, and $x_1(\lambda)$ and $A_1(\lambda)$ are analytic in D.

The function $x(\lambda)$ [or $A(\lambda)$] is entire if it is analytic in the whole plane. The orders of entire functions $x(\lambda)$ and $A(\lambda)$, respectively, are defined as the numbers

$$\limsup_{|\lambda| \to \infty} \frac{\ln \ln \| x(\lambda) \|}{\ln |\lambda|}, \quad \limsup_{|\lambda| \to \infty} \frac{\ln \ln \| A(\lambda) \|}{\ln |\lambda|}.$$

A function $x(\lambda)$ [or $A(\lambda)$] that is meromorphic in the whole plane is called a *function of finite order* ρ if in (3) we can take for $x_1(\lambda)$ and $f(\lambda)$ [$A_1(\lambda)$ and $F(\lambda)$] entire functions of order ρ, but not functions of order less than ρ.

By using Hadamard's well-known theorem it is easy to prove the following proposition: if a meromorphic function $x(\lambda)$ [or $A(\lambda)$] of order ρ is an entire function, then the order of the entire function is ρ.

We note that Lindelöf's theorem holds for the functions $x(\lambda)$ [and $A(\lambda)$]: if on the arms of the angle $|\arg \lambda| \leqslant \alpha$ the function $x(\lambda)$ satisfies the inequality $\| x(\lambda) \| < 1$, and inside the angle

$$\| x(\lambda) \| < C_\varepsilon e^{\varepsilon |\lambda|^{\pi/2\alpha}},$$

then $\| x(\lambda) \| \leqslant 1$ everywhere inside the angle.

Henceforth we denote by $A^*(\lambda)$ the operator adjoint to $A(\bar{\lambda})$. If the operator $A(\lambda)$ is analytic for $\lambda = c$, then $A^*(\lambda)$ is analytic for $\lambda = \bar{c}$.

2. We call an operator $A(\lambda)$ *completely continuous* in a domain D if it is completely continuous at each point of D.

If $A(\lambda)$ is completely continuous in a neighbourhood of an isolated singular point $\lambda = c$, then the coefficients in (2) are also completely continuous. This follows immediately from the fact that the coefficients in

(2) are determined by the integrals

$$A_n = \frac{1}{2\pi i} \oint A(\lambda)(\lambda - c)^{-(n+1)} d\lambda$$

so that by replacing the integral by finite sums we represent A_n as the limit of a uniformly convergent sequence of completely continuous operators.

A completely continuous operator which is an analytic function in D has the following property.

If λ varies on a bounded closed part F of D, and if the element x belongs to a bounded set X of a Hilbert space, then the set of elements $A(\lambda)x$ is compact.

For from any sequence $A(\lambda_n)x_n$, where $\lambda_n \in F$, $x_n \in X$, we can extract a subsequence $A(\lambda_{n_p})x_{n_p}$ so that λ_{n_p} converges to a point λ_0 of F. In a neighbourhood of the point $\lambda = \lambda_0$ the operator $A(\lambda)$ can be expanded in a series of powers of $\lambda - \lambda_0$:

$$A(\lambda) = A_0(\lambda_0) + (\lambda - \lambda_0)A_1(\lambda_0) + \ldots,$$

where all the coefficients $A_k(\lambda_0)$ are completely continuous operators; applying a diagonal process we can extract from x_{n_p} a subsequence $x_{n'_p}$ such that all the sequences $A_k(\lambda_0)x_{n'_p}$ converge; it is easy to establish that the sequence $A(\lambda_{n'_p})x_{n'_p}$ is convergent, as required.

Using an orthogonal system of coordinates we introduce a matrix representation of the operator $y = A(\lambda)x$:

$$y_i = \sum_{k=1}^{\infty} a_{ik}(\lambda)\, x_k. \tag{4}$$

Then the property of a completely continuous operator just established is equivalent to the following.

If a completely continuous operator $A(\lambda)$ is an analytic function of λ in D, then for each closed set F belonging to D we can find a sequence of positive numbers ε_n, tending to zero, for which the inequalities

$$\left\{ \begin{aligned} &\sum_{i=n}^{\infty} \Big| \sum_{k=1}^{\infty} a_{ik}x_k \Big|^2 < \varepsilon_n \sum_{k=1}^{\infty} |x_k|^2, \\ &\sum_{k=n}^{\infty} \Big| \sum_{i=1}^{\infty} a_{ik}x_k \Big|^2 < \varepsilon_n \sum_{k=1}^{\infty} |x_k|^2 \end{aligned} \right. \tag{5}$$

are satisfied. The second of these inequalities is a consequence of the complete continuity of $A^*(\lambda)$.

3. An operator $R(\lambda_0)$ is called the *resolvent* of $A(\lambda_0)$ if

$$(E + R(\lambda_0))\,(E - A(\lambda_0)) = E, \tag{6}$$

where E is the identity operator.

If the resolvent exists, then R is also a completely continuous operator, since it follows from (6) that $R = (E + R)A$. The following proposition is easily established.

If $A(\lambda)$ is a completely continuous operator, analytically dependent on λ in D, and if the resolvent of $A(\lambda)$ exists for $\lambda = \lambda_0$, then the resolvent exists over the whole of D, except for a set of isolated points, and is a meromorphic function of λ.

For consider the equation

$$x = A(\lambda)x + f \tag{7}$$

or, in an orthogonal system of coordinates,

$$x_i = \sum_{k=1}^{\infty} a_{ik}(\lambda)\, x_k + f_i. \tag{8}$$

We denote by $\mathfrak{H}_P$ the subspace $(x_1, x_2, \ldots, x_n, 0, 0, \ldots)$, by $\mathfrak{H}_{P_1}$ the subspace $(0, \ldots, 0, x_{n+1}, x_{n+2}, \ldots)$ and by P and P_1 the projection operators corresponding to these spaces. Now let Δ be an arbitrary closed subdomain of D containing λ_0, and $\varepsilon_1, \varepsilon_2, \ldots$ numbers corresponding to Δ in (5). We choose n sufficiently large so that $\varepsilon_{n+1} < 1$. We write the first n equations of (8) in the form

$$Px = PAPx + PAP_1x + Pf, \tag{9}$$

and the remaining equations (8)

$$P_1x = P_1APx + P_1AP_1x + P_1f, \tag{10}$$

on the assumption that $\|P_1A\| < \varepsilon_{n+1}$ in Δ, and therefore the operator $E - P_1A$ in Δ has the resolvent $E + S_1 = E + P_1A + (P_1A)^2 + \ldots$, which is an analytic function of λ whose norm does not exceed $(1 - \varepsilon_{n+1})^{-1}$.

Equation (10) is equivalent to the following:

$$P_1x = (E + S_1)P_1APx + (E + S_1)P_1f, \tag{11}$$

and (9) can be replaced by the equation

$$Px = PA[E + (E + S_1)P_1A]Px + P[E + A(E + S_1)P_1]f. \tag{12}$$

The relation thus obtained represents a system of algebraic linear equations in $x_1, x_2, \ldots, x_n$, whose coefficients are holomorphic functions of λ in Δ. The determinant $d_n(\lambda)$ of this system is non-zero at λ_0. For since the operator $E - A(\lambda_0)$ has a resolvent, (12) must have a solution for any f, and consequently for any absolute term. Therefore the algebraic system of equations in $x_1, x_2, \ldots, x_n$ has a solution of the form

$$Px = \frac{1}{d_n(\lambda)} SP[E + A(E + S_1)P_1]f,$$

where $S(\lambda)$ is an operator analytically dependent on λ, and $d_n(\lambda)$ is a holomorphic function of λ. For the solution of (7) we obtain

$$x = Px + P_1x = (E + R)f,$$

where

$$E+R=\frac{1}{d_n(\lambda)}\{E+(E+S_1)P_1A\}\{SP[E+A(E+S_1)P]\}.$$

In view of the fact that all the operators on the right-hand side are analytic functions of λ, we conclude that in Δ the resolvent exists everywhere, except for a finite number of points at which $d_n(\lambda) = 0$, and the roots of $d_n(\lambda)$ are poles of the resolvent. This proves the proposition.

We note that if $R(\lambda)$ is the resolvent of $A(\lambda)$, then $R^*(\lambda)$ is the resolvent of $A^*(\lambda)$.

4. If the equation $y = A(c)y$ has a non-trivial solution, then, by definition, y is an eigenelement and c an eigenvalue of $A(\lambda)$. We call y_k an *associated element of order* k to the eigenelement y if y_k is obtained as a result of solving the chain of equations

$$(13)\qquad \begin{cases} y=A(c)y, \\ y_1=A(c)y_1+\dfrac{1}{1!}\dfrac{\partial A(c)}{\partial c}y, \\ \dots\dots\dots\dots\dots\dots \\ y_k=A(c)y_k-\dfrac{1}{1!}\dfrac{\partial A(c)}{\partial c}y_{k-1}+\dots+\dfrac{1}{k!}\dfrac{\partial^k A(c)}{\partial c^k}y. \end{cases}$$

We say that $y, y_1, y_2, \dots$ form a *chain of associated elements.*

It is known that for a completely continuous operator and a given eigenvalue the number of linearly independent eigenelements is finite. We shall show below that when the resolvent of $A(\lambda)$ is a meromorphic function of λ, the order of the associated elements does not exceed the order of the pole of the resolvent at $\lambda = c$.

Note that when $A(\lambda)$ is a polynomial in λ,

$$A(\lambda) = A_0+\lambda A_1+\dots+\lambda^n A_n,$$

then

$$u(t)=e^{ct}\left(y_k+\frac{t}{1!}y_{k-1}+\dots+\frac{t^k}{k!}y\right)$$

satisfies the equation

$$u=A_0u+A_1\frac{\partial u}{\partial t}+\dots+A_n\frac{\partial^n u}{\partial t^n}.$$

Let y be an eigenelement. We denote by m the maximal order of elements associated to y. We call the number $m + 1$ the *multiplicity of the eigenelement* y.

We call

$$(14)\qquad y^{(k)},\ y_1^{(k)},\ \dots,\ y_{m_k}^{(k)}\qquad (k=1,2,\dots)$$

a *canonical system* of eigenelements and associated elements for $\lambda = c$ if it has the following properties:

1°. The functions $y^{(k)}$ form a basis of the subspace of eigenelements corresponding to $\lambda = c$.

2°. $y^{(1)}$ is an eigenelement whose multiplicity attains the possible maximum $m_1 + 1$.

3°. $y^{(k)}$ is an eigenelement, not expressible linearly in terms of $y^{(1)}, \ldots, y^{(k-1)}$, whose multiplicity attains the possible maximum $m_k + 1$.

4°. The elements (14) form a chain of associated elements.

We mention two obvious properties of a canonical system of elements.

An arbitrary element of order h is a linear combination of the elements $y_j^{(k)}$ for $j \leqslant h$.

The multiplicity of the eigenfunction $c_1 y^{(1)} + \ldots + c_\nu y^{(\nu)}$, where $c_\nu \neq 0$, is equal to m_ν.

The numbers $m_1, m_2, \ldots$ do not depend on the choice of the canonical system. We call the number $N = m_1 + 1 + m_2 + 1 + \ldots$ the *multiplicity of the eigenvalue* $\lambda = c$.

5. Let y, z be two elements of $\mathfrak{H}$. We denote by yz the operator defined by

$$Bf = (f, z)y.$$

It is easy to see that $(yz)^* = zy$, and that the trace of the operator yz is equal to (y, z).

We now note the following property of the resolvent of a completely continuous operator.

If $\lambda = c$ is a pole of the resolvent, then c is an eigenvalue of the operator $A(\lambda)$, and the principal part of the resolvent for $\lambda = c$ can be expressed in the form

$$(15)\quad \sum_{(k)} \left\{ \frac{y^{(k)}z^{(k)}}{(\lambda - c)^{m_k+1}} + \frac{y^{(k)}z_1^{(k)} + y_1^{(k)}z^{(k)}}{(\lambda - c)^{m_k}} + \ldots \right.$$
$$\left. \ldots + \frac{y^{(k)}z_{m_k}^{(k)} + y_1^{(k)}z_{m_k-1}^{(k)} + \ldots + y_{m_k}^{(k)}z^{(k)}}{\lambda - c} \right\},$$

where $y^{(k)}, y_1^{(k)}, \ldots, y_{m_k}^{(k)}$ $(k = 1, 2, \ldots)$ is an arbitrary canonical system of $A(\lambda)$ for $\lambda = c$, and $z^{(k)}, z_1^{(k)}, \ldots, z_{m_k}^{(k)}$ is a canonical system of $A^*(\lambda)$ for $\lambda = \bar{c}$, that is uniquely determined for the given system $y^{(k)}, y_1^{(k)}, \ldots, y_{m_k}^{(k)}$.

Suppose that the principal part of the resolvent for $\lambda = c$ has the form

$$\frac{R_0}{(\lambda - c)^m} + \frac{R_1}{(\lambda - c)^{m-1}} + \ldots + \frac{R_{m-1}}{\lambda - c}.$$

By (6) we have $R = AR + A$. Substituting in this the expansion of R for

$\lambda = c$ and equating coefficients of equal powers of λ we have

$$(16)\quad \begin{cases} R_0 = A(c) R_0, \\ R_1 = A(c) R_1 + \dfrac{\partial A(c)}{\partial c} R_0, \\ \dots\dots\dots\dots\dots\dots \\ R_{m-1} = A(c) R_{m-1} + \dots + \dfrac{1}{(m-1)!} \dfrac{\partial^{m-1} A(c)}{\partial c^{m-1}} R_0. \end{cases}$$

In view of the fact that $R^*(\lambda)$ is the resolvent of $A^*(\lambda)$ we can also write

$$(16^*)\quad \begin{cases} R_0^* = A^*(\bar{c}) R_0^*, \\ R_1^* = A^*(\bar{c}) R_1^* + \dfrac{1}{1!} \dfrac{\partial A^*(\bar{c})}{\partial \bar{c}} R_0^*, \\ \dots\dots\dots\dots\dots\dots \\ R_{m-1}^* = A^*(\bar{c}) R_{m-1}^* + \dots + \dfrac{1}{(m-1)!} \dfrac{\partial^{m-1} A^*(\bar{c})}{\partial \bar{c}^{m-1}} R_0^*. \end{cases}$$

Let $y, y_1, \dots, y_h$ be a chain of associated elements. We put

$$(17)\qquad \eta_h = (E - A(\lambda)) \left(\frac{y}{(\lambda - c)^{h+1}} + \frac{y_1}{(\lambda - c)^h} + \dots + \frac{y_h}{\lambda - c} \right).$$

By virtue of (13) the element η_h does not have a pole at $\lambda = c$ and it is therefore a holomorphic function of λ in D. If $y_j = y_j^{(k)}$, where $y_j^{(k)}$ are the elements of the canonical system, we denote the expression (17) by $\eta_h^{(k)}$.

We prove that the multiplicity of an eigenelement does not exceed m, where m is the order of the pole of the resolvent. For it follows from (17) that

$$(18)\qquad \frac{y}{(\lambda - c)^{h+1}} + \frac{y_1}{(\lambda - c)^h} + \dots + \frac{y_h}{\lambda - c} = (E + R(\lambda)) \eta_h(\lambda).$$

In view of the fact that $\eta_h(\lambda)$ is holomorphic and the order of the pole of $R(\lambda)$ is equal to m, we find that $h \leqslant m - 1$. We now show that

$$(19)\qquad R_0 = y^{(1)} z^{(1)} + y^{(2)} z^{(2)} + \dots + y^{(\nu_0)} z^{(\nu_0)},$$

where $y^{(1)}, y^{(2)}, \dots, y^{(\nu_0)}$ are all the eigenelements of the canonical system of multiplicity m, and $z^{(1)}, z^{(2)}, \dots, z^{(\nu_0)}$ is a uniquely determined system of linearly independent eigenelements of $A^*(\lambda)$ for $\lambda = \bar{c}$.

It follows from (16) that for any x the element $R_0 x$ is an eigenelement of multiplicity not less than, and consequently equal to, m. Therefore

$$R_0 x = c_1 y^{(1)} + \dots + c_{\nu_0} y^{(\nu_0)}.$$

By virtue of the linear independence of the $y^{(k)}$, c_k is a linear functional of x, and consequently we can uniquely determine $z^{(k)}$ so that $c_k = (x, z^{(k)})$, and therefore R_0 has the form (19). By virtue of the independence of the $y^{(k)}$ we can choose x so that $(y^{(k)}, x) = 1$ and

$(y^{(j)}, x) = 0$ for $j \neq k$. Then $R_0^* x = z^{(k)}$, and on the basis of (16*) this enables us to conclude that $z^{(k)}$ is an eigenelement of $A^*(\lambda)$ for $\lambda = \bar{c}$, of multiplicity m.

To establish the linear independence of the $z^{(k)}$ we put $y_i = y_i^{(k)}$ in (18). Equating coefficients of $(\lambda - c)^{-m}$ we obtain

$$\delta_{m-1,\, h} y^{(k)} = \sum_{j=1}^{\nu_0} (\eta_h^{(k)}(c),\ z^{(j)})\, y^{(j)}$$

and by virtue of the independence of the $y^{(j)}$

$$(\eta_h^{(k)}(c),\ z^{(j)}) = \delta_{kj}\delta_{m-1,\, h}, \qquad j \leqslant \nu_0 \tag{20}$$

the relations we have obtained prove the linear independence of $z^{(j)}$.

We establish that

$$R_i = \sum_{k=1}^{\nu_0} (y^{(k)} z_i^{(k)} + y_1^{(k)} z_{i-1}^{(k)} + \ldots + y_i^{(k)} z^{(k)}) + \sum_{k=\nu_0+1}^{\nu_1} (y^{(k)} z_{i-1}^{(k)} + \ldots + y_{i-1}^{(k)} z^{(k)}) + \ldots + \sum_{k=\nu_{i-1}+1}^{\nu_i} y^{(k)} z^{(k)}, \tag{21}$$

where

1) $y^{(\nu_{l-1}+1)}, y^{(\nu_{l-1}+2)}, \ldots, y^{(\nu_l)}$ are all the eigenelements of the chosen canonical system, having multiplicity $m - 1$;

2) for $\nu_{l-1} < k \leqslant \nu_l$ $z^{(k)}, z_1^{(k)}, \ldots, z_l^{(k)}$ is a chain of associated elements of the adjoint operator for $\lambda = \bar{c}$; the elements $z^{(k)}$ are linearly independent;

3) the following system of equalities holds:

$$\left\{\begin{array}{l} (\eta_h^{(k)}(c),\ z^{(j)}) = \delta_{kj}\delta_{m_j h}, \qquad j \leqslant \nu_i, \\ (\eta_h^{(k)}(c),\ z_1^{(j)}) + \left(\frac{1}{1!}\frac{d}{dc}\eta_h^{(k)}(c),\ z^{(j)}\right) = \delta_{kj}\delta_{m_j - 1,\, h}, \qquad j \leqslant \nu_{i-1}, \\ \cdots\cdots\cdots\cdots\cdots\cdots\cdots\cdots \\ (\eta_h^{(k)}(c),\ z_i^{(j)}) + \left(\frac{1}{1!}\frac{d}{dc}\eta_h^{(k)}(c),\ z_{i-1}^{(j)}\right) + \ldots \\ \qquad \ldots + \left(\frac{1}{i!}\frac{d^i}{dc^i}\eta_h^{(k)}(c),\ z^{(j)}\right) = \delta_{kj}\delta_{0h}, \qquad j \leqslant \nu_0. \end{array}\right. \tag{22}$$

We assume that the given situation has been established for $i < p \leqslant m$ and establish it for $i = p$. For $\nu \geqslant 0$ we put

$$R_{p+\nu} = \sum_{k=1}^{\nu_0} (y_\nu^{(k)} z_i^{(k)} + \ldots + y_{i+\nu}^{(k)} z^{(k)}) + \sum_{k=\nu_0+1}^{\nu_1} (y_\nu^{(k)} z_{i-1}^{(k)} + \ldots + y_{i+\nu-1}^{(k)} z^{(k)}) + \ldots + \sum_{k=\nu_{p-2}+1}^{\nu_{p-1}} y_\nu^{(k)} z^{(k)} + S_\nu. \tag{23}$$

If x is an arbitrary element of $\mathfrak{H}$, it is easy to verify on the basis of (16) that the elements $y_\nu = S_\nu x$ ($\nu = 0, 1, 2, \ldots, (m - p - 1)$) satisfy the chain of equations (13) and consequently

$$S_\nu = \sum_{k=1}^{\nu_p} c_k y^{(k)}.$$

By the independence of the $y^{(k)}$ the coefficients are linear functionals of x, and we can put

$$(24) \qquad S = \sum_{k=1}^{\nu_0} y^{(k)} z_p^{(k)} + \sum_{k=\nu_0+1}^{\nu_1} y^{(k)} z_{p-1}^{(k)} + \ldots + \sum_{k=\nu_{p-1}+1}^{\nu_p} y^{(k)} z^{(k)},$$

where the elements $z_j^{(k)}$ in this formula are uniquely determined.

By virtue of (23) and (24) the operator R_p can be written in the form (21) for $i = p$, and 1) is satisfied. It remains to establish 2) and 3) for $i = p$. Multiplying the $(p + 1)$-th equation of (16*) on the right by x and taking into account the fact that 2) is satisfied for $i < p$ we obtain

$$\sum_{l=0}^{p} \sum_{k=\nu_{l-1}}^{\nu_l} (x, y^{(k)}) \left(z_{p-l}^{(k)} - A^*(\bar{c}) z_{p-l}^{(k)} - \frac{1}{1!} \frac{\partial A^*(\bar{c})}{\partial \bar{c}} z_{p-l-1}^{(k)} - \ldots \right.$$

$$\left. \ldots - \frac{1}{(p-l)!} \frac{\partial^{p-l} A^*(\bar{c})}{\partial \bar{c}^{p-l}} z^{(k)} \right) = 0.$$

hence, by the independence of the $y^{(k)}$,

$$z_{p-l}^{(k)} = A^*(\bar{c}) z_{p-l}^{(k)} + \ldots + \frac{1}{(p-l)!} \frac{\partial^{p-l} A^*(c)}{\partial \bar{c}^{p-l}} z^{(k)},$$

for $\nu_{l-1} < k \leqslant \nu_l$. In view of the fact that $z^{(k)}, z_1^{(k)}, \ldots, z_{p-l-1}^{(k)}$ form a chain, we conclude from the relation we have obtained that $z^{(k)}, z_1^{(k)}, \ldots, z_{p-l}^{(k)}$ also form a chain.

We now establish 3) for $i = p$. In (18) we put $y = y^{(k)}$ and equate coefficients of $(\lambda - c)^{-m+p}$ on the two sides. Using the known expressions of R_i for $i \leqslant p$, and (22) for $i = p - 1$, we obtain the following relation:

$$\sum_{j=1}^{\nu_0} \left\{ (\eta_h^{(k)}(c), z_p^{(j)}) + \frac{1}{1!} \left(\frac{d}{dc} \eta_h^{(k)}(c), z_{p-1}^{(j)} \right) + \ldots + \frac{1}{p!} \left(\frac{d^p}{dc^p} \eta_h^{(k)}(c), z^{(j)} \right) \right\} y^{(j)} +$$

$$+ \sum_{j=\nu_0+1}^{\nu_1} \left\{ (\eta_h^{(k)}(c), z_{p-1}^{(j)}) + \frac{1}{1!} \left(\frac{d}{dc} \eta_h^{(k)}(c), z_{p-2}^{(j)} \right) + \ldots \right.$$

$$\left. \ldots + \frac{1}{(p-1)!} \left(\frac{d^{p-1}}{dc^{p-1}} \eta_h^{(k)}(c), z^{(j)} \right) \right\} y^{(j)} + \ldots$$

$$\ldots + \sum_{j=\nu_{p-1}}^{\nu_p} (\eta_h^{(k)}(c), z^{(j)}) y^{(j)} = \delta_{m-p,\, h+1} y^{(k)}.$$

By virtue of the independence of the y^j we obtain a system of equalities which, together with (22) for $i = p - 1$, give a complete system (22) for $i = p$.

The equalities, which follow from (22),

$$(\eta_{m_k}^{(k)}(c), z^{(j)}) = \delta_{jk}, \ j, k \leqslant \nu_p.$$

prove the linear independence of the elements $z^{(1)}, z^{(2)}, \ldots, z^{(\nu_p)}$.

Formula (21) proves that the resolvent can be expressed in the form (15), where $z^{(k)}, z_1^{(k)}, \ldots$ represent a chain of associated elements and $z^{(1)}, z^{(2)}, \ldots$ are linearly independent. It remains to prove that the system of eigenelements and associated elements of the operator $A^*(\lambda)$

$$(14^*) \qquad z^{(k)}, z_1^{(k)}, \ldots, z_{m_k}^{(k)}$$

is canonical. For this it is sufficient to show that the multiplicity of the eigenelements of $A^*(\lambda)$ for $\lambda = \bar{c}$ does not exceed $m_1 + 1 = m$, and that each eigenelement of multiplicity $\mu + 1$ can be expressed linearly in terms of the elements $z^{(k)}$ for which $m_k \geqslant \mu$. We put

$$(18^*) \quad \zeta_\mu(\lambda) = (E - A^*(\lambda)) \left(\frac{z}{(\lambda - \bar{c})^{\mu+1}} + \frac{z_1}{(\lambda - \bar{c})^{\mu}} + \ldots + \frac{z_\mu}{\lambda - \bar{c}} \right),$$

where $z, z_1, \ldots, z_\mu$ is a chain of associated elements of an arbitrary eigenelement of multiplicity $\mu + 1$. The function $\zeta_\mu(\lambda)$ is holomorphic and

$$(19^*) \quad \frac{z}{(\lambda - \bar{c})^{\mu+1}} + \frac{z_1}{(\lambda - \bar{c})^{\mu}} + \ldots + \frac{z_\mu}{\lambda - \bar{c}} = (E + R^*(\lambda)) \zeta_\mu(\lambda).$$

Since $\zeta_\mu(\lambda)$ is holomorphic, the order of the pole of the right-hand side is not greater than m, and consequently $\mu + 1 \leqslant m$. Equating to zero the coefficients of $(\lambda - \bar{c})^{-m}, \ldots, (\lambda - \bar{c})^{-(\mu+2)}$, calculated on the basis of (21), and using the linear independence of the $z^{(k)}$, we obtain successively the systems of equalities

$$(\zeta_\mu(\bar{c}), y^{(j)}) = 0 \quad \text{for} \quad j \leqslant \nu_0,$$

$$(\zeta_\mu(\bar{c}), y^{(j)}) = 0 \ \text{for} \ \nu_0 < j \leqslant \nu_1. \ (\zeta_\mu(\bar{c}), y_1^{(j)}) + \frac{1}{1!} \left(\frac{d\zeta_\mu(\bar{c})}{d\bar{c}}, y^{(j)} \right) = 0 \ \text{for} \ j \leqslant \nu_0,$$

. . . .

Using these relations and equating coefficients of $(\lambda - \bar{c})^{-(\mu+1)}$ in (19*), we find that

$$z = \sum_{k=1}^{\nu_0} \left\{ (\zeta_\mu(\bar{c}), y_{m-\mu-1}^{(k)}) + \ldots + \frac{1}{(m-\mu-1)!} \left(\frac{d^{m-\mu-1}\zeta_\mu(\bar{c})}{d\bar{c}^{m-\mu-1}}, y^{(k)} \right) \right\} z^{(k)} + \ldots$$

$$\ldots + \sum_{\nu_{m-\mu-3}+1}^{\nu_{m-\mu-2}} \left\{ (\zeta_\mu(\bar{c}), y_1^{(k)}) + \frac{1}{1!} \left(\frac{d\zeta_\mu(\bar{c})}{d\bar{c}}, y^{(k)} \right) \right\} z^{(k)} +$$

$$+ \sum_{\nu_{m-\mu-2}+1}^{\nu_{m-\mu-1}} (\zeta_\mu(\bar{c}), y^{(k)}) z^{(k)}.$$

This proves that z can be expressed in terms of the $z^{(k)}$ with $m_k \geqslant \mu$, since for $\nu_{i-1} < k \leqslant \nu_i$ we have $m_k = m - i - 1$.

Later we shall use the following property, which follows by simple calculations from the expression (15) for the resolvent.

Let $\xi(\lambda)$ be an element of $\mathfrak{H}$ which is a holomorphic function of λ in a neighbourhood of the pole $\lambda = c$ of the resolvent $R(\lambda)$. The principal part of $R(\lambda)\xi(\lambda)$ can be written in the form

$$(25)\quad \sum_{(j)} \left\{ a_{jm_j} \frac{y^{(j)}}{\lambda - c} + a_{jm_j-1}\left[\frac{y^{(j)}}{(\lambda-c)^2} + \frac{y_1^{(j)}}{\lambda - c}\right] + \dots \right.$$
$$\left. \dots + a_{j0}\left[\frac{y^{(j)}}{(\lambda-c)^{m_j+1}} + \frac{y_1^{(j)}}{(\lambda-c)^{m_j}} + \dots + \frac{y_{m_j}^{(j)}}{\lambda-c}\right]\right\},$$

where

$$(26)\quad \begin{cases} a_{j0} = (\xi(c), z^{(j)}), \\ a_{j1} = (\xi(c), z_1^{(j)}) + \dfrac{1}{1!}\left(\dfrac{d\xi(c)}{dc}, z^{(j)}\right), \\ \dots\dots\dots\dots\dots\dots\dots\dots\dots \\ a_{jm_j} = (\xi(c), z_{m_j}^{(j)}) + \dfrac{1}{1!}\left(\dfrac{d\xi(c)}{dc}, z_{m_j-1}^{(j)}\right) + \dots + \dfrac{1}{m_j!}\left(\dfrac{d^{m_j}\xi(c)}{dc^{m_j}}, z^{(j)}\right). \end{cases}$$

By virtue of the linear independence of the $y^{(k)}$ it is easy to verify that the expression (25) vanishes only when all $a_{jk} = 0$.

6. Suppose that the completely continuous operator $A(\lambda)$ is an analytic function in the domain D. The resolvent $R(\lambda)$ is a meromorphic function. For each pole of the resolvent we construct a canonical system of eigen-elements and associated elements. We call the totality of all chains of the form (14), corresponding to all the eigenvalues of $A(\lambda)$ in D, the *canonical system of eigenelements and associated elements of the operator $A(\lambda)$ in the domain D.*

We use a single numbering for all the chains (14) that occur in the canonical system of $A(\lambda)$.

To each chain of the canonical system $y^{(k)}, y_1^{(k)}, \dots, y_{m_k}^{(k)}$ of $A(\lambda)$ there corresponds a chain of the canonical system $z^{(k)}, z_1^{(k)}, \dots, z_{m_k}^{(k)}$ of $A^*(\lambda)$, determined by (15).

We mention the following proposition.

The chains of eigenelements and associated elements of the canonical system of $A(\lambda)$ in D satisfy the following system of equalities:

$$(27)\quad \begin{cases} (\eta_h^{(k)}(c_j), z^{(j)}) = \delta_{kj}\delta_{m_j, h}, \\ (\eta_h^{(k)}(c_j), z_1^{(j)}) + \dfrac{1}{1!}\left(\dfrac{d}{dc}\eta_h^{(k)}(c_j), z^{(j)}\right) = \delta_{kj}\delta_{m_j-1, h}, \\ \dots\dots\dots\dots\dots\dots\dots\dots\dots \\ (\eta_h^{(k)}(c_j), z_{m_k}^{(j)}) + \dfrac{1}{1!}\left(\dfrac{d}{dc}\eta_h^{(k)}(c_j), z_{m_k-1}^{(j)}\right) + \dots \\ \qquad \dots + \dfrac{1}{m_k!}\left(\dfrac{d^{m_k}}{dc^{m_k}}\eta_h^{(k)}(c_j), z^{(j)}\right) = \delta_{kj}\delta_{0h}, \end{cases}$$

where $y^{(j)}$ is an eigenelement belonging to the eigenvalue c_j, and

$$\eta_h^{(k)}(\lambda) = (E - A(\lambda))\left(\frac{y^{(k)}}{(\lambda - c_k)^{h+1}} + \frac{y_1^{(k)}}{(\lambda - c_k)^h} + \ldots + \frac{y_h^{(k)}}{\lambda - c_k}\right).$$

For $c_j = c_k$ the formulae (27) follow from (22). If $c_j \neq c_k$, then

$$\frac{y^{(k)}}{(\lambda - c_k)^{h+1}} + \frac{y_1^{(k)}}{(\lambda - c_k)^h} + \ldots + \frac{y_h^{(k)}}{\lambda - c_k}$$

is holomorphic for $\lambda = c_j$; therefore, for $\xi(\lambda) = \eta_h^{(k)}(\lambda)$ the expression (25) vanishes. Equating to zero the coefficients of (26) we obtain (27).

7. By virtue of what was stated in 5. the principal part of the resolvent with respect to the pole $\lambda = c$ is a degenerate operator, representable in the form $\Sigma\, uv$, where u and v are elements of the Hilbert space $\mathfrak{H}$. If $B(\lambda)$ is an operator that is holomorphic for $\lambda = c$, it is easy to verify that the principal parts of the operators $B(\lambda)R(\lambda)$ and $R(\lambda)B(\lambda)$ are also degenerate operators. Therefore we can uniquely determine the trace of the operators that represent the principal parts of $B(\lambda)R(\lambda)$ and $R(\lambda)B(\lambda)$. The trace of a degenerate operator $\Sigma\, uv$ is obviously equal to $\Sigma\,(u, v)$.

In the investigation of the eigenvalues the following theorem is useful.

The trace of the principal part of the operator $\frac{\partial A(\lambda)}{\partial \lambda} R(\lambda)$ *for the pole* $\lambda = c$ *is equal to* $\frac{N}{c - \lambda}$, *where N is the multiplicity of the eigenvalue* $\lambda = c$ *of* $A(\lambda)$.

By (6), we have

$$\frac{\partial A(\lambda)}{\partial \lambda} R(\lambda) = -\frac{\partial A(\lambda)}{\partial \lambda} + (E - A(\lambda))\frac{\partial R(\lambda)}{\partial \lambda}.$$

Therefore the principal part of the operator $\frac{\partial A(\lambda)}{\partial \lambda} R(\lambda)$ is equal to the principal part of $B(\lambda) = (E - A(\lambda))\frac{\partial R(\lambda)}{\partial \lambda}$. From (15) we obtain for the principal part of $\frac{\partial R(\lambda)}{\partial \lambda}$

$$-\sum_{(h)}\sum_{i-0}^{m_k}(m_k - i + 1)\frac{y^{(k)}z_i^{(k)} + y_1^{(k)}z_{i-1}^{(k)} + \ldots + y_i^{(k)}z^{(k)}}{(\lambda - c)^{m_k - i + 2}} =$$

$$= -\sum_{(k)}\sum_{i=0}^{m_k}\left[\frac{y^{(k)}}{(\lambda - c)^{m_k - i + 1}} + \frac{y_1^{(k)}}{(\lambda - c)^{m_k - i}} + \ldots + \frac{y_{m_k - i}^{(k)}}{\lambda - c}\right] \times$$

$$\times\left[\frac{z^{(k)}}{(\lambda - c)^{i+1}} + \ldots + \frac{z_i^{(k)}}{\lambda - c}\right],$$

where the sum extends over all k for which $y^{(k)}$ is an eigenfunction for

$\lambda = c$. From the last expression it follows that the principal part of $B(\lambda)$ is the same as that of the operator

$$-\sum_{(k)}\sum_{i=0}^{m_k}\eta_{m_k-i}^{(k)}(\lambda)\left[\frac{z^{(k)}}{(\lambda-c)^{i+1}}+\frac{z_1^{(k)}}{(\lambda-c)^{i}}+\cdots+\frac{z_i^{(k)}}{\lambda-c}\right].$$

The trace of the principal part of this operator is obviously

$$-\sum_{(k)}\sum_{i=0}^{m_k}\left\{\frac{(\eta_{m_k-i}^{(k)}(c), z^{(k)})}{(\lambda-c)^{i+1}}+\frac{(\eta_{m_k-i}^{(k)}, z_1^{(k)})+\frac{1}{1!}\left(\frac{d}{dc}\eta_{m_k-i}^{(k)}(c), z^{(k)}\right)}{(\lambda-c)^{i}}+\cdots\right.$$

$$\left.\cdots+\frac{(\eta_{m_k-i}^{(k)}(c), z_i^{(k)})+\cdots+\frac{1}{i!}\left(\frac{d^i}{dc^i}\eta_{m_k-i}^{(k)}, z^{(k)}\right)}{(\lambda-c)}\right\}$$

and by (27) the latter expression is equal to

$$-\sum_{(k)}\sum_{i=0}^{m_k}\frac{1}{\lambda-c}=\sum_{(k)}\frac{m_k+1}{c-\lambda}=\frac{N}{c-\lambda}.$$

8. Consider the chain (14) in the canonical system of eigenelements and associated elements of $A(\lambda)$. We define derived chains as chains of the form

$$y_h^{(k,\nu)}=\left[\frac{d^\nu}{dt^\nu}e^{c_kt}\left(y_h^{(k)}+y_{h-1}^{(k)}\frac{t}{1!}+\cdots+y^k\frac{t^h}{h!}\right)\right]_{t=0} \tag{28}$$

DEFINITION. A system of eigenelements and associated elements of $A(\lambda)$ is called *n-fold complete* if any system of n elements $f_0, f_1, \ldots, f_{n-1}$ can be represented as the limit of linear combinations

$$f_{\nu,N}=\sum_{k=1}^{N}\sum_{(h)}a_{h,N}^{(k)}y_h^{(k,\nu)}\qquad(\nu=0,1,\ldots,n-1) \tag{29}$$

with coefficients independent of ν.

In particular, in the case $n = 1$ this definition coincides with the usual definition of completeness of the system of all eigenelements and associated elements. We note that when the multiplicities of all the eigenelements are 1, the expressions (28) have the form $f_{\nu,N}=\sum_{k=1}^{N}a_N^{(k)}c_k^\nu y^{(k)}$. A method, originally due to Poincaré, enables us to establish the following proposition.

Let $A(\lambda)$ be an entire function of λ and a solution of the equation

$$y = A(\lambda)y + f$$

where f is independent of λ and satisfies the condition

$$\int_{C_j}\|y(\lambda)\|\,|d\lambda|=o(d_j^{-n+2}), \tag{30}$$

and C_j *is a fixed system of closed curves such that on* C_j *we have* $\theta d_j < |\lambda| < d_j$, *where* θ *is a constant with* $0 < \theta < 1$, *and* $d_j \to \infty$.

For n *arbitrary elements* $f_0, f_1, \ldots, f_{n-1}$ *the expansions*

$$f_\nu = \lim_{j\to\infty} \sum_{C_j} \sum_{(h)} \alpha_h^{(k)} y_h^{(k,\nu)} \tag{31}$$

hold in the sense of strong convergence, where the outer sum extends over the eigenvalues situated inside C_j.

Consider a solution of the equation

$$y = A(\lambda)\, y + g_0 + g_1\lambda + \ldots + g_{n-1}\lambda^{n-1}. \tag{32}$$

By (30),

$$\int_{C_j} \| y(\lambda) \| \, |d\lambda| = o(d_j).$$

Using the fact that $y(\lambda)$ is a meromorphic function of λ and standard arguments from the theory of functions, we obtain

$$y(\lambda) = \lim_{j\to\infty} \sum_{C_j} \omega(\lambda, c_k), \tag{33}$$

where $\omega(\lambda, c_k)$ is the principal part of $y(\lambda)$ with respect to the pole c_k. Let $f_0, f_1, \ldots, f_{n-1}$ be a system of n arbitrary elements. We put

$$\begin{aligned} g(\lambda) &= (E - A(\lambda))(-f_{n-1} + \lambda f_{n-2} + \ldots + (-1)^{n-1}\lambda^{n-1} f_0) = \\ &= g_0 + \lambda g_1 + \lambda^2 g_2 + \ldots \end{aligned} \tag{32*}$$

The solution of (32) then has the form

$$y(\lambda) = -f_{n-1} + \lambda f_{n-2} + \ldots + (-1)^{n-1}\lambda^{n-1} f_0 + \ldots,$$

since the first n terms of the expansion of the solutions of (32) and (32*) must coincide. Using (33) we obtain

$$f_\nu = \lim_{j\to\infty} \sum_{C_j} \frac{1}{(n-\nu-1)!} \left(\frac{d^{n-\nu-1}\omega(\lambda, c_k)}{d\lambda^{n-\nu-1}} \right)_{\lambda=0},$$

where on the basis of (25), by putting $\xi(\lambda) = g_0 + g_1\lambda + \ldots + g_{n-1}\lambda^{n-1}$, we have

$$\begin{aligned} &\frac{1}{(n-\nu-1)!} \left(\frac{d^{n-\nu-1}\omega(\lambda, c_k)}{d\lambda^{n-\nu-1}} \right)_{\lambda=0} = \\ &= a_{m_k}^{(k)} \frac{y^{(k)}}{c_k^{n-\nu}} + a_{m_k-1}^{(k)} \left(\frac{y_1^{(k)}}{c_k^{n-\nu}} - \frac{(n-\nu)!}{1!\,(n-\nu-1)!} \frac{y^{(k)}}{c_k^{n-\nu+1}} \right) + \ldots \\ &\ldots + a^{(k)} \left(\frac{y_{m_k}^{(k)}}{c_k^{n-\nu}} - \frac{(n-\nu)!}{1!\,(n-\nu-1)!} \frac{y_{m_k-1}^{(k)}}{c_k^{n-\nu+1}} + \ldots \right. \\ &\qquad \left. \ldots + (-1)^{m_k} \frac{(n-\nu+m_k-1)!}{(n-\nu-1)!\, m_k!} \frac{y^{(k)}}{c_k^{n-\nu+m_k}} \right). \end{aligned}$$

We put

$$\frac{a_0^{(k)}}{c_k^n} = \alpha_{m_k}^{(k)},$$

$$\frac{a_1^{(k)}}{c_k^n} = \alpha_{m_k-1}^{(k)} - \binom{n}{1}\alpha_{m_k}^{(k)},$$

$$\cdots\cdots\cdots\cdots\cdots\cdots\cdots$$

$$\frac{a_{m_k}^{(k)}}{c_k^n} = \alpha_0^{(k)} - \binom{n}{1}\alpha_1^{(k)} + \binom{n}{2}\alpha_2^{(k)} - \ldots$$

Then the preceding expression reduces to the form

$$\sum_{h=0}^{m_k} \alpha_h^{(k)} \omega_h^{(k,\,\nu)},$$

where

$$\omega_h^{(k,\,\nu)} = \sum_{l=0}^{h} y_{h-l}^{(k)} c_k^{\nu-l} (-1)^l \sum_{i=0}^{l} \frac{(n-\nu+l-i-1)!}{(n-\nu-1)!\,(l-i)!} (-1)^i \binom{n}{i}.$$

We calculate the inner sum

$$(-1)^l \sum_{i=0}^{l} (-1)^i \binom{n}{i} \binom{n-\nu+l-i-1}{l-i} =$$

$$= \frac{(-1)^l}{(n-\nu-1)!} \left(\frac{d^{n-\nu-1}}{d\tau^{n-\nu-1}} \sum_{i=0}^{l} (-1)^i \binom{n}{i} \tau^{n-\nu+l-i-1} \right)_{\tau=1}$$

The number on the right is equal to the residue at infinity of the expression

$$\frac{(-1)^l}{(\tau-1)^{n-\nu}} \sum_{i=0}^{l} (-1)^i \binom{n}{i} \tau^{n-\nu+l-i-1},$$

which in turn is equal to the residue of the function

$$\frac{(-1)^l}{(\tau-1)^{n-\nu}} \sum_{i=0}^{n} (-1)^i \binom{n}{i} \tau^{n-\nu+l-i-1} = (-1)^l \tau^{l-\nu-1} (\tau-1)^{\nu},$$

since for $l \geqslant n$ the two latter functions coincide, and for $l < n$ their expansions about the point at infinity differ only in terms with positive powers of τ. The residue of this function is equal to $\binom{\nu}{l}$ and therefore

$$\omega_h^{(k,\,\nu)} = \sum_{l=0}^{h} \binom{\nu}{l} y_{h-l}^{(k)} c_k^{\nu-l} = y_h^{(k,\,\nu)}.$$

Thus,

$$\frac{1}{(n-\nu-1)!}\left(\frac{d^{n-\nu-1}\omega(\lambda, c_k)}{d\lambda^{n-\nu-1}}\right)_{\lambda=0}=\sum \alpha_h^{(k)} y_h^{(k,\nu)},$$

which proves the proposition.

9. Next we consider the case when

$$A(\lambda)=A_0+A_1\lambda+\ldots+A_n\lambda^n. \tag{34}$$

We establish relations of biorthogonality, which in this case are satisfied by the systems of derived chains

$$y_h^{(k)},\ y_h^{(k,1)},\ \ldots,\ y_h^{(k,n-1)}.$$

First of all we note that in the case in question

$$\zeta_h^{(k)}(\lambda)=(E-A^*(\lambda))\left(\frac{z^{(k)}}{(\lambda-\bar c)^{h+1}}+\frac{z_1^{(k)}}{(\lambda-\bar c)^h}+\ldots+\frac{z_h^{(k)}}{\lambda-\bar c}\right)$$

is a polynomial in λ of degree $n-1$. For $\zeta_h^{(k)}(\lambda)$ is holomorphic for all λ, and as $\lambda\to\infty$ we have, by (34), $\|\zeta_h^{(k)}(\lambda)\| = O(|\lambda|^{n-1})$. We put

$$\zeta_{m_k-h}^{(k)}=\zeta_h^{(k,0)}+\lambda\zeta_h^{(k,1)}+\ldots+\lambda^{n-1}\zeta_h^{(k,n-1)} \tag{35}$$

and use the relations analogous to (27) for the adjoint operator, which are obtained from (27) when η, z, c are replaced by ζ, y, $\bar c$. It is easy to verify that in this way we obtain the following biorthogonality relations:

$$\sum_{\nu=0}^{n-1}(\zeta_h^{(k,\nu)}, y_l^{(i,\nu)})=\delta_{ik}\delta_{hl}. \tag{36}$$

The following proposition follows immediately from these relations.

For an operator of the form (34) there can exist only one system of expansions of the n elements $f_0, f_1, \ldots, f_{n-1}$ of the form

$$f_\nu=\sum_{k=1}^{\infty}\sum_{(h)} a_h^{(k)} y_h^{(k,\nu)} \qquad (\nu=0, 1, \ldots, n-1).$$

The coefficients in these expansions are defined by the formulae

$$a_h^{(k)}=\sum_{\nu=0}^{n-1}(f_\nu, \zeta_h^{(k,\nu)}). \tag{37}$$

NOTE. Consider the Hilbert space $\mathfrak{H}^n$ consisting of the systems of n elements $(x^{(0)}, x^{(1)}, \ldots, x^{(n-1)})$ with the norm

$$\sqrt{\|x^{(0)}\|^2+\|x^{(1)}\|^2+\ldots+\|x^{(n-1)}\|^2}.$$

It is obvious that n, the multiple completeness of the system of eigenfunctions and adjoint functions, is the completeness of the system of vectors $U_h^{(k)} = (y_h^{(k,0)}, y_h^{(k,1)}, \ldots, y_h^{(k,n-1)})$ in $\mathfrak{H}^n$. The study of the operator (34) in $\mathfrak{H}$ can be reduced to that in $\mathfrak{H}^n$ of an operator containing λ to the first degree whose components are defined by the formulae

$$(38)\ (A_0x^{(0)} + \lambda(A_1x^{(0)} + A_2x^{(1)} + \ldots + A_nx^{(n-1)}),\ \lambda x^{(0)},\ \lambda x^{(1)},\ \ldots,\ \lambda x^{(n-1)}).$$

The properties of the operator (34) mentioned above can be derived from the corresponding properties of operators depending linearly on λ. However, the results of the later chapters are closely connected with the particular form (38) or (34), and so in this introductory chapter we have preferred the present method of presentation.

We note that each linear operator $\mathscr{K}$ in $\mathfrak{H}^n$ can be written in the form

$$y^{(j)} = \sum_{i=0}^{n-1} K_{ji}x^{(i)} \qquad (j = 0, 1, \ldots, n-1),$$

where K_{ji} is an operator in $\mathfrak{H}$. We represent the operator $\mathscr{K}$ by the matrix $\| K_{ij} \|$.

Chapter II

THE COMPLETENESS THEOREM

10. In this chapter we prove a theorem on n-fold completeness of a system of eigenelements and associated elements for one class of functional equations. This theorem can be the basis of proofs of propositions on the completeness of the system of eigenfunctions of broad classes of non-selfadjoint equations.

Let H be a completely continuous selfadjoint operator. We denote the eigenvalues of the equation

$$(39) \qquad x = \lambda Hx$$

by h_i, and the orthogonalized eigenelements by x_i. If $Hx_i = 0$, then $h_i = \infty$.

If the system of eigenelements (39) is complete for $\lambda \neq \infty$, we call the operator H complete. In this case all the h_i are finite.

We call the lower bound of numbers ρ' for which

$$\sum_{i=1}^{\infty} \frac{1}{|h_i|^{\rho'}} < +\infty$$

the *order* ρ of H. It is obvious that when the order ρ of H is finite, for $m > 1/(2\rho)$ the absolute norm of the operator H^m is finite.

We denote by $H^{(\alpha)}$, $\alpha > 0$, a selfadjoint operator having the eigenelements x_i for the eigenvalues $|h_i|^{\alpha}$.

We prove the following theorem.

THEOREM. *Let H be a complete selfadjoint completely continuous operator of finite order, and $A_0, A_1, \ldots, A_{n-1}$ arbitrary completely continuous operators. For each of the equations*

(40) $$y=(A_0+\lambda H^{\left(\frac{1}{n}\right)}A_1+\ldots+\lambda^{n-1}H^{\left(\frac{n-1}{n}\right)}A_{n-1}+\lambda^n H)y,$$

(40*) $$y=(A_0^*+\lambda A_1^* H^{\left(\frac{1}{n}\right)}+\ldots+\lambda^{n-1}A_{n-1}^* H^{\left(\frac{n-1}{n}\right)}+\lambda^n H)y$$

the system of eigenelements and associated elements is n-fold complete.

The eigenvalues of equations (40) *and* (40*) *approach asymptotically the rays*

$$\arg\lambda=\frac{\nu\pi}{n}\qquad(\nu=0, 1, \ldots, 2n-1).$$

In particular, for $n = 1$ it follows from this theorem that the system of eigenelements and associated elements of the equation $y = Ay + \lambda Hy$ is complete.

We preface the proof of this theorem with a number of auxiliary propositions, which we shall also use later.

11. LEMMA 1. *Let H be a completely continuous selfadjoint operator such that H^m has a finite absolute norm, and $K_1, K_2, \ldots, K_p$ bounded operators. The operator $K = \prod_1^p K_i H^{m_i}$, $m_1 + m_2 + \ldots + m_p = m$, has a finite absolute norm $N(K)$, where*

(41) $$N(K)\leqslant\|K_1\|\,\|K_2\|\ldots\|K_p\|\,N(H^m).$$

Denote the eigenvalues of H by h_i; then

$$N(H^m)=\sqrt{\sum h_i^{-2m}}.$$

We note that if A is bounded and the absolute norm of B is finite, then

$$N(AB)\leqslant\|A\|\,N(B).$$

For if we denote by $\|a_{ik}\|$ and $\|b_{ik}\|$ the matrices of A and B in an orthogonal system of coordinates, we obtain

$$N^2(AB)=\sum_{ij}\Big|\sum_s a_{is}b_{sj}\Big|^2=\sum_{(j)}\Big(\sum_i\Big|\sum_s a_{is}b_{sj}\Big|^2\Big)\leqslant\sum_{(j)}\|A\|^2\sum_s|b_{sj}|^2=\|A\|^2N^2(B).$$

It follows immediately that the lemma is true for $p = 1$. Assuming the lemma to be true for $p - 1$ factors, we prove it for p factors. We put

$$K'=H^{m_1}\prod_{i=2}^{p}K_iH^{m_i},\qquad K''=\prod_{i=2}^{p-1}K_iH^{m_i}K_p$$

and choose as a system of coordinates the system of eigenelements of H; then

$$N^2(K) \leqslant \|K_1\|^2 N^2(K') = \|K_1\|^2 \sum_{i,j} |K'_{ij}|^2 = \|K_1\|^2 \sum_{i,j} \frac{|K''_{ij}|^2}{|h_i|^{m_1} |h_j|^{m_p}}.$$

By the theorem on the arithmetic and geometric means

$$\frac{1}{|h_i|^{m_1} |h_j|^{m_p}} \leqslant \frac{m_1}{m_1+m_p} \frac{1}{|h_i|^{m_1+m_p}} + \frac{m_p}{m_1+m_p} \frac{1}{|h_j|^{m_1+m_p}},$$

and so

$$N^2(K) \leqslant \frac{m_1}{m_1+m_p} \|K_1\|^2 \sum_{i,j} \left| \frac{K''_{ij}}{h_i^{m_1+m_p}} \right| + \frac{m_p}{m_1+m_p} \|K_1\|^2 \sum_{i,j} \left| \frac{K''_{ij}}{h_j^{m_1+m_p}} \right|^2 =$$

$$= \frac{m_1}{m_1+m_p} \|K_1\|^2 N^2(K''^* H^{m_1+m_2}) + \frac{m_p}{m_1+m_p} \|K_1\|^2 N^2(K'' H^{m_1+m_p}).$$

To each of these terms we apply the lemma for $p - 1$ factors, and thus obtain (41).

When H is a non-selfadjoint operator, the lemma remains true for $m = 1$. It is easy to verify that for $m \geqslant 2$ it is not true. For let all the elements of the matrix except $h_{2k+1,\,2k+2}$ be zero, and $h_{2k+1,\,2k+2} = 1$, and let all the elements of K except $K_{2j+2,\,2j+1}$ be zero, and $K_{2j+2,\,2j+1} = 1$; then for $m \geqslant 2$ we have $H^m = 0$ and consequently $N(H^m) = 0$. On the other hand, for example, $(KH)^2$ is a diagonal matrix whose odd elements are 1 and whose even elements are 0, and so $N((KH)^2) = \infty$.

In the proof of the subsequent theorems we use a theorem of Schur:[1] the sum of the reciprocals of the squares of the moduli of the eigenvalues of the operator λK does not exceed the square of the absolute norm of K; we also use the following proposition.[2]

If the operator K has finite absolute norm, then the resolvent of λK can be represented in the form

$$E + R(\lambda) = \frac{D(\lambda)}{\Delta(\lambda)},$$

where $D(\lambda)$ is an operator function satisfying the inequality

$$\|D(\lambda)\| < e^{\frac{|\lambda|^2}{2} N^2(K) + \frac{1}{2}},$$

and $\Delta(\lambda)$ is expressed in terms of the eigenvalues of λK by the formula

$$\Delta(\lambda) = \prod \left(1 - \frac{\lambda}{\lambda_j}\right) e^{\frac{\lambda}{\lambda_j}}.$$

---

[1] I. Schur, Math. Ann. **66** (1909).

[2] See T. Carleman, Math. Z. 9 (1921), 196–217 and also E. Hille and J. D. Tamarkin, Acta Math. 57 (1931), 1–76.

THEOREM. *Let $K = K_1 K_2 \ldots K_q$, where $K_1, K_2, \ldots, K_q$ are completely continuous operators. If the operators $K_i K_{i+1} \ldots K_{i+p}$ $(K_{q+j} = K_j)$ have finite absolute norms, then the eigenvalues of λK satisfy the inequality*[1]

$$\sum_{\nu=1}^{\infty} \frac{1}{|\lambda_\nu|^{\frac{2p}{q}}} \leqslant \frac{1}{q} \sum_{i=1}^{q} N^2 (K_i K_{i+1} \ldots K_{i+p}), \tag{42}$$

and the resolvent of K is expressed in the form

$$R(\lambda) = \frac{D(\lambda)}{\Delta(\lambda)}, \tag{43}$$

where $D(\lambda)$ is an entire operator function of order $2p/q$, and

$$\Delta(\lambda) = \prod \left(1 - \frac{\lambda}{\lambda_j}\right) e^{\frac{\lambda}{\lambda_j} + \frac{1}{2}\left(\frac{\lambda}{\lambda_j}\right)^2 + \ldots + \frac{1}{m}\left(\frac{\lambda}{\lambda_j}\right)^m},$$

where m is the largest integer for which $m < 2p/q$.

In the space $\mathfrak{H}^q$ (see 9.) we introduce the operator

$$\mathcal{K} = \begin{Vmatrix} 0 & K_1 & 0 & 0 & \ldots & 0 \\ 0 & 0 & K_2 & 0 & \ldots & 0 \\ \cdot & \cdot & \cdot & \cdot & \cdot & \cdot \\ K_q & 0 & 0 & 0 & \ldots & 0 \end{Vmatrix}.$$

The operator $\mathcal{K}^p$ will be represented by a matrix in which all the terms in the i-th row, except the term in the $(i + p + 1)$-th column, are zero, and the term in the $(i + p + 1)$-th column is the operator $K_i K_{i+1} \ldots K_{i+p}$. From this it is easy to see that

$$N^2(\mathcal{K}^p) = \sum_{i=1}^{q} N^2 (K_i K_{i+1} \ldots K_{i+p}). \tag{44}$$

We denote by $R(\lambda)$ the resolvent of the equation

$$x = \lambda K x + f, \tag{45}$$

and by $\mathcal{E}$ and $\mathcal{R}(\lambda)$ the identity operator in $\mathfrak{H}^q$ and the resolvent of the equation

$$\xi = \mu \mathcal{K} \xi + \varphi. \tag{46}$$

Splitting (46) into q equations in $\mathfrak{H}$, by putting $\varphi = (f, f_1, \ldots, f_{q-1})$,

$$\begin{aligned} x &= \mu K_1 x_1 + f, \\ x_1 &= \mu K_2 x_2 + f_1, \\ &\cdots\cdots \\ x_{q-1} &= \mu K_q x + f_{q-1}, \end{aligned}$$

[1] For $p = 1$ the convergence of the series on the left-hand side (with no indication of an upper bound for its sum) was established by S. H. Chang, Trans. Amer. Math. Soc. 67 (1949), 351–367.

we can reduce the solution of (46) to that of equations of the form (45). This gives the following expression for the resolvent of (46):

$$(47)\quad \mathscr{E}+\mathscr{R}(\mu)=$$

$$=\left\|\begin{matrix} E+R(\mu^q) & \dots & \mu^{q-1}(E+R(\mu^q))\prod^{q-1} K_i \\ \mu^{q-1}\prod_2^q K_i(E+R(\mu^q)) & \dots & \mu^{2q-2}\prod_2^q K_i(E+R(\mu^q))\prod_1^{q-1} K_1+\mu^{q-2}\prod_2^{q-1} K_i \\ \mu^{q-2}\prod_3^q K_i(E+R(\mu^q)) & \dots & \mu^{2q-3}\prod_3^q K_1(E+R(\mu^q))\prod_1^{q-1} K_i+\mu^{q-3}\prod_3^{q-1} K_i \\ \dots & \dots & \dots \\ \mu K_q(E+R(\mu^q)) & \dots & \mu^q K_q(E+R(\mu^q))\prod_1^{q-1} K_i+E \end{matrix}\right\|$$

Formula (47) shows that the eigenvalues of (46) have the form $\sqrt[q]{c}$, where c is an eigenvalue of (45). We show that if c is an eigenvalue of (45) of multiplicity N, then each value of the root $\sqrt[q]{c}$ is an eigenvalue of multiplicity N for (46).

To calculate the multiplicity of an eigenvalue we use a proposition in 7. Using the expressions for the matrices of the operators $\mathscr{K}$ and $\mathscr{E}+\mathscr{R}(\mu)$ we easily verify that the matrix of the operator $\mathscr{K}(\mathscr{E}+\mathscr{R}(\mu))$ has as diagonal terms the operators

$$\mu^{q-1}K(E+R(\mu^q)),\quad \mu^{q-1}\prod_2^q K_i(E+R(\mu^q))K_1,\ \dots,\ \mu^{q-1}K_q(K+R(\mu^q))\prod_1^{q-1} K_i.$$

Therefore the trace of the principal part of the operator $\mathscr{K}(\mathscr{E}+\mathscr{R}(\mu))$ at the pole $\mu = c_1 = \sqrt[q]{c}$ is equal to the trace of the principal part of the operator

$$\mu^{q-1}\left\{K(E+R(\mu^q))+\prod_2^q K_i(E+R(\mu^q))K_1+\dots+K_q(E+R(\mu^q))\prod_1^{q-1} K_i\right\}$$

and since the trace of the principal part of each of the operators in the brackets is equal to the trace of the principal part[1] of $K(E + R(\mu^q))$, we must calculate the trace of the principal part of $q\mu^{q-1}K(E + R(\mu^q))$, which by 7. coincides with the principal part for $\mu = c_1$ of the expression $\frac{q\mu^{q-1}N}{c-\mu^q}$, namely $\frac{N}{c_1-\mu}$. By 7. it follows that the multiplicity of c_1 for equation (46) is N.

We denote by $\mathscr{R}_p(\tau)$ the resolvent of the equation

$$(48)\qquad \xi=\tau\mathscr{K}^p\xi+\varphi;$$

[1] This follows from the equality $\mathrm{tr}(AB) = \mathrm{tr}(BA)$.

then from the expansion for $\mu = 0$ we verify that

$$\mathscr{E} + \mathscr{R}(\mu) = \{\mathscr{E} + \mu\mathscr{K} + \ldots + \mu^{p-1}\mathscr{K}^{p-1}\}\{\mathscr{E} + \mathscr{R}_p(\mu^p)\}. \quad (49)$$

By (49) the eigenvalues of (46) have the form $c_1 = \sqrt[p]{c_2}$, where c_2 is an eigenvalue of (48). We prove that the multiplicity of c_2 is equal to the sum of the multiplicities of the eigenvalues of (46) of the form $\sqrt[p]{c_2}$. For suppose that $c_1, c_1\omega_1, \ldots, c_1\omega_{p-1}$ are the different values of $\sqrt[p]{c_2}$. We denote by N the multiplicity of c_2 for (47) and by $n, n_1, \ldots, n_{p-1}$ the multiplicities of $c_1, c_1\omega_1, \ldots, c_1\omega_{p-1}$ for (46). By (49)

$$\mathscr{K}(\mathscr{E} + \mathscr{R}(\mu)) + \omega_1\mathscr{K}(\mathscr{E} + \mathscr{R}(\omega_1\mu)) + \ldots + \omega_{p-1}\mathscr{K}(\mathscr{E} + \mathscr{R}(\omega_{p-1}\mu)) = \\ = p\mu^{p-1}\mathscr{K}^p(\mathscr{E} + \mathscr{R}_p(\mu^p)).$$

We calculate the sum of the traces of the principal parts of the left- and right-hand sides with respect to the poles $c_1, c_1\omega_1, \ldots, c_1\omega_{p-1}$, noting that on the right-hand side we can define the sum of the traces of the principal parts as the trace of the sum of the principal parts. Applying the proposition in 7. we obtain

$$\sum_{k,\, j=0}^{p-1} \frac{\omega_k n_j}{c_1\omega_k\omega_j - \omega_k\mu} = \frac{p\mu^{p-1}N}{c_1^p - \mu^p},$$

from which it follows that

$$N = n + n_1 + \ldots + n_{p-1}.$$

Denoting by λ_j, μ_j, τ_j the eigenvalues of (45), (46) and (48), taking account of their multiplicity, we obtain on the basis of Schur's theorem stated above

$$\sum \frac{1}{|\lambda_i|^{\frac{2p}{q}}} \leqslant \frac{1}{q}\sum \frac{1}{|\mu_i|^{2p}} \leqslant \frac{1}{q}\sum \frac{1}{|\tau_i|^2} \leqslant \frac{1}{q} N^2(\mathscr{K}^p),$$

and using (44) we obtain (42).

To prove the second part of the proposition, we apply Carleman's theorem to (48); then

$$\mathscr{E} + \mathscr{R}_p(\tau) = \frac{\mathscr{D}_p(\tau)}{\prod\limits_j \left(1 - \frac{\tau}{\tau_j}\right) e^{\frac{\tau}{\tau_j}}},$$

where $\mathscr{D}_p(\tau)$ is an operator function of the second order.

The connection established above between μ_j and τ shows that

$$\prod \left(1 - \frac{\mu^p}{\tau_i}\right) e^{\frac{\mu^p}{\tau_i}} = \prod_{k=0}^{p-1} \prod_{(j)} \left(1 - \frac{\mu}{\mu_j\omega_k}\right) e^{\sum\limits_{l=1}^{2p-1} \frac{1}{l}\left(\frac{\mu}{\mu_j\omega_k}\right)^l},$$

where by standard properties of infinite products each factor of $\prod\limits_{(j)}$

converges and is a function of order not greater than $2p$. On the basis of (49)

$$\mathscr{D}(\mu) = (\mathscr{E} + \mathscr{R}(\mu)) \prod_{(j)} \left(1 - \frac{\mu}{\mu_j}\right) e^{\sum_{l=1}^{2p-1} \frac{1}{l} \left(\frac{\mu}{\mu_j}\right)^l} =$$

$$= \frac{(\mathscr{E} + \mu\mathscr{K} + \ldots + \mu^{p-1}\mathscr{K}^{p-1}) \mathscr{D}_p(\mu^p)}{\prod_{k=1}^{p-1} \prod_{(j)} \left(1 - \frac{\mu}{\mu_j \omega_k}\right) e^{\sum_{l=1}^{2p-1} \frac{1}{l} \left(\frac{\mu}{\mu_j \omega_k}\right)^l}}.$$

The expression obtained for $\mathscr{D}(\mu)$ shows that $\mathscr{D}(\mu)$ is a meromorphic function of order $2p$, and since $\mathscr{D}(\mu)$ does not have poles, it is an entire function of order $2p$. Thus,

$$\mathscr{E} + \mathscr{R}(\mu) = \frac{\mathscr{D}(\mu)}{\prod_{j} \left(1 - \frac{\mu}{\mu_j}\right) e^{\sum_{1}^{2p-1} \frac{1}{l} \left(\frac{\mu}{\mu_j}\right)^l}},$$

where $\mathscr{D}(\mu)$ is an entire operator function of order $2p$. By the connection established above between λ_i and μ_j we have

$$\prod_{(j)} \left(1 - \frac{\mu}{\mu_i}\right) e^{\sum_{1}^{2p-1} \frac{1}{l} \left(\frac{\mu}{\mu_i}\right)^l} = \prod_{(i)} \left(1 - \frac{\mu^q}{\lambda_i}\right) e^{\sum_{1}^{m} \frac{1}{l} \left(\frac{\mu^q}{\lambda_i}\right)^l} \tag{50}$$

where m is the largest integer for which $m < 2p/q$. We put

$$D(\lambda) = (E + R(\lambda)) \prod_{(j)} \left(1 - \frac{\lambda}{\lambda_j}\right) e^{\sum_{1}^{m} \frac{1}{l} \left(\frac{\lambda}{\lambda_j}\right)^l}$$

so that $D(\lambda)$ is an entire function of λ. On the basis of (47)

$$\| \mathscr{E} + \mathscr{R}(\mu) \| \geqslant \| E + R(\mu^q) \|,$$

hence, taking (50) into account, we have

$$\| D(\lambda) \| \leqslant \| \mathscr{D}(\lambda^{1/q}) \|, \tag{51}$$

and so $D(\lambda)$ is an entire function of order $2p/q$. This proves the assertion of the theorem regarding the representation (43).

This theorem and the lemma of the present section enable us to establish immediately the following theorem.

THEOREM. *Let H be a completely continuous selfadjoint operator such that*

$$\sum \frac{1}{|h_i|^\rho} < +\infty,$$

and K a bounded operator. The eigenvalues of the operator λKH satisfy the inequality

$$\sum \frac{1}{|\lambda_i|^\rho} \leqslant \| K \|^\rho \sum \frac{1}{|h_i|^\rho}, \tag{52}$$

and the resolvent of the operator is expressed in the form

$$E + R(\lambda) = \frac{D(\lambda)}{\Delta(\lambda)}, \tag{53}$$

where $D(\lambda)$ is an entire operator function of order not greater than ρ, and

$$\Delta(\lambda) = \prod_{(j)} \left(1 - \frac{\lambda}{\lambda_i}\right) e^{\sum_1^m \frac{1}{l} \left(\frac{\lambda}{\lambda_i}\right)^l},$$

where m is the largest integer for which $m < \rho$.

For we choose a rational fraction for which $\rho \leqslant 2p/q \leqslant m + 1$. We may assume that the operator H is positive, since otherwise it would be sufficient to prove the theorem for $(KU^{-1})(UH)$, where U is a unitary operator that transforms H into a positive operator. Applying the preceding theorem to the operator

$$\frac{K}{\| K \|} ((\| K \| H)^{1/q})^q$$

and using the lemma of the present section to calculate the right-hand side of (42) we obtain

$$\sum \frac{1}{|\lambda_j|^{\frac{2p}{q}}} \leqslant \| K \|^{2p/q} \sum \frac{1}{|h_i|^{2p/q}},$$

and $D(\lambda)$ is an entire function of order $2p/q$. Since $2p/q$ can be chosen arbitrarily close to ρ, we conclude that the order of $D(\lambda)$ does not exceed ρ and that (52) is satisfied.

More detailed arguments enable us to establish in the assumptions of the theorem that the type of $D(\lambda)$ is minimal.

We note the following consequences of the proposition just proved.

T H E O R E M. *Let H be a completely continuous selfadjoint operator such that*

$$\sum \frac{1}{|h_i|^\rho} < \infty,$$

and $K_1, K_2, \ldots, K_n$ bounded operators. Denoting by λ_j the eigenvalues of the equation

$$x = (\lambda K_1 H + \lambda^2 K_2 H^2 + \ldots + \lambda^n K_n H^n) x \tag{54}$$

and by $R(\lambda)$ its resolvent we have

$$\sum \frac{1}{|\lambda_i|^\rho} < \infty.$$

and

$$R(\lambda)=\frac{D(\lambda)}{\Delta(\lambda)},$$

where $D(\lambda)$ is an entire operator function of order ρ and

$$\Delta(\lambda)=\prod\left(1-\frac{\lambda}{\lambda_i}\right)e^{\sum_1^m \frac{1}{k}\left(\frac{\lambda}{\lambda_i}\right)^k},$$

where m is the largest integer for which $m<\rho$.

In $\mathfrak{H}^n$ consider the operators

$$\mathcal{K}=\begin{Vmatrix} K_1 & K_n & K_{n-1} & K_{n-2} \dots K_2 \\ 0 & 0 & E & 0 \quad \dots 0 \\ 0 & 0 & 0 & E \quad \dots 0 \\ \dots & \dots & \dots & \dots \\ E & 0 & 0 & 0 \quad \dots 0 \end{Vmatrix}, \tag{55}$$

$$\mathcal{H}=\begin{Vmatrix} H & 0 \dots 0 \\ 0 & H \dots 0 \\ \dots & \dots \\ 0 & 0 \dots H \end{Vmatrix}. \tag{56}$$

It is obvious that $\mathcal{H}$ is selfadjoint and each eigenvalue h_i of H generates an eigenvalue of multiplicity n of $\mathcal{H}$. The equation

$$\xi=\lambda\mathcal{K}\mathcal{H}\xi+\varphi \tag{57}$$

splits into the system

$$\begin{cases} x=\lambda(K_nHx_1+K_{n-1}Hx_2+\dots+K_1Hx)+f, \\ x_1=\lambda Hx_2+f_1, \\ \dots\dots\dots \\ x_{n-1}=\lambda Hx+f_{n-1}. \end{cases} \tag{58}$$

This enables us to express the resolvent of (57) in terms of the resolvent of (54)

(59) $\mathcal{E}+\mathcal{R}(\lambda)=$

$$=\begin{Vmatrix} E+R(\lambda)\dots & (E+R(\lambda))\sum_{k=1}^{i}\lambda^kK_{n-i+k}H^k\dots \\ \lambda^{n-1}H^{n-1}(E+R(\lambda))\dots & \lambda^{n-1}H^{n-1}(E+R(\lambda))\sum_{k=1}^{i}\lambda^kK_{n-i+k}H^k+H^{i-1}\dots \\ \lambda^{n-2}H^{n-2}(E+R(\lambda))\dots & \lambda^{n-2}H^{n-2}(E+R(\lambda))\sum_{k=1}^{i}\lambda^kK_{n-i+k}H^k+H^{i-2}\dots \\ \dots & \dots \\ \lambda H(E+R(\lambda))\dots & \lambda H(E+R(\lambda))\sum_{k=1}^{i}\lambda^kK_{n-i+k}H^k+H^{i-n}\dots \end{Vmatrix},$$

where $i = 1, 2, \ldots, n-1$ and $H^l = 0$ for $l < 0$. Using (55), (56), (59), we can easily establish that for any λ_i the trace of the principal part of $\mathcal{K}\ \mathcal{H}(\mathcal{E}+\mathcal{R}(\lambda))$ is the same as that of the principal part of

$$\frac{d}{d\lambda}(\lambda K_1 H + \lambda^2 K_2 H^2 + \ldots + \lambda^n K_n H^n)\cdot(E + R(\lambda)).$$

Therefore the eigenvalues of (54) and (57) coincide and so, by the preceding theorem,

$$\sum \frac{1}{|\lambda_j|^\rho} \leqslant n \|\mathcal{K}\| \sum \frac{1}{|h_i|^\rho}.$$

On the other hand,

$$\|D(\lambda)\| \leqslant \|\mathcal{D}(\lambda)\|,$$

where

$$\mathcal{D}(\lambda) = (\mathcal{E} + \mathcal{R}(\lambda))\,\Delta(\lambda),$$

which proves the assertion about the order of $D(\lambda)$.

12. LEMMA. *The equations* (40) *and* (40*) *have meromorphic resolvents* $R(\lambda)$ *and* $R^*(\lambda)$ *admitting representations*

(60) $$R(\lambda) = T(\lambda^n) + B(\lambda)(E + T(\lambda^n)),$$

(60*) $$R^*(\lambda) = T(\lambda^n) + (E + T(\lambda^n))\,B^*(\lambda),$$

where $T(\lambda)$ *is the resolvent of* λH, *and* $B(\lambda)$ *is a completely continuous meromorphic operator whose norm in each angle*

(61) $$\frac{\nu\pi}{n} + \delta \leqslant \arg\lambda \leqslant \frac{(\nu+1)\pi}{n} - \delta$$

tends to zero as $|\lambda| \to \infty$, *uniformly with respect to* $\arg\lambda$.

Note that in the proof of this lemma we shall not assume that the order of H is finite.

It is sufficient to prove the lemma for equation (40), since then formula (60*) for the resolvent of (40*) can be obtained by going over to the adjoint equation.

To establish the lemma completely it is sufficient to prove that the resolvent $R(\lambda)$ exists and admits a representation (60) in the angle (61) that satisfies the given condition, because then

(62) $$B(\lambda) = (R(\lambda) - T(\lambda^n))\,(E - H(\lambda^n)),$$

and since by 3. the existence of $R(\lambda)$ in the angle (61) ensures that $R(\lambda)$ is meromorphic in the whole plane, formulae (62) and (60) make sense over the whole plane.

Choosing as a system of coordinates the system of eigenelements of λH

and putting

$$f_i=(f,u_i),\quad A_\nu=\|a_{ik}^{(\nu)}\|,\quad a_{ik}^{(\nu)}=(A_\nu u_k,u_i)\qquad(\nu=0,1,\ldots,\overline{n-1}),$$

we reduce (40) to the form

$$x_i=\sum_{\nu=0}^{n-1}\frac{\lambda^\nu}{|h_i|^{\nu/n}}\sum_{k=1}^{\infty}a_{ik}^{(\nu)}x_k+\frac{\lambda^n}{h_i}x_i+f_i,\tag{63}$$

and by virtue of the complete continuity of the operators A_ν

$$\sum_{i=M+1}^{\infty}\Big|\sum_{k=1}^{\infty}a_{ik}^{(\nu)}x_k\Big|^2\leqslant\varepsilon_m^2\sum_{i=1}^{\infty}|x_i|^2,\tag{64}$$

where $\varepsilon_m\to 0$ as $m\to\infty$. We rewrite the system (63) in the form

$$x_i=\sum_{\nu=0}^{n-1}\operatorname{sign}h_i\,\frac{|h_i|^{1-\frac{\nu}{n}}\lambda^\nu}{h_i-\lambda^n}\sum_{k=1}^{\infty}a_{ik}^{(\nu)}x_k+\frac{h_i}{h_i-\lambda^n}f_i.\tag{65}$$

Taking account of the fact that the second terms on the right-hand side are the coordinates of the element $(E+T(\lambda^n))f$, and denoting by $C(\lambda)$ the operator defined by the first terms, we rewrite (65) in the form

$$x=C(\lambda)x+(E+T(\lambda^n))f.\tag{66}$$

We estimate the norm of $C(\lambda)$ in the angle (61). We choose $\beta(|\lambda|)<1$ so that $\beta(|\lambda|)\to 0$ and $|\lambda|\beta(|\lambda|)\to\infty$ as $|\lambda|\to\infty$, and let m be the first integer for which $\sqrt[n]{|h_m|}\geqslant|\lambda|\beta(|\lambda|)$; then as $|\lambda|\to\infty$ we have $m\to\infty$ and $\varepsilon_m\to 0$. We note that in the angle (61)

$$\left|\frac{\lambda^\nu h_i^{1-\frac{\nu}{n}}}{h_i-\lambda^n}\right|\leqslant\frac{1}{\sin n\delta},$$

and for $|h_i|^{1/n}\leqslant|\lambda|\beta(|\lambda|)$

$$\left|\frac{\lambda^\nu h_i^{1-\frac{\nu}{n}}}{h_i-\lambda^n}\right|\leqslant\frac{\beta^{\left(1-\frac{\nu}{n}\right)n}}{1-\beta^n}.$$

We have

$$\|C(\lambda)x\|\leqslant\left\{\sum_{i=1}^{\infty}\left|\sum_{\nu=0}^{n-1}\operatorname{sign}h_i\frac{\lambda^\nu|h_i|^{1-\frac{\nu}{n}}}{h_i-\lambda^n}\sum a_{ik}^{(\nu)}x_k\right|^2\right\}^{1/2}\leqslant$$

$$\leqslant n\sum_{\nu=0}^{n-1}\left\{\sum_{i=1}^{m}\left|\frac{\lambda^\nu h_i^{1-\frac{\nu}{n}}}{h_i-\lambda^n}\sum_{k=1}^{\infty}a_{ik}^{(\nu)}x_k\right|^2\right\}^{1/2}+n\sum_{\nu=0}^{n-1}\left\{\sum_{i=m+1}^{\infty}\left|\frac{\lambda^\nu h_i^{1-\frac{\nu}{n}}}{h_i-\lambda^n}\sum_{k=1}^{\infty}a_{ik}^{(\nu)}x_k\right|^2\right\}^{1/2}\leqslant$$

$$\leqslant n\left(\frac{\beta}{1-\beta}\varepsilon_0+\frac{n\varepsilon_m}{\sin n\delta}\right)\|x\|=o(1)\|x\|.$$

By virtue of this estimate, for sufficiently large $|\lambda|$ in the angle (46) the resolvent of the operator $C(\lambda)$ is defined by the formula

$$B(\lambda) = C(\lambda) + C^2(\lambda) + \ldots, \text{ and } B(\lambda) = o(1).$$

From (66) we obtain

$$x = E + T(\lambda^n) + B(\lambda)\,(E + T(\lambda^n))f$$

and so the lemma is proved.

We now turn to the proof of the completeness theorem stated in **10**. We note first of all that, by the lemma of this section, for sufficiently large $|\lambda|$ the resolvent $R(\lambda)$ has no poles in the angle (61). This proves the assertion of the theorem on the asymptotic behaviour of eigenvalues.

We now prove the n-fold completeness of the system of eigenfunctions and associated functions, for example, for (40). Note that the resolvent of (40) is a meromorphic function of finite order. For by the lemma of the present section the resolvent of (40*) is a meromorphic function of λ, and without loss of generality we may assume that $\lambda = 0$ is not a pole. Denoting by $E + R^*(0)$ the inverse operator of $E - A_0^*$, we reduce (40*) to the form

$$y = \{\sum_{\nu=1}^{n-1} (E + R^*(0))\, A_\nu^* H^{\left(\frac{\nu}{h}\right)} \lambda^\nu + (E + R^*(0))\, H\lambda^n\}\, y + (E + R^*(0))\, f.$$

Denoting the resolvent of this equation by $S^*(\lambda)$, we have

$$E + R^*(\lambda) = (E + S^*(\lambda))\,(E + R^*(0)).$$

By the theorem in **11**. $S^*(\lambda)$ is a meromorphic function of order not greater than ρn, where ρ is the order of H, and so the order of $R(\lambda)$ does not exceed ρn.

By the remark in **9**. the n-fold completeness of the system of eigen-elements and associated elements is equivalent to the completeness in $\mathfrak{H}^n$ of the vectors

$$U_h^{(k)} = (y_h^{(k,0)}, y_h^{(k,1)}, \ldots, y_h^{(k,n-1)}),$$

where $y_h^{(k,n)}$ are the derived chains of the eigenelements and associated elements (**8**.)

$$y_h^{(k,\nu)} = \left[\frac{d^\nu}{dt^\nu} e^{c_k t} \left(y_h^{(k)} + y_{h-1}^{(k)} \frac{t}{1!} + \ldots + y^{(k)} \frac{t^h}{h!}\right)\right]_{t=0}.$$

If the system of eigenelements and associated elements of (40) is not complete, then we can find in $\mathfrak{H}^n$ a vector $f = (f^{(0)}, f^{(1)}, \ldots, f^{(n-1)})$ orthogonal to all the $U_h^{(k)}$:

$$(f, U_h^{(k)}) = \sum_{\nu=0}^{n-1} (f^{(\nu)}, y_h^{(k,\nu)}) = 0.$$

Consider a solution of the equation

$$(67)\qquad z-\left(A_0^*+\lambda A_1^* H^{\left(\frac{1}{n}\right)}+\ldots+\lambda^{n-1} A_{n-1}^* H^{\left(\frac{n-1}{n}\right)}+\lambda^n H\right) z = \\ = f^{(0)}+\lambda f^{(1)}+\ldots+\lambda^{n-1} f^{(n-1)}=\xi(\lambda).$$

It is easy to verify that

$$z(\lambda) = R^*(\lambda)\xi(\lambda)$$

is an entire function. For by applying (25) and (26) of 9. we obtain for the principal parts of the function $z(\lambda)$ with respect to its poles $\bar{c}_j$ the expression

$$a_{jm_j}\frac{z^{(j)}}{\lambda-\bar{c}_j}+a_{jm_j-1}\left(\frac{z^{(j)}}{(\lambda-\bar{c}_j)^2}+\frac{z_1^{(j)}}{\lambda-\bar{c}_j}\right)+\ldots \\ \ldots+a_{j0}\left(\frac{z^{(j)}}{(\lambda-\bar{c}_j)^{m_j+1}}+\ldots+\frac{z_{m_j}^{(j)}}{\lambda-\bar{c}_j}\right),$$

where

$$a_{jh}=(\xi(\bar{c}_j),\, y_h^{(j)})+\frac{1}{1!}\left(\frac{d\xi}{d\bar{c}_j},\, y_{h-1}^{(j)}\right)+\ldots+\frac{1}{h!}\left(\frac{d^h\xi}{d\bar{c}_j^h},\, y^{(j)}\right).$$

Substituting everywhere the value $\xi(\lambda) = f^{(0)} + \lambda f^{(1)} + \ldots + \lambda^{n-1} f^{(n-1)}$ and using the formula that defines $y_j^{(h,\nu)}$, we easily obtain

$$a_{jh}=\sum_{\nu=0}^{n-1}(f^{(\nu)},\, y_k^{(h,\nu)})=0.$$

We now prove that on each ray other than the real axis the ratio $z(\lambda)/\lambda^{n-1}$ tends to zero. By the lemma of the present section the coordinates of $z(\lambda)$ have the form

$$z_i(\lambda)=\frac{h_i}{h_i-\lambda^n}\,\xi_i+\frac{h_i}{h_i-\lambda^n}\,(B\xi)_i.$$

Using the notation introduced in the proof of the lemma we obtain the inequality

$$\| z(\lambda)\| \leqslant \frac{\beta^n(|\lambda|)}{1-\beta^n(|\lambda|)}\,\|\xi(\lambda)\|+\frac{1}{\sin n\delta}\left(\sqrt{\sum_{n=m+1}^{\infty}|\xi_i(\lambda)|^2}+\|B\|\,\|\xi\|\right),$$

from which it follows immediately that

$$(68)\qquad \| z(\lambda)\| = o(1)\,|\lambda|^{n-1}.$$

Thus, $z(\lambda)$ is an entire function of finite order which satisfies (68) on each ray other than the real axis. By Lindelöf's theorem we conclude that (68) is satisfied uniformly over the whole λ-plane, and so $z(\lambda)$ is a polynomial of degree not more than $n - 2$,

$$z(\lambda) = z^{(0)} + z^{(1)}\lambda + \ldots + z^{(k)}\lambda^k.$$

We show that $z(\lambda) = 0$. For if $z^{(k)} \neq 0$, we find from (67) that the left-hand side of this equation is a polynomial of degree $n + k$, and the coefficient of λ^{n+k}, which is equal to $Hz^{(k)}$ by virtue of the completeness of H, is non-zero. This contradiction proves the completeness theorem.

References

[1] T. Carleman, Über die asymptotische Verteilung der Eigenwerte partieller Differentialgleichungen, Ber. Verk. Sächs. Akad. Wiss. Leipzig. Math.-Phys. Kl. **88** (1936), 119–134.

[2] G. D. Birkhoff, On the asymptotic character of the solutions of certain linear differential equations containing a parameter, Trans. Amer. Math. Soc. **9** (1908), 219–231.

[3] Ya. D. Tamarkin, Some general problems in the theory of ordinary linear differential equations, Moscow 1917.

Received by the Editors 31 March, 1971

Translated by E. J. F. Primrose

Dmitrii Konstantinovich Faddeev

Dmitrii Konstantinovich Faddeev was born in June 1907 in Yukhnov, in the district of Smolensk, in a family of a Petersburg engineer. From 1923, when he enrolled as a student of mathematics, until the end of his life all Faddeev's varied activities have been closely linked with the University of Leningrad, where for many years he held the chair of higher algebra and number theory, and where he has been Dean of the Faculty of Mathematics and Mechanics for several years. His links with the Academy of Sciences of the USSR were no less close. In 1932 he began research in the Steklov Institute of Mathematics and Physics which in 1934 this institute moved from Leningrad to Moscow and became the Mathematical Institute, and the Leningrad Division of the Steklov Institute of Mathematics of the Academy of Sciences of the USSR (LOMI) was created. Faddeev was a collaborator from the very moment of its foundation in LOMI and he has remained so. In 1964 he was elected Corresponding Member of the Mathematical Section of the Academy of Sciences of the USSR. For a long time he was in charge of the laboratories at LOMI, and head of a powerful group of well-known algebraists and number theorists. His algebra seminar was widely known throughout the whole country.

The range of Faddeev's interests is unusually broad — among more than his 150 papers some are on the theory of functions, on computational methods, on probability theory and on problems of teaching mathematics at all levels. But of course, Faddeev is known in the first place as one of the most outstanding algebraists of our time.

In his papers Faddeev touched on a wide range of algebraic problems. But there are two areas in which he began his research and to which he constantly returned. There are Diophantine equations and Galois theory. At the end of 50's Faddeev turned to problems in the theory of integral representations. Faddeev is one of the leading experts in numerical methods of linear algebra.

A distinctive feature of Faddeev's approach to the teaching of mathematics in schools was clearly expressed and logically developed chain of ideas in the courses he devised on algebra and basic analysis. Faddeev gives priority to "vivid contemplation", to awareness, to direct perception of the properties of mathematical objects, which thus reveal the content and basic ideas of mathematics as tools and as a description and the study of the regularity of the real world. His teaching over many years at the University and his general and specialized courses of lectures were always presented excellently. His numerous public speeches were a splendid example to younger mathematicians.

Dmitrii Konstantinovich Faddeev was a man of the broadest culture and intelligence. He had a wide knowledge and appreciation of classical music and was an outstanding pianist. Discussions with him of a variety of questions were highly estimated by all his friends, colleagues and acquaintances.

This text is an excerpt from the paper in Uspekhi Mat. Nauk 44:3 (1989) by A.D. Aleksandrov, M.I.Bashmakov, Z.I.Borevich, V.N.Kublanovskaya, M.S.Nikulin, A.I.Skopin, A.V.Yakovlev.

Доклады Академии Наук СССР
1953. Том XCII, № 4

МАТЕМАТИКА

Д. К. ФАДДЕЕВ

ОБ ОДНОЙ ТЕОРЕМЕ ТЕОРИИ ГОМОЛОГИЙ В ГРУППАХ

(Представлено академиком А. Н. Колмогоровым 18 VIII 1953)

1. Формулировка теоремы и некоторых ее следствий. Настоящая заметка является продолжением статьи (1). Пусть $\mathfrak{G}$ — группа; $\mathfrak{H}$ — ее подгруппа конечного индекса ν; $\mathfrak{a}$ — аддитивно записанная $\mathfrak{G}$-операторная абелева группа. Предполагается, что в $\mathfrak{a}$ возможно и однозначно деление на ν. Группа $\mathfrak{a}$ естественным образом может рассматриваться и как $\mathfrak{H}$-операторная группа. Через $C^n(\mathfrak{G}, \mathfrak{a})$, $C^n(\mathfrak{H}, \mathfrak{a})$, $Z^n(\mathfrak{G}, \mathfrak{a})$, $Z^n(\mathfrak{H}, \mathfrak{a})$, $B^n(\mathfrak{G}, \mathfrak{a})$, $B^n(\mathfrak{H}, \mathfrak{a})$, $H^n(\mathfrak{G}, \mathfrak{a})$, $H^n(\mathfrak{H}, \mathfrak{a})$ обозначаются, соответственно, группы цепей, циклов, границ и гомологий групп $\mathfrak{G}$ и $\mathfrak{H}$ в $\mathfrak{a}$. Целью работы является доказательство следующей теоремы:

Теорема 1. *$H^n(\mathfrak{H}, \mathfrak{a})$ содержит прямое слагаемое $H^n_0(\mathfrak{H}, \mathfrak{a})$, изоморфное $H^n(\mathfrak{G}, \mathfrak{a})$. Изоморфизм $H^n(\mathfrak{G}, \mathfrak{a})$ на $H^n_0(\mathfrak{H}, \mathfrak{a})$ порождается „оператором ограничения“ μ, сопоставляющим каждой цепи $F \in C^n(\mathfrak{G}, \mathfrak{a})$ цепь $\mu F \in C^n(\mathfrak{H}, \mathfrak{a})$ согласно формуле*

$$(\mu F)_{y_1, y_2, \ldots, y_n} = F_{y_1, y_2, \ldots, y_n}; \quad y_1, y_2, \ldots, y_n \in \mathfrak{H}.$$

Непосредственными следствиями из теоремы 1 являются следующие теоремы:

Теорема 2. *Если $\mathfrak{a}$ есть периодическая (в частности, конечная) $\mathfrak{G}$-операторная абелева p-группа, $\mathfrak{G}$ — конечная группа и $\mathfrak{H}$ — ее силовская p-подгруппа, то $H^n(\mathfrak{G}, \mathfrak{a}) \approx H^n_0(\mathfrak{H}, \mathfrak{a})$, где $H^n_0(\mathfrak{H}, \mathfrak{a})$ — прямое слагаемое в $H^n(\mathfrak{H}, \mathfrak{a})$.*

Действительно, в рассматриваемом случае $(\nu, p) = 1$ и, следовательно, в p-группе $\mathfrak{a}$ деление на ν возможно и однозначно.

В частности, при $n = 2$ получаем, что группа расширений $\mathfrak{a}$ посредством $\mathfrak{G}$ изоморфна некоторому определенному прямому слагаемому группы расширений $\mathfrak{a}$ посредством силовской p-подгруппы $\mathfrak{H}$ группы $\mathfrak{G}$.

Далее:

Теорема 3. *Если $\mathfrak{G}$ — конечная группа, $\mathfrak{H}_p$ — ее силовские p-подгруппы, $\mathfrak{a}$ — произвольная $\mathfrak{G}$-операторная периодическая абелева группа, то*

$$H^n(\mathfrak{G}, \mathfrak{a}) \approx \sum_p H^n_0(\mathfrak{H}_p, \mathfrak{a}),$$

где $H^n_0(\mathfrak{H}_p, \mathfrak{a})$ — некоторые прямые слагаемые группы $H^n(\mathfrak{H}_p, \mathfrak{a})$.

Доказательство. Пусть $\mathfrak{a}_p$ — p-компоненты $\mathfrak{a}$. Тогда $\mathfrak{a} = \sum \mathfrak{a}_p$, и, следовательно (см., например, [2]), $H^n(\mathfrak{G}, \mathfrak{a}) \approx \sum_p H^n(\mathfrak{G}, \mathfrak{a}_p)$. В силу теоремы 2,

$$H^n(\mathfrak{G}, \mathfrak{a}_p) \approx H_0^n(\mathfrak{H}_p, \mathfrak{a}_p) \approx H_0^n(\mathfrak{H}_p, \mathfrak{a}).$$

Теорема 3 верна не только для периодических групп $\mathfrak{a}$. В частности:

Теорема 4. *Если $\mathfrak{a}$ — единичная окружность с тождественными операторами, $\mathfrak{G}$ — конечная группа, то*

$$H^n(\mathfrak{G}, \mathfrak{a}) \approx \sum_p H_0^n(\mathfrak{H}_p, \mathfrak{a}).$$

Действительно, $H^n(\mathfrak{G}, \mathfrak{a}) \approx H^n(\mathfrak{G}, \mathfrak{a}')$, где $\mathfrak{a}'$ — аддитивная группа рациональных чисел по модулю 1. Отсюда справедливость теоремы следует непосредственно, ибо $\mathfrak{a}'$ — периодическая группа.

В частности, теорема 4 верна для мультипликатора Шура ($n = 2$).

Теорема 5. *Если $\mathfrak{a}$ есть решетка (прямая сумма бесконечных циклических групп), $\mathfrak{G}$ — конечная группа и $n \geqslant 2$, то*

$$H^n(\mathfrak{G}, \mathfrak{a}) \approx \sum_p H_0^n(\mathfrak{H}_p, \mathfrak{a}).$$

Доказательство. Погрузим решетку $\mathfrak{a}$ в линейное пространство $\mathfrak{b}$ над полем рациональных чисел и распространим естественным образом операторы из $\mathfrak{G}$ на все пространство $\mathfrak{b}$. Тогда, ввиду того, что в $\mathfrak{b}$ возможно и однозначно деление на любое целое число, $H^n(\mathfrak{G}, \mathfrak{a}) \approx H^{n-1}(\mathfrak{G}, \mathfrak{b}/\mathfrak{a})$ и $H^n(\mathfrak{H}_p, \mathfrak{a}) \approx H^{n-1}(\mathfrak{H}_p, \mathfrak{b}/\mathfrak{a})$ (теоремы 1 и 4 из [3]). Но $\mathfrak{b}/\mathfrak{a}$ есть периодическая группа, следовательно, $H^{n-1}(\mathfrak{G}, \mathfrak{b}/\mathfrak{a}) \approx \sum_p H_0^{n-1}(\mathfrak{H}_p, \mathfrak{b}/\mathfrak{a})$, что и доказывает теорему.

Отдельным рассуждением можно убедиться в справедливости теоремы и для $n = 1$.

2. Доказательство теоремы 1. Пусть $\mathfrak{G} = \sum \rho = \sum \mathfrak{H}\bar{\rho}$ — разложение $\mathfrak{G}$ по $\mathfrak{H}$, $\bar{\rho}$ — фиксированный набор представителей из классов смежности ρ, причем $\rho_0 = \mathfrak{H}$, $\bar{\rho}_0 = 1$. Обозначим, следуя [1], через $\mathfrak{j}$ группу функций, определенных на классах смежности ρ со значениями из $\mathfrak{a}$. В $\mathfrak{j}$ вводятся операторы из $\mathfrak{G}$ по формуле:

$$f^x(\rho) = [f(\rho x^{-1})]^x, \quad x \in \mathfrak{G}.$$

„Константные“ функции f_a со значениями $f_a(\rho) = a \in \mathfrak{a}$ образуют допустимую подгруппу $\mathfrak{a}_1$ группы $\mathfrak{i}$, операторно изоморфную $\mathfrak{a}$.

Укажем важный для дальнейшего гомоморфизм „усреднения“ σ группы $\mathfrak{i}$ в $\mathfrak{a}$, определенный равенством

$$\sigma f = \frac{1}{\nu} \sum_\rho f(\rho).$$

Правая часть имеет смысл, ибо в группе $\mathfrak{a}$ возможно и однозначно деление на ν. Гомоморфизм σ операторный, ибо

$$\sigma(f^x) = \frac{1}{\nu} \sum_\rho f^x(\rho) = \left[\frac{1}{\nu} \sum_\rho f(\rho x^{-1})\right]^x = (\sigma f)^x).$$

Если f_a есть константная функция со значением a, то $\sigma f_a = a$. Для дальнейшего нам удобнее рассматривать σ как гомоморфизм $\mathfrak{i}$ на $\mathfrak{a}_1$, считая σf равным константной функции, значения которой равны σf в прежнем смысле. Обозначим через $\mathfrak{b}$ подгруппу $\mathfrak{i}$, составленную из функций, для которых $\sigma f = 0$.

Ясно, что $\mathfrak{i} = \mathfrak{a}_1 + \mathfrak{b}$, причем сумма прямая и обе подгруппы $\mathfrak{a}_1$ и $\mathfrak{b}$ допустимые. Следовательно ((2), лемма 5), $H^n(\mathfrak{G}, \mathfrak{j}) \approx H^n(\mathfrak{G}, \mathfrak{a}_1) + H^n(\mathfrak{G}, \mathfrak{b})$.

Точнее, каждый цикл F из $Z^n(\mathfrak{G}, \mathfrak{j})$ есть сумма циклов из $Z^n(\mathfrak{G}, \mathfrak{a}_1)$ и $Z^n(\mathfrak{G}, \mathfrak{b})$ и $F \in B^n(\mathfrak{G}, \mathfrak{j})$ в том и только в том случае, если его составляющие принадлежат, соответственно, $B^n(\mathfrak{G}, \mathfrak{a}_1)$ и $B^n(\mathfrak{G}, \mathfrak{b})$.

Далее, в силу теоремы 1 из (1), $H^n(\mathfrak{G}, \mathfrak{j}) \approx H^n(\mathfrak{H}, \mathfrak{a})$, причем изоморфизм порождается оператором ограничения η, отображающим $C^n(\mathfrak{G}, \mathfrak{j})$ в $C^n(\mathfrak{H}, \mathfrak{a})$ по формуле $(\eta F)_{y_1, y_2, \ldots, y_n} = F_{y_1, y_2, \ldots, y_n}(\rho_0)$.

Очевидно, что применение естественного изоморфизма $\mathfrak{a}$ на $\mathfrak{a}_1$ и затем „оператора ограничения“ η равносильно применению „оператора ограничения“ μ, описанного при формулировке теоремы 1.

Тем самым теорема 1 доказана.

Замечание 1. Циклы, образующие классы гомологий, составляющие $H_0^n(\mathfrak{H}, \mathfrak{a})$, суть те и только те, которые продолжимы до циклов из $Z^n(\mathfrak{G}, \mathfrak{a})$. Обозначим их совокупность через $Z_0^n(\mathfrak{H}, \mathfrak{a})$. Из теоремы 1 следует, что каждый цикл из $Z_0^n(\mathfrak{H}, \mathfrak{a})$ продолжим, с точностью до границ, единственным способом.

Замечание 2. Нетрудно указать оператор, порождающий „проектирование“ группы $H^n(\mathfrak{H}, \mathfrak{a})$ на $H_0^n(\mathfrak{H}, \mathfrak{a})$.

Изоморфизм $H^n(\mathfrak{H}, \mathfrak{a})$ на $H^n(\mathfrak{G}, \mathfrak{j})$ порождается оператором продолжения E (см. (1), 4°). Отображение $H^n(\mathfrak{G}, \mathfrak{j})$ на $H^n(\mathfrak{G}, \mathfrak{a})$ порождается оператором σ и, наконец, отображение $H^n(\mathfrak{G}, \mathfrak{a})$ на $H_0^n(\mathfrak{H}, \mathfrak{a})$ порождается оператором ограничения μ.

Следовательно, отображение $H^n(\mathfrak{H}, \mathfrak{a})$ на $H_0^n(\mathfrak{H}, \mathfrak{a})$ порождается оператором $\alpha = \mu\sigma E$.

Нетрудно дать формулу для действия этого оператора на цепи из $C^n(\mathfrak{H}, \mathfrak{a})$, исходя из формул (2), (3) и (4) работы (1) и определений σ и μ.

Именно, если $F \in C^n(\mathfrak{H}, \mathfrak{a})$, то

$$(\alpha F)_{y_1, y_2, \ldots, y_n} = \frac{1}{\nu} \sum_{\rho} F_{\overline{\rho y_n^{-1} \ldots y_1^{-1}}\; y_1\; \overline{\rho y_n^{-1} \ldots y_2^{-1}}^{-1} \ldots \overline{\rho y_n^{-1}}\; y_n \overline{\rho}^{-1}}.$$

Из определения α непосредственно ясно, что для того, чтобы $F \in Z_0^n(\mathfrak{H}, \mathfrak{a})$, необходимо и достаточно, чтобы $F - \alpha F \in B^n(\mathfrak{H}, \mathfrak{a})$.

Ленинградский государственный университет им. А. А. Жданова

Поступило 24 VI 1953

ЦИТИРОВАННАЯ ЛИТЕРАТУРА

[1] Д. К. Фаддеев, Изв. АН СССР, сер. матем., **16**, № 1, 17 (1952). [2] З. И. Боревич, там же, **16**, № 4, 365 (1952). [3] Д. К. Фаддеев, ДАН, **58**, № 3 (1947).